D0460705

The Civil Rights Era

THE
Civil Rights Era

Origins and Development
of National Policy
1960–1972

HUGH DAVIS GRAHAM

NATIONAL UNIVERSITY
LIBRARY   SAN DIEGO

New York   Oxford
Oxford University Press
1990

Oxford University Press

Oxford   New York   Toronto
Delhi   Bombay   Calcutta   Madras   Karachi
Petaling Jaya   Singapore   Hong Kong   Tokyo
Nairobi   Dar es Salaam   Cape Town
Melbourne   Auckland

and associated companies in
Berlin   Ibadan

Copyright © 1990 by Hugh Davis Graham

Published by Oxford University Press, Inc.,
200 Madison Avenue, New York, New York 10016

Oxford is a registered trademark of Oxford University Press

All rights reserved. No part of this publication may be reproduced,
stored in a retrieval system, or transmitted, in any form or by any means,
electronic, mechanical, photocopying, recording, or otherwise,
without the prior permission of Oxford University Press.

Library of Congress Cataloging-in-Publication Data
Graham, Hugh Davis.
The civil rights era : origins and development of
national policy. 1960–1972 / Hugh Davis Graham.
p. cm.   Bibliography : p. Includes index.
ISBN 0–19–504531–9 (alk. paper)
1. Civil rights—United States—History—20th century.
2. Affirmative action programs—Government policy—United States—
History—20th century.   3. United States—Politics and
government—1945–   I. Title.
JC599.U5G685   1989
323.1′73—dc20   89–16142   CIP

NATIONAL UNIVERSITY
LIBRARY
SAN DIEGO

2 4 6 8 9 7 5 3

Printed in the United States of America
on acid-free paper

For my parents

Preface

During the academic year 1985–86 I was able to devote full time to reading, thinking, writing, and talking with colleagues about this book. For this privilege I am most indebted to the Woodrow Wilson International Center for Scholars for the fellowship that kept me ensconced during the winter and spring of 1986 high in a tower office in the Smithsonian Castle. The University of Maryland, Baltimore County granted me sabbatical leave during fall in 1985, and also provided a summer fellowship to support research in the Nixon papers during the summer of 1987. I am also grateful for the award of grants-in-aid of research from the Lyndon Baines Johnson Foundation during 1984–85 and the Gerald R. Ford Foundation for the summer of 1988, and to the National Endowment for the Humanities for travel-to-collections awards to support research in the John F. Kennedy Library in 1984 and the Lyndon Baines Johnson Library in 1985–86.

Manuscript researchers pile up debts of gratitude they can never repay to archivists who make our research trips pleasant as well as productive. My special list is dominated by the presidential libraries. It includes, at the Johnson Library, Nancy Kegan Smith, Linda Hanson, and director Harry Middleton; at the Nixon archive (not yet a presidential library), Joan Howard and deputy director James J. Hastings; at the Kennedy Library, William Johnson. Bracketing my period but providing important documents were John E. Wickam of the Eisenhower Library and David Horrocks and William J. Steward of the Ford Library. In the Washington depositories, I received generous guidance from Jerry Hess at the National Archives, Henry Guzda and Judson MacLaury at the Labor Department, Catharine Vogel at the George Meany Memorial Archives, Ronald Wartow at the General Accounting Office, Phillip Lyons at the U.S. Commission on Civil Rights, and Andrelia D. James at the Equal Employment Opportunity Commission. Zdenek V. David, librarian at the Wilson Center, provided smooth access to the Library of Congress. In mining these and related resources I enjoyed the energetic and alert assistance of intern David C. Smalley at the Wilson Center, and graduate student William Harvey at the University of Maryland Graduate School, Baltimore.

My primary reliance on documentary evidence was supplemented by the oral history interviews in the Kennedy and Johnson libraries, and also by personal interviews. I am grateful for the time provided by Lee C. White, Adam Yarmolinsky, Sonia Pressman Fuentes, Richard Graham, Aileen C. Hernandez, John Ehrlichman, William J. Kilberg, and John de J. Pemberton, Jr. I also benefitted from correspondence with George E. Reedy, Bill Moyers, Herbert Hill, Robert P. Griffin, and George P. Shultz.

Of the many colleagues who gave me critical readings of the manuscript, none was more critically demanding and yet supportive than my brother, Otis L. Graham, Jr. David R. Goldfield and Steven F. Lawson read most of the chapters in early draft form and provided detailed and instructive criticisms. Reader-critics included my long-time co-author, Numan V. Bartley, as well as my campus colleagues James C. Mohr and John W. Jeffries in history, and George R. LaNoue in political science and public policy. The Nixon chapters profitted from a careful reading by John Ehrlichman. I am also grateful for the critical assessments of legal issues I received from David S. Bogen and William A. Reynolds at the University of Maryland Law School, from Herman Belz at the sister campus at College Park, and from Charles A. Lofgren at Claremont McKenna College.

The Wilson Center gathered an eclectic group of scholars whose critical perspectives covered a wide spectrum. These included Michael J. Lacey, Louis Galambos, J. Morgan Kousser, Paul Kleppner, and Stephen Skowronek. Other colleagues whose advice I sought and freely received include August Meier (and through him, his late co-author Elliott Rudwick), Joan Hoff-Wilson, Robert A. Divine, Laura Kalman, Donald L. Horowitz, Francis E. Rourke, and James A. Sundquist.

Sheldon Meyer at Oxford University Press offered patient support when complications with the Nixon papers delayed the manuscript, and his associate Rachel Toor provided strategic guidance for cutting a manuscript that had grown very long.

Contents

Introduction 3

PART ONE   KENNEDY

I Kennedy, Johnson and the Presidency in Civil Rights   27

II Lyndon Johnson, Robert Kennedy, and the Politics of Enforcement   47

III The Civil Rights Bill of 1963   74

IV The Storm Over Racial Quotas   100

PART TWO   JOHNSON

V The Civil Rights Act of 1964   125

VI From the Civil Rights Act to the Voting Rights Act: The Johnson White House at Flood Tide, 1964–1965   153

VII The Troubled Search for Civil Rights Enforcement: Launching the EEOC, 1965–1966   177

VIII The EEOC and the Politics of Gender   205

IX From Equal Treatment to Equal Results: Transforming Civil Rights Strategy   233

X From Ghetto Riots to Open Housing, 1966–1968   255

XI From Johnson to Nixon: The Irony of the Philadelphia Plan   278

PART THREE   NIXON

XII Richard Nixon and Civil Rights Policy: "No Master Spirit, No Determined Road"   301

XIII The Philadelphia Plan Redux   322

XIV Nixon, Congress, and Voting Rights 346

XV The "Color-Blind" Constitution and the Federal Courts 366

XVI Women, the Nixon Administration, and the Equal Rights Amendment 393

XVII Race, Sex, and Civil Rights Enforcement: Culmination 1972 420

Conclusions 450

Essay on Sources 477

Notes 482

Index 571

The Civil Rights Era

Introduction

Social Movements and National Policy: The Reconstruction of the American Administrative State

This is a story about a rare event in America: a radical shift in national social policy. Its precondition was a broader social revolution, the black civil rights movement that surged up from the South, followed by the nationwide rebirth of the feminist movement.[1] The social movements took hold and spread at the grass roots, but the policy revolution that responded to them was made in Washington. This is therefore a study of national policy elites, their behavior and motives, their options and decisions and the consequences of those decisions. The combined weight of their achievements, most notably the civil rights laws of 1964, 1965, 1968, and 1972, together with supporting court decisions and administrative enforcement, broke the back of the system of racial segregation and destroyed the legal basis for denying minorities and women full access to education, employment, the professions, and the opportunities of the private marketplace and public arena.

The broader impact of these laws, and the meaning of the paradigm shift in social policy and administration that accompanied them, could not readily be discerned by their contemporaries. During the civil rights era the American administrative state reached out and embraced workaday relationships between its citizens and both public and private institutions in a way that had never before been contemplated—except, perhaps, in apologetic wartime emergencies, and in the largely failed Reconstruction.

The story's main focus—federal policy in civil rights during 1960–72—was originally conceived as centering almost exclusively on racial policy. But the evidence and especially the intensity of debate over the race-sex linkage demanded the inclusion of both major streams of civil rights challenge. The historical experience of group discrimination, the logic of civil rights theory, the politics of group protest in the 1960s, the race-sex analogy in policy debate, and the initiation of profound changes in the legal status of women,

3

all required that the black and feminist movements be considered together. Other groups of course shared the burden of historical discrimination, and civil rights theory would seem to call for the inclusion of Hispanics, Asians, and other ethnic minorities. But the nature of group protest and policy debate during the 1960s ordained that such groups would play a relatively modest role, at least through 1972. Not surprisingly, the issues of black civil rights dominate the early drama. But the policy revolution of 1960–72 involved a dual transformation that in many ways confounded the race-sex analogy. Originally driven by liberalism's classic theory of individual rights and non-discrimination, "the Movement" through its successes evoked new theories of compensatory justice and group rights. In doing so it generated new internal divisions, with the "second-wave" feminist movement providing a revealing counterpoint, invoking principles of egalitarianism that were at once radical and classic, while the black agenda began to move away from them.

Choosing the period 1960–72, like all such decisions about periodization, requires some defense. One can easily nominate plausible earlier events for a starting point, and historians are cursed by causal hindsight with a vast universe of plausible origins. Suffice it here simply to claim that the challenge of the sit-ins in the South and the presidential election of John F. Kennedy in 1960 catapulted the civil rights issue into the forefront of national consciousness. It did this with an escalating logic and force that for the first time (since Reconstruction) forced all three branches of the federal government to confront and attempt to resolve the problems of discrimination through concerted effort. As is well known, this began with John F. Kennedy's cautious executive initiative in 1961, which injected into public discourse the ambiguous phrase, "affirmative action." The momentum of reform escalated, after the Birmingham violence, into Kennedy's civil rights bill of the summer of 1963, and followed Kennedy's assassination with Lyndon Johnson's triumph in the Civil Rights Act of 1964 and, following the Selma march, the Voting Rights Act of 1965. Johnson completed his Great Society reforms in civil rights with the somewhat surprising Open Housing Act of 1968.

Thereafter both the Johnson and Nixon administrations and the Congress continued to struggle with the unresolved problems of civil rights enforcement until 1972, when Congress sent the Equal Rights Amendment to the states with resounding majorities, and Richard Nixon signed the Equal Employment Opportunity Act of 1972. Nixon confirmed the Kennedy-Johnson legacy in civil rights law in much the same way that Eisenhower had confirmed the New Deal—not by embracing it, but by accommodating to it on the margins in a way that guaranteed its legitimacy and secured its permanence in the federal establishment. After 1972, as one recent study confirmed, congressional interest in further debate over the fundamental assumptions and structures of the new civil rights laws and their enforcement mechanisms (as distinct from congressional interest in adding new groups, such as language minorities and the handicapped, to the protected-class model) dropped "virtually to zero."[2] The post-Watergate administrations of presidents Ford and Carter would grapple with Congress and respond to court decisions over policies of applying the new protections and claims, and of extending the umbrella of newly protected classes to include such groups as Hispanics and language minorities, the physically and mentally handicapped, institutionalized persons, the elderly, children, and the unborn. In the 1980s the Reagan administration would try to mount a counterrevolution, only to find in Congress, in the federal courts, and in the vast administrative apparatus of the permanent government a formidable barrier to change. Thus the events of the 1960s shattered the old mold, and by 1972 the fundamental attributes of the new order were set in place.

History and Civil Rights Policy: The Dual Transformation of 1960–72

Contemporary accounts of the civil rights movement and reforms of the 1960s, and the first wave of historical accounts during the 1970s, emphasized dramatic events like the protests at Birmingham and Selma and their causal link to the breakthrough laws of 1964 and 1965. Policy analysis similarly emphasized the formulation and enactment phases of the policy cycle, but devoted little attention to the obscure and complex phase of implementation.[3] This book broadens the focus of historical analysis to include the full policy cycle, with special attention to reconstructing from the archives the crucial process of policy implementation. But it also narrows the lens to focus more intensively on decisions and events that in the longer view have demonstrated an abiding importance, and often generated a lasting controversy, that was often not accorded them at the time. Controversies that dominated the headlines of 1960–72 have since faded in importance, many of them becoming the welcome casualties of success. These include the major battles over desegregating Jim Crow school systems, over segregation in public facilities and accommodations, the right of blacks to cast ballots in the South, the impunity with which white terrorists plied their trade. These controversies dominated the first phase of the reform era, and they were regionally focused on the South. They generally involved positive-sum rights, such as equal access to the franchise, to the public schools, to parks and hotels and restaurants, rights that all citizens could claim and the polity could accommodate. Such claims would yield belatedly to classical liberalism's Lockean demand for individual and negative rights against abusive government and private power, when the national guardians of civil authority at last commanded "Thou Shalt Not."

Lying beneath these claims, however, were conflicts that more closely approximated a zero-sum game. The perceived effect of competing individual and group claims to jobs and contracts, to appointments and promotions, to higher education and professional schools, when combined with the logic and force of rising federal efforts to rectify the "underutilization" of minorities, ultimately raised the cry of "reverse discrimination." The issue of racial quotas first arose during the national debate over the Civil Rights Act of 1964, and the apparent illogic and illiberalism of fighting discrimination with counter-discrimination produced a national consensus in 1963–64 that embedded a ban on such practices in the heart of the new law—or so it was perceived at the time. But these issues were raised again in the late Johnson years and passed unresolved into the first Nixon administration. They came to dominate the civil rights controversies after 1972, when reverse discrimination suits like *Bakke* and *Weber* split the Supreme Court, much as they divided the entire country.

These more difficult, zero-sum controversies over equal results and compensatory justice were raised in the process of resolving the problems of equal treatment following the breakthrough legislation of 1964–65. With hindsight we can seek the origins of these persistent and troubling issues in the policy deliberations of national elites who grappled with them during 1960–72, solving some and exacerbating others. The story therefore begins with John F. Kennedy's involved and perplexing executive order of 1961. Its unlikely and unhappy chief architect was Vice President Lyndon Johnson. It summoned from the obscure, statutory boilerplate of the Wagner Act of 1935 an enigmatic phrase, "affirmative action," whose ambiguous meaning would be haltingly defined in a long thread of controversy that would tie the Kennedy era to our own in ways the New Frontiersmen could not have contemplated.

Kennedy's cautious, ambiguous beginning thus launched the modern Presidency toward a triumph of classic anti-discrimination that would redeem the failures of the Reconstruction. Lyndon Johnson would fulfill the fallen Kennedy's legacy, and sign into law the watershed Civil Rights Act of 1964. But at the peak of nondiscrimination's triumph, in 1965, it confronted liberalism's paradox of history: the timeless principles of republican liberty, as they are discovered and proclaimed, are perforce applied to a timebound society. Newly discovered equal rights are applied only to the present and future. This works well with most positive-sum rights, such as nondiscrimination in hotels and restaurants, parks and schools, the voting booth. But in the more nearly zero-sum world of jobs and promotions, history has left its legacy of deep scars on excluded groups. "Freedom is not enough," Lyndon Johnson told the graduating class at Howard University in June 1965. "You do not wipe away the scars of centuries by saying: Now you are free to go where you want, do as you desire, choose the leaders you please," he explained. "You do not take a person who for years has been hobbled by chains and liberate him, bring him to the starting line and then say, 'You are free to compete with all the others,' and still justly believe that you have been completely fair."[4]

"We seek not just freedom but opportunity," Johnson said, "not just equality as a right and a theory but equality as a fact and as a result." Thus he seemed to proclaim, beyond the procedural goal of equal treatment as a necessary but insufficient precondition, a more ambitious and elusive substantive goal —"the next and more profound stage of the battle for civil rights." Lyndon Johnson confronted his battles honestly, carried them aggressively to the Congress and the public, fought them courageously, and won most of them. But the Phase II battle that Johnson launched in Congress in 1966 outlived his administration. It lasted through the first four years of Richard Nixon's administration, and culminated in the Equal Employment Opportunity Act and Title IX of the education amendments of 1972 and congressional passage of the Equal Rights Amendment. The forces that drove Phase II were broader, newer, and more complex than liberalism's vintage crusade against discrimination, which shattered the defenses of segregation in 1964–65. Otherwise the momentum of Phase II should not have survived the catharsis of ghetto riots, Black Power, political assassination, and hard-hat backlash that convulsed the nation during the late 1960s. Such forces and their consequences yield their secrets only through the perspective of time, and through the archival documents that chart the evolution of public policy.

Thus the Presidency dominates this volume. The massive paper trail of its inner deliberations, masked behind its public rhetoric, constitutes the core of the evidence. Congress in its independent role of determining the final shape of statutory law often anticipated and outflanked the President, especially during the Nixon administration. Similarly the federal judiciary, despite the shift in 1969 from the Warren to the Burger Court, asserted an expanding role in shaping social policy. Indeed the judicial rulings of the civil-rights era show a striking continuity in empowering the executive agencies and their custodians, extending the reach and reinforcing the authority of a new edifice of social regulation that was being constructed largely without blueprint or even conscious awareness. In an unconscious yet seemingly symbiotic relationship, the civil rights movement, first led by blacks and then joined by women, formed the expanding edge of the new American administrative state. Yet the civil rights thrust was only part of a broader aggregation of parallel but uncoordinated interests, including environmental, consumer, and antiwar forces, that accelerated the evolution of the new social regulation during the 1960s and beyond.

This vast and largely undirected process centered in the executive branch, where the commanding heights of federal policy are theoretically controlled by the White House. In implementing social policy the well entrenched executive establishment in the "subpresidency" has enjoyed formidable advantages in distributing the budget and designing the rules of the game. So while presidents and congresses come and go, the federal agencies abide, defining through administrative law and regulation the precise meaning of broad statutory provisions that Congress could not conceivably tailor to the nuances of America's workaday life. This then is a story not only of presidents and powerful White House aides, of prominent cabinet secretaries and attorneys general and their loyalist lieutenants among the President's battalion of senior but temporary "Schedule-C" political appointees. It is also a story of the "permanent government"—of the career civil servants in the semi-independent mission agencies and their sub-baronies in departments like Labor, HEW, Defense; of novel subagencies like the clumsily named and yet extraordinarily important Office of Federal Contract Compliance Programs in Labor; and especially of the era's fragile infant among regulatory boards, the Equal Employment Opportunity Commission.

Telling such a story necessarily involves such arcane bureaucratic entities and statutory constructions as Title VII coverage, Section 707 suits, bona fide occupational qualifications, Section 5 preclearance, cease-and-desist authority. At its heart lie such crucial statutory obscurities as Title VII's controversial Section 703(j), which meant a ban on racial quotas, and wonderfully bureaucratic inventions like "Revised Order No. 4," which when decoded meant a government requirement for minority proportionality in job distribution. Such gray contrivances of statutory law and the bureaucratic imagination lack the graphic and emotional appeal of the murders of civil rights workers in Mississippi or the ghetto riots in Watts and Detroit. But in the long run of policy continuity and its aggregated impact, they are probably more important.

Executive Orders and the Limits of Presidential Authority: A Historical Introduction

In the practical world of American government the tripartite division of powers has necessarily produced broad areas of overlap. Congressional policies have always required administrative rules and regulations by executive agencies, and the Supreme Court has long upheld certain congressional delegations of discretionary authority to the executive branch as both necessary for orderly public administration and consistent with the doctrine of separation of powers.[5] The vast majority of executive orders have in fact been routine and specific delegations of authority by Congress so executive agencies could administer statutory policy through necessary rules and regulations.[6]

It is in the not-so-routine minority of cases of apparent executive lawmaking where Congress or aggrieved private parties have been inclined to challenge presidential trespass on legislative authority. When President Truman ordered his Secretary of Commerce to seize the steel mills during the Korean War and was promptly challenged in court, Justice Hugo Black, himself a former U.S. senator, chastised the President for overstepping the proper bounds of separation of powers. Such orders "look more like legislation to me than properly authorized regulations to carry out a clear and explicit command of Congress," Black observed. "And of course," he added gratuitously, "the Constitution does not confer lawmaking power on the President."[7]

A crucial distinction between statutory and administrative law that enjoys congressional approbation on the one hand, and *ex parte* executive decrees that do not, involves the thorny matter of sanctions, or teeth, should the President order behavior that might meet resistance or evasion. Except in wartime, American presidents through their executive orders have been restricted mainly to imposing or threatening two forms of sanction without court action. The first of these is the refusal of benefits, most commonly the withdrawal or denial of government contracts. Federal contract denial or termination was potentially, to be sure, a mighty sword indeed, especially as the

New Deal and Cold War legacies were radically expanded in the 1960s. But prior to the 1960s contract deprivation had not been a very credible threat. This is generally because it pitted relatively powerless and often individual complainants against powerful firms whose goods and services were vital to the state, and whose freedom to act as entrepreneurs was sanctified by the strong traditions of America's market economy. The President's other enforcement weapon was punitive publicity. But damaging publicity was largely a function of public expectations, values, and loyalties. Franklin Roosevelt discovered this when he failed in his attempt to purge obstructionist conservatives from the Congress in the late 1930s. In the mid-century era of Jim Crow, with the Great Depression in recent and painful memory, aggressive demands for the widespread employment of Negroes (the text will reflect contemporary usage until the mid-1960s, when the term "Negro" was discarded) were often not sympathetically received by the American electorate.

It is not surprising, then, that presidential efforts to fight racial discrimination through executive orders were halting and largely ineffectual during the two decades prior to Kennedy's announcement of Executive Order 10925 on March 7, 1961. But all of these efforts contained within them the seeds of successful future attempts, eventually on a massive scale, given the rapid expansion of federal domestic programs and a radical shift of opinion concerning minority rights. Beginning with Franklin Roosevelt, the modern American presidents began to pick their way warily through the minefield.

Franklin Roosevelt's Two FEPCs

The administration of Woodrow Wilson had introduced to the federal government the patterns of racial segregation and Negro subordination that had been spreading throughout the South since the turn of the century.[8] By the 1930s, these patterns were deeply entrenched in the federal workforce. In the private economy, the southern pattern of a racially segregated labor force in a third of the states was paralleled by a northern pattern that generally relegated racial minorities to low-paying menial jobs. But the liberal spirit of the New Deal prompted both the federal executive and the Congress to enunciate new policies of racial nondiscrimination in at least a few New Deal measures.[9] Often these new nondiscrimination clauses, which typically barred discrimination on account of race, color, and creed (by which was meant religious belief, not political credo), were inserted by liberal Republicans in the Lincoln tradition, such as New York Congressman Hamilton Fish, as well as by northern liberal Democrats, like congressmen Vito Marcantonio of New York and Jerry Voorhis of California.[10]

Conservatives and southerners accepted these nondiscrimination pledges largely because such rhetoric was not self-executing. Like Prohibition, such lofty measures were toothless without the accompanying penalties of a Volstead Act. Besides, congressional conservatives were amply armed with defensive devices with which to blunt any serious liberal attempts to provide teeth for enforcement mechanisms. As early as 1933, Congress banned employment discrimination by "race, color, or creed" in the Unemployment Relief Act, and agency regulations under the National Industrial Recovery Act of 1933 prohibited racial and religious discrimination in NRA-sponsored housing programs. In 1940 a new Civil Service rule forbade such discrimination in federal employment, and discontinued the requirement for application photographs. That same year Congress itself reaffirmed this commitment in the Ramspect Act covering civilians in classified federal jobs. But all of this egalitarian precedent during the New Deal was of small immediate and practical consequence for minority employment, because the machinery and sanctions for enforcement were lacking, including definitions and standards.

This new egalitarian rhetoric was, to be sure, an important part of the symbolic outreach through which Roosevelt and the New Deal brilliantly wrenched away from the party of Lincoln the partisan loyalty of black America.[11] But when Dr. New Deal was replaced by Dr. Win the War, the crusade to defend American freedom from fascism ironically raised Jim Crow armies. The Selective Service Act of 1940 contained one of the increasingly standard new racial nondiscrimination clauses. But when the White House released a War Department policy statement in October of 1940 announcing that substantially more Negroes would be drafted—so as to reach the same proportion in the army as in the national population (10 percent)—it decreed that they would be maintained in segregated regiments.[12] To black Americans still frozen out of the defense boom, these contradictions sparked a protest that forced President Roosevelt's hand.

The story of A. Phillip Randolph and the March on Washington Movement is well known, at least through its successful first year.[13] The socialist Randolph's all-black protest and threatened march first forced a reluctant black establishment led by the NAACP to go along, then finally forced the embarrassed and irritated President to issue his pathbreaking executive order on June 25, 1941.[14] Executive Order 8802, which created the Committee on Fair Employment Practice (FEPC), was justified under the President's emergency powers in war preparedness, and this gave it a practical cachet that was nevertheless coterminous with the national emergency. It proclaimed a broad new policy of nondiscrimination by race, creed, color, or national origin in federal employment and defense contracts. To enforce the new policy it established a five-man committee authorized to receive and investigate complaints of discrimination, to take "appropriate steps" to redress valid grievances, and to recommend to the President and federal agencies steps necessary to carry out the terms of the order. On its face, the FEPC represented a dramatic new commitment of considerable promise to victimized minorities. It was responsible solely to the President himself, and operated as a de facto independent agency.[15]

But the new FEPC had four formidable weaknesses. First and most visibly, the small and part-time committee had a staff of only eight members, half of them clerical, and was funded out of the catch-all, special projects fund in Roosevelt's newly consolidated Executive Office of the Presidency.

This precluded the development of normal patterns of federal supervisory activity that were practiced by such congressionally approved agencies as the Federal Trade Commission, or even the new and controversial National Labor Relations Board (NLRB). Like the older regulatory agencies, the NLRB employed a substantial battery of staff investigators and attorneys, and it had established a network of regional offices to reflect the reality that 90 percent of federal civilian employees worked outside Washington—as of course did most defense contractors. Second, while the FEPC's coverage included federal employees and defense contract employers, the President had no direct authority over labor unions, which were not formal parties to federal contracts. Most private employment remained well beyond the President's direct reach. Third, as a creation of executive fiat, the FEPC lacked statutory enforcement powers. It therefore concentrated on persuasion, volunteerism, and public education, with pressure generated primarily through hearings. It could not initiate action without first receiving a complaint, and then could only investigate charges and issue an advisory opinion or recommendation. Actual authority to cancel contracts lay with the contracting agencies. They were understandably loath to jeopardize war production, or to yield up their independence to a tiny upstart.

Finally—and ultimately fatally—the FEPC lacked political legitimacy in public opinion and in the corridors of national power. Within the government, federal managers instinctively resented a new watchdog agency whose very inquiries seemed to insult both their competence and their sense of fairness. The FEPC threatened not only segregated facilities and practices in the South, but also the freedom of employers to hire and fire workers as they pleased. The leadership of the CIO had strongly supported FEPC, but the AFL and the rank and file of labor did not. As an executive creation unbeholden to Congress for either authority or budget, the FEPC found few friends on Capitol Hill. When the FEPC journeyed into the South in June of 1942 to hold hearings in Birmingham, it stirred congressional resentment and returned to Washington to find itself suddenly transferred to the War Manpower Commission under the directorship of Paul V. McNutt; future funding, if any, would now be channeled through the normal congressional budget and review process. In response to dismayed complaints by the civil rights lobby, Roosevelt explained that he sought "to strengthen, not to submerge" the committee by making available to it the resources of the WMC under McNutt's "friendly supervision."[16] However, when in the fall of 1942 the FEPC scheduled hearings on complaints against twenty-three resisting railroads and their unions, McNutt abruptly postponed the hearings in January of 1943 without consulting the committee. A majority of the FEPC members, including the chairman, promptly resigned, and the first FEPC essentially collapsed amid mutual recriminations.

This fiasco was especially embarrassing in the middle of a world war against fascism. So renewed negotiations centering on the mediating role of Attorney General Francis Biddle led to Roosevelt's promulgation on May 27, 1943 of Executive Order 9346.[17] This decree formally abolished the moribund first FEPC and established a second one under Roosevelt's broad authority as wartime Commander in Chief. Its jurisdiction was extended, at least by implication, more broadly to "war industries" rather than narrowly to "defense industries." Further, it declared that all employers and labor organizations had the "duty" to eliminate not only discrimination in employment, but also in "union membership." Finally, the new executive order specifically authorized the FEPC to "conduct hearings and issue findings of fact," and to "take appropriate steps to obtain elimination" of the discrimination forbidden by the order.[18] Structurally, the second FEPC enjoyed a full-time chairman, and its increased budgetary allocation ultimately provided for staff growth to 120 personnel and expansion into fifteen field offices.

Like its predecessor, the newly rechartered and enlarged FEPC operated at two levels. One was at the quiet, workaday level of complaint processing, dispute resolution, and negotiation. Concentrating on education and persuasion, the combined wartime FEPC (1941–46) was able to resolve successfully approximately one-third of the 14,000 registered complaints it received—although only one-fifth of those received from the South were resolved.[19] Given its relative powerlessness and the profound tenacity of white resistance to equal employment rights for black workers, the concrete achievements of the FEPC are all the more remarkable.[20] By the war's end, it had conducted thirty hearings involving 102 companies, 38 unions, and five government agencies. But the hearing process by its very nature invited confrontation. Under the aggressive chairmanship of Malcolm Ross, the second FEPC interpreted "appropriate steps" to authorize the issuance of "directives" to employers found to be discriminating.[21] But the FEPC's new enforcement "directives" were only appealable to and enforceable by the President. Most resisting employers and unions learned that they could simply defy and outwait the FEPC while pressing a receptive Congress to clip its wings. Ultimately, employers and unions held to be discriminating by the FEPC successfully disregarded 35 of 45 FEPC "directives" to comply. Congress, meanwhile, responded to growing resentment against the FEPC by launching a two-pronged attack.

The Congressional Destruction of FEPC

The first congressional attack took the form of an investigation in early 1944 by the U.S. House Select Committee to Investigate Executive Agencies. The committee was chaired by "Judge" Howard W. Smith, the courtly conservative who represented northern Virginia's apple-orchard country, and the committee's ostensible purpose was legitimate. It sought to reassert the traditional prerogatives of Congress against a headstrong President whose executive aggrandisements during the New Deal had been vastly compounded by the war emergency. But the Smith committee revealed its narrower ideological agenda by summoning Ross and grilling him and the FEPC's special counsel, Joseph Sarfsin, on their broad construction of Executive Order 9346, and especially on their authority for issuing "directives" to employers who had no defense contracts.[22] Smith elicited from Sharfsin a boundless claim that the FEPC's authority was as unlimited as the President's extraordinary authority in the war emergency. Having made its point that the unpopular FEPC was operating without congressional authorization under the vulnerable cover of wartime emergency, the Smith committee concluded its hearings without recommending legislation.

Later in 1944, however, Congress did move against the FEPC, this time with more artful indirection. In the Senate, Richard Russell of Georgia reflected a tightening congressional rein on the expansionist tendencies of the Roosevelt administration by sponsoring an amendment to the Independent Offices Appropriation Act of 1940. The Russell amendment prohibited federal funding of any agency that had been in existence for more than one year (including those established by executive order) without a specific congressional appropriation. It also prohibited any existing agency from using its funds to carry out the functions of an executive agency that was eliminated as a result of the Russell amendment. This would appear to bar the presidential creation of another FEPC in the absence of specific legislative authorization. After passing the Russell amendment, Congress twice appropriated money for the FEPC—the second time, in 1946, slashing its budget in half and directing the liquidation of the committee's affairs.

Thus the FEPC emerged from the wartime emergency that created it as an extremely vulnerable agency, its mighty creator gone to his grave, its congressional source of funds increasingly hostile. Its very acronym had become a devisive symbol. To its champions, FEPC signified a public commitment to fairness in the workplace. To its detractors, FEPC stood for arrogant presidential intrusion into the legislative domain, and self-seeking meddling by bureaucrats and politicians in a free market.

It would be a mistake, however, to regard the demise of the wartime FEPC as a foredoomed consequence of implacable congressional resistance, once its artificial raison d'être of wartime emergency powers was removed. As early as 1942 congressional liberals had introduced a bill to establish by statute a permanent national FEPC, and similar bills were annually introduced until the EEOC was created by the Civil Rights Act of 1964. In the first debate over congressional funding for FEPC in May of 1944, the House had rallied to the skilled but desperate pleas of New York's Representative Vito Marcantonio and approved a $500,000 appropriation bill for FEPC for fiscal year 1945. In the Senate that same year, Russell's motion to strike out the FEPC appropriation was defeated by a vote of 38 to 21. Defeated as well was the Georgian's rider to prevent the FEPC from spending more than 25 percent of its appropriations in salaries for Negro employees. In the Senate the hard core of anti-FEPC support was almost exclusively southern, and it typically included all twenty-two Democratic senators from the eleven former Confederate states except the feisty New Dealer from Florida, Claude Pepper. Moreover, a bipartisan, nonsouthern coalition of considerable prestige was formed in 1944 around the new National Council for a Permanent FEPC. Its honorary chairmen were Democratic Senator Robert Wagner of New York and Republican Senator Arthur Capper of Kansas. Leading the coalition's efforts in the Congress were Representative Mary T. Norton, a New Jersey Democrat who chaired the House Committee on Labor, and in the Senate, Democrat Dennis Chavez of New Mexico. On the Republican side, the national platform of 1944 had called for "the establishment by Federal legislation of a permanent Fair Employment Practice Commission," and the Republican members of Congress honored that Lincolnian tradition with regularity during the 1940s.

Given this array of contending forces, which tended in the Congress to be rather evenly divided numerically as the war rushed toward a close in the summer of 1945, the anti-FEPC forces repaired to their redoubt in the Senate, and especially to the filibuster. Their intransigence in the Senate forced the House in 1945 to accept a terminal FEPC appropriation for self-liquidation of $125,000 for fiscal year 1946. In 1946 itself, a Chavez-sponsored bill to create a permanent statutory FEPC could not overcome the anticipated southern filibuster. But the attempt at cloture on February 4, 1946, following eighteen days of debate, failed by a vote of 48 to 36. This was well short of the required two-thirds, but it represented a substantial majority in favor of cloture, and it came from a promising bipartisan, nonsouthern coalition in which Republicans outnumbered Democrats 25 to 22.

With the failure of Senate cloture early in 1946, the FEPC died. But the greatest tribute that marked its passing was that although it was never to be re-created in its original incarnation, virtually all subsequent debate was cast in its image and defined by its terms.

The Equal Employment Dance of Truman and the Congress

By the conclusion of World War II, FEPC had come to symbolize the ideological fault line that separated the liberal-presidential and the conservative-congressional factions of both major parties. Harry S Truman as President held steadfast to Roosevelt's FEPC legacy, keeping the second FEPC alive by his own executive order in December 1945 until the committee's congressionally ordained termination in June 1946. During 1946 and 1947, Truman repeatedly called for fair employment legislation in radio, State of the Union messages, and economic reports to Congress. In December of 1946 Truman created his own portentous and liberal-loaded advisory body, the Committee on Civil Rights, and he asked Congress to create a national FEPC in his civil rights message of 1948. On July 26, 1948, Truman issued his own executive order on fair employment, one that was timed, like his armed forces desegregation order of the same day, to boost his re-election campaign that year.

Executive Order 9980 established not another FEPC, but rather a Fair Employment Board within the Civil Service Commission.[23] But this only formalized nondiscrimination policies that had theoretically continued in federal employment after the demise of the wartime FEPC. It left the government with no claimed leverage on private employment at all, since the anti-discrimination ban in federal contracts had died with Roosevelt's second FEPC. Truman's FEB was essentially a watchdog subunit of the Civil Service Commission, and its primary function was appelate. It lacked the power and visibility of the companion Fahy committee, which was charged with military desegregation. Moreover, the FEB was dealing with two million federal civilian employees, not the 20 million workers involved in the expanding network of federal contracts and subcontracts in the private economy.[24]

During the Truman presidency the maturing civil rights coalition, centering on the NAACP and the new leadership Conference for Civil Rights, solidified around the lobbying effort for a permanent statutory FEPC. Congressional hearings were held and bills were voted out of the liberally inclined authorization committees in both houses that dealt with labor and social legislation. But the procedural defenses of the southern and conservative opposition were strong enough to prevent any serious likelihood of final passage. The committees in both chambers that were concerned with labor legislation were traditionally dominated by liberal and pro-labor legislators from urban-industrial constituencies who not surprisingly sought such committee assignments, but who scarcely reflected the broad makeup of the Congress. During the postwar period these committees would typically report out generous bills on labor and education by substantial majorities, only to be stymied or reversed on the floor of the chamber. The Senate Education and Labor Committee, for instance, reported such a bipartisan FEPC bill as early as 1945. But it was easily filibustered to death on the Senate floor. A similar bill was reported out of the same committee in the Republican-controlled Senate in 1948. But the fear of another filibuster prevented it from even reaching the floor.

In 1949 the persistent Truman again sponsored FEPC legislation and introduced it in the House, where it stood a somewhat greater chance of coming to a floor vote. That year the NAACP ranked a national FEPC law as its first priority, and the venerable civil rights organization led in the formation of an ad hoc, umbrella civil rights lobby, the National Emergency Civil Rights Mobilization. For the longer haul, the NAACP invited twenty national civil rights and labor groups to form the interracial Leadership Conference on Civil Rights. Funded mainly by organized labor, the Leadership Conference lobbied Congress from a permanent Washington headquarters. The House Education and Labor Committee in 1949 dutifully reported out the FEPC bill. But the conservative-dominated Rules Committee then bottled it up, as it was inclined to do with pro-labor and aid-to-education bills from that committee. So in a major push backed by the Truman administration, the House invoked a combination of the new "21-day rule" and the old "Calendar Wednesday" procedures to short-circuit the seniority gridlock and take the bill away from Rules. After an all-night debate on the House floor, a voluntary bill was passed by a roll call vote of 240 to 177. The exhausted House then sent it to the Senate, where it predictably died in another filibuster.[25]

Such demoralizing failures returned the initiative to President Truman.

This meant resorting to the weaker reed of executive orders, now encumbered by the Russell amendment, which the Labor Department interpreted as prohibiting the creation of an FEPC similar to Roosevelt's. So following the outbreak of the Korean crisis, Truman issued a brace of seven defense-related executive orders that he justified under the War Powers Act of 1941 and the Defense Production Act of 1950. Every order contained the policy pledge to nondiscrimination. But there remained no FEPC-style watchdog committee or enforcement machinery. On December 3, 1951, Truman by executive order created an interagency Government Contract Compliance Committee to revitalize the nondiscrimination clause in government contracts.[26]

Richard Hofstadter once observed that Lincoln's dry and legalistic Emancipation Proclamation possessed "all the moral grandeur of a bill of lading," and the straggling parade of fair employment committees during the Truman and Eisenhower administrations shared many of the same attributes. They were offered in response to a moral imperative, but they were designed to give minimal offense to a conservative-leaning Congress whose memories of Roosevelt's FEPC conjured up dark images of Reconstruction. Truman's Government Contract Compliance Committee was essentially a study committee with no enforcement powers itself, so it was not directly challenged in the Congress or in the courts. It could hold hearings but could not subpoena witnesses, could advise but not coerce contracting agencies. Its tiny staff consisted of five professionals and four clerks, and its expenses came from the regular budgets of the five large government agencies represented on the committee, especially Defense and Labor. In its single year of existence under a lame-duck administration, Truman's contracts committee could accomplish little more than a survey that clearly confirmed the need for a reinstatement of federal oversight in contractual nondiscrimination since the demise of the second FEPC in 1946.[27] Yet Truman had used his executive authority to reestablish at least a shadow of the old FEPC watchdog precedent over federal contracts, for all its relative toothlessness. If Congress would not allow the legislative establishment of such an agency, it was nevertheless unlikely to attempt to hamstring or destroy such a modest interdepartmental committee. It was also unlikely that a successor administration, even a more conservative, Republican one, would risk the appearance of backing away from the Roosevelt-Truman commitment to nondiscrimination in government contracts.

Eisenhower and the Nixon Committee

The election in 1952 of Dwight Eisenhower and a Republican Congress did indeed shift the FEPC debate in a slightly more conservative direction. But the Grand Old Party had watched three generations of Negro Republican loyalty disolve with alarming speed since the mid-1930s. In an attempt to stem this erosion, the GOP in 1944 had incorporated the FEPC plank into their party platform. But Eisenhower had demonstrated no previous interest in fair employment policy. Indeed, no major post in his illustrious career had inclined him toward the role of legislative leader or trained him in its arcane arts. He proposed no civil rights legislation during his first term, arguing that his predecessors had wasted much effort in pushing legislation that had nevertheless met defeat, and that he preferred firm executive action to the mere legislative appearance of action. Initiative in such sensitive matters was more effective when it was voluntary rather than compulsory, he said, and when it was state and local rather than national in origin. Following the Republican sweep in November of 1952, Truman's ten-man, lame-duck, contract compliance committee quickly began to disintegrate through resignations. In response, the Budget Bureau recommended that the new administration merely thank the remaining members and dismiss the lot of them. Eisenhower waited until the first session of the 82nd Congress had adjourned, and then created his own contract committee on August 13, 1953.[28]

Eisenhower's executive order abolished Truman's moribund Committee on Government Contract Compliance and replaced it with a new Government Contract Committee. The missing modifier, "Compliance," was telling. But the transposed jargon belied a few surprises. Eisenhower appointed as chairman of the new committee his Vice President, Richard M. Nixon, who had earned a progressive reputation on civil rights in California. Consistent with Eisenhower's emphasis on voluntarism and persuasion, the Vice President's leadership was designed to give the committee maximum prestige and authority. Joining Nixon on the bipartisan committee were such notables as Walter Reuther and George Meany, Congressman James Roosevelt of California, publisher Helen Rogers Reid of the New York *Herald Tribune,* businessmen John McCaffrey of International Harvester and Fred Lazarus of the American Retail Association, and Louisiana lawyer John Minor Wisdom. Joining the committee in October of 1953 to serve as its vice-chairman was Secretary of Labor James P. Mitchell, the "plumber" who joined Eisenhower's cabinet of millionaires, and who alone in the cabinet had publicly called for another FEPC.[29]

Eisenhower placed primary responsibility for compliance on the head of each contracting agency, as had Truman before him. But he charged the new nondiscrimination committee with developing systematic procedures for complaint processing and compliance review with 5,000 federal contracting and compliance officers in the twenty-seven contracting agencies. The President met personally with members of the new committee on August 19, and emphasized that he wanted a low profile and high accomplishments. At the committee's behest, Eisenhower signed another executive order a year later that required all contracting officers to ensure that every contractor posted public notice of the nondiscrimination obligation not only in hiring, but also in recruitment, upgrading, demotion, and transfer.[30]

Nevertheless, the new committee began work with a modest staff of nine and a budget of $125,000. It concentrated on educational meetings and conferences with business, labor, religious, and civil rights organizations, while the staff worked slowly through the cumbersome and often inconclusive conciliation process. In a series of self-denying ordinances, the committee ruled that it did *not* possess formal jurisdiction over labor unions, federal grants-in-aid, and federal programs in housing construction.[31] When the Warren Court's *Brown* decision of 1954 was followed by a show of surprising Republican strength among black voters in the 1956 election, Eisenhower upgraded Nixon's contract committee. An executive vice-chairman was created to manage a professional staff that by 1961 had grown to twenty-five, and small regional offices were established in Chicago, Los Angeles, Atlanta, and Dallas. Following the shock of *Sputnik* in 1957, committee chairman Nixon emphasized a policy of upgrading the positions of currently employed black workers, especially in scientific and technical fields. As the 1960 presidential election drew near, Nixon showed increasing evidence of publicly associating himself with the work of the committee. But overall the Eisenhower contracts committee represented a minimalist continuity with the Truman precedent on a very modest scale, with symbolic upgrading derived from the chairmanship of the Vice President and the vice-chairmanship of the Secretary of Labor.[32]

Similarly, Eisenhower transformed Truman's Fair Employment Board for federal jobs into his own President's Committee on Government Employment Policy. In replacing Truman's committees with his own, he had gotten rid of the words "Fair" and "Compliance" in their titles, which presumably smacked of Democratic heavy-handedness from Washington. In January 1955, Eisenhower signed an executive order on nondiscrimination in federal employment, but he issued no presidential statement to accompany the order and he made no mention of it the following day at his news conference.[33] The order removed the government employment committee from the Civil Service Commission and placed it in a direct reporting line to the President. Eisenhower wanted primary responsibility for proper personnel practices to be lodged firmly in the hands of the agency heads, and this standard arrangement clearly did not include a watchdog or ombudsman going over their heads direct to the President—at least not publicly.

In its final report in 1961 Eisenhower's federal employment committee claimed that during its six years of operation, only 225 of approximately 1,000 complaints had called for full investigation, and a mere thirty-three of these had required corrective action. Such a light traffic flow for an appellate board was consistent with the record of Truman's FEB. Eisenhower's committee compared federal employment in major cities in 1956 and 1960 and found modest gains in black hiring, especially in the lower grades. But continued black invisibility in the higher grades was highlighted by a striking irony: the major exception to the general pattern of black exclusion from white-collar federal jobs was in the post offices of *southern* cities. There the customary federal deference to local mores and social practices had in effect reserved a quota of Jim Crow clerical jobs for blacks only, much as segregated school systems had reserved jobs for black schoolteachers and principals.[34]

Until the back of southern segregation was broken, small federal advisory committees were not likely by themselves to make serious inroads in prevailing employment patterns in much of the decentralized federal bureaucracy, much less in the private sector nationwide. The Leadership Conference for Civil Rights, grown from twenty organizations in 1949 to fifty-two in 1952, and to more than seventy by 1960, was relentless in demanding nothing less than an FEPC law to create a national agency with mandatory powers. That meant, to the civil rights community, the statutory authority commonly possessed by the established regulatory agencies to issue "cease and desist" orders with judicial enforcement, should conciliation efforts fail. By 1960, seventeen states outside the South had adopted their own FEPC laws. Their experience was pointed to, although with conflicting claims and evidence, by the civil rights coalition as a minimal standard for the federal government to achieve, and as the quickest way to break the demonstrable economic subordination of America's minorities.

FEPC at the State House and City Hall: The NLRB Model

The grandfather of state FEPC laws was the New York "Law Against Discrimination," the Ives-Quinn bill passed in March 1945. It established an enforcing agency, the State Commission Against Discrimination (later renamed more positively as the State Commission for Human Rights). The political and legal background of the pioneering New York effort is especially revealing, because most of the practices outlawed by Ives-Quinn had long been against the law in New York anyway. Practices proscribed before 1945 included discrimination by race, creed, color, or national origin in jury service, in places of public accommodation and amusement, and in admission to the public schools and to tax-exempt nonsectarian schools. New York also banned such discrimination in the practice of law, access to public housing and relief, and membership in labor unions. Finally, New York laws barred job discrimination in public employment and the civil service, defense industries, utility companies, and in firms under contract to perform public works.[35] But in the absence of a specific enforcement structure, the formal law remained largely an abstraction unrelated to practice. Similar statutory and constitutional language was scattered throughout the country prior to World War II, including even some cities of the rim South. But they were seldom enforced with rigor anywhere. In the common law of tort, they placed the burden of proof on individuals claiming violation, who could bring suit in court and attempt to obtain damages. But the high cost in time and money, combined with poor odds and individual vulnerability to reprisal, effectively discouraged such suits.[36] When the war industries boom in New York brought with it charges from the Negro-Jewish civil rights coalition of widespread job discrimination against minorities, Governor Thomas Dewey appointed an investigating committee in March 1941 that ultimately led to the Ives-Quinn law of 1945. This created a commission to which individuals could bring complaints for both investigation and settlement.

The model for the new fair employment commissions, or state FEPCs, was the National Labor Relations Board. It was no accident that the first state FEPC emerged in New York, and was modeled on Senator Robert Wagner's creative instrument, the NLRB, which had been designed by the New Deal's liberal-labor coalition to remove labor law from the hostile courtrooms of the conservative judiciary. The new NLRB provided the attractive model of a single-mission administrative agency that could investigate complaints, hold hearings, negotiate and conciliate, make findings of fact, and issue orders that were enforceable in the courts. The NLRB was itself modeled after the Federal Trade Commission, which in turn derived its form and mission from the original regulatory mold of the Interstate Commerce Commission of 1887.[37] By the eve of World War II, Congress had proved willing to delegate immense powers to the "headless fourth branch of government"—the array of independent regulatory commissions and boards that was dominated by "Big Six" agencies like the ICC and FTC.[38] But Congress was willing to do so because the decision-making procedures of these quasi-judicial bodies were increasingly governed by trial-like due process, and were ultimately subject to judicial review.[39] As a result, federal appeals courts deferred increasingly to the regulatory commissions, whose officials were experts in the areas of regulation and the federal judges were not.

Administrative agencies such as the NLRB were designed to avoid the severe drawbacks of the criminal process. These included the traditionally wide discretion allowed for prosecutors, a heavy burden of proof beyond reasonable doubt, the delay and expense inherent in jury trial, and the punitive remedies of the criminal law. The NLRB was also designed by the New Dealers to avoid the traditional pattern of labor repression by conservative judges, and to create instead a third-party mechanism for identifying and eliminating unfair practices by employers.[40] The board members would be politically appointed for untenured terms, and the agency would develop expertise in employment practices and anti-discrimination remedies. In theory the board would act *in loco parentis* as guardian of the interests of victims of labor discrimination, yet it would be free to negotiate agreements with employers that would obviate the need to apply punitive sanctions against all but the most intransigent respondents.[41]

Thus the administrative mechanism and the judicial approach of New York's anti-discrimination law were new, but the legal theory of discrimination was not. Discrimination by race, color, creed, or national origin had historically been understood to mean deliberate conduct motivated by hostility and based on prejudice. Although the concept of discrimination in the law was inherently neutral, the concept of invidious discrimination had always been premised upon the *intent* to do harm to the individual. Because these basic distinctions had so long been settled and assumed in the common law of tort, neither the state fair employment practice laws nor the anti-discrimination orders of federal executives, nor indeed the various congressional bans on discrimination, had ever bothered specifically to define such discrimination. Rather, they merely specified certain discrete acts to be unlawful when based on race or creed, such as refusal to hire, promote, sell, or rent to, and the like.

In 1945, New York pioneered in developing the new machinery and approach. Procedurally, the commission would receive a complaint of discrimination and would then conduct an informal inquiry. If the investigation found probable cause to support the allegation, a conciliation process was triggered to attempt to resolve the complaint through confidential negotiations. That failing, however, the commission could repair to its more public and coercive authority by holding an open administrative hearing. Armed with subpoena power to compel evidence and call witnesses, the commission could then make findings of fact. This in turn could lead to a "cease and desist" order, backed by judicially enforced sanctions of fines and imprisonment.[42]

Cease-and-desist authority over private enterprise was the mailed fist that Roosevelt's second FEPC had ambitiously inferred from his Executive Order 9346 of 1943, and that in turn had hastened congressional destruction of the overreaching FEPC. Subsequent presidential FEP orders and committees had carefully avoided such a dangerous presumption of authority over the private sector without the statutory blessing of Congress. But the northern state FEPCs, following the NLRB model like pioneering New York, were endowed with such authority by their founding statutes. And because the threat was there, it almost never had to be used. The New York commission, which became a widely emulated model for state and municipal FEPCs, always emphasized the priority of education and conciliation, while holding its punitive sanctions in threatening reserve. The result in New York was a model fair employment law and an agency with punitive powers exercising responsible and firm enforcement practice. The model quickly spread throughout the urban-industrial northern tier of the country—first to nearby New Jersey (1945), Massachusetts (1946), and Connecticut (1947); then to New Mexico, Oregon, Rhode Island, and Washington (1949). By 1960 state FEP commissions had been added in Michigan, Minnesota, and Pennsylvania (1955); Colorado and Wisconsin (1957); California and Ohio (1959); and Delaware (1960).[43]

Several states adopted voluntary models first, without the coercive power of cease-and-desist authority—i.e., Indiana and Wisconsin in 1945, Colorado in 1951, Kansas in 1953, and Nevada and West Virginia in 1961. But by 1963 all of the more cautious, voluntary states except Nevada and West Virginia had switched to the mandatory model. The urban-industrial base of the civil rights forces also produced, not surprisingly, municipal FEPCs. Many of them arose predictably in the major cities of industrial states, such as Chicago, Cleveland, Philadelphia, and Pittsburgh. But a surprising number were in nonindustrial (including some southern) states where rural-dominated legislatures provided no state model, so the cities created one anyway.[44]

How, and how well, did the state and city FEPCs work? Most of the complaints charged racial discrimination and were directed mainly against employers.[45] They complained mostly about refusal to hire, being fired, and unequal working conditions. During its first twenty years, the New York

Human Relations Commission received a total of 8,973 complaints. The commission found probable cause to verify discrimination in 1,753 charges, or approximately 20 percent. In a typical year 60 percent of the complaints would be dismissed for lack of sufficient substance or, less often, jurisdiction. Most of the remainder would then be resolved either during the early stages of inquiry, or during "informal" investigation under the threat of public hearing and sanctions. As a result of this effective use of conciliation backed by a veiled threat, the New York commission's first public hearing was not held until 1949. Of the thirty-four hearings ultimately ordered during the commission's first two decades, only twelve were completed, and only *seven* culminated in cease-and-desist orders.[46] The restraint shown by the state and municipal FEPCs scarcely seemed to justify the fears of abuse that were invoked by the opposing coalition of conservative Republicans and southern Democrats, as they rather easily throttled the liberals' annual attempts to re-establish a national FEPC by statute. One result of this successful use of the threat of force by the state FEPCs was that court challenge and litigation was rare, and thus no comprehensive body of judicial interpretation was built up around challenged definitions and standards.

By 1960, when the Kennedy-Johnson ticket so narrowly edged out the Republicans, the civil rights coalition could point to a substantial body of experience to document the success of the mandatory FEPC approach at the state and local level. In these states and cities, both public and especially private employers had come to accept their FEPCs as a fact of modern political life. Although some employers had openly opposed FEPC at first, many had come to appreciate its enlargement of the work force. This phenomenon, after all, had traditionally pleased employers by increasing the supply and reducing the price (and also usually the bargaining power) of labor. Yet the spokesmen for organized labor also generally supported the concept of FEPC. Such a stance honored their liberal ideology, even at the considerable price of exacerbating strains with the generally unenthusiastic rank and file of blue-collar workers, and especially the all-white locals, northern as well as southern, within the more conservative craft unions of the AFL. On the whole, the the state and municipal commissioners of fair employment had functioned with responsible restraint, dismissing substantial numbers of frivolous complaints, and resolving the legitimate remainder without the glare of publicity. In their annual reports and periodic testimony before state legislative bodies, the chairs of these commissions stressed their educational function at the employer level over their complaint function at the employee level, and they laid special emphasis on the importance of voluntarism.[47]

Voluntarism especially appealed to the instincts of Vice President-elect Lyndon Johnson, whose political culture in Texas had long cherished a defiant spirit of entrepreneurial freedom. Johnson had been a member of Congress when FDR signed Executive Order 8802 in 1941, and he had watched as senior conservatives in the House hounded the FEPC into oblivion. Johnson's later mentor in the Senate, Richard Russell, had helped to consign it there. There was of course no FEPC in Texas or anywhere at the state level in the South, and in Washington, Johnson well knew the tenacity and power of congressional resistance to FEPC in particular and to coercion from Washington on race relations in general. During his 1948 race for the Senate, Johnson had mouthed the customary southern denunciations of FEPC, calling it "a farce and a sham—an effort to set up a police state in the guise of liberty."[48]

But by the late 1950s, Johnson had grown secure in his Senate tenure and in his majority leadership. Like many of his Senate colleagues, he hungered for the Presidency, and hence he needed to unshackle himself from his southern record of resistance to civil rights legislation. So Johnson achieved his own political triumph in civil rights by getting passed without a filibuster the Civil Rights Act of 1957. It was essentially a voting rights act, and as an exercise in adroit legislative leadership under difficult conditions, it was a virtuoso performance.[49] But the luster of what Senate Majority Leader Johnson called "the most important civil rights bill in history" had been quickly dimmed by embarrassing demonstrations of its inability to secure the vote for southern blacks.[50] For its enfranchising effects the 1957 law relied on the time-honored and time-consuming due process of individual court challenge, which the southern and conservative opposition had been in a position to demand. The dark specter of Reconstruction that the southerners saw in the bill's original Title III had rallied the Russell-led opposition, and they had stripped from Eisenhower's original bill the Attorney General's authority to file suit for court injunctions to protect blacks seeking to exercise their civil rights, especially the franchise.[51] Yet within two years of the act's celebrated passage, it was clear that individual litigation was not making a dent in the massive disfranchisement of southern blacks. With the presidential election of 1960 looming, another and rather patchwork voting rights bill had to be jimmied through the Congress in the spring of 1960, with Johnson in uneasy control of the Senate.[52] Once the juryrigged bill was passed, the NAACP's Thurgood Marshall acidly observed: "The Civil Rights Act of 1960 isn't worth the paper it's written on."[53]

Privately, there were few dissenters from Marshall's judgment on either side of the isle. Lyndon Johnson's reputation was that of a reasonably honest political broker with unsurpassed skill in political tactics and timing, although liberals in his party complained that he tended to compromise too early, and therefore somewhat to the right of center. Johnson had inherited from his roots in the depressed Hill Country of central Texas, however, a deep-seated resentment against the unfair handicaps that American poverty had visited so arbitrarily upon its victims. Virtually all who knew Lyndon Johnson well, including some who also feared and despised him, conceded that for all the Texan's earthy language, barnyard metaphors, and capacity for cruelty—from which no groups were spared their share of ridicule and ignominy—he seemed to have "an almost total lack of bias" toward blacks.[54] When the Kennedy-Johnson ticket was declared elected in November 1960,

President-elect Kennedy asked his new Vice President-elect to take the lead in drafting a presidential order that would launch the new administration's executive initiative in civil rights. It was a strange and novel assignment for a southern master of legislative strategy, who as Vice President would inherit the chairmanship of the executive branch's political equivalent of an FEPC.

PART ONE

Kennedy

CHAPTER I

Kennedy, Johnson, and the Presidency in Civil Rights

John F. Kennedy and the Executive Order Strategy

On March 6, 1961, John F. Kennedy signed Executive Order 10925, which established the President's Committee on Equal Employment Opportunity. It was a prodigious anticlimax. In the previous year of presidential electioneering, two ambitious young men—neither of them altogether credible in the giant shadow of Eisenhower—had postured over their ability to play tough with Khrushchev. But beyond the personal and partisan duel between Kennedy and Richard Nixon, the emotional core of the long campaign lay not in cold war abstactions and alleged missile gaps, but in the newly explosive issue of civil rights. The sit-in demonstrations of civil disobedience against racial segregation had begun spontaneously on January 31, 1960, in Greensboro, North Carolina, and had surged across the South.[1] As a presidential candidate Kennedy had endorsed the sit-ins and reached out strategically to the constituency of Martin Luther King, pledging that a Kennedy presidency would never betray the moral cause of civil rights.[2] But Kennedy's statements on the civil rights crisis ceased upon his election, and once inaugurated, his surge of priority messages to Congress summoned no action in the civil rights arena. Then on March 7 Kennedy at last broke silence, not by sending the customary legislative message to Congress, but by announcing that he had signed an executive order establishing a new interagency committee. A *committee*!

"Through this vastly strengthened machinery," the new President announced, "I have dedicated my administration to the cause of equal opportunity in employment by the government or its contractors."[3] From the perspective of a generation's remove from the Kennedy presidency, his short-lived equal employment committee is dimly recalled, if at all. At worst it is remembered as yet another futile gesture in the chain of Washington attempts to deal with the massive problem of racial discrimination through feckless committees. At best Kennedy's committee is recalled as a badly needed edu-

27

cational experience and training ground in civil rights problems and policy for its new chairman, and indeed its unlikely chief architect, Vice President Lyndon Johnson. In the long haul, it was to be much more than that. But in March 1961 it appeared to be little more than a sincere but weak gesture by a President who believed he could not get any significant civil rights legislation through the Congress, and who therefore was unwilling even to try, lest he roil the Congress and threaten his higher priorities.

Executive Order 10925 is itself recalled chiefly as the origin of the controversial phrase "affirmative action." Yet that subsequently contentious term was mentioned rather casually in the President's 4500–word decree, and then only once, together with such similar but less memorable phrases as "affirmative steps" and (twice) "positive measures." Moreover, the commitment to these unspecified positive efforts was invariably linked to the classic goal of nondiscrimination. "The contractor will not discriminate against any employee or applicant for employment because of race, creed, color, or national origin," Kennedy decreed. "The contractor will take affirmative action to ensure that applicants are employed, and that employees are treated during employment, without regard to their race, creed, color, or national origin."[4]

From its inception the notion of affirmative action in civil rights was ambiguous. Its positive obligations were undefined, although they seemed to imply more aggressive recruiting and training of minorities to broaden the pool for subsequent merit selection. But these positive measures were also tightly linked to the negative commandments of nondiscrimination: affirmative action was required to ensure that citizens were treated *without regard* to race, color, or creed. Yet the alliterative command to act affirmatively, and the tension implicit in it as a modifier for anti-discrimination, attracted little public notice at the time. What did draw attention was the breaking of the new administration's embarrassing policy silence on civil rights. The Associated Press wire story on Kennedy's presidential order of March 6 quoted Senate Majority Leader Mike Mansfield's rather lame but typically candid observation that "This explains, in part, why no reference was made to civil rights in the 16 priority matters drawn up for congressional action."[5] The United Press International wire story quoted Senate Minority Leader Everett Dirksen's complaint that the President should have asked Congress to give the committee statutory authority with its own appropriations and "not have to go around with a tin cup" to get its funds from various other government agencies.[6]

Mansfield's comment highlighted the dilemma of a new President whose negative coattail effect in the 1960 election had denied him a working program majority in the Congress (as distinct from a partisan majority). As a consequence Kennedy felt compelled, like Franklin Roosevelt before him, to defer to the powerful southern committee chairmen, who were able to hold the new administration's civil rights initiatives hostage to such higher Kennedy priorities as defense, tax, and trade legislation.[7] So that March the civil rights focus was on the gray world of executive orders—on the President's authority to act when Congress had not, would not, or might not. Dirksen's jab at Kennedy's order reflected the traditional opposition of congressional conservatives to executive intrusion into the legislative domain, and to circumvention by the White House of congressional control of appropriations. As Senate Republican leader under Eisenhower, Dirksen had resisted proposals for a fair employment committee during the pre-election negotiations over the civil rights bill of 1959–60. His junior Senate colleague, John Kennedy, had then shown little interest in the matter. Kennedy was not a lawyer, he had cultivated strategic ties to the southern wing of his party, and as a leading student of Kennedy's civil rights record observed, "Kennedy himself knew few Negroes and had no reputation as a civil rights advocate."[8] Vice President Nixon, however, was a lawyer who had chaired the Eisenhower administration's fair employment committee and who enjoyed a favorable relationship with national civil rights leaders. As the election of 1960 approached, both men shifted to the left, as did their parties, in their bids for support from the civil rights constituency.

The 1960 Presidential Campaign and the Civil Rights Legacy

During the early maneuverings of the 1960 campaign, Vice President Nixon had remained loyal to the Eisenhower administration's opposition to a statutory FEPC. Nixon had been content to echo the standard Republican claims of fealty to the Lincoln tradition by sponsoring in 1957 the first successful civil rights bill in 87 years, including its creation of the Civil Rights Commission and the Civil Rights Division in the Department of Justice. But by 1960 Nixon had every reason to press for a more aggressive stance on civil rights. Although Nixon was a card-carrying member of the NAACP, his low reputation among white liberals, which stemmed primarily from his earlier association with the House Committee on UnAmerican Activities and especially the Alger Hiss case, has obscured the relatively high regard in which he was held by black civil rights leaders.[9] He had been praised by Martin Luther King for helping pass the Civil Rights Act of 1957, and by Roy Wilkins for supporting cloture reform in the Senate.[10] In the relative obscurity of his contracts committee chairmanship, Nixon had persuaded Eisenhower to broaden the committee's scope and budget as the election approached. Biographer Stephen Ambrose affirmed that Nixon's record as Vice President on racial relations was excellent; he was "consistent in his denunciation of Jim Crow, as much so as any of his rivals for national leadership and much more so than most."[11]

During the spring of 1960 the pre-convention flanking operations by Nelson Rockefeller on Nixon's left and Barry Goldwater on his right seemed to bracket a safely centrist position on civil rights for Nixon within the Republican party, one that would accord well both with the Eisenhower legacy and with Nixon's own moderate instincts. In *The Making of the President 1960*, Theodore White puzzled over why Nixon rejected the cautious original plank on civil rights offered by the Republican platform committee for the Chicago nominating convention, and instead insisted that it be rewritten to please Rockefeller and match the aggressively liberal Democrats.[12] The rather surprising result was a Republican fair employment plank calling for "legislation to establish a commission on equal job opportunity to make permanent and expand with legislative backing the excellent work being performed by the President's Committee on Government Contracts."[13]

As a bid for the black and liberal vote, however, this poorly matched the Democrats, whose platform committee, under the leadership of freshman congressman and former Connecticut governor Chester Bowles, had drafted the most liberal civil rights plank in the party's history. The Los Angeles platform had approved the sit-in demonstrations and demanded equal access to "voting booths, schoolrooms, jobs, housing, and public facilities." It called for empowering the Attorney General to file civil rights suits on behalf of individuals (the old Title III nemesis that conservatives from both parties, with the cooperation of Senate majority leader Lyndon Johnson, had purged from the civil rights laws of 1957 and 1960). It also called for requiring all southern school districts to submit desegregation plans by 1963, and it invoked the old Roosevelt lexicon in summoning Congress to create a permanent "Fair Employment Practices Commission."[14]

Thus in 1960 both presidential parties had positioned themselves considerably to the left of their congressional counterparts in civil rights matters, and the ensuing campaign sustained the presidential bidding war until the victor the following January should confront the old congressional reality. In the second television debate in October, Kennedy attacked Nixon's contract committee for having brought only two successful actions in seven years to end employer discrimination under federal contracts. Nixon countered by urging Congress to give his committee statutory authority—a move that Senator Kennedy, in his need for southern Democratic votes, was unwilling to match. Most listeners and voters had surely never heard of the contracts committee. But the code words in civil rights were less obscure, especially to Republicans seeking to exploit Kennedy's relatively weak civil rights record and to emphasize his reliance on the Jim Crow wing of his party. So Nixon replied to Kennedy in their debate by questioning the commitment and fitness of the Texan whom Senator Kennedy had selected as his running-mate—"a man who had voted against most of these proposals," Nixon said, "and a man who opposes them at the present time."[15]

Vice President Johnson and the Civil Rights Portfolio

When Kennedy's presidential margin of victory turned out to be a tiny plurality of 112,803 votes out of more than 68 million cast, and produced a net loss in the Congress of two seats in the Senate and twenty in the House, the new political arithmetic pointed toward a backpedaling strategy that contradicted the rhetoric of the campaign. The promised civil rights initiatives in Congress would now have to be downplayed so as not to shatter the delicate

Democratic majorities on the Hill. Most political observers quickly understood this awkward reality. But Kennedy spokesmen would not formally admit that their party was so deeply divided at its congressional core, and the Kennedy administration suffered intense embarrassment for two years over its caution in civil rights policy.[16]

To the delight of the Republicans, Kennedy as President could not bring himself to exercise what he had promised as candidate—a mere "stroke of the pen" to ban racial discrimination in federally assisted housing—until after the fall elections of 1962. On their weekly "Ev and Charlie Show," Senator Dirksen and House Republican Leader Charles Halleck made a staple item of Democratic hypocrisy on civil rights. The New Frontier's maiden legislative agenda "to get America moving again" would thus contain no significant civil rights proposals, and it would avoid splitting Senate Democrats through a fight over Rule 22, the cloture requirement for shutting off filibusters. The Kennedy White House would, to be sure, work with Speaker Sam Rayburn to reshuffle the House Rules Committee so the administration's domestic legislation would stand a better chance of reaching the House floor. But this was social legislation designed to help the disadvantaged indirectly, not civil rights legislation per se. Kennedy announced just before Christmas, following a strategy session with top legislative lieutenants at Palm Beach, a five-point program of domestic legislation designed to carry out the Democrats' main campaign promises. It included medical care for the aged through Social Security, federal aid to education, an expanded federal housing program, an increased minimum wage, and redevelopment of economically depressed areas. But the legislative program was silent on civil rights.[17]

The price of avoiding congressional paralysis over civil rights legislation was an executive initiative to fill the vacuum, and during the campaign Kennedy had repeatedly called for "Executive action on a bold and large scale" to deal with civil rights issues. Specifically, he had pledged that within the first few days of his presidency he would reorganize the "do-nothing" Committee on Governmemt Contracts. He would also order all federal agencies to take "positive action" to eliminate discrimination in federally assisted housing. But if a rule-reform fight over Senate filibustering would alienate southern Democrats, Kennedy was early persuaded that a housing discrimination ban would also upset many *northern* Democrats in Congress. He therefore decided to concentrate on job discrimination through executive initiatives and announced on December 20 that Vice President-elect Johnson would chair two major committees in the new administration. One was the new National Aeronautics and Space Council. This required congressional approval, since Congress in the post-*Sputnik* scramble of 1958 had created a National Space Council to be chaired—unwisely, Kennedy thought—by the President himself. The second vice-presidential assignment was to chair the contracts committee. Because this was an inherited executive assignment, it required no congressional approval. Moreover, Kennedy could scarcely do less than appoint his Vice President to chair the equal employment committee that Nixon had chaired. For Vice President Johnson, however, such an appointment was fraught with peril. Johnson well knew and feared the miseries of the vice presidency, which historicallly had seemed to institutionalize either impotent sycophancy or, rarely but more dangerously, disloyalty fed by frustration and ambition. Unlike Nixon, Johnson was not a lawyer and had virtually no experience in the area of executive administration and contract and labor law. His vast experience fell precisely in the crucial area where Kennedy was most in need of help, in the legislative strategy and tactics of the Hill. Fearing the whiplash of the racial issue, with himself caught in the middle as head of a new quasi-FEPC, Johnson attempted twice during the post-election transition to fashion for himself a novel role of astonishing boldness—or gall.

Johnson first tried to create a de facto post as Democratic majority leader of the new administration. He persuaded his old Senate colleague and newly elected majority leader, Mike Mansfield of Montana, to propose to the Senate Democratic Caucus on January 3, 1961, that the new Vice President be invited to attend and preside over the caucus. But Mansfield's motion generated a storm of resentment in the caucus at the prospect of a lingering Johnsonian presence in the Senate chambers. Spirited constitutional objections were raised as well to this novel violation of the doctrine of separation of powers. After an intense debate in caucus, the motion formally carried by an apparently substantial margin of 46 to 17. But the strength of the dissent in the collegial Senate, given its deep tradition of deference to courtesy and consensus, was such that Johnson had in effect received a sharp rebuff for unwanted intrusion. He therefore withdrew in considerable humiliation.[18]

Johnson's second attempt to carve out a role carrying unique authority involved a draft executive order that one of his aides allegedly drew up at Johnson's request, and that the Vice President-elect approved and audaciously sent to Kennedy for his signature. According to Johnson aide George Reedy, this extraordinary document "actually proposed that President Kennedy sign a letter which would virtually turn over the national defense establishment and the exploration of space to his vice president."[19] Rumors of Johnson's power-grasping proposal appeared in the newspapers, and columnists Rowland Evans and Robert Novak reported that White House aides close to Robert Kennedy had leaked the memo to the press to embarrass Johnson.[20] The phantom memorandum was buried in a charitable silence from the Oval Office.

Reports of both of these extraordinary incidents circulated widely throughout Washington, and the twin humiliations early sharpened the pain and isolation of Johnson's vice presidency. In their aftermath, a bemused President Kennedy continued to treat Lyndon Johnson with deference, and Johnson reciprocated with a subdued but apparently genuine respect.[21] The episode appears to have considerably reinforced, however, the mutual suspicion and contempt between Johnson and Robert Kennedy, who had fought so hard to block Johnson's vice presidential nomination. In its aftermath, Johnson was pulled inward to concentrate in his profound unhappiness on the opportunities and dangers of his immediate new jurisdiction, which involved the tricky new terrain of executive authority and anti-discrimination.

The Collective Authorship of the "Affirmative Action" Order

Hobart Taylor, Jr., who was to become special counsel to the equal employment committee that Kennedy was planning to create and Johnson was going to chair, recalled Johnson's summoning him to Washington from his law practice in Detroit in December 1960 to ask his help in designing and setting up the new committee—"Because I don't know anything about it," Johnson said.[22] Taylor was a successful black lawyer with a law degree from the University of Michigan (where he was editor of the *Law Review*) and a master's degree in economics from Howard University. His roots, however, were deep in LBJ country, with a Prairie View State baccalaureate, a childhood in Houston, and a prosperous father whose business and political dealings with Lyndon Johnson extended back a quarter of a century.[23] Taylor remembered the new Vice President's giving him the famous "Johnson Treatment" by crying woe over this extraordinary new burden that President Kennedy had shunted upon him: "I don't have any budget, and I don't have any power, I don't have anything," Johnson complained. And Kennedy had replied, Johnson said, "You've got to do it because Nixon had it before, even though he didn't do anything; you're from the South, and if you don't take it, you'll be deemed to have evaded your responsibility. And so you've got to do it."[24]

As Taylor reconstructed his role, Johnson handed him a draft of the executive order and asked him to stay overnight and work it over. "I rewrote the darned thing and I was in there the next morning," Taylor recalled, "and in there were Abe Fortas and Arthur Goldberg. . . . They agreed roughly with what I was doing and made a suggestion or two. Then we went in to see [Johnson], and everybody agreed on it, and then he asked me to go up and get a draft typed because it was all handwritten." Taylor concluded with the *coup de maître*: "I went up to Abe Fortas' office, and I did it. For whatever it means to posterity, all of this talk about affirmative action, I put the word "affirmative" in there at that time."[25]

Taylor's claims to creative authorship notwithstanding, phrases like "positive effort" and "affirmative program" had become common currency in the lexicon of civil rights by 1960, especially among liberal Democrats. Kennedy's presidential campaign rhetoric had been peppered with these kinds of allusions to more positive obligations. Their meaning was commonly understood to signify a more aggressive strategy for seeking out minority applicants, as opposed to the "mere" passive nondiscrimination that liberal Democrats ascribed to the Eisenhower administration and to the Republican equal employment committees of the 1950s. But the term "affirmative action" was no invention of Hobart Taylor's. It traced its lineage at least as far back as the Wagner Act of 1935. There it was used to define the obligation and authority of the National Labor Relations Board to redress an unfair labor practice by ordering the offending party "to cease and desist from such unfair labor practice, and to take such *affirmative action*, including reinstatement of employees with or without back pay, as will effectuate the policies of this Act [emphasis added]."[26]

Using the Wagner Act and the NLRB as a model, New York's pioneering "Law Against Discrimination" of 1945 had incorporated large hunks of the Wagner language, including the relief clause that linked the commission's cease-and-desist authority to "affirmative action" for victims of discriminatory acts.[27] Thus from the beginning the concept of affirmative action was somewhat open-ended. On the one hand it was closely linked to the specific requirements of the quasi-judicial process: administrative hearings, a finding of fact identifying a victim of a discriminatory act with intent to harm, a cease-and-desist order, and "make-whole" relief in the nature of rehiring and back pay. In the history of the state FEPCs since World War II, this was the only practical or legal meaning that the phrase could claim. On the other hand, it held the potential for a broader reading in the future, perhaps to imply special efforts by government to compensate for a history of anti-union activity by employers with the support of hostile federal judges, or for a history of job discrimination against minorities.

By 1960, the liberal Democratic critique of merely passive anti-discrimination reflected a growing impatience in civil rights policy with the federal government's voluntary model of the 1950s, where weak engines of enforcement were externally driven by the filing of complaints by third parties claiming injury. The New Deal had provided unions with new protections policed by the NLRB. But the wartime FEPC was dead. In the North and West this had led to a growing list of state FEPCs. But in the southern states, aggrieved blacks could turn only to federal judges, whose local roots combined with weak and cumbersome statutes to offer little hope for substantial change. In the liberal indictment, this had been the perceived fatal flaw in the voting-rights approach of the Civil Rights Act of 1957.[28] It had relied on the traditional model of due process, with individual citizens petitioning for voting rights, claiming inability to register because of racial discrimination, and bringing suit in court, with all of the attendant procedural safeguards, delays, and expense. This is why the impatient liberals had pressed so hard and so unsuccessfully in 1957 and again in 1960 for a Title III to authorize intervention by the Justice Department. This authority symbolized for liberals the aggressive and even righteous intercession of enlightened Washington authority, in the form of the Attorney General, filing suit throughout the benighted southern hinterland on behalf of individuals who were either too poor or intimidated or perhaps even too unenlightened to do it for themselves. This is also why the civil rights leaders insisted at the minimum, in voting rights, on federal *referees* to police local registrars. But they clearly preferred federal *registrars,* who would go down South and sign up the disfranchised black voters themselves.

Nevertheless, nowhere within the liberal establishment circa 1960 was affirmative action interpreted to mean a special preference or compensatory treatment for minorities that would not be equally available to all citizens. The most strident public demands of the Americans for Democratic Action, the NAACP, the National Urban League, the Congress of Racial Equality, the Southern Christian Leadership Conference, the Political Action Committee of the AFL-CIO—indeed of the entire spectrum captured under the umbrella of the Leadership Conferernce for Civil Rights—called consistently for racially neutral anti-discrimination. The civil rights lobby, with its roots deep in Jewish philanthropy and social protest as well as black resistance, remained extremely wary of such historically unpleasant notions as racial or ethnic quotas, and job and admissions applications that required minority group identification.

The leaders of the new Kennedy administration reflected an even more cautious sense of this mission. There was wide agreement among them that in light of their political and programmatic weaknesses on the Hill, any civil rights initiative should aim toward a spectrum of authority that was fairly narrowly bracketed. Specifically, it should try to reach as far beyond the Eisenhower precedent as tactical prudence would allow, but without tempting a backlash from Congress by straying too far into the political minefields of a national FEPC, cease-and-desist orders, and fears of the regulatory entanglements and illiberal nightmares that could flow from a kind of national civil rights police.

Given these political constraints, soon after the November election Johnson had instinctively turned for advice to his close circle of advisers. Memphis attorney Abe Fortas has generally been credited with the leading role in drafting Kennedy's new executive order.[29] But what is most important about the authorship of Executive Order 10925 is that it was collective. It was constructed upon the base of an accumulated model that had evolved layer-like over the preceding two decades. Working sensibly from this inherited model, the President directed his Vice President to work with the Justice Department in proposing a draft, and Johnson, who sensed considerable danger in the charge, in turn sought a broad range of private advice. According to Reedy, it was characteristic of "the worrywart aspects of the LBJ personality" to require a "fantastic amount of checking, crosschecking, rechecking and double checking." "He was a man who could be decisive to the point of rashness once committed to a course of action," Reedy said. "But when any policy statement was under preparation, he would lash himself into a state of hysteria. In this instance, his normal fears were brought to a fever pitch by his distrust of Bobby Kennedy, who, in his mind, was laying traps all over Washington."[30]

At the formal levels of the administration, the Justice Department would necessarily be involved in drafting any presidential executive order, and Deputy Attorney General Nicholas Katzenbach worked with Fortas and Johnson aide Bill Moyers on drafting appropriate legal language and format. Arthur Goldberg, as the country's most prominent labor lawyer and the new Secretary of Labor, also played a significant role. Not surprisingly, the executive order that emerged from this collective authorship in early March shared many of the aesthetic attributes of the camel designed by a committee. It looked basically familiar, but it was unusually long and detailed. It contained no striking surprises, but rather represented incremental growth and compromise. The process of drafting the order, however, forced the new administration to confront political problems that would confirm many of Johnson's fears about the dangers it posed to his presidential ambitions and to the aspirations of his President and party.

Thorns of Jurisdiction: Labor Unions and Federal Grants

The Kennedy transition team, together with Vice President Johnson and his coterie of advisers, had early agreed that political realities required them to avoid any appearance of creating a federal FEPC by executive order.[31] This left a strengthened contracts committee as the only politically feasible model. The advisers also agreed that Eisenhower's separate committees on government contracts and federal employment should be combined. But Eisenhower's Republican administration had been able to avoid a question that the Kennedy Democrats could not: should the new committee's jurisdiction also cover the unions who worked for the contractors, but who were not themselves direct parties to the federal contracts? Eisenhower's executive order had been silent on this, and Nixon's contract committee had excluded unions from the beginning. By 1960, however, the problem of racial discrimination within organized labor was so widely recognized that unions could no longer reasonably be excluded from coverage—especially not by a new Democratic regime with such strong ties to the AFL-CIO. Moreover, the leadership of organized labor had reason to want the unions to be included. The weak federal structure of the AFL-CIO gave its leaders few methods for disciplining the independent locals, and the continued defiance of racially discriminating unions embarrassed and weakened the labor movement. On Nixon's committee under Eisenhower, George Meany and Walter Reuther had been important symbols of labor's commitment to racial fairness, at least at the top. The Kennedy administration could do no less, and was politically and morally obliged to do more.

Thus the necessity of including unions in the new committee's fair employment jurisdiction was conceded early during the construction of the executive order. But what enforcement leverage could the committee use against union discrimination? Government contracts bound the contractors, but not the unions representing their workers. The contractors, however, were also bound by their union contracts to honor union seniority in hiring and promotions. This placed contractors in a double bind, especially in the construction industry, where racial discrimination in international unions was most pronounced—outside the obvious Jim Crow unions of the South. In construction, the skilled craft locals of the building trades enjoyed de facto closed shops with a virtual monopoly over hiring hall referrals. This meant that if a builder under a union contract needed thirty electricians and twenty plumbers to construct a new post office, he took whatever workers the craft locals sent him. But any government restriction that threatened to punish the constractor could not easily reach the unions that determined his labor supply.

This system was deeply embedded in law as well as in practice. The Taft-

Hartley Act of 1947 had recognized and protected the seniority system in union procedures. Although the law had extended NLRB jurisdiction to unfair labor practices by unions, racial discrimination as an unfair union practice was expressly rejected by Congress when it passed Taft-Hartley. As a result the Kennedy administration had strong reasons to include unions in the executive order coverage. But the administration remained divided over the PCEEO's enforcement leverage on unions. Johnson insisted that the committee's sanctions must apply equally to labor as well as to management. This included the executive order's proposed Section 312 on penalties, which provided for recommendations to the Justice Department to seek court injunctions against noncomplying employers or "employee organizations." Because the prospect of court injunctions touched a raw historical nerve with organized labor, Secretary Goldberg objected. Richard Goodwin on behalf of the White House then queried the Vice President's office whether "others" couldn't be substituted for "employee organizations."[32] Johnson, backed by Fortas, refused to name management but not labor in the provisions for sanctions and penalties. He eventually compromised by accepting Katzenbach's suggestion that any "organizations, individuals, or groups" be substituted for employers and employee organizations—thereby inviting broad targets (including perhaps White Citizen Councils) for court injunctions while specifying none. But the dispute signaled the shoals ahead.

An even more complicated and politically dangerous area of fair employment jurisdiction was the huge yet obscure realm of federal grants-in-aid to state and local governments and associated organizations. The policies governing federal grants were both technically murky and politically hypersensitive, especially on the Hill where such programs were popular. Straight government contracts for supplies or services fell unambiguously within the government's procurement authority—for example, Washington could specify the performance requirements for a new battle tank and enforce them on the winning (not necessarily the lowest) bidder. But it was quite another matter to stretch this authority to cover the vast array of federal grants to state and local governments, which subsidized the construction and operation of hospitals and airports and highway systems, and especially sensitive neighborhood enterprises like schools and housing.

This question was suddenly sharpened for the new administration on January 11, 1961, when Vice President Nixon presented President Eisenhower with his contract committee's final report. It contained a broadside of politically explosive proposals for equal employment policy, now safely defused for lame-duck Republicans. They included extending the committee's jurisdiction to cover *all* federal grant programs, including federally subsidized housing programs and agreements under which the federal government contributed funds to state and local programs.[33] The report also recommended that Congress give the committee permanent statutory status—as the committee had proposed in its annual reports since 1959, and as Congress had thrice rejected.

The hostility of Congress was not surprising. Giving jurisdiction over federal grant programs to the the President's fair employment committee would potentially entangle it with virtually every popular benefit program that Congress had bestowed upon its grateful and well-entrenched beneficiaries since the Morrill Act established the land-grant colleges in 1862. The popular grant-in-aid model had grown to include the agricultural experiment and extension network; the major partnership programs for constructing water systems, highways, airports, and hospitals; and also the public health and medical research initiatives against heart and lung disease and cancer. Added to this was the growing array of organized client groups who were aided by the modern social welfare state—i.e., disabled soldiers and sailors, crippled children, the blind, the aged, dependent children and their mothers, and the mentally and physically disabled. The literature of political science and public administration has marveled at the power of the "iron triangles" that these constituent groups have formed over the years with the congressional subcommittees that authorized their grant programs and the federal agencies that administered them. Modern presidents have learned to challenge only at their peril these entitlement programs and the political alliances that protect and nurture them.[34] To the New Frontiersmen in the winter of 1961, it appeared suicidal for Kennedy's new anti-discrimination committee to threaten to ensnarl these symbiotic networks in the red tape of FEPC-type investigations and public-hearing forays into the intimate contractual relationships, all this on the questionable strength of presumed executive authority alone. The Kennedy-Johnson transition team never seriously considered including grants in the new executive order. In the wisdom of St. Matthew, sufficient unto the day was the evil thereof. The prickly issue of federal grants would have to wait for aother day. And it would have to find Congress in a radically different mood.

Constructing a "Vastly Strengthened Machinery"

Once past the inauguration on January 20, the Kennedy administration collectively shaped the new executive order into a consensual model—except for the unhappy Vice President.[35] Lyndon Johnson was anxious to replace the President as head of the National Space Council. But he was equally anxious to get himself replaced as head of the equal employment committee. He had Bill Moyers press Deputy Attorney General Nicholas Katzenbach for clarification of the Vice President's constitutional role, especially his unique legislative duties derived from presiding over the Senate. Johnson argued that these responsibilities should preclude the Vice President from directing the large enforcement effort that the administration's new commitment to equal employment opportunity would require. In early February, Johnson had his staff work up several versions of a letter to "My dear Mr. President," explaining that the Vice President had neither the facilities nor the staff to run such an executive operation. Indeed, the attachment to the vice presidency of the "large, costly, nationally-dispersed staff" necessary to transform the existing "showcase" agency into an effective instrument of enforcement would amount to a "perversion" of the vice presidency's constitutional function.[36] For these reasons Johnson recommended that he ought to be removed from the committee and replaced as chairman by the Secretary of Labor. The committee's need for enforcement staff could be provided by the Wage and Hour and Public Contracts Division of the Labor Department, he argued, with the director of the Bureau of the Budget added to the committee to beef up its muscle in light of the constitutionally prudent withdrawal of the Vice President.[37]

But Kennedy would not yield.[38] Johnson's constitutional doctrine of the vice presidency was conveniently flexible, being broadly construed for the space council, and narrowly for the equal employment committee. Even George Reedy chided as "a very weak point" Johnson's lame suggestion that "the Vice President had less time than the Secretary of Labor to exercise day to day supervision over the Committee."[39] The best Johnson could do was to have the order specify that the Secretary of Labor as vice chairman would have "general supervision and direction of the work of the Committee and of the execution and implementation" of its policies and purposes, and that the President would also designate an executive vice chairman to administer the committee staff between meetings.[40]

If Johnson could not avoid chairing the new committee, however, he was determined to direct an operation that was at least armed with some reasonable measure of responsibility and enforcement power—certainly more than that possessed by his predecessor, Richard Nixon. So first, he recommended to the President that the employment and contracts committees of the Eisenhower administration be collasped into a new committee called the President's Committee on Equal Employment Opportunity (PCEEO).[41] He further proposed that the standard nondiscrimination clause in government contracts "be revised to impose not merely the negative obligation of avoiding discrimination but the *affirmative duty* to employ applicants and to treat employees on the job on the basis of their qualifications and not because of race, creed, color or national origin [emphasis added]." Johnson also asked that contract cancellation and denial, which was only implicit in the Eisenhower order, be stipulated as a PCEEO sanction (no federal contract had ever been terminated for such a cause), and that the committee resurvey all federal agencies for an up-to-date equal employment inventory. If he was stuck with the job, Johnson wanted the tools to make it work.

By March 3, the constant whipsaw of contending suggestions and arguments was fraying the nerves of press aide George Reedy, who wrote Johnson that they should quit playing with the executive order. Reedy complained that further tinkering would be "in the realm of nit-picking," and that the latest suggestion to add women and the aged as newly protected classes for the committee's jurisdiction would "throw the committee into complete chaos and you would be charged with sabotaging the Committee."[42] Reedy did not specify the origin of these suggestions, but it is not surprising that the combination of delay and rumors surrounding the new order would prompt the national civil rights organizations to lobby for a maximal jurisdiction, using state-level precedent as a model. By 1961 several states had added age as a category newly protected from discrimination by their FEPCs. None, however, yet included sex—other than in state equal pay laws, which were prompted mainly by fears that low wages for women would undercut the earnings of men and hinder union organizing.[43]

By the beginning of March President Kennedy's patience with Johnson's crosschecking and rechecking was exhausted, and the compromise wording over labor injunctions allowed the President to pry the order out of Johnson's nervous clutches and sign it on March 6. The following day Kennedy called a press conference to issue Executive Order 10925 with the announcement that its new "vastly strengthened machinery" would grant "in specific terms, sanctions sweeping enough to ensure compliance." "I have dedicated my administration," Kennedy said, "to the cause of equal opportunity in employment by the government or its contractors. The Vice President, the Secretary of Labor, and the other members of this committee share my dedication." "I have no doubt," Kennedy said, "that the vigorous enforcement of this order will mean the end of such discrimination."[44]

The Legacy of Nondiscrimination and the Logic of Affirmative Action

As a substitute for civil rights legislation, Executive Order 10925 was presented by the White House as a forceful departure from the timid Eisenhower precedents. On its face, the 4000-word document was four times longer and far more detailed than Eisenhower's original order of 1953, and the front-page news stories and editorials that it generated emphasized the Kennedy administration's more aggressive approach. The news stories reported that the new watchdog committee, the PCEEO, combined Eisenhower's separate committees for government employment and contracts, that the order designated the main agency heads themselves as members, and that it included labor unions in its coverage. The order further provided for the Attorney General to bring suit for court injunctions against contractors, subcontractors, or employee organizations to enforce compliance, and even provided for Certificates of Merit to reward full compliance. It spoke early and repeatedly, though vaguely, of positive obligations and measures, of affirmative steps and cooperation and action.

But the rhetoric belied central elements of continuity with the 20-year fair employment legacy and the structure upon which it was built. The PCEEO was still an interagency committee without statutory authorization or budget. It thus had only an advisory relationship to the twenty-seven federal agencies where the final hiring and contractual responsibilities lay. Much of what was hailed as new only appeared so through the elaborate specificity of the package and the public relations efforts of the White House, which needed to deflect attention from its own legislative silence. The President's committee would be chaired by the Vice President, with the Secretary of Labor as vice chairman, and with an appointed executive vice chairman overseeing a modest staff that was funded through an assessment formula prorated by agency. But that had been true under Eisenhower. The new PCEEO was empowered to establish rules of procedure, receive complaints, hold hearings, receive and review reports, and sponsor conferences. It was directed to survey the federal agencies to identify employment patterns and trends. All this, however, had also been true not only under Eisenhower but under Truman as well. The new order stipulated that compliance could ultimately lead to contract cancellation or denial, and it directed the agencies to designate compliance officers. But the power to deny or cancel contracts was implicit in all federal contracting agencies, which necessarily employed contract officers to oversee compliance with the terms of their contracts. Similarly, the Attorney General's ability to file suits was implicit in the Justice Department's traditional responsibility to enforce the government's contractual terms. The Certificate of Merit was a legitimately novel incentive in this connection. But at best the certificates added a certain patina of tinsel, like points for good behavior or gold stars for regular Sunday School attendance.

What was new and auspicious about Executive Order 10925 and the PCEEO it created was, first, the concerted effort to place the full prestige of the Presidency behind the moral imperative of nondiscrimination. The presidential decrees of previous administrations had been brief and mostly technical. Indeed they had been almost apologetic or defensive, often signed into the *Federal Register* with no accompanying statement at all, and then promulgated with a cautious presidential eye cocked toward an unenthusiastic or hostile Congress. President Eisenhower had never disguised his reservations about racial desegregation as a moral imperative. But Kennedy's full press conference of March 7 launched a heavily staged public relations effort, and creating the Johnson committee was the first major shot in a campaign that was to climax so unexpectedly in the historic Civil Rights Act of 1964.

The second crucial attribute of Executive Order 10925 is customarily associated with its vague and almost casual reference to "affirmative action." Although in the long run this aspect of Kennedy's order proved to be of great significance, in hindsight it is laced with the irony of unintended consequences. Like all its predecessors, Kennedy's order was grounded in the standard model of criminal and civil jurisprudence in Anglo-American law, in which the defendant was regarded as innocent until proven guilty, and the burden of proof fell on the plaintiff and the prosecution. Also like its predecessors and like the state fair employment laws as well, Kennedy's order made no attempt to define the harmful discrimination it proscribed. Invidious discrimination was defined by convention as an intentional denial to minorities, prompted by prejudiced motives, of rights freely accorded to nonminorities. Proof of discrimination required plaintiffs to produce convincing evidence of intent after the alleged discriminatory act had occurred, even though this often required complainants to adduce proof that was uniquely possessed by the powerful organizations whose motives and behavior were the subject of the complaint. Given the dominance of the standard model, with innocence assumed absent proof of guilt, the type of activity left available for "affirmative action" was constricted in time and variety. Its meaning seemed self-defined to require more aggressive recruitment in hiring, and special training for minorities to encourage their advancement. All evidence points to these presumptions in the framing of the executive order.

A presumed state of *innocence* in criminal or civil law, however, was not the same thing as a state of *compliance* in contract law. Federal contractors had long been required to meet certain stipulations to be eligible to bid for and hold contracts. The range of these requirements was modest and mainly technical, as in minimal standards of worker health and safety, or paying the "prevailing" (meaning union-level) wage in federally funded construction. Requiring contractors to specify in advance what special recruiting efforts they might make seemed consistent with this tradition. The addition of such requirements, however, could imply a dual and potentially profound modification of the standard model. First, in fair employment matters, such requirements could shift the burden of proof away from complainants and toward the employers. Second, to the degree that the burden of proof was shifted from the complainant claiming violation toward the employer seeking to be "in compliance," the *timing* of proof would correspondingly shift from *subsequent* to a specific charge of discrimination, to *prior to* a general and even hypothetical one. Thus an employer bidding for a contract could be obligated to submit prior evidence that he was already "in compliance" as a prerequisite for future contract work. This, in turn, would reasonably require a suitable *instrument*: the Compliance Report.

Kennedy's order, while innocent of any foreknowledge or plan that contemplated any such radical transformations in the model of civil rights enforcement, nevertheless routinely provided for some systematic accumulation of employment data upon which to base assessments of compliance patterns. The order specified the submission of a Compliance Report by "bidders or prospective contractors or subcontractors," to be filed "prior to or as an initial part of their bid or negotiation of a contract."[45] But how would this crucial instrument be judged? According to what criteria might one pass or fail?

The order did not say, and the question seems not to have been raised. Affirmative action to do what? The order simply answered: "to ensure that applicants are employed, and that employees are treated during employment, *without regard* to their race, creed, color, or national origin [emphasis added]." That was classic nondiscrimination, however couched in positive new rhetoric. And classic nondiscrimination was itself a radical proposition in 1961, especially in the segregated South. That was of course the perceived heart of the civil rights problem in 1961—the South. Federal government offices throughout the region had long deferred to local customs of racial segregation and subordination in their employment patterns. U.S. military bases had historically spread throughout the South because the land was plentiful and cheap and the weather was warm. The seniority system in Congress had dis-

proportionately rewarded the Democratic South with committee chairmanships, and government installations had flowed southward accordingly. When
southern Jim Crow was combined with such nationwide but industry-specific
strongholds of job discrimination as the construction trades, then one was
dealing with brazen patterns of racial discrimination that could potentially
be isolated and ganged up on. There seemed little need for juridically novel
formulas about discriminatory intent or the burden of proof to attack such
blatant discrimination. Instead, what one mainly needed was courage and
resolve to enforce liberalism's vintage formula of nondiscrimination.

Lyndon Johnson and the Hotseat of Equal Employment

Throughout the transition period, Lyndon Johnson had necessarily steeped
himself in the complicated law and tricky politics of executive anti-discrimination. A master of consensus politics on the Hill, his instinct for wellsprings
of compromise was like a divining rod—which pointed violently away from
the specter of FEPC. On the other hand, Lyndon Johnson was determined to
be elected President in his own right, and to do that he knew he had to shuck
off the unsavory reputation he had earned in his ascent through the reactionary labyrinths of Texas politics during the postwar decade. In his biography
of Johnson's congressional years, Ronnie Dugger called this period of Johnson's Red-baiting and reaction after 1948 the "most cynical period of his
career," when "his radicalism [was] suppressed or well hidden."[46] Johnson's
biographers disagree profoundly over the depth and principled quality of his
Hill Country populism.[47] But virtually all who knew him conceded that, like
Huey Long, he seemed oddly free of the racial prejudice that was a hallmark
of his native region.[48] George Reedy, the verbose Irishman who grew up with
robust ethnic hostilities in the Chicago area, and who was scornful of Johnson's crudeness and capacity for cruelty, was puzzled by the Texan's racial
fair-mindedness: "Strangely enough, Mr. Johnson is one of the least prejudiced or biased or intolerant or bigoted men I have ever met," Reedy observed. "I don't believe there is any racial prejudice in him whatsoever."[49]
Jerry Holleman, a union leader from Texas who was appointed Assistant
Secretary of Labor, and was then appointed by Johnson as the committee's
new executive vice chairman, recalled that while Johnson was an "extremely
miserable man" as Vice President, he nevertheless met with Holleman frequently and drove him hard because the Vice President "took this task rather
seriously."[50] Despite Johnson's best efforts to avoid the task, from March 6
onward he was signed on for the duration. He quickly coined a blunt new
motto: *"We Mean Business."*

The Politics of Interagency Committees

To give Johnson's committee maximum visibility and evidence of commitment, Kennedy had stipulated membership for the heads of nine major agen-

cies (twelve when including the three armed service secretaries). For the new administration this meant cabinet secretaries Arthur J. Goldberg of Labor, Luther M. Hodges of Commerce, Robert S. MacNamara of Defense, Abraham Ribicoff of HEW, and Attorney General Robert F. Kennedy of Justice. On President Kennedy's insistence the Budget Director was dropped, but the committee included chairman John W. Macy of the Civil Service Commission, and administrators John L. Moore of the General Services Administration, Glenn T. Seaborg of the AEC, and James S. Webb of NASA. The three service secretaries, who were added at Johnsons' insistence despite President Kennedy's preference for appointing only the Defense Secretary, were John B. Connally of the Navy, Elvis J. Stahr of the Army, and Eugene Zuckert of the Air Force. Add to this Vice President Johnson as the committee's chairman, and one had an interagency committee of notables that rivaled the Cabinet itself. And that was part of the problem—it might work as badly as had the modern Cabinet, and for some of the same reasons.

Most modern presidents have begun their administrations by announcing their star Cabinet appointments and pledging a true Cabinet government of collective teamwork (though this was less true of Kennedy than most of his predecessors or his successors). Yet agency heads are intensively concerned with their own agenda rather than with the broader concerns of the President. They tend therefore to "marry the natives," immersing themselves in constituency politics and congressional committee relations. They tend also to have little in common with each other that would assist the President in collective policy deliberations, and to some degree they become natural antagonists of the President's budget priorities.[51] As new administrations mature, Cabinet meetings typically are called less frequently, and their agendas grow more contrived. But because Cabinet officers meet so conspicuously with the President, the meetings are generally well attended. Not so, however, with lesser committee assignments, where busy Cabinet secretaries tend to send deputies, and the growing roll-call of surrogates tends to reflect declining momentum and clout.

As administrative instruments for coordinating executive policy, interagency committees have earned a poor reputation. Much like the Cabinet, they suffer from member loyalty that flows primarily outward to the constituent agencies, not inward toward the common goal to be coordinated. Unlike the Cabinet, however, the chairing authority of interagency committees lacks the President's unifying command and power to reward. Such normal problems, moreover, were magnified when the purpose of the committee was to police nondiscrimination in employment. This policeman's role conflicted with a primary purpose of the mission agencies, which was to procure services and products for the government as quickly and cheaply as possible. It also conflicted with the back-scratching mutuality of the clientele-agency networks. Given these patterns of reward and loyalty, the federal government's army of procurement contract officers had little incentive to police their own agencies aggressively. As Jacob Seidenberg, a senior staff administrator in contract compliance under Truman and Eisenhower, complained, the government members of the compliance committees "must be surrogates of their agency's interests."[52] Seidenberg's preferred remedy was a clean one: to exclude *all* government members from the committee, thereby making it small and cohesive and its membership entirely "public" (meaning nongovernmental)—like Roosevelt's original FEPC, which had five commissioners, or like the newer Civil Rights Commission, which had six. But the amendment sponsored by Senator Richard Russell of Georgia in 1944, which had guaranteed the demise of Roosevelt's FEPC, had in effect decreed that all subsequent executive entities involved with job discrimination *must* be interdepartmental committees, funded in an ad hoc fashion out of the small contingency funds of participating agencies. So there were compelling reasons for the government roster of PCEEO members to be long and of senior rank.

Slightly different forces drove the "public" membership in the same direction, making the list of public members as long as the government roster, and only slightly less impressive in name recognition. First, the public members on the predecessor Eisenhower committee had outnumbered the government members, ten to six. It wouldn't do to have the Kennedy administration reverse the ratio, and swamp the representatives of business and labor and the public interest in general with senior administrative bureaucrats. Second, and more important, the politics of categorical representation, which sought a microcosm to mirror the larger society, had an escalating inner logic. Thus Eisenhower's contracts committee had internally balanced the public members, with its labor leaders not surprisingly being Democrats and its business leaders being Republicans. This eased the continuity of the Kennedy transition by making three of the hold-over public members attractive for reappointment. Two of these were the dominant figures in organized labor, George Meany and Walter Reuther. As presidents respectively of the AF of L and the CIO, they had originally been appointed in 1953 before the labor merger of 1955, and both were Democrats. The third was businessman Fred Lazarus, whose reputation for diligence as well as his Republican credentials strongly argued for his continuance under Kennedy.

To achieve religious balance, Johnson recommended and Kennedy approved a tripartite slate of a Protestant (Rev. Francis E. Sayre, dean of the National [Episcopal] Cathedral), a Catholic (Monsignor George G. Higgins, director of the National Catholic Welfare Conference), and a Jew (Rabbi Jacob J. Weinstein of Chicago). Also categorically appointed were a black (John H. Wheeler, president of the Mechanics and Farmers Bank of Durham, North Carolina), and a woman (Mary Lasker of New York's Lasker Foundation). Hispanics were not yet very visible nationally in 1961, and a representative for this minority constituency was not seriously discussed. The remaining businessmen were balanced geographically, with three southerners and three northerners.[53] This familiar process of representative accumulation led President Kennedy to appoint fourteen public members to join the fourteen government members.[54]

When the new committee first met, on April 11, 1961, in the Cabinet Room of the White House, the impressive assemblage was joined by President Kennedy. Also present were the committee's "consulting counsel," Abe Fortas; its new special counsel, Hobart Taylor; and special assistant to the chairman, George Reedy.[55] Together they almost outnumbered the committee's total support staff, which totaled forty—thirty-one of them in Washington, five in the regional office established in Chicago in 1957, and four in the Los Angeles regional office established in 1958.[56] This spartan level of staff support represented the combined staff of the two collapsed Eisenhower committees, and Johnson began by holding it at that level, with a first-year budget of only $425,000.[57] His executive vice chairman, Holleman, later claimed that Johnson had made a private deal with senior southern senators, especially Senator Russell, not to increase the committee's budget. Holleman believed that Johnson had received in return a temporary nonagression pact, and that he may also have agreed not to hold the kind of on-site public hearings that had embroiled Roosevelt's FEPC in Alabama during World War II.[58] At the committee's first meeting Johnson organized the members into five working subcommittees to cover training and apprenticeship, vocational education, promotion and upgrading, franchised industries, and religious cooperation. In his notes of preparation for the kickoff meeting, Reedy explained to Johnson that the "really important subcommittee is the one on enfranchised industries." He reasoned that the customary threat of contract cancellation was not a very useful or believable sanction when dealing with such regulated monopolies as gas companies and electric utilities, who could scarcely be disciplined by shutting off power to consumers.[59]

But Reedy's prediction was wrong. What turned out instead to be the PCEEO's only significant subcommittee was neither listed nor contemplated on the first agenda, and its chairman was not even among the fifteen members originally assigned to the five subcommittees. During the next two years public attention would focus on a controversial new program called "Plans for Progress," the special project of an enigmatic businessman from Atlanta named Robert Troutman. The dispute over Plans for Progress came to symbolize the split between hardliners and moderates in civil rights enforcement, and it caught Lyndon Johnson unhappily in the middle. It seemed to spin quite inadvertently off of an early flap over the Lockheed-Marietta plant in Georgia, and its strange evolution illustrates many of the volatile social and personal dynamics that conditioned the committee's brief and turbulent career.

CHAPTER II

Lyndon Johnson, Robert Kennedy, and the Politics of Enforcement

The Problems and Lessons of Lockheed-Marietta

Three days before President Kennedy signed his executive order on job discrimination, the Pentagon announced the nation's first billion-dollar procurement for a single military system. It was a ten-year contract negotiated with Lockheed Aircraft Corporation to build the giant new C-141 jet transport. The plane would be built at Lockheed's sprawling plant at Marietta, Georgia, near Atlanta, where one roof covered 76 acres of plant and four and a half miles of corridors. Lockheed's plant was built with federal funds on federally owned land adjacent to an airbase. There Lockheed had built the sturdy C-130 Hercules, and its $100 million payroll was the largest in Georgia. Lockheed's headquarters was in Burbank, California, but the company's strategic move into Georgia had been nurtured by Senator Russell and Representative Carl Vinson from the chairmanships of their armed forces committees. In 1961 the Marietta work force of 10,500 was still segregated by race, and its 450 black workers were confined largely to unskilled jobs.[1] The few semi-skilled workers among the blacks were segregated in an all-black local of the International Association of Machinists (IAM), where they were excluded from the apprenticeship programs under the control of the IAM's all-white local.[2] The IAM itself had been organized in Atlanta in 1888, and had formally barred Negro membership from the beginning. The race bar was removed by the international union in 1948, but IAM locals throughout the country, and particularly in the South, continued to practice racial segregation. At Lockheed-Marietta in 1961, the washrooms and drinking fountains bore "White" and "Colored" signs, as did the segregated cafeterias.

On April 7, the day Kennedy's executive order officially took effect, the NAACP's labor secretary, Herbert Hill, submitted to Lyndon Johnson's new committee 32 affadavits of complaint from black workers at Lockheed-Marietta. Hill, who was white, had a reputation for compensating for his whiteness with militancy. He had filed previous complaints against Lockheed with

Vice President Nixon's committee, but with little visibility and no successful outcome. This time he picked a moment of high visibility and professed commitment from the new Kennedy administration, and extracted from Jerry Holleman a public pledge to cancel the contract of *any* employer who refused to comply with the President's new ban on discrimination.[3] Prompt follow-up was promised by the committee's new staff director, John G. Feild, who had been prominent among John Kennedy's early supporters in Michigan.

In appointing Feild, whose roots were in the Michigan labor movement, Holleman was responding to the urgings of Michigan's two liberal Democratic senators, Pat McNamara and Philip Hart, whose ties to the AFL-CIO and to the Labor Department were close. But McNamara and Hart had never been close to Lyndon Johnson in the Senate, and the Michigan delegation had revolted against Johnson's vice-presidential selection in Los Angeles.[4] Neither Holleman nor Feild were men Johnson would likely have chosen to manage his new committee's major operations, had he been free from the political constraints that normally surround such staffing questions. Holleman had formally supported Johnson in Texas politics, but his labor forces remained solidly in the camp of Senator Ralph Yarborough, and their relationship with Johnson had never been warm. Feild's associations in Michigan's Democratic party slightly stained him with the taint of anti-Johnson credentials, although not critically enough to create problems at the beginning. Feild was, after all, a Kennedy man, and this was indeed President Kennedy's committee. Feild was also an experienced and well-connected member of the civil rights community. A former staff director of the Michigan FEPC, he was also a past president of the National Association of Intergroup Relations Officials (NAIRO). Like Herbert Hill, Feild was was a white liberal who believed in the aggressive use of sanctions in civil rights enforcement. According to George Reedy, Feild wanted to accelerate the growth of new career opportunities for the network of compliance enforcement officers represented by NAIRO.[5]

In response to Hill's complaints about Lockheed-Marietta, Feild flew to Burbank to negotiate with Lockheed's president, Courtlandt S. Gross. Within organized labor, parallel pressures to desegregate were applied from Washington by the AFL-CIO leadership, particularly IAM president A.J. Hayes, to Marietta's District 33, and especially to its all-white local. As a result, the black local voted on April 23 to integrate. When the white local made no move to accept the black machinists, Hayes ordered the district locals desegregated anyway. Supporting pressure also came from a new and previously unlikely quarter: the Pentagon. Secretary McNamara had brought Adam Yarmolinsky over from the White House as a special assistant whose troubleshooting portfolio included civil rights. Yarmolinsky, one of Kennedy's talent hunters who also hailed from Harvard, enjoyed the support of the White House as well as the strong backing of McNamara and his deputy secretary, Roswell Gilpatrick. The result was the creation in the Pentagon of a powerful new source of policy and staff support for the Johnson committee's challenge.[6] Feild was quick to acknowledge the crucial early support and close staff relationship that flowed from such senior backing at the Defense Department.

In response to these pressures Lockheed promptly removed all the "White" and "Colored" signs from the washrooms, and replaced the water fountains with paper cup taps (which according to the *New York Times* required 63,000 paper cups a day). Lockheed-Marietta also abandoned the Jim Crow cafeterias in favor of mobile, stand-up canteens. Lockheed's president Gross flew to Marietta and began long-range employment negotiations with the PCEEO staff over accelerated hiring, upgrading, and access to apprenticeships for blacks.[7] So far it looked easy.

The suddenness of the apparent turnaround at Lockheed-Marietta suggested several immediate lessons. One was that aggressive pressure by the NAACP appeared to be crucial to accelerating the momentum of Kennedy's executive order. Hill became adept at exploiting press coverage of the new committee's compliance efforts. He labeled as merely "peripheral" Lockheed's dismantling of the Jim Crow signs at Marietta, demanded immediate and substantial payoff in new black jobs, and broadened his attack to include segregated Western Electric plants in Winston-Salem, Greensboro, and Dallas.[8] Hill's barrage of complaints acted as a goad to the Johnson committee, at once accelerating its enforcement vigor while at the same time endowing it with a certain patina of judicial even-handedness in refereeing the difficult transition from Jim Crow to nondiscrimination. In the metaphor of precinct psychology, Hill's bad cop could play nicely to the committee's good cop—especially when the suspect party, a major Pentagon supplier, was so ready to reform.

A second lesson was suggested by the surprising ease with which such major national firms were willing to abandon their traditional deference to the South's biracial system, especially if the stakes and visibility were high and their competition was obliged to act similarly. Lockheed seemed particularly amenable to such pressures. With 90 percent of the company's business coming from the Air Force, and with employment at Marietta down by almost half since the peak of the mid-fifties, Gross seemed inclined to welcome the committee overtures. Besides, the nation's business and labor leaders liked the attention they were getting from and indeed *at* the White House. President Kennedy and the Johnson committee had staged a big kickoff meeting with the country's fifty largest defense contractors on May 2, and they repeated the showcase ceremonies with the leading labor leaders the following day. On May 25, Lockheed's Gross joined President Kennedy, Vice President Johnson, and Secretary Goldberg at the White House to sign what John Feild referred to as "a plan for progress," and which Johnson said might become the model or pattern for the entire defense industry.

Robert Troutman and "Plans for Progress"

Robert B. Troutman, Jr., was an ebullient southern lawyer and businessman whose reputation as a progressive entrepreneur of the "New South" in his native Atlanta had offered regional balance to the Johnson committee. Namesake son of a wealthy senior partner in the prestigious law firm of Spalding, Sibley, Troutman & Kelley, he had graduated Phi Beta Kappa from the University of Georgia in 1939 and went on to Harvard Law School, where he was a schoolmate (but not a classmate) of Joseph Kennedy, Jr. Rejected for wartime service because of a kidney ailment, he had worked during the war as an assistant counsel for Senator Walter George of Georgia. After the war Troutman joined his father's law firm in Atlanta, but he branched out into numerous business ventures, from food vending machines to shopping centers. A prominent civic booster even by Atlanta standards, Troutman was an astonishingly energetic man, rising daily at 4:30 a.m. By all accounts Bobby Troutman was also a clever and amusing man, richly endowed with southern charm, given to leaving the impression that he could best be reached in a pinch through Senator Russell's office, or Senator Talmadge's, or Congressman Vinson's, or later, perhaps even the Oval Office . . . .

Troutman had been host in 1957 to a politically exploratory reception in Atlanta in honor of young Senator John Kennedy, and in 1960 he had been one of the few Georgians actively working for Kennedy's presidential drive. This of course placed him in opposition to the Johnson forces during the primary campaign, both in Georgia and at the convention in Los Angeles. As a Harvard man, Kennedy loyalist, and a southern business professional who openly championed the Atlanta formula of progressivism in racial relations—of being "too busy to hate"—Troutman appeared to be a natural selection for Johnson's committee. But Troutman's mercurial qualities seemed always to distort and also magnify his real connections.[9] Harris Wofford, who loosely carried the civil rights portfolio among the early Kennedy aides, recalls Troutman's effective mode of operation in the White House: ". . . you had this funny, wild Georgian named Bobby Troutman, who the civil rights people all thought was a Johnson man and Johnson had never forgiven him for being a Kennedy man in the South."[10] But, Wofford, observed, Troutman "really wasn't a Johnson man."

> He was just an operation, and he operated primarily on the leverage of the fact that he'd go in, amble into Evelyn Lincoln's office and either she'd let him or occasionally he'd slip in to see the President then. He either entertained the President—and the President did enjoy him coming in, which is possible because he's a very funny fellow—or he just had the gall to keep going in. He conveyed a very close relationship to the President; I don't think it was very close at all, but I think he probably entertained the President. And with that leverage he just sort of ran with these Plans for Progress to the consternation of the civil rights people . . . .

As the civil rights people's point man in the White House, Wofford was in a strategic position to observe the growing consternation.

On May 24, the day before Lockheed's president was scheduled to sign the first of the Plans for Progress with President Kennedy at the White House, Troutman moved to claim his new project by proposing that a special subcommittee of the PCEEO be appointed under his leadership to develop the Plans for Progress program and pattern it after the Lockheed model.[11] When the Lockheed contract came under attack from the NAACP earlier that spring, Troutman had moved quickly to explore solutions. He had never denied taking part in the intensive lobbying effort mounted by Georgia's political and business community to win the lucrative C-141 contract in the first place. As Secretary Goldberg later explained, "It was down in his baliwick, so we let him handle it."[12] Hobart Taylor, who tended to defend Troutman, nevertheless believed that protecting Lockheed was one of Troutman's main motives. "I know that Bobby was motivated very largely because he didn't want that Lockheed plant stopped in Georgia," Taylor said. "It was the biggest thing in the state."[13] Troutman proposed that he be joined on the Plans for Progress subcommittee by Ribicoff and Wheeler. Johnson obliged by appointing Troutman chairman. But Johnson then cautiously broadened the subcommittee to include Ribicoff, Lazarus, Reuther, and Kaiser. On July 12, eight more major defense contractors joined Lockheed by signing Plans for Progress agreements at a public White House ceremony attended by Kennedy, Johnson, and Goldberg. Troutman was lining them up and signing them up at a crisp pace.

By August, Troutman was boasting of "startling" success, with nine of the nation's largest defense contractors signed up, and presumably covering three quarters of a million workers. In a memorandum to the parent committee outlining the objectives and assumptions of Plans for Progress, Troutman emphasized that the agreements were strictly voluntary. They were based on "surveys *the companies made themselves* without any compulsion whatever."[14] He reminded his fellow members that the committee's role was not to "create a vast Federal bureaucracy" but rather "to open the doors of job opportunity—not to become an employment agency or a policeman with a nightstick chasing down alleged malefactors." As Troutman explained his vision of the committee's overarching purpose and proper approach, there were four principal assumptions undergirding it:

(a) The Committee does NOT expect anyone to be hired for a job *because* of his race, creed, color, or national origin. But the Committee *does* seek to achieve a situation in which *qualified* people *will not be refused a job because of race, creed, color, or national origin.*

(b) Individual industries have individual problems which are based upon past experience, conditions of work and geography. Patient and understanding negotiation on the individual problems will do a lot more to achieve progress than any "crackdown."

(c) There is an amazing reservoir of goodwill that can be tapped in the employment field. However people may feel about the civil rights issue generally, most Americans will agree that *qualified* men and women are entitled to an opportunity to compete for jobs on a basis where the only tests are skill, willingness to work, and a desire to conduct themselves with decency.

(d) Where there are individual cases of injustice, the Committee, of course, will act to correct them as best it can. But the Committee does NOT intend to be a body of suspicious snoops ferreting out so-called evil doers. *The test of progress is the extent to which people are willing to open the doors to equal employment opportunity.*[15]

"Compulsion is not the thing," Troutman insisted. "I'm a lawyer. I can show you how to get around the Executive Order. It's got to be voluntary."[16] In his tireless campaign for his voluntary compact of decency and opportunity, Troutman moved the subcommittee's operation to Atlanta, installed it in rooms next to his own law office in the Peachtree Building, and paid the staff and startup expenses out of his own pocket.

Troutman's notions of voluntarism, however, were anathema to John Feild and his colleagues in NAIRO. They were equal-employment professionals who envisioned an aggressive national FEPC, wielding cease-and-desist authority unapologetically to right historic wrongs. Troutman's campaign had stolen the spotlight, but it was unclear what was meant and required by Troutman's impressive sign-ups, and how their progress would be measured. When General Electric and six other large defense contractors joined Lockheed by signing up in July in another big White House splash, Peter Braestrup reported in the *New York Times* that the plans seemed rather vague. All employers signed a pledge to observe strict nondiscrimination in hiring and dismissals. They also promised to step up recruitment of qualified Negroes, including visiting Negro colleges to interview for salaried positions. Beyond that, "several of the agreements," Braestrup observed, "contained a clause specifying that 'no specific numerical targets' would be set, but providing that the committee's evaluation of company progress would be made partly in training and occupying responsible positions."[17]

Companies signing Plans for Progress would be rewarded with special treatment, in the spirit of the section of Kennedy's executive order that provided for Certificates of Merit. They would fill out special forms designed by Troutman, rather than the PCEEO's standard reporting and compliance form. On November 24, Braestrup reported in the *Times* that 12 more leading defense contractors had agreed to sign Troutman plans at another White House ceremony. This would bring Plans for Progress coverage to 21 firms employing 860,000 workers. But Troutman's tour de force, Braestrup wrote, was stirring up a backlash among the regular committee staff and, through them, among the broader civil rights constituency. Braestrup also reported on November 24, after interviewing John Feild, that Feild expected the committee at its next meeting to approve a new compliance reporting system. It would be required of *all* contractors (holding contracts of $50,000 or more and employing 50 or more employees) and their first-tier subcontractors, *regardless* of whether they had signed any Plans for Progress.

Feild's new plan would mean that 15,000 contractors, involving 38,000 companies, 50,000 plants, and 15 million workers, would be required to submit by April 1, 1962 and annually thereafter a four-page "manpower profile" report to show the percentage of Negroes in relation to other employees in various job classifications.[18] Yet the committee's new Form 40, which was dated December 1961, stated in its instructions that "The filing of 'Plans for Progress' reports (Budget Bureau No. 44–R1174) will be accepted as a compliance report in lieu of this report form." Not surprisingly, rumors began circulating among the civil rights community that companies signing Plans for Progress agreements with Troutman would in effect be able to design and enforce their own compliance programs. Troutman himself had drawn the fire of the NAACP in early November, when he gave a banquet for 1300 white guests in honor of Senator Russell and Representative Vinson in a traditionally segregated hotel in downtown Atlanta. There local black and KKK pickets angered the guest speaker, Secretary McNamara, and embarrassed Troutman, who had pledged that the dinner would not be segregrated.[19]

By December 1961 the growing tension between the Troutman operation and John Feild's Washington staff was vented in the full committee meeting. Troutman, somewhat on the defensive, complained about lack of staff and logistical support and vowed to continue his operation by continuing to use his own personal funds and hope for later reimbursement. He explained that the data he had collected from the first nine companies to sign up, which were to be used to document the progress achieved, "were now of little use, being outmoded by the new forms."[20] When Troutman then proposed a further expansion of his program, Johnson forced him to admit that he had not first cleared his proposal with his own five-member subcommittee. Then Johnson, sensing danger, appointed a special watchdog subcommittee of himself, Robert Kennedy, and Arthur Goldberg to review any policy proposals about Plans for Progress and recommend what to do about them. In early January 1962, Holleman sent to White House aide Ralph Dungan a confidential, 30-page report from Feild. It reviewed in detail the seven-month joint effort between Feild and his 33-member committee staff in Washington, and Troutman and his four-member staff in Atlanta, to negotiate Plans for Progress with 54 major contractors, which thus far had resulted in twenty-one signed agreements.[21] Feild's memo reflected resentment at the public credit that Troutman and the Atlanta operation had received at the expense of the regular committee staff. He noted dryly that the "voluntary" aspects of "these high-minded Plans for Progress" had been "hardly voluntary," since the companies were clearly obliged to comply as federal contractors. Noting also that yet another public signing was scheduled in the White House for mid-January, Feild added: "It should be stated candidly that prolonged scheduling of ceremonial signings in the White House have within them [sic] the danger of public suspicion that these statements are mere words."

Crisis and Reorganization

During the first half of 1962 the Troutman-Feild split over voluntarism versus compulsion widened into a public exchange of charges. Johnson, caught in the middle, was a conciliator by nature who had always insisted that both approaches were necessary to induce agreement. He distrusted Troutman's soft data and resented the cultivated aura of Kennedy blessing that cushioned the Georgian's irritating independence. But he also resented Feild's apparent first loyalty to the more militant civil rights community, where anti-Troutman leaks about the committee's dissension began to surface. Feild had early adopted a pattern of going around the formal reporting structure that led through Holleman to Johnson, and cultivating sympathetic allies on the committee—most notably with the Attorney General.[22] Troutman, on the other hand, was after all a committee member overseeing a committee-endorsed program, and Feild as staff director owed loyalty to committee policy. Besides, the administration was committed to the success of Plans for Progress through the showcase White House ceremonies. So early in 1962 Johnson released what meager success reports he could get from Troutman.

The progress reports claimed that the plan signed with Lockheed had increased the company's black employment by 26 percent in just six months, and that the first six plan-signers of the previous summer had reported that "more than 2,000 new jobs were filled by non-whites, twice the number that would have gone to them under the old hiring ratio."[23] The key phrase in Troutman's system of self-reporting was "would have gone to them." Troutman labeled as "Expectable" this rather vaguely estimated number of jobs for nonwhites. He then subtracted it from reported real employment gains to generate his results. Troutman's boosting the percentages produced a growing skepticism in the senior councils of the Justice Department. Burke Marshall, head of the Civil Rights Division, complained about Troutman's inflated "expectables": "They kept using percentages in situations where percentages were a fraud. You talk about a 100 percent when you were building on the basis of one and you meant two instead of one."[24] Troutman's energy and natural salesmanship, however, were generating formidable momentum. On February 7, thirty-one more defense contractors signed plans at the White House, including such prominent firms as General Motors, Chrysler, Ford, Pan American, IBM, Goodyear, IT&T—bringing to fifty-two the number of firms participating in Plans for Progress.

In New York, the NAACP's Herbert Hill counterattacked. In response to the White House signing ceremony, he called a press conference to complain that the Johnson committee had been "powerless to effect changes" in the lowly status of black workers in the textile industry, public utilities, railroads, and the construction trades. He called instead for the committee's replacement by a new five-member FEPC. This had just been proposed in a bill sponsored by California congressman James Roosevelt and favorably reported out of the House Education and Labor Committee, which was chaired by Harlem congressman Adam Clayton Powell.[25] From the Republican flank,

Governor Rockefeller attacked the Kennedy administration's "record of broken promises" on civil rights as "one of the most cynical exploitations of minority aspirations in the history of American politics."[26]

In an attempt to quell the rumbling, Vice President Johnson in mid-February invited Roy Wilkins and Herbert Hill to his Senate office, where he was joined by Goldberg, Holleman, and Feild, but not Troutman. In a discussion that lasted two and a half hours, Wilkins and Hill argued that some major companies regarded signing the plans as guarantees of immunity against penalties for job discrimination. They complained that "symbolic breakthroughs" involving the hiring of a handful of white-collar minorities were unimpressive in the face of a black unemployment rate that was more than double the white rate. They pointed to an accelerating evaporation of semi-skilled blue-collar jobs in manufacturing, mining, and agriculture, where Negroes had gained footholds only to see automation decimate the work force.[27] Wilkins emerged from the meeting to urge continued critical support of the PCEEO, but Hill was not so easily mollified.

This dialogue was repeated in early April 1962, when Johnson released the committee's first-year report. Hill charged that the administration's fear of southern conservatives in Congress had "resulted in more publicity than progress." The "symbolic" Plans for Progress were "simply a euphemism for what a previous administration called 'voluntary compliance'," Hill said. He then announced that the NAACP was beginning to concentrate its anti-bias charges against such firms as General Electric, General Motors, and Western Electric *after* they had signed the celebrated plans—often in the expectation, Hill charged, of securing immunity from real compliance.[28]

In May, Troutman sent a letter to the President, the Vice President, and the Secretary of Labor to enclose his first-year report and to offer a novel proposal. The report contained job data only from the first six firms that had signed their plans the previous July, and the data covered only their first six months under the plans. But the report's claims of 1,820 new jobs for non-whites, as "compared with an expectable total (as per previous ratios) of 728—a net increase of 1,092 jobs," had already been released the previous February.[29] In his cover letter Troutman observed that he had worked largely alone, devoting "most of a year of my time and office without compensation (and gladly so)," making 29 trips to Washington, personally advancing almost all expenses of Plans for Progress, and requesting that he be reimbursed for $41,221 of his total expenditures (of almost $75,000) because "this total sum has become a matter of considerable personal importance to me." Finally, he proposed that a Council of 35 business leaders be formed to fund and run Plans for Progress separately in the future.

In mid-June, Peter Braestrup reported in the *New York Times* that the symbolic schism between Troutman's voluntary program and Feild's compulsory one had badly split the PCEEO. Braestrup interviewed Troutman, who explained that he wanted to kick off his new and independent council with a Washington dinner for top executives from 100 firms, to be addressed by Kennedy, Johnson, and Goldberg. He would become the group's executive secretary. It would then expand to 150 companies, including pace-setting firms that were not government contractors and hence were beyond the constraining jurisdiction of the PCEEO. He had proved that Plans for Progress could produce real jobs for qualified Negroes quickly, Troutman said—"How many jobs have the 'do-gooders' and 'talk-gooders' produced in fifty years?"[30] The major problem for industry, Troutman explained, was that "Negroes are not adequately qualified to rise to positions of dignity."[31] A passionate missionary to his fellow businessmen, Bobby Troutman insisted that the Negro must be given a stake in the free-enterprise system. An infusion of well-paid Negro white-collar workers, he reasoned, would have a calming effect on Negro communities that are too commonly dominated by extremist clergymen. He preached his evangelical message not only to his fellow businessmen, but also to the few black leaders who were willing to give him an audience, such as the NAACP's Roy Wilkins.[32] Troutman preached a gospel of conservative reform in the tradition of Henry Grady, Atlanta's 19th-century apostle of the New South. He campaigned for mutual economic progress, explaining to the *Times*' Braestrup that "I'm interested in jobs. I'm interested in improving the attitudes of big business," and that segregated plant restrooms and cafeterias were simply not worth the "fuss" of integration, as required by the Feild program.[33]

But in 1962, such a message from a paternalistic southern white was already badly dated, and it was becoming increasingly embarrassing to the administration. Senator Jacob Javits demanded on the Senate floor that President Kennedy "clarify the role of his alleged 'good friend' [Mr. Troutman] . . . in contradistinction to the work of the committee staff's director, John S. Feild."[34] Javits was especially troubled by Troutman's belief that segregated plant restrooms and cafeterias weren't worth the "fuss" of integration. He echoed previous NAACP charges that Troutman's operation provided no staff follow-up, and called attention to his own bill to establish a national FEPC, which he and thirteen other senators from both parties had introduced into the 87th Congress three days before Kennedy was inaugurated.

The following day, Herbert Hill filed NAACP charges against Socony Mobil Oil Company on behalf of 55 black workers at a refinery in Beaumont, Texas. The blacks were members of an all-black local of the Oil, Chemical and Atomic Workers International Union. Their complaint against collusion between Socony Mobil and the dominant all-white local called attention to the oil industry's Jim Crow traditions in the Southwest. Hill's press conference was designed to embarrass and deflate President Kennedy's White House sign-up the following day of thirty-three additional defense contractors, for Socony Mobil had joined the Plans for Progress spotlight at the White House only the previous February. By the Fourth of July, 1962, Hill was in full tilt against Troutman's operation. He announced a campaign of charges targeted against forty of the 85 Plans for Progress companies, denounced Troutman's program as "one of the greatest phonies of the Kennedy Administration's civil rights program," and called Troutman himself "an avowed Southern segregationist" who was "very closely involved with two of the leading racists in the United States Senate, Senator Richard B. Russell and Senator Herman E. Talmadge of Georgia."[35]

By the summer of 1962 Troutman had become a burden to the administration, and Johnson's committee was drifting in disarray. The congressional off-year elections were coming up in the fall, with all their barometric implications about the success of the new administration, and the troubled PCEEO risked becoming a net liability. In response, Johnson had moved earlier in the spring to reorder his disorderly house by commissioning an outside study of the committee's operations. The consultant he selected was Theodore Kheel, the New York labor negotiator and former Urban League president who had been one of his many advisers on the original design of Executive Order 10925. Then in May the Troutman-Feild split became acute when Jerry Holleman was implicated in the Billie Sol Estes scandal. In hindsight, Holleman's modest fund-raising sins suggest poor judgment rather than corruption.[36] But the taint of scandal forced him to resign his assistant secretaryship at Labor, which necessarily included his part-time post as the Johnson committee's executive director. Holleman was then temporarily replaced by Secretary Goldberg's special assistant, Stephen M. Schulman.

Kheel sent his troubleshooter's report to Johnson in July, and its main substantive recommendation was to place greater emphasis on the *compulsory* aspects of the committee's mission and to de-emphasize "purely voluntary approaches [which] are not likely to produce lasting results."[37] The Kheel recommendations dominated the agenda of the committee's fifth meeting on August 22, where Johnson summarized the report. Discussion centered on the damaging press play over Plans for Progress. Goldberg and Reuther received unanimous support for a public statement reaffirming the committee's "original policy of placing primary emphasis on its compliance and enforcement activities," with programs based on persuasion serving only as supplements. This amounted to a clear vote of no confidence in Troutman's operation and virtually assured that he would resign. It was regarded in the civil rights community as a victory for the committee's hard-liners, and Johnson went along with it. But it also gave Johnson the opportunity to respond to Kheel's advice about reorganizing the committee staff.

Kheel's main structural recommendation was for a full-time executive vice chairman to take firm command of the committee's turbulent operations, plus a larger budget to support the new compliance reporting system. Ironically, the position of executive vice chairman had originally been created as a full-time position in 1957 under Eisenhower. But the new post was then envisioned as a public relations liaison to persuade big business to cooperate, and it had never worked out.[38] In 1961 Johnson had agreed to Holleman's part-time position with the new committee as a political compromise, one that would allow him to avoid appointing a militant whose primary loyalty was to the civil rights coalition, yet still satisfy the Meany-Goldberg labor interests. But Holleman's sudden departure gave Johnson the occasion to act.

As for Troutman, by midsummer 1962 the Georgian had become an albatross. But the administration had grown too publicly committed to Plans for Progress to jettison it along with Troutman. Kheel rejected Troutman's proposal that Plans for Progress be separated from the Johnson committee and privately directed and funded. Instead, the committee would retain control of Plans for Progress, but would henceforth emphasize its role in "moral influence" rather than as a compliance mechanism. Feild's new compliance reporting system, on the other hand, promised to provide for the first time a method of regulatory control that was superior to the cumbersome and reactive mechanism of individual complaint-processing. It could provide the "leverage" for promoting affirmative action beyond mere anti-discrimination, as the Kheel report had insisted. But Feild's ideological rigidity and his leak-prone tendencies in the guerilla politics of bureaucracy were regarded by Johnson as both dangerous to their common purpose and intolerably disloyal. According to Hobart Taylor, during 1961 Johnson had "lost confidence in Feild's judgment. And I say, to be truthful, in his loyalty and integrity."[39] Taylor's judgment was biased, but the house-cleaning notion that both Troutman and Feild should go was congenial to Johnson, whose centrist instincts and demands for personal loyalty were violated in different directions by both men. Ideally, the new executive vice chairman should be a black, whose minority membership would be balanced by institutional and personal loyalty and by cautious instincts. Such a candidate was Lyndon Johnson's special assistant and the committee's special counsel, Hobart Taylor, Jr.[40]

Taylor's public appointment in August 1962 coincided with Johnson's giving private notice to Feild, who would leave the committee staff the following March. Troutman's first-year report coincided with his letter of resignation to the President, which was released to the public on August 22, along with President Kennedy's warm "Dear Bob" response. Troutman's letter to Kennedy summarized the statistical claims of new black jobs contained in his first-year report. Based on six months' data from thirty-seven of the 85 companies with signed plans, Troutman claimed that a total of 4,907 nonwhites were hired into new jobs, producing an increase in annual compensation of $27 million.[41] Even more significant for Troutman, who was interested in building a conservative black middle class, his data showed that the number of blacks holding salaried white-collar jobs in the 37 firms had jumped from 1 percent to 6.6 percent. But Troutman's report warned that "several invitees now express reluctance to participate," because "public denunciation of Plans for Progress by leaders of those whom these 'plans' aim to assist prompts them to hesitate to volunteer for such a role, lest by doing so they offend the nation's Negroes and thereby become subjected to undeserved racial unpleasantness."[42] Troutman also referred resentfully to the still unreimbursed personal expenses of almost $50,000 that he had advanced over fifteen months while working without salary. In his note to Johnson attaching the approved exchange of letters between himself and Troutman, Kennedy observed that Bobby Troutman's "most impressive" job figures had in effect established a measuring stick. "In short, we might well be embarrassed if the companies that have already participated pull out or if we do not continue the dynamic approach to these companies that Bobby carried out so effectively. It is not easy to find someone with the right combination of characteristics (a natural salesman with determination, intellect, charm and a bit of gall) to carry off an assignment of this type . . . ."[43]

Minority Complaints and Racial Statistics

So Troutman left, full of disappointment and bitterness. He had freely given more than a year of his expensive time and considerable energy, forfeiting his normal income and postponing his business and legal practice, and even investing his savings fairly heavily to pursue a noble vision in which he passionately believed. Plans for Progress stayed under the aegis of the PCEEO, to be absorbed, downplayed, and neutralized as much as possible of its demonstrably dangerous potential to the committee's *bona fides* as an equal employment enforcement agency.[44]

By the fall of 1962, Johnson and the committee appeared to have defused a bomb within their ranks. In the area where the committee's jurisdiction was clear—i.e., in federal contracts and in civilian employment with the U.S. government—the committee moved with an authority and effectiveness that considerably exceeded the record of all its predecessors. Within a year, the repeated public avowals of Johnson and Goldberg that contracts of violators would be cancelled and their firms blacklisted were more readily believable. The Lockheeds of American industry were so financially beholden to government contracts that they had no intention of fighting the committee, especially in light of the damaging publicity that surrounded such charges. Indeed they often welcomed such pressure as an occasion to legitimate breaking down racial job barriers, which had artificially constrained their labor supply and elevated labor costs.

Later critics of the Kennedy-Johnson committee would dismiss it as a paper tiger that never cancelled a contract. This was technically true, but failed to capture the committee's range of options and actual leverage. Most major contractors readily negotiated agreements through the committee's conciliation process. That failing, however, the committee generally proved willing to roll out its sanctions—as it demonstrated in the summer of 1962 when Attorney General Robert Kennedy announced desegregation agreements for major southern operations of several recalcitrant firms. These included Tennessee Coal and Iron in Birmingham, and the oil refineries in Lake Charles, Louisiana, run by Continental Oil, Cities Service, Cit-Con Oil, and Petroleum Chemicals, Inc.[45] At the other end of the spectrum of contractors, where second-tier subcontracting firms were many and small, the PCEEO moved with reasonable firmness to demonstrate Lyndon Johnson's claim that the committee meant business. When blacks complained of racial discrimination at five Comet Rice Mills plants in Texas, Arkansas, and Louisiana, and also at Danly Machine Specialties on Cicero, Illinois, the committee staff investigated the complaints. When the companies continued to deny the discrimination that the staff had documented, the committee announced publicly that the companies were barred from further contract work. In response to this bad publicity, both companies quickly conceded.[46]

As a result of the PCEEO's greater visibility and commitment, after only one year the staff had received and processed almost as many complaints (1,850) as had its *two* predecessor committees under Eisenhower in *seven* years (2,095).[47] Moreover, the committee had produced a much higher rate of "corrective action"—generally committee directives to employers to hire or promote or grant a petitioned transfer.[48] The PCEEO's first-year record of requiring correction in 36 percent of sustained complaints was considerably higher than the 20 percent average of the Eisenhower committees and the similar 20–year average of the New York state commission. It was also twice as high as the correction rate for federal government employees, as opposed to contract employees. Federal workers not surprisingly complained to the committee at a higher rate than employees in the private sector (and they complained most frequently over promotions). But their complaints were sustained at a lower rate because the federal government overall was far less guilty of discrimination than rice mills in Arkansas and oil refineries in Louisiana.

The PCEEO's 1963 report boasted of the superiority of its correction rates over those of its predecessor committees.[49] But in relation to the magnitude of the black unemployment problem alone, which was still more than twice the white rate (10.2 percent to 5 percent), such victories were tiny. From 2.2 million federal civilian employees and 15.5 million contract employees, there were only 4,810 complaints, and in only 1,673 did the PCEEO require corrective actions. Meanwhile 700,000 black workers remained unemployed. The committee's more vigorous enforcement measures were scarcely draconian. It is clear in hindsight, and suggestive evidence was accumulating even by the early 1960s, that the complaint process was not likely alone to produce either the dramatic breakthrough or the bureaucratic nightmare that congressional conservatives had long feared from a national FEPC operation, with or without statutory authority and cease-and-desist power.

But what the complaint process lacked in magnitude, it compensated for in precision. Filing complaints and responding to them meant filling out government forms. In processing and aggregating their new forms, the committee staff was beginning to assemble a unique national profile of the employment distribution of race (and a murkier picture of ethnicity and religion, where there was little complaint or interest) in at least government and contract employment. But when the committee moved to implement the explicit command of the executive order that a census be immediately made of minority employment in all the federal agencies, the committee discovered the surpassing irony that owing to enlightened modern practices, such records were no longer available.

At the committee's first meeting in April of 1961, Lyndon Johnson had called for prompt responses to the executive order's requirement that agencies submit statistical reports on employment patterns within 60 days. But in early May, when Harris Wofford received a copy of the racial census from the Department of Health, Education and Welfare (HEW), it explained that department policy barred the recording of employees' race, creed, color, or national origin.[50] The best they could do was to count Negro heads visually. This was a technique that agency officials conceded to be "basically inaccurate," but it did confirm that the problem of racial discrimination in HEW lay less in hiring than in opportunities for promotion. The irony was that agencies considered by liberal Democrats to be more socially progressive, like HEW and Labor, had been the first to move against the practice of recording minority information in order to prevent its possible use in discriminating against them. When a House labor subcommittee reported in August that a Labor Department employee had classified apprentice job applications by race, Secretary Goldberg told the committee that he was "shocked," and that the practice of marking applications with a code designating white or Negro was absolutely forbidden.[51] Assistant Secretary James Reynolds of Labor, in his October 1961 report to Wofford for the subcabinet group for civil rights, announced with pride Labor's recent success in requiring the D.C. General Hospital to exclude racial designations from its application for employment.[52]

Such problems continued to plague the committee's mandatory annual census. But administrators were always able to make rough visual surveys when pressed for a racial headcount. Staff reporters from *U.S. News and World Report* took the trouble to mine the PCEEO's first wave of reports, and they revealed that in the spring of 1962, the 282,600 black federal workers already represented 13 percent of the federal government's 2.2 million civilian labor force. Indeed the Defense Department, with 110,000 civilian black employees, was "probably the biggest employer of Negroes in the world." Defense also employed more than a million white civilians, and its black employment rate of 9 percent was surpassed by the General Services Administration (GSA) with 34 percent, the Veterans Administration (VA) with 23 percent, HEW (20 percent), Labor (18 percent), and the Post Office (17 percent). In all these large service agencies, however, blacks typically crowded the lower job ratings. In the smaller technical agencies, such as Atomic Energy Commission (AEC), the National Aeronautics and Space Administration (NASA), and the Federal Aviation Administration (FAA), black employment remained tiny, both in relative and absolute terms. But overall, blacks were already slightly *over*represented in federal employment.[53]

Nevertheless, at the committee's meeting on February 15, 1962, Johnson expressed his impatience with the agencies' tardy and uneven responses. He noted that even the sketchy and impressionistic results received thus far confirmed HEW's headcount impression that black employment was crowded into the lower grades. Upward mobility seemed to be slow and rare. Johnson then asked each agency head systematically to review their personnel files, paying special attention to the "under-employed resources" of "members of minority groups whose qualifications warranted shifts to new positions of usefulness."[54] By March 1963, however, despite official committee reports based on the annual June census, the agency responses remained so uneven that Johnson wrote thirty-six agency heads a terse reminder of his request of a year previous, and demanded the results in two weeks.[55] Two weeks later Hobart Taylor summarized the responses for him. Eight of the thirty-six agencies had still not responded, and the lack of statistical survey figures in most of the reports in hand made a comprehensive statistical report impossible.[56] Agency responses ranged from substantial efforts by Labor, HEW, Defense, and GSA, to the Post Office's revealing reply that no review of personnel files was made because, in Taylor's summary, the Post Office "felt that employee organizations would protest abandonment of [the] Promotion Plan."

By 1963, however, the federal agencies formally understood not only that they could once again identify employee records by race, but that they *must* systematically do so. The resulting statistics clearly indicated that federal civilian employment of blacks had increased by 17 percent in fiscal 1962 and by another 22 percent in fiscal 1963. By then the total had reached a record of 301,899 black workers, with black employment in both the higher grades (GS-12 through GS-18) and the middle or "professional" grades (GS-5 through GS-11) increasing at *twice* the rate of total employment increase in those agencies. Indeed, by the end of 1963, although no committee report ever acknowledged the fact, official PCEEO data confirmed that blacks were proportionately overrepresented in most federal agencies.[57] To elicit this information, the committee's first compliance form for contractors, in 1961, had listed columns only for "Negroes" and "Other." The "Other" column was provided optionally, "in the event that such groups (Oriental, Spanish-American, Puerto Rican, American Indian) constitute an identifiable factor in the local labor market." The bottom-line, right-hand column was labeled "Percentage Negro." The committee's June 1962 census forms for government employees added optional regional categories for Spanish speaking, Mexican and Puerto Rican origin, Oriental origin, and American Indian.

But these changes and additions represented only quiet, technical, incremental adjustments on the margin, moving toward no apparent grand design beyond generally higher minority figures and more comprehensive survey capacities. That the PCEEO was considerably outperforming its predecessors in enforcing nondiscrimination in federal employment and contracts was little appreciated and of small political consequence by late 1962, when the Kennedy administration was coming under increasing fire for its unwillingness to seek strong civil rights legislation from Congress. Moreover, in response to the glare of publicity from civil rights organizations and the Civil Rights Commission, liberal critics were learning that the federal government continued to support a massive program of grants and loans to state and local governments that in effect continued to subsidize the regional and local Jim Crow practices to which these programs had always accommodated. Lyndon Johnson and his committee could win most of the big contests with private contractors like Lockheed, and certainly the little ones with a Comet Rice Mills, judiciously brandishing the most appropriate balance of carrot and stick. But Johnson's committee had no direct authority over labor unions, or over federal grants or loans, where established patterns of deference to local customs and preferences were entrenched in normal political relationships. As the logic of the committee's charge drew it further into the political thickets surrounding these hitherto sheltered areas of federal expenditure, the limits on the committee's executive authority were increasingly revealed.

The Political and Constitutional Limits of Executive Authority

It was the Civil Rights Commission, however, more than the NAACP or Johnson's committee or the Labor Department, that crystalized the issue of the massive taxpayer subsidy of racial discrimination through federal loans and grants. Created in 1957 as a temporary investigative agency to help deal with what was hoped to be a temporary problem, the commission had been given a short lease on life and a truncated planning and budget cycle that its members and staff found unrealistic and frustrating. It reported biennially, customarily on the threshhold of statutory expiration, and its first report came in 1959, just two years after the Civil Rights Act of 1957 that had given it birth had also provided new federal tools for protecting black voting rights in the South. The commission's first report concentrated on voting rights, and it criticized the glacial pace of judicial challenge to black disfranchisement in the South.[58] Then the commission, following the logic of its charge, branched out in its investigations to include education, employment, housing, and justice. Consequently the commission's report of 1961 was issued in five volumes.[59]

The commission's 1961 report on housing embarrassed President Kennedy. It indirectly spotlighted his claim during the election campaign of 1960 that a mere presidential penstroke could desegregate federally subsidized housing and mortgages. This delighted the Republican opposition, which was attempting to make "pens for Jack" a symbol of opportunism and hypocrisy. Harris Wofford, who as one of Senator Kennedy's speechwriters had coined the hapless penstroke phrase, had left the White House earlier that fall to join Sargent Shriver in organizing the Peace Corps, and Wofford's portfolio as chief White House aide for civil rights was inherited by Lee White.[60] In late October, 1961, White received from Assistant Attorney General Katzenbach a draft executive order designed to prohibit discrimination by private parties who participated in or benefited from federally supported housing programs.[61] The order would create a President's Committee on Equal Opportunity in Housing, to be chaired by the Vice President. But Katzenbach's draft housing order of the fall of 1961 would lie fallow, while Kennedy's senior advisers debated the depressing political odds surrounding their civil rights options. Meanwhile one week later, on October 13th, the Civil Rights Commission issued its 246-page report on federally connected employment discrimination, forcing the White House to confront its dilemma in civil rights policy.

The Civil Rights Commission's report focused on the broad array of popular federal grant programs that distributed $7.5 billion annually to state and local governments to help finance highways, airports, hospitals, housing and urban renewal projects, university research programs, child welfare clinics, state employment services, public schools in federally impacted areas. The list was long, the amounts massive, and the recipients were both program-dependent and well entrenched. The Hill-Burton Act of 1946 alone had subsidized the construction of 5,390 hospitals and health centers through 1962. Construction was a crucial area in which minorities were systematically excluded—and out of reach of Kennedy's executive order of 1961. The commission's first recommendation was that Congress should endow the PCEEO or some similar agency with statutory authority to enforce equal employment policy across the board in *all* federally assisted programs. In the meantime it urged the President to issue another executive order "making clear" that all federal grant programs were subject to the same anti-discrimination ban.[62]

The commission's report and recommendations were expected by the administration. The commission's staff director, Berl Bernhard, had been a regular member of the Subcabinet Group on Civil Rights. Presidential assistant Fred Dutton had created the subcabinet group at the beginning of the administration to cut across department and agency lines and coordinate civil rights policy at the assistant secretary level. At the subcabinet group's first meeting on April 14, 1961, under the chairmanship of Harris Wofford, Dutton asked for monthly reports, with the first round concentrating on each department's grants programs. The group agreed that the PCEEO's authority should be extended to include federal construction grants.[63] By August, a core group including Bernhard, Feild, Wofford, John Doar from Justice, Hyman Bookbinder from Commerce, and James Quigley from HEW had agreed on a uniform policy. Katzenbach's office then circulated a cautious beginning draft which "would not be far-reaching," since it would apply only to grants for construction contracts that already contained a nondiscrimination clause.[64] Most contracts routinely contained such clauses, but typically the bans had not been enforced. So technically the proposed new executive order would merely make the machinery of the Johnson committee available to enforce standing policy. The subcabinet group had considered a proposal to include nonconstruction grants, but had backed away from it. Such an expanded jurisdiction would cover virtually all employment in state governments and higher education, and hence it presented far too large a new enforcement obligation for the PCEEO to assume during its shakedown cruise. Even the normally aggressive Goldberg remained nervous over including the construction grants, as this was a sensitive policy area in the Labor Department.

On November 13, 1961, White sent Kennedy an analysis of the options for future civil rights policy. He first considered standard legislative possibilities, such as a national FEPC, and Title III authority for the Attorney General to file civil injunction suits to prevent denial of civil rights. But then

White ruled them all out. He reasoned that "Any package of relatively easy items (e.g., anti-poll tax and literacy legislation) would not satisfy the civil rights groups and would still make the opponents unhappy—thus it should be a strong package or none at all." [65] But a strong legislative package seemed too risky. That left only strong executive action to counter the criticism. But Senator John Sparkman and Congressman Albert Rains, both from Alabama and chairmen of their respective chambers' housing committees, were strongly opposed to any order desegregating federally assisted housing. They might respond to such an order by defeating the proposed new Department of Urban Affairs, with Robert Weaver slated to head it as the first black member of the Cabinet. This appeared to be too high a price for the President to pay so early in his administration. So it left as a prime candidate for executive action the extension of the PCEEO's jurisdiction to cover federal grants and loans *in construction only*. White added that "the Committee (and the staff) are strong for this, but it has not yet been checked with the Vice President who it is believed will be skittish about it." [66] In early December of 1961 Katzenbach sent White his final draft of the proposed executive order banning discrimination in construction programs funded through federal grants or loans. [67]

The first session of the 87th Congress, however, had so battered Kennedy's legislative program in 1961 that he had no appetite for controversial civil rights initiatives. For the second session of the 87th Congress, Kennedy decided to continue to propose no significant civil rights legislation. He offered only Lee White's "easy items" on voting rights, the constitutional amendment outlawing the poll tax in federal elections, and a bill to establish literacy in federal elections through a sixth-grade education. [68] The noncontroversial poll tax amendment enjoyed strong congressional support. But Kennedy subsequently lost the literacy bill in a Senate filibuster. Attorney General Robert Kennedy, who resented intimations that the administration demonstrated a lack of courage and commitment by failing to press for strong civil rights legislation during the early years, irritably explained after his brother's death that prior to the Birmingham crisis of 1963, Congress wasn't remotely ready for civil rights legislation. "So, we sent up that [1962 literacy] legislation," the Attorney General recalled. "I went up and testified. Nobody paid the slightest bit of attention to me." [69]

John Kennedy was so insecure in his relations with Congress that he offered no significant executive initiatives in civil rights for almost two more years. The President's anxiety over the fate of his proposed new Department of Urban Affairs, which carried with it the promise of the nation's first black cabinet officer, was reinforced by legitimate constitutional reservations over the extent of his executive authority. These in turn centered less on construction grants and loans than on the haunting nemesis of housing. Kennedy's original "penstroke" promise had referred only to federally assisted housing. That is, it would apply primarily to Federal Housing Authority (FHA) and VA loans, which together guaranteed approximately 25 percent of mortgages on new homes. But it would not apply to commercially financed housing.

The Civil Rights Commission report of 1961 had called for a much broader outreach of executive authority, to include conventional loans and mortgages by private financial institutions regulated by federal agencies. The principal such agency was the Federal Deposit Insurance Corporation (FDIC), but also involved were the Federal Home Loan Bank Board, the Comptroller of the Currency, and even the jealously independent Federal Reserve System. Such an order would require the FDIC, for instance, to refuse to insure the deposits of banks unless they in turn required home builders to make formal pledges of nondiscrimination. But the FDIC's mission was to provide financial stability by regulating credit to banks, not to regulate housing. The FDIC claimed that it had no such authority and cited its status as an independent agency.

The Justice Department agreed, with support coming even from such civil rights champions as Burke Marshall and Archibald Cox.[70] Marshall regarded the ill-fated penstroke promise as an "awful far reach of presidential power" and a "pretty drastic step legally and constitutionally for a president to do that without, of course, any consent or approval from Congress."[71] And Congress itself had regularly rejected anti-discrimination amendments to housing bills. Besides, Marshall added, housing was too emotional an issue, especially in the North, to take such an ambitious bite so early in the administration. As the congressional elections of 1962 approached, these political worries were reflected in an eleventh-hour, election-eve panic voiced by northern Democrats in Congress over their constituents' reaction to a proposed presidential edict on fair housing.[72] So the executive order on housing was not forthcoming until safely after the fall congressional elections of 1962. Even then it was carefully buried by being sandwiched between two dominant announcements about foreign crises, and it was dumped on the news-dead Thanksgiving eve.[73]

During this period the Kennedy administration was forced into a defensive and reactive posture as racial violence exploded at the University of Mississippi in October 1962 and again in Birmingham in April of 1963. But Kennedy's political caution was reflected not only in his unwillingness to overextend his executive authority in the complex field of housing. It had stayed his hand, until halfway through his third year in office, from issuing the order that he had promised in 1960, and that a majority of his advisers had urged at the end of 1961. This was the extension of the Johnson committee's authority to include federal grants and loans for *construction only*— where the government's case for enforcing nondiscrimination through the PCEEO was strongest. Kennedy withheld the executive order banning discrimination in construction grants until *June of 1963*, by which time the turmoil over southern segregation had increased to such a pitch that his executive order scarcely attracted any notice at all.[74]

In retrospect, Kennedy's construction order of 1963 was something of a throw-away executive order, like the housing order of 1962. But *un*like the housing order, where expedient campaign rhetoric had yielded to strong constitutional arguments and legal complexities as well as to political caution, Kennedy's long-delayed executive order on construction grants represented a legally unambiguous and prudent extension of presidential authority that was both constitutionally sound and yet politically wasted. During that uncertain year and a half, the Johnson committee was bled by growing attacks on its lack of jurisdiction and enforcement teeth in fighting union discrimination, especially in construction, and also by the damaging publicity from Troutman's ill-starred campaign with Plans for Progress. It was during this difficult period also that the Attorney General, frustrated by the administration's ineffectiveness in Congress and by the lack of counterbalancing achievements in executive enforcement of civil rights, began to assert himself aggressively on the Johnson committee. No American political party had ever promised more civil rights action than the Democratic platform of 1960. As President Kennedy's legislative silence on civil rights increasingly haunted the administration, Robert Kennedy in seeming direct proportion began to attack and humiliate Lyndon Johnson.

Robert F. Kennedy and Lyndon Johnson

Lyndon Johnson once summed up his vice presidential experience during the thousand days with characteristic Texas understatement: "I detested every minute of it."[75] Johnson was both surprised and pleased by the thoughtful respect and solicitude that the President accorded him, and he generally got on reasonably well with the senior Kennedy aides who were veteran politicians and who respected Johnson's masterful political acumen—Lawrence O'Brien, Kenneth O'Donnell, David Powers.[76] But Johnson felt insecure about his weak formal education, and he was most ill at ease among eastern intellectuals.[77] Jerry Holleman, who was neither a stalwart Johnson loyalist nor an eastern intellectual, and whose executive oversight of the PCEEO gave him an apt angle of vision on the political crossfire between the White House and the mission agencies, summed up the personal source of Johnson's vice presidential miseries:

> In the first place, the Irish mafia despised him. They looked upon him as a country bumpkin. And the feeling was mutual. He didn't care much for them either. Bobby [Kennedy] hated his guts. Any time that a member of the staff could relegate Johnson to an insubordinate [*sic*], inconsequential role, they would do it. Anytime they could put him down, they would do it. He felt this. He saw it. They were not even subtle with it. He had pretty good reason to be unhappy, but he was thoroughly miserable.[78]

The bad blood between Robert Kennedy and Lyndon Johnson became a staple of press gossip, and the PCEEO became a crucible for tensions that dismayed even loyalist Kennedy lieutenants. Harris Wofford recalled that the Attorney General had early sided with John Feild and the hawks on the committee staff against Bobby Troutman and Plans for Progress, and Johnson was caught in the middle. "The Attorney General, Bob Kennedy, was critical of Johnson, I think, on all sorts of things during that period, but particularly on this," Wofford recalled, "and was not very respectful of him in the meet-ings that I saw." "Bob would just, you know, be just sort of quiet, sullen, and sulk and things like that, didn't work things out with him in advance or anything. It was not a very successful relationship."[79]

Robert Kennedy's hawkish bloc on the committee generally included the Labor secretary and Walter Reuther, in a natural alliance with John Feild and the NAIRO-oriented professional enforcement staff. When Johnson used the Kheel report in the summer of 1962 to defuse the internal warfare sur-rounding Troutman and Feild, to lower the profile of Plans for Progress, and to find a new and untarnished name for the parallel agreements to be signed by labor unions, the earlier tensions appeared to have subsided. The Attorney General was so busy driving what even his critics conceded was an excep-tionally vigorous and talent-laden Justice Department, especially during 1962 with the dangerous and frustrating negotiations with Governor Ross Barnett in Mississippi, that he attended few PCEEO meetings.

Then in January of 1963, the wounds were reopened by the publication of an attack on Plans for Progress by the Southern Regional Council (SRC), an Atlanta-based, biracial study group that opposed segregation. The fifteen-page report was based on an impressionistic survey of the twenty-four com-panies signing Plans for Progress with plants, offices, or regional headquar-ters in the Atlanta metropolitan area. It concluded that the program was "largely meaningless" because "the interpretation of the voluntary and affir-mative provisions of the program is being left to the individual signers them-selves."[80] The report's criticism was reported in *Newsweek* and prompted President Kennedy to write Johnson asking him to reply to it and to increase the enforcement pressure from the PCEEO staff.[81]

The wire service stories fixed on the SRC's assertion that the program had proved "largely meaningless," and that only seven of the twenty-four firms produced evidence of affirmative compliance with their pledges. But the report had carried the modifying phrase "except for a handful of compa-nies." It praised three of the largest—Lockheed, Western Electric, and Good-year—for demonstrating a "vigorous desire" to create new job opportunities for Negroes, and acknowledged that Lockheed had begun to respond to NAACP complaints even before it signed the Plans. In response to the report the Pentagon directed a hurried follow-up investigation by the Army, Navy, and Air Force, and found that only four of the twenty-four companies were in full technical compliance. But these four were the major employers, with 18,325 of the total of 23,084 workers in the twenty-four companies sur-veyed. The eleven companies not in full compliance employed only 4,508 workers, and nine others whose compliance status was unclear employed only 251.[82]

The *New York Times* quoted the standard "informed Washington source" in describing Vice President Johnson as being "very upset" with the contro-versy that had been rekindled by the Southern Regional Council report. With Troutman gone and Plans for Progress under renewed attack, the *Times* re-ported, "Johnson feels that he is being made the fall guy for a plan initiated by the President's friend."[83] Hobart Taylor agreed that Johnson was caught in an unfair whiplash, with Bobby Kennedy working both sides of the issue. The Attorney General, in Taylor's view, would privately advise his brother to hold up the order against housing discrimination, or to not ask Congress for civil rights legislation for fear of endangering the tax cut or higher priority economic legislative proposals. Then he would attack the Johnson committee's programs for being timid and insincere.[84] Robert Kennedy fought constantly with the Civil Rights Commission for trespassing on Justice Department turf, especially on the commission's proposed hearings in Mississippi while Kennedy and Burke Marshall were negotiating with Governor Barnett.[85] But when Lyndon Johnson, in keeping both with his own instincts and with clear White House preference, barred the PCEEO from holding regional hearings, John Feild and the staff hawks would complain, Taylor pointed out, about the "kind of adjustments that had to be made from time to time to keep the Southerners from destroying this particular committee."[86] Taylor continued,

> I might add that whatever adjustments were made were not made by Vice President Johnson but came from the White House and were frequently the product of Robert Kennedy's thinking . . . this was what frustrated Vice President Johnson so much because sometimes he was asked to do something by the White House, and then if somebody criticized it, he had to carry the brunt of it all by himself. And sometimes a lot of the criticism came from the President's brother, who was privy to the initial instruction.

The Attorney General, on the other hand, complained that the PCEEO was poorly run and not given clear direction by the Vice President. "It was mostly a public relations operation," Kennedy said, and "there wasn't any adequate follow-through."[87] Then there was "the Head of Staff . . . Hobart Taylor," Kennedy added, "whom I have contempt for because I thought he was so ineffective."

The Inner Crisis: Winter 1963

It was against this background of increasing tension in American race relations in the winter of 1963 that John Kennedy on February 28 delivered his long-awaited special message to Congress on civil rights. The President's message began by invoking Justice John Marshall Harlan's lonely dissent in *Plessy v. Ferguson* (1896) that "Our Constitution is color blind, and neither knows nor tolerates classes among citizens." But behind the President's rhetoric of anti-discrimination lay a timid set of legislative proposals. The most novel called for the mandatory appointment by federal district judges of temporary federal voting referees to register southern blacks (in counties where fewer than 15 percent of the elegible blacks were registered) while voting rights suits were pending in the federal courts.[88] In the area of school desegregation, Kennedy merely repeated Eisenhower's request for federal financial and technical aid for desegregating districts—a request that Congress had previously ignored. He also recommended extending the life of the Civil Rights Commission for four more years, with a vaguely broadened mandate that it function as a clearinghouse for information, advice, and technical assistance. But Kennedy made no mention of desegregating public accommodations. Nor was there any proposal on job discrimination, other than a notice that he had asked the Justice Department to enter pending cases before the NLRB to support ending union segregation.

Kennedy's rhetoric of February 28 was morally committed beyond any previous presidential effort. He argued that race discrimination hampered America's economic growth and world leadership, increased the costs of public welfare and crime, and marred the atmosphere of "a united and classless society." But above all, he said, "it is wrong." His legislative proposals, however, were frail. When Budget Bureau officials circulated copies of the Justice Department's draft voting bill in late February, Berl Bernhard and William Taylor of the Civil Rights Commission staff complained to Lee White. Such minor tinkering with a voting rights approach that had failed since 1957 and 1960, they said, would be greeted with "massive indifference or actual opposition" by civil rights organizations.[89] As a result it would probably fail like the literacy test bill of the previous year. Even if it passed its effects could only be marginal. Bernhard and Taylor argued instead for a bill authorizing the Attorney General to initiate school desegregation suits. But even these two senior staffers of the Civil Rights Commission made no mention of public accommodations or job discrimination, on the probable assumption that Kennedy's Roosevelt-like deference to southern power in Congress made such suggestions pointless, at least in the winter of 1963.

Nonetheless pressures for a stronger civil rights package continued to build. Early in January six liberal Democratic senators wrote Kennedy that the large Senate class of 1958 needed a much stronger legislative record on civil rights from the 88th Congress to run for re-election in 1964. Similarly, six moderate-to-liberal Republican members of the House Judiciary Committee held a press conference and issued a release calling for stronger civil rights laws. Following Kennedy's civil rights message, eight Senate Republicans introduced their own twelve-bill package. Reaching well beyond Kennedy's voting proposals, it called for all segregated school districts to file immediate desegregation plans, for Title III authority permitting the Attorney General to seek injunctive relief on his own initiative in suspected civil rights violations, and for a national FEPC.[90]

In April, Martin Luther King began leading sit-in demonstrations against segregation in Birmingham.[91] By mid-April, King and many of his supporters had been arrested and jailed. In early May, Birmingham exploded in violence. On May 24, Robert Kennedy met at the Kennedy family appartment in New York City with novelist James Baldwin and a group of black writers and artists, where he and his brother's administration were denounced for three hours. He emerged angry and embittered.[92] Just five days later, he and Burke

Marshall were scheduled to attend the first meeting of the PCEEO in 1963. Burke Marshall recalled the tension:

> Well, in May of 1963, the country was in turmoil, as you will remember, absolute turmoil because of Birmingham; it was repeating itself all over the place and everybody was on President Kennedy's neck—black people, white people, everybody was on his neck. So Robert Kennedy was trying to do all sorts of things; he was trying to persuade businessmen to open their restaurants and theaters; he was trying to get this legislation under way; he was trying to get church groups and educators and labor people stirred up about this and doing something within their own constituency. And so he was very impatient. He went to this meeting and he asked a lot of questions that were impatient, very impatient; I could see it made the Vice President mad.[93]

Marshall was referring to the tense meetings of May-July 1963, when Robert Kennedy twice humiliated Johnson in front of his entire committee.

"I was very bigshot"

Johnson had been forewarned of the liklihood of such a clash as early as the preceding February, when Berl Bernhard tipped off Johnson aide Harry McPherson that the Vice President was about to become the target of some "terrific abuse" over the committee's alleged ineffectiveness. Bernhard expected that this "wretched stuff" would have White House support, and it would unfairly hit a Vice President who "responds a great deal more sincerely and basically to the underdog than the President does."[94] At the May 29 meeting of the committee, with 38 members and proxies and senior staff attending, the Attorney General interrupted Hobart Taylor's opening report to grill him on specific statistics on black employment. Kennedy demanded to know the specific number and job levels of minority employment by southern community and plant. Taylor couldn't reasonably have been expected to produce such detailed information without advanced warning, even if then. But he responded with a bureaucratic dodge, promising to provide such information at some unspecified future date when the Budget Bureau had approved the committee's revised compliance report form. Kennedy fairly snarled his contempt for Taylor, whom he regarded as Johnson's Negro Texas lackey.

To break this tense moment, Johnson asked John Macy, chairman of the Civil Service Commission, to report on federal employment in Birmingham. Macy reported that aside from the Post Office and the VA hospital, blacks held less than one percent of the federal jobs in a city that was 37 percent Negro, or only fifteen out of approximately 2,000 federal employees. Johnson anticipated Kennedy by pressing Macy for particulars on these fifteen, and Macy responded with precise descriptions that emphasized recent federal recruiting efforts.[95] Macy was able to do this because he was directing a crash effort to place blacks in federal jobs in Birmingham. The project had received presidential and cabinet priority when Burke Marshall reported that Birmingham's civic leaders didn't see why the city's private employers should hire Negroes when the federal offices in Alabama weren't even hiring them in clerical positions.[96]

Macy's "crash job" was ordained by such high authority that the normal federal hiring rules were easily bent in order to produce the desired results quickly. When President Kennedy called his Cabinet meeting to discuss the problem, the Attorney General invited Macy to attend and report on the Birmingham situation. But when Macy discovered that the Vice President was being excluded, he called Johnson to try to bridge the chasm. Johnson smoldered over his exclusion. He especially resented the way Bobby Kennedy was able to order the emergency hiring foray to Birmingham, then turn around and blast the vice President's committee staff for ineffectiveness when it followed prescribed procedures which Macy and his special team were free to short-circuit. Macy's report at the PCEEO's tense May 29 meeting smoothed over the tensions between Johnson and the Attorney General. But the following week Johnson called Sorensen to offer his unsolicited advice on legislative strategy for civil rights, and in the process the Vice President voiced his bitterness:

> Bobby came in the other day to our Equal Employment Committee and I was humiliated. He took on Hobart and said about Birmingham, said the federal employees weren't employing them down there, and he just gave him hell and said, "We got twenty-six jobs for them." Well, obviously the President and the Attorney General can get twenty-six or twenty-six hundred if they tell them, "Put them on." But the only way *we* can tell them is take them from the civil service register and they're presidential appointees and we can't make them do anything . . . . I believe those twenty-six will cost us in the long run.[97]

The major blowup occurred at the next meeting, on July 18, when Robert Kennedy tore into James Webb. The respected head of NASA, Webb was also known as a protégé of Johnson, and Webb's highly technical agency, like the AEC, recruited all but its clerical employees from a small pool of scientific elites. Kennedy demanded to know from Webb the precise background, experience, and job description of the personnel officers with equal employment responsibilities in NASA's nine national centers. When Webb couldn't recite them at the table, Kennedy was curtly dismissive.[98] Jack Conway, who was executive assistant to the president of the AFL-CIO's Industrial Union Department and was representing Walter Reuther at the meeting, recalled the excruciating tension caused by the Attorney General's proxy assault on Johnson:

> It was a pretty brutal performance, very sharp. It brought tensions between Johnson and Kennedy right out on the table and very hard. Everybody was sweating under the armpits. . . . And then finally, after completely humiliating Webb and making the Vice President look like a fraud and shutting Hobart Taylor up completely, he got up. He walked around the table . . . shook my hand . . . and then he went on out.[99]

The effect on the committee's cohesion and morale was predictably shattering.

Robert Kennedy himself explained that his main concern for savaging Johnson's leadership at the meetings had been the election of 1964. He feared that the persistence of racial segregation and black unemployment would reveal Johnson's committee to have been largely "a public relations gimmick." "In the last analysis," he said,

> it was not going to be Vice President Johnson's Committee; it was going to be President Kennedy's Committee. The signings had taken place in the White House; and I could just see going into the election of 1964 with this great buildup, plus the internal volcano from within, plus some of the other people from without, who were dissatisfied with it and, eventually, these statistics or figures would get out; and that there would just be a public scandal.[100]

The Attorney General admitted that at those infamous meetings, where "the sharpest disputes I had with Vice President Johnson" occurred, "I was very bigshot, during that period of time. So it was unpleasant."[101] Burke Marshall observed, in the Attorney General's defense, that Kennedy was harassing everybody during that difficult period: "[Kennedy] fussed and interfered, if you want to put it that way, with almost every other department of the government in 1963 on that issue, on their employment policies, and on whether or not Negroes were allowed to participate in federally financed programs."[102]

But even Kennedy loyalists like Lee White believed that Bobby Kennedy had been far too tough on Lyndon Johnson, especially in such a public display, and for results whose potential benefits to the common purpose were far less clear than their liabilities. White, who regarded himself as "a great fan of Bob Kennedy," saw in him "a brusqueness or abruptness or roughness that would have offended anyone, especially a sensitive soul like Johnson."[103] Burke Marshall remembered that during Kennedy's harangue against Webb, Johnson grew quiet, tense, and very angry—"he didn't say anything, but I could just tell. I mean I could tell from his face. And I don't blame him. I mean, if I had been chairman of that committee, I would have been irritated by those questions, too. But the questions were justified. The question is whether it was wise to do them in the full committee or whether it could have been done in some other way."[104] The question, of course, answered itself.

Assistant Attorney General Marshall was a loyal lieutenant, but Robert Kennedy, who Harris Wofford called the Puritan who knew only black and white, was a gut fighter who knew no other way. He took his complaints about Johnson over to his brother in the White House, and reported that the President "almost had a fit. . . . He said, 'That man can't run this Committee. Can you think of anything more deplorable than him trying to run the United States? That's why he can't ever be President'."[105]

CHAPTER III

The Civil Rights Bill of 1963

The Kennedys, Johnson, and the Birmingham Breakthrough

In *John F. Kennedy and the Second Reconstruction,* historian Carl Brauer cites President Kennedy's civil rights speech of June 11, 1963, as marking "the beginning of what can truly be called the Second Reconstruction, a coherent effort by all three branches of the government to secure blacks their full rights."[1] The turmoil in Birmingham that began in April, turning quickly to violence and leading to Kennedy's June speech, had been a planned confrontation. The segregationist forces were led by the city's notorious police commissioner, Eugene T. ("Bull") Connor, who vowed to "keep the niggers in their place."[2] Connor's image as Birmingham's chief redneck seemed too archetypical to be credible, like a character from Central Casting; armed with police dogs and fire hoses and cattle prods and spoiling for a fight, "Bull" Connor as a symbol of racist intransigence had become a staple of the national news weeklies.

Less apparent at the time was the calculated nature of the black challenge in Birmingham. On the heels of the previous fall's failed protest campaign in Albany, Georgia, where the polite discipline of Police Chief Laurie Pritchett had defeated demonstrators led by Martin Luther King, the Southern Christian Leadership Conference had repaired to Savannah for a three-day review of strategy. A postmortem on strategic collapse, it quietly called into question the core of King's Christian optimism. What emerged was Project C—for *Confrontation.* To implement Project C, King accepted an invitation in the spring of 1963 to come to Birmingham and lead the protest against Connor's undisciplined defenders of white supremacy. By provoking a crisis that promised a flow of blood on the national television networks, King's "nonviolent" protest would force President Kennedy's hand.[3]

The crackle of black protest and violence that spread regionally and then northward in the wake of Birmingham convinced the Kennedy White House that its cautious legislative proposals of February 28 were inadequate.[4] The

Birmingham violence threatened nationwide turmoil. But the post-Birmingham climate of national indignation offered hope by inviting a bolder reach of presidential leadership. It offered relief from the frustrations of the first two Kennedy years, when federal protections were mocked by the bloody reception given the Freedom Riders in Alabama in 1961, and then again in 1962, when Kennedy's cautious and fitful moves during the crisis at Ole Miss had produced bloodshed without a commensurate clarification of federal policy and authority.[5] But in the spring of 1963 Kennedy needed a fitting occasion to announce his new post-Birmingham initiatives. The occasion was ironically provided by Alabama's governor, George C. Wallace. In response to a federal court order of May 5 to desegregate the University of Alabama, Wallace stage-managed a baroque posture of defiance: "I draw the line in the dust and toss the gauntlet before the feet of tyranny, and I say, Segregation now! Segregation tomorrow! Segregation forever!" But this time, unlike the Ole Miss crisis with Governor Ross Barnett, the President and the Attorney General launched a coordinated series of threats and negotiations. On June 11 Wallace was forced to capitulate. Thus the victory in Tuscaloosa gave Kennedy a fitting occasion to address the nation on the intensifying civil rights crisis.

Speaking on national television that evening from the White House, Kennedy spoke simply and eloquently. "We are confronted primarily with a moral issue," he said, "as old as the scripture and . . . as clear as the American Constitution." The moral crisis was sharpened by the painful events in Birmingham and the "fires of frustration and discord [that] are burning in every city, North and South."[6] Its resolution should not be regarded as a sectional or a partisan issue, Kennedy insisted, and on the latter assertion he was indubitably correct. This was because only Republican support could overcome the certain filibuster of southern Democratic senators in the face of a strong civil rights bill. But Kennedy's denial that the racial crisis was sectional, while statesmanlike, was also politically disingenuous—as was clear from the legislative remedies he proposed and the sectional nature of the bipartisan coalition in Congress that would be necessary to approve the remedies. The symbol of crisis was Birmingham and "Bull" Connor's police dogs and fire hoses on national television, even more than George Wallace's charade in the schoolhouse door. The emotional and political conclusions were unambiguous: the civil rights problem was exploding intolerably as a *southern* problem. Accordingly, the remedies that Kennedy proposed were sectional remedies. They included more federal muscle in voting rights and school desegregation, but they centered on a radical proposal for Congress to outlaw racial segregation "in facilities which are open to the public hotels, restaurants, theaters, retail stores, and similar establishments." There was no mention at all of job discrimination, which was a national problem.

When the Birmingham crisis first broke into the headlines in mid-April, the Attorney General had summoned the key White House aides and senior Justice Department staff to plan a new political strategy and to construct a legislative proposal to seize the moment.[7] Assistant Attorney General Norbert

Schlei, who headed the Office of Legal Counsel and was appointed chief draftsman for the new bill, recalled the focus of the decision-making process at the initial planning session: "And really the question wasn't 'Should we do it?' but, 'What's going to be in it?'," Schlei said. "We started with our existing [February] package of voting and Civil Rights Commission, something had to be done about public accommodations, something had to be done about education." "It would be nice, we thought, to do something about employment," Schlei added. "But that was very difficult, that was widely believed to have little or no chance."[8]

Also initially rejected, in addition to a fair employment proposal, was a provision for witholding funds from discriminating programs. This power was of course implicit in the executive authority of the Johnson committee. But as a legislative proposition it conjured up the nettlesome Powell amendment. Since the middle 1950s Harlem congressman Adam Clayton Powell had sought to amend federal grant-in-aid bills by banning any expenditures for segregated programs. The result, however, was often that of a Judas-kiss, which tended to endanger or doom liberal bills by requiring federally enforced desegregation of all activities they affected. The Kennedy administration had already been twice bloodied by the Powell amendment in the failed quest for federal aid to education. A vote against the Powell amendment appeared to be a vote for segregation, yet a vote for it usually sank liberal bills under the weight of the combined opposition of Democratic and Republican conservatives. So the legislative planners in the White House and at Justice wanted to steer well clear of such known congressional shoals as the Powell amendment and a national FEPC.

While the content of the new legislative package was being debated within the administration in May, the Attorney General initiated a barnstorming series of meetings with various leadership groups, beginning with himself and Burke Marshall, and by June increasingly centering on the President as his own chief lobbyist. The meetings were designed more as sales pitches than as opinion probes, and they included leaders of American business and commerce, religion, law, organized labor, and civil rights and women's groups. They began with Robert Kennedy's forays in late May (which included the acrimonious session with James Baldwin and the black artists and intellectuals in New York) and stretched into July as the President met with approximately sixteen hundred American leaders.[9] During the reassessment of legislative strategy of late May and early June, Robert Kennedy took the lead in pressing for the risky initiative in public accommodations, supported by Katzenbach and Marshall, while senior presidential aides Lawrence O'Brien and Kenneth O'Donnell and even Ted Sorensen opposed it. Throughout the intensive review of April-May, Vice President Johnson remained excluded, as usual, from the inner circle of Kennedy's policy advisers.

Perhaps understandably excluded from the President's inner circle of Irish Mafia, Johnson was occasionally excluded inadvertently even from the larger meetings, especially those including the congressional leadership, at which the President had insisted that Johnson be present. Lee White recalled with considerable embarrassment that despite Kennedy's "clear as a damned bell" instructions to "keep Johnson advised of when the meetings are, and let him know what's going on," White just frequently forgot about the vice presidency:

> President Kennedy told me, "Now look, you work with Lyndon Johnson and make sure that he knows about all of these things. I want him here. I think he can do a lot. First of all, he's pretty damned smart; and, second of all, he's a southerner and he has got a better speaking voice for a lot of these people than I have got. Third of all, I think we ought to have the whole panoply of presidential executive branch superstructure. And besides that, let's keep him involved in this."

Nevertheless, so marginal was the vice presidency that of the eight meetings, White said, "there must have been two or three when I clean forgot about the Vice President—just forgot! . . .—And Johnson didn't come. He may have known about the meeting, but if he weren't invited, he wouldn't come."[10]

Johnson was churning within, needing to be heard. He asked O'Donnell for just fifteen minutes with the President, but received instead a visit from Burke Marshall. In an oral history interview Marshall recalled the Vice President's warning that the bill "wouldn't get passed and was impossible and would cause a lot of trouble and was the wrong thing to do." Instead, the core of the problem as Johnson saw it was jobs and education, and he rambled on to Marshall about how as head of the National Youth Administration in Texas he had found jobs for Negroes during the Depression.[11] But Johnson kept pressing O'Donnell for an audience with the President. So on June 3 the Attorney General sent Norbert Schlei to Johnson's huge office in the Capitol. There the agitated Vice President "absolutely poured out his soul," Schlei remembered, even though he was clearly unhappy that his audience was with a "substitute" like Schlei (which Schlei agreed "was certainly odd") rather than a senior White House adviser or with the President himself.[12] Yet Schlei was also struck by Johnson's apparent lack of concern for the substance of the proposed civil rights bill. Instead, Johnson was concerned with strategy and timing, with psychology and symbolic meaning. He pleaded that the President appeal directly to the South in moral terms and even in religious terms, saying that racial segregation was so intolerable that we can't equivocate any longer, that we simply have to do the right thing, the patriotic thing. The newly proposed bill, Johnson said, would be "disastrous for the President's program and would not be enacted if submitted now." Johnson called instead for a seven-point legislative strategy that largely ignored substance for procedure.

First, Johnson said, call in the Republican leaders, and especially Senate Minority Leader Dirksen, putting them on the spot and demanding "promises in blood" to deliver the twenty-seven of 33 Republican votes needed for Senate cloture. Second, call in the Negro leaders and explain that the tax bill had to be passed first in order to avoid a recession in which "nobody's civil rights will be worth having," but then make an ironclad commitment to pass the civil rights bill before the end of the session. Third, seek public support from the three living ex-presidents; fourth, make major speeches based on a moral appeal in each of the main southern states; and fifth, negotiate in advance with Senator Russell. Sixth, start with the House bill and make sure it went to the Judiciary Committee, where chairman Emanuel Celler of New York would give it a friendly reception. Finally, get Senate Majority Leader Mansfield to stop campaigning for re-election in Montana and devote his undivided attention to passing the bill.[13]

Later that same day, Johnson finally got an audience by telephone with Sorensen, and in a passionately rambling, half-hour monologue he swore his loyalty to the Administration and to its newly revised bill *whatever* its content. But he warned that if it went up to the Hill cold and in its present form, Kennedy would be "cut to pieces with this and I think he'll be a sacrificial lamb."[14] Johnson was hurt that he had been excluded from the legislative planning sessions, and bitter that he had learned about the Administration's high legislative strategy from reading the newspapers. "I don't know who drafted it," Johnson complained. "I've never seen it. Hell, if the Vice President doesn't know what's in it how do you expect the others to know what's in it? I got it from the *New York Times* . . . ."[15] Johnson warned Sorensen that Kennedy hadn't done his homework, and that the chief danger was continued indecisiveness. This allowed southern whites and Negroes to agree on one point: "The whites think we're just playing politics to carry New York," Johnson said. "They're not certain the government is on the side of the Negroes." And the blacks held the same suspicion.

> Until that's laid to rest I don't think you're going to have much of a solution. I don't think the Negroes' goals are going to be achieved through legislation and a little thing here on impact area or vote or something. I think the Negro leaders are aware of that. What Negroes are really seeking is moral force and [to] be sure that we're on their side and make them all act like Americans, and until they receive that assurance, unless its stated dramatically and convincingly, they're not going to pay much attention to executive orders and legislative recommendations.[16]

Johnson was reacting to reports of a legislative strategy that had been devised four days earlier, on May 31, when the President had met with the Attorney General, Burke Marshall, and Sorensen. They had discussed an agenda that outlined three titles for the new bill, the first dealing with public accommodations and the second with school desegregation. But the third dealt with equal employment, and the agenda's attached working paper queried whether a new statutory basis for the Vice President's committee with increased enforcement powers should extend coverage to unions and to "employees not working on government contracts (FEPC?)?"[17] Its concluding query was: "Is this bill sufficiently important to add to a difficult legislative package?" And the answer was a qualified no. Instead, the President would propose a "single omnibus bill, limited to the most reasonable and urgently needed provisions . . . two additional titles on public accommodations and education."[18] Congress would be asked to give permanent statutory authority to the PCEEO, but *not* to create a new FEPC. Further, the Administration would not ask Congress to add federal grants to the labor unions or PCEEO's jurisdiction. Instead, the President would express his general support for pending fair employment legislation that had been introduced into Congress by others, but that in the past had never stood any serious chance of passing in any event.

Public Accommodations and the Universal Powell Amendment

President Kennedy's television address to the nation of June 11 on the moral crisis in civil rights stopped short of the religious appeal that Lyndon Johnson had urged—to look them in the eye and state "the moral issue and the Christian issue." But Kennedy's speech appealed early to patriotism. "When Americans are sent to Vietnam or West Berlin," he said, "we do not ask for whites only." It also appealed to a national sense of fairness, including an admixture of white guilt. "The Negro baby born in America today," Kennedy said, "regardless of the section of the Nation in which he is born, has about one-half as much chance of completing a high school as a white baby born in the same place on the same day, one-third as much chance of becoming a professional man, twice as much chance of becoming unemployed, about one-seventh as much chance of earning $10,000 a year, a life expectancy which is 7 years shorter, and the prospects of earning only half as much."[19] Kennedy's references to becoming a "professional man" and to the goal of earning an annual salary of $10,000 sound archaic to more modern ears, but no President since Lincoln had addressed the nation in such forthright moral terms about American race relations.[20] In his June 11 speech, however, Kennedy did not spell out his new legislative proposals covering public accommodations, education, and voting rights. Instead, he pledged to present them to Congress the following week. In the meantime, the White House and the Justice Department continued their bipartisan negotiations with congressional leaders. And even Robert Kennedy, who criticized Johnson's reluctance during the early planning as evidence of his faint heart and questionable commitment, later conceded that Johnson's plea for more intensive homework with congressional leaders was "very wise."[21]

On June 19, the President sent his new legislative proposal to Congress, and its controversial centerpiece was the title on public accommodations. In the parlance of legislative shorthand, this soon came to be known as Title II. In his accompanying message, Kennedy sought to take the edge off the novelty and indeed the radicalism of this federal intervention in local customer choice by pointing out that thirty states and the District of Columbia had already enacted such laws, thereby covering two-thirds of the country as well as two-thirds of its population. But Kennedy's footnoted list of precedents included no southern state except the border state of Maryland.[22] Everyone understood that the heart of the bill was aimed at the segregated South of

George Wallace and "Bull" Connor. But the President's proposal was carefully limited. It did not call for direct federal enforcement. Rather, it provided statutory grounds for aggrieved individuals to seek a court order against a discriminating business establishment. It would also permit the Attorney General to file suit in such a complainant's behalf, but only on two preconditions: (1) that the aggrieved party was unable to bring his own suit ("for lack of financial means or effective representation, or for fear of economic or other injury"); and (2) that the Attorney General had first referred the unresolved dispute to a new Community Relations Service. The President's message announced his intention to establish by executive order such a confidential, noncoercive forum for citizens to "sit down and reason together," but he also asked Congress to give it a statutory basis.[23]

But how could new federal regulation of local lunch counter patronage be justified constitutionally? In Kennedy's simple eloquence, equal access to places of public accommodation "seems to me to be an elementary right," and its denial was "an arbitrary indignity that no American in 1963 should have to endure." But that didn't make it constitutional. Eighty years earlier, in the Civil Rights Cases of 1883, the U.S. Supreme Court had gutted similar legislation passed by Congress in 1875.[24] Even though the Supreme Court by the 1960s had reversed its earlier approval of state laws requiring segregation in public facilities, such as parks and municipal swimming pools, the public accommodations section of President Kennedy's omnibus bill applied not to public acts but to the private acts of restaurant and motel owners. So the Administration derived the broad constitutional reach of its public accommodations section not from the 14th Amendment, but rather from the interstate commerce clause. Since 1937 the Supreme Court had broadly interpreted the commerce clause to accommodate the New Deal and permit the increasing federal regulation of the nation's economic life.

The Republican leadership in Congress, however, disagreed. Republican support was crucial to any hope of legislative success, especially in the Senate. But for many years Republicans had insisted that the equal protection clause of the 14th Amendment, not the interstate commerce clause, was the appropriate source of federal authority in civil rights issues. It was, after all, historically "their" amendment. Its indisputable central purpose was to provide for equality of rights among the races, whereas the central constitutional purpose of the commerce clause had no essential connection with race relations. It struck the Republicans as typical of Democratic disingenuousness to argue commerce with a straight face when the moral and constitutional problem was one of racial equality, and the obvious operable clause was the 14th Amendment's equal protection clause. As for the problem of regulating private conduct under an amendment designed to restrict state power, the Republicans pointed out that the states normally licensed restaurants and motels and the like, so public accommodations could be desegregated through the state licensing power. Besides, they said, one of the administration's main but unacknowledged reasons for pursuing the commerce route was political rather than constitutional. It derived from the embarrassing fact that the Democratic chairman of the Senate Judiciary Committee was Senator James East-land of Mississippi.

Eastland's committee had proven to be a certain graveyard for civil rights legislation. But the Senate Commerce Committee was chaired by Warren Magnuson of Washington, whose hospitality could provide a Senate vehicle to go around the Judiciary roadblock. So in his presidential message Kennedy acknowledged the secondary strength of the 14th Amendment, but had none-theless rested his case on the dim moral grandeur of the commerce clause. He claimed with somewhat strained logic that a federal public accommodations law was needed to "prevent the free public flow of commerce from being arbitrarily and inefficiently restrained and disturbed by discrimination" in business establishments. Congressional Republicans had come to despise the commerce rationale as constitutionally trite and dangerously elastic, espe-cially since Franklin Roosevelt had successfully used it to justify most of his expansive New Deal regulation of business enterprise. In Republican eyes, the commerce formula amounted to a regulatory carte blanche, designed to entangle in Washington's insatiable regulatory tentacles even the most hum-ble, mom-and-pop store at the crossroads for selling a soap bar made in Grand Rapids.

This partisan dispute, and the philosophical distance that it implied, would cause the Administration some unnecessary but not fatal difficulty on the tricky legislative road ahead. But there was little in the remainder of Kenne-dy's civil rights message that offered major cause for concern over potential Republican intransigence. There was no broad Title III, the conservative bête noir of 1957 that had appeared to give the Attorney General the authority, on his own initiative, to sue for injunctive relief over the alleged infringement of any citizen's civil rights anywhere. The section on school desegregation in the President's omnibus bill necessarily repaired to the 14th Amendment, as had the *Brown* decision. It called only for limited authority for the Attorney General to file suit against local public school boards or public colleges that were refusing to desegregate. Because the problem of school desegregation had originally flowed from the Warren Court's decrees under Eisenhower, and had since become almost exclusively a headache of southern Democratic constituencies, there appeared to be little cause here for Republican alarm—with one exception.

The exception was northern or de facto school desegregation, which was widespread but not explicitly de jure in origin. In the moral atmosphere sur-rounding the Birmingham crisis, Republicans could scarcely denounce the Democratic administration's new civil rights bill for threatening to require school desegregation outside the South. But urban Democrats from the North and West were sufficiently nervous about their vulnerability on de facto school desegregation that a quiet, bipartisan coalition suggested itself. Such a tacit agreement could later focus the bill narrowly on de jure school segregation, and hence on the South—preferably off the record, during the closed markup sessions in committee. So there seemed to be little immediate cause for non-southern alarm over Kennedy's proposed title on school desegregation.

Somewhat similarly, Kennedy's call in Title VI for a single, comprehensive congressional ban on federal funding of racially discriminatory programs found a surprisingly receptive audience on the Hill, at least among nonsoutherners. The little-noticed Title VI represented a policy reversal by the White House staff and the legislative advisers from Justice, who had feared that any enforcement proposal to cut off federal aid would be anathema on the Hill. But their fears had proved unfounded. Instead, to the surprise of Norbert Schlei and his colleagues on the drafting team, "we began to get some very interesting playback . . . all the legislators—Democrats as well as Republicans— wanted to have a Powell amendment type of feature in the bill, that turned out to be Title VI. They said, 'Please, let's get that behind us. Now, that comes up on every bill'." So the prospect of getting forever rid of Powell's Judas kiss appealed to both sides of the aisle—to Democrats frustrated by having their liberal legislation torpedoed, and also to Republicans interested in withholding federal funds as an economy measure. The serendipitous result, said Schlei, was "a practically unanimous recommendation of all the legislators that we make the Powell amendment controversy obsolete forever by having a sort of universal Powell amendment."[25]

It is one of the ironies of the civil rights era that such a casual, bipartisan consensus should form around Title VI. It was prompted by Democratic irritation over the theatrics of Congressman Powell, and Republican belief that the GOP tradition of fiscal prudence required them to support mechanisms for budget cutoff—especially if it seemed targeted exclusively against the racial discrimination of southern Democratic regimes. In the history of the modern administrative state, the armament of social regulation by Washington agencies, so traditionally loathed by Republicans, perhaps owed more to this fleeting moment of unexamined acquiescence in Title VI than to all the other titles in the landmark civil rights bill combined. This would have been true even had Kennedy included an explicit title on job discrimination—which he was careful to exclude.

Kennedy's failure to include an FEPC proposal disappointed the black civil rights organizations and angered the national labor leadership. The Urban League's analysis of the omnibus bill regretted the omission of FEPC, but judged it overall a "strong and good bill."[26] Organized labor's response, however, was surprisingly harsh. Publicly the AFL-CIO offered its customary strong support. But privately the leadership chastised the White House for failing to include FEPC. Both the CIO and even the old AFL before the merger had supported FEPC, and the reasons involved self-interest as well as social justice. Andrew J. Biemiller, the AFL-CIO's director of legislation, sent Kenneth O'Donnell a copy of labor's internal critique of Kennedy's omnibus bill. It complained that "unions are being universally blamed, often unfairly, for job discrimination. This has resulted in very strained relations between Negroes and the labor movement."[27] Thus "unions, as well as the Negro and the country as a whole, need an FEPC." It was employers, not labor, who opposed FEPC, and who shared much of the blame for job discrimination. Moreover, FEPC leverage could help the AFL-CIO leadership force nondiscrimination policies on refractory locals who were shielded by the federation's decentralized structure. As George Meany told the Senate Labor and Public Welfare Committee in July, "We need the power of the federal government to do what we are not fully able to do [by ourselves]."[28]

As for Title VI on withholding federal funds, the AFL-CIO's severe analysis of the Kennedy bill failed even to notice it. In both the public statements and the internal correspondence of the Leadership Conference on Civil Rights, the NAACP, and the Urban League, Title VI was merely acknowledged in passing.[29] The NAACP sensed the title's potential in desegregating southern schools, but still looked mainly to judicial orders.[30] The Urban League thought the President had full constitutional powers to withhold federal funds without needing to ask Congress for statutory authority, and regarded Kennedy's Title VI request as further evidence of "the moderate nature of the President's civil rights bill."[31] In the suddenly hopeful atmosphere of June 1963, most attention was focused on the controversial Title II, which would desegregate the system that Bull Connor was defending. Only later would attention shift to job discrimination, with its nationwide implications. Almost no attention was paid to Title VI, the sleeper that in time would become by far the most powerful weapon of them all—and in the process would tear at the heart of the labor–civil rights alliance.

The Fair Employment Dilemma

Lacking an FEPC proposal, Kennedy's new civil rights bill nevertheless had to address the persistent problem of employment discrimination in *some* fashion. Having decided not to propose a national FEPC for fear of losing the entire package, the Administration sought to strike a delicate balance that would not alienate the Democrats' core constituency of nonsouthern liberals. At this juncture the White House was being bombarded with outside advice, especially from the liberal-labor bloc at the heart of the New Deal coalition. One spokesman for this constituency was G. Mennen Williams, who as governor of Michigan had helped to deliver the state's critical support to Kennedy at the 1960 convention, and who was subsequently rewarded by Kennedy with an assistant secretaryship of State (for African affairs). In mid-June, Williams sent Sorensen a memo for the President that typified the liberal agenda on civil rights, especially in light of Kennedy's sudden opening to the left. Williams called for "a series of hard hitting administrative actions to be instituted every 10 days or so" while Congress was working on the new civil rights bill.[32] As for the bill itself, Williams insisted on "passage of the *whole* civil rights legislative package, including F.E.P.C. and provisions giving the Attorney General the power to take remedial action for the United States whenever civil rights of any kind are violated." Although the Administration had already decided to strike both FEPC and Title III from the new bill, Williams was calling in addition for "massive" new programs of education, housing, public works, and work relief; the establishment of a "civil rights secretariat" in the White House, to be headed by a Negro leader who "should have your ear daily"; and the presidential establishment of "a national group with subsidiaries at the state, county and city level to investigate and report on indignities suffered by Negroes and whites who are imprisoned or arrested . . . for participating in sit-ins and other such demonstrations."

Faced with such new expectations from his aroused liberal constituency, Kennedy in his June 19 message to Congress restated the case for his pending tax cut and his earlier education and training measures. The latter included not only his battered bill for federal aid to education,[33] but also increased funding and coverage for such familiar programs as manpower development and training, youth employment, vocational education, work-study, adult education, and public welfare work-relief and training—none of them especially novel or designed specifically for minorities. Kennedy had asked Labor secretary Willard Wirtz whether a three- to four-year "crash" program of these standard Democratic items would produce a breakthrough in black employment.[34] Wirtz replied that the crucial problem was not training and education but rather the sluggish economy, for which the administration had proposed the Keynesian, pump-priming tax cut. "I am forced," Wirtz said, "to the disheartening conclusion that in the present state of the economy it would be a mistake to hold out the hope that setting up a 'crash' program for Negroes would mean their being employed. It isn't true with whites, and it would be less true with respect to non-whites."[35] Kennedy agreed, and focused his fair employment efforts on Lyndon Johnson's existing committee.

Rejecting a national FEPC, Kennedy on June 19 asked Congress to give the PCEEO, under the continued chairmanship of the Vice President, a permanent statutory basis. This in effect would convert the President's committee into a national commission with a congressionally approved and appropriated budget, but would not materially affect its present structure, duties, and enforcement powers. Kennedy affirmed to Congress that "I renew my support of pending Federal Fair Employment Practices legislation, applicable to both employers and unions." But he declined to make this part of his own omnibus bill. He also pledged to issue the long-postponed executive order extending PCEEO coverage to construction jobs supported by federal grants, and he observed that he had recently directed the Secretary of Labor to enforce nondiscrimination in the federally supported apprenticeship and state employment service programs. The omnibus bill's fair employment section was fleshed out with rhetorical commitments urging both organized labor and the NLRB to end racial discrimination in unions, as the federal government itself was doing in its own conditional recognition of employee organizations. But this was relatively weak tea. Even Senator Barry Goldwater was demanding the same thing, and doing so much more forcefully than Kennedy, since it would put the hammer on unions. Kennedy's official lead, then, stopped well shy of FEPC. But he implicitly invited Congress to reach further if it could summon the will to try, and ultimately muster the consensus necessary to overcome a certain Senate filibuster.

The President, however, had no intention of establishing a White House civil rights secretariat, or empaneling a national watchdog commission on indignities, or mounting a liberal, trip-hammer campaign of "no surcease" to grind the opposition down. To the contrary, the Kennedy administration had decided to counter-balance its legislative initiative with a soft-pedaling of its parallel executive actions. This switch in emphasis offered a fitting symmetry, since the Administration during its first two years had pressed its executive authority in civil rights while asking Congress for relative trifles. But the relaxing of executive pressure was a temporary expedient, responding ad hoc to trouble spots, calculated to ease tensions and cool southern tempers in Congress during the crucial coming months of committee hearings and markup on the Hill. In order to avoid charges of hypocrisy or of selling out early, the tactic had to be more real than apparent, and the balancing was tricky.

Thus on June 4, Kennedy released a brief statement directing Secretary Wirtz to enforce nondiscrimination in apprenticeship programs. At the same time he promised to issue "shortly" the long-awaited executive order banning discrimination in federally assisted construction.[36] But Kennedy held his announcement until after he had delivered his new civil rights proposals to Congress on June 19. He then buried the new executive order by releasing it on June 22—a Saturday—without ceremony or comment.[37] In the news backwash that same Saturday, Kennedy released a potentially controversial report on race relations in the armed forces, with appreciative but noncommital letters to the chairman of the reporting committee and to Secretary McNamara.[38] But both the construction order and the Pentagon initiative provoked controversy. In doing so they demonstrated the relative futility of presidential attempts to slide controversial actions past the nation's extra-constitutional system of checks and balances, i.e., the adversarial two-party system, the leak-prone federal bureaucracy, and an alert press.

In the first instance, involving Kennedy's quietly issued executive order against job discrimination in federally assisted construction, the administration felt so vulnerable to congressional sensitivities during the committee hearings that it allowed its bluff to be called by the political heavyweights of Louisiana, and quietly backed down in private humiliation. When Louisiana highway officials complained that the new requirements would radically alter millions of dollars in properly bid and awarded construction project, Louisiana's powerful senior congressmen, most notably Senator Russell Long and Representative Hale Boggs, complained to the White House. To avoid a major blow-up and its attendant deflection of congressional attention and resources, the White House ordered the agencies to rescind their new enforcement rules. Lee White, who was normally one of the more candid and jargon-free White House aides, translated this humiliation into acceptable bureaucratese in a background memo for President Kennedy, in preparation for his meeting with the triumphant governor of Louisiana, Jimmie H. Davis: "The Governor might be interested to know that the question raised by Louisiana was considered, found to be valid, and corrective action was taken. Of course, the less that is said about this in the newspapers, the better off everyone will be."[39]

The roots of the second instance, involving the Pentagon, traced back to June of 1962. In response to the urgings of Lee White and Adam Yarmolinsky, Kennedy had appointed a President's Committee on Equal Opportunity in the Armed Forces. Chaired by Washington lawyer Gerhard A. Gesell, the ad hoc committee was advisory only. Its 93-page report, submitted on June 13, 1963, documented widespread evidence of off-base discrimination against black service personnel. This was not surprising, especially in the South. But the Gesell report's recommendations were. They included the threat of declaring off-limits those communities where discrimination remained intransigent, and ultimately the threat of closing those bases.[40] In receiving the report, Kennedy merely referred it to McNamara and asked for his recommendations within thirty days. McNamara in turn issued a strong directive on July 26 that established a new Office of Deputy Assistant Secretary of Defense for Civil Rights, and directed the military services to institute a training, reporting, and monitoring system for protecting civil rights both on-base and off-base. The Gesell report had said that the record of base commanders had "not been impressive" because as a group they did "not believe that problems of segregation and racial discrimination in the local community should be their concern." So McNamara's directive specifically charged base commanders with enforcement responsibility (the Gesell report had recommended that "officers showing initiative and achievement in this area will enhance their performance ratings and obtain favorable consideration for promotion and career advancement").[41]

Congressional conservatives complained that the Administration was using the armed forces for social reform, involving the military in political matters, and threatening local communities with economic ruin. Senator Barry Goldwater was joined in complaint by such senior southern Democrats as senators Stennis and Talmadge and representatives Rivers and Vinson. These powerful southern friends of the military were so angered at the prospect of McNamara's Trojan Horse that there was considerable overreaction, including the doomsday pronouncement of Representative L. Mendel Rivers of South Carolina that McNamara's directive represented "the beginning of the police state and the commissar program in America." Chairman Carl Vinson of the House Armed Services Committee, normally a sober sort, petulantly introduced a ripper bill making it a court martial offense for military men to carry out the directive.[42]

In his own defense, McNamara pointed out that he had rejected the base-closing recommendation of the Gesell report, and that his directive had prohibited base commanders from exercising the off-limits sanction against racial discrimination without the prior approval of their service secretary. McNamara did not back away from his directive, but he did order stopped a related recent practice of allowing off-duty servicemen to participate in off-base demonstrations against racial discrimination. Earlier that year Lee White had rather routinely approved such a request from Adam Yarmolinsky. Interestingly, the request had originated from airmen based in South Dakota, not South Carolina (the Gesell report had pointed out that off-base discrimination was not confined to the South, and that housing discrimination was often worse outside the South). But Governor George Wallace lashed out at the new practice in testimony against the civil rights bill in July before the Senate Commerce Committee, sarcastically suggesting that "perhaps we will now see Purple Hearts awarded for street brawling."[43] So McNamara quickly banned the practice, and Kennedy upbraided White for allowing it in the first place. Clearly, the Administration was walking a tightrope as the congressional hearings began on the omnibus civil rights bill.

The House-First Strategy: H.R. 7152

In June of 1963 the Administration's legislative strategists were chiefly concerned about the Senate. There a filibuster seemed inevitable if the bill was to contain any real teeth, and Minority Leader Dirksen clearly held the keys to cloture. But Dirksen had publicly declared his opposition to the heart of the bill, Title II on public accommodations. On June 29, Deputy Attorney General Nicholas Katzenbach sent Robert Kennedy a memo outlining strategic options for both chambers. But Katzenbach concentrated on the Senate and the Dirksen problem, fearing that the House would not support public accommodations "unless we can be reasonably convincing in our statements that we can get it through the Senate."[44] Senate Majority Leader Mike Mansfield feared that no such bill could pass the Senate under a Democratic President. But he agreed that success hinged on getting 67 votes for cloture, and Dirksen would hold the key to shutting off the certain southern filibuster.[45]

But the House must come first. There Katzenbach saw a need for "around 65 Republican votes." Where might they come from and who might deliver them? The liberal House Republicans, like John Lindsay, wanted a strong bill. But they were a tiny and potentially troublesome minority. To Katzenbach, Lindsay's liberal Republican minority of a minority looked like insufficient if not potentially dangerous allies; the intensely partisan House minority leader, Charles Halleck, looked highly unlikely; and minority whip Gerald Ford looked both unpromising and ineffective. Besides, both Halleck and Ford had strongly conservative voting records. So the crucial Republican dealer might turn out to be William McCulloch, a man not generally known to the public.

A respected, moderately conservative congressman from Piqua, Ohio (near Dayton), McCulloch was the senior Republican on the 35-member House Judiciary Committee, to which H.R. 7152 had been routinely referred. McCulloch had supported the moderate civil rights bills of 1956–57 and 1959–60. But he was known to have felt betrayed in both of those efforts by the Democratic congressional leadership when commitments made to the Republican leadership in the House were subsequently bartered away by Lyndon Johnson in order to avoid filibusters in the Senate. So Katzenbach urged that "McCulloch should be sounded out before he is committed to a position publicly, and we should get to work on him immediately." He specifically recommended the proven negotiator Burke Marshall for that delicate assignment, and Robert Kennedy promptly dispatched Marshall to Piqua, where Marshall and McCulloch worked out an agreement. In return for his support, McCulloch demanded two ironclad pledges from the administration. First, the President would not allow the Senate's Democratic leadership to gut the House-passed bill by compromising provisions that had been hammered out in the House. To guard against this, the Administration would give McCulloch the *sole* power to approve any changes that it would accept in the Senate. Second, the Kennedy administration would give the Republicans equal credit for passing the bipartisan bill. These were stiff terms, but McCulloch seemed to be in a position to command a high price.[46]

Later Robert Kennedy said of the pact with McCulloch, "The important thing, really, was to focus attention, first, on Congressman McCulloch and that he did it. So that nobody made any effort to take any credit for it in the Democratic Administration."[47] "[W]e wouldn't agree to any changes without his approval," Kennedy confirmed. "Robert Kennedy," Burke Marshall summarily explained, became "the lawyer for Bill McCulloch."[48] The previous January, McCulloch had introduced his own civil rights bill and, early in June, John Lindsay had submitted his version. House Republicans from the moderate-to-liberal range of the GOP spectrum flocked to these two banners. Crafting their own variations, they added two dozen bills to McCulloch's moderate lead and a similar number to Lindsay's liberal banner—eager to embarrass the paralyzed Democrats by demonstrating that Lincoln's party remained the true friend of civil rights and especially the Negro. Their party's 1960 presidential platform had pledged as much.

McCulloch's bill suggested the dimensions of leeway for negotiation. It would make the Civil Rights Commission permanent, while narrowing its jurisdiction to investigating and reporting on voting discrimination. It would also empower the Attorney General to respond to persons filing sworn complaints by seeking court injunctions against the denial of admission to public schools on account of race. But McCulloch's midwestern brand of traditional Republicanism was repelled by the prospect of expansive federal control over local merchants. So his bill was silent on public accommodations, and it proposed no extension of federal regulatory authority over private employers. But it did propose a new Commission on Equality of Opportunity in Employment, which would inherit the Johnson committee's jurisdiction and enforcement powers over government contracts and federal employment—to which would be added the significant power of subpoena. Most revealingly, and quintessentially Republican, McCulloch's bill would grant the new EEO commission explicit jurisdiction over state employment agencies and all labor unions. This would include, importantly, the grant of cease-and-desist authority *and* the power to order offending unions "to take such affirmative action as will effectuate the policies of this title"—all enforceable through the federal courts.[49] Lindsay's bill differed from McCulloch's primarily by including a public accommodations title and basing it on the 14th Amendment, and also by including "Title III" authority for the Attorney General to sue on behalf of a broad range of alleged rights infringements.

When Robert Kennedy appeared before Subcommittee No. 5 on June 26, the committee had already held eight days of public hearings on its mass of civil rights proposals stretching back to May 8. The chairman of the House Judiciary Committee, Democrat Emanuel Celler of Brooklyn, was a veteran New Deal liberal who had appointed himself chairman of his own special subcommittee, and had stacked it with six fellow Democratic liberals (including a lone southern Democrat, Jack Brooks of Texas).[50] McCulloch led the subcommittee's four Republicans, all of them regarded as conservatives.[51] Early in the Attorney General's testimony, as Kennedy talked exclusively about the contents of the Administration's H.R. 7152, he was asked by Republican George Meader whether he was familiar with any of the Republican bills that the committee had been discussing since early May. "I am not," Kennedy replied. "As I think the chairman said, there are 165 bills or 365. I have not read them all."[52] Lindsay quickly interrupted: "I am quite deeply disturbed, Mr. Attorney General, that you have never bothered to read this very important legislation." Lindsay then lectured Kennedy on the proper 14th Amendment base for the public accommodations provision, and Kennedy testily replied: "Congressman, I am sorry I have not read your bill." But the relevant question, Kennedy lectured back, was not whether Mrs. Murphy's boardinghouse was covered under the 14th Amendment or the commerce clause, but whether she was covered at all.

Mrs. Murphy was a mythical widow who had been introduced into the public accommodations debate by Republican Senator George Aiken of Vermont. Her intimate roominghouse quickly came to symbolize in the national media the potentially excessive reach of federal power. Lindsay, like most members of Congress, had no wish to cover Mrs. Murphy. But as a liberal Republican with presidential aspirations, he challenged the combative Kennedy from the left. "In view of the fact that you apparently did not consider these bills at all," Lindsay said, "I can't help but ask the question as to whether or not you really want public accommodations or not." "The rumor is all over the cloakrooms and corridors of Capitol Hill," Lindsay added, "that the administration has made a deal with the leadership to scuttle the accommodations."[53] "I don't think," Kennedy retorted, "the President nor I have to defend our good faith efforts, here to you or to really anyone else. I want this legislation to pass. I don't think, Congressman, that I have to defend myself to you about the matter."

It wasn't a good beginning. Kennedy never repaired his fences with the equally young and ambitious Lindsay, whom he disliked and distrusted. But Kennedy's impatience and his partisan and filial loyalty had reinforced his reputation for arrogance, and his ham-handed performance on June 26 had also irritated McCulloch and the Republican centrists, whose support could prove crucial. In his subsequent testimony the Attorney General better observed the exaggerated decorum that protocol had long demanded on the

Hill. More important, the following week the diplomatic Burke Marshall was sent to Piqua, Ohio as the Kennedy emissary to negotiate with McCulloch. Meanwhile in the Senate, a dual set of committee hearings was gearing up to reflect the peculiar nature of the Administration's Senate strategy, especially in light of that chamber's southern-dominated committee chairmanships, and the crucial position of Senator Dirksen.

Dual Hearings in the Senate: The 14th Amendment versus the Commerce Clause

The Senate picture was clouded not only by the prospect of filibuster, but also by the southern domination of the Senate's committee structure. Both circumstances suggested the wisdom of a House-first strategy, through which a moderately strong bill might be reported out of Celler's sympathetic Judiciary Committee, then overwhelm Judge Smith's Rules Committee, and thereby offer bargaining chips with which to negotiate for Dirkesn's support for cloture in the Senate. Meanwhile the civil rights coalition could build up a groundswell of national support for a bill aimed primarily at the dramatic southern abuses, with the ironical cooperation of the South's Bull Connors. But because the Senate Judiciary Committee was chaired by James Eastland of Mississippi, who could be counted on to block any significant civil rights measures, the Administration worked out a special arrangement with Mansfield and Dirksen. First, Mansfield would introduce the Administration's civil rights bill, S. 1731, and this would routinely be referred to Eastland's Judiciary Committee. Mansfield would then be joined by Dirksen in introducing S. 1750, which was the same bill *minus* the public accommodations title to which Dirksen objected. The Mansfield-Dirksen bill would also be referred to Judiciary.[54] Finally, Mansfield would be joined by Democratic Senator Warren Magnuson of Washington in introducing S. 1732, which dealt *only* with public accommodations. Because S. 1732 (which was essentially Title II of the Administration's bill) derived its constitutional authority primarily from the interstate commerce clause, it was referred to the Senate Commerce Committee, which Magnuson chaired. This artifact of the congressional seniority system presented a strong tactical argument for resting Title II mainly on the commerce clause. But that was not the main reason for the commerce rationale, as the Administration would soon explain.

The Commerce Committee was a sympathetic forum for Title II. Its 17 members included only five Republicans, none of them from the former Confederacy. Two were moderates from the border states (Thurston Morton of Kentucky and J. Glenn Beall of Maryland), but none of the Republicans was inclined to defend Jim Crow. The committee's twelve Democrats included only two southern senators. One of these was former Dixiecrat Strom Thurmond of South Carolina. But Thurmond's intransigence was well known, and was balanced by the unusually strong liberalism of the other southerner, Ralph Yarborough of Texas. In eleven days of hearings that demonstrated the routine openness of debate in the American legislative process, the Commerce Committee heard seventy-nine witnesses, including half of the nation's governors.[55] Opposition spokesmen were given full voice, as eight southern governors (including George Wallace of Alabama and Ross Barnett of Mississippi) were joined by three southern attorneys general, plus the segregationist journalist James J. Kilpatrick of the *Richmond News-Leader,* and the arch-conservative nonsouthern publisher, William Loeb of the Manchester (New Hampshire) *Union Leader.*

The opponents based their case less on a defense of Jim Crow per se than on the hoary rhetoric of neoConfederate state-rights, the mythic memories of Black Reconstruction, the specter of federal tyranny as a betrayal of the Founders, and a plea for gradualism and voluntarism. But those too-familiar pleas were mocked by the televised visions of Birmingham. August witnessed the massive and dignified March on Washington for Jobs and Freedom, highlighted by King's moving sermon, "I Have a Dream." The heart of King's dream was that black Americans could "one day live in a nation where they will not be judged by the color of their skin but by the context of their character." Arrayed against this dream, the state-rights and property-rights arguments that the conservatives advanced in Senate testimony were potentially powerful, but they rang hollow in the manifest absence of state responsibilities and the palpable presence of state brutality. Largely failing to rally to a more credible defense even of Mrs. Murphy's boardinghouse, the conservatives seemed implicitly to defend the constitutional rights of "Bull" Connor. They were politely heard, and utterly overwhelmed by the weight of the moral argument. The best states rights defense against a federal Title II was equivalent state efforts at reforming the humiliating abuses of the biracial caste system in public commerce, and in this the southern conservatives manifestly failed. Even the commissioners of baseball and of the National and American football leagues, who were no strangers to the halls of Congress, testified in the hearings to the irrelevance and inefficiency of Jim Crow in a free market society.

The state-rights argument in constitutional debate had always been in defense of regional minorities, including early New England. By invoking it the white defenders of the South's peculiar folkways, like their Confederate forebears, denied the theory's implicit protections to their own regional minority. By default, then, the serious constitutional arguments concerning the form, limits, and rationale of Title II's radical federal intrusion into the sanctity of local retail prerogatives fell to the ideological and partisan proponents of the 14th Amendment rather than the commerce clause—that is, to the Republicans. Equal protection was such a loftier banner than interstate commerce. And it was here, in making his case for the commerce clause against deep-seated Republican instincts, that Attorney General Robert Kennedy in his Senate testimony performed superbly. In arguing for the primacy of the commerce clause, he was aided not only by the experienced battery of legal minds in the Justice Department, and the dutifully supporting written testimonials from nine federal agency heads, but additionally by the knowledgeable testimony of Commissioner Erwin N. Griswold of the U.S. Commission on Civil Rights, who was also dean of the Harvard Law School. Kennedy's argument was supported most impressively, however, by Griswold's learned colleague on the Harvard law faculty, Professor Paul A. Freund, whose brief was prepared at the request of the committee.

Because the Republican congressional leadership had resisted the commerce clause rationale out of instinctive resentment at its historic use by Democratic administrations to justify broad expansion of federal regulatory power, their arguments appeared to be consistent with the strict-constructionist safeguards of principled constitutional conservatism, especially as it bore on protecting the freedom of private enterprise from strangling federal encroachment. But Freund's constitutional brief reversed the equation. He argued instead that the commerce clause approach was the cautious and prudent one, and that a 14th Amendment approach would open up a broad new class of constitutional claims against private enterprise.[56] Freund cited overwhelming Supreme Court support for congressional authority to regulate private enterprise, stretching back a half-century to the white slave laws. This long regulatory tradition included legislating for specific moral purposes, such as protecting child labor from abuse and protecting vulnerable consumers from securities fraud and adulterated food and drugs. It also included legislating against various forms of discrimination, such as discrimination against consumers through price-fixing (the Robinson-Patman Act) and against unions through unfair labor practices (the National Labor Relations Act)—even when the context of the proscribed activity seemed purely local. Because such congressional regulation based on the commerce clause was indisputably constitutional, Freund reasoned, the question was not one of constitutional power but rather one of legislative policy. The great and indeed conservative advantage of such a legislative grant of power to Congress was precisely that it could then "be exercised in large or small measure, flexibly, pragmatically, tentatively, progressively." This was quite unlike a new class of constitutionally guaranteed rights, which "if they are declared to be conferred by the Constitution, are not to be granted or withheld in fragments."[57]

For this reason, Freund explained, the Supreme Court had never explicitly "overruled" its narrow interpretation of the 14th Amendment in the Civil Rights Cases of 1883. First, it had been unnecessary to do so in order to secure federal protection against racial discrimination in public accommodations, because the commerce clause and similar constitutional grants of authority to Congress (e.g., defense, spending) were more specifically tailored to the ends to be achieved. Second, it was unwise and even potentially dangerous to do so. Because the 14th Amendment was "spacious in its guarantees (equal protection and due process), and is cast largely in terms that are self-executing," such a broad, 14th Amendment-based protection would carry much larger constitutional implications, creating an unclear range of new rights.

Republicans who called for a 14th Amendment basis for a public accommodations law typically argued for a narrow application, not a broad one, by linking its enforcement to the state licensing practices. According to this 14th Amendment logic, then, the states would be prohibited from participating in racial discrimination against Negroes through their licensees, the hotels and restaurants that were a main target of Title II. But the states varied widely in their licensing requirements, Freund observed, which were designed to promote and protect public safety and sanitation and the like, as well as incidentally to raise tax revenue. Moreover, the states could change these practices in order to avoid the reach of a 14th Amendment-based Title II. Furthermore, would such a novel constitutional logic then also include all corporations operating under state charter? Would it newly encompass all licensed professions—all lawyers and medical practitioners, for instance? All private schools and colleges and charitable institutions? And how could such a new constitutional guarantee make desirable exemptions for size, as in Mrs. Murphy's boardinghouse? Clearly, Freund warned, such a 14th Amendment approach promised to "open up new areas of direct constitutional relationships which will call for judicial creativity on a formidable scale."

In his opening testimony before both the Senate Commerce Committee and the Judiciary Committee, Robert Kennedy skillfully pressed this brace of arguments.[58] Before Judiciary, he faced not only the courteous opposition of chairman Eastland, but also, on public accommodations, the coy opposition of the ranking Republican, Everett Dirksen. In historically legitimizing the commerce clause approach, Kennedy cited the full pantheon of the nation's major commerce-based, landmark laws, littered as they were with the names of distinguished and historic senators: Sherman, Clayton, Wagner, Taft-Hartley. They sought fair labor standards, pure food and drugs, meat inspection, water pollution control, truth in securities. They sought also moral reform of child labor, gambling, prostitution. But a 14th Amendment approach through state licensing would confront a fascinating array of state differences, Kennedy said. In Alabama (a favorite Kennedy example), it would necessarily include architects, embalmers, sleight-of-hand artists, and even the state's curiously licensed feather renovation industry. Yet in Minnesota it would *exclude* department stores, and in Pennsylvania, hotels and motels. Kennedy acknowledged that the 14th Amendment could be cited as backup authority. But his argument that primary reliance must rest on the commerce clause was compelling, and the Commerce Committee hearings constituted an admirable case study of a focused and effective forum for reaching reasonable legislative consensus in a controversial area of public policy. Besides, the Commerce Committee was loaded in favor of some effective form of Title II, and as Katzenbach observed in his frequent legislative status reports to Kennedy, this was especially important because the "Title II reported by the Commerce Committee should be a strong one since it will set the floor for the House."[59]

"Committee Filibuster" in Judiciary

The same constructive assessment could not be made for the Judiciary hearings, however. There Kennedy's chief antagonist was neither Eastland nor Dirksen, but committee member Sam Ervin of North Carolina. Ervin's frequent protestation that he was nothing but a "poor old country lawyer" was confirmed by his folksy, piedmont demeanor, but belied by his Harvard Law credentials and by his experience as a justice on North Carolina's Supreme Court.[60] A younger generation with a short memory would later recall Ervin as the learned constitutional scourge of Nixon and Watergate. But like the other, and mostly lesser, southern defenders of the South's biracial caste system, Ervin was forced ultimately to repair to the weak reed of states rights gradualism, which certainly was constitutionally defensible in theory in 1963, but was clearly not politically defensible in practice.[61]

From his privileged chair among the committee majority, and fortified by chairman Eastland and the elaborate protocols of senatorial courtesy, Ervin conducted what journalists soon called a "committee filibuster." As the Judiciary hearings, going nowhere, dragged on into August and September, bringing the long-suffering Attorney General with them, Ervin hammered away at the threat of federal tyranny, appealing grandiloquently and at great length to Magna Carta and the Founders and Daniel Webster. But he repaired less to the 10th Amendment, that traditional and antiquarian shrine of state-rights confederalism, than to the due process clause of the 5th Amendment. Ervin argued with dissembling self-deprecation—"in my feeble way"—less for state-rights than for separation of powers, and especially for constitutional restraints and legislative resistance to executive law-making. He therefore moved beyond public accommodations to Title VI on fund cut-off and especially to a House-proposed Title VII on fair employment, complaining that the "discrimination" that was proscribed was never even defined in the proposed law. Thus the definitions of what behavior was deemed illegal, Ervin said, would perforce be left to the considerable discretion of the President and his agents, who would also apparently determine the punishments. Violators would therefore not be told precisely what behavior was prohibited, Ervin warned, and hence would be held accountable to a kind of floating bill of attainder, with both the crime and the punishment to be decided, *post hoc,* by federal bureaucrats.

A decade later, Ervin would distinguish himself with the same arguments, and Americans confronted with shocking presidential and executive abuses would resonate to Ervin's stern indictment of Nixon and his co-conspirators. But in 1963, Ervin had no smoking gun. So he was forced to imagine one. And in relentlessly pressing this logic, he was reduced to the *argumentum ad horrendum*—that the omnibus bill would cover, under its equal employment provisions for federal contracts, the ordinary American housewife and her maid. By hiring a domestic servant and paying social security taxes on her wages, the American housewife would thereby become a federal contractor! As Kennedy wearily but politely observed, "Senator, I just do not think that makes any sense."[62] And in the political context of the summer of 1963, it didn't.

Origins of Title VII: From FEPC to EEOC

Throughout the prolonged hearings in the Senate Judiciary's "committee filibuster," then, Senator Ervin found it hard going to pin the Attorney General down by demonstrating the horrors that could be expected to follow passage of the administrations's omnibus civil rights bill. There were several reasons for this. First, Kennedy was too adept at deflecting Ervin's *ad horrendum* logic by returning attention to the simple moral purpose of the bill, which was to eliminate the unfair discrimination against blacks that was so massively demonstrable, especially in the South. Also, Ervin had to rely principally on hyperbolic imagination because he could find so few genuine examples of federal tyranny in the field. Ironically, this was partly because he and his like-minded predecessors had successfully thwarted all attempts at creating a statutory FEPC whose outrages he could document. Also—although Ervin would not concede this—the state and local FEPCs had been, by and large, models of responsible restraint in wielding their cease-and-desist authority. Furthermore, Lyndon Johnson's EEO committee had similarly avoided any heavy-handed abuse of its more limited jurisdiction and circumscribed authority.

Finally, the Administration bill that Kennedy defended proposed no new federal FEPC, but only a statutory recognition of the Vice President's EEO committee. This later point was helpful to the Attorney General, but it was distressing to the civil rights coalition. The well-organized civil rights community—the NAACP, the Urban League, CORE, Dr. King's SCLC, together with such white-led liberal groups as the ACLU, the Anti-Defamation League, the ADA and the AFL-CIO, most of them cooperating closely under the Leadership Conference for Civil Rights—was long accustomed to this combination of presidential timidity and congressional resistance, and they had done their homework well. Necessity had forced the civil rights lobby to become practiced manipulators of the pluralistic system of American democracy, and early in the Kennedy regime they had seized an effective opportunity to forge an instrument to serve their purposes.

When Harlem's Adam Clayton Powell in 1961 replaced Carl Perkins of Kentucky as chairman of the House Labor and Education Committee, he established a special subcommittee on labor and charged it with a singular mission. Powell appointed Congressman James Roosevelt of California to chair the new subcommittee, and dispatched it on a highly visible, transcontinental tour of hearings to inquire into the condition of equal employment opportunity across the land.[63] But the tour was confined to the upper, nonsouthern tier of the United States. Powell and Roosevelt wanted to demonstrate that racial discrimination in employment was not a peculiarly southern problem, and that even states with vigorous FEP laws and commissions were in need of a strong federal statute and commission to enforce it. So they held their subcommittee hearings during the fall of 1961 in Chicago, Los Angeles, and New York, and consolidated them back in Washington in January 1962.[64] Not surprisingly, the Roosevelt subcommittee's list of witnesses was loaded in favor of their proposed federal commission, as was the membership of the subcommittee and its parent committee.

The Roosevelt subcommittee concentrated its hearings on loosely structured areas of employment that the state FEPCs found difficult to reach, such as waiters and bartenders and hotel bellmen in Los Angeles, and nonsalaried commission salesmen in New York. In smoking out persistent discriminatory practices in such areas, they summoned reluctant witnesses by subpoena, and on at least one occasion their zeal overreached their responsibilities. At the New York hearing, where Powell joined the subcommittee, the staff had sent handpicked blacks to apply to the Wearever Aluminum Company for jobs as commissioned salesmen of household stainless and china. When the committee's applicants were not offered jobs, the committee staff summoned Wearever's hapless district sales manager to explain why the subcommittee's carefully qualified applicants were rejected and why the manager's force of 550 salesmen included only one recently hired Negro.[65] In response to this unusual episode of traveling committee entrapment, Congressman Roman Pucinski complained that however deplorable was the discriminatory practice, the subcommittee nevertheless was not New York's State Commission for Human Rights. Representative Charles Goodell joined Pucinski, objecting that "this committee has no jurisdiction in enforcing this State commission . . . We are not an enforcement body."[66]

On the whole, however, Roosevelt's subcommittee conducted itself with responsible restraint and credibly documented what it set out to demonstrate. It sought from witnesses like New York's Will Maslow, executive director of the American Jewish Congress and a veteran of Franklin Roosevelt's original FEPC, knowledgeable advice about the strengths and weaknesses of the state FEPCs, and the advantages and disadvantages of a NLRB model for a federal FEPC. Such informed testimony educated the subcommittee and its public on the functional and legal distinctions between the quasi-judicial, administrative agencies like the state commissions on the one hand, and on the other such federal enforcement instruments as the Vice President's interagency committee. A chief and troublesome distinction was that state FEPCs ultimately enforced their cease-and-desist orders through the courts. But a national FEPC as a federal agency would run afoul of the doctrine of separation of powers, and could scarcely be effective by trying to sue other federal agencies in court.[67] Yet because agencies like the Johnson committee had ultimate access, through the President, to command authority over government employees and contracts, they enjoyed sanctions in those areas that were potentially more effective than cease-and-desist orders, which seemed better tailored to the private sector.[68]

It was, indeed, unusual for congressional hearings on legislative matters of such widespread popular concern to focus so constructively on the substance and language of the law. Most congressional hearings served primarily to allow groups arrayed on both or all sides of an issue to express their sentiments, customarily in lay rather than technical language. In the Washington hearings in 1962, the Roosevelt subcommittee briefly examined the implications of including age and sex discrimination, with spokesmen for organized labor showing considerable enthusiasm for protecting age and very little for gender. Before the committee was finished, it gave opposition spokesmen, most notably fellow congressmen from the South, their opportunity to explain why the subcommittee's proposal amounted to a "Star Chamber Employment Bureau."[69] But throughout the entire EEO hearing process stretching back to Franklin Roosevelt's original FEPC, the proliferation of aggressive proponent groups contrasted starkly to the sameness and defensive scattering of opposition spokesmen, who reflected a static group base of southern political constituencies and officeholders.[70]

From H.R. 405 to Title VII

By 1963, the Powell committee had refined its own full proposal for a fair employment title, and entered it into the House hopper as H.R. 405. Under normal circumstances, such liberal proposals from the Labor and Education committee stood little chance of surviving the Rules Committee, where their aid-to-education efforts, for instance, and been ritually slaughtered during the Kennedy administration. But these weren't normal times, at least not for civil rights. In 1962 subcommittee chairman Roosevelt had gotten Labor secretary Arthur Goldberg to testify, implicitly for the Administration as well as for his department, that he supported the subcommittee's efforts "in principle." Roosevelt had resisted the frequent efforts of the more zealous civil rights spokesmen to arm the proposed commission with the powers of both prosecutor and judge. He was constantly looking over his shoulder at the hostile Rules Committee, and he wearied of complaints from angry civil rights activists that his committee's emerging proposal was too timid.

On June 6, 1963, the Roosevelt subcommittee completed ten days of final testimony on H.R. 405, and the full committee reported the bill out on July 22. On the face of it, H.R. 405 created nothing very much resembling a "Star Chamber Employment Bureau." It declared it an unlawful practice for an employer, labor union, or employment agency "to fail or refuse to hire or to discharge any individual, or otherwise to discriminate against any individual with respect to his compensation, terms, conditions, or privileges of employment, because of such individual's race, religion, color, national origin, or ancestry."[71] There was no mention of sex discrimination anywhere, but age was added as a protected category (excluding, however, the operations of a bona fide seniority system). The proposed enforcing instrument was a five-member Equal Employment Opportunity Commission, which had the advantage of replacing the controversial acronym FEPC with the at least then neutral acronym EEOC. The commission's six itemized "powers" testified to its ostensible conciliatory spirit: to cooperate with similar state and local agencies, to pay the expenses of witnesses, to offer technical assistance, to assist employers in conciliation upon request, to make technical studies, and to create an advisory and conciliation council. Nothing very frightening there.

The proposed federal EEOC, however, was modeled on the state FEPCs. It thus would be similarly armed with the authority, after a structured investigation and a finding of probable cause, and also only after the failure of conciliation, to order the offending respondent "to cease and desist from such unlawful employment practice and to make such affirmative action, including reinstatement or hiring of employees, with or without back pay . . . as will effectuate the policies of the Act." The proposed EEOC's guarantees of due process to potential respondents (meaning employers charged with discrimination) were substantial, including time and geographical limits on claims and rights to confidentiality and to judicial review. But the EEOC could also compel witnesses, take testimony under oath, investigate complaints, enter and inspect premises and records, and require all employers to maintain stipulated records and to make periodic reports. As for jurisdiction over federal employees and government contractors, H.R. 405 recognized the inappropriateness of endowing such a quasi-judicial, administrative agency with court-enforceable authority over other federal agencies. So the bill yielded this authority to presidential discretion.

H.R. 405, however, was no part of the Administration's bill. Moreover, Roosevelt's special subcommittee had always been a sideshow in relation to the real powers of Congress. By August 5, Celler's Subcommittee No. 5 had completed 22 days of public hearings on the administration's omnibus civil rights bill, and compiling in the process 1,742 pages of printed testimony. Celler then took the committee into private markup sessions. President Kennedy asked Celler to stall the markup sessions, however, out of a fear that rumors of progress on that front would upset the seven southerners on the House Ways and Means Committee (including chairman Wilbur Mills of Arkansas) and thereby threaten the President's higher priority tax cut bill. So Celler and McCulloch agreed to restrict the first seven markup sessions to bland discussions with no amendments, and Congress broke for the Labor Day recess on the heels of the March on Washington for Jobs and Freedom on August 28.

That evening, the President received the civil rights leaders in the White House, where he rattled off the political arithmetic of the Congress, and discouraged their demands for Roosevelt's FEPC-style commission and their perennial call for strong Title III authority for the Attorney General. Kennedy's listeners included Martin Luther King, A. Philip Randolph, Roy Wilkens, Whitney Young, Walter Reuther, and also Lyndon Johnson. Kennedy warned them that the strategically placed Congressman McCulloch had said that "If I wanted to beat your bill, I would put FEPC in. And I would vote for it, and we would never pass it in the House."[72]

Celler would not reconvene his subcommittee for serious markup until September 10, when the President's tax bill had been safely voted out of the

Ways and Means Committee. In a confidential letter to Celler of August 13, Katzenbach had spelled out the Administration's cautious negotiating strategy on H.R. 7152, concentrating on titles I on voting (which had been a chief interest of Celler's) and II on public accommodations.[73] Katzenbach had been leery of the draft of Title III on school desegregation because it included the problematical term "racial imbalance." This was intended to help justify some forms of federal financial assistance to northern school districts with de facto segregation. But to Katzenbach it carried the risk of sparking a controversy over "quotas, bussing children across town, etc. These are not intended to be endorsed by the act." Katzenbach's six-page letter to Celler devoted only two sentences to the committee's job discrimination proposals, and then only to confirm that the Administration had no objection to explicitly including unions in the PCEEO's jurisdiction. Celler, however, would ignore the Administration's cautious strategy and Katzenbach's letter, for he had long entertained a private strategy of his own. Soon he would spring this on an angered Robert Kennedy and a dismayed William McCulloch, and threaten to unglue the entire package. The results of this imbroglio would fashion the basic shape of the Civil Rights Act of 1964. And while that congressional debate was occurring, the nation would become involved in a corollary debate on the hitherto latent notion of racial quotas, which seemed to lurk fugitively but implicitly in such vague normative concepts as "underutilization." As a result, the Civil Rights Act of 1964 would bear the unambiguous imprint of that debate.

CHAPTER IV

The Storm Over Racial Quotas

The Economic Crisis of Black America

The televised brutality in Birmingham triggered a national revulsion that led with surprising speed to a frontal attack on segregation in the South. But the new sense of breakthrough in black civil rights that quickened in the summer and fall of 1963 masked a more sobering economic reality: because black workers were not sharing proportionately in America's great postwar economic surge, their relative economic well-being seemed to be deteriorating. The leaders of the major civil rights organizations were aware that by 1958 black unemployment had reached two digits and generally doubled the white rate, especially among young males.[1] During the early 1960s civil rights leaders pressed their case with compelling evidence, strongly supported by the northern philanthropic foundations and the national leadership of the AFL-CIO. But they lacked the gripping symbol of nationwide economic discrimination that was provided in the South by the black martyrs in the fight against Jim Crow. They also lacked, during the early Kennedy years, a prominent champion in the Administration. The continued absence in Kennedy's white male Cabinet of a Department of Urban Affairs, headed by a black secretary who might symbolize the emerging national crisis, meant that the logical candidacy for leadership in the fight against job discrimination fell largely by default to an unlikely source: the Department of Labor.

Labor secretary Arthur Goldberg was a forceful leader whose political acumen was widely respected. But as labor's superlawyer architect of the AFL-CIO merger, Goldberg poorly concealed his ambition for a "labor" seat on the Supreme Court, and his tenure as secretary was too brief and too crowded with other agendas to accommodate such a major role. So the potential mantle fell to his deputy secretary and heir, the Stevensonian liberal W. Willard Wirtz. An academic labor economist, Wirtz carried a professorial lineage and rather formal demeanor. But behind it lay a willful independence. These qualities tended to generate among the senior White House staff both admiration and irritation.[2] Samuel Merrick, an old sailing partner of Jack Kennedy's who was appointed director of legislative liaison for the Labor Department, admired Goldberg as a "completely political animal" who had fought his way up through the hurly-burly of local politics, who knew the congressional pulse intimately, never complained, and had "plenty of armor-plate."[3] Wirtz, on the other hand, was by temperament and training an intellectual. Unlike Goldberg, Merrick said, Wirtz "doesn't like the blarney process on the Hill, and he thinks it's a waste of time to spend time up there." But Merrick conceded that Wirtz was usually convincing in formal congressional testimony.

To inform the secretary's testimony, the Labor Department contained a fine repository of historical evidence and a relatively precise monitor of current economic trends in the Bureau of Labor Statistics. By 1963, Wirtz had hit full stride in mining these data. He argued before congressional committees that the American economy was employing its workforce neither effeciently nor democratically, and that the crippling matrix of penalties against black workers that was historically embedded in its structure was in many ways widening rather than narrowing the gap between white and black standards of living. In the years immediately following World War II, Wirtz said, the black unemployment rate had averaged about 60 percent higher than the white rate. But by the early 1960s this had climbed to a plateau that doubled the white rate—in June of 1963 the official white jobless rate was 5.1 percent and among blacks it was 11.2 percent.[4] The early 1960s was an era of economic staleness that buzzed with fears of automation and the potentially radical displacement of workers through "cybernetics." In congressional testimony Wirtz emphasized the grow vulnerability of black workers in an economy that was rapidly moving toward a more technical and service-oriented base. In 1963, in an economy that had flattened following the robust 1950s, nearly 15 percent of employed blacks still worked on farms, but only 5 percent of whites did so. Fifteen percent of employed blacks but only 2 percent of whites worked in private households; 14 percent of nonwhites outside agriculture held unskilled jobs, as against 4 percent of whites. Only 17 percent of nonwhites held white collar jobs, compared with 47 percent of whites.[5]

At midsummer in 1963 the Census Bureau released a study by economist Herman P. Miller that documented an alarming pattern of relative black deterioration since World War II. In 1963 black income hovered at about 55 percent of white income, Miller reported, and in the South the earnings of nonwhites were only one-third of those of whites of similar occupations and schooling.[6] While blacks had raised their occupational levels faster than whites since 1940, most of this gain was accounted for by black migration from the rural South rather than through major national improvements in job opportunities. "In most states," Miller concluded, "the nonwhite male now has about the same occupational distribution relative to whites that he had in 1940 and 1950."[7] During his frequent testimony before congressional committees over the summer of 1963, Wirtz emphasized the importance of the Kennedy tax cut bill, repeating the Administration's admonition that a rising tide raised all boats, and that "Unless we have more jobs, the cost of eliminating discrimination will mean the loss of a job by someone else." He warned Celler's Subcommittee No. 5 of House Judiciary that "It will be a hollow victory if we get the 'whites only' signs down, only to find 'no vacancy' signs behind them."[8] Wirtz's message was a dual one. Racial discrimination in the workplace must be stopped because it was an imperative of both moral philosophy and economic efficiency. But without the Keynesian tax cut to stimulate growth, anti-discrimination could become a fratricidal, zero-sum game.

The Double-Edged Sword of Racial Quotas

Organized labor was historically quite familiar with zero-sum games and racial quotas. Indeed, both union-busting and closed shops were essentially zero-sum conflicts, and the racially segregated unions were essentially zero-quota membership organizations. But even in some of the unsegregated unions, especially in vintage areas like the railroad brotherhoods of the old AFL, officially sanctioned quotas had long protected the desirable jobs of senior white workers, or had been used to strip away traditionally Negro jobs. Early in the century, for example, the job of shoveling coal into engines was hot, dirty, "Negro work," and as a result, by 1910, 6.8 percent of all rail firemen and 41.6 percent of southern firemen were black.[9] But as diesel engines made the fireman's job more attractive, and the solidifying seniority system made service as a fireman a prerequisite for promotion to engineer, whites began to displace black firemen. They achieved this by first excluding new union members who were "unpromotable" (i.e., Negro), then by making quota agreements with rail carriers that would get around the awkward fact of senior black firemen by placing a two-thirds limit on jobs for "nonpromotable" firemen. The euphemisms veiled a devastating process of job-snatching by white labor. By 1950 the vulnerable black firemen had been reduced to only 4 percent (2,130 of 53,310) of all firemen (with none at all listed in New York and New Jersey). Yet this was not rapid enough to satisfy the demands of white workers. So in 1941 the Brotherhood of Locomotive Firemen and Enginemen (BLF) worked out the "Southeastern Carriers Agreement" with twenty-one railroads, and this established an explicit quota limit on Negro firemen.

The BLF's agreement backfired, however, when a displaced black fireman in Alabama sued. This led to the crucial *Steele* decision of 1944, in which the Supreme Court broadened the doctrine of fair representation to include racial equity.[10] During the decade following World War II, and especially since the AFL-CIO merger of 1955, the national leadership of organized labor under George Meany and Walter Reuther struggled to end the anachronistic and embarrassing tradition of racially segregated unions. By midcentury, despite the strength of the segregationist tradition in the transportation industry—as symbolized in mirror image by A. Philip Randolph's all-black Brotherhood of Sleeping Car Porters—the locus of segregationist defensiveness had shifted to the construction trades. By 1963, the national leadership of the AFL-CIO had been able to eliminate segregated international unions completely, and to reduce the number of segregated locals, out of the federation's total of 60,000 locals, to about 170. These remained mostly in the South (although in the construction trades the pattern was more national) and considerably on the defensive. They feared less the pressure from the AFL-CIO brass, which might threaten and cajole but would not expel them, than they feared the NLRB and the threat of union decertification contained in *Steele*.[11] During 1963 a trial examiner for the NLRB heard charges filed by an all-black local on the Texas Gulf Coast against the International Longshoremen's Association, where an all-white local continued to profit from a 1950 agreement that allocated a maximum quota of 25 percent of the work to a Negro local. Early in 1964, the trial examiner ruled that the racial quota violated the fair representation doctrine. The following June the NLRB itself finally dropped the other shoe, as implied in *Steele* in 1944, and ruled in the *Hughes Tool* case, which had also originated in Texas, that racial discrimination was grounds for stripping unions of NLRB recognition as official bargaining units.[12]

By 1963, then, the racial quotas against blacks in the workplace, like the earlier hidden quotas against Jews in the Ivy League colleges and the prestigious medical schools, were being flushed out and exposed to an emerging national censure. Such racial and ethnic quotas were uniformly seen as an embarrassing relic of the past, a last-gasp, often covert defense by selfish and conservative elites against true integration. They found no public defenders. In such an atmosphere, it scarcely seemed conceivable in the early summer of 1963 that racial quotas in "benign" form might be advanced by the political left as a desirable reform.

The Birmingham crisis had unleashed a complicated new mood that combined contradictory elements of optimism and anger, hope and urgency, and it endowed the debate over the civil rights bill with a carrot-and-stick quality. Against a backdrop of sporadic black violence that was scattering throughout the non-South, and facing the challenge at last of a serious congressional push for a civil rights breakthrough, the 13-year-old, 52-member Leadership Conference on Civil Rights organized an intensive lobbying drive in early summer. The conference formed an Ad Hoc Council for United Civil Rights Leadership with the special mission of galvanizing the support of the major, national, white-dominated religious organizations—the (Protestant) National Council of the Churches of Christ, the National Catholic Welfare Conference, and the Synagogue Council of America—to apply moral pressure, especially in those states where blacks constituted a small and politically weak minority. Meanwhile the major black-dominated civil rights groups—the NAACP, the National Urban League, the Congress of Racial Equality (CORE), the Southern Christian Leadership Conference (SCLC), the NAACP Legal Defense and Education Fund, the National Council of Negro Women, the new Student Nonviolent Coordinating Committee (SNCC), and Raldolph's Negro American Labor Council—met in New York to announce an emergency-fund drive to raise a $1,500,000 war chest and to coordinate plans for the March on Washington for Jobs and Freedom in August.[13] The Urban League's director, Whitney Young, was asked to head the fund-raising drive in recognition of the League's successful record in eliciting corporate generosity. At its annual convention in Chicago that July, the NAACP reflected the impatient new mood, attacking President Kennedy for the weaknesses of a civil rights bill that would have thrilled them just months before, and warning that the dangerous surge of racial unrest would continue until full Negro demands were met.[14]

CORE and the Logic of Compensatory Discrimination

The NAACP, however, with its long commitment to legal initiatives and its base in the black middle class, was in no position to control the new grassroots forces. Nor was the Urban League, with its close financial ties to the business establishment and the Community Chest. But the Congress of Racial Equality had been essentially a grassroots organization since its founding to protest northern racism in World War II. James Farmer, its national director, was by 1963 therefore riding a wild horse. A black native of Texas who had studied for the ministry at Howard University under the Methodist pacifist Howard Thurman, Farmer had taken his divinity degree to Chicago in 1941 as a full-time field worker for the Fellowship of Reconciliation. The FOR in turn had been founded during World War I, and by World War II had added to its legacy of Christian pacifism and commitment to racial equality a new interest, under the leadership of the radical reformer A. J. Muste, in Gandhian nonviolent direct action. As a veteran of CORE's founding at the University of Chicago in 1942, Farmer frequently reminded his listeners that the O in CORE stood for the Congress *of*, not *on*, Racial Equality, because CORE was a racially mixed group. "We do not think it is possible," Farmer had explained in testimony before James Roosevelt's special subcommittee on labor, "to fight racial discrimination through a segregated weapon."[15] By 1963, however, CORE had moved far beyond its wartime origins in the Christian-pacifist Fellowship of Reconciliation.

CORE had pioneered in the freedom ride against segregated bus transportation in the upper South in 1947. But it had disintegrated in the 1950s, only to be revived by the galvanizing Freedom Rides into the Deep South in 1961, when Farmer returned to CORE as national director.[16] During the next two years CORE chapters proliferated throughout the country, concentrating on voting and public accommodations in the South, and on housing and job programs in the North and West. Like their co-workers in SNCC, the CORE workers in the South resented the failure of the Kennedy administration to protect them from violent white reprisals, and this was a shared prelude to their common radicalization after 1964. Between 1961 and 1963 CORE's northern chapters also grew more militant, escalating their picketing and boycotts from retail chains to banks, the construction industry, and even to manufacturers. Ultimately they escalated their demands from nondiscrimination to preferential employment policies.

The modern boycott as a weapon against northern job discrimination had gained prominence in Philadelphia during the early 1960s. There a loosely organized group of almost four hundred black ministers used their pulpits effectively to coordinate single-product consumer boycotts against such firms as Pepsi-Cola, Esso, Gulf Oil, and Sun Oil.[17] The Philadelphia boycotts of 1960–1962 persuaded twenty-four firms to agree to specific hiring goals for blacks, and the tactic quickly spread to Boston, New York, and Detroit. CORE chapters found themselves attracted to the boycott as a galvanizing and recruiting device. This threatened the more traditionalist local groups like the NAACP. But CORE itself was threatened on its left by newly emergent community groups who lacked a restraining ideology or national organization, and hence who tended to outbid CORE for local militant support. By early 1962 these pressures had forced an internal debate by CORE's national council, which was split over the payoff promised by a racial quota system and the attendant price of reverse discrimination. But by the end of 1962, CORE's militants had won national endorsement of racial employment preferences.

During the winter of 1962–63, New York's CORE chapter led a boycott of Sealtest Milk and negotiated an agreement in which Sealtest pledged to give Negroes and Puerto Ricans "exclusive exposure" for at least a week when hiring their next fifty employees.[18] As national program director Gordon Carey explained CORE's new guidelines on preferential hiring to the Denver chapter, "CORE has begun recently to change its 'line' on the national level. Heretofore, we used to talk simply of merit employment, i.e., hiring the best qualified person for the job regardless of race." But now, Carey said, "CORE is talking in terms of 'compensatory' hiring. We are approaching employers with the proposition that they have effectively excluded Negroes from their work force for a long time and that they now have a responsibility and obligation to make up for past sins."[19]

Thus by the summer of 1963, CORE found itself in intensifying competition, at the grassroots level in the volatile northern inner cities, with local black protest groups whose aspiring leaders were themselves in competition, like Gandhi, to rush to the head of their angry followers in order to lead them, and who were unrestrained by the policies of national organizations, much less by historic nonviolent inhibitions. The rapid expansion of CORE brought in an influx of new members, many of them inner city blacks who identified increasingly with the black poor and with black separatist tendencies, and who were uninterested in CORE's Christian tradition of nonviolence. Farmer tried to crack down early on escalating tactics that violated CORE discipline, such as blocking traffic, dumping garbage, and foot-stomping disruptions. But such strictures only seemed to hamper CORE's local chapters in their competition for members and for attention from the media.[20]

In New York City, CORE launched its July 1963 protests with sit-downs and picket lines at Jones Beach, at the state headquarters of the New York Human Relations Commission, and at the boards of education in Brooklyn and Harlem. But CORE was upstaged by a bolder Harlem group, led by the Reverend Nelson Dukes of the Fountain Street Baptist Church, who patterned his style after Adam Clayton Powell. The Dukes-led delegation descended on Mayor Robert Wagner's office (Wagner managed to be out of town, leaving Deputy Mayor Edward Cavanaugh to catch the flak) on June 24 and demanded that Negroes be given 25 percent of all jobs on city contract—or else "the dikes will break."[21] The following week pickets from the Dukes group successfully halted work on a city construction project in Harlem, whereupon Dukes upped the ante of his demands to 25 percent of all *state* construction contracts. The stymied contractor in Harlem, interestingly, seemed to have no fundamental objection to such hiring quotas as long as his work could resume. But public officials and the civil rights establishment in New York certainly did. A joint civil rights committee that represented the city's major civil rights organizations promptly disavowed the Dukes group. But the *New York Times* and the wire services seized upon the bizarre bidding war involving racial quotas, and similar phenomena were soon reported in New Jersey, Boston, and San Francisco.[22]

By August, news of the new demands by militant blacks for racial job quotas had swept across the nation, in vast disproportion to the phenomenon's highly local and infrequent occurrence. At the White House press conference of August 20, President Kennedy was asked his opinion of Negro leaders who demanded "job quotas by race."[23] Kennedy replied that he didn't think that was the general view of the Negro community. "[W]e ought not to begin the quota system," Kennedy said, "not hard and fast quotas. We are too mixed, this society of ours, to begin to divide on the basis of race and color." "I don't think we can undo the past," Kennedy concluded. "In fact, the past is going to be with us for a good many years in uneducated men and women who lost their chance for a decent education. We have to do the best we can now. That is what we are trying to do. I don't think quotas are a good idea. I think it is a mistake to begin to assign quotas on the basis of religion, or race, or color, or nationality. I think we'd get into a good deal of trouble."

Racial Quotas and the Congressional Hearings of 1963

The new nationwide debate over racial quotas found its way into the congressional hearings on the civil rights bill. In the marathon Senate Judiciary Committee hearings on the omnibus bill, Senator Ervin tried to lead Robert Kennedy toward a testimony that embraced the quota doctrine, or at least that acknowledged entanglement with its relentless logic. In his quest for a smoking gun from the Kennedy administration, Ervin had found one promising candidate. It was an equal employment opportunity directive to personnel officers, issued by the New Orleans district of the Corps of Engineers on January 24, 1963. The regulation was issued in response to pressure by the Johnson committee, and it was couched in the customary federal bureaucratese:

> In any case where a Negro is known to be within the top three on a list of available eligibles for a vacancy and such Negro is not selected, prepare and direct to the district engineer, through the civilian personnel officer and the deputy employment policy officer a formal letter citing the reason for nonselection. Included therein will be a comparative résumé of the qualifications, including education and experience, of the three available eligibles. In such cases, no appointment will be made without the written approval of the district engineer.[24]

Ervin translated that into "plain English" to say that if a list of eligible applicants under civil service procedures contained the name of a Negro, the personnel officer was expected to then hire the Negro without giving any reason. But if a person of another race was chosen, the hiring authorities had to prepare an elaborate written rationale for higher review, and the appointment could not be made until it received the written approval of the district engineer. Didn't the Attorney General think, Ervin asked, "that provision is calculated to cause discrimination in reverse?" "Senator," Kennedy replied, "I would not personally issue that kind of regulation." "[N]obody is being hired because he is a Negro," the Attorney General insisted, "but also they are not being discriminated against because they are Negroes. But we have found discrimination in the past and we are trying to rectify that situation."[25]

Ervin persisted. He had also unearthed another new regulation, this one issued by the Federal Housing and Home Finance Agency, in connection with a federally assisted low-rent housing project in Elizabeth City, North Carolina. It attempted to define job discrimination quantitatively by providing that

> . . . if the contractor pays to the Negro skilled labor at least 20 percent of the total amount paid in any period of 4 weeks for all skilled labor under the contract (irrespective of individual trades) and pays Negro unskilled labor at least 79 percent of the total amount paid in any period of four weeks for all unskilled labor under the contract, it shall be considered as prima facie evidence that the contractor has not discriminated against Negro labor.[26]

Ervin pounced: "Under this provision, even though 99 percent of the available unskilled labor might be white and 1 percent Negro, the contractor would have to hire at least 79 percent from the 1 percent in order even to establish a case of prima facie nondiscrimination." The population of Pasquotank County, which contained Elizabeth City, was 40 percent black (and 41 percent of the county's families earned less than $3,000 annually), so it was unlikely that the available unskilled labor force would be 99 percent white.

But Ervin had stumbled upon at least a smoking popgun—an apparent new bureaucratic hiring quota based on some implicit notion of proportional representation by race in the labor force. And he found this even in the absence of any Title VI on federal fund cut-off for discrimination, since that legislative proposal was what the hearings were about in the first place.

"Do you approve of that stipulation?" Ervin asked Kennedy.

"I would not issue those regulations," Kennedy replied. "I don't think that they are wise regulations." "I am sure there have been instances like that," Kennedy conceded, and "I am sure if you make the search, you will find more. But I am sure there have been many instances of discrimination against Negroes by those employed by the Federal Government or those employed in Federal programs or those employed in State programs. You never speak out about that."[27] The senator from North Carolina, however, was in a poor position to exploit his opening. He expressed outrage at the "two glaring examples of discrimination being practiced against white people by agencies of the Federal Government." But he was a formally committed defender of racial segregation—as, less politely, were Bull Connor's storm troopers.[28] Ervin's indignation that Elizabeth City's whites might face new forms of racial discrimination seemed remote and abstract, especially in light of Connor's televised demonstrations of more customary racial arrangements in the South. In the national mood of 1963, a white southerner like Sam Ervin lacked the credibility to shift the focus of opprobrium from racial segregation to a Kennedy administration that so consistently denied any intention of permitting racial quotas.

But in the House Judicary Committee hearings, James Farmer had an equally difficult time in struggling, from quite the opposite perspective, with the new quota nemesis. There his interlocutor was Congressman Peter Rodino, a white Democrat from majority-black Newark who was second in seniority on the subcommittee only to chairman Celler, and who was a long-committed partisan of strong civil rights legislation. Farmer's position was precarious at the helm of the fragile national CORE, which was beset by local storms of competitive militance. And in the House hearings, Farmer got into trouble through excessive initial candor. He was criticizing the weakness of Title VII in Kennedy's omnibus bill when Rodino suddenly asked him what he thought of racial job quotas: "We are not one of the organizations that believe in a quota," Farmer replied. "We do believe, however, in aggressive action to secure the employment of minorities, but not in terms of a quota."

"That is fine," Rodino said, "but do you still believe that it should be based on education and opportunity? "Yes," Farmer answered, "but if two people apply for a job and are equally qualified and generally or roughly have the same qualifications, one is Negro and one is white, and this is in a company which historically has not employed Negroes, I think then that company should give the nod to a Negro to overcome the disadvantages of the past."

RODINO: Well, isn't this then preferential?

FARMER: Well, you could call it preferential, you could call it compensatory, but sir, we have been seeking . . .

RODINO: Isn't that discriminating against a white who may have been innocent of any discrimination against anyone else in that time?

FARMER: You see none of us are really innocent because we are caught in a society, the social system which has tolerated segregation. Negroes have received special treatment all of their lives. They have received special treatment for 350 years. All we are asking for, all I am asking for now is some special treatment now to overcome the effects of the long special treatment of a negative sort that we have had in the past.

I am not asking that any white person be fired. We do not want Negroes to displace whites.[29]

Rodino remained troubled. "I think it would bother many people if this were then construed that people such as you and I who support civil rights and equality in every phase, are then going to do so to the point where the rights of others are being denied," Rodino said. "You said as between two people, white and a Negro, you should give preference to the Negro?" Farmer hedged slightly, but stood his ground. He replied that if that particular employer had a history of excluding Negroes from his work force, "and now comes an opportunity for him to repair the 'imbalance,' to use that term again, and two persons, one white and one Negro of the same qualifications. He has to choose among them somehow and wouldn't it be fairer and wiser, indeed, for him to choose among them in a way that will overcome the balance which past discrimination has created?"

Before Rodino could reply, however, the subcommittee's general counsel, William Foley, interrupted Farmer to ask about the recent demand in his former home of Brooklyn that 25 percent of the employees working on state construction jobs must be Negroes. Farmer quickly backed off: "CORE does not make that demand, that 25 percent of the people working there should be Negroes." Foley pressed harder, to the "next question, since we are talking of equality of opportunities, should it be 25 percent Negroes, 25 percent Puerto Ricans, 25 percent Italians and 25 percent Indians?" And what, Foley demanded, "do you do with a Puerto Rican who is a Negro?" Farmer was routed. In retreat he thrice denied that CORE had any interest in racial quotas. Despite the recent but unpublicized commitment of CORE's national council to just such a policy, the argument for compensatory racial quotas flew in the face of an overwhelming consensus of American opinion, and hence was politically suicidal. And Farmer knew it.

And so it went. If civil rights opponents like Ervin could no longer credibly defend the biracial caste system in 1963, national civil rights leaders could and would make no credible public case for racial quotas per se to redress historic "imbalance." Bayard Rustin, the deputy director of the August 28 March on Washington, publicly rejected preferential job hiring for blacks or any other minority group because it would "create psychological and economic problems for the country."[30] Walter Reuther said that "If the Negroes were asking for something more than equal opportunity, I think that would be crazy."[31] George Meany agreed, saying in his characteristic vernacular that "We cannot visit injustice on the white boy to make up to the black boy for injustice done to him in the past."[32] Vice President Lyndon Johnson also agreed. He told a Mexican-American group in Los Angeles that "We are not going to solve this problem by promoting minorities. That philosophy is merely another way of freezing the minority group status system in perpetuity."[33]

Compensatory Justice in the Public Arena

Yet Johnson's insistence that the problem of job discrimination was not going to be solved by a philosophy of promoting minorities had an odd ring to it. It seemed in a general way to contradict the enforcement posture of his EEO committee and his own occasional rhetoric about "underutilization." More specifically, it seemed inconsistent with the main purpose of his West Coast visit, which was to kick off a five-state southwestern conference that his committee had targeted on the region's large Hispanic minority. The day-long conference, held at the Ambassador Hotel in Los Angeles on November 14, attracted 2,000 persons, an estimated half of them of Mexican ancestry. They were responding to an implicit challenge issued by Johnson himself the previous August, when he observed that most complaints about job discrimination came from black Americans and few from Hispanics. Mexican-American leaders therefore obliged the Vice President and his committee by gathering in Los Angeles and demanding special Spanish-oriented programs in bilingual education and apprenticeship training programs for Hispanics (Oriental Americans were little in evidence at the conference). But the Hispanics made no public quarrel with Johnson's objections to minority preferences.

Similarly, the dominant black leaders and civil rights organizations steered well clear of the preferential argument. Like Lyndon Johnson, Bayard Rustin had rejected preferential quotas based on race. But he had done so as a balancing caution in his pitch for a massive federal public works and job retraining program to help "the needy—the poor, the unemployed, the unemployable, and youths—black and white." Rustin was, after all, a democratic socialist, and he had always insisted on the primacy of economic class over the politics of race in social discourse. The NAACP, on the other hand, represented a middle-class black constituency that had remained steadfast in its classic call for racial nondiscrimination. During the spring and summer of 1963, the NAACP's interracial leadership was embarrassed by the demagoguery of Adam Clayton Powell, whose flirtation with the emerging black separatist mood in Harlem had led him to denounce the NAACP for having a white president—although Powell well knew that the ceremonial post had traditionally symbolized historic Jewish support for the NAACP's interracial efforts. In 1963

Powell began to call for a black boycott to protest white leadership in predominantly black civil rights organizations. In response, Roy Wilkins battled back through the editorials of *The Crisis,* and no rhetoric of racial radicalism or separatism could be heard from that quarter.[34]

It remained, surprisingly, for Whitney Young, Jr., whose National Urban League had historically mirrored the more cautious instincts of the business community, to raise the volatile question of racially compensatory damages as reparations for historic discrimination. A native of Kentucky with an academic background in social work, Young was named executive director of the League in 1961. At the League's national meeting in Grand Rapids, Michigan, in September 1962, Young persuaded the board of trustees to support the inclusion of racial identification in government statistics, and in announcing the new policy in New York he called for "a decade of discrimination in favor of Negro youth" to help close the gap left by "300 years of deprivation."[35] The following January, in preparation for the board's February meeting in Columbus, Ohio, Young sent the trustees a confidential position paper on "Special Consideration To Close the Gap." Its preamble stated that "Our present concept of equal opportunity is not sufficient," and called instead for a compensatory, preferential Marshall Plan for black America.[36]

Young's plan held that the "simple act of granting 'equal opportunity' to Negro citizens" and the mere disappearance of old barriers could not compensate for 300 years of discrimination. Strict impartiality was "unrealistic at this moment in history," which required placing a "higher value" on the "human potential when it comes incased in a black skin." Specifically, this meant moving "above and beyond providing equal schools and equal teachers." "Token integration and pilot placement in business, industry and government is not enough." Instead, "*qualified Negroes*—because they are Negroes" must be placed in entrance jobs in all types of employment and in supervisory and policy-making positions. Thus employers "who throughout the years have never considered a Negro for top jobs in their institutions must now recruit qualified Negro employees and give preference to their employment." The draft cited as precedent the preference on civil service examinations given to veterans following World War II.[37]

The response of the League's trustees was consensual: Whitney was right in his indictment, but such a stance could be disastrous for the Urban League. Wendell G. Freeland, president of the Urban League of Pittsburgh, replied that while he basically agreed with the compensatory logic, "I have serious reservations about stamping the concept as official League policy."[38] "What we are actually saying is that our definition of 'equal opportunity' has changed," Freeland said, but he foresaw "adverse reactions to the pronouncement as such." The League's public would ask: "What in blazes are these guys up to? They tell us for years, that we must buy [nondiscrimination] and then say, 'It isn't what we want.'" Lawrence Lowman of the New York League reacted more strongly against "seeking special privileges" when "We [already] have much law on our side."[39] Lowman objected to the "misplaced sarcasm" implicit in the draft's assertion that "For more than 300 years Negroes have received special consideration of exclusion." He objected also to the "bitterness of such words as 'incased in a black skin.'" But most objectionable was "the heart of it"—the "business of employing Negroes 'because they are Negroes.'"

As is the normal fate of strong first drafts in collective review, Young's draft was successively winnowed and honed toward a more rounded consensual model. The League's final and official statement, published on June 9, was purged of its rhetoric of preferential discrimination.[40] It still called for a "massive 'Marshall Plan' approach" and a "crash attack." But gone were all the references to compensatory treatment, as well as all explicit or implied criticisms of the traditional concept of equal opportunity. The plan's ten-point program of implementation was unexceptional. It listed the customary items: greater federal investment in jobs, schools, housing, and health care, with needy blacks chiefly but not exclusively in mind.

Young persisted in drafting and circulating strongly worded versions of his reparations plan, which he called a compensatory effort based on "the *concept of indemnification.*"[41] By indemnification Young meant "realistic compensation (not necessarily in money alone) and realistic reparation for past injuries." But Young's indemnification plan never saw the light of day. Instead, Young was obliged to transform the logic of black reparations into a broader summons for an antipoverty coalition similar to Bayard Rustin's. In public, Young talked not just about 22 million American Negroes, but also about "40 million citizens locked in the cycle of poverty," and nearly 40 million others who lived in deprivation.[42] He called for a National Works Corps in the image of the New Deal's WPA. But the Urban League's new Marshall Plan stopped well short of any serious hint of a conscious reparations policy. In his public role as spokesman for the League, however, Whitney Young could not or would not entirely mask his private convictions. The *Wall Street Journal* quoted him as urging that industry should "hire the Negro when two job applicants, Negro and white, are equally qualified."[43] Young acknowledged that he was receiving phone calls from corporate executives who were angry at the implications of preferential treatment for Negroes. But he replied that many employers were now quietly doing it anyway.[44] So there remained considerable public ambiguity about what affirmative action in support of anti-discrimination meant and what it did not mean.

In an attempt to clarify the ambiguity, the *New York Times* asked Young and Kyle Haselden, the managing editor of the *Christian Century* and author of *The Racial Problem in Christian Perspective,* to debate in the pages of the *New York Times Magazine* the "so-called doctrine of compensation" that had recently emerged from the civil rights campaign. The exchange in the *Times* was published on October 6. Young began his argument by documenting the dimensions of the racial gap. Black mean family income was only 45 percent of white income, and had actually *fallen* two percentage points in the past decade. Seventy-five percent of Negro workers were found in the three lowest-paid occupational categories, such as domestics and unskilled farm workers, compared with 38 percent of whites. Twenty-one percent of high school dropouts but only 7 percent of high school graduates were Negroes. Young then outlined his proposed "crash program," a ten-year domestic Marshall Plan in education, employment, housing, and health and welfare as a "deliberate and massive effort to include the Negro citizen in the mainstream of American life."[45]

Young was vague, however, about his plan's specific components, and he attached no dollar cost to it. His definition of compensation remained at the macro-level of greater aggregated expenditures to benefit the Negro population generally. But it made no clear connection at the micro-level of specific programs, where the zero-sum mechanism of preferential or compensatory quotas would invite resistance from racially excluded nonblacks. Moreover, Young insisted that the Urban League was "asking for a special effort *not* [emphasis added] for special privileges." Such terms as "preferential treatment," "indemnification," and "compensatory activity," he said, were "scare" phrases that "go against the grain of our native sense of fair play." Employers should consciously seek to hire "*qualified* Negroes," Young explained. But that did "not mean the establishment of a quota system—an idea shunned by responsible Negro organizations and leaders."[46]

Because Young had purged his own "scare" phrases from his formal position statement, his sanitized case for compensatory justice left little for Haselden to disagree with. Indeed, Haselden also proposed a "crash program," and like Young he borrowed an analogy from Truman's foreign policy to make his case. Haselden summoned a domestic Point Four program based not on race but on economic need. This was required because even were American society immediately desegregated, he said, "most Negroes could not, in a free and impartial society, compete on equal terms with most white people for jobs and preferments." But Haselden condemned the compensatory approach, which he claimed would inevitably legalize, deepen, and perpetuate a subtle but pernicious form of racism that must be eliminated from the social order, not confirmed by it. Racial preference would be unfair to other minorities handicapped by their history, and hence would "penalize the living in a futile attempt to collect a debt owed by the dead." It would thereby compound the ironic tragedy of exploitation: "It leaves with the descendants of the exploiters a guilt they cannot cancel and with the descendants of the exploited a debt they cannot collect." To Haselden, the new theory of compensation failed all the crucial tests of a moral struggle for racial justice, which "is valid only if it honors the moral ground on which the Negro makes his claim for justice, preserves in the human relationship values which are equivalents of justice, and promotes rather than prevents the Negro's progress." In 1963, the civil rights movement derived its moral authority from liberalism's core value of equal treatment for individuals. Haselden's elegant phrases, like Martin Luther King's vision at the Lincoln Memorial, appealed to the ethical equilibrium of ends and means in human affairs. In such a contest, demands for racial preferences echoed unsavory practices from the past under a color-conscious Constitution.

The Practical Ambiguity of Preferential Discrimination

A gap remained, however, between public rhetoric and the workaday world of jobs and pressure groups. The notion of an accelerated catch-up carried certain practical attractions to agency and corporate officials who had inherited racially unfair enterprises. But this was balanced against the overwhelming public rejection of a race-triggered theory of compensatory justice that would necessarily penalize the innocent for the sins of their forebears. During the latter half of 1963, while Congress debated the omnibus civil rights bill, there occurred two revealing episodes of quota-linked behavior. One was in the federal government, the other in private industry. Both illustrated the momentum and appeal of compensatory preference as a response to racial inequity as well as its political and ethical awkwardness.

The first instance involved the Labor Department's apprenticeship programs. The Civil Rights Commission in its 1961 report on employment had singled the apprenticehsip programs out for special criticism as being both racially discriminatory and too small even to replace the mostly white workers who retired. Labor's Bureau of Apprenticeship and Training (BAT), which funded training programs for 150,000 apprentices in twenty-two states, worked closely with the Construction Industry Joint Council, which represented the eighteen national building trades unions and the major national contractor associations. BAT cherished its collegial program relationships with industry and union officials, which in turn hinged on a spirit of voluntarism and mutually compatible goals. BAT had evolved historically as a promoter, not a policeman. The Bureau's director even testified to James Roosevelt's subcommittee in 1962 that he did not *want* the enforcement power to eliminate alleged discrimination in apprenticeship training.[47] BAT thus participated symbiotically with the construction unions and the major contractors in an "iron triangle" bound by customary log-rolling relationships.[48]

This comfortable arrangement began to change for BAT in the early 1960s, however. The source of the change was the triangle's third member—the congressional authorizing committee that created and monitored the program with which the agency serviced its constituency. For BAT, this was Adam Clayton Powell's House Education and Labor Committee. Like the Kennedy administration generally, Powell's committee was becoming more responsive to the civil rights constituency than to the trade union constituency.

Early in June of 1963, both in responding to PCEEO pressures and in implementing President Kennedy's pledge to move against discrimination in the apprenticeship program, Secretary Wirtz issued a press release spelling out new selection standards for apprentices. These were promulgated in a BAT circular on July 17 and immediately provoked a howl of protest from the construction industry that they would impose a racial quota system.[49]

The chief objections were to new selection provisions that did not, on their bureaucratic face, reveal the mailed fist that BAT's veteran clients immediately perceived in their obliquely worded codicils. The first required that future selection of apprentices must be made according to merit standards alone— *"provided that,* where there are established special applicant practices," such apprenticeship lists "must be *disregarded* to the extent necessary to provide opportunities for current selection of qualified members of racial and ethnic minority groups for a significant number of positions [emphasis added]." [50] The second required the "taking of whatever steps are necessary, in acting upon application lists developed prior to this time, to offset the effect of previous practices under which discriminatory patterns of employment have resulted." A third provision simply—and ironically—required nondiscrimination in all phases of the apprenticeship program. In enforcing its contract provisions, the Labor Department's main enforcement sanction (shy of outright contract cancellation) was program deregistration. Deregistered contractors would no longer be permitted to pay apprentices lower wages than the prevailing rates for journeymen. Furthermore, apprentices in deregistered programs might lose their draft deferments. Potentially, the Johnson committee might even terminate the contracts of deregistered construction companies.

Because the third regulation simply required straightforward anti-discrimination, it produced no formal objection. But the first, translated from its bureaucratic code, obliquely referred to and directly threatened the construction industry's widespread "sponsorship" system, which favored relatives or friends or existing members of the building trades in the recruitment of apprentices. This was the engine of the infamous father-son unions in the skilled crafts, whose historic origins extended back to the medieval guilds. In response to such a threat, the national president of the powerful Plumbers and Pipefitters union, Peter Scheomann, rose to defend the threatened sponsorship system with becoming candor: "[S]ponsorship and favoritism are phenomena of American political and business life," Schoemann explained. "Indeed, one may wonder whether they are not necessarily inherent in a free and democratic society, in which men derive much of their motivation from a desire to accomplish something for their families and friends, and where they have a free choice of selection of people, in government or private employment." [51]

It was the second new requirement, however, that drew the most fire, because it raised the specter of racial quotas for unions. The Building and Construction Trades Department of the AFL-CIO joined in an industrywide objection to the phrases "significant number" and "to offset." What, precisely, was a "significant number" of jobs for minorities? What was the meaning of a requirement that contractors must "offset" the effects of previous discrimination? These troublesome phrases implied the adoption of a "veiled quota system," union officials charged, and the new language clashed with the administration's rhetoric about consistent nondiscrimination. Secretary Wirtz denied the quota charge. But the storm of labor protest was so intense that the regulations were withdrawn, revised, tentatively reissued in October, and officially promulgated in December. In order to satisfy such a united constituency, especially during the sensitive congressional debate over the civil rights bill, the Labor Department's revised regulations eliminated entirely the offending phrase "significant numbers," and also substituted "to remove" for the original "to offset."[52] *Sic transit quotum.*

The second incident of quota-linked behavior in late 1963 involved private industry. As Whitney Young had pointed out, some businesses were turning to racially preferential hiring practices while officially denying it. In October the Plans for Progress companies claimed to have hired 60,000 new workers in the previous summer quarter, of whom almost 25 percent were black.[53] But one substantial company announced at the beginning of December that it would adopt a *formal* policy of preferential hiring of Negro workers. This was Pitney-Bowes of Stamford, Connecticut, a major manufacturer of postage and mailing equipment with 7,400 employees. The quota-hiring announcement from Pitney-Bowes followed within days of President Kennedy's assassination in November 1963, and the timing was no coincidence. Its president, John O. Nicklis, was so troubled by the "mindless murder of President Kennedy and the dark horror of that weekend [that] lies heavy on our hearts," that he told the annual meeting of the National Social Welfare Assembly in New York that a national purging of the soul was in order. "Now is the time to do the purging," Nicklis said, "and the Negro and his struggle for justice and opportunity should be the place for us to begin."[54]

Ninety-five percent of Pitney-Bowes employees, however, were not black, and they and their potential nonblack co-workers were not pleased to be volunteered as a class for the purging. President Nicklis was quickly so informed. Thus chastened, he explained one week later to the *New York Times* that he had been misunderstood. His company was not making "a direct comparison of two applicants, one white and one Negro, and chosing one because of his color."[55] Instead, Pitney-Bowes was only attempting to find and employ more Negroes, who after a period of training would make "good" employees. The company would "continue to employ qualified white people, of course." Like Secretary Wirtz, the president of Pitney-Bowes learned that in the zero-sum game of job discrimination, "benign" quotas appeared malignant to workers facing displacement because of the color of their skin.

The Inner Transformation of the Civil Rights Vision

During those watershed months of 1963–64, two parallel but somewhat independent dialogues were occurring, with little connection at the elemental level of assumption and definition. One dialogue dominated the public debate and centered on the moral imperative of nondiscrimination. It represented the accumulating momentum of a century of liberal thought and reform in racial relations, and it sought belatedly to fulfill the failed promise of the Reconstruction. Since World War II the racial attitudes of the white majority had turned against the hypocrisy of the racial caste system;[56] the intellectual underpinnings of racial segregation had collapsed, and its institutional barricades were cracking. Because the chief offender was the southern region, this dialogue took the familiar political form of pitting desegregation against Jim Crow. But philosophically the debate was national, and in the national policy arena of equal employment policy the debate set state-enforced nondiscrimination against traditional employer freedoms in a dispute over fair procedures and the limits of government intrusion. In the classic liberal vision, racial discrimination was traditionally defined as a conscious act of prejudice. It was widespread and deeply rooted in American society, but would yield to the vigorous enforcement of anti-discriminatory measures by new regulatory commissions modeled after the New Deal's NLRB and Roosevelt's FEPC. Its opponents were state-rights conservatives who opposed the FEP commissions with the same arguments of free-market capitalism that they had always marshalled against the engines of the regulatory state. The conservatives had won the first round against Roosevelt's FEPC. But the liberals had countered with New York's Ives-Quinn Act in 1945, and they had swept the northern half of the country with their FEP commissions by 1963.

This "classic" public debate, however, was paralleled by an underground dialogue on the political left that was more of a Hegelian dialectic than a debate. Its parameters were implicitly more radical than those associated with liberalism's standard FEP formula. It was also mostly a private and internal discussion, still inchoate and theoretically immature. It was driven by a perception of fundamental contradiction grounded in the brute fact that the relative well-being of black Americans had at best remained static and in some cases was continuing to deteriorate *even in* those enlightened northern states, like New York, where anti-discrimination had been the operative law for a generation, and where FEP commissions were armed with full cease-and-desist authority.[57] The logic of catch-up suggested some form of compensatory preference. But pursuit of this means caught up its advocates in the moral paradox of justifying the means of preferential racial discrimination to achieve the ends of racial nondiscrimination.

Men like James Farmer were caught in the middle, pushed by a radicalizing constituency from below, yet knowing that a candid public expression of CORE's new private credo would be politically suicidal. So in public discourse Farmer backed off quickly from the liberal Rodino's challenge, just as Whitney Young had blurred the edges of his reparations program until it was scarcely distinguishable from the WPA. But privately, Farmer and many of his colleagues in the civil rights leadership saw in New York-style FEP commissions little potential solution to the massive maldistribution of resources between white and black America. Farmer later claimed in his memoir that he had told Vice President Johnson as early as 1961 that the traditional FEP codes were "obsolete," and that Johnson had seemed to agree.[58]

As black leaders of national civil rights organizations with integrated memberships and nondiscrimination principles, Farmer and Young had muted their sentiments during the great debate of 1963–64. But these constraints did not effectively apply to Herbert Hill, the ranking white radical within the orthodox NAACP. In 1963 Hill had completed a study of the effectiveness of state and local FEP commissions since World War II.[59] He concluded that the status of Negro labor in northern FEPC states had declined since World War II, and that the complaint-based, fair-procedures approach of the FEP commissions had proven unable to cope with changing occupational patterns and structural unemployment. Hill concentrated on macro-level aggregations, comparing the weak national record of black employment progress since the boom war years with the typical state commission record of high levels of rejection of individual complaints and low levels of finding probable cause and issuing cease-and-desist orders. His data broadly supported a contention that was weakened, however, by his partisan affiliation and reputation.

During the debate of 1963–64 a larger academic study of the state and local FEP commissions was published by labor economists Paul Norgren and Samuel Hill. In *Toward Fair Employment,* they defended the achievements of the New York commission but were more critical of the record of the FEP commissions in general. Norgren and Hill concluded that "the FEP laws in effect in most Northern and Western states and in several major Northern cities have up to now resulted in only a very modest and spotty decrease in discriminatory employment practices."[60] But neither the Herbert Hill nor the Norgren and Hill study had analyzed the data intensively at the state level. No major and intensive case study of a single state FEP commission had been produced since Morroe Berger's landmark study of New York in 1952.[61] Berger's book had defended New York's State Commission Against Discrimination (SCAD) as a worthy liberal innovation, and he generally praised the commission's successful balancing of the need to protect minorities while buttressing the commitment to merit-based fair competition of "the moral order." Berger's support for SCAD reinforced liberalism's sustained assault on William Graham Sumner's dictum of conservative Darwinism that "stateways cannot change folkways." But it also reflected liberalism's defensive mood of the postwar decade.

By the early 1960s, however, a minority of frustrated liberal reformers was beginning to question the adequacy of the FEP model. In 1963, Rutgers law professor Alfred W. Blumrosen studied the effectiveness of the enforcement division of New Jersey's Civil Rights Commission at the invitation of its chairman. Blumrosen was a committed "plaintiff's lawyer" who would spend a sabbatical year in 1965–66 helping the new EEOC organize its enforcement procedures. Like Herbert Hill, he brought to his study a "tough-minded" model of "maximum enforcement," but with "a lawyer's bent"—a "mental image, a model of how the state agency *should* operate in order to have maximum impact on the problems of discrimination."[62] Blumrosen concluded that the New Jersey commission's enforcement patterns "typified administrative caution and ineptness at every turn; its procedures were incredibly sloppy; it narrowly construed a statute which the courts were prepared to construe broadly; it did not secure relief for the complainants, or for the general class of victims. It was a failure."[63]

Blumrosen's aggressive advocacy in a law review article still did not constitute an objective and comprehensive state study. But Leon Mayhew's Harvard dissertation in sociology did. A doctoral student of Talcott Parsons,

Mayhew's interest in the relationship between law and society had been stimulated in seminar with Gordon Allport, and his dissertation analyzed the structure and functioning of the FEP commission in Massachusetts, a state with strong liberal traditions, like New York. Mayhew collected his data, conducted interviews, and observed the commission's operations during 1961–62. His study was a Parsonian analysis of the social organization produced by the Massachusetts Commission Against Discrimination (MCAD) since its creation in 1946. But it concentrated on the years since 1959, when MCAD's jurisdiction was extended to include housing, and it focused more on the complexities of institutional processes than on the efficacy of results. Mayhew found that MCAD was skilled at co-optation. It was less effective in combating job than housing discrimination (because the latter was inherently more blatant and provable), and it treated anti-discrimination law as essentially a system of private law.[64] Herbert Hill had flatly dismissed MCAD's record as "dismal." But Hill's study had simply summarized MCAD's typically low rate of public hearings and cease-and-desist orders, and let that record presumably speak for itself.[65]

Mayhew, like Hill, was sympathetic to the purpose of the FEP commissions. But unlike Hill, Mayhew was bound by the canons of scholarship enforced by a dissertation committee. In his research he had closely followed the processing in Massachusetts of 118 complaint cases. The pattern he found was surprising. The evidence showed that "complaints developed by individuals whose structural position provides limited perspective are objectively poor," Mayhew concluded. "They tend to be based on mere suspicion, they are quite likely to eventuate in a finding of 'no probable cause,' and they tend to be made against the very firms that do not discriminate."[66] Because the ordinary operations of the job market "regularly produce experiences that could be interpreted as discrimination," Mayhew wrote, the complaints often "turn out to be mistaken." This ambiguity "permits Negroes to blame discrimination for their troubles. Hence, some complaints represent a projection of one's own deficiencies on to the outside world."[67]

To Mayhew, merely defining discrimination as the proportional difference between white and black employment rates begged the important and complicated questions of employability and motivation and fairness that the FEP commissions' quasi-judicial procedures were designed to balance and weigh. To result-oriented critics like Hill and Blumrosen, however, the fairness of due process in the individual complaint model was largely irrelevant. Equal treatment was desirable, but not where it produced unequal results. Both critics indicted the state FEP commissions as, in Hill's language, "ineffectual agents of social change."[68] Blumrosen was more optimistic than Hill about the redeeming possibilities of successful enforcement through liberal construction of the existing FEP statutes. The more cynical Hill tended to view the primal fault as inherent in the "inadequacy of FEPC" itself as a model "to deal with broad patterns of employment discrimination." But the analysis of both men reflected a growing conviction among fair employment activists that the individual complaint model's deliberate due process in determining discriminatory intent was irrelevant to the root problem of "institutional racism."

In the newly evolving view of institutionalized racism, individual intent was at best a secondary consideration. Instead, employment discrimination should be defined and attacked statistically as a differential, rather than traditionally as an invidious and injurious act of prejudice. Its measure was simply the gap between the white and minority employment rates. This presumptive new definition in turn rested on an implicit normative theory of proportional representation in the workforce, absent the discrimination that institutional racism had built into the employment structure. The chief political weakness of this theory was that it violated the American creed that rights inhered in individuals rather than in groups, and that immutable factors like race and ethnicity should be irrelevant as employment criteria. Its chief political strength lay in its practical utility as an implicit and self-justifying formula for equity. This was captured in the workaday concept of "underutilization," a term that Vice President Lyndon Johnson and his associates on the PCEEO had rather casually accepted as early as 1961. But the distance between the vague notion of undertilization and a conscious theory of proportional representation was politically vast in the fall of 1963. The ensuing debate over the civil rights bill was to widen rather than narrow it.

Clearly that fateful fall, the national mood was receptive to the need not only for an end to racial segregation in the South, but also for a serious attempt to end racially discriminatory barriers in the nation's economic life. But the debate over racial quotas elicited a virtually unanimous public condemnation of the notion of racial preference, however allegedly benign. Overall, the burst of national debate over preferential discrimination had been rather shallow, truncated, and one-sided. Few of the serious arguments that would later be deployed to rationalize and sustain a doctrine of preferential discrimination were raised in its defense in 1963 and 1964—beyond occasional references to veterans' preferences, to the inertia of historical lag, and to the analogous unfairness of an ostensibly even start in a footrace between the sturdy and the starved. Clearly, too, in the great battle for the civil rights bill, the stakes were far too high to risk alienating liberal support by calling for a purportedly benign new form of racial discrimination. But more important, the evidence suggests that the traditional liberalism shared by most of the civil rights establishment was philosophically offended by the notion of racial preference. Such pillars of the liberal establishment as the *New York Times,* Gunnar Myrdal, and Senator Hubert Humphrey rejected it. Despite the maverick dissent of Herbert Hill and Whitney Young, the NAACP and the National Urban League disassociated themselves from it, and continued their alliance with Martin Luther King's SCLC in support of classic, liberal nondiscrimination. The chief exceptions were SNCC and CORE. But SNCC and CORE by 1964 were accelerating in their radicalization, soon to hurtle down the path to black nationalism, isolation, and violence that was to lead to their self-destruction.

Given the rising national sense of urgency in the wake of Birmingham, it would be up to the Congress to construct and enact an effective instrument for ending racial segregation in the South and constructing a fair employment policy nationwide. The crucible of decision, in the early fall of 1963, was in the white male hands of the House Judiciary Committee's Subcommittee No. 5—meeting, as the great national debate flourished so openly, in closed executive session.

PART TWO

Johnson

CHAPTER V

The Civil Rights Act of 1964

Running the Double Gauntlet in House Judiciary

The Kennedy administration's House-first strategy for the omnibus civil rights bill faced six known obstacles. The first was Emanuel Celler's Subcommittee No. 5, from whence a successful bill would then face the full, 35-man (and 35-lawyer) House Judiciary Committee. Then on to Judge Smith's Rules Committee, and finally to the full House. The penultimate hurdle—never previously accomplished in a civil rights bill—was cloture against a certain Senate filibuster. The final barrier was the Senate-House conference committee, which was potentially treacherous because the conference agreement, if there was one, would have to be ratified by each chamber. In the volatile atmosphere following the Birmingham violence, any misstep could be fatal for the bill, and potentially disastrous for the Administration and the nation's race relations. The House-first strategy meant that most of the major substantive decisions would be made in the closed markup sessions of the House Judiciary Committee. There Chairman Celler had a private strategy which he did not entirely share with the White House.[1]

Celler's legislative strategy was not inherently unreasonable. It had twice worked for him in previous civil rights hearings, in 1957 and again in 1959. On both occasions he had begun by emerging with a strong civil rights bill from a subcommittee that he had stacked with liberal Democrats. Then he had traded off concessions to the more conservative Republicans in the full Judiciary Committee, in order to report out a moderately strong bill that centered on voting rights and that commanded sufficient bipartisan support to pass the House. In the Senate the Democratic leadership then compromised further in order to stave off a southern filibuster. This in effect had meant that Majority Leader Lyndon Johnson had weakened the bills enough to convince Richard Russell that the bill was not threatening enough to justify a full filibuster. The final compromise was then presented to President Eisenhower as a statesmanlike achievement by the Democratic-controlled

Congress. In 1963, however, senior House Republicans like William Mc-Culloch were determined not to be thrice used and ultimately betrayed by these Democratic bargaining tactics, and especially by what they regarded as a sell-out in the Senate. They were unwilling, without further guarantees, to support unprecedented enforcement provisions under a Democratic President who could so easily blame them if the bill failed. Moreover, by the fall of 1963 the political and emotional climate was so hypercharged that the forces that added urgency to the liberal agenda were also unsettling the customary bargaining milieu on the Hill.

On September 10, the House Ways and Means Committee approved Kennedy's tax cut bill, as did the full House on September 30, thereby removing the restraining fear that committee progress on the civil rights bill might endanger the tax cut. So Celler began serious (and closed) markup sessions as soon as the tax cut was safely out of committee. Then on September 15, a bomb blew up the Sixteenth Street Baptist Church in Birmingham, killing four small black girls and injuring twenty other children, and sparking protest riots in which two more children died. On the last day of September, the Civil Rights Commission issued its third biennial report to the President and Congress, and for the first time its report and recommendations were unanimous. The commission's two white southerners—Robert G. Storey, former dean of the Law School at Southern Methodist University, and Robert S. Rankin, chairman of the department of political science at Duke University—concurred in recommendations that called for broad new federal guarantees of uniform national voting standards (including federal registrars if necessary). Reaching beyond voting rights, the commission endorsed a new fair employment practices statute that would cover private businesses in interstate commerce and would be enforced by the Labor Department.[2]

In response to these pressures, Celler began markup in an initially bipartisan spirit. This was suggested early in the markup when the subcommittee agreed to an amendment by McCulloch that signified a powerful if rather quiet bipartisan consensus, at least in the North. McCulloch moved to strike from the Administration's title on school desegregation a proposal to extend financial aid and technical assistance to school districts that attempted "to adjust racial imbalance."[3] This was code for northern de facto school segregation, and the nonsouthern congressmen clearly wanted to strip from the bill a potential nonsouthern application. It left the bill's school desegregation title applying to southern de jure segregation only. Southern congressmen denounced the maneuver as hypocritical because it made the bill essentially a one-region attack on a national problem. In their own hypocritical way, the southerners were quite right. But they were also heavily outnumbered on the subcommittee, so they were ignored.

That accomplished, Celler then began to override the subcommittee's four-man Republican minority. McCulloch was especially upset when Celler supported an amendment offered by Robert Kastenmeier (D.-Wisc.) that would extend Title II on public accommodations to cover virtually every private form of business licensed by the states. This would exclude Mrs. Murphy's sacrosanct boardinghouse, but it otherwise would cover private schools, law firms, medical clinics and similar enterprises that had explicitly been excluded in McCulloch's earlier agreement with the Justice Department, and that Celler had promised he would support. Then in rapid succession, the subcommittee's seven-Democrat majority rammed through a brace of strengthening amendments that reflected virtually the full demands of the Leadership Conference on Civil Rights.[4] Chief among them was a Title III amendment by Byron G. Rogers (D.-Colo.) to allow the Attorney General to sue on behalf of individuals to protect virtually any civil right anywhere, and also an amendment by Peter Rodino (D.-N.J.) that replaced the Administration's weak FEP title with the full EEOC provision of H.R. 405—including cease-and-desist authority.

By October 2, Celler's partisan demarche had radically transformed the bill. The voting provisions were to cover all state and local elections, not just federal contests. The public accommodations section now covered all state-licensed activities—presumably including Alabama's feather renovation industry, and also McCulloch's local bar association in Piqua, Ohio. The public education section in effect exempted all northern schools from desegregation requirements. The Civil Rights Commission was to become permanent. The cut-off provision for federal funds would allow the Attorney General to sue state and local governments to stop programs like Hill-Burton hospital construction. Vice President Johnson's PCEEO was now to become an FEPC—or rather, an EEOC that was a quasi-judicial administrative agency with both prosecutorial and cease-and-desist authority. Finally, Celler's compliant Subcommittee No. 5 had newly added to the omnibus bill the Leadership Conference's old and symbolic standby, Title III.[5]

McCulloch, furious, felt betrayed again—just as in 1957 and 1959. "It's a pail of garbage," snapped the normally mild-mannered ranking minority member.[6] But the Leadership Conference was delighted. And many southern Democratic congressmen were both—but in that order. They were initially angry at being railroaded in subcommittee by their liberal colleagues. But then they warmed to the prospect of first voting *for* Celler's liberal Democratic *coup de main* in the Judiciary Committee, then joining the Republicans to sink it on the House floor. To the more conservative southerners and Republicans, Celler's liberal new package on further reflection had many of the enticing earmarks of a super-Powell amendment.

Liberals are "in love with death"

Robert Kennedy did not conceal his contempt for "professional liberals."[7] The Attorney General associated the Eleanor Roosevelt–Adlai Stevenson wing of the Democratic party, where he and his brother had rarely felt welcome, with a Kamikaze purity. To Bobby, such liberals had a "sort of death wish, really wanting to go down in flames. . . . Action or success make them suspicious; and they almost lose interest. I think that's why so many of them think that Adlai Stevenson is the second coming," Kennedy said. "But he never quite arrives there; he never quite accomplishes anything. . . . They like it much better to have a cause than to have a course of action that's been successful."[8] To Burke Marshall, Celler's subcommittee that September "ran away and closed out this impossible bill," which McCulloch confirmed would never pass the House. So the thirty-seven-year-old Attorney General summoned the seventy-five-year-old Celler and upbraided him for allowing doctrinaire liberals like Kastenmeier to railroad through a liberally sublime but politically doomed bill.[9] "We'd lost him," Kennedy said of Celler, "and he wasn't giving any leadership. He'd indicated that he'd come along with us and then hadn't."[10] Kennedy lectured the resentful Celler that he was "no good to us," that consequently "the bill was going to go down the drain," and that McCulloch and the cooperating Republicans felt stabbed in the back.[11]

The result of this crisis of mid-October was a White House-led salvage operation, in which Celler agreed to meet with Katzenbach, Marshall, and McCulloch to fashion a substitute bill that was acceptable to the Republican minority. Meanwhile, House Speaker John McCormack was meeting with Katzenbach and House Minority Leader Charles Halleck, asking the House Republican leadership to help bail the Democrats out. In return, McCulloch demanded that Kennedy's earlier pledge of constancy be converted into an "ironclad oath." The Republicans insisted not only that Celler recant his support of the subcommittee's new Leadership Conference version of H.R. 7152, but also that the Attorney General himself come testify before the full Judiciary Committee and reinforce the disclaimer. At the risk of further alienating his party's liberal core, Robert Kennedy agreed, and testified in executive session on October 15 and 16.

His performance was a tour de force, an impressive reversal of his poor showing of June 26. Some Republicans even called "brilliant" his mastery of the details of the bill, as he never seemed to refer to his notes (beyond his initial formal statement) or to confer with the ever present duo of Katzenbach and Marshall beside him.[12] Kennedy concentrated his arguments on the subcommittee's revised Title II on public accommodations and the new Title III. The public accommodations section now carried Kastenmeier's "catch-all" inclusion of all private businesses operating under state or local authorization, permission, or license. This represented to Kennedy an unnecessary overreaching, possibly beyond the full limits of the 14th Amendment's constitutional power; its impact would be both unclear and widely uneven because it depended upon state licensing practices that varied widely and were easily changed. "It would seem to extend Federal regulation to law firms, medical partnerships and clinics, private schools, apartment houses, insurance businesses," Kennedy claimed, and "potentially, to all businesses which a State does not affirmatively ban."[13]

Kennedy's most eloquent testimony, however, was directed against Title III's major expansion of the powers of his own office. He briefly traced the history of Title III, which had originated in 1957 as a tool designed to enable the Justice Department to accelerate the glacial pace of southern school desegregation in an era of "massive resistance." The Administration's 1963 omnibus bill, however, now contained specific federal provisions tailored to combat the denial of the vote, of equal access to public accommodations, and of equal education and employment opportunities. But the new Title III was "not concerned with those matters," Kennedy said. Instead, it was at once a post-Birmingham symbol of impatience and urgency, and a weapon to use against southern police abuse of civil rights demonstrators. And as a weapon it was a blunderbuss. Its court injunctions, Kennedy claimed, could not prevent or punish bombings or isolated acts of brutality by individual police officers. It would necessarily involve the federal courts in advance in police functions that had historically been exercised by local officials. If such a federal intrusion led local police to abdicate their law-enforcement responsibilities, it would risk the creation of a national police force. Yet so broad was the reach of Title III that it extended federal power potentially far beyond mere control of police abuses against racial protests. Its language would extend to "claimed violations of constitutional rights involving church-state relations; economic questions such as allegedly confiscatory ratemaking or the constitutional requirement of just compensation in land acquisition cases; the propriety of incarceration in a mental hospital; searches and seizures; and controversies involving freedom of speech, freedom of worship, or of the press." [14] In the name of civil rights, in effect, it would endanger the delicate Madisonian balance that had historically protected civil liberties.

Coup and Counter-coup in Fair Employment

In ticking off his response to the revised bill's titles and in concentrating his criticism on Title III, the Attorney General had brushed quickly by the subcommittee's revised fair employment title. This seemed to indicate that the Administration had no fundamental problem with the substitution of H.R. 405 for the Administration's original proposal to give merely statutory recognition and budget authority to the Vice President's EEO committee. But the Republican minority had such a problem, and they had good reason. In February 1962, when the Powell committee had originally reported out its EEO bill (H.R. 10144—which few students of Congress thought had any chance to pass), a cohesive bloc of seven moderate Republicans on Powell's committee had demanded one basic change as the price of their support. [15] Led by Robert P. Griffin of Michigan, they insisted that the proposed new EEOC, *un*like the state FEPCs, must be *denied* the quasi-judicial power to hold hearings and issue cease-and-desist orders. Instead of in effect holding a trial and declaring a judgment, then, the federal EEOC would be empowered only to investigate and prosecute, leaving the judicial decisions about guilt and remedy to a federal district court. [16]

Griffin and his Republican colleagues argued that the American Bar Association's long-standing principles of American jurisprudence required that final determinations be made by the judiciary rather than by an investigative, prosecuting agency. This position reflected the great battle over administrative reform of the 1940s, in which a coalition of Republicans and southern Democrats had attacked the regulatory abuses they associated with the New Deal.[17] Their prime target had been the NLRB, and their chief instrument had been Howard Smith's special House investigating committee (Roosevelt's wartime FEPC had drawn Smith's special ire because it lacked even the statutory authority that the Wagner Act had conferred on the NLRB). Despite its own conservative bias, the Smith investigation uncovered damaging evidence of pro-labor prejudice on the part of the early NLRB and its hearing examiners. These disclosures had produced attacks on the labor board not only from the American business community and the American Bar Association, but also from the AFL, which complained of heavy-handed bias by the NLRB in favor of the CIO. The conservative counterattack also unearthed evidence of communist infiltration, to an extent that was alarming even when heavily discounted for the manifest witch-hunting of the Dies Committee.[18] The chief result under a Republican Congress was the Administrative Procedures Act of 1946, which sought to "judicialize" the procedures of the quasi-judicial regulatory agencies. This in turn had been followed by the Taft-Hartley Act of 1947, which sought to balance the conscious pro-labor tilt of the Wagner Act.

The Administrative Procedures Act rested on the principle of Madisonian separation between the police function of investigation and prosecution on the one hand, and the judicial function of rendering judgments and determining relief and penalties on the other. Its administrative reforms concentrated on mandating procedural safeguards in such areas as rules of evidence, burden of proof, and rights of appeal, safeguards that tended to place judicial fairness above administrative efficiency. Structurally, it provided for the replacement of agency hearing examiners with administrative law judges, who were deemed less likely to be captured by constituency clienteles.[19] The Republicans on the House Education and Labor Committee had inherited a strong commitment to this conservative tradition of administrative reform, and in 1962 Griffin and his colleagues had even secured a memorandum of agreement from subcommittee chairman James Roosevelt that quoted, of all the unlikely people, his father Franklin D. Roosevelt, on the potential evils of mixing judicial and administrative functions in the independent federal regulatory commissions![20]

The Republicans pointed out that their *prosecutorial* model was the standard method through which the Labor Department enforced the wage-and-hour provisions of the Fair Labor Standards Act, the unfair labor practices provisions of the Landrum-Griffin Act, as well as the new provisions of the Equal Pay Act for women. Griffin and his colleagues had expressed a common Republican complaint against NLRB-type administrative tribunals, which was that despite the limited reforms of the Administrative Procedures Act of 1946, they "have acquired a well-deserved reputation for ignoring the rules of evidence." What the Republicans meant was that in practice, the federal appeals courts so rarely overturned the decisions of such administrative tribunals that the normal burden of proof was reversed, and the accused on appeal found that "he must bear the burden of proving his freedom from guilt."[21]

In addition to this appeal to judicial principle, the Republicans in 1962 had added a practical argument against judicial authority in fair employment regulation. This was that in the experience of the state FEPCs since their Ives-Quinn beginnings in New York in 1945, the cease-and-desist orders and accompanying court enforcement almost never had to be used. In support of this assertion they appended the following summary of state FEPC activity from their origins through the end of 1961:

State	Cases	Hearings	Cease-and-Desist Order	Court Action
California	1,014	2	2	2
Colorado	251	4	3	1
Connecticut	900	4	3	3
Massachusetts	3,559	2	2	0
Michigan	1,459	8	6	4
Minnesota	184	1	1	1
New Jersey	1,735	2	2	2
New York	7,497	18	6	5
Ohio	985	2	1	0
Oregon	286	0	0	0
Pennsylvania	1,238	19	0	0
Rhode Island	286	0	0	0
Total	19,394	62	26	18

This was, at first glance, a rather odd supporting argument. It seemed to support the conclusion that the state FEPC model worked quite well—or at least that it functioned without any abuse of cease-and-desist authority. Indeed, in less than 0.3 percent of the cases had formal hearings even been necessary. In only 0.1 percent had cease-and-desist orders been issued. Even less frequently had such judgments been adjudicated in state courts—only eighteen out of a total of 19,394! The Republicans argued that because the cease-and-desist orders of quasi-judicial bodies had no inherent force to punish offenders, state FEP commissions had to seek court enforcement anyway if their desist orders were defied. So on principle the Republicans insisted on divorcing the judicial from the prosecutorial role as their price for supporting the fair employment bill in 1962, and that year they had prevailed.[22]

The Powell committee's Democrats, however, were unhappy with that compromise. In 1962, when the House Education and Labor Committee could never seem to get an important bill past the Rules Committee, perhaps it wasn't worth fighting about. But in the post-Birmingham atmosphere of 1963, the committee's Democrats sensed a new political opening. So on July 11 they reversed the 1962 decision by a straight partisan vote of 13–7.[23] When the committee reported out H.R. 405 on July 22, 1963, they had returned their proposed EEOC to the original quasi-judicial model based on the NLRB.

In impressive tribute to the "power of the first draft," this model had been familiarly precribed in legislative boilerplate since as early as the 78th Congress of 1944, and certainly by the 81st Congress of 1950. In dissent, Griffin and his Republican colleague, Peter Frelinghuysen (R.-N.J.), objected in July 1963 that "the historic safeguard of trial before an impartial judiciary would be abandoned in this bill." But their objections were ignored by the Democratic majority. So when Celler's Subcommittee No. 5 in October incorporated H.R. 405 as the new Title VII of H.R. 7152, back came the EEOC's judicial role with cease-and-desist orders. Then came the White House salvage operation, which allowed the Republicans to counterattack under the leadership of McCulloch and Halleck. Out again went the quasi-judicial model, in again came the prosecutorial model, like a yo-yo.[24]

Such a prosecutorial EEOC would neither decide cases nor issue orders. Instead it would seek relief in federal district court, and then in a trial de novo, rather than in an appeals-level review of a commission decision. The *New York Times*'s Anthony Lewis, who was close to the Kennedys, reported that the Administration preferred the Powell committee's quasi-judicial model in H.R. 405.[25] But as Burke Marshall was fond of recalling, Robert Kennedy was Mr. McCulloch's lawyer and press agent. So the Administration embraced and pressed the compromise. The partisan whipsaw with the EEOC and cease-and-desist authority, however, had stirred up ideological suspicions in the House, and this would make successful compromise difficult.

The House Leadership Compromise of 1963

So there the matter stood, on the brink of indicision and disagreement amounting to near chaos in late October in Chairman Celler's normally iron-ruled Judiciary Committee.[26] To counter privately the intensive public lobbying of the Leadership Conference, the President personally entered the fray. On October 23 he summoned to the White House an unsettled council of notables that included Vice President Johnson, Speaker McCormack, Majority Leader Carl Albert (D.-Okla.), Minority Leader Halleck, Minority Whip Les Arends (R.-Ill.), Chairman Celler, and McCulloch.[27] The goal of this meeting and a profusion of subsequent spinoff meetings was to forge a bipartisan compromise bill that could attract a minimum of 17 votes in committee. This would be enough to allow Celler to cast a tie-breaking vote in favor of substituting the compromise package for the liberals' subcommittee bill. The showdown came on October 29, in a series of three votes that produced unusual political coalitions. On the first ballot, over a motion to approve the subcommittee bill, most of the southern conservatives joined the northern liberals and voted "aye," in hopes of sending to the House an ultra-liberal bill they could then defeat on the floor. But in a classic cliff-hanger of roll-call uncertainty, the Administration's centrist coalition held and the motion failed 15 to 19.[28] Then a motion to substitute the compromise bill was carried by a margin of 20 to 14. Finally, with the issue no longer in doubt, the committee voted 23 to 11 to send the Administration's preferred compromise bill to the House.

The House leadership compromise of 1963 represented a strategic victory for the Kennedy brothers, although it was purchased at the cost of considerable resentment by the civil rights forces allied under the umbrella of the Leadership Conference. Most attention was focused on the bill's first three titles—on voting, public accommodations, and the symbolic Title III. On voting rights, the compromise limited to federal elections its prohibitions against subjective literacy tests and similar disfranchising technical devices.[29] On public accommodations, the compromise dropped Kastenmeier's sweeping inclusion of all state-licensed establishments, and accepted McCulloch's specific exclusion of personal service establishments (e.g., barber shops and shoeshine parlors) and retail shops without eating facilities. But the public accommodations language basically retained the broad coverage of the original Kennedy bill. This was easily its most controversial provision, and arguably its most radically transforming one for the South.

Finally, the broad language empowering the Attorney General in Title III was deleted. Instead of authorizing him to sue to protect *all* constitutional rights everywhere, the compromise empowered him only to enforce equal protection of the laws "on account of race, color, religion or national origin." The loss of Title III was a hard blow to the Leadership Conference. The NAACP's chief Washington lobbyist, Clarence Mitchell, called it a "sell-out" (during Judiciary's tense negotiations in closed session, Mitchell pointed out that "Everybody in there is a white man").[30] But the compromise kept the bill alive, and it was indubitably stronger than the original Kennedy omnibus proposal of June 19. With the exception of the new EEOC title, the remaining elements of the House leadership compromise were substantial but did not reflect major changes. The school desegregation title retained its exclusion of northern de facto segregation; the Civil Rights Commission was made permanent while the Community Relations Service was dropped; Title VI on fund cut-off remained essentially the same potentially powerful weapon, while attracting little legislative or public attention; and titles VIII through XI reflected relatively minor consensual adjustments.

During the summer of 1963 most attention was fixed on the civil rights bill's provision to desegregate southern hotels and stores and restaurants. But by the fall the novelty had somewhat worn off the edge of Title II, and with the demise of Title III, increasing attention was turned to the omnibus bill's greatly strengthened FEP provision, Title VII. The compromise bill now contained, in by far its most complicated and lengthy title, a federal FEPC newly christened as the EEOC. It provided for a five-member commission with powers over most private employers to receive complaints and investigate charges of discrimination, including authority to subpoena witnesses and require record-keeping and periodic reporting. The compromise had stripped the proposed new agency of cease-and-desist authority. This in turn had stripped away much of the structural apparatus of quasi-judicial agencies that typified the proposed EEOC's administrative models—the NLRB, the Federal Trade

Commission, and the state FEPCs. But the EEOC retained its prosecutorial role. The compromise bill's EEOC could still file civil suits in federal district court for injunctive relief against future violations, including reinstatement and the potential recovery of back pay for victims.

These distinctions, however, were not understood by the nation at large. Both the national media and the congressional debate had done a poor job of identifying and explain them. Half the states possessed their own FEP agencies, most of them modeled on the respected New York initiative of 1945, and between them encompassing a rich variety of powers, coverage, and structure from which an educational comparison might well have been derived.[31] The other half of the states, of course, did not, and this implicit control group was dominated by southern states wherein political debate had long consigned the FEPC to an opprobrium unworthy of serious discussion. All the more reason, then, for the enlightenment of informed national debate. But such enlightenment largely failed by default. The penetrating explorations of the congressional hearings, floor debate, and media coverage, which had effectively illuminated such fundamental controversies as those involving the 14th Amendment versus the commerce clause in public accommodations, or the reach of federal power in Title III, found no parallel in the arcane world of quasi-judicial agencies. Perhaps this should not be surprising, given the pressure of time and events, of emotion and escalating violence, of guilt and fear and anger, and also given the inherent complexities of the subject. But the consequences of this failed dialogue would risk a resolution that fell between two stools, thereby badly confusing and mismatching the relationship between ends and means in fair employment. In November of 1963, however, H.R. 7152 still had a long way to go.

Only three weeks after the Judiciary Committee had cleared the compromise civil rights bill, John Kennedy was assassinated in Dallas. Five days later, President Lyndon Johnson told a joint session of Congress that "We have talked long enough in this country about civil rights. It is now time to write the next chapter and to write it in books of law."[32] No eulogy, Johnson said, "could more eloquently honor President Kennedy's memory" than the "earliest possible passage of the civil rights bill for which he fought so long."

Judge Smith, the House Rules Committee, and the Irony of Sex Anti-discrimination

In the fall of 1963 the House Rules Committee had refused to allow the House Education and Labor Committee's H.R. 405 to proceed to the House floor, much as it had regularly and rather easily bottled up such FEP legislation in the past. But once Congressman Griffin and his minority colleagues on the Powell committee had negotiated their compromise with McCulloch and retransformed H.R. 405 into a "Republicanized" Title VIII in the Judiciary Committee, Judge Smith and his normally dominant conservative cohort on Rules faced a broad bipartisan compromise that they could delay but stood little chance of stopping. In an attempt to push Rules toward some early action on H.R. 7152, Celler filed a discharge petition on December 9, and the Republican leadership joined in by threatening to invoke the Calendar Wednesday procedure. Both of these short-circuiting procedures were difficult to achieve and neither was pressed with determination. But they carried the message that a bipartisan majority for H.R. 7152 was building even on Chairman Smith's bastion in Rules, and that further delaying tactics (such as the chairman disappearing into his apple orchard in Virginia) only risked the humiliation of having his committee formally take the bill away from him. The ranking Republican of the fifteen-member Rules Committee, Clarence Brown of Ohio, represented a district adjacent to McCulloch's and he shared McCulloch's moderately conservative Republican convictions, including a belief that Republicans should push rather than block the revised H.R. 7152. Brown made it clear to Smith that he controlled enough Republican votes on Rules to join the northern Democrats, if reluctantly forced to do so, and take the bill away from the chairman.[33] So, on December 19, Smith grudgingly scheduled the Rules hearings for ten days between January 9 and 30, 1964.

The hearings were, as it turned out, anticlimactic. This was partly because Kennedy's martyrdom and Johnson's pledge to redeem Kennedy's civil rights promise had set a newly expectant tone to the debate. The conservative coalition was less free than in the past to savage the bill in Rules. More important, because Celler and McCulloch were united behind the compromise, they could count on a supportive majority on Rules, and these circumstances effectively postponed to the House floor any serious battles over major amendments. The result on Rules was a sort of low and folksy theater, with Smith needling Celler for having railroaded his bill through Judiciary in closed session, then failing to produce the printed testimony of its sole witness, the Attorney General. Celler's excuses were weak on both counts. But in such a contest of political muscle that was largely irrelevant.

The Rules hearing's elaborate rhetorical courtesies contrasted sharply with the determination of the combatants, and occasioned much guffawing from observers. Most discussion centered on integrating public accommodations, and this tended only to echo rather perfunctorily the conservative objections of the summer and fall in defense of state rights. William Colmer of Mississippi worried aloud that the barbers he knew "tell me that they're not equipped or trained to cut the hair of the opposite race," which was probably true. Smith expressed dismay at the new plight of a hypothetical podiatrist whose office might be in a hotel: "If I were cutting corns," the octenagarian committee chairman said, "I would want to know whose feet I would have to be monkeying around with. I would want to know whether they smelled good or bad."[34] Smith later had second thoughts about the taste of that remark, and had it expunged from the record. During the Rules hearing Celler and McCulloch faced down all hostile amendments that were offered, and accepted several technical and clarifying amendments of no significant impact, beyond one that returned the tenure of the Civil Rights Commission to the four-year renewal originally proposed by President Kennedy. On January 30, to no one's great surprise, the resolution granting a rule for H.R. 7152 carried by the comfortable margin of 11 to 4.

Once the bill was before the full House, however, Howard Smith sprang his one-word surprise. On February 8 he interrupted debate and simply moved to add "sex" to the list of classes protected by Title VII.[35] "This bill is so imperfect," he asked, "what harm will this little amendment do?" Actually, the proposed addition of sex protection was not such a surprise to the Administration. Nor should it have been. Like so much of the statutory language and legislative debate during 1963–64, the "friendly" southern amendment barring sex discrimination had been prefigured in the donnybrook over FEPC in 1950. That year the House had degenerated into a bizarre caricature of parliamentary gamesmanship, and Democrat Dwight Rogers of Florida had won voice-vote approval of a similar—and similarly mischievous—sex-discrimination amendment to the substitute, voluntary FEPC bill that ultimately passed the House, only to die later in the Senate.[36]

Katzenbach, in his capacity as chief shepherd of H.R. 7152, had warned supporters against the maneuver, and had asked Representative Edith Green of Oregon to help deflect its mischief. A senior Democrat on the Labor and Education Committee, Green had been a chief sponsor of the Equal Pay Act of 1963, which had enjoyed strong union backing because it would reduce the ability of employers to hire low-wage women to displace union-wage workers. But Green believed that discrimination against blacks was much more severe than against women, and she feared that adding sex as a protected category might sink the civil rights bill.[37] Otherwise why were southern congressmen pushing it? The House leadership had in fact rather easily beaten back several attempts by Democrat John Dowdy of Texas during the early floor debate to add sex discrimination bans to several titles of the bill, and thereby to divide its supporters. But Dowdy lacked Smith's unique committee leverage, and Smith shrewdly confined his amendment barring sex discrimination to Title VII.

The courtly Virginian, however, was also curiously sincere in his unlikely feminist egalitarianism. He had been a congressional sponsor of the Equal Rights Amendment since as far back as 1945—almost as far back as he had been a scourge of FEPC. Ever since then he had maintained loose political ties with the National Women's Party, which had been founded in 1913 in the suffragist campaign under the aggressive leadership of Alice Paul, and had pushed single-mindedly for the ERA since 1923. For two score years Paul and the NWP had engaged in a running and basically losing battle with the League of Women Voters and the national establishment of liberal women's groups that came to be dominated by Eleanor Roosevelt.[38] Yet the radical, all-female NWP symbolized an abiding paradox of the women's movement. It generally represented the views of elite, white, affluent, professional, and highly educated women whose demand for the franchise had included plausible arguments that woman's suffrage would double the vote of the white middle and upper class. Such an argument appealed to both progressives and conservatives who feared being overwhelmed by the swarms of strange and swarthy new immigrants in America, as well as by the enfranchised black masses.

For forty years the NWP's campaign for the ERA was opposed by vintage liberal Democrats like Eleanor Roosevelt and Emanuel Celler, largely because it threatened their proud legacy of progressive legislation to protect women in the workplace.[39] It was also opposed by the American Association of University Women (AAUW), who combined a vigorous advocacy of black civil rights with public opposition to the insertion of "sex" into Title VII. It is difficult today, looking back at these events through the prism of the modern feminist movement—especially since the founding of the National Organization of Women in 1966—to appreciate the logic that wed the elitist and sometimes racist NWP to the ERA, and that set the vanguard of traditional liberalism so firmly against the ERA's egalitarian feminism for so long.[40]

The divisiveness of this historic and paradoxical split in the women's movement, especially within the Democratic party, had led President Kennedy in 1961 to create the President's Commission on the Status of Women, to which he appointed Eleanor Roosevelt as chair (until her death in 1962).[41] But the commission's membership also reflected the opposition to the ERA of Esther Peterson, whom Kennedy had appointed director of the Women's Bureau and also an assistant secretary of labor.[42] When the commission, after troubled debate, recommended against the ERA in its final report of 11 October 1963, Alice Paul and her NWP colleagues pressed Judge Smith to sponsor their egalitarian amendment to Title VII. On December 16, the NWP's annual convention unanimously resolved that protection against sex discrimination should be added to the civil rights bill. Otherwise, the resolution observed, the bill would not offer "to a *White Woman*, a *Woman of the Christian Religion*, or a *Woman of United States Origin* the protection it would afford to Negroes."[43]

On the House floor, opposition to the Smith amendment was led by Celler and Edith Green. But Green was isolated in her opposition by the other women members. Martha W. Griffiths of Michigan, who was the first woman to join the House Ways and Means Committee, led the bipartisan and ideologically strange fight for the Smith amendment.[44] Griffiths was a warrior who took no prisoners; she declared that "a vote against this amendment today by a white man is a vote against his wife, or his widow, or his daughter, or his sister."[45] "It would be incredible to me that white men would be willing to place white women at such a disadvantage," Griffiths said. Under the bill as reported, "you are going to have white men in one bracket, you are going to try to take colored men and colored women and give them equal employment rights, and down at the bottom of the list is going to be a white woman with no rights at all." In response to the nervous ripples of male laughter that occasionally stirred through the House during the debate, Representative Katherine St. George, a genteel Republican from Tuxedo Park, New York, hurled defiance in the teeth of her male colleagues: "We outlast you. We outlive you. We nag you to death. So why should we want special privileges?" "We want this crumb of equality," she said, "to correct something that goes back, frankly, to the dark ages." Her Republican colleague, Catherine May of Washington, was a less strident petitioner, yet quaintly echoed the sentiments of the NWP: "I hope we won't overlook the white native-born American woman of Christian religion."[46] In their rather lame and chiefly tactical arguments against Smith's ban on sex discrimination, northern liberals Green and Celler and Lindsay could only plead that the amendment's timing was "inopportune," while offering so substantive arguments against it.

The long congressional battle over the civil rights bill had created many unusual political coalitions, but the brief and decisive row over the Smith amendment created even more ironical pairings than had the left-right coalition in Judiciary the previous October. Emanuel Celler and Edith Green battled the unlikely combination of Howard Smith and Martha Griffiths, and Alice Paul got her posthumous revenge on Eleanor Roosevelt. Republicans and southern Democrats had made common cause before, but such solidarity in defense of radical feminism was a novelty. After only two hours of debate, Smith's amendment passed on a teller vote of 168 to 133, with most of the senior, white, male, conservative southern Democrats apparently voting for it. By February 10, a total of 124 amendments to H.R. 7152 had been offered, and thirty-four had either been accepted by the Celler-McCulloch leadership or passed on the floor. Most of these were technical adjustments, although the left-right, egalitarian-chivalric vote for "the ladies" had been followed by a voice vote to permit bona fide fraternal and religious organizations to employ "their own kind," and also by a vote of 137 to 92 to deny protection to atheists and communists. Having failed on his own accord in overburdening the entire bill with sex-protection clauses, Congressman Dowdy tried to do the same thing with a blanket amendment to add age as a protected class. But this too was defeated, by a standing vote of 94 to 123. The black civil rights coalition, which had generally steered clear of the floor fight over sex discrimination, more actively resisted complicating Title VII by adding age as a protected category—"age" being a much less clearcut variable than sex, as the few state FEPCs that included it were discovering.

On February 10 the exhausted House, trying to adjourn before the Lincoln holiday, voted to approve H.R. 7152 by the lopsided vote of 290 to 110. Favoring the bill was a bipartisan phalanx of 152 Democrats and 138 Republicans. Opposing were 96 Democrats and 34 Republicans, with 86 of the Democrats and 10 of the Republicans representing the states of the former Confederacy. During the floor debate, one of the amendments that the bipartisan, nonsouthern coalition had rejected, by a standing vote of 83–137, was a proposal by former Governor William Tuck of Virginia that the racial voting statistics that Title VIII required the Census Bureau to collect be kept for the entire country. The bill provided that racial voting statistics be kept only for areas recommended by the Civil Rights Commission, which was expected to confine its coverage largely to the South. Clearly the civil rights bill was aimed mainly at the benighted South, which was both morally culpable and politically vulnerable. So the bill's measures to protect voting rights and to desegregate schools and public accommodations had all been carefully regionalized by the bipartisan coalition. This had not been done, however, for the bill's fair employment provisions in Title VII. As the bill moved to the Senate, major attention shifted to the two constituencies that, in the heat of the debate over racial quotas, were most likely to be worried about the bill's possible dangers. One was the employer community, which feared bureaucratic intrusion on its traditional freedom to hire and promote and fire on merit. This traditionally Republican group would look primarily to Senate minority leader Everett Dirksen to protect their interests. The other was organized labor, which supported both the Johnson administration and a national FEPC. But now that the House bill with its new equal employment provisions was headed for the Senate, labor would have occasion to review its commitment.

Organized Labor and the Specter of Racial Quotas

In mid-January, Senator Lister Hill of Alabama, a Democrat of New Deal vintage and loyalties who by southern standards had been a friend of the labor movement, warned in a speech on the Senate floor that the civil rights bill would undermine labor's seniority system by requiring racial preferences in hiring and promotions.[47] In a pained response, the AFL-CIO explained why labor supported the civil rights bill and why Senator Hill's alarming prediction was impossible under its provisions.

Spokesmen for the AFL-CIO gave two reasons for supporting H.R. 7152. The first was the moral objection that racial discrimination in employment "dulls the public's sensitivity toward all injustice, which is the wellspring of the humane kind of welfare legislation that laboring men and women want."[48] The other reason was more practical: racial discrimination hurt union membership. It undermined the solidarity of working people, and the persistence of segregated locals mocked the loose, federationist structure of the AFL-CIO. So as George Meany had testified in the House the previous summer, the unions needed the added federal muscle to make a small minority of their craft locals do what was right. The crux of the problem, however, lay not with labor but with nonunion employers. Since the *Steele* decision of 1944 the federal courts had held that unions must fairly represent all workers irrespective of race. Thus in the absence of the proposed Title VII, federal law would continue to ban discrimination by both unions and unionized employers, but not by *un*organized employers. Only Title VII could terminate their unfair advantage. Not only were they free from federal constraints, they were in fact *rewarded* for discrimination by hiring cheap, docile, fearful, nonunion workers.

Senator Hill did not address these issues. Instead he warned that under Title VII the new EEOC would find that "underutilization" of minorities required compensatory racial preferences. Like Senator Ervin in his joust with Robert Kennedy the previous summer, Hill pointed to hidden dangers of Title

VI's short and ostensibly simple ban on discrimination in federally assisted programs. Hill feared that federal civil rights bureaucrats might find violations on a federally aided construction job "because there were less [*sic*] carpenters, proportionately, of a given race than of another race, and that the job was not racially balanced." The AFL-CIO response branded this as an "utterly false" premise that led to a "long tale of horrors" over job preferences for Negroes. AFL-CIO lobbyist Andrew Biemiller insisted that the civil rights bill "does not require 'racial balance' on a job."[49] "It does not upset seniority rights already obtained by any employee. It does not give to any race the right of preferential treatment in hiring or terms of employment," Biemiller said. "It does not prevent an employer or union from relying on genuine gradations in skill or experience or similar qualifications in deciding whom to hire or promote or refer to a job." Neither did it force a union "to favor nonunion men over union men." What it did do was to lay down just one overriding rule regarding employment: "starting just one year after the bill is effective, unions and employers alike are forbidden to judge a man by the color of his skin, or the faith he professes."[50]

The AFL-CIO's reply to Hill in January did not anticipate the addition of sex to Title VII in February. Labor's posture of demanding equal treatment by race while defending special protective laws for women was confounded by the sudden addition to Title VII of gender discrimination. Senator Hill's warning to labor implied that the civil rights bill might invert labor's logic by requiring its opposite: equal treatment for women and special treatment for blacks. Union spokesmen assumed that racial balancing would be impermissible because Title VII prohibited racial discrimination of any kind. Its sections "only forbid discrimination in employment 'because of' race or religion," Beimiller replied to Hill, "and do not affect an employer's right to hire whomever he wants for whatever other reason."

> If such nondiscriminatory hiring results in "racial imbalance," because whites are better qualified, have greater seniority, or apply in greater numbers, there is still no violation of the equal employment guarantees of the bill. Accordingly, no violations could be grounded on the more general provisions of the bill dealing with federally assisted programs, merely because whites hired on a nondiscriminatory basis were receiving a larger number of jobs from that particular federally aided program.[51]

A former congressman from Wisconsin, Biemiller concluded that because "it is well-established that the specific provisions of a statute govern the general," Title VII's specific ban against racial discrimination would also ban any use of racial quotas in Title VI. And because any fund cutoff under Title VI would be subject to judicial review, "even if the Senator's nightmare" about "overzealous civil righters" in the EEOC ordering racial balancing should come true, "the judges of the nation's courts could be counted on to insist that legal standards be observed." It seemed inconceivable to Biemiller that Congress or the federal courts could approve of "righting ancient wrongs by perpetrating new ones," or that "white workers who possess hard-earned seniority should be discriminated against in the future because Negroes were discriminated against in the past."

The AFL-CIO's interpretation was consistent with every judicial precedent coming into the Senate debate that winter and spring, and also with every pronouncement of the bill's sponsors on the Senate floor. Lister Hill of Alabama, like Sam Ervin of North Carolina, appeared merely to be whistling Dixie. With organized labor securely behind the Johnson administration and the civil rights bill, major attention shifted to Senate Minority Leader Dirksen. His price for supporting cloture was expected to hinge on the potential reach of Title VII beyond the South, to affect the behavior of private employers whose interests the Republican party had traditionally defended.

The Courting of Senator Dirksen

Early in January, 1964, while the Rules Committee was gearing up to review H.R. 7152, Lyndon Johnson made his pact with Robert Kennedy. "I'll do on the Bill just what you think is best to do on the Bill," Johnson told the Attorney General. "We'll follow what you say we should do on the Bill. We won't do anything that you don't want to do on the legislation. And I'll do everything you want me to do in order to obtain the passage of the legislation."[52] Kennedy revealed this to Anthony Lewis in an oral history interview the following December, and Lewis replied that the President's pledge was "rather extraordinary." Lyndon Johnson "didn't want President Johnson to be the reason, to have the sole responsibility," Kennedy explained. "If I worked out the strategy, if he did what the Department of Justice did, said, recommended, suggested—and particularly me—then, if he didn't obtain the passage of the Bill, then, he could always say that he did what we suggested and didn't go off on his own." "He had a particular problem being a southerner," Kennedy conceded. "If we decided that he should follow a particular line of strategy, and then it didn't work, it could be very, very damaging to him. So, I think that for political reasons, it made a great deal of sense. Secondly, our relationship was so sensitive at the time that I think that he probably did it to pacify me . . . ."[53] "[S]ince January, I haven't had any dealings with Johnson," Kennedy admitted. Johnson "used to tell Kenny [O'Donnell] and Larry [O'Brien] and all the others that he thought I hated him and what he could do to get me to like him, and whether he should have me over for a drink," Kennedy said. "I thought that an awful lot of things that were going on that President Kennedy did, that he [Johnson] was getting the credit for and wasn't saying enough about the fact that President Kennedy was responsible."[54]

Both men understood that their ambition and their obligation to the murdered President required them to submerge their enmity and get the civil rights bill through the Senate. "Where are you going to get the votes?" Johnson asked his Attorney General. "The person who you're going to get the votes from," Kennedy said, "was Everett Dirksen."[55] Would Johnson, in the bargaining process, allow the Senate conservatives to strip away the most controversial provisions of the House bill, as his critics said he had done as Senate majority leader in 1957? In early February Johnson was asked at a press conference whether he expected a filibuster and whether, in order to get the bill passed in the Senate, it would "have to be substantially trimmed." Johnson replied with a laconic smile that the answer was "yes" to the first question and "no" to the second.[56] Burke Marshall emphasized the Johnson administration's united front against the suspicions of early cave-in, insisting on the record that Title VII "was not put in there as a trade-off" to bargain for cloture. But hard bargaining for Senate cloture dominated the next three months, and necessarily involved some trade-offs in response to Dirksen's artful probes.

Johnson's consistent public and private refusal to bargain away major elements of the civil rights bill, and his deference to the strategic leadership of his Attorney General, narrowed the maneuvering room that was available to Dirksen. Bargaining space was further reduced by special circumstances in the Senate that winter and spring. These ranged from such specific constraints as Robert Kennedy's ironclad oath of 1963 to McCulloch, to the newly impatient national mood of the post-Birmingham era.[57] The latter was captured in a typical Dirksenian flourish by the minority leader's "Sermon on the Mount" of May 19, when he announced that the Senate leadership had reached a bipartisan compromise, and quoted Victor Hugo's pronouncement that no army was stronger than "an idea whose time has come."[58]

The forces converging on a civil rights consensus in 1964 included two unusual combinations. One was a delicate balance between carrot and stick, partly attributable to superior leadership and partly to luck. Its positive incentives included vigorous moral suasion by the nation's prestigious religious leaders, the constitutional blessings of the dominant legal establishment, and both the martyrdom of President Kennedy and the impatient victimhood of black America—especially Martin Luther King's God-fearing southern legions. On the negative side stood the rnational anger that grew in response to southern intransigence and violence, ranging from Bull Connor's defiance in Birmingham to the murder of three civil rights workers in Mississippi after their disappearance on June 21. Beneath this lay a muted anxiety over the potential for retaliatory black violence. In March, King warned that if the filibuster lasted more than a month, he and his followers would "engage in a direct action program here in Washington and around the country." In April, when the Brooklyn CORE chapter threatened to paralyze the opening of the New York World's Fair with a massive "stall-in" traffic jam, they were restrained with great difficulty by Farmer and the worried civil rights leadership.[59] Ghetto rioting did indeed explode that year, in Harlem in July, thus heralding the extraordinarily destructive string of mostly northern and increasingly violent "long hot summers" that would trigger white backlash and would escalate through 1968. But the Harlem riot occurred just *after* the House voted final passage of the civil rights bill on July 2.

The second unusual combination leading toward consensus was the superior organization of the Senate leadership, and the fatal strategic blunder of Senator Russell's southern bloc. To fill the vacuum posed by the aloof, statesmanlike posture of Majority Leader Mansfield, the majority whip, Hubert H. Humphrey (D.-Minn.) teamed up with the liberal Republican whip, Thomas Kuchel of California, to organize a group of disciplined anti-filibuster teams to combat the cadres of experienced southerners. Humphrey channeled his boundless energy toward an uncharacteristic path of patience, moderation, tolerant understanding, and bargaining flexibility.[60] For their part the southerners, despite serious internal debate, ultimately followed Russell's standfast script to their strategic doom.

When the Senate formally took up the House-passed bill on March 9, Katzenbach sent O'Brien a memorandum outlining the Justice Department's legislative strategy. In it he posited a cluster of self-serving motives that he expected to determine three ranked principles of Republican strategy. The Republicans' first interest, Katzenbach assumed, was to block the bill. Otherwise its passage was "bound to benefit President Johnson and the Democratic Party and spike the myth that civil rights can only be enacted in a Republican administration."[61] Second, they would try to cut a deal for a considerably weaker bill. Third, they might offer strengthening amendments to the House bill, thereby forcing the Democrats into the embarrassing posture of opposing the amendments in order to save the bill (and also to honor the promises to McCulloch and Halleck which were not on the public record). Meanwhile the Republicans would complain that the Administration was plotting a deal with the southerners, that the Democrats didn't really want a strong bill, and that the Republicans weren't being consulted. Katzenbach was confident that the southern Democrats would block and delay while mischievously supporting both strengthening and weakening amendments—and profiting either way, "No mater what these amendments do." He therefore concluded that the Administration must stand behind the House bill and oppose any major amendments. This would require the hard discipline of taking heavy liberal criticism while involving the Republicans in direct private negotiations and publicly identifying them with the decisions made. This was, after all, only a logical and indeed honorable confirmation of the bargain they had struck with McCulloch and the House leadership some ten months earlier.

Katzenbach was right about the strategy, although his premises were flawed. He was too pessimistic about the partisan and spoiler motives of the Republicans. He overestimated the potential bargaining leverage of the GOP's northeastern liberal wing, which was led in the Senate by Javits and Kenneth Keating of New York and Clifford Case of New Jersey. Katzenbach overestimated, also, the political wisdom of the southern Democrats. They stood dogmatically by their record-breaking filibuster, generally refusing to exercise their potentially enormous bargaining power in amendment maneuvers until it was too late. Then they lost the war totally.[62] The Republican leadership, on the other hand, had far less room to maneuver than Katzenbach assumed. Sitting on a national volcano, they could not risk appearing to combine with the southerners and kill the bill. Moreover the Lincoln tradition was genuine. So Dirksen did what he had to do, and what he loved to do, playing coyly with the intrigues of backroom negotiation and floor posturing, whittling down the areas of disagreement toward a compromise that would allow him to blunt the bill's most objectionable features and bask in the reflected glory of Victor Hugo's moment of destiny—and, in the tradition of Lincoln, save the Republic.[63]

The Senate debate fell into three phases. The first lasted from March 9 to March 26, and centered on an important procedural question: should H.R. 7152 be referred to the Senate Judiciary Committee, as was customary? On February 26, Mansfield had intercepted H.R. 7152 on its way over from the House and had placed it directly on the Senate calendar. Lyndon Johnson as majority leader had done the same thing with the civil rights bills of 1957 and 1960, and for the same reason—to avoid a potentially fatal delay. But the arguments for normal committee referral were powerful. As had so often been the case with ad hoc coalition-building in the long history of civil rights legislation in the American Congress, the strategic logic linked constituencies that on other grounds held few interests in common. Least surprising was the insistence of southern Democrats that all civil rights proposals belonged in the jealous protective custody of committee chairman James Eastland of Mississippi. Dirksen agreed. He argued as a procedural conservative that to do otherwise was highly irregular. Furthermore, H.R. 7152 had been reported out by the House Judiciary Committee after closed negotiations, and hence had received no proper hearings in a substantive committee at all. Surprisingly, Senator Wayne Morse of Oregon also agreed. A liberal Democrat and former law professor (and also a former Republican) whose independent cast of mind frequently bordered on irascibility, Morse said he supported referral to Judiciary because he supported the bill, knew that if it passed it would generate a flood of litigation, and hence called for hearings to document fully and properly its legislative history and to explicate the intent and meaning of its many and complex titles.[64]

But Eastland's committee had a jaded past. Humphrey pointed out that during the previous decade only *one* of 121 civil rights bills had ever emerged from Eastland's lair—and even that one had been a fluke. Indeed, Eastland's committee had already held eleven days of hearings on the original bill during the previous summer, but had heard only *one* witness, Attorney General Robert Kennedy. Kennedy's long-suffering nine days of testimony during that hearing had been monopolized by the running colloquy with Ervin—which was one reason why fellow committee member Dirksen complained that he hadn't had a proper crack at the bill yet in committee. Senator Thomas Dodd (D.-Conn.), in defending a bypass of Judiciary, pointed to H.R. 7152's cumulative record of 83 days of congressional hearings in six committees, involving 280 witnesses and 6,438 pages of printed testimony. Meanwhile, the clock ticked toward the 1964 elections, with a third of the Senate scheduled to face the voters that fall. On March 26 the Senate first agreed to take up Morse's motion to refer the bill to Judiciary, then rejected the motion by a vote of 34–50. Then on March 30 the contest entered its second and more substantive phase. This lasted into mid-May. It was dominated by Dirksen's intricate dance of negotiations, and culminated in the crucial Senate leadership compromise of 1964.

The Senate Leadership Compromise of 1964

The second phase of Senate debate mirrored, in collapsed form, the original omnibus bill's hearing process of 1963. Before the Senate could turn to the hard negotiations over the kind of zero-sum dilemmas that were posed by Title VII's concern for the racial distribution of jobs, the Senate felt compelled to first review the moral debate. These exchanges on the floor centered on the more obvious and almost exclusively southern, positive-sum questions of racial fairness. They were starkly posed, in the nationally televised new format ushered in by the sit-ins of 1960, by the massive southern denial of the right to vote, to eat in restaurants and sleep in hotels, and to go to neighborhood public schools irrespective of one's race. This early, South-centered debate was reinforced frequently by news stories that dramatized the South's racial unfairness, such as the early spring's southern foray by Mrs. Malcolm Peabody of Boston. The Brahmin 72-year-old mother of Massachusetts governor Endicott Peabody got herself successfully thrown into jail in St. Augustine, Florida, for trying to dine with Negroes at a segregated motel. As the *New York Times* editorially explained in response to Mrs. Peabody's septuagenarian raid, "the real test of the Dirksen program will be in its proposals for changes in the provisions forbidding racial bias in access to hotels, restaurants and other places of public accommodations."[65] Such regional attacks invariably invoked southern charges of northern hypocrisy, as when Eastland accused Javits of demanding massive busing to desegregate southern schools, while New York City alone contained at least 165 all-Negro schools. Javits volunteered that the number of predominantly Negro schools in New York was probably closer to nine hundred, but he denied that the city's de facto racial school separation amounted to de facto racial school segregation. Eastland replied that this procedural distinction without a substantive difference continued to elude him, but the colloquy was wearily familiar.[66]

By the spring of 1964, however, this stale scenario would no longer suffice. The senators were long in the habit of rehearsing the same old rhetoric, but something fundamental had changed in the atmosphere of expectations. Javits's self-righteousness had long galled the southerners. But Eastland could no longer bring himself to announce, as he had done so confidently and so incredibly in 1950 during the debate over the House-passed FEPC bill, that "In all the hearings which are talked about, there will not be found a scintilla of proof that there is discrimination in Mississippi or any other state in the deep South. I deny most emphatically that there is discrimination in my state based on race or color."[67] Indeed, the public humiliation wrought by the Jim Crow system was so palpable by 1964 that southern politicians in Washington no longer seemed to have real heart for its defense. Even Dirksen, who had warned since the previous summer that he opposed Title II's infringement on business freedom, sought a graceful way out of the impasse. So in his negotiations with the Justice Department team and the Democratic leadership, Dirksen found a reasonable compromise that would permit him to yield to the moral imperative of Title II while still constraining federal intrusion in commerce.

The formula that Dirksen agreed to was devised primarily to limit the EEOC in Title VII cases. It would allow the Attorney General but not the commission to file anti-discrimination suits, and to do so only where he could document a "pattern or practice" of systematic discrimination. By preventing third-party suits filed by groups like the NAACP, such an arrangement could avoid a sea of unnecessary litigation against businesses while still providing for some certain measure of enforcement by federal authorities. On the other hand, Dirksen well knew that the Justice Department was a relatively small, elite cabinet agency, in comparison with the more typical and large program-running departments like HEW, and so prided itself on enforcement through key case selection rather than through massive litigation. As a result the Justice Department posed a smaller threat of potential harassment to employers than would a new mission agency like the EEOC, which reminded Dirksen and his more conservative colleagues uncomfortably of its crusading early model: the NLRB. To Dirksen, desegregation of public accommodations had essentially nothing to do with northern constituencies like Illinois in any event. So Dirksen continued to protest Title II's potential abuses until the Senate's penultimate act. Then Dirksen—like Lord Byron's Julia, whispering "I will ne're consent"—consented.

During the five weeks following the first of April, Dirksen floated dozens of trial amendments, most of which further refined, moderated, clarified, or restricted the reach of federal enforcement power over private enterprises and citizens. He proposed amendments that would variously delay and stagger implementation, restrict record-keeping requirements, require confidentiality of proceedings, exempt educational institutions, curtail the EEOC's power of subpoena, and require that only individually aggrieved complainants rather than third parties could bring suit (and then only in the districts wherein they resided). Dirksen further sought to restrict the new EEOC's budget, which called for $2.5 million in the first year and $10 million the second. He was reminded unpleasantly, as ever, of the fledgling NLRB, which had jumped from a budget of $659,000 and a staff of 189 in 1936 to a budget of $21 million and a staff of 2,056 by 1963.[68] Many of his proposed amendments were relatively minor, technical, and often helpful to the Administration, such as deleting the House floor amendment concerning atheists, and adding coverage of union hiring halls. Others represented playful Republican harassing fire, such as abolishing the PCEEO by congressional mandate, and these were generally not pressed with great seriousness.[69]

Some of Dirksen's more substantive amendments, however, would eventually lead to major litigation. Their subsequent debate on the Senate floor would provide a legislative history that would fuel later charges that the courts were ignoring and even contradicting the unambiguous intent of Congress. One of these was the requirement in Title VI that federal funds could be cut off only from the specific program against which the complaint of discrimination was lodged, and not from the entire institution or umbrella program.[70] Another was the addition of the word "intentionally" to Section 703(g) of Title VII, to make it clear that discrimination could not be legitimately inferred from statistical distributions in employment patterns. Still another of Dirksen's amendments protected differences in compensation or conditions of employment that resulted from bona fide seniority or merit systems, and that were not based on intention to discriminate on account of race, color, religion, sex, or national origin. Dirksen also reinforced Representative McCulloch's original ban on racial balance as a legitimate concern of federal enforcement authorities by extending it from Title VI to include Title VII and job distribution.

During the first two weeks in May, Dirksen met five times in his Senate office suite for culminating negotiations with Mansfield, Humphrey, Kuchel, and the Justice team of Kennedy, Katzenbach, and Marshall, plus key staff aides and occasionally other senators.[71] By May 13, the essential elements of the compromise that would lead to cloture had been determined. The package of 70-odd amendments required the complete rewriting of a substitute bill for H.R. 7152. But the significant changes in fundamental approach boiled down to only two. Here Dirksen had insisted on his bottom price. Yet the minority leader's two crucial demands did *not* include the stipulations that later became so controversial in the courts, such as the narrow definition of fund cut-off in Title VI, or, in Title VII, the ban on racial quotas, the requirement that discrimination be intentional, and the protection of bona fide seniority and merit systems. That was because Dirksen did not have to exert his unique leverage in these areas. Outside the South, the Senate consensus on the classic understanding of anti-discrimination was unambiguous and overwhelming. As Humphrey and the Senate leadership took great pains to reiterate in floor debate, the amendments of the compromise package only clarified and codified the original intention of the administration and the congressional leadership. Dirksen's two basic changes extended beyond this to further reduce the authority of the EEOC, which he regarded as a potential bureaucratic monster, like the early and runaway NLRB. For the new EEOC, Dirksen would mandate deference to state and local FEP agencies where they existed, and, more important, strip the EEOC of its prosecutorial role.

The first change was designed to further isolate the North and West from the impact of the new law, although this was not acknowledged as its goal. The civil rights bill was, after all, targeted primarily against the intransigent South, as symbolized by Bull Connor. This alone made it politically possible in 1964, because the southern senators could not sustain a filibuster without their Republican allies, and the latter would not join a last-stand defense of the state rights of Bull Connor. The second change, however, seemed to reflect a growing awareness outside the South that while Jim Crow schools and segregated restaurants and black disfranchisement were essentially southern problems, job discrimination was a national phenomenon. When the *New York Times* editorially endorsed the Dirksen compromise as both a sound bargain and yet another congressional lesson that the perfect was the enemy of the good, it conceded that job discrimination had become "a national, not a Southern, problem."[72] Thus Dirksen's two key changes reflected a tension between, on the one hand, the nonsouthern and bipartisan consensus that the new law should mainly reform the wicked South, and on the other a fear that the bill's two job discrimination titles, VI and VII, might impinge heavily upon their nonsouthern congressional constituencies. Hence they must be carefully constrained.

Dirksen justified both changes by a higher principle. The first was the Republican (and conservative Democratic) principle that in government-business relations, local primacy must prevail over Washington-knows-best. Responsible fair employment "starts back home," Dirksen said; his amendment requiring EEOC deference to state FEPCs retained the "local spirit."[73] This would both recognize and reward the kind of responsible local initiative that the state FEPCs represented, he argued, and it would also greatly reduce the administrative burden on the new federal agency. But the brute problem, as Humphrey and Keating and other Senate liberals pointed out, was that such a provision would seem to invite the instant creation of phantom or paper FEPCs by the defiant southern state legislatures. The House bill, recognizing this danger, had provided for the EEOC to yield to a local FEP jurisdiction only if the EEOC determined that the local agency was "effectively exercising such power." But Dirksen and the Republicans instinctively distrusted the fairmindedness of Washington regulators, especially in light of the early behavior of the EEOC's regulatory model, the NLRB. So the Dirksen compromise gave state or local FEP agencies exclusive initial jurisdiction over complaints, but limited this to sixty days, beyond which the complainant could then turn to the EEOC. This Dirksen compromise, however, together with the Senate's further reinforcement of McCulloch's earlier House exclusion of de facto segregation or racial balance in the North, predictably drew Richard Russell's complaint that the "bill now has been stripped of any pretense . . .[and] stands as a purely sectional bill."[74] While Russell's charge was not entirely true, the point was telling, and the political value of Dirksen's compromise was clear.

The principle that underpinned Dirksen's second major change, which stripped from the EEOC the power to bring suit, was drawn from the regulatory reform tradition that had produced the Administrative Procedures Act of 1946, and had earlier led Congressman Robert Griffin and his Republican colleagues on the House Education and Labor Committee to strip away the EEOC's judicial functions. Dirksen wanted to make the EEOC "largely a voluntary agency."[75] He argued that under his amendments the law would encourage states lacking fair employment agencies to create them, because his amendment would double to 120 days the period of exclusive jurisdiction enjoyed by such new agencies during their first year. Moreover, every major study had shown that the state FEPCs were so effective at negotiating voluntary compliance that they almost never had to resort to the courts. To counter Dirksen's argument, the NAACP's Herbert Hill in late April released his own study of the state FEPCs.[76] Hill pointed out that despite almost two decades of the Ives-Quinn model of FEPC in New York and other major industrial states, virtually *all* of which were armed with the cease-and-desist power, black unemployment still stood at twice white unemployment, and blacks continued to stand at the bottom of the union scale. Hill attacked the "insensitivity" of "timid political appointees" who ran such inefficient FEP agencies as those of Michigan and Wisconsin, and warned that if such feckless behavior continued, the state FEPCs would have to be regarded as "obsolete, for the rising Negro mass movement will proceed to the attack in its own way."[77] Such a prominent public attack from the left on northern liberalism's proud tradition of model state FEPCs was unprecedented. But it coincided with a sudden development out of Dirksen's own home state of Illinois, one that involved behavior that was quite different from the kind Hill pilloried. The episode was picked up in mid-March by Arthur Krock's syndicated column in the *New York Times,* and it sent alarms through the employer community across the land.

Cease and Desist at Motorola

In the fall of 1963 a twenty-eight-year-old black man named Leon Myart, who had dropped out of high school and enlisted for a tour in the army, applied to the Motorola Corporation in Chicago for an assembly line job checking for flaws in television sets. He took the company's standard multiple-choice, 28-question general ability test, and his performance led Motorola to reject him. Myart then complained of racial discrimination to the Illinois Fair Employment Practice Commission, which appointed a black attorney, Robert E. Bryant, as hearing examiner. Bryant held the hearing on Myart's case on January 27, 1964, and on March 5 he declared against Motorola, holding that the test itself was unfair to "culturally deprived and disadvantaged groups" because it did not take into account "inequalities and differences in environment."[78] Bryant then ordered Motorola to hire Myart and to cease giving the test. The Illinois FEPC's unprecedented cease-and-desist order on testing was attacked and appealed to the full commission by Motorola, whose outcry was soon joined by the 1400-firm Employers Association of Chicago, the Illinois Manufacturing Association, and the *Chicago Tribune.* In New York, Arthur Krock took up the case in his *Times* column of March 13. Krock warned that Title VII of the pending civil rights bill threatened "to project the rationale of the Illinois F.E.P.C. ruling throughout the free enterprise system of the United States." "Then a Federal bureaucracy would be legislated into senior partnership with private business," Krock said, "with the power to dictate the standards by which employers reach their judgments of the capabilities of applicants for jobs, and the quality of performance after employment, whenever the issue of 'discrimination' is raised."

The Motorola case quickly became a *cause célèbre* among conservative critics of the proposed EEOC. Senator John Tower (R.-Tex.) placed the decision and the order of the examiner in the *Congressional Record* and used it to attack the civil rights bill. On June 13, Tower won passage on the Senate floor of an amendment, which ultimately became Section 703(h) of Title VII, stipulating that it would not be an unlawful employment practice for an employer "to give and act upon the results of any professionally developed ability test" unless it was "designed, intended, or used to discriminate because of race, color, religion, sex, or national origin." [79] Humphrey agreed to Tower's amendment, which the leadership had found to be "in accord with the intent and purpose" of Title VII. The amendment's equation of design, intent, and use reflected the law's traditional standard that proof of discrimination required evidence of intent to harm and that statistical imbalances or disproportional distributions would not suffice. [80]

Dénouement on the Senate Floor

In the spring of 1964 the national debate over racial quotas was focused on the Senate floor during the intense six-week courtship of Everett Dirksen, and the Motorola case brought this concern to a fine pitch. [81] In response, the civil rights bill's sponsors repeatedly clarified their determination to ban discrimination against *any* citizen on account of his or her race, color, religion, national origin, or sex. Humphrey thus defended the as yet unamended H.R. 7152 in the March debate:

> Contrary to the allegations of some opponents of this title, there is nothing in it that will give any power to the Commission or to any court to require hiring, firing, or promotion of employees in order to meet a racial "quota" or to achieve a certain racial balance.
>
> That bugaboo has been brought up a dozen times; but it is nonexistent. In fact, the very opposite is true. Title VII prohibits discrimination. In effect, it says that race, religion and national origin are not to be used as the basis for hiring and firing. Title VII is designed to encourage hiring on the basis of ability and qualifications, not race or religion. [82]

In April, the floor managers for Title VII, Joseph S. Clark (D.-Pa.) and Clifford P. Case (R.-N.J.), submitted a joint memorandum that responded directly to the complaints of Russell and his southern colleagues that the discrimination that was prohibited was nowhere defined. "The concept of discrimination . . . is clear and simple and has no hidden meanings," Clark and Case explained. "To discriminate means to make a distinction, to make a difference in treatment or favor . . . which is based on any five of the forbidden criteria: race, color, religion, sex, and national origin."

> There is no requirement in title VII that an employer maintain a racial balance in his work force. On the contrary, any deliberate attempt to main-

tain a racial balance, whatever such a balance may be, would involve a violation of title VII because maintaining such a balance would require an employer to hire or refuse to hire on the basis of race. It must be emphasized that discrimination is prohibited to any individual.[83]

As for the Motorola case, which raised the specter of biased public bureaucrats dictating employee choice, Clark and Case reassured the Senate:

> There is no requirement in Title VII that employers abandon bona fide qualification tests where, because of differences in background and education, members of some groups are able to perform better on these tests than members of other groups. An employer may set his qualifications as high as he likes, he may test to determine which applicants have these qualifications, and he may hire, assign, and promote on the basis of test performance.[84]

By June, when Humphrey was presenting and defending the Senate's compromise with Dirksen, he explained that Section 703(j) on racial balance was added to "state the point expressly" and as "clearly and accurately" as the power of language would permit what the leadership had maintained all along about the bill's intent and meaning: "Title VII does not require an employer to achieve any sort of racial balance in his work force by giving preferential treatment to any individual or group."[85] The *New York Times* similarly reassured its readers that the "misrepresentations by opponents of the civil rights legislation are at their wildest in discussing this title." "It would not, as has been suggested, require anyone to establish racial quotas," the *Times* editorially insisted. "To the contrary, such quotas would be forbidden as a racial test. The bill does not require employers or unions to drop any standard for hiring or promotion or membership—except the discriminatory standard of race or religion."[86]

Humphrey reiterated his assurances with regard to the new amendment in Section 706(g), which required a showing of discriminatory intent: "This is a clarifying change. Since the title bars only discrimination because of race, color, religion, sex, or national origin it would seem already to require intent, and, thus, the proposed change does not involve any substantive change in the title." "The express requirement of intent is designed to make it wholly clear," Humphrey explained, "that inadvertent or accidental discriminations will not violate the title or result in entry of court orders. It means simply that the respondent must have intended to discriminate."[87]

With that, the leadership compromise was honorably sealed. On June 10, the Senate voted 71–29 to close off the record-shattering southern filibuster that had consumed, according to Russell's calculation, 82 working days, 63,000 pages in the *Congressional Record,* and ten million words. On June 17 the Senate approved substituting the leadership compromise bill for H.R. 7152 by a vote of 76–18. Two days later it voted 73–27 to pass the bill—with Barry Goldwater ominously joining the southerners in voting against it. The bill then returned to the House, where Halleck and McCulloch, who had kept in close touch with Robert Kennedy's team at Justice throughout the negotiations, blessed the Dirksen compromise as consistent with their original grand bargain. In response to the difficult Senate triumph, the House leadership avoided the customary conference committee and agreed to the Senate amendments on July 2 by a roll-call vote of 289–126. That same day President Johnson signed the epochal bill into law, transforming H.R. 7152 at last into Public Law 88-352—a name by which no one would remember it.

The Civil Rights Act of 1964 was by any comparative measure a spectacular accomplishment. It easily ranked, as Allen Matusow proclaimed, as "the great liberal achievement of the decade."[88] In response to one of its boldest provisions, Title II, the destruction of Jim Crow in public accommodations would occur with surprizing speed and virtually self-executing finality. By shattering southern defenses that had heretofore seemed impregnable, it paved the way for the effective Voting Rights Act of 1965, which would enfranchise not only the mass of southern blacks but, ironically, even greater numbers of southern whites, and in its uniquely contradictory way would permanently transform the classic gridlock of southern politics that V.O. Key so brilliantly described in 1949.[89] The Civil Rights Act of 1964, following upon a decade of paralysis in school desegregation since the *Brown* decision, would within another decade help transform the South into the most desegregated region in the country. Within that same decade, however, the Civil Rights Act's attack on job discrimination would lead to a metamorphosis of executive and judicial behavior throughout the nation that was both unanticipated and unprecedented. The implementation of the Civil Rights Act would provoke from many of the moderate and conservative architects of the compromise of 1964 cries of dismay that the law in practice was betraying the spirit and even the letter of the triumphant liberal compact that gave it birth.

CHAPTER VI

From the Civil Rights Act to the Voting Rights Act: The Johnson White House at Flood Tide, 1964–1965

The Missing Civil Rights Task Force

Because Lyndon Johnson's legislative victories in civil rights occurred so early in his regime, the civil rights issue never quite fit the model for Johnson's famous task force operation. For two decades scholars have been intrigued by Johnson's secret task forces for planning a uniquely Johnsonian legislative agenda for the Great Society. This scholarly interest may be partly explained by the unusual prominence of university professors in the task force deliberations. But the task forces commanded attention as a creative innovation in the way presidential policy was formulated.[1] Johnson's initial legislative agenda was inherited from Kennedy, and during the winter and spring of 1963–64 the new President concentrated successfully on enacting such major Kennedy initiatives as the tax cut, civil rights reform, and the increasingly Johnsonian war on poverty. But in the late spring of 1964, Johnson began to build his own agenda for the fall presidential election and beyond. In his "Great Society" speech at the University of Michigan commencement on May 22, Johnson announced: "We are going to assemble the best thought and the broadest knowledge from all over the world." "I intend to establish working groups to prepare a series of White House Conferences and meetings," Johnson said, "on the cities, on natural beauty, on the quality of education and on other emerging challenges. And from these meetings and from these studies, we will begin to set our course toward the Great Society."[2]

The task force idea was not new under the sun, and variations of it had been employed by most modern presidents. But Johnson's task forces were unique in their number and intensity, their involvement of nongovernment advisers, their beguiling secretiveness, their comprehensive scope across the spectrum of domestic policy, and their legislative productivity.[3] The project had initially been urged on Johnson by Budget Director Kermit Gordon and by Walter Heller, chairman of the Council of Economic Affairs. White House aides Richard Goodwin and Bill Moyers directed its implementation.[4] Johnson announced the task force operation to his Cabinet on July 2, correctly anticipating that the fall election against the vulnerable Barry Goldwater (who had won the Republican nomination two weeks later) would provide him with the kind of political mandate and program majorities in Congress that Kennedy had so sorely lacked. On July 6, Moyers sent out his marching orders to the policy elite in the subpresidency.

The initial task forces were to be small, secret, focused on major domestic issues and problems, composed mostly of "outside" (i.e., nongovernmental) experts, and charged "to think in bold terms and to strike out in new directions."[5] They would report the week following the fall election, and their recommendations would then be winnowed by small, interagency task forces, where the Budget Bureau would cost out the program proposals and senior White House aides would assess their practical and political feasibility. Moyers's initial ad hoc model was subsequently refined under Joseph Califano into an elaborate system. It involved an annual rhythm of outside task forces with a one-year reporting deadline (usually midsummer), each followed by a three-month (fall) interagency task force, and culminating in the pre-Christmas decisions at the LBJ Ranch. Then came inclusion in January's annual state-of-the-union, budget, and special congressional messages for legislative enactment the following year. The volume of Great Society legislation that this system produced was extraordinary.[6] But its timing precluded inclusion of the major civil rights laws of 1964 and 1965, which were forced through the new Johnson administration before the task force system was fully geared up.

Moyers's original directive of July 1964 listed fourteen task forces, and a civil rights task force was duly included. Lee White was listed as the White House liaison, and the Justice Department was identified as providing the executive secretary and the logistical support. In preparation for the planned civil rights task force, the Budget Bureau prepared an "issue paper" to guide the initial discussions, as it did for all the task forces. Dated June 17, the Bureau's issue paper assumed that the pending civil rights bill would pass. Yet it was deeply pessimistic. The veteran, elite civil servants who staffed the Budget Bureau prided themselves on being broad "generalists" rather than mere budget experts; they cherished their reputation for possessing a uniquely historical grasp of the complexities of the governmental process. As the loyal keepers of the institutional memory of the presidency since the Bureau's creation in 1921, they were aware of the gap between expectations and delivery and between program enactment and implementation; operating out of the cockpit of the Executive Office, they knew the political costs to the presidency of such disappointments. Their issue paper on civil rights began by positing the "fundamental premise [that] we are doubtful that enactment of the pending civil rights bill will stave off the pressure for further formal and binding action for more than a brief period. Much will depend on how the law is implemented, and much will depend on the Negroes' acceptance of inevitably slow and limited gains."[7]

The process would be slow and the disappointments would be inevitable, the issue paper predicted, because the new reforms ultimately relied on the courts to enforce civil rights. This would necessarily require "a great deal of local cooperation and compliance cooperation from local job and law enforcement agencies, compliance by the thousands of those who are in a position to break the law." One example was in voting, where Kennedy's original proposal in 1963 for federal registrars had been dropped from the bill that passed in 1964, and where the fall elections in 1964 would likely demonstrate a continued pattern of massive black disfranchisement. The education title of the bill faced the prospect that the present pace of school desegregation would integrate the southern schools by the year 2063. Worse, the growing problem of de facto school desegregation in the North was a "diffuse, difficult, and dangerous problem" that was side-stepped when the House Judiciary Committee dropped from the 1963 civil rights bill the section on technical assistance to reduce "racial imbalance." Moreover, the problem of de facto segregation hinged less on education than on housing policy, which Congress had also avoided. As for job discrimination and economic advancement, the issue paper warned, "We do not believe that the FEPC title of the bill will provide shelter for the Federal Government for very long. The FEPC title may suffer from a fatal flaw of providing a legal solution for what is essentially a social and economic problem." Title VII would require at the minimum an enhancement of economic leverage that had previously been exercised only through defense contracts.

On the eve of the triumphant passage of the Civil Rights Act of 1964, then, the Budget Bureau's issue paper was sober to the point of gloom. Its authors knew that the new law would raise expectations about results that the labyrinth of federal enforcement machinery was poorly designed to deliver. But in response to the President's charge to the task force to think boldly, the Budget Bureau's experts suggested some novel approaches. These included federally sponsored model new cities on the urban periphery that *required* integration by both race and economic class, and federal sponsorship of "voluntary 'bridge' organizations where people would sign up to go to church together, exchange visits at home, and play together." New federal programs might reduce the geographic concentration of blacks ("Quotas? Housing requirements? Relocation allowances?"). Perhaps a joint, federal-state foster home program could encourage whites "to take in and raise children from minority groups?" From the hard-headed realists of the Budget Bureau, such was the groping sincerity and yet touching naïveté of 1964.

But the civil rights task force was stillborn. Johnson and his senior aides knew that the fundamental challenge of the new civil rights law lay in its executive administration, not in summoning a group of university professors, foundation executives, and civil rights leaders to brainstorm for more novel proposals to Congress. On the heels of the Civil Rights Act, leaders of American business complained that the new requirements were unclear and even contradictory, and that the government's administrative jurisdictions were jumbled. The Budget Bureau quietly agreed and took the lead in a high-level effort to coordinate federal enforcement.

The Search for Enforcement Coordination

Although Title VII was by far the longest and most complicated section of the new civil rights law, its effect was additive, layered upon a mosaic of accumulated requirements that confronted employers. Reflecting this uncertainty, an analysis in the Commerce Clearing House's *Labor Law Journal* observed that management's employment practices now came within the multiple scrutiny of "municipal ordinances, state fair employment laws (both civil and penal), union contracts, Executive Orders, state and federal registered apprenticeship requirements, Plans for Progress, Title VII of the Civil Rights Law, the Wage-Hour Law and the National Labor Relations Act."[8] The proliferation of agencies involved "staggers the imagination," the article said. In New York City, for example, an employer faced regulation from

> . . . municipal civil rights bodies, the New York State Human Rights Commission, local district attorneys, the state Attorney General, the Industrial Commissioner, the President's Committee for Equal Employment Practices [*sic*], federal and state contracting agencies, the Commission for Equal Employment Opportunity, the Wage-Hour Administrator, the Secretary of Labor, the National Labor Relations Board, the United States Attorney General, Arbitrators under union contracts, state and federal courts, the Civil Rights Commission and that roving linebacker of the Civil Rights Act, the Community Relations Service. The list is probably not complete.[9]

Management was faced with a "jungle of great expectations" in which "the allocation of jurisdiction between these agencies ranges from hazy to nonexistent." An employer would typically find himself "clearly within the jurisdiction of perhaps six of these agencies and colorably within the jurisdiction of probably an additional half dozen." The web of legislation and regulation "covers the entire ambit of the employer-employee relation and embraces within its protection all employees from the company president to the lowest grade janitor."

At the heart of the government's enforcement problem lay a troublesome dualism. Its tensions were posed most immediately by the differences between Titles VI and VII. Title VI gave statutory authority to the powerful enforcement tool of fund cut-off that Vice President Johnson's committee had coordinated under Kennedy's executive orders on federal contracts and federally assisted construction. But the new civil rights law had now broadened the purview of Johnson's original PCEEO to include virtually all federal grants and loans as well as contracts. This would require new and coordinated regulations in at least twenty-four agencies. Many of these, moreover, were large mission agencies, like HEW, that had little previous involvement with procurement, and hence had little working experience with contract compliance in equal employment law. Such agencies had little experience in policing their own grants, and even less enthusiasm for developing a policing relationship to their constituencies. Title VI thus posed new and potentially disruptive challenges to the mission agencies. But the law had greatly strengthened the

President's hand, and the executive watchdogs in the Budget Bureau were jealous guardians of the governing authority of the presidency.

Title VII, on the other hand, would soon bring under the authority of the new EEOC virtually all significant employers and unions in the nation's private sector, including government contractors who previously had dealt only with their funding agencies under Johnson committee requirements. The complicated new title· was to be administered, moreover, by a new agency that did not yet exist. Because the new EEOC lacked its own enforcement powers, its efforts would necessarily involve court proceedings in unpredictable ways. Despite its considerably defanged status as essentially a voluntary agency, the EEOC was a strange hybrid creature—a national FEPC of unknown potential for amelioration or mischief. Unlike the PCEEO and the mission agencies involved in contract compliance, which were directly responsive to the President's wishes, the new EEOC was *in* the President's executive establishment but not really *of* it. Fashioned after the quasi-judicial model of the federal regulatory agencies—the "headless fourth branch of government"—the EEOC like all independent regulatory agencies stood potentially beyond the direct reach and control of the President. The Civil Rights Commission, for instance, had asserted its independence from Administration policies since its creation in 1957. Indeed this was partly by design for such a purely investigatory and recommending agency. But the nascent EEOC shared with the Civil Rights Commission the same increasingly militant civil rights constituency, and as a regulatory agency it possessed far more potential than a mere advisory commission as an instrument of policy control. Hence this new and somewhat oddly constructed agency deserved both careful nurture in light of its formidable new responsibilities, and thoughtful planning toward integrating its role into a coherent enforcement effort. It received neither—at least not in sufficient measure and timeliness to avoid a nearly disastrous beginning.

In late July of 1964, the head of the Budget Bureau's Office of Management and Organization (OMO), Harold Seidman, recommended to Budget Director Kermit Gordon that an interagency working group be created under BOB leadership to design a new structure to coordinate enforcement efforts.[10] Seidman's proposals for Title VII and the EEOC pictured the agency as essentially a public information office. The EEOC would educate the public about the new anti-discrimination law and assure the minority communities of the government's commitment to equal employment opportunity. It would also assure the business community that the government would administer Title VII "carefully and fairly, as it has administered the Executive Orders," and it would reassure businessmen that "they will not be harassed."[11] The five new EEOC commissioners should be selected by the President for their ability to "lead the nation toward *merit hiring* [emphasis added] and toward overcoming such causes of disadvantages as poverty and poor education, training, housing and health." The commissioners should be good explainers and persuaders who could weigh evidence and devise sound policy recommendations for the Attorney General. "However," the Budget Bureau advised, "the judicial requirements for these positions will be minor."

Seidman's office feared that "the public would not understand the necessity for two organizations having the same general responsibility." Both would police job discrimination in private employment, but one would apply the leverage of contract compliance under the executive orders and Title VI, while the other would use Title VII. So Seidman posed the proposition that the government's equal opportunity policy should be established by only one organization. The affirmative case was logically appealing: government contractors would now be covered by both the executive orders and Title VII, and dual regulation posed problems for a nervous business community. But both the legal and practical circumstances argued against this, or at least against assigning the dominant role to the EEOC. The coverage and the requirements of the executive orders, administered by the PCEEO, and Title VII, administered by the EEOC, were quite different. Whereas the EEOC could seek final enforcement authority only through the slow and uncertain means of court orders in adversary proceedings, the established PCEEO had "largely been successful because of the contractor's relationship to the government, because of his basic interest in pleasing a customer," and because this buyer-seller relationship had enabled the PCEEO and the agencies to "move quickly and decisively . . . . It is unlikely that a five-man Commission, or that a Chairman reporting to a Commission, could act in this way." The PCEEO's success had flowed from compliance reviews required of the agencies, and from voluntary company activities under Plans for Progress, neither of which were authorized by Title VII. Were the EEOC to undertake their administration, "it would be supervising activities foreign to its other activities under Title VII."[12]

Moreover, the nature of these "other activities" and responsibilities that might derive from Title VII remained unclear. These included sorting out the relationship between Title VII policy and NLRB policy, and also the Labor Department's apprenticeship policy. Additionally, there were the unknown and unexplored complications of age and sex discrimination. Although Congress had postponed any decision on age discrimination, Section 715 of Title VII required the Secretary of Labor to study the problem of age discrimination and report recommendations to the Congress by 30 June 1965. One of Lyndon Johnson's early acts as President was to issue an executive order in February 1964 "Declaring a Public Policy Against Discrimination on the Basis of Age." But because the order created no program or enforcement mechanism or even any clear policy implications, it was little more than a statement of sentiment, a political gesture to a constituency that was not well organized or aggressively led.[13]

As for sex discrimination, a series of congressional committee hearings had at least probed the field of age discrimination. But none had explored the uncharted complications of the new ban on sex bias. If race discrimination might touch the lives of 30 million Americans, sex discrimination might involve more than 100 million. A close working relationship with Labor would be required to harmonize Title VII's prohibition against sex discrimination with the Equal Pay Act of 1963 and also with the activities of the department's Women's Bureau, which had opposed including sex discrimination under Title VII in the first place. This was a sensitive area, because the Women's Bureau, created in 1920, was a product of the progressive reforms that had historically nurtured special protective legislation for women. Given all these ambiguities, and given the Budget Bureau's conviction that the executive orders provided "better remedies than Title VII in the case of government contractors," the Bureau's staff believed that the executive orders and the PCEEO should be continued, and that the PCEEO should remain dominant. To make sure that the EEOC followed the President's policy in civil rights enforcement, Seidman concluded, "Probably the chairman of the EEOC should be appointed to the PCEEO." [14]

The Politics of Equal Employment Coordination

Seidman's staff of managmeent experts was politically sophisticated, but as the politically ticklish question of coordinating civil rights enforcement bubbled upward toward the White House, the senior staff surrounding the Budget Director grew cautious about the Bureau's involvement. The Bureau's general counsel, Arthur B. Focke, wrote Kermit Gordon that he agreed with Seidman's recommendation that high-level coordination of civil rights enforcement was needed. But he "seriously question[ed] whether the Bureau should either assume or accept leadership in bringing it about." [15] Focke argued that the crucial questions were primarily legal, and these could best be resolved by the Justice Department. After all, Justice had drafted the original civil rights bill and coordinated its amendments. It had prepared the executive orders that established and governed the PCEEO. Furthermore its Civil Rights Division was the only organization in any regular agency that was devoted solely to civil rights questions. The Attorney General, Focke noted, had urged Lee White to direct all agencies to prepare regulations under Title VI, stipulating that the Budget Bureau would review them "with the assistance of the Justice Department." [16] But the reverse had occurred, and the BOB had ended up assisting the Justice Department. "I very much fear," Focke warned, "that we will find the agencies working directly with Justice in disregard of any coordinating role of the Bureau." Focke therefore urged that the matter first be "thoroughly discussed with Lee White, and if he deems it appropriate, with the President." The Bureau's position should be that overall responsibility for coordination be assigned to Justice. But if the Bureau must do it, then the assignment should come from the White House, not from the Bureau's own invitation. William D. Carey, the Bureau's assistant director, was similarly cool to the "appearance of our thrusting the Bureau into the center of a civil rights problem." [17] Carey advised Gordon to urge President Johnson to make the White House the focal point, with Lee White or Hobart Taylor heading up the effort.

While the bureaucratic jockeying and polite buck-passing was building pressure for a presidential decision, Lee White was already leading the hurried effort to coordinate the agencies' Title VI regulations. In September he began a series of fall meetings by summoning an enlarged version of the Subcabinet Group on Civil Rights.[18] In the three years since its inception under Fred Dutton and Harris Wofford, the subcabinet group had steadily decreased in influence. As it had grown larger its members' institutional loyalty to their agencies had grown more protective. Further, as centrally involved departments like Justice and Labor and the coordinating PCEEO had refined their responsibilities, the other agencies assumed a more passive role. The PCEEO, however, was essentially headless in 1964, with no Vice President to direct it. Labor Secretary Wirtz officially presided. But under the uncertain conditions of a post-assassination transition, compounded by both a new civil rights law in prospect and an imminent presidential election, the PCEEO fell understandably into a holding pattern. The subcabinet group still existed as a policy sounding board, but White's autumn meetings had to culminate in a presidential decision. No decision was likely, however, until after the November 3 election. When Johnson and Humphrey swamped the hapless Goldwater-Miller ticket, White immediately convened a high-level meeting on civil rights policy at the White House.[19]

The agenda for White's mid-November meeting was dominated by the need to issue coordinated Title VI regulations before the newly elected 89th Congress convened in January. The second priority for discussion was appointment of the EEOC commissioners. Four months had passed and President Johnson still had not decided upon a chairman. And because the chairman had to be appointed first, no commissioners at all had been chosen. The third item on White's agenda stipulated that "The proliferation of groups with responsibilities in this area requires a new and more sophisticated coordination *somewhere above departmental level* [emphasis added]." The Budget Bureau was represented at the meeting by Phillip S. Hughes from Legislative Reference. Hughes reported to Gordon the following day that final presidential approval of Title VI regulations was set for December 4; that everyone agreed that the EEOC commissioners must be appointed as soon as possible; and that the civil rights organizations were not pushing for new civil rights legislation—at least not yet.[20] Because no new legislative proposals were under consideration, the discussion centered on enforcement coordination. Here Lee White agreed with the new Acting Attorney General, Nicholas Katzenbach (whom Johnson had appointed the previous September 3 to replace the resigning Robert Kennedy), that the solution to the vexing problem of coherent interagency enforcement lay outside the normal purview of Justice, and would have to be found above the departmental level. No one at the meeting—i.e., representatives of the major Cabinet departments, the Budget Bureau, and the senior White House aides—wanted to take on the thorny problem. In the time-honored tradition of assigning undesirable committee chores to members not in attendance, a consensus began to converge as if by default on the vacant post soon to be occupied by a driving force behind the new civil rights law, Senator Hubert Humphrey. As Hughes wryly reported to

Kermit Gordon, "some consideration is being given (very quietly) to the Vice President doing this job." The new coordinator of federal enforcement in civil rights would inherit a hornets nest, yet someone had to do it. No one knew better than Lyndon Johnson how vulnerable to such assignments a new Vice President was. But unlike Johnson, this particular new Vice President seemed eager for such assignments.

Hubert Humphrey and the President's Council on Equal Opportunity: 1965

The White House consensus was confirmed by President Johnson's letter of December 2 to Vice President-elect Humphrey. In it Johnson asked Humphrey to "undertake during the next weeks to consider how best we should coordinate the functions of the various federal agencies in the area of civil rights."[21] Humphrey responded with his customary energy, and unlike Johnson in 1960, with enthusiasm. He consulted widely with the relevant officials both in and out of government, recommending to Johnson on January 4 that the President establish by executive order a President's Council on Equal Opportunity, to be chaired by the Vice President.[22] Humphrey attached to his letter of self-nomination a 36-page report. Most of the report described the government's jumble of inherited structures and responsibilities in civil rights. These ranged far beyond equal employment opportunity, and included the Civil Rights Commission, HHFA and the President's Committee on Equal Opportunity in Housing, HEW and Title VI in education, the Community Relations Service and the anti-poverty efforts of the new Office of Economic Opportunity. But Humphrey saw the employment problem as "the area presenting the greatest diversity in agency commitment and responsibility, and the most obvious potential for needless duplication and overlapping of effort, both within and without the Federal government."[23] Humphrey concluded that *"present circumstances do not appear to require the creation of a new civil rights agency or the appointment of a single 'czar' with overriding authority to compel or direct specific agency action."* Instead, primary reliance must be placed upon each constituent agency using its own resources and retaining operational responsibility.

What, then, was the function of the proposed new council? Humphrey explained that it should not carry an operational burden. Rather, it should offer "leadership, support, guidance, advance planning, evaluation, and advice to foster and increase individual agency effectiveness, cooperation, and coordination." To these laudable buzzwords were added equally vague pledges to maintain "simplicity, flexibility, and a pragmatic relationship to existing agencies." And who would sit on the new council? Not surprisingly, Humphrey's long list looked strikingly like the roster of the PCEEO. It contained the cabinet secretaries of Defense, Agriculture, Commerce, Labor, and HEW, plus the Attorney General. Also included were the chairmen of the Civil Service Commission, the Commission on Civil Rights, the President's Committee on Equal Opportunity in Housing, and the new EEOC. Add to this the administrators of the GSA and the HHFA, the Commissioner of Education, and the directors of OEO and the Community Relations Service. Indeed, Humphrey even included the chairman of the PCEEO—which was to say, himself—and then proposed his own designation as Council chairman as well. Thus the Council was to be superimposed upon the PCEEO. It would recommend coordinating measures to the President, recruit an executive secretary and staff, appoint subcommittees and interagency working groups, receive reports and collect data, and hold conferences. Just like the PCEEO.

It all appeared very mushy and redundant. But the senior veterans of the executive establishment had passed the buck to the ebullient new Vice President, and he must be given running room. Johnson asked Bill Moyers to study Humphrey's report and recommend what to do. On February 3, Moyers wrote Johnson that he was mindful of the President's wish *not* to "create another agency," and also to avoid issuing another executive order.[24] Moyers was pleased that the report had not leaked to the press. But that luck might not hold, and it was time to decide between two alternatives: create the Council by executive order as Humphrey had recommended, or simply charge the Vice President with coordinating responsibility by presidential letter.

Moyers explained that Lee White joined him in urging the latter course, especially in light of Johnson's reluctance to establish a new presidential entity by executive order. But Moyers also summarized the case *for* the executive order, reiterating Burke Marshall's argument that "it is virtually meaningless to ask any Vice President to take on a coordinating role without providing the necessary mechanism or structure for the coordination." Moyers explained unnecessarily to his boss that "The limitations inherent in the Vice Presidency are such that he would find himself subsequently unable to fulfill the responsibilities placed upon him by the President." Moyers also recalled how Katzenbach rightly had insisted, back in 1961, on such a supporting structure for Vice President Johnson's chairmanship of the National Space Council. So Moyers included a draft of both an executive order and a letter for Johnson to choose and sign. Mindful of his own miseries of powerlessness in the vice presidency, Johnson reluctantly signed the executive order.[25]

The Election of 1964 and the Origin of the Voting Rights Act of 1965

The creation of Humphrey's council on February 5 drew little public notice, largely because civil rights attention had once again shifted to explosive developments in the South. The new focus was Selma, Alabama, where Martin Luther King was mounting a major voting-rights offensive. But Johnson had taken the voting-rights initiative earlier. Immediately following his election victory on November 3, he directed Katzenbach to draft "the next civil rights bill—legislation to secure, once and for all, equal voting rights."[26] Johnson had crushed Goldwater with a stunning 16-million vote margin, carrying 44 states and the District of Columbia to Goldwater's six states, and in the process achieving a net gain of 38 seats in the House and two in the Senate. But Goldwater had carried the five Deep South states of Alabama, Georgia, Louisiana, Mississippi, and South Carolina. In the process Goldwater's coattails had added five new Republican representatives in Alabama and one each in Georgia and Mississippi—in all instances these were the states' first Republicans elected to Congress since Reconstruction. Black Americans had voted 9 to 1 for the Democrats. But in much of the South they remained massively disfranchised, while in the Deep South most whites were suddenly voting Republican.[27] These tea leaves were not difficult to read: the vote of southern blacks had spelled the margin of victory for Johnson in such Rim South states as Arkansas, Florida, Tennessee, Virginia, and probably North Carolina as well. Johnson was determined to use his new mandate and congressional majorities to enact a voting-rights law that worked and worked swiftly.

But no voting rights law had really worked since Reconstruction.[28] This of course was part of the problem, because the federal manipulation of the postbellum franchise had stained the memory of Reconstruction with partisan corruption, and this legacy of "Black Reconstruction" had long provided conservatives with powerful practical as well as constitutional arguments for hedging the voting rights provisions of 1957, 1960, and even 1964 with elaborate judicial due process. The voting (and nonvoting) patterns of 1964, however, had demonstrated the bankruptcy of this good-faith approach in the face of so much blatant bad faith. So, too, had the intimidation and violence that greeted the voter registration efforts of the combined civil rights organizations throughout the Deep South, and especially in Alabama and Mississippi, during the "Freedom Summer" of 1964.[29] Even Katzenbach, who was inclined to be something of a stickler for proper constitutional restraints on national power in a federal system, reluctantly concluded that the national government would have to go beyond "the tortuous, often ineffective pace of litigation."[30]

Accordingly, on December 18 he recommended to Johnson three ranked alternatives for enfranchising southern blacks. His first preference was a constitutional amendment that would prohibit *all* states from imposing *any* voter qualifications beyond age, a short period of residency (60 to 90 days), conviction of a felony, or confinement in a mental institution.[31] To Katzenbach, this was the most effective way to eliminate the typical southern literacy tests and their accompanying subjective requirements for "understanding" various constitutional provisions, vouchers of testament to "good moral character," and the like. The 24th Amendment, which banned the poll tax in federal elections, had passed Congress in 1962 under its sponsor's argument that a statutory ban on any qualifications set by the states for voters would violate Article 1, Section 2 of the Constitution and also the 17th Amendment. It was ratified by the required three-fourths of the states in January 1964, which followed congressional passage by only 17 months. The amendment affected elections for federal office only, and hence was defended as a legitimate concern of Congress and the presidency. But Katzenbach conceded that a constitutional amendment was the "most drastic" alternative. It was slow and cumbersome, vulnerable even after congressional approval to blockage by as few as thirteen states, and even more immediately vulnerable to "opposition from sources genuinely concerned about federal interference with a fundamental matter traditionally left to the States."

The Attorney General's second choice was legislation to create a new federal commission that would appoint federal registrars for federal elections only. Such an approach, however, risked a two-pronged Republican attack. Republican conservatives would raise a state-rights objection to its potential for "federal dictatorial control of the electoral process." Republican moderates and liberals would complain that it would still not enfranchise blacks in state and local elections where their political powerlessness was most painful. (Oddly, Katzenbach did not explain whether such a commission would cover the entire country, or, if it would focus only on areas of demonstrated disfranchisement, how it would do so.) Katzenbach's third choice was legislation granting an existing federal agency "the power to assume direct control of registration for voting in both federal and state elections in any area where the percentage of potential Negro registrants actually registered is low." This approach promised the quickest payoff. It had been recommended by the Civil Rights Commission in its first report on voting in 1959, and it had been proposed by President Kennedy as part of his omnibus civil rights bill in 1963. But it had also been rejected then "because of the opposition of Cong. McCulloch and others," and Katzenbach himself entertained doubts about the constitutionality of a statutory takeover by the national government of state and local election procedures.

Martin Luther King and the Crisis at Selma

While Katzenbach's poposals were germinating toward an uncertain conclusion, King returned from his Nobel triumph at Oslo to launch a carefully planned voting-rights offensive in Selma. In his Pulitzer Prize-winning biography of King, *Bearing the Cross,* David Garrow has described the internal debate within King's SCLC, and also involving the restive young recruits in SNCC, that led to the climax at Selma.[32] The strategic paradox was the realization by King and his lieutenants that nonviolent protest would fail unless it triggered violent repression. Perferably this should occur on prime-time national television, leading to bloodshed and martyrdom, and generating thereby a wave of revulsion that would carry Congress and the Johnson administration before it. What this radically revised strategy of nonviolent protest lacked in philosophical consistency and purity, it gained in proven political leverage. King had failed miserably in his major desegregation campaign in Albany, Georgia, in 1961 and 1962. There police chief Laurie Pritchett had politely arrested and jailed all civil disobeyers, thereby providing little martyrdom and small inducement for television news cameras. But not so in

Birmingham in 1963, where Bull Connor played the perfect foil, and the chief result was the Civil Rights Act of 1964. King needed a similarly inhospitable environment to dramatize the need for radical federal intervention to enfranchise southern blacks, and he found it in Selma.[33]

King picked Selma because it promised to provide three crucial ingredients. The first was a demonstration of indefensible racial disfranchisement. In 1960 the population of Dallas County, Alabama, of which Selma was county seat, was 57.6 percent black. But only 2.1 percent of its registered voters were listed as Negroes. Of its voting-age population, 9,542 of 14,400 whites were registered, but only 335 of its 15,115 blacks. The second requirement was a demonstration of the ineffectiveness of the traditional federal reforms of 1957, 1960, and 1964, operating glacially as they did through litigation. Despite a sustained black voting drive in Selma between May 1962 and August 1964, only 93 of 795 blacks were successfully registered there, as against 945 of 1,232 whites. This was a racial differential in registration of 11 percent to 76 percent. Then in February of 1964, the all-white government of Dallas County had further stiffened its already draconian registration requirements by reducing registration to only two days per month, and lengthening the process by requiring applicants to respond to 68 questions about constitutional provisions and government procedures. Alabama's registration requirements were the most severely comprehensive in the nation, followed closely by Mississippi (although Alabama's registration of 19.3 percent of eligible blacks in 1964 far surpassed Mississippi's 6.7 percent).[34] Alabama's model was devised by the state supreme court, and included the ability to read and write any article of the U.S. Constitution in the English language, to understand and correctly interpret constitutional passages, and to produce a voucher from already registered voters that testified to the applicant's "good moral character." Attorney General Robert Kennedy had filed a voting-rights suit against Dallas County as early as April of 1961, and the local federal district court had imposed a permanent injunction on the registrar in November of 1963. But such protracted and expensive efforts placed few black names on the voter lists.

Finally, King needed the ironical and inadvertent cooperation of defiant southern white officials. He found it not only in Governor George Wallace, but most superbly in the county's corpulent sheriff, James Clark—who appeared as if on cue from Central Casting, his racial animosity worsened by an uncontrolled temper, his deputies armed with electric cattle prods, his lapel sporting a "Never!" button.[35] The result was violence and death. A young Selma black, Jimmie Lee Jackson, was gut-shot by a state trooper on February 18 (he died on February 26). A Unitarian clergyman from Boston, the Reverend James J. Reeb, was clubbed on the head by local whites on March 9 (he died on March 11). The bloodletting culminated on March 7, "Bloody Sunday," when 500 demonstrators who were marching toward Montgomery were charged on the Edmund Pettus Bridge by Governor Wallace's state troopers, leaving forty injured in the chaos, with the news networks' television cameras whirring. King's protest against such a demonstrable civic outrage was at once physically dangerous, philosophically somewhat disingenuous, and politically brilliant. The nation was enraged.[36]

The Justice Department and the Voting Rights Bill of 1965

President Johnson had early set in motion the machinery to produce a new and effective voting rights bill, and in his State of the Union message on January 4, he proposed to eliminate "every obstacle to the right and the opportunity to vote."[37] But he did not commit himself to press for another civil rights bill so close on the heels of the 1964 law. King and his allies, however, quickly provided the occasion for the new bill and the emotional catalyst that would drive it through Congress. The new 89th Congress was so lopsided in its Democratic skew and so inspirited by the massive mandate over Goldwater, who had conspicuously voted against the Civil Rights Act of 1964, that there was little doubt that the Administration's new voting-rights bill would pass in some form. The Democratic margin over the Republicans in the House was swollen to 295–140, the largest since 1936. In the Senate, the modest net Democratic gain of two seats (yielding a ratio of 68 Democrats to 32 Republicans, the largest since 1940) was less important than the re-election of 13 Democrats from the bumper class of 1958—liberals whom the Goldwater Republicans had hungrily targeted. The conservative coalition in Congress of Republicans and southern Democrats had been seriously weakened—ironically, in the House, by the Goldwater-Republican replacement of so many senior southern Democrats from the Deep South. Emanuel Celler's liberal hold was tightened on the House Judiciary Committee, and Judge Smith's weakened grip on the Rules Committee could only delay the inevitable. In the Senate, the conservative coalition was weakened by the increased seniority of the re-elected class of 1958, and Senator Eastland's conservative cohort on Judiciary was now outnumbered by a dominant bloc of nine liberals. The new Congress awaited Johnson's pleasure.

The Selma crisis accelerated the Justice Department's already hurried efforts to draft a bill radical enough to be quickly effective, yet still reasonably respectful of the long traditions of due process and federalism that had made all its predecessors both passable and yet largely ineffective. Katzenbach's favored constitutional amendment was by definition the most constitutionally sound approach. Yet it was also the slowest and most uncertain. Johnson rejected it. Similarly rejected was Katzenbach's second alternative, the new federal commission to register voters for federal elections only. Because it would not intrude on traditional state responsibility for state and local elections, this approach reflected Katzenbach's compunctions over unconstitutional infringements on traditional state responsibilities. But as Katzenbach conceded, it also risked a partisan Republican whiplash against a halfway measure that offended both the left and the right. Such a commission approach would leave intact, although temporarily dormant, the subjective literacy test devices that were the heart of the problem. By leaving unaffected the state and local elections where black political voice and power were weakest, it would risk a resurgence of disfranchisement when the federal heat was off. It would also require the maintenance of a dual system of voter lists in the states and localities wherein it applied. Moreover, President Johnson was leery of creating yet another quasi-independent, bipartisan federal regulatory commission, especially if its task of full enfranchisement would be needed only *temporarily* to redress a historic wrong. What Johnson appeared to have in mind was both radical and yet short-lived: a one-shot quick-fix that by permanently enfranchising southern blacks would provide them with the one essential tool wherewith to protect themselves—like everybody else.

This process of elimination led to the Attorney General's third alternative, a law giving an existing agency the power to register voters for state and local as well as federal elections in areas of substantial disfranchisement. It promised to be fast and effective. But it might not be constitutional. This had been the problem with its predecessor laws from the Reconstruction era, at least in the hands of a conservative judiciary. It necessarily would require some novel arrangements that explored terrain that was unfamiliar both to the Justice Department and to Congress. But otherwise the problem of black disfranchisement in the Deep South seemed both intolerable and intractable. A radical new formula would have to be devised that would achieve five objectives. It must (1) apply to all elections, (2) identify the target political subdivisions without violating the Constitution, (3) suspend the literacy tests and related devices and freeze the electoral governance machinery therein, (4) provide for federal voter registration in the covered areas, and (5) provide some escape mechanism for a return to normal federal-state relations once the offending practices had been safely purged. At least in the aftermath of the Civil War, the national government enjoyed the extraordinary practical and legal sanctions of a conquering army over a treasonous (white) populace that had been largely disfranchised by the victors. In 1965, there was little precedent to guide such a hurried quest, beyond those iron-fisted arrangements of the previous century's Reconstruction.

Yet Civil War metaphors were not entirely inapt. Joseph Rauh, counsel for the Leadership Conference on Civil Rights, sent Katzenbach a draft bill that would authorize the President to order the "nullification" of voter registration lists in areas where blacks were substantially underrepresented. Rauh would identify the affected states as those that maintained *any* racially segregated or "otherwise unequal" educational facilities in recent years, or that made *any* change in their registration qualifications (other than age or residence) since the Civil Rights Act of 1957 was enacted. Rauh clearly had the offending South in mind. But such a loose definition could conceivably have encompassed every state in the nation. Even a definition that safely confined the target areas to the South, however, would also require some formula for identifying the specifically affected counties and municipalities so the President could nullify their electoral laws and establish therein a "Federal Enrollment Office."[38] The Deep South's abused literacy tests were the correct target. But literacy tests per se were not. They found historic protection in constitutional law and in naturalization requirements. As voting requirements for reading and writing the English language, they were widely used outside the South—i.e., in Alaska, Arizona, California, Connecticut, Hawaii, Idaho, Maine, Massachusetts, New Hampshire, New York, Oregon, Washington, and Wyoming. The trick was to link literacy tests to massive racial disfranchisement, and to do so through a triggering formula that was automatic.

In drawing up a bill for the President, Katzenbach was chiefly assisted by Solicitor General Archibald Cox. Cox in turn was assisted by Louis F. Claiborne, a constitutional lawyer on his staff; by Harold Greene, chief of the Appeals and Research section of the Civil Rights Division; and by Sol Lindenbaum, an attorney in the Office of Legal Counsel.[39] Cox sent Katzenbach a rough first draft on February 23, and during the next three weeks the lawyers in Justice struggled intramurally with an Administration bill that was toughened by lawyers in the Civil Rights Division, and that left the Solicitor General and several of his associates worried about its constitutional vulnerability. The essential ingredients of the model they converged upon boiled down to three. The crucial and thorniest item was the "trigger" provision to identify the disfranchising jurisdictions. Through this formula the Attorney General could determine whether a literacy test or similar device was used as a voter qualification in the presidential election of 1964, *and* the Director of the Census could determine whether less than 50 percent of the voting-age population in that jurisdiction was registered or voted in the 1964 election. This determination would then automatically suspend all literacy test devices, and this in turn would trigger the second ingredient. This was the Attorney General's authority to dispatch federal examiners to the area identified to compel the registration of qualified applicants. Finally, the federal registration would cover state and local as well as federal elections.[40] The automatic triggering device was on its face a racially neutral formula; its 50 percent threshhold was arbitrary but implicitly majoritarian in identifying an unacceptable level of citizen disfranchisement.

The office of the Solicitor General, however, which represented the Administration before the federal courts, was worried that a bill constructed in such haste and through such patchwork accretions, and also built upon such a thin body of precedent, would be constitutionally vulnerable. Louis Claiborne sent Cox a memorandum in mid-March arguing the unconstitutionality of much of the bill, and attacking the arbitrariness of a formula that would include counties in such states as Alaska, Arizona, Idaho, and Maine, where Indian populations could trigger the formula. Cox complained about a formula that could demonstrate "the lack of any but a coincidental relationship between racial discrimination and the facts to be determined by the Attorney General and Director of the Census." He suggested that greater precision could be achieved through a straightforward bill of attainder: "One might equally well make the Act applicable to any State whose name begins with Vi or Mi or Lo or Al or Ge or So. Indeed, since even this description covers Alaska as well as Alabama, it has exactly the same effect as the determinations now required to be made."[41]

Cox urged stipulating as a minimum requirement for covered states that "more than 20 percent of the persons of voting age [be] of the Negro race." This would eliminate Virginia as well as states like Alaska, he said, which seemed "reasonable and proper since we have very little reason to suppose that the Fifteenth Amendment has been widely violated in Virginia." Indeed, it would eliminate all states except the five Deep South states plus North Carolina. Cox was able to purge an earlier draft provision that would authorize federal intervention based on a single act of discrimination by a single official anywhere in the state. But he was anxious to avoid implications of attainder, and he wanted the triggering formula to provide "only rules of thumb giving rise to a presumption." Such a presumption ought to provide the states or political subdivisions identified by the formula "a genuine chance to dispose of the inference that the statute draws." But "at present it is so tightly drawn as to defeat the purpose," and "renders the statute subject to still further attack for arbitrary vindictiveness."

The Solicitor General also objected to several "unsound" provisions in Section 5, the "preclearance" provision. This section was added by enforcement-minded lawyers in the Civil Right Division, who had cut their teeth doing combat with local southern officials during the era of "Massive Resistance." They had dealt with the most defiant southern officials, and their distrust was palpable. This included some segregationist federal district judges, and the experience prompted the Civil Rights Division attorneys to add the extraordinary proposal that judicial preclearance could come *only* from the federal court in the Dictrict of Columbia—by long reputation the most liberal in the land. Section 5 provided that if any covered state or subdivision subsequently sought to make *any* change in its election laws or procedures, it would either have to obtain the prior approval of the Attorney General, or else obtain from the U.S. district court for the District of Columbia a prior declaratory judgment that such proposed change "does not have the purpose and *will not have the effect* [emphasis added] of denying or abridging the right to vote on account of race or color." Section 5 was added in practical anticipation that covered jurisdictions would seek to sabotage the remedial process through constant obstructive and subversive changes. An ostensibly sensible precaution, it was a sleeper provision that in time would create a substantial monitoring bureaucracy in the Justice Department, and would generate a flood of resentment and litigation.[42] To Cox, such a "prohibition would forbid substituting of voting machines for ballots, changing the hours at which polls were open, changing the location of the polling places, creating new precincts, and anything else having to do even with the bare mechanics of either registration or voting." As such it was "totally unnecessary and probably unconstitutional," and was not only inappropriate but "exceedingly dangerous."[43]

Cox and Claiborne were troubled by many of the constitutional concerns that Senator Ervin would later trumpet in the Senate Judiciary hearings. But they shared little of Ervin's philosophy—Cox was a Harvard law professor of impeccable liberal credentials, and Claiborne was a career constitutional lawyer from New Orleans with philiosophical roots in the minority tradition of liberal dissent by southern aristocrats. Both men sought an effective voting rights law to crush the evil of racist disfranchisement. But they wanted one that could pass an honorable constitutional muster.

There simply wasn't time, however, to draft a carefully considered bill and straighten out all the kinks. The violence in Selma had ignited the nation's indignation against the Deep South's wholesale disfranchisement and repression of blacks. President Johnson wanted to present the receptive 89th Congress with an effective voting rights bill immediately. But the previous decade of congressional committee hearings had concentrated on the traditional, judicially enforced methods of voting-rights enforcement that stood some chance of enactment, and neither the Congress nor the Administration was prepared for the sudden moment of possibility that Selma provided. Katzenbach's hurried draft was after all a more circumspect bill than the kind demanded by Rauh and the Leadership Conference, and its formula satisfied Lyndon Johnson. Johnson in turn had been persuaded by Martin Luther King, when King conferred with Johnson at the White House on February 9 (freshly bailed out of the Alabama jailhouse), that the voting law should rest on an automatic formula rather than on litigation in court, and that the hated literacy tests in such areas must be suspended.

Congress was hungry to act, and the Republicans were chiding the Democratic administration for having fobbed off a series of ineffectual voting rights laws in deference to the powerful southern Democrats in Congress. The Warren Court, moreover, had given every early indication that the boldest reforms of the Great Society were constitutionally acceptable. On December 14, 1964, a unanimous Supreme Court in *Heart of Atlanta Motel v. U.S.* had quickly upheld the controversial public accommodations section of the Civil Rights Act of 1964.[44] Then on March 8, 1965, the Supreme Court gave further indication of its hostility toward the South's traditional battery of voting restrictions by unanimously ruling unconstitutional Louisiana's understanding clause, by overturning a lower federal court's ruling against a similar Justice Department suit in Mississippi; and by agreeing to review a lower court's ruling that approved poll taxes for state and local elections in Virginia.[45]

Thus committed, Johnson journeyed to Capitol Hill on March 15 to address a joint session of Congress on voting rights—it was the first such personal appearance by a President on a domestic issue since 1946. There Johnson delivered his forceful "We Shall Overcome" address on the Negroes' right to vote: "Every device of which human ingenuity is capable has been used to deny this right," he said, and "It is wrong—deadly wrong." "Their cause must be our cause too," Johnson pledged. "Because it is not just Negroes, but really it is all of us, who must overcome the crippling legacy of bigotry and injustice. And we *shall* overcome."[46] Two days later the Administration's hastily assembled bill, which was still in the process of spirited debate within Justice, was sent to the Hill.

The 89th Congress and the Voting Rights Act of 1965

Unlike the Civil Rights Act of 1964, the Voting Rights Act of 1965 has still not found its historian.[47] A full generation after the act's passage, this is surprising, especially because the law was both radical in design and extraordinarily effective in enfranchising the southern blacks who were its intended beneficiaries. Most of the literature has concentrated on the law's implementation and the battles over its periodic renewals (in 1970, 1975, and 1982), rather than on its original construction and enactment.[48] In the spring of 1965, the political challenge for the Johnson administration was to move quickly and to steer a middle course between Senate liberals who sought a stronger bill, and House efforts by the loose conservative coalition of Republicans and southern Democrats to substitute a weaker compromise bill. There appeared to be little possibility that a voting-rights bill could be blocked in the 89th Congress, and this essentially reversed the circumstances of 1964. With neither Eastland's Judiciary Committee nor a southern filibuster posing once formidable obstacles, the Administration could avoid the delay of a House-first strategy and push the bill simultaneously through both chambers.

The bipartisan Mansfield-Dirksen leadership sent the Administration's bill, S. 1564, to the Judiciary Committee on March 18, but with tight instructions to report it back by April 9. In Judiciary a dominant bloc of nine liberals further strengthened the bill in three ways. First, they added a ban on the poll tax in state and local elections. Second, they added a second automatic trigger that would send federal voting examiners to districts where fewer than 25 percent of voting-age racial minorities were registered. Finally, they exempted from the triggering formula all political subdivisions that according to the 1960 census had a nonwhite voting-age population of less than 20 percent. This provision would exclude from the bill's coverage almost all of the nation's political subdivisions outside the South.[49] But the Mansfield-Dirksen leadership shared Katzenbach's fear that a poll tax ban by statute was unconstitutional and would endanger the entire bill. So they refused to accept it, and on May 11 in a narrow 45–49 roll-call vote they defeated an attempt led by Edward Kennedy to retain it. The leadership provided instead for the Attorney General to test the constitutionality of the poll tax in court. On May 25, the Senate voted 70–30 to break a desultory southern filibuster, and the following day the Senate passed the bill by a 77–19 roll-call vote and sent it to the House.

In the House the Administration's bill, H.R. 6400, was sent to Celler's Judiciary Committee, where first Subcommittee No. 5 and then the full committee strengthened it and reported it out on June 1. Like its Senate counterpart, the House Judiciary Committee was dominated by liberals who viewed the poll tax ban as a litmus test of commitment to civil rights. So Judiciary's amended version included the statutory poll tax ban that the full Senate had narrowly dropped. Then the Rules Committee delayed the bill for the month of June. But when H.R. 6400 reached the floor of the House on July 9,

Congressman McCulloch and House Minority Leader Gerald Ford offered their substitute bill (H.R. 7896). It would delete both the poll tax ban *and* the Administration's automatic trigger as both unconstitutional and unnecessary. Like the Senate bill, McCulloch's substitute would direct the Attorney General to bring suit against the poll tax. To replace the blanket triggering formula, which McCulloch called a "pure fantasy—a presumption based on a presumption" (echoing the strong reservations of Cox and Claiborne in the Solicitor General's office), it would potentially cover *all* political subdivisions in the nation by authorizing the appointment of voting examiners when the Attorney General received twenty-five or more meritorious complaints of voter discrimination.

McCulloch's substitute was an attractive alternative for conservatives and moderates who believed that Congress had no power to ban state and local poll taxes as election requirements, and who also regarded the state-based automatic trigger as a crude and disingenuous device for attacking voter discrimination, which usually was specific to individual political subdivisions, not to entire states. Moreover, McCulloch's county-by-county approach seemed superior because it would necessarily include the southern states of Arkansas, Florida, Tennessee, and Texas, all of which were excluded by the automatic trigger because they lacked literacy tests. Also, it would render irrelevant the trigger's idiosyncratic inclusion of three counties in Arizona and one county in Idaho. But when southern House Democrats, led by former governor William Tuck of Virginia, rallied to the Republican substitute as less "objectionable" than H.R. 6400, many northern Republicans grew fearful of voting with the southerners in the shadow of Selma. On July 8, the McCulloch-Ford substitute was defeated in a 166–215 teller vote. The following day, H.R. 6400 was passed by an overwhelming roll-call vote of 333–85.

The ensuing Senate-House conference committee, however, got hung up on the question of the poll tax ban, which H.R. 6400 had embraced and S. 1564 had rejected. Yet the poll tax question was far more symbolic than real, for the device had long since ceased to be a serious source of racial discrimination in voting. It was true that all eleven former Confederate states had originally adopted the poll tax around the turn of the century as part of their disfranchising armament, which was directed rhetorically against blacks but was also designed to decimate the voting ranks of the lower economic order generally.[50] But by 1965 seven of these states had dropped the levies. This left them as voter qualifications only in Alabama and Mississippi, where they were quite minor adjuncts to the literacy test battery, and in Texas and Virginia, which historically had disfranchised blacks and poor whites almost equally, and where by the 1960s a discriminatory racial impact could not readily be identified.

If the poll tax question by 1965 was largely irrelevant to racial disfranchisement, however, it provided a rallying symbol for liberals to stand up for "tougher" civil rights enforcement. This, too, represented a reversal of previous roles and symbolic behavior in Congress. In previous civil rights debates, Congress had typically centered its debate on issues set by the conservative opposition, like jury trial rights in criminal and civil contempt in 1957, and Mrs. Murphy's boardinghouse in 1964. These marginal but symbolic issues reflected the strong bargaining position of the conservatives prior to the Johnson landslide of 1964. They symbolized tempering the laws in the direction of less federal intrusion on traditional and valued prerogatives of state and local self-governance and private freedom. Defending Mrs. Murphy and the sanctity of jury trials was far more palatable than defending an odious nuisance tax on the franchise, however, and by 1965 southern politicians had so abused the original republican principles of federalism that their currency was debased.

But in the conference committee in July of 1965, with the members of Congress already on record in their speeches and votes, the Administration and the Senate conferees stood firm against the poll tax ban, and the Leadership Conference privately conceded that the poll tax question was not worth endangering the bill. So the House conferees yielded to the Senate on the poll tax, and in the August 2 conference report, the Senate conferees returned the favor by dropping both of the Senate provisions that were geared to specific racial percentages. These were the 25 percent level for a second automatic trigger, and the exemption for areas with less than 20 percent racial minorities. The conference report agreed to many such trade-offs on secondary and technical matters, but the compromise bill that emerged closely resembled the original Administration bill in its basic provisions.

Despite all the fencing and posturing over the poll tax, Congress added only a largely redundant provision for the Attorney General to challenge the tax in court—a path that he was pursuing anyway. Congress also authorized the federal courts as well as the Attorney General to suspend tests and appoint examiners in voting-rights suits brought by the Attorney General. Finally, the conference accepted the Senate's "American flag" amendment, which senators Robert Kennedy and Jacob Javits had proposed to enfranchise Puerto Ricans living in New York who were illiterate in English, but could demonstrate completion of at least the sixth grade in a school "under the American flag" where classes were taught in a language other than English.[51] By and large, in the 1965 conference committee on voting rights the two houses cancelled out each other's distinctive changes, and gave the Administration essentially what it had asked for in the first place.

On August 3, the House adopted the conference report by a thumping vote of 328–74, with 37 southern Democrats voting for the bill. Most of the latter, like Charles Weltner of Atlanta, who had also voted for the Civil Rights Act of 1964, represented urban areas where the black vote was substantial. The Senate approved the report the following day by a similar margin, 79–18 (including six southern Democrats in support), and President Johnson made a special gesture by traveling to the Capitol on August 6 to sign the bill into law.[52]

President Johnson's Howard Commencement Address:
"Equality as a Fact and as a Result"

On June 4, with the Senate having just passed S. 1564 and with ultimate success in the House a virtual certainty, the confident President delivered a historic address before an audience of 14,000 persons attending the graduation ceremonies at Howard University. In a speech suggested by Bill Moyers and written by Labor assistant secretary Daniel Patrick Moynihan and White House aide Richard Goodwin,[53] Johnson celebrated the imminent passage of the Voting Rights Act as the latest in a long line of victories, in which freedom was only a beginning:

> But freedom is not enough. You do not wipe away the scars of centuries by saying: Now you are free to go where you want, do as you desire, choose the leaders you please.
> You do not take a person who for years has been hobbled by chains and liberate him, bring him up to the starting line of a race and then say, "You are free to compete with all the others," and still justly believe you have been completely fair.[54]

It was not enough just to open the gates of opportunity, Johnson said, because "All our citizens must have the ability to walk through the gates." Then the President proclaimed "the next and more profound stage of the battle for civil rights. We seek not just freedom but opportunity—not just legal equity but human ability—not just equality as a right and a theory but equality as a fact and as a result."

When Johnson announced that "equality of opportunity is essential, but not enough," he was, to be sure, voicing without change the liberal prose of two gifted intellectuals he had inherited from Kennedy.[55] But the effect was electric on his immediate and mostly black audience, which "sat in stunned silence," and then finally "applauded out of shock and self-identification."[56] The *New York Times*'s Tom Wicker compared Johnson's speech favorably in historic importance to the Supreme Court decree in *Brown v. Board of Education*. The Congress was then debating, and would soon pass, a provision in the voting-rights bill authorizing federal courts in voting suits brought by the Attorney General to suspend literacy tests and similar devices if they were used "with the effect" of discriminating, even if this was not their purpose. The preclearance provision of the bill's Section 5 would also authorize the Attorney General to disallow any electoral changes in affected areas if in his judgment the proposed changes would "have the effect" of racial discrimination. Such hints of a radical shift from procedural to substantive criteria in civil rights law, from intent to effect, from equal opportunity as a right to equality as a fact and as a result, seemed to be gathering momentum. But so were contrary trends. It was a time of crucial transition, but the countercurrents were powerful and their direction was unclear.

Five days after the President signed the Voting Rights Act of 1965 into law, a spectacular riot led by angry blacks exploded in the depressed Watts section of Los Angeles. An estimated 7,000 to 10,000 persons took part; thirty-four were killed, and 864 were treated in hospitals for injuries. In retrospect, the Watts riot marked a fateful watershed for the Johnson administration and the Great Society. In the euphoria of signing the Voting Rights Act, Johnson could bask in the knowledge that the momentum of the Great Society was producing a cornucopia of achievement for 1965: the Voting Rights Act, Medicare, federal aid to education, urban mass transit, the new Department of Housing and Urban Development, clean air and water pollution programs, immigration reform, national endowments for the arts and the humanities. Johnson's popularity ratings were soaring, and the media had grown highly respectful of the political genius of the former "cornpone" Vice President. Johnson had achieved a White House intimacy with Martin Luther King, Roy Wilkins, and Whitney Young, and he even seemed to enjoy a mutually surprising rapport with the university-based community of intellectuals with whom he had always been uncomfortable. In his Howard speech the President announced that he would convene a summit White House conference on civil rights, and invite the nation's leading scholars and experts, civil rights leaders, and top government officials. Its theme would be "To Fulfill These Rights," echoing the founding promise of the Declaration of Independence, and specifically intended to signify the fulfillment of the report of President Truman's Committee on Civil Rights: *To Secure These Rights*.

But the euphoria and consensus would not last. In July, Johnson committed the country to a major combat role in Vietnam. Then in August, the racial explosion in Watts seemed to mock the recent victory for black voting rights in the South. Beneath these unsettling events, black nationalist sentiment was radicalizing SNCC and CORE, deflecting them from their interracial and nonviolent origins toward "Black Power" and a self-destructively violent path of racialist separatism.

Within the Administration, Johnson's White House conference moved unknowingly toward a nationally divisive fiasco over the Vietnam war and the Moynihan Report on the black family.[57] The President's puzzling delay in appointing the EEOC commissioners raised troubling questions about his commitment to the new agency, and Vice President Humphrey's efforts to pull together a coherent program of civil rights enforcement were quietly foundering. In September, Johnson would issue a pair of executive orders to restructure the government's entire enforcement operation, stripping his Vice President of his short-lived role with a disingenuous public explanation, and leaving Humphrey in a posture of loyal humiliation. Johnson would abolish both the PCEEO and Humphrey's fledgling coordinating council, and thus would begin the Administration's sustained and largely obscured process of enforcing the major new commitments of 1964–65 through the permanent government. It was a normal maturing of the implementation process, with enforcement moved out of its unnatural, ad hoc environment in the White

House cockpit and the Executive Office of the presidency, and into the career bureaucracies of the subpresidency. There out of the spotlight, the new EEOC, and an even newer agency with an especially clumsy title, the Office of Federal Contract Compliance Programs (OFCC), would struggle to work their own quiet revolution.

CHAPTER VII

The Troubled Search
for Civil Rights Enforcement:
Launching the EEOC, 1965–1966

The Missing Commissioners of Equal Opportunity

Lyndon Johnson could not fairly be held accountable for many of the domestic events and controversies that so quickly soured his honeymoon following the triumphal summer of 1965. Especially unanticipated were the shock of the Watts riot; the ferocity of the sudden storm over the Moynihan Report on the deteriorating black family, which linked growing impatience in civil rights to liberal dissent over the expanding war in Vietnam; and finally the radical cries of "Black Power" that would lead to political "backlash" and sharp Democratic reverses in the congressional elections of 1966. But Johnson's abortive experiment with Vice President Humphrey's layer-cake coordinating council and his unconscionable delay in appointing the EEOC commisioners were foolishly self-inflicted wounds.

There is no evidence, however, that the precious months that Johnson wasted with a leaderless EEOC were prompted by hidden motives.[1] Johnson was proccupied with the grand designs of his Great Society, as well as with responding to overseas crises in Guantanamo Bay, Panama, the Tonkin Gulf, the Dominican Republic, and increasingly Saigon. The occasional prodding of Johnson's aides through the winter and early spring of 1965, working against these priorities, produced no apparent movement toward identifying the new EEOC commissioners.[2] As early as December 1964, John Macy had urged Johnson to appoint Lee White to the EEOC chairmanship.[3] But Johnson wanted to retain the proven and loyal senior aides he had inherited from Kennedy, and White had demonstrated steady judgment in a volatile policy area. Probably more important, Johnson wanted for the EEOC a responsible "name" chairman to signify the prestige and consensus that he sought for the commission. He had in mind the centrist model of LeRoy Collins, whom he had tapped the previous year to head the Community Relations Council. But by the spring of 1965, with still no commissioners to lead the commission, civil rights leaders were beginning to complain to the press about the delay, and this was echoed by liberals in Congress. Senator Joseph Clark, who had been floor captain for Title VII in 1964 and who chaired the employment and manpower subcommittee of the Senate Labor Committee, sent a sharp note to Johnson in early May demanding action after almost a year's inexplicable delay.[4]

In response to these pressures, Macy and Lee White proposed to Johnson on April 29 a full slate of candidates for the commission. They noted the difficulties inherent in balancing the demands of labor and management, race and sex, geography and party affiliation (no more than three of the five commissioners were allowed by Title VII to be of the same political party—which was to say that two must be Republicans).[5] Once Johnson's attention was commanded by the rising chorus of complaint, he had no difficulty in approving recommendations for the four commissioners beyond the chairman (both the slate of nominees and Johnson's early approvals assumed that the chairman would be a white male Democrat). To fill these four posts Macy proposed three names for the two Democratic slots. First was Luther Holcomb, a Baptist minister long loyal to Johnson, who in 1961 had arranged John Kennedy's successful encounter with the Protestant ministers in Texas, and who was executive director of the Greater Dallas Council of Churches and chairman of the Texas advisory committee to the U.S. Civil Rights Commission. Second was Fred K. Koehler, Jr., a professor of labor relations at the University of Michigan, who enjoyed strong AFL-CIO support. The final Democratic nominee was Marjorie McKenzie Lawson, a black juvenile-court judge in the District of Columbia and a "public" member of the PCEEO.

Johnson picked Holcomb and Lawson, balancing a southern white male moderate with a black female veteran of Washington's political culture. When Lawson proved unwilling to sacrifice her recent move into private law practice, however, Johnson turned not to Professor Koehler but to another labor nominee, Aileen C. Hernandez. Born Aileen Clarke in New York City in 1926, daughter of Jamaican immigrants, her Hispanic surname came from a four-year marriage in 1961 to a cutter in the Los Angeles garment industry. She was a Democrat with strong union ties and had worked for eleven years as educational director for the International Ladies' Garment Workers Union in California. Appointed assistant chief of the California FEPC in 1962, Hernandez was the only EEOC commissioner with state experience in fair employment law.

For the two Republicans, Johnson agreed to both of Macy's nominees. One, Richard A. Graham, was then heading the Peace Corps program in Tunisia, and was sponsored by Bill Moyers, who had worked with Graham during his Peace Corps days under Kennedy. Graham had an engineering degree from Cornell, and before joining the Peace Corps in 1961 he had been an executive of Graham Transmission in Menominee Falls, Wisconsin. The other Republican was Samuel C. Jackson, a black attorney who was president of the NAACP branch in Topeka, Kansas, and who worked as a social welfare lawyer for the state government of Kansas.

The commission's chairmanship, however, remained the sticking point for

Johnson. Macy, looking for the Collins model of a former southern governor with political savvy, Democratic affiliation, and moderate persuasions in race relations, suggested a brace of three such candidates: Ellis Arnall (Georgia), Burt Combs (Kentucky), and Terry Sanford (North Carolina). Then he added Moyers's candidate: Morris B. Abram. Moyers sent Johnson his own memorandum nominating Abram (and also Richard Graham) on the same day. He described Abram as a native of Georgia, a lawyer, and a liberal southerner who was chairman of the American Jewish Congress, and who "*at my request* helped stem the tide of Jewish opposition to the Education bill."[6] "He is your appointee to the tentative post of representative to the Human Rights Commission of the United Nations," Moyers wrote. "He is a superb speaker . . . and a skillful negotiator. He would be neither harsh to the South nor a tool of the civil rights extremists. As a close friend of mine, he would be a vital link for us to the FEPC." Johnson approved asking Abram to chair the EEOC. But Abram had left Atlanta to join a New York law firm only two years earlier, and he cited that commitment and his partners' pressures in declining the President's offer for the full-time EEOC chairmanship (Abram did, however, agree to serve as "white" co-chairman, with William Coleman of Philadelphia as the black co-chairman, for the ill-fated White House Conference on Civil Rights). So Johnson, under intensifying pressure, turned to a proven "name," if not to a proven administrator: Franklin D. Roosevelt, Jr.

In 1965 FDR's namesake son was serving as Under Secretary of Commerce, having been appointed by Kennedy in 1963. He also had represented New York's 20th congressional district from 1949 to 1955, where he had become familiar with legislative issues in civil rights. The fifty-year-old Roosevelt was also a lawyer (like Robert Kennedy, he had graduated from Harvard and his law degree was from Virginia), and these attributes plus his magic name made him at least ostensibly a plausible choice as chairman of the EEOC. But in reporting the New Yorker's appointment, the *New York Times* observed that the surprise naming of Roosevelt causes "something of a stir" in Washington, where Roosevelt was believed to be preparing to run either for mayor of New York City or, more likely, for governor of the state in 1966.[7] Roosevelt had been seeking a platform more visible than Commerce undersecretary, and only recently had failed to win appointment as U.S. representative to the Economic and Social Council of the United Nations. Roosevelt later admitted that he had accepted the EEOC appointment on less than twenty-four hours' notice.[8]

When chairman Roosevelt and his four fellow commissioners were sworn in on June 2, they had no staff, no office space, and only one month to prepare to enforce the longest and most involved title in the Civil Rights Act of 1964. They faced the daunting task of immediately hiring a staff director and a general counsel, recruiting a staff of approximately three hundred, and creating regional offices in New York, Chicago, Kansas City, Atlanta, New Orleans, Dallas, Albuquerque, Los Angeles, and San Francisco. More important in the long run, they would have to draw up regulations to clarify the many ambiguities in the complex law. In doing so, however, they would enjoy a broad range of interpretive discretion within which to exercise their collective political judgment and ideological predispositions.

Chairman Roosevelt responded to this challenge by promptly sailing off for a week of yachting. He appointed Luther Holcomb as vice chairman, leaving the Dallas clergyman in the capital to struggle with weak support in an unfamiliar environment. Meanwhile the President retained White House control of the major elements of civil rights policy. He authorized a loose group of White House aides to plan for the White House Conference on Civil Rights, and directed his new White House chief of staff, Joseph A. Califano, to work with Deputy Attorney General Ramsey Clark to respond to the Watts riot. Roosevelt did testify before the House Labor subcommittee on July 21, briefly reporting on progress in organizing the EEOC, but mainly supporting chairman Powell's proposal for cease-and-desist authority for the commission. In early August, however, when Congress was considering the next fiscal year's budget for the EEOC, Roosevelt, astonishingly, was off sailing again. Johnson had requested $3.2 million for EEOC operations in fiscal year 1966. But the Senate Appropriations Committee combined conservative uneasiness over the EEOC with general irritation at Roosevelt's unavailability for testimony, and so approved only $2.25 million—a cut of $950,000 that would prevent the opening of regional offices. Vice President Humphrey joined Senator John O. Pastore (D.-Rh.I.) in restoring $500,000 of the cuts on the Senate floor. But Chairman Roosevelt remained at sea on his yacht for most of early August, a circumstance that was noted in the press, considerably irked the Congress, and embarrassed the White House.[9]

All this prompted Republican congressman William Brock of Chattanooga to denounce on the House floor Roosevelt's frequent absences. The new chairman's "fondness for yachting," Brock said, "handicaps [the EEOC's] operation, inconveniences Members of Congress needlessly, and makes a mockery of the administration's emphasis on equal opportunity for all Americans."[10] Moyers sent Brock's statement in the *Congressional Record* to Johnson, who returned it with the note: "Bill M—Send to FDR & ask for comments?"[11] Such a shoddy beginning for the EEOC was inexcusable. Moreover, the President's new scheme to coordinate the government's hydra-headed effort through Vice President Humphrey's council was foundering.

The Collapse of Hubert Humphrey's Coordinating Council

Humphrey's ill-starred council lasted only eight months. It's slap-dash evolution and rapid demise was a failure not only for Humphrey but, deservedly, for President Johnson as well. The President was poorly served by his senior advisers (excluding perhaps Bill Moyers, who argued against issuing the executive order); unwisely advised, he also ignored his own better instincts, and in the process needlessly humiliated the Vice President. Ironically, the council's aborted career was terminated by an executive order in September 1965 that was greeted with surprising hostility by the civil rights community, yet that has since been virtually enshrined as the lodestar of the federal government's affirmative action policy. Both of these reactions, however, misjudged the context and meaning of the short and unhappy life of the President's Council on Equal Opportunity and the central purpose of Executive Order 11246.

When Johnson created Humphrey's coordinating council on February 5, 1965, the Vice President surveyed the administrative realities of his new co-ordinating domain and was distressed by what he found. Lacking any commissioners or staff, the new EEOC existed only on paper. The PCEEO was functioning in its workaday routines, but without initiative. It was presided over but not driven by Secretary Wirtz, and it was staffed by a defensive cadre of Johnson loyalists led by Hobart Taylor. Most disconcerting, the PCEEO was also essentially broke.[12] Humphrey tried to shift to the PCEEO unexpended funds from the EEOC budget, but he was advised by the Budget Bureau that this would be illegal. So Humphrey was forced to send out a pro-rated quarterly "billing" to the cabinet departments for $225,000 to cover the PCEEO's main activity, which was its contract compliance oversight of 27,000 contractors representing 11 million employees.[13]

The Vice President's reputation for genial bumbling was reinforced by the miserable half-year episode with the council, and this was not altogether un-justified. But Humphrey was a veteran politician with far more victories than losses under his belt. He was shrewd enough to see that the new council was his potential instrument for building a staff and for having a portfolio that brought great personal meaning to him, and that also brought some respon-sibility to the limbo of the vice presidency.[14] On the other hand, the inherited PCEEO had become something of an albatross. It was tainted, however un-fairly, in the eyes of much of the civil rights community by the Johnsonian vice presidency, the symbolic battles with Robert Kennedy over Plans for Progress, and its general lack of muscle as a mere interagency committee that never seemed to use its ultimate weapon of contract cancellation. At the heart of Humphrey's problem was the persistent dualism in the government's en-forcement traditions. The American public, expectant in the wake of the Civil Rights Act and its newly created EEOC, understood that the new fair em-ployment agency would now lead the attack on job bias. Citizens claiming discrimination could now file complaints with the EEOC, and the Attorney General could bring suit in their behalf against employers. But the political and legal traditions of the federal government in fair employment were rooted in executive orders and contract compliance. Popular expectations and un-derstanding were poorly matched against reality. So was the task of Hum-phrey's coordinating council.

Johnson's executive order had charged Humphrey's council with recom-mending to him "such changes in administrative structure and relationships, including those for merger, combination, or elimination within the Federal establishment, as may be necessary." In his unwelcome March billing letter to the participating agencies, Humphrey said he intended to "begin to phase down and ultimately eliminate the President's Committee on Equal Employment Opportunity."[15] His billing was an interim measure, he explained, and he hoped that May 1 would bring a smaller and *final* billing for the PCEEO. By late June, Humphrey had prepared his basic recommendation to Johnson for realigning the confused and overlapping structure of federal responsibilities in employment discrimination. It contained a proposal advanced by his council's task force on employment (chaired by Wirtz, and including Katzenbach, Macy, Hobart Taylor, and by June, also EEOC chairman Franklin Roosevelt), that the PCEEO be abolished.[16] Responsibility for nondiscrimination in federal employment would be transferred to Macy's Civil Service Commission, and the PCEEO's lesser functions, such as education and community relations, labor liaison, public information and surveys, would be transferred to the EEOC. But the PCEEO's major authority over contract compliance would be transferred *not* to the EEOC, but instead to Humphrey's council. Humphrey saw a major task of his council as "figuring out how Title VI would be carried out," which was now the statutory source of the PCEEO's authority for fund cut-off.[17]

But why not have the EEOC oversee *all* equal employment programs? Humphrey agreed that "Ideally, responsibility for Government programs directed toward non-discrimination in private employment ought ultimately to be centered in the Roosevelt Commission." But he and his employment task force were unanimous (and this presumably included Roosevelt himself) in opposing a transfer of authority over contract compliance "at this time." Foremost among his reasons was his knowledge, as the Senate whip-hand behind the Civil Rights Act of 1964, that because Title VII "carefully limits" the EEOC's authority to conciliation and mediation functions, such a transfer "would be contrary to the clearly expressed Congressional intent." Second, such a transfer "would likely provoke negative reactions, not only on the Hill, but among civil rights groups as well, mainly because the Commission is still new and has not yet had a real opportunity to 'prove itself.' "

Third, the EEOC clearly had "more than enough to do in simply getting established." Thus Humphrey proposed to redelegate to his council's Task Force on Employment, chaired by Wirtz, direct responsibility for running the contract compliance program. In order to pay for it, and yet avoid the continued uncertain financing and pro-rata billings that had plagued this unusual interdepartmental committee, Humphrey airily observed that "We anticipate seeking a direct appropriation for discharge of the contract compliance function." To implement his plan, Humphrey attached for Johnson's convenience "a simple housekeeping Order" to transfer the functions. On August 12, Lee White sent Humphrey's recommendations to Johnson with his own observation that "The proposal makes sense to me and is one that could easily be accomplished by executive order."[18]

But the proposal didn't make sense to Lyndon Johnson. He was in effect being asked to sign a new executive order that once again reshuffled the government's admittedly messy and ineffective arrangements in civil rights enforcement, shifting crucial and sensitive responsibilities from one interdepartmental committee to yet another one. Title VI was already proving to be exceedingly volatile. Humphrey's recommendations reached the White House just as new HEW guidelines on Title VI were about to spark a political explosion over de facto school segregation in Mayor Richard Daley's Chicago. Moreover, with Humphrey's proposed rearrangement, jurisdiction over 27,000 government contractors and their 11 million employees would fall to a subcommittee of the larger interagency committee. These politically sensitive issues would be presided over by a chairman whose loyalty was above reproach, but whose unrestrained loquacity was legendary, whose prudence was increasingly suspect in the White House environment, and whose star was clearly plummeting in the presidential firmament. To support this dubious transfer, Johnson was expected to ask Congress for a direct budget allocation for a presidential committee—with all the attendant implications of congressional inquiry, oversight, and control. Shades of the Russell amendment!

In August of 1965 the White House mood was already beginning to retreat defensively from the euphoria of the spring. Bill Moyers had suddenly been shifted to press secretary, prompting puzzled press rumors that Johnson was worried about his slipping image. To replace Moyers as chief of the White House staff, Johnson raided McNamara's "whiz kid" cardre in the Pentagon and brought in the aggressive Joseph Califano to build a new core staff for domestic policy. Marvin Watson, the literal-minded, politically inexperienced, and jealously loyal Texan whom Johnson had brought in from the Lone Star Steel Company to replace Kenneth O'Donnell as appointments secretary, was stationed as guardian of the President's door. Externally, the Watts riot coincided with growing public uneasiness over Vietnam policy, especially on the nation's campuses and among the intellectual community. The planners of the White House Conference on Civil Rights were encountering hostility over Vietnam and charges that the Moynihan Report on the Negro family insulted blacks and blamed the victims of racial oppression.[19]

As these pressures grew, Humphrey and his staff drew increased criticism from the White House, and Humphrey himself lost access to the President. On August 26 he sent a pleading memorandum to Watson, listing four recent examples of how he and his cabinet-level council had been excluded from White House initiatives that they had also been working on. These White Houe-directed operations included Assistant Attorney General Ramsey Clark's task force on the Watts riot, Secretary John Gardner's report on school desegregation and Title VI fund cut-off, a recent White House report on progress in implementing the Civil Rights Act of 1964, and planning for the fall civil rights conference. Humphrey complained to Watson that his council had scheduled a high level meeting for August 18 to discuss the Los Angeles situation, only to have it suddenly cancelled by the White House staff. It was all quite humiliating, and Humphrey was reduced to petitioning Watson for permission to discuss it with the President—"because I do want to be sure of my ground and what is expected of me."[20] Instead Watson, presumably doing the President's bidding, bucked Humphrey's plea to White. Watson observed that there was too much overlapping of effort in Humphrey's proposals, and asked White to suggest a better way to coordinate the groups working on the same problems.

The Katzenbach Plan for Civil Rights Enforcement

By the end of September, Humphrey's grand design had been shattered, at least insofar as his own future role and prestige in civil rights was concerned. To replace the Humphrey operation, White had instead worked with Califano and Katzenbach to design a structural solution that not only would abolish both the PCEEO *and* Humphrey's council, but in doing so would force Humphrey to recommend his own removal from civil rights enforcement. The chief architect of the new arrangement was Attorney General Katzenbach, who prepared for Johnson an eight-page reorganization plan. In it he agreed with Humphrey, as he had on the council's employment task force, that the PCEEO had well served its original purpose, but that under the new circumstances it should now be abolished. Katzenbach, however, urged Johnson to abolish Humphrey's council as well. Responsibility for federal employment would then be assigned to the Civil Service Commission, as Humphrey had urged. But contract compliance would be transferred *not* to Humphrey's disappearing council, nor to the shaky new EEOC, but rather to the Labor Department. (Ironically, this had been the original argument of Vice President Lyndon Johnson in February 1961, when he so desperately tried to escape chairing the new PCEEO.) "Putting the Government contract function in the Department of Labor is, I believe, preferable to assigning that function to a multi-headed commission," Katzenbach wrote. "Furthermore, the Roosevelt Commission has not yet gained the confidence of civil rights groups which I believe the Department of Labor and Secretary Wirtz enjoy." [21] Enforcement of Title VI would be formally assigned to Justice, which had been coordinating the development of Title VI guidelines for the past year anyway. This arrangement was consistent with Katzenbach's belief that high level interagency committees had not only proven ineffective in enforcement, but in the process had diluted direct agency responsibility. As for the Vice President, he could be used as a "trouble shooter or evaluator" under presidential assignment. [22]

Katzenbach was less sure of what to do with three other loosely hanging civil rights entities. First, the President's Committee for Equal Opportunity in Housing, created by Kennedy's executive order in 1962 and chaired by former Pennsylvania governor David Lawrence, was near mutiny. "They insist quite rightly that the Executive Order signed by President Kennedy is ineffective," Katzenbach wrote Johnson. Many of the housing committee's members had been threatening to resign unless the order was extended to cover mortgage loans made by all banks regulated by the federal government. But the legal basis for such an order was "very difficult," Katzenbach said, as President Kennedy had learned in 1962. An attempt to work out a legislative alternative would be preferable, although it would doubtless face hard going in the Congress on such a sensitive issue.

Second, the Community Relations Service was a puzzle. Originally designed to deal with problems anticipated in desegregating public accommodations throughout the South, the service had instead been met with massive peaceful compliance. So it had busied itself with the more pressing trouble spots, which were in the northern cities. But this in turn seemed both unintended and unwise to Katzenbach. He even questioned "the need or wisdom for continuing" the CRS, which "has no muscle nor is it responsible to any department or agency directed to carry out a particular program," such as in housing or education or jobs, where the specific black grievances seemed to lie. Katzenbach saw conciliation as an inherent function of enforcement. He admitted that he was "not familiar in detail with what the Service does," and recommended a reorganization plan that would transfer its conciliation function to the Justice Department—which was to say to his own office.

Finally, there was the Civil Rights Commission, which was set to expire in 1968 unless renewed by Congress. Katzenbach reflected his predecessor's general unhappiness with the commission. It had irritated Robert Kennedy by seeming to stir up disputes that Justice was trying to negotiate and resolve in the South. It also relied on staff work that Kennedy had called "deplorable," and that even its own staff director had privately admitted was often sloppy and biased.[23] Katzenbach thought the commission should instead be encouraged to study the problems of de facto school segregation, discriminatory housing, and the success of voter registration in the South. "At the moment," he wrote, the commission was "beset by a certain amount of uncertainty and needs direction. During such a period, it has performed the function of educating Negro citizens as to their rights and has encouraged them to assert these rights. In my judgment, continuation of this function only makes enforcement as well as compliance more difficult."

So Katzenbach remained somewhat vague about the proper placement and relationships of such advisory civil rights entities as the housing committee, the Community Relations Service, and the Civil Rights Commission, none of which had program responsibility. But he was firm on the need to abolish both the PCEEO and Humphrey's council, and to farm out enforcement responsibilities to the mission agencies that ran the programs.

On September 20, White sent Katzenbach's proposal to the President. It called for Humphrey to "show his 'bigness' by recommending the dissolution of a group that he heads which has performed its assignment and no longer needs to remain in existence."[24] White reminded Johnson that "although there is no written record," it was "perfectly obvious that you signed the Executive Order creating the Council with considerable reluctance and indicated to the Vice President that it was to be a temporary body." White even volunteered that requiring Humphrey to recommend abolishing his own council "should pose no problem" because the Vice President's proposal to abolish the PCEEO was, "with slight modification," what Katzenbach was recommending. But the slight modification was of course the destruction of Humphrey's council itself:

> This may afford an opportunity to urge that the Vice President work directly with the White House staff on this matter, rather than his own group, because of their demonstrated inability to work quietly, keeping things out of the newspapers (If this is a problem for him, he might ask only John Stewart, an able and discreet fellow, to work with him but not his other people.) His memorandum or letter to you making the various proposals could be answered by a very warm letter from you indicating your great pleasure at his recommending a dissolution of a body he heads.

So the new executive orders were drawn up, and the Vice President dutifully claimed authorship of the new recommendations. He received in return not a "very warm letter," but only a press-release memo from Johnson that concurred in his Vice President's recommendations and rather perfunctorily acknowledged Humphrey's energy and dedication to the cause of equal opportunity. Bill Moyers then worked up a contrived scenario for a press conference that was called to explain all these maneuvers to the White House press corps on September 24.

The Irony of Executive Order 11246

With Humphrey, Katzenbach, and Wirtz in attendance at the press conference, Moyers orchestrated a media blitz. He ticked off seven structural changes to "strengthen and streamline" federal responsibilities in civil rights policy, all in response to the Vice President's recommendations. These were, first, both the PCEEO and the President's Council on Equal Opportunity would be eliminated. Second, nondiscrimination in federal employment would be shifted to the Civil Service Commission, and third, contract compliance would come under the Department of Labor. Points four through six dealt with relatively secondary functions: Plans for Progress would be continued on a "private voluntary basis," the Community Relations Service would shift from Commerce to Justice, and its data gathering and clearinghouse functions would be taken over by the Civil Rights Commission. Finally, the Justice Department would assume responsibility for coordinating federal enforcement policy under Title VI.

The reporters' questions reflected a weak grasp of the issues involved, and most of them focused on the relatively inconsequential role of the Community Relations Service.[25] But Humphrey was questioned about his loss of two coordinating chairmanships, and he gamely deflected the questions. When queried why the Roosevelt Commission was scarcely mentioned in the new executive order, Humphrey replied that the EEOC's responsibility for nondiscrimination in private employment remained unaffected, and that the commission was plenty busy getting staffed and established.

Humphrey handled his demotion with manful loyalty. But in his memoirs the bitterness showed. He chiefly blamed Johnson's new coordinator for domestic programs, Joe Califano, for his humiliation. "Joe was smart and able," Humphrey said, "liked power, and did not want to share it." "He convinced the President that direction of federal civil rights efforts should flow from him."[26] But Califano was too new to the White House to have defrocked the Vice President. The decision was Lyndon Johnson's, who had been persuaded to create Humphrey's redundant council in February against his sounder instincts.[27]

In rushing through the press conference, Moyers was interrupted at one point by a perceptive reporter: "Bill, in going over these points, several times you mentioned discrimination by reason of race, religion or color, but not sex?" Moyers preemptorily replied: "Sex also." But the question was pertinent, and hinted at one of several ironies that attended the two executive orders that were issued that day. One, E.O. 11247, was narrow in scope. It formally assigned coordinating authority for Title VI enforcement to the Attorney General, and revoked Executive Order 11197 of the previous February 5, which had established Humphrey's coordinating council.[28] But the main order was E.O. 11246. Like Title VII, it was quite long for the genre, and six times repeated the standard prohibition against discrimination on account of "race, creed, color, or national origin." But it *nowhere* mentioned sex.

President Johnson did not bother to rectify this omission until the fall of 1967, when he issued Executive Order 11375 to add sex discrimination retroactively as a prohibited activity under E.O. 11246. It is true that the PCEEO had never been charged with monitoring sex discrimination, and that Title VI on contract compliance nowhere prohibited it. But Title VII unequivocally did. Thus the great majority of federal contractors, all of whom were governed by Title VI, were also large enough to fall under Title VII's ban on sex discrimination. The Civil Rights Act's ban on sex discrimination of 1964, however, was still regarded by the White House men of power in 1965 as a fluke, and even a distracting nuisance. They were long accustomed to associating civil rights and discrimination almost exclusively with racial problems, and the parallel problems of sex discrimination seemed rarely to occur to them. There were no ranking women among them, nor any cohesive women's lobby to remind them of its presence. But the ineluctable logic of sex anti-discrimination would soon present the fledgling EEOC with a major and unanticipated test of its theory of discrimination, and with rather surprising results all around.

This leads to a second irony of Executive Order 11246. It concerns the order's reputation, as it has been passed down through subsequent administrations, as the standing source of the federal mandate for affirmative action. This is of course technically correct, since Johnson's E.O. 11246 formally superseded and abolished an entire string of preceding orders on anti-discrimination. The series began with Eisenhower's two contract committee orders of 1955 and 1957, and included Kennedy's order of 1961, plus Kennedy's directive on construction grants-in-aid of 1963, and also a minor order of

Johnson's in 1964 adding the Postmaster General to the PCEEO. Indeed, so hurried was the construction of Johnson's September 1965 order that it formally abolished Eisenhower's two executive orders for a *second* time—both had been specifically abolished by Kennedy's order in 1961. Moreover, E.O. 11246 mentioned "affirmative action" only once. Even then it did so only in the boilerplate language of the mandatory contract for all federal contractors that was taken directly from Kennedy's original order of 1961. The latter had also alluded to the need for "additional affirmative steps" and "positive measures" to enforce nondiscrimination in federal employment, and its collective design had involved prolonged discussion, although reaching no clear final agreement, on what such positive requirements might mean.

The internal debate leading to Johnson's new executive orders in September 1965 focused almost exclusively on structural arrangements, administrative jurisdictions, bureaucratic politics, and to some degree on questions of leadership and staff personalities. Affirmative action had simply not been fundamentally at issue in the tortured evolution of E.O. 11246. What the order did require in this regard was that in farming out direct enforcement responsibility to the mission agencies, "each executive department and agency shall establish and maintain a positive program of equal employment opportunity" consistent with a prohibition of discrimination "because of race, creed, color, or national origin." The order's requirements for federal contracts was lifted verbatim from Kennedy's order of 1961: "the contractor will take affirmative action to ensure that applicants are employed, and that employees are treated during employment, *without regard* to their race, creed, color, or national origin [emphasis added]." What this might mean was left up to the Secretary of Labor. His future rules and regulations would bind all contractors through Compliance Reports that would be required "*prior to or as an initial part of* their bid or negotiation of a contract [emphasis added]." The potential sword was mighty, as it had always been. But the language was vintage 1961.

A final historical irony is that Johnson's order of September 1965, which re-created the enforcement sword that others would subsequently wield, was greeted with surprise and growing dismay by the liberal civil rights community.[29] Liberals were alarmed that Humphrey was so suddenly relieved of his civil rights command. It is true that Humphrey's rhetoric and reputation as a civil rights stalwart contrasted sharply with the less known Katzenbach's reputation as a careful moderate, or with the Labor Department's historic reputation for accommodating local segregationist practices in its apprenticeship and state employment services programs. But the Vice President was inexperienced and ineffective in executive administration, and he was poorly served by his staff.[30] More important, Johnson's new order provided the Labor Department with the occasion to create, under the provenance of Secretary Wirtz, a new enforcement subagency. Created with the clumsy title of Office of Federal Contract Compliance Programs (OFCC), it potentially possessed the awsome power of the federal contract purse. Moreover the OFCC lacked the constraining bureaucratic environment of established clientele relationships. Meanwhile Johnson's order left intact the independent and more visible EEOC to find its way toward a policy consensus on enforcement guidelines. To this task the fledgling EEOC promptly turned, with an uncertainty about its role and its permissible range of behavior that only mirrored the contradictions in the Civil Rights Act that gave it birth.

Public Enforcement of Private Rights: The Retail Model of EEO

The EEOC's Deputy General Counsel, Richard K. Berg, joined the new commission from the Office of Legal Counsel in the Justice Department, where he had assisted in drafting Kennedy's omnibus bill and had subsequently followed the progress of Title VII through the Congress. Following its enactment, Berg published in the *Brooklyn Law Review* a close analysis of its evolution and the enforcement implications of the compromises that shaped it.[31] As Berg observed, at Title VII's heart lay a contradiction between competing models for the EEOC. The model originally envisioned by H.R. 405 in 1963, which was initially incorporated into H.R. 7152, was the familiar or standard model of regulatory agencies like the FTC, and more specifically the NLRB. It envisioned a quasi-judicial body with authority to hear cases, compel witnesses, and to issue orders enforceable through the courts. When in the fall of 1963 the House compromise stripped away the proposed EEOC's cease-and-desist authority, it still left the commission a prosecutorial role to sue for compliance, and thereby attempt to enforce the *public's* interest in nondiscriminatory employment. But when the prosecutorial role was then itself stripped away in the Dirksen-engineered Senate compromise of the spring of 1964, it left the EEOC with a core responsibility of enforcing a *private* right to nondiscrimination by responding administratively to individual complaints.[32] Liberals complained that this would reduce the largely toothless agency to a reactive posture. It would respond to the initiatives of others on a case-by-case, "retail" basis, rather than take the initiative to attack broad "wholesale" patterns of discrimination in large firms, unions, and entire industries. Given such a constricted function, the EEOC's five-member board, which had been appropriate for a quasi-judicial commission, made less administrative sense than would a single chief administrator. "The leadership compromise turned the NLRA pattern inside out," Berg observed, "and the emphasis is now shifted to the resolution of individual grievances and away from obtaining broad compliance."[33]

By the summer of 1965, on the eve of the EEOC's launching, the needless temporizing of the Johnson administration had squandered so much of the good will generated by passing the Civil Rights Act that the major civil rights organizations were starting to condemn the hobbled EEOC even before it began. Their anger prompted public vows to prove its *un*workability by inundating it with a tide of complaints. Jack Greenberg, head of the NAACP Legal Defense and Education Fund, flayed Title VII's provisions as "weak, cumbersome, probably unworkable." "We think the best way to get it amended," Greenberg said, "is to show that it doesn't work."[34] James Farmer threatened nationwide CORE demonstrations to protest the EEOC's cumbersome procedures, which Farmer said would mean that "before an aggrieved person can get a remedy, he may have found another job or starved to death."[35]

During the commission's first week of meetings in mid-July, chairman Roosevelt announced that he did "not want any organization or agency out to drum up complaints" for the EEOC.[36] But the NAACP had launched an effective national drive, led by Herbert Hill, to flood the EEOC with complaints. On July 15, Hill personally delivered to Roosevelt the NAACP's first batch of job discrimination complaints. They were arrayed against a broad range of employers and unions, mostly in the South. The targets included Sears, Southern Bell, Kroger of Memphis, four large department stores in New Orleans, and Darling Chemical of East St. Louis, together with Darling's associated and segregated Local 127 of the International Chemical Workers Union, AFL-CIO.[37] Hill was accompanied by NAACP General Counsel Robert L. Carter, and they demanded at a press conference that Roosevelt invoke the commission's "full power not merely to resolve individual complaints but to eliminate overall patterns of discrimination in employment."[38]

By mid-September, the NAACP alone had recruited 374 such complaints.[39] By mid-December the EEOC had received 1,383 formal complaints in its first one hundred days, although it was only budgeted to handle 2,000 during its entire first fiscal year.[40] By April 1966 it had received 5,000 complaints, apparently heading toward 9–10,000 for the year. Thus the EEOC started from behind, and continued to fall further behind in processing complaints and in pursuing the lengthy and manpower-intensive conciliation process. Roosevelt wrote Budget Director Charles Schultze to plead for a budget supplement. "We are being overwhelmed," Rooevelt said, with an "uncontrollable complaint workload."[41] In asking for an additional $625,000 to process the complaint backlog, Roosevelt explained that what the commission urgently needed was an affirmative action program of "persuasion, education, and promotional efforts" on a nationwide basis, which would "in the long run reduce the cost of Commission operations which are now devoted almost exclusively to investigation of individual complaints."

The EEOC As a Subversive Bureaucracy

If the EEOC was to move beyond the tight limitations of the complaint model— beyond putting out fires with voluntary persuasion on a case-by-case, retail basis, or beyond "mere" passive nondiscrimination toward an affirmative program that attacked broad patterns of discrimination—then four fundamental transformations would have to be made, either in the language or the interpretation of the commission's founding law. This seems reasonably clear from historical hindsight. But it was by no means clear in the middle 1960s. One should be wary of any imputation of grand strategic design, by projecting backward upon the early participants motives that matured subsequently. One could not reasonably expect any such clear and early consensus on the part of a commission whose members ranged the ideological spectrum from Luther Holcomb to Aileen Hernandez, or a staff that ranged the spectrum of aggressiveness, albeit more narrowly, from Richard Berg and Tom Powers to Alfred Blumrosen and Herman Edelsberg. The minutes of the commission meetings during the EEOC's first year reflect not a grand design but rather an honest groping among the conflicts between nondiscrimination and affirmative remedy, color blindness and minority consciousness, individual complaint processing and pattern detection, confidentiality and publicity, race versus gender, even Title VII versus Title VI.

The first of the four transformations would necessarily center on a definition of what discrimination was. To move radically beyond the complaint model, the definition of what constituted discrimination would have to be extended beyond the *intent* standard of the common law tradition, which was stipulated by Congress in Title VII, toward an *effects* standard. This would require a shift in criteria from invidious intent on the part of discriminators to harmful impact upon members of the affected class. Such a radical shift was implicit in the newly current metaphor of "institutional racism." By this was meant the deeply embedded patterns of historic preference through which minorities had been excluded from institutions and their power structure so long that discriminatory intent was no longer required to maintain the pattern of majority dominance. Indeed, such institutionalized racism could theoretically be maintained even in the absence of racial prejudice. Thus it was argued that "mere" passive nondiscrimination would not change such patterns, and that some interventionist form of compensatory affirmative action was therefore required.

But such a basic switch from an individual to a group-based concept of rights would be far too radical to expect from Congress. So the civil rights organizations in general and the EEOC in particular concentrated on the more promising courts to forge a new body of case law that led in this direction.[42] The EEOC, however, could not itself file suit. The Justice Department could file "pattern-or-practice" suits (or "707" suits in the jargon of the civil rights bureaucracy, after Section 707 of Title VII, which provided for suits by the Attorney General). But Justice could be expected to exercise its traditional caution in screening and pressing such cases—as it had historically done in anti-trust and similar areas, where carefully selected test cases were slowly nurtured by a small and elite staff toward bellwether victories. So the EEOC would have to explore such promising terrain slowly, filing amici briefs in selected private test cases, shopping the federal courts for sympathetic judges.

Second, an evolving effects standard would have to be based on a statistical demonstration of unequal impact in the distribution of jobs and rewards. This would require a comprehensive, national data base to demonstrate such distributions. But such an information bank did not exist, and there were strong historical as well as theoretical reasons why civil libertarians as well as civil rights leaders had long opposed the maintenance of such records. Third, such a statistical demonstration would then require an appropriate model of fair distribution against which to compare actual distributions for rendering judgments. Such a model would necessarily be based on some variation of a theory of proportional representation. That is, equal opportunity would be defined as, absent discrimination, an equal chance to represent one's group in the work force relative to the group's size. The proportion of blacks or women or other minorities in certain job categories would therefore be expected to equal their proportion in some appropriate labor pool—e.g., the general U.S. population, the regional or local population, the workforce or skill population, the applicant pool. This would pose obvious conflicts over which pool was selected and why. It would conflict with traditional standards of individual merit divorced from the group's historic context. But without such a model of proportional representation, there could be no measurable benchmark of "underutilization" with which to justify coercive remedies. Fourth, given an effects definition of discrimination, a comprehensive data base with which to compare distributions, and a proportional model of appropriate distributions, the EEOC would need the authority to require specific measures of hiring and promotion sufficient to rectify discrepancies with the model.

In 1965, however, despite increasing impatience on the part of the civil rights lobby, the EEOC was feeling its way cautiously into its new and circumscribed environment. Moreover, the commission's founding law quite deliberately barred it from adopting *each* of the four paths. Indeed the congressional conservatives of 1964, alarmed by the talk of compensatory quotas and by the Motorola case, had demanded as the price of passage that such a model of compensatory affirmative action be prohibited by the statute. Thus on the question of prejudicial intent versus discriminatory effect, Section 706(g) of Title VIII required that the courts find that a respondent "has intentionally engaged" in an unlawful employment practice. Humphrey had explained to his Senate colleagues: "The express requirement of intent is designed to make it wholly clear that inadvertent or accidental discriminations will not violate the title or result in entry of court orders. It means simply that the respondent must have intended to discriminate."[43]

As for a proportional model of fair employment, Section 703(j) of Title VII prohibited any requirement of "preferential treatment to any individual or to any group" because of their "race, color, religion, sex, or national origin," or on account of "an imbalance which may exist" in their numbers or percentages relative to nonminorities. This language reflected an amendment offered by Senator Gordon Allott (R.-Colo.) and incorporated into the Dirksen-Mansfield compromise. In floor debate, Harrison Williams (D.-N.J.) grew angry at opposition charges that H.R. 7152 could permit quota hiring or require racial balance. Reinforcing the like-minded disclaimers of Title VII's bipartisan floor managers, Senators Clark and Case, Williams testily explained on the Senate floor that "to hire a Negro solely because he is a Negro is racial discrimination, just as much as a 'white only' employment policy."[44] Again, Humphrey had explained the new subsection: "The proponents of this bill have carefully stated on numerous occasions that title VII does not require an employer to achieve any sort of racial balance in his work force by giving preferential treatment to any individual or group." "Since doubts have persisted," Humphrey explained, "subsection (j) is added to state this point expressly."[45]

Only a subsequent Congress, or possibly the federal courts on constitutional grounds, could fundamentally change such an unusually firm congressional assignment of functions, duties, and limits. Also, only Congress could endow the embarrassingly defanged EEOC with the customary cease-and-desist authority that had long been enjoyed by the state FEPCs, where traditional definitions of intent and anti-discrimination had historically held sway. So in the first month of the EEOC's existence, Roosevelt testified before chairman Powell's sympathetic House Education and Labor Committee that Congress should give the EEOC cease-and-desist authority. Beyond seeking cease-and-desist authority, however, Roosevelt and his colleagues showed little interest during the EEOC's formative early months in loosening their founding charter's prohibitions against an effects-based definition of discrimination, or adopting a group-rights theory of proportional representation in employment. There was, however, one area remaining in the four that might permit some creative interpretation. That was the area of record-keeping and reporting.

Creative Administration: The National Reporting System

All regulatory agencies necessarily required record-keeping by and periodic reports from the organizations whose activities they regulated. Even the nonstatutory PCEEO had routinely required this from government contractors. In recognizing this reality, the Senate in 1964 originally adopted the House version of Title VII's provision on records and reporting, Section 709(c). This provided for the EEOC's adoption, after a public hearing as required by the Administrative Procedures Act, of "reasonable, necessary, or appropriate" record-keeping and reporting requirements. But Senator Dirksen added a further restriction in a new subsection (d). This exempted organizations already reporting to state or local FEPCs, and also those complying with Executive Order 10925 on government contracts. Dirksen's technical proviso was consistent with the conservatives' general stance of hostility to the red tape and bureaucratic imperialism of federal regulatory agencies in general, and certainly in particular to the bureaucratic clutches of federal regulatory officials who were wedded to a cause of fire-breathing liberalism like racial and sexual equality in the workplace. Dirksen's avowed purpose was to prevent the EEOC from harassing businessmen by requiring duplicative reporting in states with their own EEO agencies (like Illinois). But subsection (d) was also intended to shorten the new agency's bureaucratic reach. Humphrey explained to the Senate that the new provision would prevent duplication. "Where the employer, agency, organization, or committee is also subject to a State fair employment practice law," Humphrey said, "the Commission may not prescribe general record keeping requirements."[46] Clear enough.

But the EEOC was, after all, a government agency, with an inescapable need for minimal information. Like the proverbial camel's nose in the tent, this *sine qua non* provided an entrée of bureaucratic imperatives against which Dirksen's statutory constraints were simply no match. Thus Dirksen's subsection (d) contained a reasonable codicil that was fatal to its purpose. It provided that "the Commission may require such notations on records as are necessary because of differences in coverage or methods of enforcement between State or local law and this bill."[47] This was Dirksen's gesture of recognition toward federalism's complexities, as a way of partially balancing his insistence that the EEOC yield primary jurisdiction to FEP states. But it was also his undoing. For such an ambiguous opening was a challenge for a zealous and inventive young bureaucracy.

Alfred W. Blumrosen, who took leave from the Rutgers Law School to become the EEOC's chief of liaison for federal, state, and local agencies, claimed to have figured out how to subvert Section 709(d)'s limitation after only two weeks with the agency. In the summer of 1965 Blumrosen had early discovered the "gold mine" of Form 40 reports that the PCEEO had required for several years from all government contractors, and from which they had derived "zero lists" of employers with no Negro employees and "under-utilization lists" of employers with very few Negro employees.[48] But the PCEEO staff did not want to share their data files with the upstart EEOC—and especially not with the naive academician, Blumrosen, who almost got himself fired for "borrowing" the PCEEO's zero-list printouts anyway.[49] So Blumrosen convinced his colleagues on the EEOC staff, including the hesitant deputy general counsel, Richard Berg, to exploit the nonduplication rationale of Section 709(d). They sought to do this by demonstrating that the scattered and inconsistent patterns of state FEP reporting hardly constituted the systematic data base that the EEOC would need to enforce Title VII. Therefore the EEOC, by establishing a uniform national standard for reporting, would not really be requiring duplicate reporting. This claim, in addition to Dirksen's codicil allowing for adjustment to state-federal "differences," provided the EEOC staff with a plausible justification for the agency to develop a national reporting system of its own.

At the beginning the EEOC commissioners were divided on the issue. As a NAACP lawyer, Commissioner Jackson shared the NAACP's bitter memories of racial record-keeping. He opposed questions from organizations that might require individuals to incriminate themselves.[50] But Aileen Hernandez disagreed. She had worked at California's EEO agency, where racial reporting was required on some forms, and she defended its logic in monitoring equal employment efforts.[51] When Secretary Wirtz wrote Roosevelt on September 22, 1965, asking support for a reversal of the Labor Department's policy against racial identification, Jackson objected that the NAACP had fought for years to force the Labor-funded state employment services in Alabama and Mississippi to *stop* listing race on their application forms.[52] Indeed, it had taken the civil rights coalition until 1962 to convince Labor to adopt a color-blind policy in the first place! Not surprisingly, the white commissioners let their black colleagues take the lead in hammering this awkward issue out. But because the EEOC as a newborn agency was constantly forced to the cutting edge of equal employment policy, it lacked the NAACP's institutional memory, and early discarded the old inhibitions.

That the EEOC was able to reinterpret its founding law this way was crucial to its ultimate transformation. Yet the failure of the conservative opposition to challenge the agency on its self-aggrandisement is, like many nonevents, difficult to explain on the basis of the extant evidence. The EEOC's circumlocution was certainly open to challenge. The candid Blumrosen admitted that his "creative" reading of the statute was "contrary to the plain meaning," and that the language of subsection (d) indicated that the nonduplication rationale applied only to the exemption for government contractors.[53] Berg believed that the intent of the Senate compromise was to prevent the EEOC from requiring record-keeping and reporting in FEP states—although like most liberal critics of the congressional compromise, he believed that such unusually severe restrictions helped to hamstring the agency and defeat the purpose of the statute. "While the nominal purpose of this provision is to prevent duplication of record keeping requirements," Berg explained, "the provision is applicable whether or not the state or local authority actually imposes such requirements, and in fact, few, if any, do."[54] Like Berg, EEO authority Michael Sovern interpreted Section 709(d) as broadly "exempting all those covered by a state fair employment practice law, without regard to whether that law contains record-keeping provisions."[55] In his widely praised study for the Twentieth Century Fund, Sovern speculated that the EEOC might try to exploit Dirksen's vague codicil. "But its efforts seem certain to encounter resistance," Sovern wrote, "and it could well decide that the game is not worth the candle, especially since the quoted language *cannot possibly be stretched* [emphasis added] to permit the Commission to insist on the filing of reports."

But if Berg and Sovern were right about legislative intent and the putative restrictions of the founding law, the professional staff of the young EEOC was right about exploiting the advantages of the permanent bureaucracy in making and interpreting its own rules and regulations. It was entirely normal behavior for a government agency to seek maximum leverage and freedom of manuever by broadly interpreting its enabling legislation. This was especially critical (but also therefore potentially dangerous) for a new regulatory agency where the stakes were high and passions were easily aroused. To begin the process of building specific procedures and requirements from general statutory directives, Roosevelt appointed Jackson to chair a task force to draft rules and regulations. Jackson's expanded his task force by borrowing Richard Berg and Howard Glickstein from Justice, Eric Feirtag from Labor, Bruce Hunt from the NLRB, and added from the EEOC staff Blumrosen from Rutgers and Frank Reeves from Howard Law School.[56] As the infant EEOC's brains trust, they began the process of maximizing agency power by subverting the congressional restrictions on reporting employment statistics. During its first few months the EEOC also jousted with the resistant Budget Bureau over the form and extent of its proposed reporting system, held public hearings in December 1965, and ultimately emerged with a master EEO-1 reporting form that would be required from 60,000 of the nation's employers beginning the following March.[57] The victorious EEOC owed a large debt, ironically, to the voluntary precedent of Plans for Progress, whose companies had to submit employee data by race and ethnicity largely in order to *avoid* having the government require them to do more. So approximately 50,000 of the 60,000 large employers were already in the habit of submitting such annual reports to the old PCEEO or to Plans for Progress.[58] Bobby Troutman's source of new (and voluntary) minority hires over "expectables" had badly boomeranged on him. Once the EEOC's authority for requiring the reports was thereby established, its subsequent expansion was difficult to challenge.

In winning the battle of the reports, the EEOC provided itself with what it described in its administrative history as a "calling card" that "gives credibility to an otherwise weak statute."[59] It also thereby demonstrated the advantages over Congress that accrue to permanent bureaucracies in a contest of administrative aggrandizement. The EEOC had precedent on its side, both in the general sense of the self-evident need for performance data by regulatory agencies, and specifically in the recent practice of the PCEEO and the Plans for Progress companies. Congressional conservatives remained opposed. But the congressional opposition, having dutifully taken their stands on record in 1964, was poorly positioned to exercise close oversight as the new agency fashioned its rules and regulations. In 1965 Congress was distracted by debates over voting rights and Vietnam and Watts and inflation and scores of other issues more pressing than agency records. Balancing such opposition was the EEOC's powerful national network of allies, as reflected in the demonstrated power of the Leadership Conference, and the crushing of formerly invincible southern filibusters in both 1964 and 1965.

Outside of Congress and beyond the capital, the state and local FEPCs were potential contestants over the regulatory turf. But they were also ideological allies, and the EEOC early discovered that it could co-opt the state FEPCs through another creative interpretation. This was a broad construction of Section 709(c), which allowed the EEOC to reimburse state agencies for help in implementing the law. The EEOC interpreted this to permit outright grants to such state agencies as reimbursements in kind for processing the states' own complaints. By 1968 the EEOC's state grant budget of $700,000 was considerably sweetening the cooperative instincts of the state agencies, and was effective in cementing their cooperation in supporting and enforcing EEOC policy guidelines.

Within the federal government's executive branch, the EEOC's chief inherited bureaucratic rival for claims to turf, the PCEEO, was abolished just two months after the EEOC opened for business. The EEOC's main new rival, the Labor Department's OFCC, was even newer to the terrain. An obscure entity with an awkward title, it was nestled within a traditional, constituency-captured, program-running agency that was instinctively uncomfortable with the OFCC's policing mission. Finally, on the issue of reporting and record-keeping, the commission also had some advantage in the appearance of reasonableness. Given its public mission, and the smothering backlog of accumulated complaints, the new commissioners and staff were rapidly concluding that the only effective path to equal employment was through a wholesale attack on broad patterns of discrimination. Thus the national reporting system was viewed as essential for locating the most likely targets and then following up on the results.[60] But before the EEOC could convince its external public of the need for such racial record-keeping, it had to convince its more immediate civil rights constituency—many of whom were deeply distrustful of an earlier legacy of racial and ethnic record-keeping, and bore deep scars to prove it.

Entangled Myths of Affirmative Action: Racial Quotas and Racial Records

Part of the array of activities planned for the White House Conference on Civil Rights, as President Johnson had pledged in his Howard University commencement address of June 4, was a White House Conference on Equal Employment Opportunity. It was scheduled early, for August 19–20, 1965. Like most such issue-centered conferences, its main purpose was less policy formulation than constituency-building, consensus-molding, and networking—and, in the case of the new EEOC, staff recruitment. Also like most such widely advertised White House conferences, this one was too big for the White House. So the 600 conferees convened instead at the nearby State Department. Faithfully reflected there were the assumptions of the key policymakers on the cutting edge of implementation, as well as the general tone of current debate on civil rights enforcement. One of the conference's seven panels discussed the meaning of affirmative action.[61] There Herman Edelsberg, of Washington's Anti-Defamation League of B'nai B'rith (who would shortly join the EEOC as staff director, serving from October 1965 to March 1967), captured for the conferees a consensus that "if you try to achieve equal employment opportunity by the case by case litigation or even conciliation [method] you are trying to fill the bucket with a medicine dropper."[62] Edelsberg discussed the concept of institutional racism. Despite Title VII's statute of limitations, which started the clock the day the Civil Rights Act took effect, Edelsberg explained that true pursuit of job equality meant that one could not "look solely at what has happened after July 2, 1965." Instead one had to look "inside the plant or union," at "conditions in the community, pre-existing institutions of discrimination, the climate of hostility in the community."

Later in the crowded session on affirmative action, when panelist Charles Wilson of the California FEPC was describing recent affirmative action negotiations with the Bank of America, a questioner finally uttered the magic, or dreaded, word:

> VOICE: May I use a bad term, do you set a quota?
>
> WILSON: That's illegal.
>
> VOICE: I am putting that in quotes.
>
> WILSON: No Sir, we don't.
>
> VOICE: What criteria will you use?
>
> WILSON: We try to get away from setting any kind of standards to be obtained. When we get down exactly to what we want we get very vague, something like "substantial," words of that kind. Mainly because we don't want to be caught in the bind establishing quotas and neither does the bank. . . .[63]

Wilson's honest disclaimer was to become a standard refrain of EEO bureaucracies: affirmative action had nothing to do with racial quotas. That was illegal. It was important for civil rights spokesmen, at least in the mid-1960s, to reassure themselves, their public, and their critics that affirmative action did not require quotas or racial preferences, that indeed the former was as morally imperative as the latter was morally opprobrious—as well as being unambiguously prohibited by Title VII. As Edelsberg reassured the conferees in his summary of the conference panels, "We are agreeably surprised that there was almost no suggestion that the various methods of achieving affirmative action might be construed as a kind of preferential treatment."

That affirmative action necessarily excluded racial quotas or preferences was one of two rationalizing myths that emerged with the new concept in the 1960s, and that temporarily conditioned its meaning. The other was its corollary, that racial and ethnic record-keeping was inherently perverse. History had demonstrated that such indicators constituted a dangerous invitation to institutional officials to exercise their inherited prejudice. This kind of nefarious record-keeping was most notably associated in recent American memory with the maintenance of Jim Crow institutions in the South and with discrimination against Jews in the North. The recent war against fascism had seared into the consciousness of that generation a horror at the consequences of official recognition by the state of racial, ethnic, and religious distinctions among the citizenry, and of state policies of favoritism and disability that flowed from these distinctions. Both myths were thus grounded in historical reality as well as in moral philosophy, and both were linked by American liberalism's repugnance at the use of racial or religious-ethnic indicators in pursuit of public policy. Hubert Humphrey symbolized traditional liberalism's animus against racial or ethnic quotas, as when he told the Senate in 1964 that "Title VII prevents discrimination," because "it says that race, religion, and national origin are not to be used as the basis for hiring and firing." Title VII's central purpose, Humphrey explained, was "to encourage hiring on the basis of ability and qualifications, not race or religion."[64] Thus

Edelsberg and his colleagues were "agreeably surprised" that affirmative action did not mean preferential treatment.

Liberal opposition to requiring minority identification in institutional records was symbolized at the EEO conference by the NAACP's chief Washington lobbyist, Clarence Mitchell. At the panel session on record-keeping and reporting requirements, Mitchell explained: "The history of the reason why we do not include this is sadly and surely proven, that the minute you put race on a civil service form, the minute you put a picture on an application form, you have opened the door to discrimination and, if you say that isn't true, I regret to say I feel you haven't been exposed to all of the problems that exist in this country."[65] When challenged that affirmative action required racial identification, Mitchell's long memory fueled his anger: "It seems to me incredible that the Government of the United States, recognizing that there is a nasty, underhanded little system for keeping track of people through a cute little code system . . . would make it easy for discrimination by saying 'Oh, no, you don't use obscure little marks. You put a nice big thing which shows this is a Negro so you don't have to put on your glasses to find out."[66] Mitchell had been fighting the ancient battle against the southern humiliations of Jim Crow for decades and he was cynical about the pretensions of beneficent racial discrimination. "I am just amazed that you people who come from these northern states, where you are exposed to this problem," Mitchell said, "would fall into the crevasse which has no bottom, of keeping racial statistics." He warned them that by reintroducing such statistics into the Armed Forces and the Civil Service Commission, they would "put us back 50 years and nullify all the work we have been doing to correct this thing."

But Mitchell, who had helped persuade the Eisenhower administration to ban the practice of racial record-keeping in federal records in 1955, and who had also helped persuade the Kennedy administration to reaffirm the prohibition in agency employment records in 1962, rather quickly lost the renewed battle during the first year of the EEOC. The practical logic of racial identification to enforce desegregation was powerful. In March of 1966, John Macy announced that the Civil Service Commission would henceforth *require* each federal agency to maintain records of their employees' race.[67] The employees themselves, however, would make that designation on special forms, and these documents would be kept outside agency personnel offices. Nevertheless, the federal schizophrenia on this question made for awkward policy directives. Macy explained that the purpose of his order was to help identify any patterns of discrimination in agencies and to guide corrective action. He then added the customary bureaucratic disclaimer: while the resulting statistics would be used as "group data in studies and analyses which contribute affirmatively toward achieving the objectives of the equal employment opportunity program," they would *not* result in preferential treatment of minorities, Macy said, because "This is a program of equal opportunity, not preference."[68]

It remained for Willard Wirtz to explain liberalism's rapid reversal on the issue. On May 19, 1966, eight months after he had first raised the notion with the EEOC, the Labor Secretary announced at a luncheon meeting of the NAACP Legal Defense and Education Fund in New York that he was reversing the department's 1962 order against racial identification in Labor's personnel records. The reasons were eminently practical: the approximately 2,000 offices of the federal-state employment service, which had come under attack for its history of routinely accommodating segregationist job preferences in the South, would henceforth be required to produce racially specific "statistical tools" to show where "concentrated action" was required.[69] Wirtz said he was "sick and tired of the false piety of those who answer inquiries about racial aspects of their employment or membership practices with the bland, smug answer that 'we don't know because of course we don't keep records about that kind of thing.'" Besides, Wirtz could scarcely demand racial accounting from the Labor-funded SES if he did not also require it throughout the agency itself.

The Labor Secretary's practical argument was weighty. And like the Civil Rights Act, it pointed southward in its corrective rationale. It paralleled the judgment of the federal courts that monitoring the progress of school desegregation was impossible without knowing the race of the school children. But the surprisingly rapid destruction of the myth that affirmative action was possible without what Clarence Mitchell called the "dangerous" and "scurrilous" system of record-keeping by race, also undermined liberalism's corollary myth of the 1960s, that affirmative action was possible without racial preferences or de facto quotas. California's Wilson had explained to the EEO conference in August 1965 that his FEP negotiators had made safe progress by persuading Bank of America officials to pledge "substantial" improvement in minority hiring, while remaining purposely vague about the specific measures that might be required to remedy underutilization. But such a vaguely suspended tension was unnatural in the numbers-driven world of employment negotiations and regulatory agencies. As had long been the case in labor negotiations, built-in pressures for "more" tended to become explicit and measurable.

Thus the EEOC's victory in the national reporting system threatened eventually to contradict the operational myth that affirmative action had nothing to do with illegal racial preferences and quotas. In 1965–66, however, that day was not yet near at hand. What was immediate and pressing was the EEOC's need to define its role in relation to its anomalous structure as a new regulatory agency with almost no formal regulatory power. Circumstances conspired to launch the EEOC poorly and at an awkward moment. Its first year began with the explosion in Watts, and coincided with the Johnson administration's rapid escalation of the American combat role in Vietnam, and with the attendant inflation and breakdown of the liberal consensus—the last including the radicalizing surge of black nationalism that found its slogan in the summer of 1966 in Black Power. All of these disruptive forces were reflected in the politics of "white backlash" that news commentators used to explain the sharp Republican gains in the congressional elections at 1966. In the middle of all this Chairman Roosevelt, never very attentive of the helm of the EEOC, left the agency to run for governor in New York.

The Hardening Political Mood of 1966

When Franklin Roosevelt, Jr., informed the White House in early April 1966 that he would not even complete the first year of his two-year term as EEOC chairman, the angry President told Harry McPherson to refuse Roosevelt's request for the customary exchange of letters. "Get a good successor immediately," Johnson scrawled at the bottom of McPherson's memo—"A Man of judgment."[70] But having implicitly ruled out the female half of the population in the immediate search for a chair, and probably the black tenth as well, Johnson was once again distressingly slow to appoint a successor. He did decide early, partly on the recommendations of Roosevelt and John Macy, that none of the sitting commissioners was suitable as chairman. Vice chairman Luther Holcomb was a Johnson loyalist and caretaker who seemed inappropriate for the chairmanship. He also claimed not to want it; he had no substantial political ties to the important black and labor constituencies; and he was actively opposed by the Hispanics, who lobbied against him in Texas through their Democratic ally, Senator Ralph Yarborough.[71]

In his search for chairman candidates, Macy concentrated on white male Democrats from the South and the East, in order "to maintain the present political and geographical balance of the commission."[72] But the commission lacked ethnic balance in one rather obvious category: there was no apparent Jewish commissioner.[73] The commission's senior staff, to be sure, was heavily Jewish—Edelsberg, Blumrosen, Berg, Markham, and others. Such a pattern was not surprising, given the pioneering of Jews in the FEPC enterprise. Indeed, there had been some early black criticism of the staff dominance of a group representing only 2 percent of the population. Moyers had commented the year before, during the first round of recruitment (but after Morris Abram had turned down the chairmanship), that no Jew sat on the commission.[74] Macy's first choice this second time around was Abram again. His second was Sol M. Linowitz, chairman of the executive committee at Xerox. Macy described Linowitz as "a Johnson Democrat [and] an unflagging liberal" whose attractive attributes included "support of your policy in VietNam."

Johnson again preferred Abram, but Abram again declined. For reasons unexplained, Johnson showed no interest in Linowitz, and instead suggested another "name" Democrat, Adlai Stevenson III. Macy inquired, but Stevenson, a candidate for state treasurer in Illinois, was not interested. This process dragged on repetitively through the summer, with Johnson frequently rejecting suggested Democrats who were somewhat better known (Linowitz, Theodore Kheel, David L. Cole, R. Peter Strauss, Robert Law, Orin Lehman) in favor of others who turned out not to be available (G. William Miller, Courtney Smith, George A. Spater, Patrick F. Crowley).[75] Few of these names were household words in 1966. All were white males. In light of Roosevelt's early departure, the obvious teething troubles within the commission, and the hardening political mood surrounding it, persuading well-known Democratic men with multiple career options to accept the hotspot chairmanship was going to be difficult.

In the meantime, Richard Graham's one-year term expired in July, and he was not reappointed. It is not clear why, although circumstances suggest a White House wariness at Graham's independence, especially on the troublesome new issue of gender discrimination (see Chapter Eight). Macy had alerted Johnson to Graham's expiring term in April, reporting that "Roosevelt has no serious fault to find with Graham. He believes Graham more capable of identifying problems than finding solutions, but believes it might be difficult to find a liberal Republican to replace Graham."[76] By August, the vacancy at the helm of the three-member EEOC was attracting unfavorable publicity, including a critical article in *Jet* magazine charging the White House with crippling the new agency. The Budget Bureau complained that the continued vacancy was "creating competition among the three remaining Commissioners," who were clearing staff appointments with the White House and congressional politicians "in order to gain political favor in their bid for the Chairmanship."[77]

With the situation becoming critical, Macy went to see Johnson on August 22 with a list of five fresh candidates. Topping the list was Stephen N. Shulman, general counsel for the Air Force.[78] Shulman was a lawyer-technocrat who was unknown to the general public. But at age thirty-three, his credentials included a Yale law degree, a clerkship to Justice Harlan, private practice in Washington with Covington and Burling, and experience in labor and civil rights law both as special assistant to Secretary Goldberg at Labor, and subsequently at the deputy assistant secretary level in Defense. If Johnson was determined to appoint a "safe" and loyal as well as competent white male Democrat to chair the troubled EEOC, especially during a time of increasing domestic turmoil and political tension, then Shulman's candidacy was plausible. And both time and acceptable candidates were running out. The agency's administrative and budgetary disarray could certainly profit from the firm hand of an experienced administrative technician. So Johnson told Macy to check Shulman out "with all the Negro and labor groups"—and by Negro groups he specified those represented by Wilkins, Young, and Martin Luther King *"in that order."*[79] On September 21, Johnson introduced Shulman to the press as the new EEOC chairman.

That fall the nation's political mood was turning sour on two prime commitments of the Johnson administration. First, rising inflation combined with growing antiwar sentiment to challenge Johnson's insistence on a policy of "guns and butter." Second, public support for the civil rights laws of 1964 and 1965 was dimmed by resentment at the black rioting in Watts. A Harris poll of mid-1966 showed that 75 percent of white voters believed that the Negroes were moving too fast—up sharply from 50 percent in 1964.[80] The congressional elections of November were looming against the backdrop of a collapsed consensus. The Administration's civil rights proposals met defeat in 1966 for the first time, and the House Democrats began the process of stripping Adam Clayton Powell first of his committee chairmanship, then of his House seat itself. As the *Congressional Quarterly* observed, "The pro-civil rights coalition which had operated so effectively in previous years—Republicans and Northern Democrats in Congress and civil rights, labor, and church groups outside Congress—fell apart in 1966." [81]

James Farmer, whose replacement at CORE by Floyd McKissick paralleled Stokeley Carmichael's takeover from John Lewis at SNCC, agreed that by the fall of 1966 "the movement was reeling." With fundamental disagreements not on tactics but on goals and objectives, Farmer said, the major civil rights organizations "didn't know where they were going or what to do at that point." [82] The November elections ratified the popular disenchantment with the Johnson administration's tutelage. The Republicans, so disastrously defeated in 1964, rallied to capture eight new governorships, three new Senate seats, and 47 new House seats. [83] The Democrats would of course still control the White House and Congress. But in the new atmosphere they could not be expected to expend major political resources on behalf of the EEOC or new civil rights legislation while the more conservative 90th Congress sat in 1967–68.

The relatively unknown Shulman inherited the warm chair of the EEOC just as congressional resistance was stiffening. The recoil of a nervous Congress was the EEOC's most palpable threat, given the unsteady new agency's need to win congressional support for expanded authority and a budget to support it. In the internal budget hearings for the fiscal 1968 submissions during September and October, the Budget Bureau came down hard on the EEOC for a host of administrative shortcomings. In its first year the EEOC had received 8,854 complaints. But it had processed *none* of these cases within the statutory limit of sixty days, and in only one year it had accumulated a six-month backlog of 3,000 cases. Worse, the EEOC was devoting only one-third of its operational budget to its major statutory function of investigating and conciliating complaints. More than half of its staff resources were devoted to its Washington headquarters, where the staffing pattern showed a top-heavy ratio of one clerical to 1.3 professional positions. [84] In light of the evidence of mismanagement and accumulating EEOC fat in headquarters, the Budget Bureau recommended that the agency's request of $8,870,000 for FY 1968 be cut by $2 million and that it redirect its resources from special headquarters projects to processing the huge complaint backlog and to staffing its undernourished field offices.

Finally, the Budget Bureau's severe review of the new commission's performance contained one criticism that seemed typical and even innocent enough at the time, but that foretold an ironical new source of conflict and misery for the EEOC. In response to the EEOC's request for a supplemental budget amendment of $430,000 to attack its growing backlog of complaints, Jack Buchanek of the Bureau's Education, Manpower and Science Division recommended to Director Schultze that the request be approved—but with revealing comments. In exploring additional ways for the EEOC to reduce the backlog, Buchanek noted that one-third of the complaints charged sex discrimination. He recommended that "less time be devoted to sex cases since the legislative history would indicate that they deserve a lower priority than discrimination because of race or other factors."[85] During its troubled first year of operation the EEOC, struggling with the backlash of racial politics, would be blindsided by the sudden explosion of the gender issue. The Budget Bureau's Buchanek had been correct: Title VII's legislative history indicated a much lower and indeed almost an inadvertent concern over sex discrimination, at least on the part of the overwhelming majority of men who participated in constructing and passing the historic Civil Rights Act of 1964.

CHAPTER VIII

The EEOC and the Politics of Gender

The Janus Face of Sex Discrimination

Historically, the issue of gender equality in America has been far more complicated by the cross-pressures of economic class than has been the issue of racial equality. The American black community, to be sure, has always reflected conflicting, class-based dualisms, ranging from the classic split between Booker Washington and W.E.B. DuBois to the social and programmatic chasm between Martin Luther King and Malcolm X. But three centuries of racial oppression cemented the dominance of racial over class concerns for black Americans, for whom the color line threatened survival and blocked class solidarity with other Americans. The primacy of racial over class subordination has been symbolized in the 20th century by the pre-eminence of the NAACP. As an elite-led organization of the black middle class, the NAACP has nevertheless consistently fought against racial discrimination without making basic class distinctions. As Roy Wilkins and George Meany well understood, racial divisions were the common enemy.

In American women's history, however, the role of class has dominated gender issues. Turning discussion so abruptly from race to sex discrimination therefore requires a brief historical shifting of gears. Because the institutions of marriage and the family dictated a distribution of women throughout the American class structure that paralleled the male distribution, women, unlike blacks, have lived intimately with "the oppressor" and have shared his class identification. As a result the politics of female equality has reflected that division.[1] In the 20th century, this class division within the women's movement has persisted, and between 1920 and 1965 it largely neutralized the feminist impulse.[2]

The victorious suffragist coalition celebrated the ratification of the 19th Amendment in 1920 by splitting into two warring factions that mirrored the frustrating ambiguities between sex and class.[3] The dominant National Women's Suffrage Association, having led the coalition that won the franchise, was essentially transformed into the League of Women Voters. Declaring victory, the League concentrated on civic-minded voter education. Meanwhile the progressive women associated with the trade union movement pursued special protective legislation for women and children, and found new voice and leverage in the creation of the Women's Bureau in the Department of Labor in 1920.[4] On the other side of the split, the ostensibly radical-egalitarian National Women's Party, led by the irrepressible Alice Paul, in 1923 dedicated itself exclusively to crusading for the newly proposed Equal Rights Amendment. This was the same year in which the Supreme Court announced in *Adkins v. Children's Hospital* its apparent hostility to the concept of special protective legislation, now that women had the vote.[5]

In the ensuing four decades of intramural feminist conflict, the anti-ERA, protectionist wing held dominance. It was led by such Democratic luminaries of trade-union liberalism as Mary Anderson, Eleanor Roosevelt, Frances Perkins, India Edwards, and Esther Peterson. Its achievements were anchored in the New Deal triumphs for organized labor, most notably the Wagner Act of 1935 and the Fair Labor Standards Act of 1938. But the progressive legacy also included the enactment of protective legislation in most states, laws that generally limited the working hours and physical burdens of work for women, and that also granted them such special considerations as rest and maternity benefits. Opposing this labor-oriented group in the women's movement were the upper-class, often Republican, "tennis-shoe ladies" of the NWP, later in alliance with the elite clubwomen of the National Federation of Business and Professional Women's Clubs and the General Federation of Women's Clubs. Being highly educated and generally affluent, the ERA women embraced a brand of radical egalitarianism that emphasized procedural freedom over substantive equality. In demanding equal freedom to compete with men in professional life and financial transactions, the NWP's egalitarianism was more libertarian than liberal. It thus attracted the support of such bastions of conservatism and otherwise unlikely feminist allies as the National Association of Manufacturers, the National Association of Chambers of Commerce, and the American Retail Association.

This feminist dichotomy with its curious twists of alliances persisted into the Kennedy administration.[6] It set the stage at the beginning of the Johnson administration for the unlikely alliance between the ERA's "tennis shoe ladies" and Judge Howard Smith that unexpectedly popped sex into Title VII in 1964.[7] The sexual politics of this period have been ably reconstructed from the Kennedy and Johnson archives by Cynthia Harrison in *On Account of Race: The Politics of Women's Issues 1945–1968.*[8] Harrison describes the central role of Esther Peterson, a veteran Democratic labor lobbyist who served as director of the Women's Bureau and also as assistant secretary of Labor under Kennedy and then Johnson, in guiding the work of the President's Commission on the Status of Women (PCSW), which Kennedy established by executive order in December 1961.[9] Such study commissions had historically been used by union-based Democrats to blunt drives for the ERA and contain the damage from the abiding feminist split, and Kennedy's PCSW was no exception.[10] Nominally chaired by the infirm Eleanor Roosevelt, the PCSW reflected the consistent opposition to ERA of Peterson's dominant wing of the Democratic party. Under Peterson's determined prodding, the Johnson Democrats in 1964 yielded to labor demands and dropped support for ERA in their presidential platform.

Peterson and her PCSW allies concentrated their lobbying on the Equal Pay Act of 1963, which the Women's Bureau and Secretary Goldberg had successfully engineered. But the trade-union movement had supported the principle of equal pay for the same work since World War I, when the large-scale movement of women into the industrial labor force threatened to depress wages for male family heads. Thus the new law of 1963 was no ERA-style, constitutional abstraction. Rather, it was a statutory amendment to the Fair Labor Standards Act of 1938, and its coverage *excluded* business and professional women. But the political compromises necessary to get it passed would also initially exclude almost two-thirds of working women from its provisions, especially the low-paid women in agriculture and domestic service. The Equal Pay Act nonetheless represented another small pragmatic step, consistent with Peterson's trade-union and Labor Department credo of "specific bills for specific ills."[11]

Thus when the Civil Rights Act of 1964 was passed with women suddenly included, American women were neither united, effectively organized, nor psychologically prepared to press effectively for its enforcement. The ostensibly victorious leaders of the NWP and the business and professional women's clubs had historically been committed to ERA as a symbolic abstraction. They lobbied the Congress to pass the Smith amendment, but they were neither organized nor politically interested in lobbying the Johnson administration and especially the new EEOC over the longer haul to interpret broadly and enforce vigorously the country's new ban on sex discrimination. On the other hand, Eleanor Roosevelt and Esther Peterson and their labor allies had kept the ERA out of the Democratic platform in 1960, and they had won a recommendation against the ERA in the women's commission report of 1963.[12] Because the Women's Bureau under Peterson's leadership had opposed Smith's sex amendment as an albatross that would endanger the original civil rights bill and the state protective laws, the Bureau had worked with representatives Emanuel Celler and Edith Green to defeat it. They had been quietly encouraged in this effort by many old-line, labor and civil rights liberals of the Leadership Conference, who didn't want race nondiscrimination to have to compete for resources against sex nondiscrimination.[13] Once the Smith amendment passed, however, the national network that Esther Peterson had forged around the PCSW was confused and split by the sudden emergence of sex in the nation's new anti-discrimination law.[14] Because no national debate over gender discrimination had preceded its inclusion in the Civil Rights Act, a consensus remained among the public that the problem of civil rights was not only mainly a racial concern, but was also essentially a *male* problem—especially insofar as it involved jobs for minority breadwinners.

The Male World of Anti-discrimination: The Moynihan Report and Playboy Bunny Jokes

Since the feminist revolution in consciousness of the 1970s, it has become a commonplace to observe that a male bias has historically dominated the social thought and public policy of American political life. Little distinction in this regard has been discerned between liberal Democratic and conservative Republican regimes. It is nonetheless striking how candidly the modern liberal reformers against prejudice during the 1960s voiced their conscious and unconscious male prejudices. This was true even of the two major organized constituencies for the modern civil rights movement, both of them male dominated: the labor unions (meaning chiefly the AFL-CIO) and the black civil rights groups (dominated by the NAACP). Since World War II the NAACP had quietly discouraged the potential encumberment of FEPC bills with new categories of rival claimants, like women and the elderly. Like the Urban League, the NAACP was careful to avoid antagonizing potential allies in the civil rights coalition by openly attacking their inclusion. But there was little ambiguity in the higher echelons of black civil rights groups that the core of the problem was the unemployment and underemployment of black *male* breadwinners and putative heads of families.[15]

As for organized labor, its political and financial support for FEPC since the AFL-CIO merger had been consistent and financially generous. But so had been its support for protective legislation and its opposition to the Equal Rights Amendment. The correspondence of former Wisconsin congressman Andrew J. Biemiller, the AFL-CIO's chief congressional lobbyist, was peppered with snide and fraternal allusions to the ERA. "[I]f the ladies who are now worrying about [the ERA] did not have this to worry about," Biemiller wrote a colleague in 1960, "they would find something else. Vive le (sex) difference!"[16] When Congressman James Roosevelt was beginning to fashion what was to become the heart of Title VII in his House subcommittee hearings in 1962, George Meany wrote him to "strongly urge you not to include a prohibition of discrimination based on sex in the proposed Equal Employment Opportunities bill."[17] After Title VII nevertheless became law with sex included, the AFL-CIO continued to oppose the ERA, even in the face of snowballing sentiment in its favor. Organized labor, having lost the great postwar campaigns to organize the South and repeal Taft-Hartley, or at least to repeal its Section 14(b) authorizing state "right to work" laws, continued to stand by its guns in defending women's protective legislation. As late as 1970, Biemiller reminded Bayard Rustin, who chaired the Leadership Conference's executive committee, that "the AFL-CIO is now and always has been on record against the Equal Rights for Women [sic] Amendment, and strenuously objects to the Leadership Conference taking a position on this question."

Within the executive establishment of the Kennedy-Johnson administration, the white male liberals who forged breakthrough policies in civil rights were not inclined to include women within their normal civil rights purview.

From the beginning of the PCEEO to the launching of the EEOC, these politically astute men had been leery of the encroachments of sex discrimination on the already hypersensitive domain of race discrimination. This fear extended from George Reedy's plea to Vice President Johnson in March of 1961 that adding sex to the PCEEO's jurisdiction would "throw the committee into complete chaos," to the perfunctory acknowledgment of EEOC staff director Tom Powers to Secretary Wirtz in September 1965 that "The Commission is very much aware of the importance of not becoming known as the 'sex commission.' "[18] Even Hubert Humphrey, in his ill-fated plea to Johnson in August 1965 to retain vice presidential command of the short-lived Equal Opportunity Council, explained to the President that his proposed "simple housekeeping" executive order

> . . . does not cover discrimination on grounds of sex. Careful consideration was given to this matter and on the basis of strong recommendations from Secretary Wirtz and Chairman Macy, it was concluded that the Government's program of non-discrimination in Federal employment on grounds of sex should not be combined with its program of non-discrimination on racial and other grounds.[19]

Humphrey reassured the President that he had been advised by both Macy and Wirtz that "the liklihood of adverse political reaction from women's groups is small." Not surprisingly, sex discrimination was omitted entirely from Johnson's famous Executive Order 11245 of September 1965—despite Bill Moyers's nimble-tongued reaction at the press conference at which it was announced.

That same summer, 1965, saw the release of the explosive Moynihan Report on the Negro family. Moynihan's alarm over the disorganization of the black family structure was unapologetically male-centered. It emphasized the systemic weakening of the position of the black male and the self-defeating encouragement, by the federal government's welfare policy as well as by broader economic and social factors, of female-headed families. In a preview memorandum to Bill Moyers in January 1965, Moynihan warned that the deteriorating stature and authority of the black male had created a "pathological matriarchal situation which is beginning to feed on itself."[20] In May, Wirtz sent Johnson a summary memorandum by Moynihan outlining his alarming thesis, which Wirtz called "nine pages of dynamite."[21]

Moynihan's memo observed that "the master problem is that the Negro family structure is crumbling." Illegitimacy rates had soared past *one-quarter* of all black births (the white rate was 3 percent). The "main reason for this," Moynihan explained, "is the systematic weakening of the position of the Negro male." "By contrast," Moynihan noted, "Negro women have always done and continue to do relatively well." They had higher rates of college graduation and were getting "better jobs, higher salaries, more prestige." Moynihan told Moyers that when he had asked a distinguished Negro sociologist what could be done for the Negro man, the sociologist had replied that he was sure of only one thing: "Anything that could be done to hurt the Negro woman would help. He was not smiling."[22]

When the official version of Moynihan's report was released by the Labor Department in July, its tone was considerably softened from the summary that Moynihan had sent Johnson (indeed it lacked even the name of the assistant secretary for policy planning who had authored it). As a clinical analysis of a sensitive area of social policy, it contained no policy recommendations for government. But Moynihan told Moyers that the time had come for the federal government to adopt a "master strategy" to deal with the deteriorating Negro family. Accordingly, Moynihan's May memo to Johnson had recommended seven steps for action. Some were general, calling for more housing and birth control and data collection. But two were sex-specific, and they were unambiguous in their message about gender and role. First, "Men must have jobs." "We must not rest until every able-bodied Negro male is working," Moynihan said, "[e]ven if we have to displace some females." "In the Department of Labor," Moynihan added, "after four years of successful effort to increase Negro employees, we found that in non-professional categories 80 percent were women!" Moynihan's second gender-related recommendation was that many of the new jobs for black men would have to be taken from black women. "More can be done about redesigning jobs that are now thought to be women's jobs and turning them into men's jobs." "This is a great problem for the Negro male," Moynihan added. "[H]is type of job is declining, while the jobs open to the Negro female are expanding."

The subsequent uproar over the Moynihan Report saw him publicly ostracized and quickly hounded into resignation by much of the liberal academic and civil rights community. But the liberals were incensed by Moynihan's allegedly patronizing attitude toward American blacks and their culture, not toward women. Liberal critics likened Moynihan's report to a Brutus-kiss, his alarm to crocodile tears; they indicted his analysis for blaming the black victims of white discrimination for their plight.[23] So effective was the intellectual community's ban on discussing such a forbidden topic that public debate was silenced for two decades—by which time the ignored pathology had worsened by a factor of two, and impoverished female-headed families with fatherless children had become the norm in the black underclass.

The outcry against Moynihan's report in 1965 merged with the swelling protest over Johnson's Vietnam policy. But it included no significant voice of protest from women's groups over its male-centered policy implications, beyond little-noticed complaints from Women's Bureau director Mary Keyserling (who had succeeded Peterson in 1964 when Peterson moved to the White House as special assistant for consumer affairs) and from Pauli Murray, a black senior fellow at the Yale Law School.[24] Moynihan's thesis was indeed politically insensitive, and in some ways his report was legitimately vulnerable to the technical criticisms it received from the social work establishment within the permanent government. But in hindsight his alarum of 1965 about the hemorrhaging black underclass has been sadly vindicated. Not so his male-centered social analysis, which he shared with the leading sociological scholars of his generation.[25]

Small wonder, then, that the American public mind at large, as mirrored in a "masculinist" press coverage and political discourse, did not seem to take the new issue of sex discrimination seriously, and indeed found in it much occasion for comic relief. When the *Wall Street Journal* in a special report asked corporate executives how they felt about Title VII, a perplexed airline personnel executive said, "We're not worried about the racial discrimination ban—what's unnerving us is the section on sex." "What are we going to do now," he asked, "when a gal walks into our office, demands a job as an airline pilot and has the credentials to qualify? Or what will we do when some guy comes in and wants to be a stewardess?" [26] An electronics executive in Nashville, whose nonunion firm hired only nimble-fingered women to assemble the delicate electronic components, complained that he would have to "hire the first male midget with unusual dexterity if he shows up." A telephone company official similarly worried about the threat of male operators, although he dismissed as "fantasy" the possibility that some women might want to become linesmen.

Even EEOC chairman Franklin Roosevelt played to this gallery at the commission's kick-off news conference on July 2, 1965. "What about sex?" a reporter asked Roosevelt. "Don't get me started," Roosevelt replied with a laugh. "I'm all for it." [27] Shortly thereafter the *New York Times* ran a major story on the newly discovered "bunny problem." This would arise when a male applied for a job as a Playboy Bunny—or a woman to be an attendant in a male Turkish bathhouse, a man to clerk in a corset shop, a woman to be a tugboat deckhand. [28] The following day the magisterial *New York Times* editorialized that Congress ought to "just abolish sex itself." [29] This was because in light of the new federal policy against sex discrimination, everything was going to have to be neuterized—housemaid, handyman, Girl Friday, even the Rockettes. "Bunny problem, indeed!" the *Times* despaired, "This is revolution, chaos. You can't even safely advertise for a wife anymore." It was of course inconceivable that the *Times* would use such a tone in an editorial over black civil rights. But behind the gleeful hyperbole of such popular speculations as the many varieties of the bunny joke, lay a series of hard questions that Title VII now raised and that the new EEOC would be expected to address—although virtually no systematic thought had been given to the answers.

The New Feminists and the EEOC: The Question of Race-Sex Linkage

In her study of women and public policy during the Kennedy-Johnson years, Patricia Zelman argues that two especially frustrating encounters with new programs of the Great Society drove a divided community of women leaders toward a more militant feminism. [30] The first and lesser of these encounters was with the early war on poverty. The Job Corps had been envisioned by its male planners as an exclusively male enterprise to provide job training for male heads of families. OEO director Sargent Shriver told Congress that he and his fellow antipoverty warriors naturally assumed that the Job Corps should concentrate on training "primary breadwinners"—meaning of course, men. But this angered even the redoubtable Congresswoman Edith Green, who had opposed both the ERA and Judge Smith's sex amendment, and who was given to explaining that "I do not consider myself a suffragette." Green extracted from Shriver and the Administration grudging concessions toward a Women's Job Corps. But they amounted to little more than pacifying gestures, and seemed to confirm that the Great Society was less hostile than blind to the problems of the female third of the nation's workforce.

That experience left a sour taste with Democratic women who had been pleased by President Johnson's eagerness to appoint women to important posts.[31] It primed them for a second and more sobering surprise, which was with the new commission on equal opportunity itself. Zelman called this "the first major political battle for the new feminists." More than any other single catalyst it sparked the formation in 1966 of the National Organization of Women (NOW), soon followed by the Women's Equity Action League (WEAL) in 1968—and onward to the feminist revolution of the 1970s.[32]

When the EEOC opened for business in the summer of 1965, virtually all political indicators pointed the commission toward caution in interpreting the new ban on sex discrimination. Lacking substantial authority to coerce employers, the EEOC was expected to nurture a cooperative atmosphere of voluntarism, especially in light of its need to avoid offending a suspicious Congress, from which it must seek greater funding and authority. Moreover, the commission's dominant constituency was overwhelmingly concerned with black civil rights, not with sex discrimination. This consensus was captured in one of the EEOC's early staff studies by consultant Frances R. Cousens, who had been research director of Michigan's FEPC. She observed that the addition of sex discrimination to Title VII "was to seriously weaken the Commission. Intergroup relations professionals have long believed that although discrimination according to sex needs consideration, it differs substantially from that based on race, religion, or national origin; to include sex as a provision of Title VII does, in fact, undermine efforts on behalf of minority groups."[33] Cousens complained that from the beginning the EEOC staff was "inundated by complaints about sex discrimination that diverted attention and resources from the more serious allegations by members of racial, religious, and ethnic minorities."

Furthermore, the perennial split in the women's ranks between the protectionists and the proponents of ERA had effectively neutralized the women's collective voice. But because there had been no congressional hearings on the Smith amendment, the EEOC needed to hear from women's groups. So early in 1965 the Women's Bureau urged the cabinet-level Interdepartmental Committee on the Status of Women (ICSW), which President Kennedy had established in the fall of 1963 to follow through on the recommendations of the PCSW, to lobby the EEOC on behalf of women's rights.[34] But the ICSW soon deadlocked over the irrepressible issue of state protective legislation, and this lack of consensus left the EEOC during its critical shakedown cruise without consistent advice on sex discrimination policy from the affected constituency.

When the EEOC opened its doors for business on July 2, 1965, only one of the five commissioners was female, and there were no women among the supergrade staff. Commissioner Hernandez complained that a meeting between the commissioners and employers in California was arranged in a private club that barred women. "The message came through clearly that the Commission's priority was race discrimination," she recalled, "and apparently only as it related to Black *men*."[35] When Johnson announced the White House Conference on Equal Employment Opportunity for August 19–20 to showcase the new agency, only nine of the seventy-five speakers and panelists were women, six of them on the panel dealing with sex discrimination that Hernandez chaired. Chairman Roosevelt called the question of sex discrimination "terribly complicated," and indicated that the commission would formulate policy slowly on a case-by-case basis.[36]

In the summer of 1965 the disagreement among the women's groups, together with the priority both in legislative history and public pressure for a main thrust against race discrimination, pointed the EEOC toward great caution in defining and enforcing sex discrimination. The commission's immediate problem was how to apply Section 708 of Title VII. This provision was originally designed to supersede segregationist state laws in the South. But its clumsy lawyer language made no specific mention of the offending Jim Crow laws. Rather, it simply asserted that state laws that conflicted with Title VII were now invalid. This stricture, however, pointed in turn to the heart of Title VII, which in Section 703 listed what acts were now prohibited. These were: to refuse to hire or to discharge any individual or to limit, segregate, or classify employees in any way which would deprive them of employment opportunities or otherwise adversely affect their status as employees, because of such individual's race, color, religion, sex, or national origin. But what did this require of the EEOC if a woman in Arizona filed a complaint against an employer who refused to hire her because state law prevented her, but not men, from working more than eight hours a day?[37] Was California's 50–pound limit on employee-lifting a legitimate ceiling for women but not for men, especially in light of higher pay for heavier work? Was the 15–pound limit in Utah? The years of congressional testimony on what became Section 708 had uniformly assumed its racial focus against southern Jim Crow laws. And the lack of hearings and committee reports on Smith's sex amendment had prevented the AFL-CIO and the dominant women's protectionist groups from testifying to exempt women's protective laws.

Thus in August of 1965 the applicability of Section 708 remained unclear and the women's groups remained at loggerheads. In light of the policy paralysis of the ICSW, its chairman, Willard Wirtz, wrote to Roosevelt in August and urged the EEOC on behalf of the Labor Department (and especially Mary Keyserling's Women's Bureau) to preserve the women's protective laws. Meanwhile the states themselves, assisted by their new status-of-women commissions, could re-examine them in light of current developments.[38] This advice posed the EEOC no real problem. There was little pressure on the new EEOC to invalidate the women's protective laws—other than mere logic. But what did cause interpretive and communicative problems, and led with surprising speed to a public rupture between the EEOC and the women's groups, was disagreement over the meaning of Section 704(b) concerning segregated classified ads.

The Problem of Jane Crow in the Classified Ads

Section 704(b) prohibited job advertisements or referrals "indicating any preference, limitation, specification, or discrimination, based on race, color, religion, sex, or national origin . . . ." Like most of the language of Title VII, it was designed with the South's racial caste system in mind, and the single word "sex" had simply been added to the list by Judge Smith's amendment. Thus the provision was targeted against Jim Crow listings in the want-ads that effectively segregated blacks into menial jobs. Most southern newspapers had traditionally listed separate columns for "Help Wanted—Colored." But in recent years many of the more cosmopolitan southern dailies had dropped the practice, and by 1965 resistance to abandoning it had been reduced, like resistance to integrated restaurants, to marginal redoubts in the small-town and rural South.

Most American newspapers, however, continued to publish separate classified advertisements for men and women. On Friday, July 2, 1965—the day the Civil Rights Act of 1964 took effect—the *New York Times* published its customarily segregated columns of help-wanted ads for males and females. The listings were typical of the national press, with the dominant category of female jobs falling into the file clerk/steno/typist/secretary bloc, and the next largest bloc calling for "Girl Friday." The men's section, which was roughly twice as long, carried the male equivalent to the low-paying female jobs— "mail boy," driver, guard, building superintendent, administrative assistant, bank teller. But here, additionally, were the jobs with futures, separating white-collar from pink-collar New York: accountancy, auditor, draftsman, engineer, exports, management, production. There was no third, sexually neutral column. Racially separate ads would have been unthinkable outside the South in 1965, and they were rapidly disappearing within the South. But the only indication of a coming challenge to the long tradition of single-sex segregation in the want-ads was the *New York Times*'s insertion of a special notice, in a small box atop the help-wanted section (surely where job seekers rarely looked). It explained that in observance of the passage of the Civil Rights Act of 1964, "Qualified job seekers of either sex are invited to consider job opportunities in either the Male or Female help wanted columns . . . ."[39]

The anomaly was by no means obvious to Americans in 1965, and it illustrates the vast distance between the race and sex analogies in civil rights, both in theory and in perception. State FEPCs were enforcing equal employment rights in fully half the states when Title VII became law, and forty-three states had protective laws for women. But only three states—Hawaii, New York, and Wisconsin—had added a ban on discrimination based on sex.[40] In September 1965 in the pioneer FEPC state of New York, the *Wall Street Journal* ran a special report on another leading FEP state, Wisconsin. The FEP law in progressive Wisconsin had been amended in 1961 to cover sex discrimination. Since then, the *Journal* reported, only sixteen complaints had been filed, and only one had required a public hearing.[41] In that short time, a male advantage in schoolteacher salaries in Wisconsin had been removed, the face amount of employee pension benefits had been equalized for the sexes, and a male former medic in the army was ordered admitted to a state vocational school's course on practical nursing. But no complaint had challenged the single-sex classified advertisements in politically liberal Wisconsin's newspapers.

Because the public debate over sex discrimination remained confused, and was still marked by deep divisions among political active women, the Congress had added exceptions clauses to Title VII in the aftermath of the Smith amendment. The chief exceptions clause for women exempted them from coverage when sex was "a bona fide occupational qualification for employment." The latter phrase quickly translated into the acronym BFOQ in the lexicon of government, and this BFOQ potentially exempted men also, since it simply wouldn't do to have restroom attendants of the opposite sex in either restroom. But again no committee hearings or report explicated just what such a BFOQ might mean. For blacks, the ban on racially segregated want-ads was accorded widespread understanding, and the highly regionalized practice virtually ceased overnight. But for women, the ban produced bunny jokes.

When Wirtz wrote Roosevelt in August of 1965, he urged a narrow construction of Section 704(b) on job advertising. This was "not a provision," Wirtz said, "that permits a general labeling of jobs as 'men's jobs' or 'women's jobs.' "[42] At the White House conference a week and a half later, the EEOC's deputy counsel, Richard Berg, warned that the commission was going to put the burden on the employers to make the case for BFOQ exceptions. "If they can't think of any reason not to do it [hire women for jobs traditionally held by men]," Berg cautioned, "then they'd better do it."[43] Shortly thereafter, in the EEOC's first ruling on its first job complaint (the *Wall Street Journal* reported that "the honor went to the ladies"), Roosevelt announced the commission's ruling that any labor-management agreement or corporate policy that required dismissing females when they marry would be regarded as a violation of Title VII.[44]

But the preponderance of signals the EEOC was receiving emphasized the differences between sex and race, not their similarities. The commission was urged to impose a total ban on special state laws for Negroes and on "colored" job ads. But at the same time great caution was urged on behalf of women's protective legislation, and there was no evidence in public opinion of significant support from either sex for a ban on single-sex ads. Furthermore the American Newspaper Publishers Association, which offered no defense of racially segregated want ads, defended the business necessity for separate-sex classified ads. The ANPA argued not only that both their readers and their advertisers preferred the traditional and market-efficient single-sex ads, but also that nothing in Title VII gave the EEOC authority over the publishers of classified advertising, as opposed to the help-seeking employers themselves. Whatever the merits of the ANPA's economic and cultural arguments, their legal argument about the limited reach of the EEOC was a serious one. And implicit in their complaints was a powerful political argument—i.e., politicians traditionally accorded great weight to the views of newspaper publishers.

The voice of caution on policing sex discrimination made sense to most of the males who dominated the EEOC as well as the ANPA. Loud in dissent was Aileen Hernandez. "It is the advertiser's responsibility to insert in his ad that he does not discriminate and the job is open to males and females," she argued. "If there is a B.F.O.Q. he should come to the EEOC and get a ruling, not make the determination himself."[45] But EEOC General Counsel Charles Duncan—also a black—disagreed. "I have never heard it suggested that the Commission had the power to pass on in advance what was and what was not B.F.O.Q," Duncan replied. "I have never heard it said we could not," Hernandez countered. Staff director Tom Powers warned: "If the Commission lets itself get into the position where it has to pass on every single instance in which any covered employer in the United States believes that Sex is a bona fide occupation [BFOQ], you are going to load yourself down with so many questions in the Sex area that you are not going to have time to do anything else of significance."

In response to this debate, Roosevelt on August 18 announced a flat ban on racially separate want-ads, but he appointed a seventeen-member advisory committee to study the question of single-sex ads.[46] He appointed Richard Graham, the commission's only business executive, to work with the group, and Graham reported to the commission in late September that "the newspaper people were strongly of the opinion" that while racially segregated want ads violated Title VII, sex-segregated ads did not because the purpose was not to express preference by sex but rather their desire to reach the greatest number of readers.[47] Graham offered a compromise proposal requiring advertisers to publish a disclaimer with their classified ads. Roosevelt directed Graham to rewrite his statement to make sure "that the EEOC will not become a police operation in every case where an employer claims a B.F.O.Q.," and circulate it for publication in the *Federal Register*.[48] Commissioner Hernandez, who had objected so vigorously at the previous meeting, was absent at the decision meeting.

In October the *Wall Street Journal* carried a special report on the young EEOC's "shadow boxing" with the varied permutations of job discrimination against Title VII's newly protected classes, searching for a defensible balance of policy. Summoning a prizefighting metaphor, the *Journal* predicted that the "youthful pugilist" EEOC would demand preferential treatment for Negroes under the euphemistic guise of affirmative action. But "on behalf of the ladies" the EEOC would only produce "a succession of light jabs against unchivalrous treatment, just enough to qualify as effort though far short of enthusiasm."[49] "We're not going out on our charger to overturn patterns," an unnamed commissioner was quoted as admitting. Thus "the ladies, at most, will be given the letter of the law," the *Journal* story predicted. "But the Negroes will get the spirit. . . ."

The *Journal's* prediction of a softer line and milder tone on sex bias was at least partially verified rather quickly. In October the EEOC ruled that racially segregated local unions, seniority lines, and promotion lists violated Title VII. Then in November the commission announced its first guidelines on sex discrimination (which it had formulated the previous September). Consistent with its stance on race discrimination, the commission led off by ruling several common business practices as illegal acts of sex discrimination. First was a refusal to hire a woman based on "assumptions of the comparative characteristics of women in general," or of "preferences of co-workers," or on "stereotyped characteristics of the sexes." Second, it was impermissible to classify jobs as "light" or "heavy" or "male" or "female," or to maintain separate seniority lists if this were done to adversely affect women. Third, employers could not fail to hire or promote women solely because they were married.[50] So far the race-sex linkage was close.

When it came to classified advertisements segregated by sex, however, Roosevelt acknowledged the commission's belief that "Culture and mores, personal inclinations, and physical limitations will operate to make many job categories primarily of interest to men or women."[51] So the commission would not require unified want ads by gender, as it had done by race. Instead, publishers could maintain, "for the convenience of readers," separate classified sections labeled "Jobs of Interest—Male" and "Jobs of Interest—Female." In doing so, the EEOC would require only that advertisers specify in the ad that the job was open to both males and females (for example, by indicating "M & F")—if indeed it was (that is, if it didn't qualify as a BFOQ)—and that publishers print a notice on each help-wanted page that their listings were not intended to exclude persons of the opposite sex. The EEOC's cautious guidelines on sex discrimination, as compared with its aggressive stance on race discrimination, were explained by a male environment so dominant that even the EEOC's new executive director, Herman Edelsberg, could explain to the press in the fall of 1965 that "There are people on this Commission who think that no man should be required to have a male secretary—and I am one of them."[52]

The EEOC formally announced its different treatment of race and sex distinctions in classified ads at the same time it announced a similar stance on state protective laws. While urging state legislatures to update archaic and irrelevant provisions in such laws, the commission said it could not assume that Congress intended to strike them down. So it would continue to "consider qualifications set by such state laws or regulations to be bona fide occupational qualifications and not in conflict" with Title VII.[53] But here the EEOC hedged by saying that in the future it nevertheless would judge such complaints on a case-by-case basis. If, for example, a state's restriction on the weight a woman could lift were set "unreasonably low," and thereby excluded her from otherwise desirable benefits, then the EEOC would rule it invalid. Implicit in this hedge was of course an assumption that the EEOC possessed the authority to rule against such restrictions.

So the EEOC itself was resisting the analogy that equated sex and race in civil rights enforcement. This generally pleased the status-of-women lobby when it came to protective legislation. But it began to anger them in the matter of classified advertisements. The inconsistency between the EEOC's guidelines on sex and race in job advertisements created a symbolic rallying point to unify women leaders who for so long had been paralyzed by their divisions over protective legislation. Powerful sources of women's sentiment for decades had pressed serious arguments in defense of the protective laws. But virtually no women had defended the sexually segregated classified ads. The EEOC's inconsistency and timidity in enforcing women's rights would likely have provoked the divided women toward greater unity of purpose in fairly short order. But the internal politics of the EEOC soon produced a colossal blunder which by the spring of 1966 had managed to goad the status-of-women groups into a unified attack on the reeling agency.

The EEOC's Strategic Retreat of 1966

In hindsight, the EEOC's weak compromise of November 1965 that acquiesced in the continuation of single-sex classified advertisements seems short-sighted and philosophically inconsistent at best, and a sexist double standard at worst. But the political and ideological environment of 1965 included no reported sentiment in public opinion against single-sex ads. It did include, however, a continuation of divided counsels in the women's groups; legitimate uncertainties over congressional intent; the EEOC's initial rulings against other areas of sex discrimination; and its artful hedging on future review of the range of permissible single-sex classified ads (and protective laws as well). In this context the compromise guidelines of November 1965 were not inherently unreasonable as a point of departure for a new and unsteady agency just finding its way in a volatile environment.

Instead of prudently feeling its way forward from such a cautionary beginning, however, the EEOC soon *retreated* in its want-ad policy. The result was a political fiasco for the new agency. But it was an unanticipated bonanza for the new feminists, as it triggered a unifying burst of feminist "consciousness raising" among the status-of-women groups. The protagonist in this pivotal episode was the Reverend Dr. Luther Holcomb, whose subsequent role as a feminist goat was vaguely similar, in an inadvertent, catalytic kind of way, to Sheriff Jim Clark's crucial role for black civil rights in Selma.

What happened was that the American Newspaper Publishers Association, unhappy with even the compromise guidelines of November 1965, and confident of the fabled political weight of the press lords, decided to press for a policy rollback. The ANPA had a long history of challenging potential gov-ernment intrusions on the First Amendment rights of the press. They were anxious to emphasize that this new regulatory agency should not try to tell *publishers* what to print and where to print it—as opposed to requiring *em-ployers* to carry certain stipulations in *their* advertisements. Such an argu-ment could have represented a consistent defense of important philosophical principles. But this was not the argument that the ANPA most forcefully pressed on the Johnson administration and the EEOC. In light of the mischief that flowed from the ANPA's failure to find sufficient satisfaction in the com-promise of November, its failure to base its objections on First Amendment rights is puzzling.

In seeking to roll back the November compromise, the ANPA enlisted the support of EEOC vice chairman Holcomb. By Dallas standards Holcomb was a liberal integrationist who had worked within the city's ministerial alliance during the early 1960s to accelerate school desegregation. In 1961 Holcomb had lobbied Vice President Johnson to stiffen the Justice Department's de-mands that at least forty rather than a token dozen Negro children be en-rolled in integrated schools in Dallas—otherwise "the Negro community will feel betrayed."[54] But by the Washington standards of 1965, Holcomb was the most conservative of the commissioners, and also the most closely asso-ciated with the White House. Holcomb was approached by Stanford Smith, the ANPA's executive secretary, who informed Holcomb that the EEOC had handled the want-ad issue in a "manner irritating" to the ANPA. In mid-December, Holcomb sent to President Johnson through Marvin Watson a message that the ANPA had become "unusually alarmed" by the EEOC's new policy. But Holcomb's summary to Johnson of ANPA objections made no mention of First Amendment complaints. Instead, his report to Johnson suggested that the publishers were irritated chiefly by inconvenience, loss of business, and the general impertinence of governmental meddling.

> . . . when private individuals call to insert ads in the classified section of newspapers, these individuals are compelled by certain members of my Com-mission to answer specific questions as to possible discrimination pertaining to both race and sex. In many instances the newspapers claim that they have lost ads as a result of being compelled to ask these questions of the private individuals who wish to insert the said ads.[55]

Holcomb feared that the EEOC was "overly anxious and eager to move with a 'big stick' at business," and he worried about how this "affects the rela-tionship of the business community to the President" and "the influence that this could have upon newspapers generally in regard to the President."

Holcomb's memo, in fact, read like an espionage report. It is odd to read from the new commission's vice chairman that he had learned from a "reli-able source" that the commission's chairman, executive director, and general counsel were soon going to Louisiana to meet with representatives of labor and management at Crown Zellerbach's plant in Bogalusa, there to attempt to negotiate a conciliation agreement in a dispute over job seniority. This was, after all, what the EEOC was supposed to do—attempt to conciliate such complaints without the glare of publicity. Holcomb concluded his confidential memorandum to Johnson, as he typically did, with a note of personal fealty that was far too convincing for merely cloying sycophancy: "I appreciate more than words can express the opportunity of working with Marvin in your behalf. I have followed everything that pertains to your welfare with prayerful concern. Please know that my admiration and devotion to you is akin to a religious dedication."

Holcomb's December memo proposed that Johnson authorize him to negotiate secretly with the ANPA in an attempt to frame a revised policy that would protect the President's interests. Johnson was noncommittal, but he instructed Watson to have Holcomb report his progress to Bill Moyers. By March 1966, Holcomb was able to send Watson the text of an agreement that he worked out "to the complete satisfaction of the Publishers."[56] It effectively nullified even the weak compromise requirements on dual-sex classified ads of the previous November. Armed with this apparent presidential preclearance (no written evidence of Johnson's approval has surfaced), Holcomb on March 24 persuaded Graham and Jackson (Hernandez was absent) to agree at least tentatively to what he but not they knew was the ANPA-sponsored revision. It read as follows:

> Help-wanted advertising may not indicate a preference based on sex unless a bona fide occupational qualification makes it lawful to specify male or female.
>
> Advertisers covered by the Civil Rights Act of 1964 may place advertisements for jobs open to both sexes in columns classified by publishers under "male" or "female" headings to indicate that some occupations are considered more attractive to persons of one sex than the other. In such cases, the Commission will consider only the advertising of the covered employer and not the headings used by publishers.[57]

Like the ANPA, Graham believed that Title VII gave the EEOC legal authority over employers placing the classified ads but not over newspapers who published them.[58] But he insisted that single-sex advertising nevertheless violated Title VII and he wanted to stop the practice. So he tried to attack it from another direction by urging that Executive Order 11246 be revised to include sex discrimination. That would add the enforcement muscle of the Labor Department and the OFCC to the campaign against sex discrimination. It might not immediately reach newspaper publishers, but it would greatly accelerate the lagging campaign against sex discrimination, and focus public attention on the double standard in the race-sex analogy.

This mood explains why that spring Commissioner Graham, a Republican businessman and egalitarian feminist, responded so critically to one of the EEOC's great (and rare) early victories. On March 30 the EEOC's compliance chief, Alfred Blumrosen, briefed the commissioners on a conciliation triumph that the EEOC, the OFCC, and the Defense Department had jointly reached with the giant Newport News Shipbuilding and Drydock Company. Graham mixed his congratulations, however, with a complaint. He pointed out that the landmark agreement to upgrade Negro employment at the Virginia shipyard had omitted any reference to sex discrimination as an obligation under the Civil Rights Act. Blumrosen replied that sex discrimination had never been raised as an issue in the Newport News conciliations. But Graham was not finished. He further offered his "violent opposition" to part of the agreement because it reached beyond affirmative action to require reverse discrimination. It provided that if "a Negro employee has not moved up through the grades within the classification in which he is presently employed as rapidly as the norm or standard derived from the sample for white employees, he shall forthwith be assigned the first grade in his job classification."[59] "If we are going to say because of his race," Graham said, "that considerations of the employee's skills and ability are not germane," and that "regardless of this—his ability, he must be moved up now," then "we are going to lose."

Graham's fellow commissioners generally voiced agreement with his reservations about reverse discrimination. But they also agreed that the Newport News conciliation was a great breakthrough for blacks, and it only indirectly raised the awkward question of sex discrimination. As Blumrosen explained, "when we have a case, which is a 'race' case, no overtones of sex discrimination in it, [we] do not give them the vagueness, uncertainty and haze that surrounds the 'sex' provision for nobody knows what it means." Blumrosen continued:

> They are beginning to understand what the "race" provision is, but they don't know what the "sex" provision is. To inject any item into the conciliation that is new and difficult makes the achievement of the settlement that much more difficult. These people had not been accused of "sex" discrimination. The fact that "sex" is not referred to in the settlement agreement, in no way privileges these people to engage in "sex" discrimination, but it does mean that to handle the "sex" aspects of the case would involve a separate conciliation.[60]

On April 20, with Hernandez in attendance, the commission agreed unanimously to release the revised guidelines on sex discrimination in want ads.[61] Hernandez, like Graham, felt frustrated by her apparent legal inability to reach newspaper publishers directly.[62] She said she opposed the new policy, but agreed not to obstruct its publication in the *Federal Register*.[63] The EEOC announced the revised policy on April 22, and it left the advertisers essentially where they were in November. But it now completely exempted the publishers, who were free to continue running their single-sex ad columns with impunity.

The EEOC and the Revolt of the New Feminists

While the EEOC had been backpedaling, however, the resentment of the status-of-women groups had been accelerating. When the *New York Times* published its too-clever, bunny-joke editorial on women's equality following the White House EEO conference the previous August, Esther Peterson wrote a letter of protest to the *Times*. She chided the editors for featuring a tone of ridicule and suggested that the *Times*'s worries over male Playboy bunnies and Rockettes were misplaced. They should ask instead: "How many jobs are 'women's jobs' merely because they are menial, routine, monotonous and, of course, low-paying?"[64] Peterson's letter reflected an abandonment of her earlier opposition to the Smith amendment, and this in turn signaled the imminence of a sea change in feminist thought. At about the same time the ICSW, which had been unable to agree on a common policy to urge on the EEOC, formed a special committee chaired by EEOC commissioner Graham (who sat on the ICSW as Roosevelt's deputy), and charged it with building public support for a more vigorous enforcement against sex discrimination.[65]

In a parallel development, the Citizens Advisory Council on the Status of Women (CAC), a presidentially appointed, twenty-member group or private citizens chaired by *Ladies' Home Journal* editor Margaret Hickey, overcame its own paralysis from the disagreements over protective legislation. The CAC members did so not through the weight of superior logic or evidence—indeed, such arguments had rarely budged the historic logjam. Instead, they followed Lincoln's advice about the stubborn green tree stump in the cornfield: they simply plowed around it. In effect, they adopted the same pragmatic position that Secretary Wirtz was urging on the EEOC, which was to ask the states to review their protective laws in light of new circumstances, to meanwhile judge the complaints as they arrived on their individual merits, and otherwise to get on with the rest of the equal employment agenda.[66]

Such a pragmatic stance finally freed the status-of-women groups to quit exercising their ancient divisions and to attempt instead to combine their voices. An intellectual benchmark of this internal women's dialectic was the publication in December of a law review article by Pauli Murray and Mary Eastwood, entitled "Jane Crow and the Law: Sex Discrimination and Title VII."[67] Like many such essays that mark and help shape rather than trigger major turning points in public discourse, Murray and Eastwood's "Jane Crow" article broke no new theoretical ground. Indeed, as law review articles go, it was little more than a broad overview of recent developments in sex discrimination law. But the law reviews were scarcely full of legal assessments of gender discrimination in 1965. More important, the Murray-Eastwood essay accelerated the metamorphosis in feminist legal thought by finessing the old debates over protective legislation and by pressing hard, as the "Jane Crow" title indicates, the case for a race-sex analogy in EEO law that would grow toward a race-sex equivalence for enforcement purposes. Having at last begun to get the debilitating old protectionist wars behind them, the leaders of the status-of-women groups were freeing their energies to convert first themselves and then other women leaders into a more cohesive and effective bloc of feminist lobbyists. Their ultimate target was of course American public opinion, male and female alike, which was still tolerating bunny jokes.

So the EEOC wasn't the real enemy. But by May of 1966 it was the immediate and proximate offender. Given its unique egalitarian charge, it presented a splendidly ironical target for the ire of the new feminists. This time the chief castigator was a member of Congress, Representative Martha Griffiths of Michigan. In 1964 Griffiths had led the drive that forced the Smith amendment through the House—against a background of masculine giggles and guffaws. Her sensitivities had been well riled by such indignities, and in May of 1966 her rage at the EEOC's new step backward was palpable.[68] Griffiths sent a blistering letter to acting EEOC chairman Holcomb, with a copy to President Johnson, noting her disbelief "that this is what equality of opportunity means."[69] On June 1, Holcomb replied that Section 704(b) properly applied only to the individual advertisement of the employer, which could indicate no gender preference, but not to the layout decisions of the publisher, which could. Such decisions were based, Holcomb explained, on publishers' desire "to obtain a maximum reader response and not on a desire to exclude applicants of a particular sex."[70] He informed Griffiths that advertisers who in the wake of Title VII had moved their single-sex ads to the male *and* female column had suffered a marked drop-off in responses. "Thus, it is primarily the reading habits of job seekers which presently dictate the placement of ads." But beyond this, Holcomb advised Representative Griffiths, "as you know, section 704(b) is not applicable to the policies of the newspapers in classifying advertising."

On June 20, Griffiths took the House floor and blasted the EEOC in an attack that consumed six pages of the *Congressional Record*.[71] She began by echoing the resentment voiced by Esther Peterson and the CAC that the media's concentration on various odd and hypothetical cases of sex discrimination demeaned the seriousness of the problem. Even the *EEOC Newsletter* of May-June 1966, Griffiths observed, had discussed the reasonableness of women serving as dog wardens and men as house mothers. She accused the EEOC's executive director, Herman Edelsberg, of (1) insisting at his first press conference (when he had replaced Tom Powers the previous fall) that he and his male colleagues at the EEOC were "entitled" to have female secretaries, and (2) of explaining to an audience at New York University in April 1966 that the sex provision in Title VII was a "fluke" that was "conceived out of wedlock."[72]

Beyond her *ad hominem* arguments, however, Griffiths grounded her case in an equation of race and sex discrimination that would require the same legal remedies from the EEOC. Such a flat equation carried problematical implications in the realms of legal philosophy and jurisprudence, quite apart from biology and physiology. But it had a compelling ring in the immediate arena of politics and ethics. As a case in point, Holcomb's defense of the Jane Crow want ads had repaired to the intent doctrine. By arguing that advertisers intended no gender exclusion when they placed their job notices in one column rather than the other, Holcomb in effect echoed the rhetoric of Jim Crow's defenders that racial segregation was a mutually preferred social arrangement and was intended to inflict no harm or insult on either party. Griffiths retorted that "If even a single employer uses sex-segregated ads for the purpose of discrimination, that is prohibited by law." This was certainly true. But it was also true in connection with job interviews, employment tests, and myriad other standard items and normal procedures in the job applicant screening process. These were procedures that would be illegal if used for the *purpose* of discrimination, but that according to Title VII were not illegal per se. Griffiths's main reply to Holcomb's intent defense, however, was to counter with a *results* doctrine. "In any event," she continued, "even if a particular employer has no intention or desire to exclude applicants of a particular sex, the law prohibits *any action* which *tends to accomplish such discrimination* [emphasis added]."[73]

Any action? Which merely *tends* to accomplish alleged discrimination? Griffiths did not dwell on the radical implications of such a results-centered definition of discrimination. It represented, in hindsight, not the antithesis of the EEOC's own intentions, to the degree that these were discernible in the confusion of 1966, but rather the main strategic direction that the EEOC would ultimately take once it survived the current turmoil. Thus the EEOC's clumsy retreat of 1965–66 on gender advertising was an anomaly, a cautionary hedge to the right that immediately boomeranged—to the ultimate benefit of the angry feminists. In the process of responding both to a congressional mandate for some unclear measure of different treatment of race and sex, and also to short-term political pressures from the publishers, the EEOC had inadvertently accelerated the social and political momentum of the civil rights movement toward a results-centered strategy that would equate race and sex in EEO enforcement. The new strategy would seek to determine the extent of both race and gender discrimination, and potentially the extent of any other categories of alleged discrimination, not by attempting to ascertain intent on a case-by-case basis, but rather by broadly applying a proportional model of statistical representation in the workforce.

In the summer of 1966, however, such a portent could not be reasonably guessed. So Congresswoman Griffiths scored more heavily against the EEOC by making more practical arguments. One telling argument for abolishing Jane Crow help-wanted ads was the most practical one of all: it had recently and successfully been tried—by the *Phoenix* [Arizona] *Gazette*, the Honolulu *Star-Bulletin*, and in Toledo by both the *Blade* and the *Times*. Moreover Griffiths, as the acknowledged if self-appointed congressional whip who successfully drove the Smith amendment through the House, took umbrage at the "whining" excuse of EEOC officials that the sex provision was a "fluke." "Since when is it permissible," she asked, "for an agency charged with the duty of emforcing the law, to allude to the assumed motive of the author of legislation as an excuse for not enforcing the law?"[74]

Historically, a strong case could be made that the passage of the Smith amendment had indeed been something of a political fluke. Its legislative history was so devoid of the customary hearings and committee reports that congressional intent was poorly documented. But Griffiths's indictment was on target in a deeper political sense: regulatory agencies often exploited such murky areas in their congressional charge in order to expand their authority and please their constituencies, or to create new constituencies. It was, in fact, a staple of Weberian theory in public administration that bureaucracies would normally behave in such a self-aggrandizing and self-perpetuating fashion.[75] So when regulatory agencies complained in public that their guiding legislation did not mandate a certain policy stance, it was often because they weren't interested in that policy stance. Griffiths acknowledged as much when she attacked the "shilly-shallying" EEOC for "wringing its hands about the sex provision" when its real excuse was that "sex discrimination cases take too much time and thus interfere with the EEOC's 'main' business of eliminating racial discrimination."[76] Certainly the future EEOC, in its more secure and aggressive days of the late 1960s and the 1970s, would advance some interpretations of congressional intent concerning specific Title VII clauses in 1964 that were quite bold in their revisionist inventiveness.

Women's Liberation and Administration Resistance

In *It Changed My Life*, Betty Friedan recalled how during the EEOC's first tumultous year of interpreting Title VII, Commissioner Richard Graham sought her help in organizing independent women's groups to lobby the EEOC in the same way the effective black groups had done:

> Richard Graham, a real fighter for women who had emerged on that commission, told me that he personally had gone to see the heads of the League of Women Voters, the American Association of University Women and other women's organizations with national headquarters in Washington, to get them to use pressure to have Title VII enforced on behalf of women. And they had been appalled by the very suggestion. They were not "feminists," they had told him.[77]

Friedan heard a similar plea from a young attorney on the EEOC's legal staff, Sonia Pressman, who requested anonymity for fear of losing her job.[78] These sympathetic, internal appeals to the new feminists, asking for an independently organized source of counterbalancing pressure on the EEOC, reflected the inherent limitations of the status-of-women groups. They were a new, thin, elite network of top-down reformers, not representatives empowered by a grass-roots social movement. Their leaders could scarcely create independent lobbying groups, because so many of them were either government officials or represented established women's organizations that had traditionally avoided such active lobbying.[79]

The turning point for the feminist lobby, but not yet for the Administration or the EEOC, came in June of 1966, when several events coalesced to trigger the new feminist phenomenon that NOW both reflected and accelerated. First, during June, Congresswoman Griffiths took to the House floor to berate acting chairman Holcomb and the EEOC staff. Second, Richard Graham's initial one-year term as commissioner was allowed to expire, even though the Administration gave no reason and had no replacement in mind. Third,

June also saw delegations from the state status-of-women commissions gather in Washington for a convention, with Friedan in official attendance as a credentialed writer, and unofficially in attendance as a feminist organizer. Out of this fertile sea of discontent NOW was conceived. On October 29, when NOW was officially christened, EEOC castoff and proto-feminist Richard Graham became a founding vice president.

So, too, did Aileen Hernandez (in 1970 she would succeed Friedan as NOW's president). With Roosevelt having resigned in May to run for governor in New York, and with Graham's one-year appointment allowed to expire without renewal or replacement in July, Hernandez found herself blocked by acting chairman Holcomb, whose vote was required to produce a working majority of the three remaining commissioners. Hernandez suddenly announced her resignation from the EEOC on October 10, to take effect on November 10, having participated in the EEOC's decisions on the new gender guidelines, and also having announced that she would accept the position of executive vice president of NOW.[80] But for the growing feminist coalition, NOW's gain was the EEOC's loss, because the commission was now stripped of its two lonely feminist voices. Fittingly, the commission's two ERA partisans were a black woman Democrat and a white male Republican.

When John Macy informed President Johnson of the surprise resignation of Hernandez—"not in protest or disappointment," Macy explained, "but because she is not personally happy in Washington"—Johnson scrawled on the memo: "See Connally list of Mexicans at once . . . ."[81] The previous summer Johnson had concentrated on white male Democrats to replace Franklin Roosevelt. Although he had seriously though unsuccessfully pursued such prominent female Democrats as Marjorie Lawson and Patricia Roberts Harris for the EEOC, in the fall of 1966 Johnson moved to shore up his troubled Hispanic flank, where his senior appointments had been far less visible than his female or black appointments.

Johnson was responding to stirrings of discontent among the restive Hispanic leaders. They were voicing their resentment at being generally ignored, and at being putatively represented on the EEOC by a Negro woman with a Spanish surname derived from her former spouse. In the spring of 1966, Mexican-American groups had signaled their dissatisfaction by ambushing the unoffending Richard Graham in a meeting in Albuquerque (Graham believed that the ambush had originally been set for Holcomb, who got wind of it through his Texas connections and sent Graham as a last-minute substitute).[82] In response to Johnson's request, Governor Connally searched his Hispanic network in Texas and identified as a candidate for the commission a World War II bombardier, Vincente T. Ximenes. Then serving as an AID official in Panama, Ximenes was a native Texan with training in economics from the University of New Mexico. A former staff member for the Democratic National Committee, he had deep roots of loyalty and patronage in the southwestern Democracy.

As a Democrat, Ximenes replaced Aileen Hernandez on the EEOC, thereby swapping a Hispanic male for a black female. Joe Califano confirmed for

Johnson Ximenes's assigned role on the EEOC with unmistakable clarity: he "will actively take charge of the Mexican American problem and keep it away from the White House." [83] In the aftermath of the conference fiasco on black civil rights, Califano confirmed a new strategy that combined containment and appeasement for the Mexican-Americans. He urged Johnson to appoint Ximenes head of a "Cabinet-Level committee to make sure the Mexican-American communities are getting their fair share of federal programs." A national conference would be necessary to achieve maximum visibility and credit for the gesture, but it would best "be held in Albuquerque sometime in the future, thus keeping any conference out of the White House." [84]

So Johnson moved swiftly to repair damaged or strained relations with dominant racial and ethnic groups in the Democratic coalition. But his administration seemed blind to the new feminist reaction that was building. The President had first sweated out the ill-starred White House Conference on Civil Rights during 1965–66, with the capable help in damage control of Ben Heineman's rescue mission. He then had successfully used the EEOC to shore up his relations with the disaffected Hispanics, or at least with the dominant Mexican-American bloc. Despite his relatively strong record of appointing women to intermediate posts, however, Johnson left curiously unattended his crumbling relations with the feminists. The liberal Republican and proto-feminist Richard Graham was essentially replaced by nobody until the eve of Johnson's retirement from the White House. [85] Indeed, following Graham's unexplained termination, no Republican was appointed to join the commission's other Republican, Samuel Jackson, although such a minimum partisan balance was required by law, until Jackson himself resigned early in 1968, thus unacceptably leaving the commission with no Republicans at all. So in October 1968 Johnson nominated William H. Brown III of Philadelphia, who like Jackson was a black male Republican. Somewhat similarly, following the resignation of Hernandez in October 1966, Johnson chose no woman to replace her until the appointment in March 1968 of Elizabeth Jane Kuck, a Wisconsin social worker and Democrat with strong labor ties. Thus when Graham and Hernandez disappeared from the commission in 1966, no identifiable feminist voice replaced them on the EEOC at all.

Furthermore, the tidal shift of June 1966 toward a new feminist sentiment and mood in the status-of-women groups did not sweep the established women's groups before it. The new EEOC chairman, Stephen Shulman, reported to President Johnson his continuing meetings through 1966–67 with a stream of group lobbyists who urged him not to interfere with state protectionist laws. [86] These included the National Consumers League, the National YWCA, the National Council of Negro Women, Amalgamated Clothing Workers, ACLU, International Union of Electrical Workers, Communications Workers of America, and the American Association of University Women. Shulman told Johnson that all of these groups were stepping up their lobbying to counterbalance the new efforts of NOW, which had thrown its weight on the side of the Business and Professional Women's Association and the ERA forces in general.

To the Women's Bureau coalition, NOW represented a new threat on the egalitarian left, reaching across the ideological spectrum to embrace the out-numbered libertarians who had carried the ERA banner for two generations. Alarmed by NOW's strident new voice, the protectionist leaders appealed to Wirtz, in his dual capacity as secretary of Labor and ICSW chairman, to resist the growing threats to what they called the states' "labor standards laws."[87] The Washington lobbyist for the National Council of Jewish Women shared her anxieties over NOW with her NAACP counterpart: "As you know, Mr. Celler wrote to the Commission last year and expressed his opinion that Congress never intended to invalidate the State laws. However, there are some misguided souls who insist that these laws are discriminatory and should be invalidated in one fell swoop."[88] The chief culprit was the "National Orga-nization of Women—NOW," whatever that means, [formed] for the sole purpose of harassing the Commission on this subject. There are also a few misguided souls within the government who are active in the Interdepartmen-tal Committee on the Status of Women where they manage to create a good deal of confusion."

Patterns of Women's Dissatisfaction

Yet the removal of the feminist commissioners from the EEOC, the policy rollback engineered by the ANPA, the apparent inattentiveness of the Presi-dent, and increased lobbying by the protectionist coalition failed to deflect the EEOC from pursuing the race-sex equation. Following the Holcomb de-bacle of the spring and summer of 1966, the agency moved steadily toward adopting policies that strengthened enforcement of women's employment rights.[89] This was partly because the logic of the race-sex analogy in anti-discrimination kept asserting itself from the EEOC's unique accumulation of evidence. This evidence took three main forms. One was the flood of newly required EEO-1 forms on which the covered employers were annually obliged to report employee statistics. Even in the days of the PCEEO and Standard Form 40, the government routinely had provided blank data columns for each sex, as in a census report—although the PCEEO had neither interest nor authority in gender discrimination. The EEOC of course possessed both. But its early interest was so preponderantly racial that the avalanche of EEO-1 forms, which were to produce such powerful data in documenting patterns of racial discrimination in 1967 and 1968, were simply not analyzed for gen-der patterns until the very end of the Johnson administration. In any event the massive EEO-1 data were not usefully digested for any sophisticated an-alytic purpose until Shulman's computerized operations began to take hold in 1967.

But the individual complaints filed with the EEOC, the second source of systematic evidence, immediately produced a surprising pattern. The EEOC's first annual report revealed that only 60 percent of the complaints were ra-cial, and 37 *percent* charged sex discrimination![90] Unlike the black complainants, where the dominant charge during the EEOC's first year was over hiring discrimination (23.7%), only 5.5 percent of the women complained about hiring.[91] Almost none complained that they were barred from the corporate executive suite or blocked from career success in such professions as law and medicine. Instead, their main complaint was over unequal employee benefits (30%), closely followed by discriminatory seniority lines (24%). Only 12 percent of the women complained about the unfair restrictions of state protective laws, and the preponderance of these objected to the common restrictions on their hours, especially the right to earn extra pay through voluntary overtime, like the men. While 37.3 percent of the charges filed by blacks complained about hiring and firing, only 5.6 percent of the women did so. So while the feminist case for a race-sex equivalence in EEO enforcement was reinforced by the surprising number of complaints against sex discrimination in general, the striking differences in the racial and gender patterns of complaint argued against a simple race-sex equation. This, however, was a distinction that attracted little attention in 1966–67.

The third source of evidence on gender discrimination was from special EEOC staff studies. These also tended to reinforce the race-sex analogy, although they identified important differences. In August 1966 the EEOC reported a staff study of 18,000 help-wanted ads in 21 newspapers. The study was done by staff under Commissioner Hernandez, and found few violations concerning race, national origin, and religion. But sex-segregated want ads remained common. Acting chairman Holcomb himself announced in August that "The Commission is launching a campaign against classified ad violations of Title VII."[92] As a result of the study the EEOC brought commissioner charges against 75 advertisers for running unlawful ads (although by law it could not identify those charged until after conciliation). The 21 newspapers that published the ads, however, were not so charged, consistent with the commission's reinterpretation of April. Some of the offending ads specified a preferred race—e.g., a "white attendant" or "Anglo carhops," and these were ruled illegal per se. But most of the offending ads indicated a gender preference—e.g., "Executive sales positions for men," "Lady in charge, northside shop," "Insurance trainee, man age 22 to 25"—and these the EEOC regarded as unredeemed by the requisite BFOQ.

The Belated Feminist Conversion of the EEOC

During the last two years of the Johnson administration, all three of these sources of hard new evidence combined with the pressure of the status-of-women groups, their supporters in Congress and in NOW, and also in the states to drive a recalcitrant EEOC toward implementing policies that equated race and sex discrimination.[93] The agency did so in fits and starts at first, but then with increasing momentum.

Indicative of this trend was the EEOC's yo-yo pattern on the controversial protective laws. On August 19, 1966, the EEOC confronted a challenge to California's rather typical limit for women of eight hours a day and 48 hours a week. In its original guidelines on sex discrimination issued in November 1965, the EEOC had affirmed its belief that Congress had not intended for Title VII to supersede state protective laws. Its language had invited challenges that it would decide on a case-by-case basis, with the burden of demonstrating a BFOQ presumably resting on the states, and with the authority to disallow them clearly resting, at least by implication, with the EEOC. When in August 1966 the commission was confronted with the "two competing values" of anti-discrimination and special protection for women, the commission "refrained from ruling squarely" on the California Labor Code. "The Commission cannot rewrite state laws according to its own views of the public interest," the EEOC observed in judicious self-restraint, adding that it had "no authority by such an interpretation to insulate employers against possible liability under state laws."[94] In backing off from the broader implications of its initial guidelines of 1965, as it had done in the spring of 1966 on the question of single-sex classified ads, the uncertain commission nevertheless advised the complaining women of their right to bring suit against California in federal court within 30 days under Section 706(e) of Title VII. The commission also urged state lawmakers to revise their laws, and state administrators to provide more flexibility in their interpretations.

With the EEOC waffling between the contending pressures, the contenders on both sides increased their pressures.[95] Feminist groups found that they could command the news by challenging baroque relics of state discriminatory laws, as a useful softening-up and attention-grabbing prelude to going after the seriously crippling laws. One price they seemed invariably to pay during this transitional period, however, was in the coin of cute condescension that was accorded their reform efforts by the media.

News reporters and editors, most of them male, found it difficult to resist coy leads that played on sexual stereotypes. *Wall Street Journal* reporter Wayne E. Green, for example, began his otherwise serious roundup article of May 1967 on the women's legal and legislative assault by exclaiming: "Shades of the suffragettes! The ladies are up in arms again." "They're demanding equal rights," Green wrote.[96] Then he reported that women were making headlines by attacking state laws that seemed bizarre, vulnerable, or unfair. These included a Texas law against women dancing in tents, a Nebraska law keeping women off juries unless the judge approved of their restrooms, laws in seven southern states making fathers the presumptive legal guardians of their minor children, a Washington state law preventing a married woman from filing suit in a state court unless her husband joined in the suit, and Louisiana's Napoleonic code granting men most property rights.

Many of these laws were crippling residues of the ancient code of *feme covert*. Some were long unenforced, but usefully bizarre—such as the Texas law of 1856 that recognized the legal right of a husband to kill his wife's paramour if he caught them in *flagrante dilecto* (the wife's equivalent reprisal was simply regarded as murder). But the feminist reformers in the states were also going after the protective laws of the Progressive Era and the liberal-labor legacy, and in December 1965 they had succeeded in wiping out in one swoop *all* such laws in Delaware. In pursuit of such objectives, the state federations of business and professional women's clubs was especially active in Texas, where communal property laws were rather brazenly male-dominated, and in Ohio, where historically strong union forces had helped shape a body of industrial law that barred women from nineteen job categories. In response to this snowballing threat from the business and professional women on the hustings, Mary Dublin Keyserling counterattacked from the Women's Bureau, summoning to renewed arms the tiring protectionist coalition.[97]

Caught in the middle, the EEOC gave feminists a modest but bellwether victory early in 1967 in a conciliation agreement over seniority involving the Dubuque Packing Company and the meat cutters local of the AFL-CIO.[98] The agreement opened up potentially hundreds of new job opportunities for women, but it was binding only on Dubuque Packing. In response to the stepped-up pressure, the EEOC then called a major hearing for May 2–3, 1967 on the three most controversial issues involving sex discrimination: state protective laws, classified advertising, and different pension benefits for men and women.[99] These were full hearings and gave the EEOC needed breathing room. But they were inconclusive, and they were not soon followed by a commission verdict, as would be expected of a court. Following these hearings, however, the EEOC slowly began to flow with the changing tide of opinion and argument on the major disputed issues of gender discrimination. By 1968 the agency had managed to reverse virtually all of its earlier cautionary retreats, and in 1969 it began to rule almost as aggressively on gender as it had from the beginning on race.[100]

This transition was paralleled by a transformation of the commission membership itself, from one dominated by white males to one dominated by members of the "affected class." By the end of the Johnson administration, Commissioner Holcomb was often reduced to a lone minority dissent, as in the flight attendant case, typically against a bloc consisting of two blacks, a Hispanic, and a woman. On the issue of protective legislation, for example, the EEOC had invited case-by-case review in November 1965. It then retreated in August 1966, when acting chairman Holcomb presided over a Thermidorian reaction, in favor of court review.[101] But the EEOC switched back again in February 1968, when it used its new effects standard to strike down certain state protective laws. By August 1969 the commission announced that such protective laws by their very nature inherently "conflict with Title VII of the Civil Rights Act of 1964 and will not be considered a defense to an otherwise established unlawful employment practice or as a basis for the application of the bona fide occupational qualification exception."[102]

On single-sex classified ads, the commission repeated the cycle. On August 6, 1968, it reversed its precedents of November 1965 and April 1966, and ruled in a 3–2 decision that separate want-ad columns by gender were unlawful per se.[103] In the protracted dispute over whether airline stewardesses constituted a BFOQ, the agency was initially paralyzed both by its internal divisions after the departure of Roosevelt, Graham, and Hernandez, and by successful legal challenges by the Air Transport Association.[104] But new hearings were held in September 1967, and on February 13, 1968—nearly 27 months after the first complaints were filed—the EEOC denied the BFOQ request on flight attendant jobs.[105] In that 3–1 decision, Holcomb was outvoted by two blacks and a Hispanic. The commission later struck down airline rules barring married stewardesses and limiting their age. The EEOC was somewhat more cautious on the controversial question of pension rights. Following the omnibus gender hearings of May 1967, the commission ruled unanimously in February 1968 that employers may not set different retirement ages for men and women.[106] But the commission met stiff congressional resistance to interfering with private pension arrangements, and so avoided ruling on different levels of survivors' benefits for the sexes.

Overall, however, by the end of the Johnson administration the EEOC seemed settled on a path that would tightly link race and sex in EEO enforcement. The commission would seek to build a powerful coalition of constituent groups, adding women to blacks, Asians, Hispanics, and American Indians as members of officially protected classes—a "minority" coalition that in the 1970s approached three-fourths of the United States population.[107] But the EEOC would achieve this in the face of stiffening congressional resistance. The explanation for such a contradictory development lay in an unusual combination of circumstances. This included certain unique historical elements—black ghetto riots, the concommitant surge of black nationalism, rising antiwar protest and campus turmoil, and in 1968 the assassinations of Martin Luther King and Robert Kennedy. It also included certain bedrock continuities—i.e., the powerful tendency of clientele groups in the modern administrative state to capture the bureaucratic apparatus of the agencies that were designed to regulate their access to benefits and entitlements and their participation in the American economy.

CHAPTER IX

From Equal Treatment to Equal Results:
Transforming Civil Rights Strategy

The Flow and Ebb of the Great Society

The conservative backlash that fueled the Republican comeback in the congressional elections of 1966 owed much to the Watts riot of 1965 and the Black Power movement of the summer of 1966. In the resentful eye of the white middle class, "We Shall Overcome" had been replaced by "Burn, Baby, Burn." But the uniqueness of these events should not obscure the normality of such losses by an incumbent regime in the off-year elections. The Democrats' off-year battering in 1966 was made more severe by the abnormality of Johnson's electoral triumph in 1964. Newly elected in his own right, Johnson knew that he must ram most of his Great Society proposals through the 89th Congress in 1965–66 while his landslide majority held. During Johnson's post-assassination honeymoon of 1964, he had realized the Kennedy legacy of the tax cut, the Civil Rights Act, and the Economic Opportunity Act. Beginning with a rush in 1965 he would add his own stunning array. This included Medicare, aid to elementary and secondary and higher education, the Voting Rights Act, a new Department of Housing and Urban Development, immigration reform, national endowments for the arts and the humanities, environmental initiatives against air and water pollution, and a host of supporting entering-wedge social programs.[1]

This momentum carried through the second session of the 89th Congress in 1966 and produced Model Cities, the new Department of Transportation, Teacher Corps, and expanded safety and consumer legislation.[2] But the weakening of support in the 90th Congress in 1967–68 was substantial, when Congress produced a modest record of federal innovations, including public broadcasting, college work-study, and an almost unnoticed ban on age discrimination. In civil rights, however, 1968 produced an unanticipated open housing act, which was curiously coupled in mood and congressional tactics with get-tough laws on crime and riot control.

The Great Society's flood of new laws and social programs was unprecedented since the New Deal, and underlying both bursts of liberal Democratic reform were similar patterns of flow and ebb in presidential mandate, initiative, and power. Within this pattern were two basic variations in the way major social legislation was accepted after enactment by political constituencies and the broader public. One variation is the dominant success model, like the familiar New Deal initiatives in labor law and social security. For Johnson's Great Society this successful model embraced Medicare, federal aid to education, and to a lesser extent environmental and consumer protection. In such breakthrough areas of social innovation, new coalitions would typically overwhelm the opposition in their flood tide of legislative victory during the 89th Congress. They would then consolidate their gains through the 90th Congress, and extend them into the Nixon administration and beyond—much as the New Deal's major social initiatives survived Eisenhower's Republican administration and prospered with enhanced legitimacy. The second model involves major initiatives that withered. Following initial success they quickly fade, like a meteor—as in Franklin Roosevelt's alphabet-soup programs for recovery and national planning, or the institutional core of Johnson's war on poverty.[3]

Lyndon Johnson's crusade in civil rights shared with the success model the creation of a newly empowered coalition that subsequently commanded respect and fear from politicians. In the process it shattered the invincibility of the southern filibuster and the conservative hammerlock on key congressional committees. But like the war on poverty, the great civil rights reform harbored a vulnerability in the mercurial image of its chief clientele. During 1965–66 the Negro image in America was sharply transformed, although probably in opposite directions for whites and blacks. In the spring of 1965 the dominant symbol was a petitioning black voter being brutalized by Sheriff Jim Clark in Selma. By 1966 this had been countered, if not displaced in the volatile world of *Time* and *Newsweek*, by the rampaging ghetto rioter in Watts, or the black racist harangues of an H. Rap Brown.[4]

The EEOC in Resentment and Disarray

Given this political stand-off in the 90th Congress, the EEOC entered 1967 in a state of disrepair. The wobbly new agency began the year with its third chairman in as many years (including the five-month acting chairmanship of Commissioner Holcomb during the summer of 1966). It also was embarrassed by *two* empty commissioner chairs, one consequence of which was an all-male commission weakly enforcing the nation's new ban on sex discrimination. Moreover, Congress had rejected the Administration's legislative proposal of 1966 to grant the EEOC cease-and-desist authority. Then Johnson, in his State of the Union address of January 1967, surprised most of Washington by proposing a merger between the departments of Labor and Commerce to provide the core of a new superdepartment of Business and Labor.[5] The proposed consolidation would include, and deeply submerge, the already demoralized EEOC. The new commission understandably regarded the novel (and as it turned out, congressionally doomed) proposal as a threat to its independence and shaky prestige.[6]

By early 1967, EEOC chairman Stephen Shulman had remolded the senior staff by bringing in a cadre of young experts in managerial effeciency and data processing, which the press quickly identified as a manifestation of the McNamara "whiz kid" phenomenon. Part of the new senior staff's task was to convert the complaint processing system from its archaic base in manual files to computers, and also to analyze the data from the new national reporting system. But the agency's complaint backlog seemed almost hopeless, even with computers. There were several reasons for this, although the most telling reasons pertained more to the realms of spirit and morale and ultimately to ideology than to the sheer crush of paper and logistical demands. The latter burden was real enough, and its arithmetic for efficiency experts was depressing. By the 1st of January 1967, for example, the EEOC had received almost 15,000 complaints. Of these 6,040 had been earmarked for investigation, and the tiny agency's overwhelmed investigators had completed inquiries on only 3,319. These then faced a major middle bottleneck at the commission level, where determinations of probable cause had to be made by the commissioners themselves. Like their veteran counterparts in the state and local FEPCs, the new EEOC commissioners took these quasi-judicial duties seriously (conciliation agreements worked out by the staff, however, were almost routinely approved by the commissioners).[7] The third and most severe bottleneck was in conciliation. There only 110 cases (involving 330 complaints) had been concluded by January 1967. This modest achievement was claimed by a national staff of only *five* conciliators, including chief of conciliation Alfred Blumrosen.[8] No amount of Shulman's whiz-kid staff and computerized efficiency could speed the flow past such a bottleneck.

Why only five conciliators for the whole country? The Budget Bureau had wondered as much, when it called for internal reassignments of EEOC personnel to address the growing backlog problem in the fall of 1966. The answer in part lay in limited resources. The commission's $5.2 million budget in 1967 was smaller than that of the Office of Coal Research, and its staff of 314 was smaller than that of the Federal Crop Insurance Program. But the answer also lay in the resentment felt at the EEOC, and shared by the more aggressive elements of the civil rights community, over the symbolic castration of their national FEPC. The EEOC's complaint-response model symbolized to the EEO community the trivialization of the nation's most pressing domestic and moral concern. To them it rang of the voluntarism of a largely neutered agency. They were stung by the impracticality as well as the moral inappropriateness of the studied, case-by-case, retail approach to an outrageously obvious and massive continuation of wholesale job discrimination against minorities.[9]

Lacking the threat either of cease-and-desist authority or even of damaging publicity in complaint processing, the EEOC had little leverage in such voluntary conciliations. Its meager results showed not only how ineffectual the agency was in producing retail conciliations on the bottom line, but also how little heart they had for the whole feckless, nickel-and-dime, due-process-ridden, complaint-processing enterprise. The early EEOC thus functioned as quiet co-conspirators with the agency's critics on the left, deploring its manifest weaknesses as a "poor enfeebled thing." Its staff in effect invited inundation by men like the NAACP's Herbert Hill, who would drown the small agency in a paper sea of complaints to dramatize the need for reforms to put teeth into the law. "We are out to kill an elephant," Chairman Shulman acknowledged, "with a fly gun." [10] Shulman had pledged to Johnson, as he and his successor were to tell Congress, that the complaint process "must remain the Commission's primary concern." [11] But to the degree that this was taken to mean merely processing individual complaints, such a prosaic commitment did not speak for the agency's soul.

Vicarious Lawsuits

The EEOC's record of successfully conciliating only 110 cases in its first twenty-one months, then, was pitiful. While the agency struggled under its tiny budget to get organized and mount a serious attack—meanwhile merely minding the store in complaint processing and conciliation—it needed outside help to endow its threats with credibility. Unable to sue, the EEOC could recommend that the Justice Department do so. It could also encourage suits by private organizations like the Legal Defense Fund. Between July 1965 and January 1967, when conciliation had failed in cases the EEOC judged important to pursue, it had referred fifteen cases to the Justice Department and recommended that the Attorney General bring suit. In response, however, Justice did not file a pattern-or-practice (Section 707) suit on the recommendation of the EEOC until mid-December 1966, when it filed against a segregated union in New Orleans. The first Section 707 suit by Justice against an employer was not filed until March 1967, against the Dillon Supply Company of Raleigh, North Carolina. [12] What explained such caution?

The Attorney General's slowness to file 707 suits was criticized by the Legal Defense Fund, which had already filed 35 private suits (under Section 706 of Title VII, which provided for private litigation) where EEOC attempts at conciliation had failed. LDF spokesmen speculated that Justice had selected such a small and obscure supplier as Dillon because a suit against a major government contractor would embarrass the Labor Department's OFCC. If an allegedly discriminatory employer were large and important enough, they reasoned, like U.S. Steel or General Motors, the threat of litigation would be quashed by the White House. [13]

The EEOC, however, had never recommended that the Attorney General sue such a major government supplier. Moreover, the Justice Department had its own honest problems. The Civil Rights Division was one of seven major divisions in the department, and in 1967 it employed only 87 lawyers; the division's fiscal 1967 budget was only \$2.5 million, which was less than 1 percent of the total budget for Justice. Yet the small division had to cover seven broadly defined and increasingly active areas of civil rights jurisdiction and litigation. Of these, the President had assigned top priority to voting rights, school desegregation, and criminal law enforcement cases involving violence, local defiance, and intimidation. These were the hot areas of pressing public and political concern, where blowups were frequent and dangerous and Congress was quick to blame the President. Lyndon Johnson had insisted that the Attorney General stay on top of them. By Katzenbach's own admission, during 1964–66 less urgency had attended the slow-moving equal employment cases, which had been accorded lowest priority more by default than by design.[14]

The Perennial Search for a New EEOC Chairman

On top of all these troubles, on May 4 Shulman wrote President Johnson that he would be leaving the commission at the expiration of Franklin Roosevelt's original term on July 1.[15] As John Macy once again geared up to renew the perennial search for an EEOC chairman, he reminded Johnson that the last time around they had considered "21 different persons before you prevailed upon Steve Shulman to serve."[16] Schulman's notice had prompted Macy to review that list, but only one potential candidate had emerged with the requisite qualifications: the EEOC's first staff director, Tom Powers. Macy reminded the President that a year earlier Johnson had been prepared to turn to Powers had Shulman declined. But Powers was now in private law practice and he was "highly pleased," Macy reported, with the "remuneration therefrom." Shulman himself had suggested another competent, young, male, white, mid-level executive, an assistant postmaster general named Richard J. Murphy. Macy suggested yet another: promising young John Douglas in the Justice Department—whose father, incidentally, was a prominent liberal representing Illinois in the Senate.

But this time the search for chairman would reveal that the notion of an elite white male Democrat at the helm of the EEOC was an idea whose time had passed. Roy Wilkins and Whitney Young were upset over the revolving-door succession of short-term chairmen, and they told Macy that conditions demanded a chairman who would promise to remain for a full five-year term. Moreover, in the interest of continuity, the new chairman ought to be someone already on the commission, preferably a Negro. That could only mean Sam Jackson, Macy observed—and Macy was surprised. "Jackson is too closely involved in the politics of the NAACP," Macy wrote the President, "and has not demonstrated the necessary qualities of leadership."[17] Besides, Jackson was a Republican who bore no ties of loyalty to the Johnson administration, and he represented the aggressive wing of civil rights reformers.

But if not Jackson, then who? Macy early discovered that the troubled

EEOC chairmanship was as unattractive to prominent black candidates as it was to white.[18] Federal district judge A. Leon Higginbotham, Jr., was unwilling to leave his tenured bench in Philadelphia. Patricia Roberts Harris preferred the independence of the Howard law faculty. Marjorie Lawson would not give up her private law practice. Other black names surfaced on Macy's running lists,[19] and other white possibilities as well.[20] Early in his explorations, Macy asked White House aide Clifford L. Alexander, Jr. for his suggestions. Alexander was a young black lawyer whose rapid rise from the fringe of the White House staff was reflected in the escalated rank of his typically vague White House titles—from Deputy Special Assistant in 1964 (when he was only 31), to Associate Special Counsel, then to Deputy Special Counsel. Macy told Johnson that, to his considerable surprise, Alexander was "very interested in this appointment himself"—indeed he was "an eager candidate."[21]

Alexander's credentials were indeed gold-plated. But initially that was a big part of the problem: Johnson did not want to lose him from the White House. Unlike Eisenhower's E. Frederic Morrow or Nixon's Robert Brown, Alexander was uniformly regarded as exceptionally bright, energetic, and capable, rather than as merely the requisite marginal black face on the White House staff. He was, moreover, a strikingly attractive man, whose confidence in his ample talents left him unusually free of the uncertainties that hampered black front men like Morrow and Brown, whose status was derived more from symbolic and representational than personal qualities.[22]

A native of New York City, Alexander had graduated *cum laude* from Harvard in 1955, and earned his law degree from Yale in 1958. Then after a tour in the army he worked as an assistant district attorney for New York County and did youth development work in Harlem. In 1963 he left private practice in New York to join the National Security Council staff of McGeorge Bundy, who had known Alexander when he was student council president at Harvard and Bundy was arts and sciences dean. Alexander was spotted and recruited to the White House staff to work with Ralph Dungan and Lee White, mostly in a head-hunting capacity for potential black appointees.[23]

Johnson initially refused to consider losing his lone black aide to the EEOC, and Macy agreed.[24] But as the witching hour of July 1 approached, and with it the embarrassing prospect of yet another lacuna under acting chairman Luther Holcomb, the attractions of Alexander's appointment began to grow on Macy. He was soon joined by Harry McPherson in arguing to Johnson (correctly, as it turned out) that Alexander's appointment would bring the President praise, not criticism for stripping the White House of its ranking black aide.[25] So on June 27 Johnson announced Alexander's nomination for Senate confirmation, and following a brief summer interim under acting chairman Holcomb, Alexander was sworn in as the EEOC's first black chairman on August 4, 1967.

From Retail to Wholesale Enforcement

To the EEOC's well-advertised troubles with the selection of chairmen and commissioners, and its ill-defined role in the jumbled network for federal enforcement and regulation, were added increasingly strained budgets and adversarial relations with a more conservative Congress. As the complaint backlog grew with the expanded coverage, the commission's senior staff concluded that complaint processing was a losing game.[26] They sought wholesale leverage, not retail enforcement. Behind the musical chairs of commissioners and chairmen, they planned investigations into entire industries. While Congress complained about increasing delays in responding to individual complaints, the EEOC staff was creating a nationwide system of mandatory employment reporting that would far better serve their more ambitious purposes.

The new commission's resentment at the elaborate limits of its congressional mandate set it early at odds with Congress. After the elections of 1966, this distance grew, especially in the more conservative House. Despite Shulman's computerized system, the commission fell further behind in its backlog of complaint processing. By the summer of 1967 the backlog had stretched processing delays to fifteen months, and a year later it had reached two years.[27] This made a public mockery of the EEOC's statutory duty to process all complaints within sixty days, however unrealistic such a tight legislative stipulation may have been in the first place. Congressional conservatives wanted the EEOC to remain primarily or even exclusively an educational and promotional agency. The civil rights lobby, on the other hand, preferred the model of an activist investigatory agency with broad police powers—a kind of Civil Rights Commission with ombudsman authority. The early EEOC chairmen, however, took a middle position, agreeing that the EEOC was primarily a complaint-processing regulatory agency, like the NLRB, but asking Congress for cease-and-desist authority and larger budgets as the necessary tools.[28] Congress resisted.

In 1966 the House cut $670,000 from the Administration's already lean EEOC budget proposal for FY 1967. The following year's $6.5 million budget represented little change, especially for a new agency that for its first three years had an automatic escalator built into its workload. On July 1, 1968, Title VII would reach its full coverage of all employers and unions with at least 25 workers, and this would add 195,000 companies (a 250% increase in the number of firms covered) and almost six million workers to the EEOC's official purview. So despite the growing pressures of inflation and Vietnam, the Johnson administration asked for a doubled EEOC budget, to $13.1 million, for FY 1969. The House promptly slashed this in half, to $6.9 million. Only eleventh-hour intercession by senators Mansfield, Javits, and Pastore in conference committee got it increased to $8.7 million.

Given this congressional standoff on the one hand, which denied the EEOC both a substantial budget increase and cease-and-desist authority, and on the other the EEOC's preference for wholesale modes of enforcement through identifying and attacking patterns of discrimination rather than merely reacting to individual complaints, the agency turned increasingly to the new national reporting system as its most promising tool for investigations and enforcement. The commission used this instrument with ingenuity and, in light of the danger of congressional backlash, also with considerable risk. Title VII's stairstep increase in coverage would bring in all but the smallest commercial employers by July 1968, and all such employers would be required to submit EEO-1 forms annually. The Chamber of Commerce would not welcome the new burden of regulatory paperwork and surveillance on its small business constituency, and Congress would hear from the Chamber.

Many important segments of private employment, however, were not labor-intensive enough to fall automatically under EEOC jurisdiction. For example, only 2,000 of the nation's 14,000 banks employed more than fifty workers, which was Title VII's cut-off point in 1967. So in February 1967 the EEOC repaired to the logic of the old PCEEO and required that all banks that accepted federal deposits must also submit EEO-1 reports, irrespective of number of employees.[29] This was Title VI and executive order authority, and exceeded the authority of Title VII. But the EEOC was flooding the country's employers with its new reporting forms, and getting away with it. Then in March of 1967, 7,500 apprenticeship committees, which were jointly run by construction companies and unions and were funded by the Labor Department, received directives from the EEOC to submit EEO-2 statistical reports. In the summer of 1967, the commission added 52,000 union locals, representing 16 million workers, to the reporting pool, thus massively inaugurating form EEO-3.[30] By the spring of 1968 the EEOC had added to its huge data pool all private employment agencies, and thereby rounded out its bureaucratic armamentarium with form EEO-4. Thus the EEOC's net was both increasingly wide and fine. What would it produce, and how might it best be used?

The key to the success of the commission's national reporting scheme lay in its mandatory provision of precise employment data at the *middle* level. For the first time this would provide data for comparisons between geographic areas and within specific industries. The individual complaints, to be sure, already provided *micro*-level data. But their generation was arbitrary, unsystematic, small in number, and episodic. It therefore allowed only crude generalizations. On the other hand, the Bureau of Labor Statistics for years had regularly provided macro-level data on employment trends, which it derived from national surveys of a sample based on 52,500 households. The longitudinal BLS data could reveal, for example, that the number of employed blacks had increased from 6.4 million or 10.2 percent of the U.S. labor force in 1955, to 7.7 million or 10.7 percent in 1965. Or the BLS data could show that in 1966 the proportion of whites in managerial jobs was 15 percent and in technical/professional jobs it was 12 percent, whereas for blacks it was only 5 percent for both categories.[31] But such surveys could not provide that revealing middle level of aggregation, such as the distribution by race and ethnicity and sex of workers in the electrical and aerospace industry in Los Angeles, or for railroads in Cleveland, airlines in Atlanta, or retail trade and medical services in New York. The EEOC reports could. And beginning in 1967, they did—with telling effect.

Hearings and Conciliation Agreements

In 1966, before this mass of unique data had been collected and analyzed, the EEOC planned its first set of "hearings." Lacking formal subpoena power for hearings, and committed by law to confidentiality in processing complaints, the commission euphemistically called its first such public foray a "forum."[32] It was originally scheduled for the fall of 1966 in Charlotte, N.C. The commission chose the Carolina piedmont because North Carolina generated more complaints than any other state; the area's dominant textile industry had historically been a hotbed of labor animosity, and the EEOC wanted to expose the industry's discriminatory labor practices.[33] Not surprisingly, the Carolina textile executives were hostile and uncooperative, and the forum was postponed until January 1967.[34] The American Textile Manufacturing Institute boycotted the forum, as did eight of ten companies invited to testify, and the forum's exchanges were tense and characterized by anecdotal charges. EEOC officials privately conceded that while the textile industry made a vulnerble target, it was a low-wage, contracting, technically backward industry that scarcely typified the country's present or future employment problems or opportunities.[35]

When the preliminary results from analysis of the EEO-1 returns began to come in mid-1967, the precision of these findings permitted the EEOC to zero in on crucial target areas. The textile forum in January 1967 represented a crude and not very productive stab at exposing southern backwardness. But the EEOC's next such initiative was a halfway step toward its goal of a model full hearing based on hard and comparative statistics on job distribution. This was a closed Washington hearing in October 1967 on the pharmaceutical industry. A consultant study had identified consistently low minority employment in the drug industry, and early EEO-1 returns confirmed the pattern.[36]

By the fall of 1967 the EEOC's research staff had analyzed the data contained in the first batch of two-page EEO-1 forms and had broken it down for nine major metropolitan areas. The data were drawn from the 1966 reports of 43,000 employers covering 26 million workers, and the EEOC's selected cities were Atlanta, Chicago, Cleveland, Kansas City, Los Angeles, New Orleans, New York, San Francisco, and Washington, D.C. These data revealed with unprecedented precision that blacks were disproportionately represented in blue-collar jobs throughout the nation. This was especially true in the South. In Atlanta, for example, blacks were 23 percent of the population but held 33.8 percent of the blue collar jobs. In New Orleans, blacks were 30.7 percent of the population and 44.8 percent of the blue-collar work force.[37] In Chicago, blacks were 14 percent of the population but held 20.6 percent of the blue-collar jobs.

The most striking finding of the EEOC report was its corollary documentation of black underrepresentation in white-collar jobs, where the economy's future job growth lay in technical and service enterprises. A double penalty against minorities was involved. First, blacks were underrepresented according to their area populations in white-collar employment generally. For example, blacks in Kansas City were 11.2 percent of the population and only 2.1 percent of the white-collar work force; in Cleveland the corresponding figures were 13 percent and 3.2 percent. Second, within those white-collar categories the great majority of blacks worked either in low-paying clerical and sales positions or as hospital orderlies in white-collar growth businesses such as health services. In Chicago, for example, where blacks held only 4.7 percent of white-collar jobs, 79.6 percent of that tiny group of black white-collar workers fell into this low-status category. In certain industries, blacks held a strikingly low share of white-collar jobs: 2.8 percent in the aerospace industry in Los Angeles, 1 percent in air travel in Atlanta, 0.7 percent in Cleveland railroads.[38]

Armed with this new information, the EEOC planned a major hearing on white-collar employment and scheduled it for New York City in January 1968.[39] Exclusive of Washington, where blacks were disproportionately represented in federal employment, New York had the highest percentage of white-collar blacks. But that share was still only 5.7 percent in a metropolitan area that was 11.5 percent black. Both New York state and City had long boasted of grand-daddy FEPCs. Yet of the 4,249 New York firms reporting employment statistics to the EEOC in 1966, 1,827 reported no black employees at all, and 1,936 reported no Hispanics.[40] Only 1,207, or 3.5 percent, of the 35,819 employees in New York's securities and brokerage business were black, and blacks held only 2.1 percent of white-collar jobs in the city's rapidly expanding air travel business.

The New York hearings of 1968 were successful because, unlike the Carolina textile forum of 1967, their inquiries were based on the self-reports of the employers themselves; they focused on a broad sector of enlightened growth industries rather than on a sick and regionally defensive industry with a siege mentality; and they benefited from Clifford Alexander's crisp image as a firm but fair new chairman. Together with subsequent West Coast hearings in the spring of 1969 in San Francisco and Los Angeles, the hearings demonstrated what Alfred Blumrosen had early observed—that the EEOC's national reporting system would be more crucial in the North than in the South. This was because the main problem in the North and West was black entry into skilled trades and especially into the white-collar growth sectors.[41] But in the South, where black labor had always been exploited to do low-skilled, low-wage, blue-collar work, the main problems were promotion and transfer out of such dead-end jobs. Because such a segregated system required tightly sealed job compartments that were both racially obvious and vulnerable to assault under Title VII, the bulk of the EEOC's early and landmark conciliation agreements involved southern employers. These were the easy pickings. The

EEOC thus benefited from forging its most effective weapons in the South's Jim Crow environment, where outrage against the palpable unfairness of traditional racist practices deflected attention from the potential consequences of later employing these same weapons outside the South.

The federal government's Newport News settlement of March 1966 illustrated the deceptive ease with which the larger southern transgressors might yield. Builder of nuclear submarines and aircraft carriers, the company ran one of the world's largest shipyards, and it was easily the largest commercial employer in Virginia. In 1965, some 5,000 of its 22,000 workers were black, and 41 of these black workers, supported by the NAACP, had filed charges of race discrimination with the EEOC. In response the company was attacked by an extraordinary coalition of government agencies. Because 75 percent of the shipyard's business came from naval contracts, the EEOC was joined in its formal charges by the departments of Justice (Civil Rights Division), Labor (OFCC), and Defense (Navy). Like Lockheed-Marietta during the early 1960s, Newport News during the middle 1960s seemed to welcome such external pressures to change. Indeed, the shipyard had never maintained typical southern patterns of rigid job segregation. Its ties to the U.S. military establishment were far too crucial economically to risk defending the region's Jim Crow traditions, and clearly it made for poor public relations to launch the nuclear submarine *George Washington Carver* in the middle of a televised demonstration against racial injustice. So Newport News agreed to promote 3,890 of its black workers, and also to designate 100 blacks to become supervisors.[42]

This concerted effort by the otherwise poorly co-ordinated federal establishment against such a vulnerable target, however, was distinguished during the middle 1960s by its singularity. The EEOC's other conciliation agreements were relatively few and far more modest—for example, integrating segregated union locals in a small Alabama foundry, or producing the first black truck drivers for a paper board company in Richmond.[43] The EEOC still had no power to coerce agreement, so even its national reporting data could only yield identifiable gaps for publicity and pressure, at least in the short run.

In the long run, however, the EEOC's hopes for forcing significant change in the nation's discriminatory patterns of employment seemed to lie less in aping the state FEPCs, with their complaint-response model of retail enforcement, and even potentially with their cease-and-desist authority, than in finding a way to exploit the new national commission's unknown potential for wholesale enforcement. This meant, first, that increased regulatory muscle would emanate most immediately from the binding, rule-making authority that regulatory agencies traditionally enjoyed in setting definitions of permissible conduct (although in the case of gender discrimination, the initially timid EEOC had been confronted with rule-making demands in novel territory rather earlier than it would have liked). Second, and more gradually, the commission's leverage for change would heavily depend on the ease with which it could identify and prove broad patterns of discrimination—which for the early EEOC meant *racial* discrimination—and thereby encourage the courts to require widespread remedies.

Establishing discriminatory patterns would hinge in turn on the standards of proof of discrimination that the EEOC could successfully urge on the federal courts in defining the meaning and reach of Title VII. Convincing the courts to uphold the commission's various new rules, regulations, and guidelines would be eased by the federal courts' modern tradition of according "great deference" to the expert judgments of the regulatory agencies. More difficult would be the EEOC's task of exploiting its new employment data to win court approval of an aggressive wholesale attack on employment discrimination. This was partly because the agency had no direct access to the courts as litigant. But adequate vicarious access could be gained through amicus representation in private suits, in coordination with the Legal Defense Fund and the entire panoply of civil rights organizations that constituted the EEOC's aggressive constituency.

The commission's major difficulty lay in the historic legal definitions of invidious discrimination, which centered on the required proof of harmful motive. To meet this challenge, the EEOC's comprehensive new data bank was documenting employment patterns that were so massively skewed, especially by race, that they intuitively demanded an inference of systematic job discrimination, and hence invited a judicial remedy. How could the EEOC legally use these incriminating data to destroy the nation's discriminatory employment patterns?

Intent, Statistics, and the Burden of Proof

In the spring of 1966, the EEOC's general counsel, former Howard law professor Charles T. Duncan, requested a staff analysis of the status of EEO law and the relative promise of the options available. On May 31, Duncan received a heavily documented, 21-page memorandum from staff lawyer Sonia Pressman.[44] A Jewish refugee of Hitler's Germany, Pressman had graduated Phi Beta Kappa from Cornell in 1950, then graduated first in her law class from the University of Miami, and came to the EEOC with a background in labor law from the NLRB.[45] Pressman knew from her NLRB experience that discriminatory intent was extremely difficult to prove. As the New York Court of Appeals had observed in a frequently cited case in 1954, "One intent on violating the Law Against Discrimination cannot be expected to declare or announce his purpose. Far more likely is it that he will pursue his discriminatory practices in ways that are devious, by methods subtle and elusive . . . ."[46] This difficulty was widely documented in postwar labor law, where arbitrators had stressed the need for affirmative proof of motive in discrimination cases. In 1948, for example, a NLRB umpire found it "indeed a surprising fact" that the Consolidated Steel Corporation had hired many whites but no Negroes during a six-month period, and had rejected all thirty Negroes during a three-day period when whites were hired.[47] But the arbitrator rejected the charge of discrimination as unproven because the nature of the steel industry and the qualifications of the Negro applicants made the assumption of discrimination only a *possible* and not a *necessary* result. As another NLRB arbitrator observed in a widely cited case in 1955, when he rejected charges of discrimination against a meat-packing company for passing over five black applicants while hiring whites, "discrimination cases are invariably difficult cases because they usually involve questions of motive, intent, and credibility that are not susceptible to objective or quantitative standards of proof."[48]

Despite the barrier posed by the common law tradition that evil intent be proven, however, Pressman's memo to Duncan identified several promising precedents to guide future lines of attack. All of them hinged on the potential but problematical use of employment statistics. The goal would be to document such large disparities in employment patterns that discriminatory intent might legally be inferred. The classic line of statistical attack had been in jury discrimination, and was anchored in a tension between two vintage decisions of 1880. These were *Strauder v. West Virginia,* which had affirmed the 14th Amendment's ban on state juries from which Negroes were explicitly excluded, and *Ex parte Virginia,* which nevertheless approved all-white juries if the racial exclusion was merely de facto.[49] But in 1935 in *Norris v. Alabama,* the Supreme Court overturned the conviction of the Scottsboro Boys by an all-white jury because black jurors had systematically been excluded, and in 1954 the Court affirmed that such de facto patterns "constitute prima facie proof of the systematic exclusion of Negroes from jury service."[50] Although judges had been historically reluctant to render substantive judgments in areas remote from their expertise, as lawyers they had been quicker to infer discrimination from statistical patterns in familiar areas like jury selection and legal education. Pressman cited more recent cases in which the courts had relied on similar statistical evidence to infer racial discrimination in voter registration and teacher pay. She even found occasional supporting rulings in labor law, where the early NLRB had found too suspicious in unorganized plants the strikingly high correlations between laid-off workers and union members.[51]

But the EEOC in 1966 faced several peculiar obstacles that had not been encountered in the courts' gradually broadening use of statistical evidence in areas like jury selection and voter registration. In such cases, racial discrimination where proven had always violated the 14th Amendment. Plaintiffs could document damaging patterns of discrimination by joining statistical evidence of current disparities to past illegal practices. But racial discrimination in private employment was not a violation of federal law prior to 2 July 1965. In 1964 the Motorola case had fueled business fears that statistical evidence might support a prima facie case requiring the preferential hiring of minorities. To allay these fears the Senate had added in Section 703(j) a prohibition against requiring "preferential treatment . . . on account of an imbalance" of the "total number or percentages of persons of any race, color, religion, sex, or national origin . . . ." As evidence of congressional intent, Section 703(j) seemed unambiguous. Typical of this understanding was an interpretation published in the *University of Chicago Law Review* in 1965, explaining that any "explicit attempt to prefer a racial or religious group, or to hire in accordance with quotas, would be contrary to congressional intent and would undoubtedly be held unlawful by the courts."[52] Thus according to the received law and judicial doctrine, historic patterns of employment were not relevant, and even subsequent workforce patterns would not necessarily indicate discrimination practiced since 2 July 1965. Pressman conceded that "Under the literal language of Title VII, the only actions required by a covered employer are to post notices, and not to discriminate subsequent to July 2, 1965. By the explicit terms of Section 703(j), an employer is not required to redress an imbalance in his work force which is the result of past discrimination."[53]

The language of Title VII seemed to leave little room for broad statistical interpretations. But the EEOC's statistical dilemma had been worsened just months earlier when the Supreme Court delivered a major new ruling on racial discrimination in jury selection. In *Swain v. Alabama,* the Court in a 6–3 decision affirmed the conviction of a black man for raping a white woman, even though the defense showed that no Negro within recent memory had served on any petit jury in any civil or criminal case tried in Talladega County, Alabama.[54] In ruling to protect the historic unchallengeability of trial counsel's peremptory challenge, the majority in *Swain* struck a major blow against the agrument for statistical proportionality. Speaking for the majority, Justice Byron White observed that "a defendant in a criminal case is not constitutionally entitled to demand a proportionate number of his race on the jury which tries him nor on the venire or jury roll from which petit jurors are drawn."[55] "Neither the jury roll nor the venire," White ruled, "need be a perfect mirror of the community or accurately reflect the proportionate strength of every identifiable group."

Sonia Pressman concluded from the relevant law and precedent that the language of Title VII explicitly protected employers from any requirement to redress an imbalance in their workforce, even if it was the result of past discrimination. Furthermore, the language in *Swain* reaffirmed "the fact that a minority group does not have the right to proportional representation." It indicated to Pressman that "an employer cannot be found in violation of Title VII simply because his use of minority groups does not mirror their representation in the community."[56] Nevertheless, Pressman argued that the EEOC should attempt to exploit circumstances where the employment statistics show gross disparities, and then apply the doctrine of the prima facie case. She observed that the Supreme Court had affirmed in 1957 that "the ordinary rule, based on considerations of fairness, does not place the burden upon a litigant of establishing facts peculiarly within the knowledge of his adversary."[57] Thus the plaintiff, Pressman said, need only supply "a certain quantum of evidence, after which the burden of going forward, and of offering rebuttal to that which has been presented, shifts to the respondent."[58]

Surely the common patterns of southern industrial employment would present abundant opportunities to identify the measure of this "quantum" of evidence that was necessary to make a prima facie case, and thereby throw the burden of proof upon the employers.

Pressman conceded the difficulty of trying to stretch the logic of statistical disproportion from conscripted juries, where standards were relatively simple and juries were short-lived, to the complexities of the employment market, where such factors as job requirements, turnover, and the competence of applicants would require other evidence beyond employment statistics. She conceded as well that such litigation would "always be open to the danger that in a particular case a personnel decision was made not for unlawful reasons," and that the free choice of employers legitimately provided for some necessary range of "permissible discrimination." Nevertheless, Pressman urged the commission to pursue in the courts an affirmative theory of nondiscrimination, one that the employer cannot satisfy "merely by silently and passively ceasing his past discrimination and posting EEOC notices." Under such a definition "the active pursuit of an equal opportunity policy," Pressman said, would mean that "Negroes are recruited, hired, transferred, and promoted in line with their ability and numbers."[59]

And numbers. Sonia Pressman and her colleagues in the federal government's struggling new civil rights bureaucracy felt boxed in by the very terms of their founding charter. Title VII not only leaned over backward to disallow most potentially compensatory measures that might narrow the racial gap. It also explicitly barred as corroborating evidence the nation's long history of employment discrimination prior to July 1965. Thus the new EEOC faced elevated hopes with modest tools indeed. The logic of this dilemma drove civil rights lawyers toward a model of proportional representation, yet one that seemed to require the disguise of euphemism, because it was statutorily proscribed in the enabling legislation. Pressman herself suggested, somewhat lamely in her memo, that an EEOC requirement that Negroes be advanced "in line with their ability and numbers" could be considered "not so much the redress of past discrimination as the application of a nondiscrimination policy to the current work force." This was tricky logical terrain for anti-discrimination, and the most promising targets for such a politically high-risk strategy clearly lay southward. For if the Civil Rights Act of 1964 had protected job discrimination prior to July 1965, it had also outlawed racial segregation, and in doing so had left a massive legacy of blocked opportunity that seemed ripe for statistical challenge, now that the old excuses were dead.

From Harmful Intent to Disparate Impact

Thus the EEOC's more ambitious vision of its social mission and legal leverage was circumscribed by the prohibitions against ratios and quotas in Title VII and by the recent language of *Swain*. But the commission nevertheless enjoyed an abundance of targets throughout the South that fairly begged for a linkage between statistical disparities and certain highly suspect practices of the workplace. High on such a list were two formidable barriers to black advancement: employee tests and seniority systems. Both tests and seniority could claim worthy traditions. They had theoretically claimed and historically demonstrated their utility in protecting worker rights while upgrading the workforce and also assigning and rewarding employees according to merit. But job tests and seniority also lent themselves to abuse as arbitrary barriers to employment and advancement. As in the Newport News case, the most obvious abuses were found in the South. But after July 1965 employers could no longer simply hide behind Jim Crow ordinances, and black workers seeking promotions could now demand a better answer. In going hunting where the ducks were, however, the EEOC engaged two of the most complex issues in modern labor relations. It did so, moreover, in areas that had been explicitly protected by the civil rights law that had founded the agency.

Clearly employers had a rational interest in applicant and employee testing to maximize efficiency and productivity. Similarly, workers and unions had a rational interest in the job protection that seniority had historically afforded. But when both practices produced results that were racially disproportionate and punitive by virtually any measure, as was so obvious in many southern workplaces, they raised two questions that reached the core dispute in EEO policy that was to dominate the next decade. The first and most important of these questions was to what degree, if any, could discriminatory intent be inferred from "disparate impact," or disproportionate results. The second (depending upon the answer to the first) was the question of the appropriate legal remedy.

The EEOC legal staff was aware from the beginning that a normal, traditional, and literal interpretation of Title VII could blunt their efforts against employers who used either professionally developed tests or bona fide seniority systems.[60] The EEOC's own official history of these early years records with unusual candor the commission's fundamental disagreement with its founding charter, especially Title VII's literal requirement that the discrimination be intentional. "Under the traditional meaning," which was the "common definition of Title VII," the EEOC's first *Administrative History* observes, an act of discrimination "must be one of intent in the state of mind of the actor."[61] Senator Dirksen's amendment to Title VII in Section 706(g) required a finding that "respondent has intentionally engaged in or is intentionally engaging in an unlawful employment practice charged in the complaint." In his accompanying remarks to the Senate (when he originally proposed the term "willfully"), Dirksen explained that his purpose was "to denote an intentional act . . . as distinguished from an accidental act," so as to avoid a situation wherein "Accidental, inadvertent, heedless, unintended acts could subject an employer to charges under the present language."[62] But by the end of the Johnson administration the EEOC, by its own self-description, was disregarding Title VII's intent requirement. "unlike state FEP agencies which continue to rely on intent," the agency's *Administrative History* explained, "the Commission has begun to rely on the constructive proof of discrimination."

But how would the EEOC square such a novel interpretation of Title VII with the statute's prohibitions, especially as they were explained by the elaborate iterations of classic nondiscrimination theory that the Mansfield-Dirksen compromise had invoked on the floor of the Senate in 1964? The answer was a counter-appeal to a kind of higher-law doctrine. In this view Congress really had in mind in 1964 a broader purpose than its narrow legislative strictures would indicate. Thus Commissioner Jackson told members of the NAACP that the "EEOC has taken its interpretation of Title VII a step further than other agencies," and concluded that the law banned not only racial discrimination per se, but also employer practices "which *prove to have a demonstrable racial effect.*"[63] "The underlying rationale for this position," Jackson explained, "has been that Congress, with its elaborate exploration of the economic plight of the minority worker, sought to establish a comprehensive instrument with which to adjust the needless hardships resulting from the arbitrary operation of personnel practices, as well as purposeful discrimination." Jackson admitted that "[t]his approach would seem to disregard intent, then, as crucial to the finding of an unlawful employment practice."[64]

Thus a broad, global purpose was imputed to Congress in 1964 and was then invoked to override the limitations that Congress, in defining what it meant to achieve, had set upon itself and upon the instruments it created. A strong case could be made that, in light of subsequent events, Congress *should* have taken a more global view of the social context of discrimination in 1964. Certainly the federal regulatory agencies have traditionally preferred to read in their founding statutes a global charter that entrusts them to guide the public weal in their domain. But Congress in 1964 was politically driven by a national wave of anger at southern segregation, especially as it was so brutally manifested in Birmingham. So Congress began Title VII by prohibiting employers from either refusing to hire minorities or segregating or classifying them into second-class status in the workplace. The dominant targets were the whites-only workforce and the Jim Crow workplace.

These twin goals led Congress to declare it an unlawful practice under Section 703 (a) (1) for employers "to fail or refuse to hire or to discharge any individual, or otherwise to discriminate against any individual with respect to his compensation, terms, conditions, or privileges of employment, because of such individual's race, color, religion, sex, or national origin."[65] The second subsection then declared it unlawful for an employer "to limit, segregate, or classify his employees in any way which would deprive or tend to deprive any individual of employment opportunities or otherwise adversely affect his status as an employee, because of such individual's race, color, religion, sex, or national origin." Civil rights lawyers like Sam Jackson sought to exploit the dualism in Section 703 (a) by aligning the first prohibition with the standard harmful intent doctrine, but by realigning the second with their new doctrine of harmful effects or disparate impact. As Jackson explained, "[e]ven in the absence of 'disparity of treatment,' *per se,* as it concerns Negro workers compared with white workers, the person aggrieved is harmed by an act of omission or commission that has a racial consequence which manifests the harm."[66]

Given the determination of the EEOC's professional staff—and by 1967 of the chairman and a majority of the commissioners—to mount a "wholesale" attack on institutionalized racism, the agency was prepared to defy Title VII's restrictions and attempt to build a body of case law that would justify its focus on effects and its disregard of intent. But commission lawyers acknowledged that such a course would set it at odds with the compromise language that was the key to Title VII's passage. Such a course might set it at odds with the courts as well, the commission conceded, at least insofar as the courts continued to interpret EEO law in the traditional manner:

> When the word "intentionally" was added by the Mansfield-Dirksen compromise package, the then Senator Humphrey accepted [Dirksen's intent amendment] as being without substantive change. However, the courts cannot assume as a matter of statutory construction that Congress meant to accomplish an empty act by the amendment. When this application is made in a given case, then the Commission and the courts will be in disagreement as to the basis on which they find an unlawful employment practice.[67]

When that occurred, someone would have to blink. "Eventually this will call for reconsideration of the amendment by Congress," the commission's administrative history concluded, "or the reconsideration of its interpretation by the Commission."

The EEOC lawyers, in consultation with civil rights attorneys at the NAACP Legal Defense Fund and elsewhere, developed a plausible rationale for abandoning the intent test. They reasoned that EEO law should be detached from its historic grounding in criminal law, where intent was customarily a matter of evil or improper motivation. Instead, standards of proof in equal employment law should be shifted toward tort law, where it could be argued that intent was merely an awareness of forseeable consequences of one's conduct, and hence could be presumed. To the EEOC lawyers it seemed irrational that the standards of proof and penalty in EEO law had historically evolved in such a mismatched fashion, with the daunting standard of evil intent from the criminal law paired with the timid penalties of the civil code—often at the punitive level of mere misdemeanors. So at the heart of the EEOC's radical legal strategy was an attempt to build a body of case law in the lower federal courts that would replace the traditional intent test with an effects test. This in turn would allow the agency to construct prima facie cases based on statistical data irrespective of intent, and through this device to throw upon employers a burden of proof that, in light of the damaging statistics, would be difficult to sustain.

Testing, Seniority, and Discriminatory Effect

When Title VII finally became fully effective in its coverage on 2 July 1968, the Legal Defense Fund sponsored a major conference in New York City to review the status of its fifty-four EEO lawsuits, and to reaffirm its decision to concentrate on the abuse of testing and seniority in the South. These two devices were "the most frequently used means of discriminating against minority-group workers," said conference co-sponsors Gabrielle Kirk and Robert Belton.[68] Yet because their utility to both employers and employees (especially labor unions) had been long established, they seemed invulnerable to assault under the intent test. They became, therefore, prime targets for statistical demonstration of disparate impact.

Of the two, litigation over testing was slower to develop. Unlike the effect of seniority in reinforcing white preference in segregated workforces, the effect of testing was difficult to isolate in dramatic fashion. Moreover, the Tower amendment (Section 703(h) of Title VII) had protected professionally developed testing quite explicitly. The EEOC *Administrative History* summarized the agency's early internal debate over that awkward circumstance: "The controversy over testing caused much discussion. Many members of Congress were concerned about this issue because the court order against Motorola was handed down during the debates. The record establishes that the use of professionally developed ability tests would *not* be considered discriminatory [emphasis added]."[69] As a result of these protections for employee testing, the early EEOC's most controversial legal offensive, in pursuit of its core strategy of displacing intent with an effects test, concentrated on seniority.

The seniority issue offered a more promising challenge to the EEOC because unlike the testing issue, seniority offered a target that was flushed into the open by the Civil Rights Act. Testing was protected by Title VII, but segregated workplaces and unions had to be dismantled forthwith. In southern plants where racially segregated departments or unions had to be abandoned, the result left the whites with all the seniority for all the more desirable jobs. This was true even though veteran black workers often held seniority with the firm in their low-paying, unskilled labor units that exceeded the seniority of many whites in skilled departments.[70] In American labor practice, seniority had evolved in a dual fashion. "Benefit seniority" had evolved on a plant-wide or company-wide basis, and it determined employee rights to pensions, vacation pay, sick leave and similar fringe benefits. It customarily was transferred with the employee from one department to another. But "competitive-status seniority" had evolved around job and skill (and hence pay) classifications and was traditionally department-specific. Unions cherished such seniority because it avoided employer favoritism, and it also removed union leaders from the dangers of political crossfire inherent in having to make merit judgments among their competing members. In segregated southern plants, blacks and their status seniority were typically confined to unskilled classifications with the lowest pay, while whites were protected in the higher-paid skilled departments. Thus the desegregation that Title VII demanded immediately unmasked seniority's crucial *effect* in perpetuating racial discrimination, whatever its original intent.[71]

To attack seniority this way carried major risks, however. It meant taking on an uncommon alliance of union chiefs and businessmen. Employers shared with labor a stake in maintaining the predictability and labor peace in the workplace that flowed from seniority's regularity. The prospect of merging desegregated lines of seniority offered a nightmare to employers and union leaders alike. For labor it especially exacerbated the strains within the civil rights movement between black leaders and their labor allies. Furthermore, while the discriminatory impact of such seniority arrangements appeared to make a compelling case for EEOC assault, the seniority problem was probably the most complex in all of EEOC law. Even the Legal Defense Fund's Jack Greenberg agreed: "The problem of seniority is how to unravel threads of discrimination which existed for years. Separate seniority lines are a clear violation of the [civil rights] act," Greenberg acknowledged, "but in many cases it is not clear what is the best way that segregated seniority lines can be merged or connected without destroying the seniority system."[72]

In the middle 1960s, seniority law in such cases was governed by the *Whitfield* decision of 1959, involving Armco Steel Corporation and the United Steelworkers. *Whitfield* had held that in a merger of formerly segregated seniority lines, it was legally sufficient to tie the black seniority to the bottom of the white seniority list. *Whitfield* represented nondiscrimination without affirmative action. As Judge John Minor Wisdom explained for the 5th Circuit majority in *Whitfield,* the steelworkers and Armco had made a fresh start for the future, with a new contract "that *from now* on is free from any discrimination based on race."[73] Wisdom conceded that this arrangement left the black workers at a disadvantage. But he explained that "This is a product of the past. We cannot turn back the clock."

Within a decade both Judge Wisdom and the 5th Circuit were to revise considerably their unwillingness to redress the wrongs of past generations. But the *Whitfield* formula was accepted by many unions and companies in the early and middle 1960s as a liberal incentive to change, especially in growth industries, where more redistributionist approaches would stiffen resistance. The *Whitfield* formula was therefore incorporated into a few of the EEOC's early conciliation agreements. In industries with static or declining employment, however—which characterized much of the South's extractive and labor-intensive industries—the formula meant that blacks would shift from a protected status in their old seniority line to a vulnerable status at the bottom of the white line. They thus would remain still junior in seniority to much less veteran co-workers in the firm, ripe to be laid off before most of the company's less senior white employees.

The manifest unfairness of this practice led the EEOC early in its enforcement probes to escalate its originally modest *Whitfield* demands. The commission first pushed "bumping rights." This meant a right, for blacks wishing to transfer to higher paying job classifications from which they were previously barred by race, to displace whites with less plant-wide seniority in their new department. The agency reasoned that black workers should get some credit for their years of plant seniority, especially since it was earned while suffering from low wages and dead-end jobs. Commission lawyers argued that the critical determinant of a bona fide seniority system was the system's *effect* on job distribution to minorities, *not* its purported intent, or the fact that it was the result of collective bargaining, or was adopted prior to July 2, 1965.

This stance, however, once again placed the EEOC at odds with the restrictive statutory language in Title VII's Mansfield-Dirksen compromise. The Dirksen language in Section 703(h) protected the use by employers of "different standards of compensation, or different terms, conditions, or privileges of employment pursuant to a bona fide seniority or merit system . . . provided that such differences are not *the result of an intention* to discriminate because of race, color, religion, sex, or national origin [emphasis added]. . . ." Union spokesmen, moreover, replied that departmental seniority had been the fruit of a long and hard-fought battle in modern labor history; it protected workers from the insecurity of "bumping" threats through departmental transfer, and in the process it also increased industrial productivity by improving the efficiency of technical specialization.[74] Spokesmen for both U.S. Steel and the United Steelworkers defended the *Whitfield* formula by observing that transfering one's seniority into a new line had *never* been permitted to an employee in their contracts, white or black. "The NAACP," a U.S. Steel spokesman said, "is trying to turn back the clock. They want super-rights for Negroes to make up for past discrimination."[75] Essentially, the EEOC quite agreed.[76]

Because the EEOC could not participate directly in litigating these questions, the issues rather slowly percolated upward in the lower federal courts in suits filed by the Legal Defense Fund or the Justice Department or various private groups, with the EEOC providing support through amici briefs. The controversial, quota-war implications of the EEO challenges to testing and seniority would not fully surface until 1970–71, when parallel lines of litigation produced both the *Griggs* decision on testing and the Crown Zellerbach decision on seniority—both circuit court decisions that were upheld by the Supreme Court.

In the meantime, the EEOC would face annual congressional battles over its proper jurisdiction and authority. In 1966, the second session of the 89th Congress would actually pass a bill in the House, by a deceptively wide margin, that granted cease-and-desist authority to the EEOC. But the bill died when the Senate adjourned without acting on it. The more conservative 90th Congress, elected in the aftermath of the Watts riot of 1965 and the Black Power summer of 1966, stiffened its resistance to the civil rights measures that President Johnson continued to propose and lose. During the last two years of the Johnson administration, civil rights debate would center on the

President's controversial proposal for open housing legislation, with equal employment measures relegated to a secondary status. The dispute over EEO enforcement would be inherited, still unresolved, by Richard Nixon in 1969.

To the considerable surprise of most congressional observers, Johnson's failed fair housing bill of 1966 and 1967 would ultimately pass in 1968—in a strange, carrot-and-stick marriage with anti-riot legislation, and bitter-sweetened by the martyrdom of Robert Kennedy and Martin Luther King. Thus during 1966–68 the EEOC would quietly and effectively pursue its long-range objectives in the administrative and judicial arenas. But these important advances were obscured by the agency's more visible struggles with organizational discontinuity and disarray, an embarrassing growth of case backlog, and repeated defeats in the Congress.

CHAPTER X

From Ghetto Riots to Open Housing, 1966–1968

The Hawkins Bill and the Katzenbach Task Force: 1965–66

On April 27, 1966, the House passed a bill granting cease-and-desist authority to the EEOC by a deceptively lopsided, roll-call vote of 300–93.[1] This was H.R. 10065, or the Hawkins bill, after California Democrat Augustus Hawkins, who was a senior member of the House Education and Labor Committee and a black veteran of California's FEPC effort. Yet the next day President Johnson, in a special message to the Congress on civil rights, proposed a four-part bill that made no mention at all of the EEOC or equal employment needs. By August, the House had passed an amended version of the administration's bill, which featured a controversial provision for open housing, and the Hawkins equal employment bill had essentially disappeared. Then in September Senate opponents of the administration's House-passed civil rights bill, H.R. 14765, rather easily filibustered it to death (the opposing combination of Dirksen and Ervin symbolized the strength of the forces arrayed against the bill), and the 89th Congress adjourned to face the November elections. There would be no Civil Rights Act of 1966.

The apparently abrupt turnabout of late April 1966, however, when Hawkins's strong EEO bill yielded to Lyndon Johnson's dead-end open housing bill, was less sudden than it appears. It signaled the Congress's post-Watts shift from civil rights bills targeted against southern racism to an ambivalent national mood that mixed anger and guilt in unstable measure. The results in civil rights policy were paralyzing. But the April shift also reflected the Johnson administration's determination to press for new civil rights legislation despite the stiffening new resistance in Congress and the inability of the fracturing civil rights coalition to agree on a central focus, as they had in 1964 and 1965.

The sudden demise of the Hawkins bill is puzzling at first blush. The bill had initially appeared to enjoy considerable momentum on the Hill, and even modest support from the White House. It had started out in the summer of

1965 as the Powell bill, H.R. 9222. The mercurial Powell, as chairman of the general subcommittee on Labor as well as chairman of the parent Education and Labor Committee, had greeted the official launching of the EEOC that July by inviting chairman Roosevelt and the various spokesmen of the Leadership Conference coalition to help construct a bill to replace the EEOC's missing teeth.[2] The result was a bill that would arm the EEOC with cease-and-desist authority, like most of the state FEPCs. It would also extend coverage to include employers with eight or more workers—also like most state FEPCs—rather than the bottom limit of 25 employees or union members that the Civil Rights Act of 1964 had scheduled for 1968.

The bill was reported out of the liberal-dominated Education and Labor Committee on August 3, 1965, as H.R. 10065. But its new House sponsor was the committee's other black member, Representative Hawkins. Powell's erratic behavior had already alienated enough constituencies—namely, organized labor, the Johnson administration, the House Democratic leadership, southern members of Congress, and Republicans—to doom any bill carrying his name. But the newly christened Hawkins bill, cosmetically shorn of Powell's sponsorship, could still not get past the Rules Committee. So it was carried over to the second session of the 89th Congress in 1966. Budget Bureau officials then routinely invited agency comments on the bill, especially from Justice and Labor. The Bureau found no major opposition to the bill's two key provisions, which were cease-and-desist authority for the EEOC and extending coverage to employers with at least eight rather than twenty-five workers. There was, however, resistance from Justice and even from the EEOC to the Civil Rights Commission's call for extending Title VII coverage to employees of state and local governments.[3]

Despite the Administration's general agreement with the Hawkins bill provisions, there was no mention of them in the President's State of the Union message of January 12, 1966. This was partly because Attorney General Katzenbach, who had led the dethronement of Hubert Humphrey as chief coordinator of the Administration's enforcement policy, was urging a different approach. Katzenbach was leading an interagency task force (formed at his own behest, as distinct from Califano's highly organized and White House-centered task forces of 1966–68) that included officials from Labor, HEW, Defense, the OEO, and, revealingly, the Housing and Home Finance Agency. Katzenbach's group was charged by the President with recommending a new legislative program in civil rights in light of the triumphs of 1964–65, the subsequent ghetto rioting and the surge of black nationalism, and the unfinished agenda in civil rights enforcement.

Katzenbach's task force, which began work in late August 1965, generally shared the Budget Bureau's belief that the Hawkins bill's two main provisions "may be controversial and their essentiality has not been demonstrated."[4] Moreover, the EEOC's worsening backlog argued against extending its coverage to state and local governments. The problem of continued job discrimination against black state and local employees in the South was admittedly severe, as the Civil Rights Commission had complained. But this would add yet another eight million workers to the struggling EEOC's current coverage of 30 million. Worse, it raised for conservatives the specter of further eroding federalism by giving one highly suspect new federal agency the authority to overrule the personnel decisions of *all* state and local government agencies. So Katzenbach's report to Califano called not for new EEOC authority, but rather for new provisions to remove the exemption for state and local governments in the Civil Rights Act of 1964, and then to empower the Attorney General to sue them as necessary in federal court to eliminate discrimination.[5]

As for the problem of the EEOC's "very modest powers," Katzenbach's task force considered two strengthening options. One was the muscular model of the FTC, which was favored by the Leadership Conference and the civil rights community generally. This would expand the EEOC into a "full-blown enforcement agency" by giving it general administrative subpoena and investigative powers, substantive rule-making powers having the force of law to prescribe certain practices, and cease-and-desist authority not just over individual complaints, but also broadly over pattern and practice cases. Katzenbach conceded that the EEOC likely would favor such a vast enpowerment. But he added that "Needless to say, it represents much more serious political problems. I cannot conceive of Senator Dirksen's going this far."[6] Such a radical change from the compromise model of 1964 would "substantially reduce the protection which the court-oriented procedures of Title VII now provide for covered employers," the Attorney General observed. In any event the EEOC experience had been far too brief to make such a premature judgment. The alternative, which Katzenbach told Califano he preferred, was to invest the EEOC with the power it lacked to conduct evidentiary hearings, make factual findings, and issue cease-and-desist orders to redress the grievances of *individual* complainants only. Such a proposal would leave intact the Attorney General's authority to seek broad remedies in court to eliminate general patterns of discrimination.

Katzenbach's group had worked all fall to shape the civil rights section of the President's State of the Union message in January 1966, and also the special and more legislatively detailed message to Congress that would follow. But given the breakthrough laws of 1964 and 1965, what additional legislation was needed that might be both passable and workable? As Katzenbach reported to Califano, his task force had initially concentrated on "schools, housing, employment, and the administration of justice."[7] The last area referred to such continuing problems of southern discrimination as black exclusion from state and federal juries and criminal attacks on civil rights workers. Here Johnson agreed, and he made jury nondiscrimination and protection of civil rights workers the first two proposals in his State of the Union message on January 12, and also the first two items in his special message to Congress on April 28.[8] In response to Katzenbach's recommendations on schools, Johnson made the third goal of his special message a broadening of the Attorney General's authority to bring suit for the desegregation of southern schools and public facilities.

But in response to Katzenbach's recommendations on employment, both Johnson's State of the Union and special civil rights messages were silent. It is not clear from the archival record precisely why, especially since the Administration seemed mildly in favor of the Hawkins bill. Indeed, both the Kennedy and Johnson administrations had always seemed mildly in favor of congressional proposals for cease-and-desist authority at the EEOC. *New York Times* reporter John Herbers speculated that the Administration was not actively supporting the Hawkins bill because it did not expect the bill to pass the Senate, and it doubted the EEOC need or capacity to handle such increased powers.[9] Herbers further speculated that the Hawkins bill was too closely associated with Powell, who the previous year had blackmailed the Johnson administration by holding hostage to his civil rights proposals the Administration's high-priority labor legislation on common-situs picketing and repeal of Taft-Hartley 14(b)—the latter being the Democrats' standard legislative commitment to reward labor support. For the President to so punish the arrogant Powell was not inconsistent with the vintage LBJ style (House Democrats would soon strip Powell of his committee chairmanship, then of his House seat itself). But it remains unproven.

The real surprise of Johnson's January message, however, was his call for "legislation, resting on the fullest constitutional authority of the Federal Government, to prohibit racial discrimination in the sale or rental of housing." In his April message he asked Congress "to declare a national policy against racial discrimination in the sale or rental of housing." Not content to rest with the Civil Rights Act of 1964 and the Voting Rights Act of 1965, Johnson now targeted discrimination in the inner sanctum of middle-class America—housing. And unlike the first two targets, housing discrimination was less a southern than a northern phenomenon.

Furthermore, it already had a tortured history within the executive branch. Why was Lyndon Johnson reaching so far so fast?

Origins and Politics of Johnson's Fair Housing Bill

When Katzenbach first reported to Califano on the progress of his legislative task force for civil rights in October 1965, he described a preliminary proposal that would prohibit racial discrimination in both the sale and rental of housing units. It would cover apartment houses, but not sales of single homes (except in development tracts), and it would probably include a "Mrs. Murphy" provision to exclude rentals where the owners lived on the premises. But Katzenbach had warned: "There are many constitutional problems here."[10] His approach to the housing issue was cautious and vague, yet it was listed second only to school desegregation in the Attorney General's preliminary agenda of civil rights initiatives. School desegregation commanded first attention by virtue of its newly explosive quality, for it had recently broken its southern confinements in a showdown between HEW and Chicago's Mayor Daley. This imbroglio embarrassed the Johnson administration, and quickly nationalized the school segregation issue by linking the new billion-dollar program of federal school aid to de facto patterns of racial distribution.[11] A new federal proposal to ban racial discrimination in housing would at least promise to attack the roots of de facto school segregation in the North and West. It would represent a radical departure as a presidential request to Congress (its only precedent was the truly Radical civil rights law of 1866, which an earlier southern President named Johnson had vetoed).[12]

Meanwhile, pressures were growing for Lyndon Johnson to take some strong initiative against housing discrimination quite independent of the controversy over school desegregation and the storm over the HEW guidelines of 1966. The pervasiveness of racial discrimination in housing was widely conceded, and contemporary scholarship was beginning to document the historic complicity of the federal government in fostering housing segregation through the FHA, the HOLC, and the Federal Home Loan Bank.[13] The Watts riot of August 1966 had underlined the explosive consequences of ghetto housing. In response, Martin Luther King began to shift the focus of his protest from voting in Alabama to northern slum conditions in Mayor Daley's Chicago.[14] Several hundred civil rights experts had converged on the White House in late November 1965 in a planning session for the promised White House Conference "To Fulfill These Rights." Their long list of "urgent" recommendations called for presidential intervention in the housing market, largely on the strength of executive authority alone.

Moreover, the federal government's existing fair housing policy, dating from Kennedy's executive order of November 1962, had proven ineffective. Because it had applied only to the sale of *new* housing under the FHA-VA mortgage insurance programs, it reached less than 3 percent of the nation's total housing supply of 60 million units. In 1965 the FHA and the VA had helped finance only 17 percent of the country's 1.5 million new private housing units for that year.[15] Moreover the median income of black families in 1964 was only $3,724 a year, about half of white family income. The new housing that was going up in the suburbs was thus beyond the financial reach of most blacks, who in any event had demonstrated more interest in purchasing resale housing closer to familiar neighborhoods or, most commonly of all, were financially limited to renting.

Finally, the great civil rights victories of 1964 and 1965 had peeled away the long-dominant and primarily southern issues of public accommodations, de jure school segregation, and racial disfranchisement, thereby isolating housing discrimination as a bedrock problem as yet untouched by the Great Society. By early 1965 Johnson was coming under increasing pressure to expand the scope and force of Kennedy's housing order of 1962. Pushing Johnson to act on executive authority were the various spokesmen of the Leadership Conference, Democratic liberals in his own administration whose spokesmen was Vice President Humphrey, and liberal congressional Republicans like senators Javits and Case.[16] In March 1965, Governor David Lawrence's near-mutinous President's Committee on Equal Opportunity in Housing had forced the issue by recommending that Johnson extend Kennedy's 1962 order to cover housing financed by federally insured banks and savings and loan associations. By confronting Johnson with a choice between decisive action or public resignations, this crisis led to a series of top-level meetings over the next seven months that culminated in mid-November 1965 in an agreement that further expansion of executive authority could not be justified on legal grounds, and would probably be ineffective anyway.[17] Johnson's senior advisers therefore agreed to switch priority from executive strengthening of Kennedy's order to seeking congressional action—and to do so promptly.

Califano agreed with the emerging consensus. But he and Katzenbach feared moving too quickly. Califano told Johnson on November 20 that "Such legislation is so complex and difficult to put together that it is unlikely to be ready for consideration by you and presentation to Congress before January, 1967."[18] Then Katzenbach, in his 12-page progress report on proposed new civil rights proposals to Califano of December 13, 1965, listed recommendations for desegregated juries, federal protection for civil rights workers, cease-and-desist authority for the EEOC in individual complaint cases, school construction to ease desegregation, and expanded employment training programs. But he omitted any mention of the housing issue. "I would confine the civil rights message," the Attorney General advised Califano, "to jury legislation, personal security legislation . . . and employment legislation."[19] But Johnson, once committed to the legislative consensus, saw no virtue in further delay, especially while Republican partisans like Jacob Javits were hammering away at him for failing to strengthen Kennedy's executive order, while enjoying congressional freedom from having to vote on such a controversial measure themselves.

So in 1966 Johnson shelved the equal employment recommendation from Katzenbach's task force as premature and sent the Congress in late April an omnibus civil rights bill. Title IV of Johnson's bill would place Congress as well as the Presidency on record in declaring that nondiscrimination in housing was a national policy. Johnson's original bill was surprisingly comprehensive in its coverage, applying to almost *all* sale and rental housing, old as well as new. Its enforcement, however, was more cautious and followed the judicial model. It provided for private suits in federal or state and local courts for injunctive relief and damages, the latter limited to a modest $500. The Attorney General, however, would be empowered to bring pattern-or-practice suits in federal courts, as he was in Title VII cases on employment discrimination. Procedurally, the Administration followed its customary House-first strategy in pushing the omnibus civil rights package. There the sympathetic Celler controlled the Judiary Committee (as Eastland still did in the Senate), and the Administration could negotiate its early compromises with dependable Republican moderates like McCulloch, whose bargains tended to command bipartisan respect and hence to stick.

Yet oddly, the compromising in the House during the summer of 1966 flowed in contradictory directions. Dilutions in coverage were matched by a potentially radical transformation in enforcement machinery. Because the bill dealt with the citadel of the voter's home, moderate Republicans under the leadership of McCulloch and a liberal young Republican from Maryland, Charles Mathias, bargained for a compromise that would exempt home owners and sales agents, while still covering persons fully engaged in the real estate business, such as builders, apartment-house owners, mortgage lenders, and real estate brokers.[20] But the Administration's judicial model of enforcement, which would make nondiscrimination in housing a private right, like Title VII's right to nondiscrimination in employment in 1964, was turned on its head by the Judiciary Committee. On June 29, 1966, Celler's full committee adopted the compromise Mathias amendment on coverage, which the Administration supported. But then it adopted an enforcement amendment offered by John Conyers, a black congressman from Detroit. The Conyers amendment, which was opposed by the Administration but pushed by the Leadership Conference, would create a potentially powerful new watchdog agency called the Federal Fair Housing Board. It would be patterned after the NLRB, and similarly armed with cease-and-desist authority, with its orders enforceable through the federal courts of appeals.

House conservatives were horrified at the prospect of a new federal agency dictating to private citizens just who could buy what house or rent what apartment. But the Goldwater debacle had so decimated their ranks in 1964 that they were powerless to defeat the bill in the House, especially since the Mathias compromise had protected Mrs. Murphy, at least for the time being. Conservatives had little doubt that the sweetener of limited initial coverage would quickly erode, like the proposed drop in EEOC coverage from twenty-five to eight, once the enforcing agency was established. So the House conservatives of both parties were forced to rely, as they had done so often in the past, on their colleagues in the Senate. The Senate was by nature better armed with defensive devices to protect minorities (including propertied and conservative minorities in the original Hamiltonian sense), and hence to prevent the creation of muscular and therefore dangerous new executive agencies that were endowed with a social mission and staffed by its ideological devotees. In the late 1940s, Senator Taft had performed this role for the conservatives by leading the Senate counterattack on the Wagner Act and the NLRB, and Senator Dirksen had similarly defanged the EEOC in 1964.

Nevertheless the House, even in passing the omnibus civil rights bill on August 9, 1966, by a roll-call vote of 259–157,[21] also voiced its dawning disaffection with the increasing tendency of civil rights issues to spill over the southern borders. The House-passed bill denied new authority for the Attorney General to bring school desegregation suits without first receiving a written complaint from someone who had actually been turned away from a school. Although this provision was sponsored by a North Carolina Democrat, and it essentially returned the Attorney General's authority to the level set by the Civil Rights Act of 1964, it found great appeal as a northern hedge against potential attacks from the Justice Department on de facto school segregation in order to achieve racial "balance." This in turn was reinforced by a specific new ban on withholding federal school-aid funds from any school district because of "racial imbalance"—a pointed rejoinder against HEW's

recent and politically maladroit attempts to attack de facto school segregation in Mayor Daley's Chicago. Finally, the House lodged in H.R. 14765 a
pointed complaint against the growing practice of social protest by riot and
arson. It voted 389–25 to support Florida Republican William Cramer's
criminal ban on interstate travel with the intent of committing violence or
inciting a riot.

The hardening mood of the public and hence of the Congress, as the civil
rights issue moved north with the fires of black protest, was not lost on Sam
Ervin and his southern colleagues in the Senate. Ervin greeted the omnibus
bill by observing that its housing provision "is not aimed solely at the South.
Its impact may be felt more strongly in other areas of the country." "It should
be most interesting," Ervin said, "to watch the politics or the debate now
that others' oxen are being gored." [22] It was indeed interesting. But not for
very long—at least not in the Senate, where the *Wall Street Journal* described
the Senate's response to the House-passed bill as a desultory process of spending
"only two listless weeks discussing the bill before its backers dropped the
fight." [23] Majority Leader Mansfield blamed both Negro rioters and white
rock-throwers, images that were increasingly associated with northern and
western cities like Los Angeles and Chicago, not with southern cities. Senators Javits and Case blamed Lyndon Johnson for having dumped the prickly
housing issue on the Congress in the first place. The Johnson administration
in turn tended to blame Everett Dirksen for failing this time around to tease
the drama into its delicious eleventh hour, and then pull one of his famous
switches to save the day and reap the statesman's acclaim (Dirksen had early
announced his opposition to Title IV on housing, and, not sensing the kind
of national groundswell that had prompted his dramatic reversals in 1964
and 1965, he stuck by his guns). [24] No one much bothered to blame the southern
Democratic senators, like Ervin, who had so easily talked the bill to death.
But the very ease of that feat bore ominous tidings for the bill's future chances—
especially when in the November elections, the Democrats lost 47 seats to
the Republicans in the House and three in the Senate.

Califano's Civil Rights Task Forces and the Combative 90th Congress: An EEOC for Housing?

When the 90th Congress convened in January 1967, Lyndon Johnson would
welcome them with a full slate of legislative proposals to swell the inventory
of Great Society programs and budgets. His distinctive instrument for legislative program formulation, the task force device, was cranked up to a fever
pitch during 1966–67 by Califano and his rapidly growing domestic staff. In
the late summer and early fall of 1966, Califano appointed a record array of
11 outside and 34 interagency task forces (there had been only 4 outside and
11 interagency task forces in 1965 to work on the legislative agenda for
1966). [25] The interagency task forces of fall 1966 were asked to recommend
new legislation for the 90th Congress, and also to propose administrative

measures not requiring new legislation that would maximize the efficiency and impact of Great Society programs already on the books and in operation. Their recommendations were to receive the close scrutiny of the Budget Bureau, where Budget director Charles Schultze felt squeezed by the camel's-nose implications of the proliferating new, "entering-wedge" programs of the Great Society, especially such expensive items as Medicare and Medicaid, aid to education, and the various efforts to reduce poverty and clean up the environment.[26] But new measures to enforce anti-discrimination were not regarded as carrying such major cost implications. So Califano's civil rights task forces were generally spared the Administration's increasing tensions between the forces represented by Schultze, who worried about the impact of spiraling Vietnam costs and inflation, and Califano and his program-boosting allies in the mission agencies.

Because by late August the administration's proposed Civil Rights Act of 1966 seemed destined not to pass, Califano on September 5 asked Katzenbach to chair a new interagency task force for civil rights.[27] That September, however, Katzenbach was stepping down from the hot spot atop Justice to replace George Ball as Under Secretary of State. So a new and acting Attorney General, Ramsey Clark, inherited the task force. Clark promptly bucked the group's working chairmanship down to the Civil Rights Division's first assistant, Stephen J. Pollack. Pollack's unusually large task force involved 34 representatives from 12 agencies.[28] On November 30, Clark sent Califano his recommendations in response to the 58 proposals generated by Pollack's task force.[29] Most of these were minor and administratively technical, and paralleled proposals from the prolific White House Conference of the previous June.[30] Such long lists of proposals, especially on the major areas of housing (16 proposals), equal employment (15), and federal organization and coordination (6), made for a fat document. But the important and controversial proposals were few, and they demonstrated the relative narrowness of the Administration's room to maneuver.

Housing was by far the most controversial policy area for the task force, and the sixteen proposals dealt with essentially three policy options. In light of the Administration's recent failure to pass civil rights bills, one group of proposals explored stronger varieties of executive action, such as extending Kennedy's 1962 order beyond federally financed housing to include federally insured lenders for mortgages and construction. In the past, however, the complexities of the federal regulatory structure and the private housing market had conspired to defeat all such proposals. The battery of lawyers from Justice on the Pollack task force reviewed the tangled debate and the task force concluded that the President "lacks the authority to order the bank regulating authorities to impose such a requirement." Furthermore the banks themselves "probably lack the authority to issue such regulations even if ordered to do so by the President."[31]

A second potential source of leverage was to tie new strings to federal grants-in-aid. This had recently been attempted in the field of education when the Elementary and Secondary Education Act and the Higher Education Act of 1965 had been linked to Title VI of the Civil Rights Act of 1964. The most notable result of this connection had been the HEW guidelines on school desegregation of 1966. But this had also produced an explosion by Mayor Daley in Chicago and a hurried retreat at HEW. The President was unlikely to follow such a political fiasco by ordering, for instance, on his own authority and without congressional approval the construction of low-income public housing projects in every American suburb that received any form of federal housing assistance. So the Pollack task force proposed new *legislation,* not an executive order, "to condition all federal housing assistance to any local government entity" on its agreement to construct a "reasonable number" of low income public housing units and to take "adequate steps to permit" the construction of private housing for low-income families.[32] Attorney General Clark, however, recommended against this proposal because he believed that Congress would not enact it.

Having rejected the stick, Clark also rejected a proposed carrot. The task force proposed financial inducements to suburbs to build low-income housing by paying local governments as an incentive a set dollar amount of aid per low-income family unit, such as $500. The model was similar to Title I of the billion-dollar ESEA of 1965, which had distributed federal school aid to school districts through an antipoverty formula based on their percentage of low-income students. But in the case of the new forms of education aid, the students were already resident in the school districts seeking the additional money to serve them. The purpose of the new housing incentive, however, would be to lure into suburban communities low-income families from elsewhere who were not customarily welcome, and indeed from whom the suburban residents had presumably fled in the first place. Clark reasoned that a housing inducement as low as $500 per low-income family would not even cover a community's real costs, and that in any event "suburban communities probably do not shun the poor because of costs anyway, but because of the unfortunate tendency of many persons to view the poor as undesirable neighbors."

So the process of elimination had reached past executive orders to regulatory agencies and carrot-and-stick inducements on public housing, and eventually passed on to the Congress. The Pollack task force therefore contemplated asking not for a presidential order but rather for new statutory authority. One obvious possibility was a law requiring all federally insured lending institutions to require nondiscrimination from builders and tract developers. But even if such a law could be passed, it would reach only about 60 percent of the new housing market, and only 15 percent of the total market. Like the proposed executive order, it would also be easy to evade, and would probably encourage builders and developers who wanted to discriminate to take their business to lenders who did not carry federal deposit insurance.

A second legislative proposal was modeled on the Voting Rights Act of 1965. It would contain a "trigger" mechanism to make it operate only when and where conditions of serious housing discrimination were determined to exist. But the task force could identify no feasible formula for such a trigger. Third, the task force considered the Administration's failed bill of 1966, which had called for judicial rather than administrative enforcement. But this option was rejected because it "would be a repudiation of the compromise made this year. Sufficient reasons for repudiation have not been found."[33]

This was a peculiarly cryptic observation. The "compromise" referred to had been the Administration's rather silent acquiescence in Representative Conyers's proposed new Federal Fair Housing Board—a potential EEOC for housing. By process of elimination, then, the Pollack task force had arrived at a consensual Proposal No. 1, and the Attorney General agreed. This would combine Title IV of Johnson's 1966 open housing bill with an administrative rather than a judicial mode of enforcement. The task force observed that the previous spring the House Judiciary Committee had amended the original Title IV to substitute administrative for court enforcement, and commented that those new provisions had "provoked surprisingly little opposition or even discussion on the floor." The task force therefore recommended a bill that would "ban discrimination in housing by tract developers, apartment house owners and mortgage lenders but not in sales of owner-occupied one- and two-family homes, room rentals in a home, and rentals by religious and fraternal organizations, and with enforcement responsibility in an administrative agency." Proposal No. 1 did not, however, suggest which agency would do the enforcing, or how.

Whither the EEOC?

The task force's other tough policy area was equal employment. Its first three proposals captured the hard questions, all of them involving expanded authority for the EEOC. Two of these reflected the Hawkins bill's two main provisions: extending coverage to all employers or unions with eight or more employees or members, and granting the EEOC cease-and-desist authority, including the power to award back pay. The third proposal would add state and local government agencies to the EEOC's jurisdiction. This would specifically include educational institutions, all of which were exempted in 1964. Expanding coverage by including smaller employers and unions with only eight workers would dramatically increase the number of firms or unions covered by the EEOC from 258,000 to 786,000. This would add an estimated 6.5 million more workers to the 30.6 million already covered. But that promised a modest 21 percent increase in workers covered, at the hefty price of a 305 percent increase in the number of firms and unions that would be required to enter the unwelcome maze of federal paperwork. The marginal potential payoff from such increased bureaucratic hassle was not very promising, especially in light of the EEOC's growing backlog with its original coverage.

On the other hand, the proposal to add state and local government employees to the EEOC's jurisdiction would add another 8 million workers.

These, however, were scarcely associated with mom and pop stores or tiny union locals. Nonfederal public employment was an area characterized by both rapid growth and a special symbolism. "Exclusion of members of minority groups from positions as visible representatives of public authority," observed the working paper of the Office of Legal Counsel in Justice, "may contribute to the alienation of these groups from authority and thus to disregard of such authority."[34] Finally, granting the EEOC the same sort of cease-and-desist authority as was enjoyed not only by older regulatory agencies like the FTC, SEC, and NLRB, but also by most of the state and local FEPCs, seemed logical and desirable to the task force and to Clark. But they conceded its political vulnerability, like that of the two coverage-extension proposals, and speculated that a passable compromise alternative might be to amend Title VII to give the EEOC authority to bring civil actions in the courts. This had been the original McCulloch compromise on the Kennedy civil rights bill of 1963.

A fourth proposal revealed just how murky the role of the EEOC was, even in the fine collective legal minds of the Justice Department. As a kind of bastard compromise between a quasi-judicial regulatory commission, an administrative agency, and an educational and conciliation bureau, the EEOC was essentially *sui generis.* Thus it was proposed that Title VII might be amended to give the EEOC the power to issue substantive as distinguished from merely procedural rules. This echoed Katzenbach's distinction of the year before, which contrasted the weak EEOC with a "full-blown regulatory agency." But the Attorney General's Office of Legal Counsel, in attempting to place the EEOC in that framework, revealed a generalized confusion of role:

> In its first year of operations, the EEOC has written seven hundred fifty opinion letters, has issued a number of General Counsel Opinions, and has published guidelines on such matters as sex discrimination, advertising, religious discrimination and professional testing. The precise meaning and scope of these matters are left open by Title VII. The opinions and guidelines may be relied on by employers, unions, employment agencies and individuals, but do not have the status of substantive rules of law . . . .[35]

Yet the Justice lawyers noted that a federal court in northern Mississippi had recently held in *International Chemical Workers v. Planters Corp.* that the interpretations of Title VII by the EEOC are entitled to judicial deference. Amid all the confusion, the Justice lawyers had noticed early indications that despite the EEOC's statutory confinement to voluntary and conciliatory efforts, the federal courts were likely to defer to its presumed substantive authority much as they had with the FTC and the other established regulatory agencies. Such judicial deference would seem to empower the EEOC, retrospectively, to make substantive rulings even in the absence of the normal statutory provisions. Indeed, Title VII's ill-defined exemptions for BFOQs and state protective laws in gender discrimination, and for employment testing and bona fide seniority and merit systems as well, seemed to imply at least some substantive rule-making authority for the "poor, enfeebled thing."

President Johnson, however, having failed for the first time to pass his civil rights proposals in 1966, wanted neither to retreat from his prior proposals, nor to add controversial new items to the burden. So in his State of the Union message to the first session of the 90th Congress on January 10, 1967, Johnson asked Congress to legislate in the same four areas as he had the year before. And in his special civil rights message of February 15, he explained that "The Act I am proposing this year is substantially the same as last year's bill."[36] Two of the four measures, nondiscrimination in jury selection and federal protection of civil rights workers, were indeed mere repeaters. But for employment discrimination, Johnson for the first time formally proposed giving the EEOC cease-and-desist authority. Johnson had appeared to support such new authority, at least implicitly, the year before by tacitly accepting the Hawkins bill. But once the Hawkins bill had cleared the House Judiciary Committee with Congressman Conyers's novel and potentially draconian enforcement club of a Federal Fair Housing Board, the administration had allowed it to die quietly while it pursued its own omnibus bill. Johnson's 1967 bill included the cease-and-desist proposal. But it omitted any reference to increasing the coverage for the beleaguered EEOC.

The main emphasis of the Administration's 1967 civil rights bill, however, remained on fair housing. Here Johnson compromised between his previous year's call for judicial enforcement through private suits, and the Conyers model of a new Federal Fair Housing Board that was so alarming to conservatives. Instead, the Administration's omnibus bill for 1967, H.R. 5700 and S. 1026, struck a rather inventive compromise on the enforcement machinery. Instead of relying on court enforcement of an essentially private right, as in 1966, the 1967 proposal would empower the Secretary of Housing and Urban Development to hold an administrative hearing and issue cease-and-desist orders of its own in response to individual complaints. As in 1966, the Attorney General would be authorized to sue in pattern-or-practice cases.[37]

Congressional Backlash and White House Pessimism: 1967

In the first session of the 90th Congress, Johnson's carefully crafted omnibus bill went nowhere. A major reason for this was widespread resentment among the House members that they had been required to cast a vote on the bill's most controversial provision, open housing, just prior to the fall election, but the Senate had not. Emanuel Celler, the sponsor of H.R. 5700, explained that the House, having "fought with moments of great bitterness for the open housing section [in 1966] only to have it killed in the Senate," would take no action this time around until the Senate passed an open housing bill first. This was understandable. But because the Senate's perfunctory hearings on open housing in 1967 never even got beyond the subcommittee stage, neither chamber took action on open housing even at the committee level that year. A similar fate befell the equal employment proposals in the omnibus package, which received no committee recommendations in either house.

Indeed, Johnson's omnibus bill was essentially ignored in 1967. The package was broken up into separate bills, only the most minor of which became law in 1967—i.e., a bill to extend the life of the Civil Rights Commission for five years. Neither chamber took any action on nondiscrimination in state jury selection, although the Senate did pass a similar bill for federal juries (it emerged from Senate Judiciary with the qualified support even of Sam Ervin, who insisted that it wasn't a civil rights bill at all), and sent it to the House with voice-vote approval on December 8.

The tough-minded 90th Congress was primarily interested in cracking down on the nationwide escalation of violence, including both ghetto riots and southern violence against civil rights workers. A bill to increase federal protection for the latter passed the House on August 16 by a vote of 327–93, and by the end of the first session it had narrowly emerged from Senate Judiciary by a margin of 8–7. In the House, Representative Cramer reintroduced his anti-riot amendment of 1966 as H.R. 421, and it thundered through the anti-riot House by a vote of 347–70 on July 19. This occurred just days after 23 people had died in Newark's ghetto riot, and just before the Detroit ghettos exploded with a savagery that left forty-three dead. The destruction of life and property in Detroit was so severe that Governor George Romney called in the Michigan National Guard, and President Johnson ultimately dispatched U.S. paratroopers and tanks. Senate action on the House anti-riot bill was delayed only because the Judiciary Committee held extended hearings on the rioting, and dickered over how best to strengthen the House bill.

Meanwhile, Califano had appointed yet another round of task forces, eleven of them outside and twenty-four interagency. The civil rights task force was led by Ramsey Clark, and under his direction 67 staff officers representing 18 federal agencies and the White House worked intensively during September and October to propose 35 new measures. But as the Attorney General observed in his mid-November memo of transmittal and recommendation to Califano, the Administration's main agenda must remain the enactment in 1968 of the legislative proposals left hanging in 1967.[38] His task force's proposals were therefore supplemental to that main effort, and they concentrated on administrative rather than legislative measures (partly on the reasonable assumption that there might not be any new civil rights legislation). The largest group of proposals concerned housing and these were premised on the assumption that the open housing bill could not pass. Clark conceded this liklihood at the White House meeting on November 24, where the proposals were assessed for final recommendations to the President.[39]

That White House session in late November represented political triage of a hard-nosed variety. The President's senior advisers sought to protect him (and themselves) from endorsing wishful proposals that the Congress would not pass, and that might damage his own and his party's re-election prospects in 1968. At this penultimate level of winnowing, the gates of political and budgetary feasibility were therefore greatly narrowed, and most of the proposals that had proliferated so creatively out in the task forces were rejected, deferred, "passed over," or remanded for further clearance with specific agencies with a presumed interest and possible policy conflict. Politically, this translated into a gauntlet that few liberal initiatives survived. Task force chairman Ramsey Clark was rooted in the liberal-populist wing of his party—his sympathetic response to the frustrations of ghetto protesters in Watts and elsewhere had led press critics to refer to him less as the nation's chief law enforcement officer than as its Public Defender. Yet at the White House showdown meeting, Clark urged the adoption of few of the task force's proposals. He did support, however, a proposal that President Johnson amend Kennedy's original housing order to eliminate the exemptions for one- and two-family, owner-occupied dwellings. This would add 500,000 housing units to the 300,000 covered by the FHA and VA regulations, and it called for enforcement through "affirmative compliance investigations by trained civil rights specialists."[40] But there was virtually no likilhood that Johnson could be persuaded to abolish such a politically crucial exemption by executive order.[41]

Similarly, none of the task force's inventive proposals in the education area were accepted. These included grants by the Commissioner of Education "to reduce racial isolation in public schools" and federal payments to local schools based on a formula to reward "the number of students entering racially integrated classes from racially isolated schools" (the Budget Bureau was horrified at an estimated cost that ranged, on the high-incentive end of the scale, up to $17 *billion* by 1971). Also rejected were Attorney General suits to integrate de facto segregated schools even in the absence of written complaints, and denial of tax exemptions to private schools that avoided desegregation or discriminated on the basis of race. In short, given the narrowing political and budgetary constraints of 1967–68, there seemed little hope in asking a recalcitrant and riot-scarred 90th Congress to support more expensive new civil rights programs.

There also seemed to be too much risk in pressing for stronger use of the government's administrative muscle in such sensitive areas as housing, schools, or Title VII authority on jobs. The best the group could do was to agree, "as rapidly as feasible by quiet administrative arrangements," to require "assurances" of nondiscrimination in employment from state and local agencies receiving federal assistance.[42] That is, if Congress wouldn't extend the EEOC's coverage to the employees of state and local governments by amending Title VII (and the Johnson administration had declined to request such a move from Congress), then the federal agencies should start whittling away faster at the uncovered state and local agencies by applying the leverage of Title VI—as HEW was not-so-quietly doing with the nation's public-school teachers and administrators. When at the close of the disappointingly negative meeting, Califano asked the participants whether there should even be a presidential message on civil rights for 1968, the group agreed that "there should be but that the emphasis should be on what will be done under existing authorities and that accomplishments to date would be emphasized."

The Strange Career of the Open Housing Act of 1968

The pessimistic sentiments reflected in the derailment of most proposals from Clark's civil rights task force mirrored the broader public anticipation that Congress was not likely to pass any significant civil rights legislation in 1968. Congress-watchers agreed that the coalition of northern Democrats and moderate Republicans, which had been essential to passing the laws of 1964 and 1965, was by 1968 in disarray.[43] Senator Dirksen continued his stout opposition to open housing, and the House was clearly in a more conservative and sullen mood. Similarly, the public seemed in no mood to "reward" the Detroit rioters of 1967, and congressmen in an election year were not expected to rally to such volatile issues as open housing and equal employment enforcement, where the redeeming southern focus of the crusade of 1964–65 was lacking. On January 24, President Johnson did indeed greet the returning Congress with a special message on civil rights. But in it he merely reiterated his request of 1967 for the 90th Congress to protect civil rights workers, require nondiscrimination in federal and state juries, grant cease-and-desist authority to the EEOC, and pass the open housing bill—"because it is decent and right."[44]

Then in a rapid and surprising chain of events, the Senate took up the fair housing bill in early February, invoked cloture on a southern filibuster on March 4, and sent the bill to the House. On April 10 the House agreed to accept the entire Senate bill in order to avoid the dangers of a conference, and President Johnson signed the bill into law the following day as Public Law 90–284. It was an astonishing turnaround. The surprising Civil Rights Act of 1968 has not yet found its historian, but a recapitulation of the major steps in the moribund bill's resuscitation identifies the main protagonists in this unlikely drama.[45] Its equally unlikely vehicle was H.R. 2516, the bill to protect civil rights workers that the Senate Judiciary Committee had favorably reported on November 2, and that had nothing whatever to do with housing.

On December 15, Majority Leader Mansfield had placed H.R. 2516 on the Senate's agenda of pending business when the chamber reconvened after the holiday recess on January 15. This procedure was irregular but not uncommon with House-passed civil rights bills, because it avoided the possibility of a filibuster over the mere motion to bring the bill up. Once the bill was the Senate's order of business, proposed amendments could not be procedurally filibustered. Then on February 6, Senator Walter Mondale (D.-Minn.) proposed a major open housing amendment to the relatively minor H.R. 2516. Mondale's amendment was in fact identical to the Administration's open housing bill of 1967, S. 1358, except for its new "Mrs. Murphy" exemption for owner-occupied dwellings that housed up to four families. Mondale's open housing amendment was co-sponsored by Republican Edward W. Brooke of Massachusetts, the Senate's only black member, and was supported by the usual bipartisan coalition of liberal northern senators, but not by Dirksen. The amendment produced the expected filibuster on the merits, led by Ervin, and as late as February 19 the *Wall Street Journal* was referring to "the almost certainly doomed bill banning racial discrimination in housing."[46]

On February 20, however, a surprisingly close cloture vote showed the Senate's Republicans to be evenly split (they had twice voted reliably and heavily against cloture, at Dirksen's behest, in the 1966 open housing debate). This in turn reflected an important twist to the more conservative 90th Congress, one that had gone largely unnoticed following the elections of 1966, when most attention had focused on the 47-seat loss by the Democrats in the House. The Democratic losses of 1966 had obviously strengthened the conservative Republican bloc in the House, now newly led by Representative Gerald Ford of Michigan. But in the Senate, the modest Republican gains had not correspondingly increased the conservativism of the Senate. Rather, they in effect had shifted the considerable strength of the civil rights bloc toward the Republican side of the isle. The net impact of these new and more liberal Republicans—Charles Percy for Democrat Paul Douglas in Illinois, Robert Griffin for Democrat Pat McNamara in Michigan, Mark Hatfield for Democrat Maurine Neuberger in Oregon, and Howard Baker for Democrat Ross Bass in Tennessee—was not lost on Dirksen, who promptly shifted into his negotiating mode.[47]

The result was another famous Dirksen switch and bipartisan compromise to save the civil rights bill. Dirksen offered an amended version of the bill that would cover an estimated 80 percent of all housing, compared with Mondale's 91 percent. The Dirksen version's more tender solicitude for Mrs. Murphy—a Republican hallmark—would exempt single-family, owner-occupied housing *if* it was sold or rented by the *owner* rather than through real estate brokers or agents. Despite its relatively modest Republican trimming on coverage, Walter Mondale called the Dirksen compromise "a miracle." Dirksen explained his sudden and widely heralded discovery that a federal open housing law wasn't unconstitutional after all by observing modestly: "One would be a strange creature indeed in this world of mutation if in the face of reality, he did not change his mind."[48] On March 4 the Senate voted cloture by the precise two-thirds requirement in a 65–32 roll-call vote, and on March 11, following a lively amending process on the floor that concentrated on anti-riot provisions, the Senate passed the bill by a roll-call vote of 71–20.[49]

This plunged the House into a debate over whether to vote directly on the Senate bill, which the Administration and the House Democratic leadership favored, or to send it to conference. Republican Minority Leader Ford called for the latter course, observing correctly that the House in the 90th Congress had never considered the legislation (a quite different version had passed the House in the 2nd session of the 89th Congress in 1966). But both of the major Republican presidential candidates, Nelson Rockefeller and Richard Nixon, urged accepting the Senate version. This debate intensified during a procedural delay caused by the closely divided House Rules Committee, but that was shattered when an assassin gunned down Martin Luther King in Memphis on April 4. In the emotional aftermath of King's murder, the House accepted the Senate version without conference on April 10, by a roll-call vote of 250–172.[50] The House bill then became Public Law 90–284 the following day with President Johnson's signature.

In explaining the surprising turnaround on the open housing bill that spring, Lyndon Johnson's consistent and courageous support is easily overlooked, since it was a background constant. The President had sprung his open housing surprise in January 1966, when it was widely judged as overreaching if not quixotic, and he had kept the pressure on right up until the House accepted the Senate's amendments without a conference 27 months later. Dirksen's leadership switch of course cannot be missed. But other important contributors included House Republicans Charles Goodell of New York, Albert Quie of Minnesota, and especially Ohio's William McCulloch, whose role as senior conservative negotiator in the House was similar to Dirksen's in the Senate—only much steadier. The *Congressional Quarterly* gave especially high marks to NAACP lobbyist Clarence Mitchell for coordinating the efforts of the Leadership Conference.[51]

On the negative side of the lobbying effort, the National Association of Real Estate Boards waited too late to mobilize for opposition the 85,000 members of local real estate boards. The NAREB assumed that the controversial open housing provisions would remain safely bottled up in the Senate, and they were caught napping by the early March cloture vote.[52] Accidents of timing were elsewhere important, especially the April 4 murder of King. Although this occurred *after* the crucial decision in the Senate, it created a national mood of remorse that even the massive urban rioting that followed could not erase. The postmortem on this surprise victory for the civil rights reformers should not minimize the arresting fact that the majority decision in Congress to pass the open housing bill was crystalized and affirmed while Washington burned.

Clearly such extrinsic factors as the ghetto riots of 1964–68 cut both ways. Most contemporary attention was given to the backlash components, and they explain why the open housing bill was balanced in 1968 by tough anti-riot provisions. Indeed, what is commonly called the Open Housing Act of 1968 was in fact a three-part civil rights bill. Its core was the bill to protect civil rights workers, and this was balanced by two comprehensive amendments, one on nondiscrimination in housing, the other to punish the perpetrators of riots. Sophisticated Congress-watchers like the *New York Times* and the *Wall Street Journal* were struck by the rather naked role-swapping that this carrot-and-stick package seemed to require of both liberals and conservatives, with law-and-order conservatives demanding federal intrusion into traditional state jurisdiction over rioting and arson, and with liberal senators suddenly discovering the sanctity of state rights.[53]

Members of the 90th Congress on both sides of the isle vented their frustrations against the racialist demagoguery of Stokeley Carmichael and H. Rap Brown. But the National Advisory Commission on Civil Disorders, which was appointed by Johnson in the wake of the Detroit riot of 1967, released its report on February 29 with strong expressions of sympathy for the ghetto protestors, and also with well-timed support for the open housing bill as an essential means of reducing the explosive pressure.[54] Dirksen explained his switch in part by observing that the riots of 1967 "put this whole matter in a different frame." "I do not want to worsen," Dirksen said, "the restive condition in the U.S."[55] In his study of the political impact of the 1960s riots, James W. Button concludes that despite the public rhetoric of backlash, the policy response of the federal government toward redistributionist social programs shifted in a strongly *liberal* direction throughout the Johnson administration, and basically shifted rightward only after the outbreak of student and antiwar violence in 1968–69 that led into the Nixon administration.[56] Button emphasizes not only that the open housing law of 1968 was passed after the assassination of King and the rioting that followed, but that the less widely publicized but landmark National Housing Act was passed even later in 1968. As Harry McPherson observed, "Martin Luther King's death provided the final impetus for one more civil rights law, just as the abuse he bore during his life had helped to persuade the nation, and then the Congress, that the earlier laws should be passed."[57]

The Lessons of 1968

If this was a generally cheering conclusion to draw from the turmoil of 1968—and it was—then some balancing observations were also in order. One was that quite unlike the landmark civil rights laws of 1964 and 1965, none of the titles of the civil rights legislation of 1968 was to have a major impact, at least not directly, through enforcement of the new statutes themselves. Both the anti-rioting and civil-rights protection provisions were passed largely as symbolic responses to violent phenomena that had already peaked—the "long, hot summers" of ghetto rioting from 1964 through 1968, and the parallel attacks on civil rights workers in the Deep South. Federal authorities would employ some of the new anti-riot authority of 1968 to the subsequent wave of campus and antiwar violence from the more radical left, such as Weatherman and the Symbionese Liberation Army. But the statutory authority of 1968 was less crucial to their efforts than the anti-conspiracy laws of 1870–71. As for housing, the patterns of cause and effect are complicated, but a reasonable conclusion appears to be that while the open housing law of 1968 had an important indirect and longer-range impact on discrimination, its direct and immediate impact was quite modest—although not so modest as its liberal detractors complained.[58] Both points bear more elaboration than can be accommodated here, but such judgments hinge heavily on the question of enforcement, a crucial element that was almost ignored in the housing debate of 1968.

Despite the many broad differences between the civil rights battles in Congress of 1964 and 1968, including the lasting importance of their results, one common ingredient was a scenario that hinged on the fabled Dirksen switch and compromise. These courtship dances surrounding Dirksen had the effect of muzzling serious analysis of enforcement problems and machinery.

Dirksen's compromises tended to balance relatively liberal provisions for coverage (such as exemptions for "Mrs. Murphy" and religions organizations) with conservative models of enforcement (such as the largely voluntaristic EEOC). They were secretly negotiated, but with maximum publicity over the brewing deal. Then they were announced to a breathlessly expectant public—whereupon they were then stoutly defended by the Administration and the congressional liberals. This shielded them in congressional debate from the customarily vigorous scrutiny of the left, which otherwise was quick to condemn "weak" and "toothless" social reforms. Dirksen's compromise package for 1968 thus closely resembled his model of 1964. It asserted a right to nondiscrimination in housing, as in jobs, as essentially a private right. Aggrieved complainants might seek injunctive relief and damages through the courts, wherein the "burden of proof shall be on the complainant."[59] They might also file written complaints with the Secretary of HUD. But the secretary's authority, beyond receiving subpoena power to call witnesses in hearings, was limited to seeking voluntary compliance through the conciliation process. Also as in 1964, the federal enforcement administrator could not bring suit, although the Attorney General was authorized to bring suit in pattern-or-practice cases.

Again as in 1964, but much worse, the way Congress passed the bill prevented a full deliberation on its contents. In both 1964 and 1968, the strangely irregular nature of the power-play politics, which the Administration and the congressional liberals found necessary to short-circuit the conservative defenses, also thereby circumvented the standard hearing process. The normal legislative model provided for negotiation and bargaining in the mark-up sessions of well-staffed committees, where the legislators necessarily grappled with important but often technical questions of definition and enforcement authority, with majority and dissenting views captured in the committee reports. Not so with the open housing bill in the 90th Congress. Instead, in this vital respect it more closely resembled Judge Smith's sex discrimination provision of 1964. Passed over by the House committee in 1967, the Administration's housing bill was rammed backwards through the full House in 1968 without a conference committee or even floor amendments. Introduced subsequently in floor debate in the Senate, it produced no clarifying committee or conference reports at all.

Thus the level of debate during the 1968 Senate discussions of February 6 through March 4 remained at the highly generalized level of discourse that the senators thought appropriate for the public toward whom they were aiming their remarks. The result was too much of the customary, philosophical posturing by conservatives like Ervin, who compulsively lectured his colleagues and the bemused galleries on the constitutional wisdom of Burke and the Elder Pitt. This stylized genre of debate was accompanied by lengthy exchanges over the disputed percentages of housing to be excluded under the varying Mrs. Murphy provisions. Liberal sponsors of the housing amendments, like Mondale and Hart, countered that their slightly more modest protection of Mrs. Murphy's privacy better balanced the public interest in broad coverage for a moral imperative. But the difference between 80 percent and 90 percent coverage did not easily translate into matters of highest principle.[60] The alert public understood quite well the nature of the line-drawing that the senators were arguing about, and on which side of that line their home seemed to fall.

The spirited debate over coverage, however, was not matched by any serious debate over enforcement. The Mondale amendment of February 6 was essentially Johnson's moribund fair housing bill of 1967. It had provided for enforcement by the HUD Secretary, rather than by an unknown new entity like Representative Conyers' ominous-sounding Federal Fair Housing Board, which had actually passed the House in 1966. But Johnson's bill had called for empowering the secretary with a circumscribed yet potentially creative form of cease-and-desist authority. This prudent new compromise merited a full congressional hearing that it never even began to receive. Instead, Mondale offered it as a floor amendment without hearings, and Ervin predictably denounced it as giving the Secretary of HUD the "functions of prosecuting witness, investigator, prosecutor, jury, judge, and executioner."[61] Then when the Dirksen compromise stripped the HUD Secretary of the roles of judge and executioner, the liberals rallied to defend the compromise as the best they could get (and a "miracle" at that), and thus silenced any countervailing debate on their left.

Small wonder, then, that when the bill was passed, Congress funded its enforcement quite modestly. In response to the Administration's request for $11.1 million for FY 1969 to employ 850 enforcement officers, Congress appropriated only $2 million. Small wonder, too, that the Civil Rights Commission would continue to criticize the law's weak enforcement throughout the next decade.[62] But throughout the open housing debate of 1966–68, contemporary authorities had cautioned that such laws were unlikely to bring about rapid change in patterns of racial segregation. Oliver Hill, director of intergroup relations for the FHA, explained to the *Wall Street Journal* that even cities with the most active fair housing groups, such as in the District of Columbia, San Francisco, and St. Louis, worked primarily "to get higher-income Negroes into higher-priced neighborhoods."[63]

The *Journal* had predictably opposed the open housing bill. But when the bill passed, the *Journal* editorially greeted its passage with surprising warmth. The *Journal* editors acknowledged that the new law would "bring some real benefits" as well as create some risks. They argued in its behalf that, like the massive desegregation of southern public accommodations in 1964, the 1968 open housing law would relieve the developers of large tracts of their fears of competitive disadvantage that had hitherto sustained so much underhanded resistance to integrated housing.[64] A decade later, the director of the National Committee against Discrimination in Housing told the House Judiciary Committee that while the enforcement machinery of the 1968 open housing law had been ineffective, the law had nevertheless made open housing a national policy, and discrimination in most of the nation's housing a crime. "This, in itself, is a measure of progress," he concluded, "because resort to elaborate concealment makes discrimination less effective."[65] Later studies using census data indicated that the rapidly growing black middle class of the 1970s and 1980s found outlets in the suburbs, but that unlike upwardly mobile Asians and Hispanics, black Americans continued to live in racially identifiable neighborhoods.[66] How much of this mixed pattern of middle-class growth and mobility for blacks, coupled with both racial and class isolation for the black underclass, can be attributed to the open housing compromise of 1968 remains debatable—like so much of the reform of the Great Society. But that it was even attempted in 1968 was a surprising testimony to the strength of the reform impulse even in the aftermath of Black Power and ghetto rioting.

Congress, then, in 1968 fought through to its last great victory in civil rights in its inimitable 1960s style. It then apparently regarded the prickly matter as largely dealt with, and removed the question from its active agenda for another generation.[67] On the other hand, in 1968, Congress for the first time entirely ducked the equal employment issue. Johnson had asked in January, as he had in 1966 and 1967, for a congressional grant of cease-and-desist authority for the EEOC. But Dirksen remained adamantly opposed, and neither chamber took any action. The postponed battle over enforcement power under Title VII would return under the Democratic-controlled 91st and 92nd Congress, to engage the Republican administration of President Richard Nixon. In the meantime, the EEOC would continue its low-profile campaign of bureaucratic aggrandizement, as indeed had all regulatory agencies before it—comforted, like them, in the knowledge that the agency's interest and the national interest remarkably coincided, and that Congress poorly understood the importance of their mission. For relief from the Congress's narrow construction of the EEOC's authority in Title VII, the agency would have little success persuading Congress. But it would turn with increasing vindication to the federal courts.

In 1968, however, the open housing debate had combined with another destructive wave of ghetto rioting to underline the Kerner Commission's conclusion that white society was conspiring in the Negro's entrapment in the inner city slums.[68] With open housing now established at least on paper as national policy, black jobs remained the key to escape from the ghetto. The key to providing those jobs, in turn, was destruction of the historic impediments to minority hiring, and access to the education and training required to qualify miniorities for advancement. As the proliferating new grant-in-aid programs of the Great Society extended the hand of federal assistance to most major areas of school, hospital, and urban renewal construction, the Johnson administration turned increasingly to its Title VI authority from 1964, and especially to the contract-enforcing machinery created by Executive Order 11246, in an attempt to find an effective way to transform that leverage into jobs for unemployed or underemployed blacks. In this quest, the Labor Department under Secretary Wirtz would invent the Philadelphia Plan. Its birth in 1967 and its meteoric career in 1968 would add a new and complicating dimension to the political and bureaucratic warfare over the direction and force of civil rights policy. This new battle would center on the unlikely figure of the Comptroller General of the United States, whom few knowledgeable observers outside the Washington beltway could even name prior to 1968.

CHAPTER XI

From Johnson to Nixon: The Irony of the Philadelphia Plan

Origins of the Philadelphia Plan: The Nettle of Construction

One of the more striking ironies of the era of civil rights originated in the city of Philadelphia. There a controversial plan to integrate racially the city's construction industry was created under the nurturance of the Johnson administration. When the plan's requirements for racial hiring brought attacks from builders and from congressional conservatives, the Johnson administration rescinded the plan. Following Richard Nixon's presidential victory in 1968, however, his putatively more conservative Republican regime revived and strengthened the Philadelphia Plan. The irony seemed especially painful for Philadelphia, where the normal urban pathology of the industrial Northeast defined its own irony in the historically Quaker "City of Brotherly Love." So it had been during World War II, when 5,000 army troops were required to put down the racially triggered Philadelphia transit strike of 1944. So again, in 1963, black protest against white union monopolies in the construction trades in Philadelphia triggered an outburst of racial violence that followed on the heels of the violence in Birmingham. When angry pickets led by the Philadelphia NAACP sought to close down a school construction site in June 1963, thirty-nine people were injured in the turmoil.[1]

Philadelphia's AFL-CIO Council was embarrassed by the racial clash over construction. The labor council had been working since the local AFL-CIO merger in 1960 to integrate the city's construction unions. In February 1963, however, the council's 23-member Human Rights Committee had conceded their failure. In a bitter report the union committee condemned with "vigor the failure within our own family" to "break down the pattern of segregated locals."[2] The report noted that C.J. Haggerty, the national president of the AFL-CIO Building and Construction Trades Department, had made "strong speeches" against union discrimination, but the national federation possessed little leverage on the union locals beyond jaw-boning rhetoric. "We have been

278

unable to get Mr. Haggerty to take an interest in our local problem," the report complained, even to the point of answering committee letters seeking his assistance.[3]

The outburst of violence in Philadelphia that June produced little change in Philadelphia itself.[4] But in Washington, where the Birmingham crisis was transforming President Kennedy's civil rights bill, Lee White used the Philadelphia incident to help persuade Kennedy to issue Executive Order 11114, which extended the PCEEO's authority to include federally assisted construction.[5] Within six months John Kennedy was dead, and President Johnson drove the civil rights bill he inherited through the Congress the following June. In September 1965, Johnson issued the confusing brace of executive orders that abolished the PCEEO and established a decentralized enforcement network, and the result was a jerrybuilt structure that faithfully reflected the realities of political power in the subpresidency. But in doing so, it purchased bureaucratic peace at the usual price of weak coordination, program overlap and duplication, and unclear accountability. The civil rights lobby had shouted its dismay at Johnson's complicated rearrangement of the enforcement machinery in the fall of 1965. In doing so they unfairly impugned Johnson's motives, but the result scarcely constituted a policy for coherent implementation. No longer was there a titular czar for civil rights, in the vicarious form of the Vice President. Indeed, there was no pretense of a civil rights czar—just as there was no education czar, and no urban policy czar. The pluralistic network of American government just didn't work that way—for better and for worse.

With the PCEEO now abolished, President Johnson in 1966 appointed yet another interagency committee at the White House level to monitor the effectiveness of his new, decentralized enforcement machinery. The committee was announced as a promised follow-up to the White House Conference on Civil Rights; it was chaired by Harry McPherson, with staff provided by the Budget Bureau.[6] One major consequence of this rather typical attempt to coordinate the civil rights-related activities of so many otherwise unrelated functions and agencies was that there seemed to be no place in the scheme for the unique construction industry. There were three main reasons for this. The first was that federal construction assistance (meaning tax dollars) was usually given not directly to construction companies themselves, but rather to "applicants," which typically meant units of state and local government. Cities and counties and state agencies asked for federal money to help build roads, bridges, dams, hospitals, schools and colleges, public housing. Thus Part III of Johnson's executive order of September 1965 had dealt separately with federally assisted construction, and it did so in elaborate language that underscored, to the practiced eye, the considerable weakness of attempting to apply such direct sanctions as contract cancellation to such an indirect flow of federal aid.

The second reason why construction couldn't fit into the normal assignment of agency jurisdictions was the inherently fluid, ephemeral, project-centered nature of the construction enterprise. Typically, prime contractors formed ad hoc clusters of subcontractors to bid for projects, and the winners (theoretically the low bidders) hired their workers not directly from the street or employment agencies, but rather from the union hiring halls under prior and exclusive union contracts. Thus the contractors typically did not really hire their own workers, and the workers were not permanent employees of the construction companies. Federal agencies could well use their burgeoning contract muscle to bludgeon a heavily dependent Lockheed-Marietta or a Newport News Shipbuilding and Dry Dock Company into changing their employment policies. But no agency could effectively order a construction company to change the work rules of a standing union contract.

The third reason was a corollary of the second: admission to the union hiring halls was in turn controlled mainly by the feeder apprenticeship programs. This was a bottleneck of entry that was guild-like in its design and in its role as jealous guardian of the craft.

The Apprenticeship Game

Since its creation under the New Deal in 1937, the federally funded apprenticeship program had served the dual function of controlling the supply of union labor in the skilled crafts and guaranteeing quality control through rigorous training. But it had worked all too well. By 1968, with the robust economy driving a building boom with inflationary pressures, the Bureau of Labor Statistics projected a need for four million additional craftsmen during the decade 1965–75. But the apprenticeship program was producing only about 30,000 completions a year—less than 10 percent of the new craftsmen needed annually. In response, the Labor Department called for the apprenticeship program to grow to one million annually. But this was not well received by organized labor. Union leaders, especially in the skilled crafts, tended to respond to such proposals much as the American Medical Association responded to demands for a massive building of new medical schools to flood the country with doctors. As a result the federal-state program supported only about a quarter of a million apprentices annually, with slightly more than half of them in the construction industry. Each year about 75,000 new apprentices would begin training. But more than half that number would quit. With wages increasing at the rate of 15 percent a year in the skilled construction trades, and with employment of craftsmen and foremen increasing at almost half that rate annually, the low-producing apprenticeship program seemed a self-defeating bottleneck. But there was a self-conscious, guild-related reason for it.[7]

The national apprenticeship program, especially in the construction trades, had evolved into what Willard Wirtz's successor, labor economist George P. Shultz, called a "craft elite" program. Completion often required as many as four years of intensive, high-level training—usually far in excess of what was normally required for journeyman labor. Accordingly, the majority of journeymen did not acquire their skills by completing an apprenticeship program.

Instead, the program produced a disproportionate number of the construction industry's work supervisors, foremen, and self-employed contractors.[8] It had become the equivalent of a business college education translated to a craft setting. This is what provided the long-range incentive for apprentices to work so long at subjourneyman wages (plus the shorter-range incentive of deferment from military conscription). Relatives of journeymen received preferences through a kind of white old-boy network. In fact the apprentice system performed its "elite craft" function of upward mobility quite well. It did so at unusually low cost compared with other manpower programs, and it maintained consistently high standards of quality. But at the same time it reinforced, especially in construction, the white dominance of the inherited social order. Small wonder, then, that so few minorities had cracked their way into the ranks of the skilled construction trades.[9]

In 1960 there were only 2,200 active nonwhite apprentices. This amounted to only 2.5 percent of all apprentices—the same percentage as in 1950. In 1963 Secretary Wirtz had issued anti-discrimination guidelines for the program. But the employer-union constituency that jointly operated the programs had insisted on maintaining high standards of quality for competitive entry. The guidelines stipulated that applicants would be chosen from the top of an "ability ranking list," whereon they would be "selected and ranked solely on the basis of qualifications alone in accordance with objective standards."[10] Final screening often included oral interviews. By 1966, Wirtz was able to report that nonwhite apprentices had increased only to 4.4 percent.[11] But by then, the cities were burning, and the old admonitions would no longer suffice.

The upshot of all this was increasing frustration from all quarters with the post-1965 machinery that Johnson had established to enforce nondiscrimination. Like its predecessors, it still lacked a promising mechanism for making any fundamental changes in a federally assisted boom in construction, one that seemed to mock the Administration's promises by subsidizing the maintenance and growth of employment patterns it was supposed to change. By the late 1960s, federal funds were reaching 225,000 contractors involved in $30 billion in annual construction. This gave the federal government a direct economic impact on 20 million workers, or almost one-third of the entire U.S. labor force. While the potential federal leverage was enormous, however, the exercise of direct EEO influence had thus far been negligible. The challenge that faced the Johnson administration was to devise a strategy of implementation that would transform a residual and intractable headache for civil rights enforcement into a unique opportunity to use federal pressure to break down the racial barriers.

By 1966, the Johnson administration's assignment of enforcement responsibilities under the executive order, construction projects excluded, seemed to be sorting themselves out. Nondiscrimination in federal employment itself, which had formerly been watched by the now defunct PCEEO, was entrusted to John Macy's Civil Service Commission.[12] Minority employment in the federal government continued to accelerate, although mostly at the lower grade levels. The new EEOC was responsible for nondiscrimination in private employment under Title VII of the Civil Rights Act of 1964. Title VI, on the other hand, required nondiscrimination in all federally *assisted* programs. This meant, in effect, grant-in-aid programs like federal aid to education rather than the standard procurement contracts of agencies like the Defense Department and its purchase of tanks and bombers. Because so many of the Great Society's new grant-in-aid social programs dispatched their funds directly to state and local governments, and thus necessarily involved delicate legal and political questions of policy control, Title VI enforcement was coordinated by the Justice Department.

Title VI, then, was the prickly one. Brief and straightforward on its face, with its simple threat of fund cutoff for not complying with federal contracts, it was a legal and poliltical minefield. The storm over cutting off federal school aid in Chicago in 1965–66 had quickly reinforced the instinctive deliberateness of Justice. Consequently throughout the remainder of Johnson's presidency the Budget Bureau reiterated, in its frequent evaluations for McPherson's oversight committee, its standing criticism that Title VI was being enforced too timidly.[13] Justice coordinated Title VI *policy,* but Justice could not monitor the EEO compliance of hundreds of thousands of contractors in procurement and construction annually. With the PCEEO abolished in 1965, enforcement of nondiscrimination in federal contracts fell naturally enough to the Labor Department, which had backstopped the government's various contract committees since 1953. But no Secretary of Labor, not even a secretary as respected and as liberally inclined as Willard Wirtz, could order the other cabinet officers and their agencies about. This had always been a prime weakness of interagency coordinating committees, irrespective of which agency may have been given a "lead" role. Yet in the case of the construction industry, this arrangement inadvertently worked out to be a unique source of strength.

Edward Sylvester and the Office of Federal Contract Compliance

When Johnson in September 1965 designated Labor as the lead agency in coordinating contract compliance, Wirtz promptly established a new coordinating and enforcement entity. Unfortunately it was given an awkward bureaucratic name with an unpronounceable acronym: the Office of Federal Contract Compliance Programs (OFCC).[14] Wirtz appointed as OFCC director a black administrator in Labor's international bureau, Edward C. Sylvester, Jr. A graduate of Wayne State University in engineering, Sylvester had worked as a civil engineer for the city of Detroit during the 1950s; he joined the Labor Department in 1961 after working in the presidential campaign of Senator Stuart Symington. Early in 1966 the OFCC began implementing Johnson's executive order of 1965 by assigning to the various agencies what the order called "primary responsibility" for EEO contract compliance. This was determined by simply assigning responsibility according to dollar volume. That meant, not surprisingly, that the largest share of procurement contracts went to Defense. But it also meant that the upstart OFCC had inherited a political problem it was ill-equipped to handle.

During the era of the PCEEO, Defense had taken a strong and surprisingly effective lead in attacking racial discrimination as practiced by such major suppliers as Lockheed-Marietta and Newport News Shipbuilding. But the giant agency was not about to reprogram its basic contracting procedures in order to conform to the different priorities and procedural preferences of a tiny new office in another agency that was itself dwarfed by the DOD. Neither was the Pentagon, or any other agency with normal Weberian instincts, interested in submitting its vital contract-making power to an implicit veto by a network of compliance officers whose ties to the parent agency were strained by competitive loyalties of ideology, and often also loyalties of race.

Since 1961 the Pentagon had required explicit commitments to nondiscrimination in its standard contracts, and it stood ready to enforce them with vigor upon complaint and a showing of cause. Then in 1966 came the bureaucratic challenge, when the new OFCC circulated a draft of its proposed new rules and regulations. At their heart was a requirement for OFCC clearance on nondiscrimination grounds *prior* to any agency's award of contract. The Pentagon replied on November 25 that this was quite impossible. "Several new requirements of the [OFCC] regulation," wrote Deputy Secretary of Defense Paul Nitze, "could not be accommodated by this Department."[15] When the OFCC persisted, Nitze wrote directly to Secretary Wirtz to explain in unusually blunt language that the proposed new pre-award procedures would be "crippling" to the Pentagon's ability to "accomplish its procurement mission in providing for the national defense." The delays involved would be "totally impracticable in the Department of Defense," Nitze said, "which handled over 15.1 million contract actions in the last fiscal year."[16] Such a contest was highly uneven, so Defense simply ignored the OFCC's pre-award maneuverings.

The OFCC experienced similar frustrations of inadequate leverage in many of its other initiatives. These included attending first to cleaning up its own backyard by increasing the number of minorities in Labor's own apprenticeship programs.[17] It also required this new subagency with a clumsy title, an unclear mandate and authority, and only 28 employees to respond with the customary due process to the trickle of formal complaints (averaging only 375–400 a year for the entire country) that reached such an obscure new office, and to pursue the protracted process of barring noncooperative contractors from future contracts.[18] Early in his tenure as head of the OFCC, Sylvester talked freely to the press about contract cancellation and debarment.[19] But while the OFCC's posture was more aggressive than that of Lyndon Johnson's PCEEO, its behavior did not fundamentally differ from its predecessor's in regard to policing problem companies.[20] Like the PCEEO, the early OFCC mainly went after either southern segregationist targets, like the Armstrong Rubber plant in Natchez, Mississippi, or after northern in-transigents like Timken Roller Bearing, a company in Canton, Ohio, known for its rock-ribbed Republican philosophy and its president's sponsorship of conservative causes.[21] When the OFCC teamed up with the EEOC, Justice, and Defense to force a settlement on Newport News Shipbuilding, the results were impressive.[22] All four agencies enjoyed claiming major credit for the big Newport News settlement. But such major showdowns were rare, mostly be-cause the major southern defense employers rather quickly ended their old segregationist behavior. So the OFCC was left to grind down slowly such small-fry recalcitrants as the right-wing Republicans at Timken Roller Bear-ing.[23]

The Virtues of Necessity: Exploiting Pre-award Leverage

If the Labor Department was still struggling with unimpressive results merely to desegregate its fully subsidized programs in apprenticeship and state em-ployment services, how could it expect to make a major dent in the discrim-inatory traditions of the thrice-removed construction trade unions? Unlike Title VII, which gave the EEOC direct jurisdiction (but no direct power) over all but the smallest unions, Title VI and Executive Order 11246 provided for direct government leverage only on contractors, not on unions. Furthermore, the construction industry presented unique problems because manpower was organized on an area and skill basis rather than on an employer basis; job assignments were generally controlled through strong union hiring halls; and employment was intermittent. Thus the Labor Department's standard model of assigning government contractors to their dominant funding agency, such as the Pentagon or HEW, could not be effective in the mercurial construction industry. Unlike airplane and ship builders, construction companies hustled bids and built structures for all federal agencies, especially the thousands of state and local government agencies, but they rarely worked primarily for a single agency that might be given the "lead" role in compliance monitoring.

For these reasons the Labor Department as early as April 1965 had begun to establish for the construction industry a system of "area coordinators" on a metropolitan or labor market basis.[24] Then in May of 1966, the new OFCC established its crucial *pre-award* program for construction only. Its strategy was to take advantage of the government's greater leverage *prior* to awarding contracts. To do this the OFCC began to require, initially on contracts of $1 million or greater, that the potential recipient's employment *system* must *first* be reviewed and approved for compliance with the executive order. Final contracts, then, could not be awarded without a finding by the compliance officer that the contractor's employment system and data were "in compli-ance."[25]

Sylvester enjoyed a reputation for energy and competence. But such a pre-award system required no special insight or strategic brilliance. Indeed, its development was virtually invited by the language of the executive order itself, which called for the submission by bidders of "Compliance Reports *prior to* or as an initial part of their bid or negotiations of a contract [emphasis added]." The upstart OFCC, however, lacked the muscle and reach to force the giant mission agencies to police their client relationships in such a radical new fashion. The OFCC's frustrating contest with the Defense Department over pre-award authority reflected a lop-sided equation of bureaucratic power that would defeat the pre-award strategy under normal circumstances. In the standard system of "lead" agency assignment, the contract compliance officers worked for and reported to their own agencies, where they normally directed their loyalties and found their career rewards. By 1966 the mission agencies were rapidly dismantling the vestiges of outright racial segregation, and they could effectively police the normal requirements of nondiscrimination. But they were in no position to inject a racial numbers game into the calculus of employment, and thereby risk the paralysis of important procurement routines that would surely flow from such a radical departure.

In the amorphous world of construction, however, where contractual relationships seemed more ad hoc and transitory, the OFCC was able to dilute and even transfer to itself normal patterns of employee loyalty. Compliance officers remained on the payroll of their funding agency. But in the OFCC's new scheme of enforcement for the construction industry, agency compliance officers were assigned to report directly to the OFCC's assistant director of construction, Vincent G. Macaluso.[26] The OFCC enjoyed another special advantage in construction: much of the Great Society's urban renewal construction was funneled through HUD, and the OFCC enjoyed a special relationship to the new housing agency. Lacking the established ties of clientele interest that characterized the veteran mission agencies (certainly including Labor), HUD was also generally staffed by a younger and more socially sensitized cadre of civil servants, and was headed by the Administration's lone black cabinet officer, Robert P. Weaver.

By 1967 the OFCC, like the EEOC but on a smaller scale, was accumulating a staff of activist blacks and white liberal reformers whose zeal for enforcement mirrored that of their young, attractive, and aggressive leaders— Edward Sylvester at the OFCC and Clifford Alexander at the EEOC. The OFCC was then ready to apply its new, coordinated, area-based, pre-award system to construction in selected experimental cities. In the first three cities the OFCC tried—St. Louis, San Francisco, and Cleveland—the experimental process of trial and error produced more knowledge about what didn't work and why than about what did. But on the fourth try, in inhospitable, unlikely Philadelphia, the OFCC found its elusive model.

The OFCC's first experimental attempt, in St. Louis, was almost immediately short-circuited by a Justice Department suit that threw the entire question into the courts for several years. The Labor Department had begun attempts in 1965 to desegregate the construction trade locals in St. Louis, where the National Park Service was building the soaring visitor center in the Gateway Arch Park. On January 6, 1966, the OFCC officially launched its pre-award program by providing data sheets that instructed builders to "actively recruit minority group employees for work in the trades where they are not now frequently represented."[27] On the following day in St. Louis, the Hoel-Steffen construction company, under heavy pressure from the Labor Department, hired a black plumber and two black helpers who were not members of AFL-CIO Plumbers Local 35. This triggered a walkout by five locals of the Construction Trades Council, which in turn triggered the first pattern-or-practice suit filed by the Justice Department under Title VII.[28] The Justice brief charged that the combined union membership of the pipefitters, electrical and sheetmetal workers, and plumbers in St. Louis was more than 5,000, of whom only *three* were black. The suit by Justice spun out into a series of complex and lengthy trials, and in such a legal morass, the OFCC's new pre-award program had no chance for a normal test. So Sylvester's initiative shifted to San Francisco.

In the San Francisco Bay area, the OFCC early in 1967 launched a slightly different plan, this one centering on the huge federal investment in the Bay Area Rapid Transit system (BART). The plan required contractors to provide detailed plans addressed to such specific union items as joint apprenticeship committees, pre-apprenticeship training, and journeyman referrals. But the San Francisco plan specified no standards by which employer performance could be judged. Moreover, the Bay Area was such a labor stronghold that the contractors produced, according to the Labor Department's solicitor for civil rights, "a program of paper compliance" based on "beautiful affirmative action programs which resulted in few minority placements on the job."[29] There were so few, in fact, that in 1967 the Civil Rights Commission held extensive hearings in San Francisco on union discrimination in the construction trades.[30]

The OFCC's third attempt concentrated during 1967 on the seven-county Cleveland area, where Sylvester announced in December 1966 that it would require pre-award conferences affecting $125 million in hospital and school construction.[31] On March 15, 1967, Sylvester further announced that all pre-award affirmative action plans must "have the *result* of assuring that there was minority group representation in all trades on the job in all phases of the work [emphasis added]."[32] Sylvester lamented to Wirtz that the government's basic weakness was applying pressure, either directly or indirectly through a local government agency, "on contractors to produce results notwithstanding the inherent problems of collective-bargaining agreements, hiring hall arrangements, recalcitrant unions, or the availability of qualified Negroes." "Further," Sylvester complained, "there has been no demonstration by the Government that it will shut down contracts. There is too much carrot and not enough stick."[33]

But Sylvester was starting to lay on more stick. In June 1967 a major and carrot-hungry NASA contractor offered, in response to the suggestive coaching of pre-award negotiations, to provide a "manning table" for a Cleveland job. This crucial piece of paper would list the specific *number* of minority workers to be hired in each trade. It amounted to a contractor's promissory note that if he was awarded the job, he would hire X number of minority workers (meaning, in Cleveland and similar industrial cities, black male workers). The logic of the numbers game pressed relentlessly toward such a quota requirement. The contractors resisted this evolution, fearing to be caught between their government paymasters and the deeply entrenched unions. But there was a certain iron inevitability to the suggestive invention of the "manning table" by that first hungry contractor, angling to beat out his rivals for the juicy NASA contract. And once he signed on, his competitors could scarcely afford to remain far behind. Once the low-bid protocol was breached, contractors would scramble to meet the new criteria.

So the OFCC pounced on the well-coached offer of the NASA contractor, guaranteed his contract award, and then delayed awarding $83 million in Cleveland area contracts while pressing the manning table device on the other bidding contractors. By mid-November, the contractors had taken the hint and reinforced their bids by promising to hire 110 minority craftsmen out of total crews of 475 in the mechanical trades and for operating engineers.[34]

Wirtz, Sylvester, and the Philadelphia Pre-award Plan

Sylvester's memo to Wirtz in March 1967 had mentioned only the first three experimental cities, not Philadelphia. In May, Wirtz issued a related order requiring written "assurances" from *all* bidders on federal contracts (not just in construction) exceeding $10,000 that they did not maintain segregated facilities.[35] This seemed odd at first blush, since such old-fashioned segregation had been federally banned for several years, and would seem to require no new directives. But Wirtz had a hidden agenda. He wanted to lay the legal groundwork for applying OFCC sanctions quickly, in the form of criminal prosecution against contractors who filed fraudulent assurances.[36] The noose was tightening. But the OFCC's coordinating and enforcing mechanisms were still third-party and confusing. Cleveland, for example, was a prime example of what Richard Nathan called "some truly unique interagency relationships," with the area coordinator for construction being an employee of HUD, sharing a field office with Labor's Bureau of Apprenticeship and Training, and yet reporting directly to the OFCC.[37] This messy model, however, was soon to yield in a fourth city to a leaner and cleaner one, where local rather than Washington-based initiative would force the coordination.

On June 14, Sylvester informed Wirtz about a new proposal in Philadelphia.[38] It differed from the previous three in that it was locally proposed by Philadelphia's Federal Executive Board (FEB). These rather obscure interagency boards were first established in 1961 to coordinate better the disjoint programs of federal agencies in the field, rather than through the central agency nexus in Washington. The FEBs were similarly structured. Philadelphia's FEB was composed of the top federal officials in each contracting agency in the five-county Philadelphia area. But its moving spirit and chairman was Warren P. Phelan, the aggressive regional administrator for HUD. Phelan had forged an agreement among his FEB colleagues that federal muscle must be used to break the lily-white grip of the construction trades in the area. Racially explosive Philadelphia must not burn like Watts or Detroit. Each FEB member had obtained prior approval from their home agencies in Washington to support the Philadelphia initiative.[39] Their instrument of enforcement was an aggressive compliance committee composed of representatives from the EEOC, the OFCC, the Community Relations Service, and the U.S. Attorney's office.

The FEB's goal was simply to force integration on the construction trades unions by concentrating irresistable eonomic leverage on all the area's major builders. While the FEB worked through the summer of 1967 toward unveiling their plan by mid-fall, Sylvester alerted all federal agency heads to the Philadelphia initiative and pledged Labor's strong support. John Macy similarly alerted all chairmen of FEBs across the country, notifying them that a promising new model of EEO enforcement was being forged in Philadelphia, with the little-known FEB as its central mechanism.[40]

On October 27, 1967, FEB chairman Phelan sent out his operational plan.[41] To justify his get-tough measures, Phelan first surveyed the patterns of union membership in the Philadelphia area. In light of the subsequent controversy over the Philadelphia Plan, the surprising result of Phelan's survey was lost in the turmoil over the racial exclusiveness of the skilled construction locals. But the FEB survey revealed that the white-union problem was largely confined to a small but highly visible minority of craft unions that held only about 4 *percent* of the area's 225,000 union members. In all the subsequent furor over the Philadelphia Plan, the significance of its context—that the plan was targeted against only a 4-percent problem of blatant union discrimination—was quickly lost.

It was not, however, an insignificant 4 percent. These locals attracted the FEB's ire through a combination of high visibility and pay, lily-whiteness, and defiant resistance to change. Of the 75,000 unionized construction workers in the Philadelphia area, approximately 28,000 were members of 22 building-trades locals. But the majority of these locals nevertheless had what Phelan agreed were "sizable minority memberships." Phelan and his colleagues had originally included the 5,000-member Local 542 of the Operating Engineers, for example, on their target list of eight discriminating construction locals. But the FEB survey revealed that Local 542 had made rapid recent progress in response to the combined pressures of the national AFL-CIO, federal contracting officials, and local black organizations. By the summer of 1967 the operating engineers, formerly all-white, included 800–900 minority workers and 150 minority apprentices. So Phelan dropped them from the hit list. But several highly skilled—and highly paid—locals remained.

By the mid-1960s, black workers in the American construction industry nationwide were typically concentrated in the laborer locals, with substantial representation in the lower skilled and lower paying "mud" or trowel trades. Philadelphia's FEB identified seven offending locals in Philadelphia that remained virtually all-white. These were the high-paying electrical and sheet-metal workers, plumbers, roofers, structural ironworkers, steamfitters, elevator constructors, and stone masons. Phelan estimated their combined membership at 8500–9000, including 700 apprentices. But his survey revealed that this included only 25–30 minority journeymen, and perhaps 15 minority apprentices and helpers/tenders.

Phelan's plan of October 1967 dryly sketched the damning statistics for the offending 4 percent. The structural ironworkers had the "best" record. But this produced only 5–10 minority members out of more than 800. Sheet-metal Workers Local 19, with 1300–1400 members, had *no* minority journeymen. In the past three years Local 19 had tested 17 Negro sheetmetal journeymen who had applied, some with experience in Philadelphia's naval shipyard. Yet *none* of the blacks had passed. Elevator Constructors Local 5, with 500 members and 148 helpers, not only had no minority journeymen, but also no minority apprentices. Indeed, Local 5 had no apprenticeship program at all, and no apparent prospect of getting one. So no Negroes need apply. The same was true of Stone Masons Local 3, with 435 members—all white, and getting elderly.

Phelan conceded that stonemasonry was one of the declining crafts, and the FEB eventually dropped the stonemasons' local from the high-priority target list as not worth the deflected energy. But this small minority of lily-white, high-wage trade unions symbolically dominated the conspicuous work sites in Philadelphia, such as the new U.S. Mint, where the FY 1968 federal budget had earmarked $550 million in construction for the five-county area, and $250 million for the city. Approximately 30 percent of Philadelphia's population of two million was black, and the city's race relations, always tense, by 1967 were explosive. The Philadelphia Plan was hammered out, instructively, while Detroit burned from the ghetto rioting.

As for the mechanism of the plan itself, it was simple. The contract compliance committee would set no firm target numbers, because this would suggest illegal quotas. But contracts would be held up until bidders submitted detailed manning tables that listed by trade in all phases of the work the specific number of minority workers they pledged to hire. But the term "pre-award" led to some confusion. Standard procurement practice in government contract work required that all potential bidders receive the same pre-bid specifications on the work to be done. Their bids were then officially opened and compared, with the contract normally going to the lowest bidder. But in the Cleveland and Philadelphia plans, compliance officials would intervene *after* the opening of bids, and then insist on negotiations over the adequacy of the numbers of minorities listed in the manning tables before finally approving or officially awarding the contracts. Thus the key to the government's leverage, and therefore to the contractors' complaints, lay in the irregular and uncertain period of negotiations following the opening of bids. Under the new arrangement, contractors who emerged as the low bidders were now held in the limbo status of "apparent low bidders" while their compliance posture was examined. In this sense the "pre-award" program required ad hoc, post-award negotiations before the award could be completed and the contract signed.

Under these new plans, the contractors were responsible for producing the promised minority workers by craft category. Thus for federal contracting officials, the unions with whom the builders had previously signed hiring-hall contracts were the *contractors'* problem. Phelan observed that "Although affirmative action is criticized as ambiguous, the very lack of specific detail and rigid guideline requirements permits the utmost in creativity, ingenuity, and imagination." He simply demanded that the affirmative action programs "achieve equal opportunity results." To ensure that they did, the compliance committee would follow up the pre-award negotiations with regular on-site project checks and evaluations, including periodic head counts against the manning tables.

Counterattack: 1968

Building contractors were important clients and constituents in departments like Defense, HEW, Transportation, and HUD. But the builders' only significant client relationship with Labor had been in the joint apprenticeship programs, where their long relationship with BAT had generally been a cozy one.[42] Building contractors tended therefore to resent Labor's sudden crackdown from the side door in Cleveland, where the builders' contracts were signed with local government units, which in turn were being funded from agencies like HEW and NASA. As a result, some builders brought lawsuits to protest what they regarded as Labor's radical and extraneous new EEO requirements. Thus the litigation that would ultimately test the Philadelphia Plan was in fact initiated in Cleveland.[43]

But organized labor was of course the backbone of the Labor Department's constituency, and the rapid evolution of the Philadelphia Plan in 1967 produced cries of protest from the union movement that quickly shattered the historic but strained solidarity between the black civil rights organizations and the national leadership of the AFL-CIO. That fall Secretary Wirtz tried to calm and reassure his major constituency. But he did this somewhat disingenuously, in view of his internal support for Sylvester's drive and also in light of subsequent developments. In November, Wirtz claimed in a speech to the AFL-CIO's restive Building and Construction Trades Department that the circumstances in Cleveland and Philadelphia were exceptional. The numerical requirements of the manning tables there, Wirtz said, would not become a general policy:

> In at least two cases—in Cleveland and in Philadelphia—the government contract situation had gotten so bad, with antagonism and recrimination piled on top of each other, to the point where symbolism was more important than substance, evidence more important than equity, that there was probably no effective alternative to that kind of ruling. But it isn't right as a general policy, and it won't work. Even if it drags someone who worships his prejudices into line, it demeans somebody else who has done the right thing for the right reason.[44]

Similarly John Macy, who had alerted the government's FEB heads to the imminence of the new Philadelphia model, sought in January 1968 to reassure anxious members of Congress that the Philadelphia Plan was experimental. "There are no intentions to implement such a plan on a nationwide scale," Macy said.[45] The anxious legislators included men as divergent in party and ideology as South Dakota's two senators, Democrat George McGovern and Republican Milton Young, and this was an early indicator that the Philadelphia Plan was going to create strange political combinations. The plan's opponents featured an alliance of builders and construction unions, and ultimately this common opposition linked the Chamber of Commerce and the National Association of Manufacturers to the AFL-CIO. Thus both conservative Republicans and liberal-labor Democrats complained that the Philadelphia Plan's de facto racial quotas violated the Civil Rights Act of 1964.

The builders were first to feel the heat, and first to complain. When Sylvester's office in November 1967 distributed drafts of the Philadelphia Plan to interested parties and invited comment, as was customary, the director of the Associated Contractors of America, William E. Dunn, fired back strong objections. With its quota-like allusions to "headcounts," Dunn said, the proposal was "drastic," and its implementation was scheduled on a "crash basis."[46] Dunn demanded prompt public hearings to review Labor's radical new approach in all four experimental cities. When the OFCC declined, Dunn appealed directly to Wirtz. Contractors were firmly and honorably bound by their hiring-hall agreements with the unions, Dunn insisted, so Wirtz should "place the response for discrimination under the hiring-hall system where it belongs."[47]

In response, Wirtz not surprisingly demurred, since the OFCC had jurisdiction over contractors, but not over unions. So early in January 1968 Dunn wrote to all chapter presidents of the Association of General Contractors, summoning them to a war council in Washington on January 31.[48] Like most Washington-based associations, the contractors were adept at lobbying Congress and the mission agencies, in the normal workaday routines of their "iron triangle" relationships. But this time around, the agencies seemed politically united to fend them off. The aroused contractors came to discover, however, a strange new ally, in the form of the General Accounting Office—a watchdog agency they usually feared because it so jealously guarded the taxpayers' dollar, on which they normally fed.

Comptroller General Elmer Staats and the Weberian Imperative

The storm over the Philadelphia Plan of 1968–69 confounded many of the traditional alignments of ideology and partisanship that had coalesced to encourage or resist the expansion of the Great Society. It shattered the historic solidarity between the national leadership of the AFL-CIO and the major civil rights organizations. It set the Labor Department against its own union constituency. And ultimately—and quite oddly—it sent a conservative Republican administration to the rescue of a radical civil rights proposal that had been abandoned by a liberal Democratic regime. But at a deeper level that usually remained hidden, it triggered a seismic jolt along a historic fault line in the substructure of government. In doing so it flushed out briefly into open warfare a cluster of bureaucratic animosities that long had simmered. The flap centered on the unlikely figure of Comptroller General Elmer Staats, a highly regarded civil servant of the old school whose main interests in the affair had little to do with civil rights policy.

A midwesterner of Scandanavian descent, Staats had earned his doctorate in public administration from the University of Kansas in 1936, and most of his long, distinguished, and nonpartisan service as a career civil servant had been spent in the Budget Bureau. Joining the Bureau in 1939 and rising through recognized proficiency, Staats reached the penultimate level of deputy Budget director in 1958. In 1966, President Johnson responded to the wishes of Congress and appointed Staats Comptroller General of the U.S. As the federal government's chief accountant and a gray bureaucrat of vintage invisibility, the Comptroller could trace his office all the way back to the creation of the Treasury Department in 1789. Since 1921 he had been head of the General Accounting Office, which was created as the fiscal and budgetary watchdog of Congress by the Budget and Accounting Act of that year. This in turn threatened the traditional autonomy of the mission agencies, who feared that the GAO's cost-conscious reviews of their contracts would chill their client relationships and inhibit their programs. Since this was a rather naked political objection, the agencies also objected on the constitutional grounds that as a legislative department the GAO would improperly be exercising judicial functions over executive agencies. The separation-of-powers argument was a serious and difficult one. But the agencies' concern for their own congressional appropriations kept them from objecting too overtly or strenuously. Like the proliferating federal regulatory agencies in the "fourth branch of government," the GAO did not fit neatly into the tripartite logic of separated powers that the Founders had envisioned.[49] Thus James Madison was a less useful guide than Max Weber to the era of sustained but customarily muted intergovernmental warfare that followed the GAO's creation.

Not surprisingly, over the years the Department of Justice took the lead, on behalf of the executive branch, in resisting the GAO's claims of legal jurisdiction over contract disputes. Periodically the Attorney General would directly challenge the Comptroller's assertion of authority to rule on the finality of government contracts. The GAO stood at some disadvantage in these disputes, because the federal courts had not generally accorded the GAO the kind of judicial deference enjoyed by the Big-Six regulatory agencies. On the other hand, the GAO enjoyed the special solicitude of the House of Representatives, because both institutions jealously guarded the power of the purse. So the years of legal and bureaucratic skirmishing had left the lines of jurisdiction and authority still tangled, and scattered in the law review journals are arcane analyses of the ambiguous legal basis of this normally obscure warfare over separation of powers.[50] But the fight over the Philadelphia Plan suddenly stripped away the veils.

In 1966, Elmer Staats, newly appointed as Comptroller and armed with a quarter-century of intergovernmental experience in the Budget Bureau, moved briskly to reinforce the GAO's disputed claims to review agency contracts for finality. Staats began by asserting for the first time the GAO's authority to review contractual disputes of fact as well as law. The occasion he selected for expanding the GAO's review authority involved a contract let by the Atomic Energy Commission, in which Staats reversed a minor payment concerning the construction of a test basin. Like John Marshall in the Marbury case, Staats had picked his target well, because the substance of the decision was so minor (or, in the case of Chief Justice Marshall, technically moot) that Staats could quietly establish his precedent without provoking a major fight. Then he reached out further in 1968 and took on the Pentagon, ordering the Air Force to reexamine a disputed decision.[51] This outreach, however, brought Attorney General Ramsey Clark into the lists in defense of executive authority, especially his own, and Clark was exchanging broadsides with Staats literally up until the Attorney General's last week in office in January 1969.[52] Thus when the contractors' associations complained to Washington in late 1967 and early 1968 that the new manning table requirements in Cleveland and Philadelphia were violating the government's own long-established bidding procedures, the GAO welcomed with unusual alacrity the contractors' complaints against such agency irregularity.

The GAO, in fact, was already ahead of the complaining contractors—and was well in advance of the complaints from southern congressmen who would figure so prominently in the news accounts in 1968. When at Sylvester's urging, Wirtz sent out his May 1967 order requiring written "assurances" from all contractors that they did not maintain racially segregated facilities—a requirement that had seemed so odd at the time because such Jim Crow arrangements had been illegal since 1964—the GAO's alert director of procurement regulations dispatched an early objection to the GAO's deputy general counsel.[53] In defending their claims of turf, the staff lawyers at the GAO circulated position papers arguing that Labor's new EEO requirements were proliferating in a direction that was "inconsistent with the GAO's historic position" that low bids may not be rejected merely for failure to furnish information. By burdening the procurement process with extraneous requirements, the GAO claimed, such practices threatened to decrease competition and increase bid prices. Furthermore, there was "no statutory authority for such requirements."[54] One staff attorney obliquely attacked the OFCC in a revealing *obiter dictum*: "In addition I feel that as a matter of policy we should register a protest against the creation of a new sub-empire in the DOL without a shadow of authorization."[55]

Against this bureaucratic background, the congressional complaints that were generated early in 1968 by the intensely lobbying contractors began to be felt. Meanwhile the building-trades unions were busy with a parallel effort to head off direct attacks like the Philadelphia Plan by joining in a broader (and vaguer) effort of the Great Society to meet the crisis of the cities. Under the urging of President Johnson, they were negotiating a national pact with contractors that represented a classic Johnsonian consensus: builders would rebuild the cities, labor would train minority workers in exchange for robust employment, and the federal government would pay for it all.

The key to the pact was the multi-billion dollar fund planned for the new Model Cities program to rehabilitate the slums. The Model Cities law of 1966 required that ghetto dwellers be given "maximum opportunities" as indigenous workers for these rebuilding jobs. The construction unions were persuaded by the prospects of such a building boom to agree to train under-qualified slum dwellers in new, government-funded "trainee" categories. Herbert Hill thundered that the new pact was a "fraud" because unlike the normal apprenticeship programs, the Model Cities pact for construction would assure successful trainees neither job preferences nor union membership.[56] But C.J. Haggerty was enthusiastic, and even plumbers' union president Peter Schoemann, who in the past had denounced similar efforts as "wild schemes" leading to socialism, came around to order his locals to stop "pussyfooting and bellyaching" over integration and instead sign on with Model Cities to "help our underprivileged minorities."[57] This was vintage LBJ politics, and the divisive Philadelphia Plan, with its explosive potential for splitting the Democrats' labor–civil rights alliance, was not. Sensing the Democrats' vulnerability on this issue, especially in a presidential election year in which Johnson had suddenly (on March 31) withdrawn from the race, the Republican congressional leadership moved to confront the Administration with its own internal contradiction. At the center of this effort was a southern Republican, Congressman William Cramer of Florida.

A Phi Beta Kappa graduate of Chapel Hill with a law degree from Harvard, Cramer was first elected to the House in 1954 from St. Petersburg. At the age of thirty-two, he had been the first Republican congressman elected from Florida since 1875. A conservative who had opposed the Civil Rights Act of 1964 as well as most Great Society programs, Cramer nevertheless had supported the Voting Rights Act of 1965. In 1968 he even voted for the open housing bill, presumably because his anti-riot provision was in it. A seven-term veteran by 1968, Cramer was also the ranking Republican on the House Committee on Public Works. The committee's minority counsel, Robert May, had been meeting with the complaining contractors since January, as the Public Works committee was a natural lobbying point. In early April, Cramer sent Staats a letter objecting to the Cleveland Plan and re-emphasizing the long-standing congressional requirement that "In order for the bidding to be truly competitive, all bidders must compete on the same basis with no allowance for negotiation on particular aspects of the program after the bids are opened."[58] Cramer referred to the statutory code on competitive bidding, and requested the Comptroller's official judgment.

In response to Cramer's letter, the GAO sent a staff attorney, Wilbur Allen, to the Hill to examine the complaints that the building contractors had sent to May. The contractors had complained not about the disruption of their hiring-hall agreements, but rather about the increased and unpredictable costs that they and therefore the taxpayers faced from the new pre-award negotiations. Operating like regional real estate barrons, the builders preferred either no unions at all, or very strong ones that bound equally all major contractors to their high wages and hiring-hall agreements. They did not relish a fight with the unions, and they feared the prospect of committing themselves to low bids in the face of such unpredictable factors controlling future costs. But Allen reported back to the GAO that at that early point in the skirmishing over the manning table device, he did "not feel [the contractors complaints] prove that contractors suffer unanticipated costs."[59]

But Allen was not a senior staff counsel, and the political culture of the GAO, especially under the expansive leadership of Staats, was reinforced by its professional culture, which reflected the accountant's insistence on strict regularity in bidding and contractual procedures. Accordingly, on April 22 Staats sent Cramer a memorandum of opinion, with an official copy to the Secretary of Labor. The Comptroller's opinion formally dealt only with the experimental programs in St. Louis, San Francisco, and especially Cleveland, since technically the Philadelphia Plan was still circulating in draft form for comment. Staats's ruling observed that the GAO records contained no guiding precedents that might govern such novel negotiating requirements after the bids were opened. Therefore he reaffirmed the GAO's basic principles of competitive bidding, which require awards "only on the basis of a low bid, *including* any additional specific and definite requirements set forth in the invitation, and [stipulating] that the award will not thereafter be dependent upon the low bidder's ability to successfully negotiate matters mentioned only vaguely before the bidding."[60] The Comptroller's opinion therefore held that the Cleveland Plan was invalid because the manning table requirements failed to "include a statement of definite minimum requirements to the bidder's program, or any other standards or criteria by which the acceptability of such prerequisites will be judged."

But what about the Philadelphia Plan? It imposed even more forcefully a vaguely defined burden upon contractors to pledge to hire specific numbers of minority workers in each trade. Yet it contained no criteria for determining what was acceptable and what was not, or why. The Philadelphia *Evening Bulletin* had run a major story on the controversy on April 26, and following the Comptroller's ruling against the Cleveland Plan in May, *Bulletin* reporter Gene Meyer began calling the GAO to discover whether the Comptroller was also going to rule against the Philadelphia FEB's contracts for the $30 million U.S. Mint, or the Presbyterian–University of Pennsylvania Medical Center, or new buildings for Drexel Institute of Technology or the Spring Garden Institute. Staats preferred to stand pat on his Cleveland ruling of May. But, in October, Cramer again wrote Staats and asked for a specific ruling on Philadelphia. Citing with approval the Comptroller's insistence on bidding specificity in the Cleveland opinion, Cramer attached a copy of the Philadelphia FEB proposal and noted that chairman Phelan had actually bragged about the creative ambiguity of the FEB's demand for affirmative action.[61]

Staats replied to Cramer on November 18. His letter had been drafted by the GAO's general counsel, then shared by the GAO's legislative liaison staff with minority counsel May of the Public Works committee. May in turn had cleared it with Cramer, in the time-honored mode whereby veteran officials, who were required to deal with one another in permanent institutional relationships, therefore learned to avoid springing nasty surprises on one another.

On November 18 the Comptroller not surprisingly ruled that the Philadelphia Plan, too, was illegal, and for the same reason as the Cleveland Plan. Staats pointed out that he was not ruling that pre-award affirmative action programs were illegal per se. But he insisted that "the invitations for bids contain or incorporate a statement of minimum requirements to be met by the bidder's program." Where federally assisted contracts are required to be awarded on the basis of publicly advertised competitive bidding, Staats ruled, "award may not properly be witheld pursuant to the Plan from the lowest responsive and otherwise responsible bidder on the basis of an unacceptable affirmative action program, until provision is made for informing prospective bidders of definite minimum requirements to be met . . . ."[62]

Thus Wirtz and Sylvester were caught up in a classic dilemma, like Joseph Heller's Catch-22. If their affirmative action plans required explicit numbers of minorities to be hired, they ran afoul of the prohibitions of Section 703(j) of Title VII of the Civil Rights Act of 1964, which prohibited preferential treatment on account of race, sex, ethnicity, and so forth. If on the other hand the OFCC's model for affirmative action remained carefully, even artfully, vague, as in Phelan's FEB plan for Philadelphia, then they ran afoul of the GAO's insistence that bidders be specifically informed in advance of all job requirements. No rational planner, to be sure, would design a system of public bidding that included anything like the coy guessing game fashioned by the OFCC in Cleveland or the affirmative action committee in Philadelphia. But the Philadelphia Plan was less a public works construction document than a political document, forged in the subpresidency for social purposes that had little directly to do with building college dormitories. Staats and his careerist associates at the GAO had a quite different political agenda, and it served their purposes to call the shot according to their accountants' rulebook. Lacking any backing from the wounded, lame-duck President Johnson on the eve of a presidential election, the sponsors of the Philadelphia Plan looked defeated. In late November, the Labor Department quietly announced that the Philadelphia Plan was rescinded.[63]

Thirteen days before Staats ruled that the Philadelphia Plan, like the Cleveland Plan and for the same reasons, was illegal, Richard Nixon was elected President of the United States. Within a month, Califano had passed along to the Nixon transition team a recommendation from Ramsey Clark's interagency task force that the ineffective OFCC be transferred from the Labor Department to the EEOC, where the tiny office might find more collegial support for a coordinated enforcement of both Titles VI and VII.[64] That December the NAACP urged the incoming Nixon administration to transfer the

OFCC to the Justice Department, claiming that the Department of Labor was a captive of "bigoted labor unions."[65] Well before Nixon's inauguration, then, the Philadelphia Plan seemed permanently scotched by Republican rectitude and high professional standards—an early casualty of the zealous overreaching of LBJ's Great Society. That the new Republican President would intercede to rescue it seemed scarcely within the realm of political possibility.

PART THREE

Nixon

CHAPTER XII

Richard Nixon and Civil Rights Policy: "No Master Spirit, No Determined Road"

The Enigma of Richard Nixon

The legacy of Richard Nixon remains a cipher. For more than a decade and a half our understanding of his first administration has been filtered through the lens of Watergate and the American defeat in Vietnam. Over that span we have accumulated a shelf of books on a lost war and a failed president that dwarfs the serious literature on the Nixon presidency. The Nixon lieutenants have left us with the customary clutch of aide-memoirs; despite the self-justifications that are common to the genre, some of them are quite full and helpful on domestic policy, particularly those by John Ehrlichman, Raymond Price, and William Safire.[1] Nixon's presidential papers remained closed by the litigation associated with Watergate until late 1986, when documents on domestic policy during Nixon's first administration began to be opened to researchers.[2] As a consequence, scholarly assessments of Nixon's tenure, beyond the long shelf on Watergate, have remained few and evidentially thin. Furthermore, beyond these problems of limited evidence and perspective, it has seemed especially difficult for political writers and scholars to be fair to Nixon.[3]

Most post-Watergate attention has focused on Nixon's innovations in foreign affairs, especially the opening to China and detente with the Soviet Union.[4] In domestic policy, there has been some recognition of Nixon's surprisingly creative contributions in planning and government organization, beginning with Otis Graham's *Toward a Planned Society*.[5] Books like Moynihan's *The Politics of a Guaranteed Income* have called attention to Nixon's curious presidential record of "Tory socialism" or Disraeli liberalism, of which his surprising proposal for a national family assistance plan remains the prime example.[6] But a list of parallel domestic initiatives and reforms would be substantial, including creation of the Environmental Protection Agency, the Occupational Safety and Health Administration, and related new programs for clean air and water, child and consumer protection.[7] Oddly enough, it would also include a wage-price freeze and cost-of-living indexing for Social Security recipients—both policies violating the heart of traditional Republican economic doctrine.[8]

Nixon's pre-Watergate record of legislative and program achievement as President is surprisingly rich in reform initiatives, attempts at strategic planning, fresh reassessments of federal-state relations, and far-reaching executive reorganization. Substantively, however, the domestic achievements of the first administration are a strangely mixed lot, and the mix reflects Nixon's own lack of any coherent and internally consistent ideology, especially when he is compared with a liberal incrementalist like Lyndon Johnson on his left, and a conservative ideologue like Ronald Reagan on his right. Unconstrained by the kind of ideological boundaries that defined a Rockefeller or Goldwater within the Republican party, and little interested in the substance of domestic policy beyond its political repercussions, Richard Nixon was free to tailor his policies on civil rights to maximize their political payoff. As Rowland Evans and Robert Novak described Nixon's view: "The President must run foreign affairs and let his Cabinet handle the humdrum of domestic affairs."[9] The result of this Faustian bargain was a trade-off of political advantage for policy incoherence. Lacking any internally consistent model of civil rights theory, the Administration was free to pursue contradictory policies for short-term tactical gains. One major consequence was Nixon's smashing electoral victory of 1972. But this result, like the White House entourage that savored its triumph, was short-lived. The consequences of the policy incoherence, however, were profound.

Nixon and the Civil Rights Legacy

In his public persona as a political candidate or incumbent in office, Nixon prior to his presidency had avoided directly engaging the volatile race issue in a negative way. He had learned to hedge his bets carefully, however, in the direction of his naturally more conservative Republican constituency. In 1960 Nixon had captured 32 percent of the black vote in his narrow contest with Kennedy; had he come closer to Eisenhower's impressive 39 percent of 1956, he would have beaten Kennedy. Writing two years later in *Six Crises*, Nixon recalled how Kennedy's public intervention after the arrest of Martin Luther King had apparently cost him the election: "this one unfortunate incident in the heat of the campaign served to dissipate much of the support I had among Negro voters because of my record."[10] As a private citizen of California in 1964, Nixon had called the pending civil rights bill "a step forward if it is administered effectively."[11] But this was balanced by criticism of the "hate engendered by the demonstrations and boycotts" and a summons to "responsible civil rights leaders to take over from the extremists." Nixon accused Rockefeller of "political demagoguery" for trying to outpromise the Johnson administration and "raising hopes that can't be realized." Once the civil rights bill was passed, however, Nixon supported it and distanced himself from the direct racial appeals of the Goldwater Republicans. In 1966, Nixon warned a meeting of the party faithful in Jackson, Mississippi, that there was "no future in the race issue" for the Republican party.[12] Biographer Stephen Ambrose defined Nixon's mainstream Republicanism on civil rights as safely bland: "he was for progress, but it should be at the state level and voluntary . . . . He opposed the poll tax, was in favor of antilynching legislation. He opposed compulsive [sic] legislation by the federal government, arguing that it would set back race relations by fifty years."[13]

As the leading Republican presidential candidate in 1968, Nixon had danced cautiously around the dangerous open housing issue, generally avoiding the topic while the bill was pending in Congress. Then after the assassination of King and the passage of the bill, he told delegates at the Republican convention in Miami that the prudent course had been to "vote for it and get it out of the way . . . to get the civil rights and open housing issues out of our sight so we didn't have a split party over the platform when we came down here to Miami Beach."[14] Nixon could not compete with liberal Democrats in a bidding war for black votes, and he wanted to avoid the polarization that had doomed Goldwater. Lacking the ideological conservativism of Republican challengers like Goldwater and California governor Ronald Reagan, whose commitment to creed moved them to confront civil rights issues as matters of principle, Nixon was free to adjust to the political winds and tides. (On women's rights, Nixon was to prove tractable on the feminist litmus-test of the Equal Rights Amendment, which was subsequently to drive his party in quite the opposite direction.)

Nixon knew that the racial issue was dangerous, that played with a heavy hand it could split the coalition of moderates and conservatives on which his presidential ambitions hinged.[15] Yet he owed his nomination in 1968 to southern Republicans who preferred him to northern liberals like Rockefeller, Romney, or Percy, and who feared that they could not win with their real favorite, Ronald Reagan.[16] Political commentators reached an early consensus that Nixon's campaign manager, his law partner John Mitchell, had engineered a "Southern Strategy" with Senator Strom Thurmond of South Carolina, and that Nixon had pledged to repay the debt by curbing federal pressures for school desegregation and by appointing southern conservatives to the Supreme Court.

Nixon's "Southern Strategy," however, was really a border strategy. His overtures were directed not toward the Deep South which Goldwater had carried, but rather toward the Rim South as symbolic of Middle America. He enjoyed a centrist position between Humphrey and Wallace, and inherited almost by default the legions of what Richard Scammon and Ben Wattenberg called the "unyoung, unpoor, unblack."[17] While campaigning in 1968, Nixon was asked by the Republican head of the Philadelphia Chamber of Commerce to schedule a "publicized" private meeting with local black Republican leaders. Observers reported that Nixon refused, asking "What's the use?"[18] "I'm not going to say anything. I am not going to campaign for the black vote at the risk of alienating the suburban vote," Nixon said. "All we can do is not say anything or do anything which will cause the Negroes to lose confidence in me, because I am going to be President of the United States, and no President can do anything without having the confidence of the black community." "If I am President," he added, however, "I am not going to owe anything to the black community."

In 1968, Nixon ran a campaign that was richly financed, measured in pace and tone, and largely issueless. Although his chief interest and claim to expertise lay in foreign affairs, on the most burning issue of the day, the war in Vietnam, Nixon said little, claiming that to do otherwise might jeopardize the peace negotiations being conducted in Paris. Stressing the advantages of his experience in the presidency rather than the substantive issues of debate, he appealed in broad terms to the "forgotten Americans," citizens who paid their taxes and did not riot. To Nixon this included middle class Negroes with a stake in society, and it was consistent with his rather generalized campaign pledges to foster black capitalism.

On November 5, Nixon won by a surprisingly narrow plurality, with 43.4 percent of the three-party vote as against 42.7 percent for the fast-closing Humphrey, and only 13.5 percent for the fading George Wallace. This represented, perhaps, endorsement by a conservative majority of 57 percent, while nine out of ten black votes went to Humphrey. But it was scarcely a substantive mandate. With the Vietnam issue scrambling the ideological spectrum, and with candidate Nixon proposing little more than opposition to crime and inflation, there seemed to be no clear domestic mandate at all.

The Burns Task Forces: Civil Rights in a Policy Vacuum

As President-elect, Nixon during the transition assembled a group of senior appointees whose ranks contained few men (and no women) of strong ideological predispotitions.[19] Nixon even claimed that his Cabinet was slightly to the left of his own centrist position.[20] His almost exclusive interest in foreign affairs led him to abandon the customary practice of outlining the State of the Union and his new legislative program in January 1969. Instead he planned a post-inaugural trip to Europe to try to arrange a face-saving end to the Vietnam war. Meanwhile, immediately after the November election Nixon had asked Columbia University economist Arthur Burns to direct a task force operation on domestic policy issues, most of them focusing on the economic implications of policy choices.[21] Burns's fellow economist, Paul McCracken, coordinated a score of task forces whose recommendations were winnowed in turn by a small team headed by Burns. On January 18 Burns gave Nixon his summary report.[22]

The agenda for the Nixon planning task forces, especially as they were filtered and transmitted to the President-elect by Burns, generally steered well clear of what Scammon and Wattenberg called the Social Issue. There was no task force on civil rights per se, none on equal employment opportunity, or on school desegregation or voting rights. Instead Burns concentrated on applying the principles of tighter management and orthodox Republican economics to discipline and reshape the organizational sprawl and fiscal disorderliness of the Great Society. A few symbolic and politically vulnerable programs of the Johnson legacy were scheduled for the chopping block, like Model Cities and the Job Corps. But the Burns agenda represented no Republican counter-revolution, much as the Eisenhower agenda had carried no repudiation of the New Deal. The chief Burns recommendations, like revenue-sharing and block grants and reorganizing the federal executive structure, marked a politically moderate and fiscally prudent program of Republican reform that could be hurt by a major campaign on the volatile social issues.

The Burns Report addressed 100 separate topics, many of them ranging beyond domestic policy to involve international economic policy and even defense policy. Several touched tangentially on race relations issues—for example, revenue-sharing, federal aid for inner-city schools, an all-volunteer army, manpower training, repeal of the Davis-Bacon prevailing wage. They included the controversial recommendations of Richard Nathan's task force on public-welfare reform (which the conservative Burns dutifully reported but to which he also briskly objected).[23] But only two touched directly on civil rights policy. One recommended that the Nixon administration support the status quo and oppose cease-and-desist authority for the EEOC. The other was an endorsement of the vague recommendations of the manpower task force, headed by Secretary of Labor-designate George P. Shultz, that the new administration should "press forward with new programs relating to (1) ending discrimination in minority hiring, (2) job opportunities, and (3) introducing off-site fabrication in the construction field."[24] Presumably the construction unions would pay lip-service to the first goal, embrace the second, and resist the third. But none of the Shultz recommendations appeared to require major new legislation from the Democrat-controlled 91st Congress.

The Nixon White House shared a policy consensus on civil rights that, with the Great Society laws of 1964, 1965, and 1968 newly on the statute books, and with fresh battles over their interpretation already welling up through the enforcement agencies and the courts, there was little need for any creative burst of new legislation.[25] As candidate Nixon had explained to the voters of Boise, Idaho, in April 1968, "we've had for ten years marches for civil rights and petitions for civil rights and the Congress has passed civil rights legislation in all fields."[26] But "we've now reached a watershed," Nixon said. "[S]peaking as somewhat an expert in this field and somewhat an expert in law itself . . . I do not see any significant area where any additional legislation could be passed that would be helpful in opening doors that are legally closed." "Whether it's housing, or whether it's education, or whether it's voting rights, or whether it's jobs," candidate Nixon said, "civil rights legislation now covers the field." As President-elect in December 1968, Nixon explained that the decade of the 1960s had witnessed a "needed revolution" in civil rights. But, what was now needed was "something that no Congress and no law can provide . . . preparing people to walk through the doors that have been opened."[27]

The Nixon White House and Domestic Policy: 1969

Most of the books on the Nixon presidency sketch a similar picture of the spirited but confused contest over the direction of domestic policy during the formative months of 1969.[28] Having outlined no domestic agenda during the presidential campaign, Nixon—like Johnson but unlike Kennedy—kept his task force operation secret. In his inaugural address on January 20 he asked Americans to "stop shouting at one another" and give him a chance to honor America as peacemaker—a task he pursued by planning trips to Europe in February-March and Asia in July and August.[29] Unlike Lyndon Johnson, Nixon sent no special message on civil rights to Congress. Indeed, compared with his predecessors, Nixon during his 100 days seemed to have no substantive domestic agenda for Congress at all.

Despite his lack of interest in the substance of domestic policy and legislation, Nixon was nonetheless a keen student of the governmental process, and of theories as they related to structures, organizations, functions, and above all to power. Nixon preferred to surround himself with young staff aides who were rewarded for partisan loyalty, campaign success, administrative efficiency, and personal deference. He had collected since 1960 a youthful but proven cadre of campaign veterans, men like H.R. Haldeman and John Ehrlichman, whose tough staff reputations veiled a diffidence toward their boss that included a fear of his temper and the fickle quality of his favor.[30] Nixon took special pleasure, however, in the substantive ideas and intellectual glitter of his new team of academic advisers. Henry Kissinger provided a model of strategic policy leadership as National Security Advisor. The unexpectedly strong George Shultz had taken a firm grip on the customarily thankless Republican helm at Labor. The veteran economist Burns added experience from the Eisenhower White House. The most intriguing anomaly of them all was the renegade Democrat from the Kennedy-Johnson brain trust, Daniel Patrick Moynihan.

Organizationally, Burns was appointed Senior Counselor to the President, the only presidential assistant designated with cabinet rank. But Burns also headed a new White House entity of unclear jurisdiction and authority called the Office of Program Development. Moynihan became head of the somewhat vague Urban Affairs Council, which the Burns Report had endorsed as a "highly promising instrument." But Burns apparently felt that way because the task force on urban affairs had called for the new council to "subject the 400-odd existing urban programs to cold, hard scrutiny, eliminating all that can be pared and consolidating those that should be saved," and also to "enforce a rule of restraint upon the bureaucracies whose natural tendency is to magnify their callings."[31] Moynihan not surprisingly had a more liberal vision of the council's possibilities. He exploited his proximity to the President (operating out of the West Wing, while the lofty Burns was quartered next door in the elegantly rococo Old Executive Office Building) and profited also from his arresting pen, his roughish wit and charm, and his continuity of experience in serving three presidents. Meanwhile Nixon ran about the globe, preaching of a new era of cooperative international relations and world peace.[32]

By mid-March, however, when Nixon had returned from his trip to Europe to reinvigorate the Atlantic Alliance and seek support for a Vietnam settlement, grumbling was increasing from Republicans in Congress that the new administration had no clear legislative agenda. Nixon heard these complaints at congressional receptions in early March, and he asked Ehrlichman to try to "feed into our own executive organization a little more imagination and ingenuity," because "we seem to be pretty short on such ideas. . . ."[33] The White House had dispatched a presidential memo to all agencies on February 12, directing them to respond to the 100 policy recommendations contained in the Burns Report. But this broadside of initiatives from the top necessarily triggered a time-consuming dialectic within the bureaucratic layers of the mission agencies. When, in early March, Nixon asked Bryce Harlow, the Oklahoman who headed congressional liaison, to summarize the legislative agenda for the administration's first 100 days, Harlow described an unimpressive and even haphazard pattern.

The White House had sent Congress only six domestic proposals, Harlow reported, none of them major (e.g., postal reform, coal mine safety, increased debt ceiling). The Administration was avoiding the controversial issues, but this included the more conservative ones that the Republican legislators preferred. "Our troops across the nation are grumbling about the Administration's 'liberal' direction," Harlow told the President.[34] What was needed before the end of April was a coherent package containing proposals for federal revenue-sharing, block grants, welfare reform, tax and "clean election" reform, and an attack on organized crime. Harlow worried that if the Administration postponed grasping the nettle on the sensitive civil rights issues— i.e., the HEW guidelines for school desegregation and extending the Voting Rights Act, which was scheduled to expire in the summer of 1970—the congressional Democrats would seize the policy initiative. "Arthur Burns," Harlow volunteered, "is the only person positioned to review the progress on Presidential directives to produce legislative recommendations."

In response to the storm warnings, Nixon directed Haldeman to have some kind of domestic program pulled together by April 2, when Congress adjourned for Easter recess. So Haldeman sent Ehrlichman an "URGENT!" message: "[the President] urges that you get Burns cracking on this." Haldeman's command captured Nixon's astonishing inattention to the substance of his own domestic policies during his first 100 days: "He at least has to know what is the program," Haldeman wrote, "what is going down, etc."[35] Nixon's political anxiety was quickened in early April by reading reports in his daily news analyses about the abrupt decline in the administration's domestic activities and news stories, thus giving "a virtual open field to the opposition."[36] Ehrlichman, however, was involved in a testy exchange of memos with Burns over their relative authority in formulating legislative policy, and this clash in turn reflected jurisdictional confusion and rising tension between the Burns shop and Moynihan's Urban Affairs Council.[37]

The result of these hurried efforts was Nixon's dispatch to Congress on April 14 of a slapdash agenda for domestic legislation. He explained to Congress that his search for peace abroad and executive reorganizations at home had delayed his legislative proposals.[38] In the field of social legislation he had inherited a "hodge-podge of programs piled on programs," he said, which had led to "the growing impotence of government." Thus his first task had been to reorganize the executive branch, creating an Urban Affairs Council, a Cabinet Committee on Economic Policy, overhaul the Office of Economic Opportunity, and (although he did not mention it on April 14) create a task force under Litton Industries executive Roy Ash that would redesign the instruments governing domestic policy and administration.[39] Substantive priorities included initiatives in crime control, new strategies toward welfare, and a new kind of federal-state partnership centering on revenue-sharing. Now that he was getting around to his legislative agenda, Nixon explained, it would include such issues as revenue-sharing, new programs in air and mass transit and in manpower training, indexing Social Security payments against inflation, and reforming tax abuses and providing new tax credit incentives. On civil rights his congressional message remained vague, alluding only to "a program to increase the effectiveness of our national drive for equal employment opportunity."[40]

The analysis and planning on behalf of government reorganization during the first Nixon administration, and especially the creative efforts of the Ash task force, was arguably unmatched since FDR's effort with the Brownlow committee. Probably no President since Wilson could match Nixon's knowledge of governmental structure and capacity for functional analysis.[41] But the resulting inattention by the President to the substance of domestic policy and the organizational incoherence of the White House's domestic policy machinery had produced an incoherent domestic agenda. Lacking any substantive theory of domestic policy, Nixon possessed no philosophical gyroscope or ideological predisposition with any identifiable core—certainly not the kind of conservative commitment that the Republican congressmen resonated to, and grumbled to Bryce Harlow about in its absence. As Evans and Novak concluded after the Administration's first year, "If anything distinguished the President's conduct and habit in foreign as contrasted to domestic policy," it was that "in one he had a theme, in the other he did not."[42] In civil rights policy, even more than in domestic policy generally, the new Nixon administration would reflect Wordsworth's complaint: "No single volume paramount, no code/No master spirit, no determined road."

As the political demands on the President for greater clarity in domestic policy intensified through the spring, Haldeman asserted his presence as dour keeper of the President's door. By the summer of 1969, Nixon's pleasure in the formidable erudition of Burns and Moynihan was fast eroding under the grind of their competition for his policy preference. Their jockeying for advantage began to intrude into Nixon's memo traffic, and, in June, Ehrlichman told Burns that the program planning between "your office and Pat Moynihan's" was in danger of becoming "dis-jointed and uncoordinated."[43] Nixon began "dreading his appointments with the antagonists," Ehrlichman said, with Moynihan pressing his novel Family Assistance Plan, and the orthodox Burns flaying its costs and its moral irresponsibility.[44] Even Vice President Agnew chimed in, volunteering his services as the final referee short of the President.[45]

The system wasn't working and the internal feuding was beginning to leak to the press. So Haldeman suggested to Nixon that his own chief lieutenant, Ehrlichman, might usefully screen the President as coordinator of the Burns-Moynihan controversies. A zoning lawyer from Seattle with strong credentials as a loyal, firm-handed partisan in the campaign organization, but a man whose substantive expertise was limited to local policy issues like land use and environmental regulation, Ehrlichman qualified by his very inexperience as a disinterested referee, and he qualified by his nature as a tough-minded one.[46] Like Haldeman, he was a pragmatic political operator with strong loyalty to party and President, but little driven by ideological commitments. Nixon approved Haldeman's proposal, and by October the President had appointed Burns to replace William McChesney Martin as chairman of the Federal Reserve Board. Then in response to the recommendations of the Ash task force, Nixon created the Domestic Council on the NSC model and put Ehrlichman, in the new post of Assistant to the President for Domestic Affairs, in charge of its staff.[47]

With Burns gone, Moynihan was appointed Counselor to the President with cabinet rank. But the announcement stipulated that his Urban Affairs Council, together with all agencies and all other White House working groups, would report *through* Ehrlichman's Domestic Affairs staff. To William Safire, the end of 1969 signaled the replacement of the competitive reign of Moynihan and Burns in domestic policy by the rise of Ehrlichman and George Shultz— "Shultz became the senior domestic man, Ehrlichman the junior, but Shultz was his own man and Ehrlichman was Nixon's man."[48] Moynihan was left to concentrate on shepherding his welfare reform through the Congress, and to function as "the President's Counselor and resident thinker," Ehrlichman approvingly concluded, free "to rove from subject to subject as he wished, stimulating our intellects and crying alarms."[49]

The Other Moynihan Report: Race, Class, and Hostages

Two weeks before his inauguration Nixon had asked Moynihan to "sit back" about once a month and give him a general view of a particular area of public policy.[50] On March 19, Moynihan dispatched a 28-page "Report to the President" on the state of race relations. In it he pictured the Nixon presidency as a turning point between stark alternative futures. One was grim:

a "calamitous interlude during which the incapacity of 'middle America' to govern was for once and all established." The other was glowing: a Nixon administration that was "an extraordinary and wholly unexpected success out of which grew the miraculous disappearance of race as a central problem of American life."[51]

Moynihan found some encouraging evidence for the latter possibility in recent black gains in income and education. But hidden behind the race problem, he warned, was a class problem. That is, there was no longer *a* black community in America—there were two. The black poor was a too-well known entity—heavily southern in their crippling pathology, abnormally dependent, demographically under siege, unusually self-damaging in their behavior, soaring toward a rate of illegitimacy that was approaching half of all black births. Their level of educational achievement was appalling, but the topic was "extremely sensitive."

Black apologists and white ideologues attempted to explain away this crippling learning gap as entirely the product of external discrimination and suppression. But, as Moynihan cautioned, "the facts would seem to be otherwise."[52] Recent research in urban IQ decline had showed an alarming trend: "a Negro intelligence distribution sharply skewed toward incompetence." Moynihan alluded to the work of Arthur R. Jensen of Berkeley, which he had recently shared with the President, and with whom he was then in correspondence in a collegially courteous but substantively noncommittal fashion.[53] Moynihan included a table showing that the percentage of Negro children with IQs under 75—the point of retardation—was 42.9 percent compared with 7.8 percent for whites in the lowest quintile of socio-economic class. In the two highest SES categories, the black-white ratio in retardation was a devastating 13.6 to 1. Moynihan acknowledged the recent scholarly revival, "in impeccably respectable circles," of genetic explanations for these persistent racial discrepancies. He volunteered his own opinion that "I personally simply do not believe this is so." But he acknowledged that as a matter of scientific dispute in light of divided evidence, "it is an open question."[54]

Moynihan had been sending the President notes since late January commenting on the disturbing black-Jewish split. His essay-memos had concentrated on the bitterness of New York Jews over the recent teachers' strike, which had featured black-power drives, under the banner of community control, to oust tenured Jewish teachers and administrators from their jobs in black-majority schools. In early March he sent Nixon a copy of an article in *Commentary* entitled "Is American Jewry in Crisis?" by Milton Himmelfarb, a staff member of the American Jewish Committee. In his cover note Moynihan quoted two chilling sentences from Himmelfarb's penultimate paragraph: "In New York our remaining years in the civil service, and above all in the schools, are not many: if policy does not drive us out, terrorism will. (It has started.)"[55]

In his March 19 report to the President, Moynihan summarized the dilemma in New York. The origins of the strike lay primarily in the failure of the New York City school system, which had "transformed two generations of Jewish immigrants into the intellectual elite of the world's most powerful nation, to be able even to bring its black students, now almost a majority, up to grade level at any point in the school sequence."[56] Black parents had been assured by white ideologues that it was altogether within the power of the school system to do this. So the black parents naturally enough assumed that when it was not done it was because "some white persons—the Jewish principal or whomever—*desires* that it not be done. It is difficult to conceive a better formula for inducing mass paranoia!"

The chief danger of this situation, to Moynihan, was that the politicized and violent Negro poor "have given the black middle class an incomparable weapon with which to *threaten* white America." This had been "an altogether intoxicating experience," because the rising black middle class had created a new *caste* problem, a psychological demand for "equality of *self valuation*." Behind the courteous optimism of the black middle class lay "volcanoes of hate and rage"—including "self hate as much as anything." Thus the existence of a dependent, alienated, angry, black urban *lower* class "has at last given the black middle class an opportunity to establish a secure and rewarding power base in American society—as a provider of social services to the black lower class." They would therefore demand the Jewish teachers' jobs. More generally they would demand civil service jobs, proportionate in all ranks to their population, throughout the federal service and especially in the rapidly growing halls of state and municipal bureaucracy. Their demand for Black Studies programs at universities was "essentially a demand that black professors be hired to teach black students to read books by black authors, thereby displacing the white—typically Jewish—professors and writers who have gained the rewards of these roles up until now."[57]

"The era of equal opportunity, nondiscrimination, integration and such," Moynihan concluded, "is coming to an end." But Moynihan seemed to be of two minds about this. In the universities, he said, there "is no true Negro intellectual or academic class at this moment. (Thirty years ago there was: somehow it died out.)" Books by black authors were "poor stuff for the most part," and the new Black Studies programs tended toward "the worst kind of ethnic longings-for-a-glorious-past." The logic of these distortions was relentless: "before long blacks will be demanding Eleven Percent of all public places and services—in universities, civil services, legislatures, military academies, embassies, judges. This is not what the civil rights movement expected to come about, or hoped to see, but it does appear to be the outcome nonetheless."[58] On the other hand, given his own impeccably Irish, Hell's Kitchen credentials, Moynihan historically observed: "What building contracts and police graft were to the 19th century Irish, the welfare departments, Head Start programs, and Black Studies programs will be to the coming generation of Negroes. They are of course very wise in this respect. These are expanding areas of economic opportunity. By contrast, black business enterprise offers relatively little."

This last gratuitous jab was directed at Nixon's half-hearted efforts at stimulating black capitalism through his Minority Business Enterprise program. And in light of that modestly launched program's subsequent slide into near invisibility under the crushing weight of muddled design, incompetent administration, and the Administration's anti-inflationary constraints, Moynihan's dismissal was prescient—if impolitic. But it was also self-interested. Moynihan's answer to the racial crisis was radical welfare reform, specifically his own plan for a guaranteed annual income for the poor. Thus he closed his extraordinary report to the President by returning to his optimistic (and Nixon-flattering) scenario and abruptly concluding that "the present course of events could bring about a final transformation of the race problem in America into an ethnic problem." Within a generation these black/white distinctions might be reduced to the significance of Protestant/Catholic, Irish/ Yankee, North/South distinctions. "The essential fact," Moynihan wrote, "is that what even the most militant blacks at this moment are demanding is that they be co-opted into the system. Made judges, professors, congressmen, cops." But such an outcome cannot be assured, he concluded, unless "*the nation sets out in earnest to dissolve the great black urban lower class (and the rural slums that feed it) that the militant middle class now uses as a threat to the larger society.*"[59] That meant using government to rescue the disintegrating lower class family, especially the poor urban black family. To Moynihan, this was a familiar crusade. But its politics had changed dramatically.[60]

So the black middle class was holding the black poor hostage, threatening white America with urban violence. To make the newly militant black bourgeoisie safe for black capitalism and even for Republicanism, their hostage would have to be freed. Thus Moynihan was providing Nixon with an attractive rationale for a radical intervention in welfare policy, one that violated virtually all the canons of traditional Republicanism. Lyndon Johnson's rationale for his own war on poverty had been simple and direct, a self-evident combination of social justice tempered by social control. Nixon wanted to defang if not dismantle that Democratic legacy, so quickly grown unpopular through its encouragement of takeover by community militants—in the contemporary argot of gonzo journalist Tom Wolfe, of *Mau-Mauing the Flak Catchers.*

Would the Republicans not appear heartless, though, by replacing it with nothing? Nixon had only offered the shibboleth of black capitalism, which scarcely seemed to touch the heart of the ghetto's pathology. Comes now Daniel Patrick Moynihan, with his intriguing, hostage model of class and caste. He had rejected the Great Society's *services* strategy in Lyndon Johnson's failed war on poverty (his indictment, *Maximum Feasible Misunderstanding,* was even then being rushed to press).[61] Now he proposed a more radical *incomes* strategy, and offered it to Richard Nixon as the key to determining whether his administration would be a "calamitous interlude" or an "extraordinary and wholly unexpected success" which produced "the miraculous disappearance of race as a central problem of American life."

Moynihan would find ironic vindication in Richard Nixon for the passionate concerns that led the former assistant secretary of Labor to be hounded out of the Johnson administration. But this vindicating crusade for welfare reform would also narrow his options. He was both too politically astute and too liberal a Democrat to presume to guide the Republican administration's overall policy in civil rights.[62] Prudently, he would save his chits and use them in the service of welfare reform. He would direct his secondary and (unlike his campaign for a guaranteed income) largely successful efforts toward expanding federal aid for higher education and moderating the Administration's control of campus unrest.[63] Arthur Burns would have even less of a role in shaping civil rights policy, as he directed his short-lived and fading efforts as cabinet-ranked aide toward economic issues like textile quotas, the investment tax credit, the SST.

Both Moynihan and Burns were struck by the *tabula rasa* nature of Nixon's philosophical beliefs—that is, he seemed to have none. Moynihan remarked that "the president had no view" on the welfare question, "at least none he disclosed."[64] But Moynihan saw this as an opportunity to persuade, and he kept his gifted pen busy. Burns worked hard to persuade the President also. But Nixon's lack of a philosophical keel bothered the systematic and scholarly Burns, even when his persuasion was successful: "The president would sometimes accept my arguments without hearing them out," Burns complained to James Reichley. "This bothered me. I was able to persuade the president to go ahead with revenue sharing in two minutes. This was an achievement of which I was not very proud. It was wrong for the president to make decisions without being fully acquainted with the problems."[65] When Burns lost out to Moynihan in the battle over the Family Assistance Plan, he complained to Ehrlichman that the FAP was not in agreement with President Nixon's philosophy. Ehrlichman had laughed: "Don't you realize the president doesn't have a philosophy?"[66]

Thus the policy vacuum remained, and this invited an ad hoc pattern of interesting but essentially uncoordinated and incoherent policy initiatives (or inactions) by top Administration leaders within their various domains. During the second half of 1969 these were dominated by the congressional brawl over Shultz's revival of the Philadelphia Plan and the Senate snarl over Nixon's nomination of Judge Haynsworth. In such a policy sprawl, the Administration's signals on civil rights could scarcely have been more contradictory. But at the beginning of the Administration's first year, attention had focused on candidate Nixon's sole campaign commitment of 1968 in response to the racial crisis: his frequently iterated support for black capitalism. On March 5 the new President issued an executive order creating an Office of Minority Business Enterprise in the Department of Commerce, thereby redeeming his campaign pledge to foster "black capitalism," but broadening its embrace through a more suitable acronym.[67] Nixon's initiative in fostering minority small business was something of a meteor in the spring of 1969. It illustrates the baleful consequences of combining promising rhetoric with a hastily assembled program in an environment of policy incoherence.

The Perils of Black Capitalism: Cynicism and Hair Grease

Richard Nixon brought with him to the White House in January 1969 a *policy* vacuum in civil rights, but the Office of Minority Business Enterprise (OMBE) he created on March 5 did not operate within a *program* vacuum. Nurturing business enterprise had been a hallmark of Eisenhower Republicanism during the 1950s, when the Small Business Administration was created in 1953 to speed economic recovery in areas hit by natural disaster. In 1958 the SBA was made permanent and given a major role in providing new equity capital to support small business investment companies (SBICs). In the 1960s the Democrats added on their multi-layered anti-poverty law of 1964, with special attention in Title IV to ghetto rehabilitation by stimulating small business initiatives. Anti-poverty amendments in 1967 had boomed the SBA aid budget from $650 million to $2.65 billion, and required that at least half of the SBA loans go to help the poor (of whatever race) in the slums.

At the same time, however, Congress demanded a crackdown on the SBA's chronic record of mismanagement and defaulted loans. Congress even granted the SBA cease-and-desist authority to use against the swarms of incompetent and corrupt companies that joined the more deserving petitioners for the SBA's generous grants and low-interest loans.[68] During its first decade the SBA had drawn congressional criticism for the opposite reason: its lending policies were too cautious. But the SBA had also earned a reputation for maladministration by second-rate political appointees. And long before the advent of the SBA, the risky world of small business had been a frequent graveyard for naive hopes and underfunded dreams. Against this not very promising background, Richard Nixon pledged to extend the dream to aspiring black entrepreneurs—but without any investment of new resources.[69]

When Nixon created the OMBE by executive order on March 5, 1969, he broadened the target spectrum from "black capitalism" to include "blacks, Mexican-Americans, Puerto-Ricans, Indians, and others."[70] Presumably the "others" included Asians, but not women. In any event, to most Americans "minority" mainly meant black in 1969. But the new OMBE had no program budget and no authority. Instead it would "coordinate" the efforts of 116 existing programs in 21 different federal agencies. Its tiny staff of ten professionals would be buried in the Commerce Department. Nixon proposed no transfer of federal programs or agencies and asked nothing in support from Congress.

The gap between the OMBE's purpose and its resources and authority invited a troubled beginning. The *Wall Street Journal,* which had editorially blessed the new gesture, was reporting by midsummer 1969 that "behind-scenes battling between [OMBE director Thomas. F.] Roeser and Hilary Sandoval, head of the Small Business Administration, has been furious."[71] The OMBE was trespassing on the SBA's turf, and Sandoval was a Hispanic who resented Roeser's concentration on blacks.[72] On the other hand, so many complaints began to pour into aide Leonard Garment, who handled the OMBE program in the White House, about "SBA's managerial and personnel [in]ability to get their part of the show on the road" that Garment considered asking Dwight Ink at the Budget Bureau to form a management team and try to straighten out Sandoval's operation.[73] Roeser announced a new $1 million grant (with money borrowed from other federal agencies) to establish an institute for minority business education at Howard University, and to help the Reverend Leon Sullivan of Philadelphia export his black shopping-center idea to thirteen other cities. But to black critics this was a pathetic payoff. "Nothing much is happening," deadpanned Clarence Mitchell; "Black capitalism is a shambles," said Whitney Young.[74]

Commerce Secretary Maurice Stans struggled to piece together some sort of believable program with no budgeted support. Roeser, who had been a public-relations man for Quaker Oats, sought to enlist "big brother" companies to sponsor new minority businesses. He and Stans urged the auto industry to sponsor 100 new minority dealerships, and they also sought franchises from the shoe, oil, and grocery industries. But the potential "big brothers" were cautious, fearing the high financial risk of failure in their tightly competitive markets. It could not be done merely with good will and few resources. Commerce was a relatively small and weak department, and the tiny and upstart OMBE was trying to exercise the kind of coordination through persuasion that had hobbled Lyndon Johnson's old PCEEO, yet doing so not from the White House but from the remote corridors of Commerce.

In August 1969, Secretary Stans asked Budget director Robert Mayo for an additional $20 million in net outlay ceiling so the SBA could fund 100 new minority SBICs.[75] But Mayo got a strong negative recommendation from his budget and program analysts. They told Mayo that the SBA currently had a backlog of $33 million in loan applications from existing SBICs, and the political implication in such a new program "of there being 'black' money available but no 'white' money is just too strong."[76] Then, in December, Nixon asked his department and agency heads to cooperate better with the OMBE, the SBA, and a special task force on procurement coordinated by Leonard Garment.[77] The Defense budget was expected to fund 80 percent of the minority procurement contracts, but the Pentagon was leaking complaints that the no-bid minority contracts cost 11 percent more than those bid for competitively. Defense Secretary Melvin Laird complained to Garment that the Pentagon was "vulnerable to Armed Services Committee charges that the premiums paid to minority firms are a 'load' which should be authorized in separate legislation."[78]

The rhetoric surrounding Nixon's minority enterprise program generated at the grassroots expectations that ironically seemed to combine with a deep reservoir of cynicism on the part of the civil rights leaders. Between them they tended to catch up Nixon's civil rights initiatives in a left-right whiplash. Berkeley Burrell, a black businessman from Washington who was vice chairman of the Nixon-appointed Advisory Council on Minority Enterprise, offered only sarcasm for the President's efforts to involve OMBE in the procurement process. "I'm thrilled to death," Burrell said, "and the reason I'm so thrilled is that the companies they're talking about haven't even been born yet. Let's face it, right now we don't make anything but hair grease."[79] During the summer of 1969 the Senate Select Committee on Small Business began hearings that would lead to charges of incompetence against the OMBE.[80] This was perhaps to be expected from liberal Democrats in the Senate. But the presidentially appointed Advisory Council on Minority Enterprise produced a similarly damaging report on President Nixon's ill-starred initiative.[81] Darwin W. Bolden, a black business executive from New York who sat with chairman Sam Wyly on the 83-member council's executive committee, described the OMBE as "one of the poorest-managed federal agencies I know." The OMBE was a "tragic failure," Bolden said, that "cannot be saved."[82] Salvageable failure or not, the OMBE was a clumsy government acronym that translated for journalists and the political public into "Black Capitalism." No one seemed to mention Brown Capitalism. The omission was revealing.

Gestures for Hispanics: The Fragmentation of Brown Power

The Nixon administration did not create a distinct civil rights program and policy for Hispanics, as it did for blacks and women. In its relationship with Hispanics the Nixon administration continued the Johnson administration's pattern of gestures in political symbolism. Like Lyndon Johnson, Nixon milked the Hispanic constituency of its political payoff, but otherwise did not attempt to forge a self-consciously Hispanic program. Nixon and Johnson followed the same path with Hispanics because unlike the black civil rights organizations within the Leadership Conference, and after 1968 the bipartisan coalition of feminists united behind the Equal Rights Amendment, the Spanish-speaking constituency was too fragmented and quarrelsome to create a united front.

Hispanic Americans shared many minority grievances with American blacks, such as historic discrimination and its consequences of disproportionate poverty, illiteracy, and political weakness. But unlike American blacks, Hispanic Americans were set apart from the American majority by a cultural bond of a separate language. Also unlike blacks, the Hispanics remained internally divided by national origin and its regional subcultures. By the 1960s the nine million Mexican-Americans or Chicanos were concentrated in five southwestern states; their low levels of income inclined them toward Democratic economic policies, but Republicans saw potential appeal in their strong ties with family and church and their respect for hierarchical order. The two million mainland Puerto Ricans were concentrated on the northeastern seaboard, where like blacks they generally provided low-income clients for the urban Democratic machines. Finally the 700,000 Cuban Americans were clustered around Miami and the New York-New Jersey megopolis, where their entrepreneurial abilities and anti-communist passions provided fertile ground for Republicanism. Reflecting this cultural diversity, Hispanic organizations were generally fragmented by regional cultures and problems, local loyalties, and leadership feuds.[83]

In 1967 Lyndon Johnson had responded to Joe Califano's political proddings by creating the Inter-Agency Committee on Mexican American Affairs. Not taken seriously at the time, other than as a precautionary move in a pre-election year, the committee not surprisingly accomplished nothing of consequence. But when the Democrats lost the White House in 1968, the Democratic 91st Congress in 1969 gave the committee a statutory basis, and renamed it the Cabinet Committee on Opportunity for the Spanish Speaking (CCOSS)—to indicate that it now represented all Hispanics, not just Mexican-Americans.[84] President Nixon routinely signed the CCOSS bill in December 1969. But the Democratic origin of CCOSS and the disorganized nature of its constituency gave it a low priority in the Nixon White House. A year later neither a permanent chairman nor the required advisory council had been appointed, and the new "Cabinet" committee had never met with the members of the Cabinet. Indeed, a year and a half after its statutory creation, this symbolic new high-profile committee for Hispanics, which was required by statute to meet quarterly, had never met at all! By 1970 such lassitude would have been inconceivable on the part of a top-level black or women's committee. Because the wheels of the nation's scattered Hispanic organizations squeaked faintly, and rarely in unison, they got little oil.

The black civil rights movement, however, had helped created federal agencies that would naturally turn, in their normal program-building and their recruitment of supportive clientele, to a constituency as large as 13 million citizens. During the spring of 1970 the Civil Rights Commission released a report that criticized the unequal justice accorded to Mexican-Americans in the Southwest, and the EEOC held hearings in Houston to dramatize job discrimination against Tex-Mex Americans.[85] Ehrlichman warned the President that certain Hispanic leaders were reportedly "fast approaching the stage of confrontation."[86] But Nixon sensed more opportunity than danger from the splintered Hispanic organizations. Mere symbolic gestures seemed to produce political payoff at low cost, while Hispanic infighting seemed to deflect complaints. Looking toward the 1972 election, Nixon in 1970 brought Charles Colson to the White House to woo ethnic voters away from the Democrats. In July 1970, Nixon gave Colson and Peter Flanigan "an absolute demand that we do more for Italians, Cuban Americans, Poles and Mexicans," insisting that "these ethnic groups be included on the names of every commission formed hereafter."[87] Nixon was not entirely cynical. He told Dwight Chapin that he wanted his aides to generate a list of "brilliant Black, Spanish-speaking, and Indian achievers—physicists, historians, medical students, school deans, symphony conductors, business leaders, lawyers, etc."[88] By the fall of 1970 the White House was geared to lure Democrats from the New Deal coalition by building ethnic coalitions for Nixon in 1972.

In November 1970, Nixon announced a 16-point program to bring more Hispanics into the federal civil service, where they constituted only 2.8 percent of federal employees. Also, that November, Garment sent Nixon a list of twenty-none initiatives for wooing minorities.[89] Most of these were symbolic overtures toward blacks, who represented 11 percent of the American population, and only two dealt specifically with Hispanics, who represented an estimated 6 percent.[90] Nixon told Ehrlichman he was willing to meet with the CCOSS once it was formed and had a chairman. But when the White House had still produced no CCOSS chairman or advisory council by the summer of 1971 (although the phantom Hispanic committee had a staff of thirty and an annual budget of $860,000), Republican senators Percy and Javits criticized the Administration for ignoring the statute that created the committee in 1969.[91] Garment finally got the committee to meet for the first time on June 8, 1971. Despite a Nixon campaign promise of 1968 to call a White House conference on Hispanic issues, White House aides persuaded the CCOSS to schedule regional meetings instead—for fear of "factionalism within the brown community . . . [which] could end in disorder."[92] On August 5 Nixon finally nominated for Senate confirmation as CCOSS chairman a Hispanic staff member from the Civil Rights Commission, Henry M. Ramirez, who was a Republican Chicano from Nixon's hometown of Whittier, California.

The Lebanon-like pattern of politics that the White House feared seemed irrepressible, however, among Hispanic leaders. Nixon appointed a CCOSS advisory council consisting of four Chicanos, three Puerto Ricans, and two Cuban-Americans. But one of Ramirez's first acts as chairman was to fire two of the committee's staff, both of them Puerto Ricans. This produced a counterattack from Democratic congressmen Herman Badillo, a Puerto Rican from New York, and Edward R. Roybal, a Mexican-American from Los Angeles, who demanded that a Puerto Rican be appointed as the committee's executive director.[93] Colson admitted to Ehrlichman in late 1971 that the CCOSS was a "neglected step-child" and an "ineffectual mendicant." But the Hispanic communities had produced no strong leaders of national stuture, other than Cesar Chavez, who seemed no friend of Republicans. It was symptomatic of the Hispanic political weakness that in the Nixon White House as late as the early 1970s, Hispanic issues would be handled by men with names like Malek, Marumoto, Garment, Colson, and Finch. The Hispanics would not find a unifying theme until the mid-1970s, during the presidency of Gerald Ford, when Hispanic groups would link voting discrimination to the new issue of language discrimination. But in the first Nixon administration, voting rights was a black issue. Hispanic discontent would remain largely unfocused and intramural.

Nixon and the 91st Congress: Contradictory Models of Equal Opportunity

Black suspicion of Nixon's intentions in civil rights policy began with the campaign rhetoric of the Southern Strategy and increased during the transition when Nixon appointed no black to his cabinet. Nixon himself admitted at a press conference in early February 1969 that he was aware that he was not considered "a friend by many of our black citizens." But he vowed that "by my actions as President I hope to rectify that."[94] During the spring of 1969, while Nixon was traveling in Europe and the White House was struggling to sort out both its domestic agenda and its internal divisions of authority, a series of events external to the White House increased the anxieties within the civil rights establishment about the Administration's uncertain intentions. During Senate hearings in March over the contract compliance programs, Everett Dirksen threatened to use his influence with the White House to fire EEOC chairman Clifford Alexander, whom an increasingly ill-tempered (and profoundly ill) Dirksen accused of conducting "carnival-like hearings" that symbolized his agency's "punitive harassment" of businessmen.[95] On April 9, Alexander resigned as chairman with a blast at the Nixon administration for "a crippling lack of support."[96] In early May the Civil Rights Commission released a study by Richard Nathan, *Jobs and Civil Rights,* which the commission had funded two years earlier. Although all of Nathan's data predated the Nixon administration, the commission used the occasion of its release to attack the new administration as "seriously deficient" in enforcing the civil rights laws.[97]

By the end of the summer, however, the Administration had launched initiatives in three arenas of civil rights policy: school desegregation, job discrimination, and voting rights. Each of these addressed a set of political needs in an internally consistent fashion. In combination, however, they were logically incoherent, and rested on contradictory models of civil rights policy. The contradictions are clearer in our hindsight than they were then, especially because the policy area that received the most attention—school desegregation—blurred the contradictions in a muddle of charges over busing and Supreme Court nominees. The turmoil over Nixon's Southern Strategy and the politics of school desegregation commanded the domestic headlines of his first administration, and Nixon's school- and court-centered policies have dominated the analytical literature in the generation since. The issues these policies addressed, which centered on desegregation in the South, have largely faded with time. The consequences of Nixon's contradictory policies in equal employment and voting rights, however, have grown while their origins have remained obscure.

Nixon's move on school desegregation came first. On July 3, Attorney General Mitchell joined HEW Secretary Robert Finch in announcing that henceforth HEW would shift its enforcement emphasis on school desegregation from fund cut-off to bringing suits in federal court.[98] Accordingly, on August 19, federal lawyers switched sides and opposed NAACP lawyers on a desegregation suit. It was the first time the Justice and NAACP lawyers had differed fundamentally on a major question of judicial policy since the Justice Department began participating in such public litigation in the 1950s. At Nixon's insistence, Secretary Finch then formally intervened in a school suit to withdraw approval and request further delay in the desegregation plans of 33 school districts in Mississippi that HEW had earlier approved. The new,

Nixon-appointed Chief Justice, Warren Burger, on behalf of a unanimous Supreme Court was to disapprove the unusual request later that fall. But Mitchell was bent on shifting the onus of forced busing from the presidency to the courts.[99]

The Attorney General's call for enforcing school desegregation through Justice Department lawsuits rather than through HEW fund cut-offs was on its face a plausible alternative. It had long been advocated by John Doar, the assistant attorney general for civil rights under Kennedy and Johnson, whose liberal credentials and civil rights commitment were mainstream for the Democratic regimes of the 1960s. Doar had argued that Justice lawsuits as an enforcement tool were more equitable than fund cut-off because they did not automatically penalize the victims of discrimination along with the perpetrators.[100] More important for Mitchell, judicial rather than administrative enforcement might shift the blame for school desegregation in the eyes of southern whites from the presidency to the federal judiciary. Nixon could neither control nor defy the federal courts. But he could fill them with strict constructionist judges, especially if he occupied the White House for eight years. In 1972, Wallace was expected to run again, and resentful southern whites must not direct their blame primarily at President Nixon. As for apprehensive northern whites, they must see the President as a protector who could confine school desegregation to the historically de jure systems of the South.

The Nixon administration's rather muddled challenge to "forced busing" engaged both the 91st and the 92nd Congress in a running debate over legislative and even constitutional attempts to stop the busing. Although this turmoil ultimately produced no significant legislation, it did lead to an ironical dual conclusion. First, Nixon presided with quiet responsibly and surprising success over a sustained effort to ease the path of school desegregation throughout the South.[101] The proportion of black children in the South attending all-black schools plunged from 68 percent in 1968 to 8 percent in 1972. By the end of Nixon's first term, the southern school systems led the nation in racial integration. Second, the Supreme Court, by a narrow margin that was determined by Nixon Court appointees, ultimately stopped the busing for racial balance before it reached the northern suburbs.[102] By 1974, the share of black children attending schools that were 95 percent or more black had declined in the South to 20 percent, but had *risen* to 50 percent in the northern and western states.[103]

Nixon is remembered far more for his Supreme Court nominations of judges Clement Haynsworth and G. Harrold Carswell than for his quietly successful desegregation of the South's schools. But while the controversies over racial busing and judicial appointments captured public attention, the Administration was developing two parallel policies that were based on contradictory models. One of these involved voting rights. The issue was raised in a rather routine fashion by the scheduled expiration in August of 1970 of portions of the 1965 Voting Rights Act. In the spring of 1969 the House Judiciary Committee held hearings designed to support a straight five-year renewal of the act, and this created an immediate dilemma for the Administration, because a simple extension would perpetuate the pattern of regional enforcement that southern whites resented.

In response the Administration argued that the act should be amended to apply equally to all fifty states. The implicit model was classic procedural egalitarianism, based on a core doctrine that affirmed individual rights against all public and most private discrimination. It would expand the Kennedy-Johnson legacy, as embodied in the civil rights laws of 1964 and 1965, by extending their anti-discrimination provisions to cover all individuals equally throughout the entire country. This would be done on the theory that specifications and limits of coverage might appropriately be applied to scale, as in Mrs. Murphy's boarding house, but not to geography, as in the regional design of the Voting Rights Act. The motives behind such an even-handed defense of individual rights were not necessarily pure, but the logic was consistent.

At the same time an opposing theory was pushed by the Nixon administration in another policy arena that was based on an implicit theory of group rather than individual rights. Its core model was one of proportional representation of racial and ethnic groups, and it emphasized substantive rather than procedural equality. By the end of the Johnson administration the proportional or equal-results model was coming to dominate the enforcement strategies of the EEOC and the OFCC. The legislative history and language of the 1964 law, however, argued against such a novel interpretation, and both the EEOC and the OFCC remained small and relatively powerless. Then the Nixon administration in its first year suddenly revived the moribund Philadelphia Plan. When conservative Republicans and southern Democrats in Congress reacted in dismay that autumn by trying to strip away the Plan's provision for racial job quotas, the Nixon administration hurled the full force of its lobbying muscle against them. From the viewpoint of some sort of consistent civil rights policy, it was all very bewildering.

CHAPTER XIII

The Philadelphia Plan Redux

George P. Shultz and the Rebirth of the Philadelphia Plan

In his memoirs, John Ehrlichman described Richard Nixon as being convinced that "the majority of Americans did not support open housing, affirmative action, busing to achieve racial balance, Model Cities, the Equal Employment Opportunity Commission and the other federal civil rights activities."[1] Nixon, Ehrlichman said, believed that such programs "simply would never do any good." Why, then, did such a man, who would appeal to southern and suburban whites on the busing issue, and who would enlist Charles Colson in the task of wooing Catholics and hard-hats away from the Democrats, begin his new administration by reviving the liberal Democrats' explosively controversial Philadelphia Plan? The racial preferences of the Philadelphia Plan violated Nixon's long-standing and, in 1968, successful strategy of avoiding a policy stance on the race issue that wavered very far from the center-right of the political spectrum. It also seemed palpably inconsistent with normally conservative Republican principles, which had led Everett Dirksen in 1964 to demand in Title VII's Section 703(j) a ban on preferential quotas in the first place. By reputation Nixon was often moved by motives no loftier than a tactical desire to confound his enemies, or no more principled than a politician's practical instinct to balance off one thrust with a counterthrust—in this case to counterbalance his attack on busing and his conservative Supreme Court nominations with a positive gesture toward providing black jobs and thereby building a black middle class. But Nixon was also responding to a powerfully argued initiative that came from an unanticipated source: the new Republican Secretary of Labor, George P. Shultz.

On December 30, 1968, as Secretary of Labor-designate, Shultz appeared in Chicago at the annual meeting of the Industrial Relations Research Association to deliver his presidential address. In covering the story, the *Wall Street Journal* played as the lead Shultz's surprising priority on the issue of race and employment—which Shultz ranked as even more pressing than the nation's problems of productivity, inflation, and industrial conflict.[2] A labor economist, Shultz argued that until quite recently the research in his profession had mostly ignored the "appalling unemployment experience of black teen-agers," who were dangerously excluded from the American mainstream.[3] He talked about the exclusion of ghetto Negroes from the informal, relatives-and-neighbors network of job information. He alluded opaquely to the need for "special measures," for wholesale job redesign that would match "hiring standards and actual job requirements," and for government incentives to persuade employers that under the "explosive" new circumstances of racially skewed unemployment, "supervisors cannot conduct business as usual."

An experienced labor negotiator and dean of the University of Chicago Graduate School of Business, Shultz did not attack the unions directly in his maiden speech as Secretary of Labor-designate. He was a Republican academic who worked comfortably within the intellectual webb of Chicago's conservative, market-oriented school of economic thought. The "Chicago school" was dominated by Shultz's monetarist colleague and friend, the more famous Milton Friedman. But Shultz was no doctrinaire conservative or market-libertarian, and he defended the need for a variety of government-sponsored insurance and welfare programs to cushion fluctuations in the job market.[4] In his IRRA speech, however, Shultz criticized the American steel industry's inflationary, postwar *pas de deux* with the unions, and in private he deplored the unions' role in driving wages irresponsibly upward in a self-destructive inflationary spiral.[5] Such union hammerlocks were inherently unhealthy for a competitive market economy, Shultz believed. But at least the industrial unions, under the prodding of George Meany and the AFL-CIO leadership, were rapidly moving to racially integrate their membership.

Not so the skilled trades of the construction unions, Shultz said. Their AFL craft traditions outrageously excluded minorities, and in such matters of local tradition and fraternal preference they ignored the national leadership. Local craft monopolies, enjoying Davis-Bacon guarantees of prevailing wage rates, effectively protected the skilled construction trades from wage competition, and this encouraged a soaring wage inflation that spread, in Shultz's view, like regional cancers.[6] Shultz knew that as Secretary of Labor he could not realistically take on such a sacred cow as Davis-Bacon, which Democratic Congresses had periodically strengthened since its enactment for organized labor in 1931. But Shultz could challenge the refractory construction unions on their sorry record of racial exclusion in the skilled trades, and try to bend their will to socially constructive purposes.[7]

Like many highly educated Republicans, Shultz was conservative on economic issues but more liberal on social issues—just the opposite of the hard-hat construction workers who had inherited their Democratic loyalties. But Shultz was also what Nixon later was to call a "bulldog"—an academic intellectual whose Buddha-like demeanor disguised an iron determination, a Princeton graduate who had joined the Marine Corps to fight in World War

II.[8] After the war Shultz returned to academe, earning his doctorate in industrial economy from MIT in 1949. Nixon, like most Republican presidents, cared little about the Labor Department, with its overwhelmingly Democratic constituency, and he welcomed the recommendation of Arthur Burns that his fellow economist, Dean Shultz, be appointed Labor secretary. Somewhat like his pipe-smoking mentor, Burns, Shultz was a rather stolid figure, an avuncular "Dutchman" who was not much given to showing his emotions. He combined the practical temperament of a labor negotiator with a stubbornly Calvinistic sense of moral rectitude. Shultz had served under Burns in the Eisenhower administration as senior economist for the Council of Economic Advisors during 1955–57, then left to join the faculty at Chicago, becoming the business school dean in 1962.

As both a professional consultant in labor economics and a close colleague of Willard Wirtz, who had taught labor law at Northwestern prior to becoming Labor secretary, Shultz knew his way around the Labor Department. During the 1968 presidential campaign he had chaired a task force for Burns on manpower training that had focused on Labor Department programs. Ehrlichman characterized Shultz with rare approbation as a quiet man of action who controlled his bureaucracy and delivered on his commitments. Like a good Marine, Ehrlichman said, Shultz demanded loyalty up and down his chain of command; he was blunt with painful truths in private councils, and in his public remarks he was invariably dull, straight, and boring.[9]

When Shultz assumed his cabinet duties in late January 1969, he told Nixon's newly appointed Budget director, Robert P. Mayo, that he agreed with the recommendation of Ramsey Clark's 1968 task force that the OFCC should be transferred out of Labor and moved to the EEOC.[10] Such a move would undoubtedly have eased the complications of Shultz's mediating role with organized labor. But the available White House and Labor Secretary records do not clarify when and why Shultz changed his mind, and decided instead not only to keep the OFCC in Labor, but also to revive the Philadelphia Plan.[11] Mayo wrote Arthur Burns on February 13 that the question of what to do with the OFCC was complicated because the recommendation of Ramsey Clark's task force to transfer the OFCC to the EEOC had leaked to the press. Furthermore, Clifford Alexander had refused in public to tender his resignation as EEOC chairman to President Nixon, and Shultz did "not want to get the Department into a position which could be interpreted as pulling it out of the anti-discrimination field." Then Mayo added a handwritten note at the bottom: "Arthur—This is getting hot and we should get together tomorrow if possible. Harlow's office and Packard are both desirous of straightening this out as soon as we can."

Bryce Harlow's concern of course meant that members of Congress were beginning to ask questions, and the Packard allusion was to Deputy Defense Secretary David Packard. What was "getting hot" was the spotlight thrown by Senator Edward Kennedy's subcommittee of the Senate Judiciary Committee investigating a $9.4 million Pentagon contract signed with three southern textile mills, with only oral assurances on affirmative action plans.[12] This was the hearing where Dirksen attacked both the EEOC and the OFCC and threatened to get Alexander fired. The spotlight seems to have been hot enough to change Shultz's mind about getting rid of the OFCC, and thereby appearing to back away from enforcing an anti-discrimination commitment against such notoriously anti-labor employers as the J.P. Stevens Company. Shortly after Dirksen lashed the EEOC and OFCC, Shultz told the National Press Club that ensuring equal employment opportunity was a "top priority" of the Nixon administration, and therefore he had decided to strengthen rather than transfer the OFCC.[13] Having decided to take the offensive with the OFCC, Shultz pressed his argument on Nixon, Ehrlichman recalled, by appealing to the President's bourgeois solicitude for jobs, to the Quaker virtue of honest work, and to the prospect of creating black conservatives by giving them a stake in the middle class.[14] Not lost on Nixon, however, was the delicious prospect of setting organized labor and the civil rights establishment at each other's Democratic throats. Ehrlichman recalled Nixon's rather rapid conversion by Shultz to the Philadelphia Plan: "Nixon thought that Secretary of Labor George Shultz had shown great style in constructing a political dilemma for the labor union leaders and civil rights groups."[15] "The NAACP wanted a tougher requirement; the unions hated the whole thing," Ehrlichman said. "Before long, the AFL-CIO and the NAACP were locked in combat over one of the passionate issues of the day and the Nixon Administration was located in the sweet and reasonable middle."

In his *Memoirs*, Nixon recalls that *he* came into office with a stout determination to do something about racial discrimination in jobs and promotions by the major labor unions. "I asked Secretary of Labor George Shultz to see what could be done [about unemployment]," Nixon explained. "He proposed a plan which would require all contractors working on federally funded construction projects to pledge a good faith effort toward the goal of hiring a representative number of minority workers."[16] Nixon recalled that as Vice President he had chaired Eisenhower's contracts committee, which had used "persuasion and publicity to encourage companies with government contracts to hire more minority workers." The plan Shultz devised, Nixon said, "would require such action by law [and] was both necessary and right. We would not fix quotas, but would require federal contractors to show 'affirmative action' to meet the goals of increasing minority employment."

Shultz's major reason for reviving the Philadelphia Plan may well have flowed from professional and personal convictions about the social irresponsibility of the construction-trades unions and the economic irrationality of racism in the American political economy. But Ehrlichman conceded that Shultz was not above appreciating, with a twinkle in his Republican eye, the partisan virtues as well as the moral splendor of linking the Democratic-voting black laborers and the lily-white construction craft unions in intimate dialogue by "tying their tails together."[17]

Arthur Fletcher and the Revised Philadelphia Plan:
A Mandate for Proportional Representation

In the spring of 1969, Shultz reorganized the Labor Department. This was not unusual behavior for a new agency head. Such structural rearrangements provided senior slots for the new management team, and often also provided the new leaders a quicker way to make their mark on the agency than by changing the substance of institutional policy against the natural inertia of the permanent government. Secretary Wirtz had shown less interest in Labor's workaday management than in his special projects, which he tended to keep close to his own office, and one result was that the new OFCC had reported directly to the Secretary. But Shultz, who was a more systematic manager of the organizational lines and boxes, created a new assistant secretaryship for Wage and Labor Standards, and filled it with a black Republican entrepreneur, Arthur A. Fletcher. Fletcher had lost a race for lieutenant governor in Washington state the previous year; a former journeyman football player for the Baltimore Colts and the Los Angeles Rams, Fletcher in effect replaced the departed Sylvester as Labor's black point man on EEO contract compliance.[18] Under the new arrangement, the OFCC reported to Fletcher, and was headed by Fletcher's deputy, John A. Wilks, who was black also.

During the late spring of 1969 Fletcher presided with Shultz's blessing over the redesign of the Philadelphia Plan. The chief obstacle to be overcome was Comptroller Staats's objection to post-award negotiations of affirmative action plans without clear guidelines for bidders. The Department of Labor under the previous administration had already committed itself to the development of "specific goals and timetables," which had first been required in Labor's general affirmative action regulations of May 1968.[19] But these had required that the detailed goals themselves, with target numbers and percentages of minority employees, had to come from the bidding contractors, not from the Labor Department. Otherwise, the OFCC would appear to be dictating racial quotas, and hence would collide squarely with Dirksen's ban in Section 703(j) of the Civil Rights Act. If in the aftermath of the Comptroller's ruling, the OFCC could no longer arrange for the reluctant contractors to come up with these specific goals and timetables through post-award coaching, then the OFCC would have to devise a scheme that would somehow avoid the appearance of setting quotas and requiring racial preferences in construction hiring.

The result of the spring retooling was the Revised Philadelphia Plan, which Fletcher announced to the heads of all agencies on June 27, 1969. The plan took the form of an administrative order, which Fletcher signed at a public ceremony in Philadelphia. There he announced that "under my administration" the new plan "will be put into effect in all the major cities across the Nation as soon as possible."[20] He explained that specific "goals or standards for percentages of minority employees" were necessary because America's history of segregation had been forcefully imposed upon her minorities. In that historic experience, Fletcher said, "quotas, limits, boundaries were set." He agreed that "It might be better, admittedly, if specific goals were not required—certainly the black people of America understand taboos." But the brute fact of historic discrimination meant that "Visible, measurable goals to correct obvious imbalances are essential."

What Fletcher meant by "imbalances" was clarified by the revised plan itself.[21] At its heart was a model of proportional representation. This had been a vaguely implicit model ever since Roosevelt's FEPC in 1941, and Vice President Lyndon Johnson had seemed to imply as much when he had complained during the early 1960s, as head of the PCEEO, of minority "underutilization." The notion of underutilization implied an unfair shortfall in the representation of certain groups, whose presence under normal circumstances presumably would reflect their proportion in the total population of the relevant worker pool, absent invidious discrimination.

But the Civil Rights Act of 1964 had seemed with equal clarity to ban any government requirement that would coerce employers toward proportional representation. So the revised Philadelphia Plan sought to avoid this apperance by providing for the OFCC area coordinators to assess local conditions and then establish, for each designated construction trade, not a specific numerical target but rather a target *range*. The range in turn was expressed as a percentage rather than as a numerical goal—although the percentage translated into a number whenever it was applied. Thus for the seven mechanical trades originally targeted in the Philadelphia Plan, the Labor Department held hearings in Philadelphia during August 26–28 to gather information for setting the new standards, and on September 23, Fletcher published a 20–page memorandum of implementation with "findings" that set five-year target ranges. The plumbers and pipefitters, for example, with what Fletcher's figures claimed were only 12 minority workers among 2,335 union members in Philadelphia (or .51 percent), were given a minority goal for 1970 that ranged from 5 to 8 percent. This escalated annually to reach a range of 22–26 percent for 1973. Similarly escalating ranges were set for the other trades. These target ranges in turn represented a rough splitting of the difference between Philadelphia's 30 percent black population in the metropolitan work force, and its 12 percent black representation in the skilled construction trades in 1969.[22]

Fletcher thus sought to finesse what Hubert Humphrey had referred to as the "bugaboo" of quotas by establishing, in the invitation for bids, suggested ranges of minority employment within which bidding contractors themselves could presumably choose targets. The revised plan acknowledged Staats's objection by precluding any negotiations after bids were opened. As a disarming gesture toward the anti-quota language of Title VII's Section 703(j), the revised plan also contained a formal disclaimer of uncertain import: the requirement of specific goals "is not intended and shall not be used to discriminate against any qualified applicant or employee." Fletcher insisted that the goals were not rigid quotas, and that employers who subsequently failed to meet them would be given a chance to demonstrate that they had made "every good faith effort" to meet their goals.[23] On the other hand, putting the blame for discrimination on the unions was ruled an unacceptable excuse for contractors. The revised plan stated that because employment discrimination was prohibited by the Civil Rights Act of 1964, contractors could claim no excuse simply because their hiring-hall unions had failed to refer any minorities to them.

But which groups qualified as minorities? Obviously blacks did. The Revised Philadelphia Plan defined the covered minority groups as "Negro, Oriental, American Indian and Spanish Surnamed American."[24] Oriental was not further defined. Spanish Surmaned American was defined merely as "all persons of Mexican, Puerto Rican, Cuban or Spanish origin or ancestry." This rather casual enumeration of protected groups, which was buried in an appendix, essentially repeated the enumeration contained in the old PCEEO's Standard Form 40, which had since evolved into EEO-1.[25] But there was one significant omission, and also one interesting addition. The *omission* was women. Like the Johnson administration, Richard Nixon's administration simply forgot about women when they contemplated EEO enforcement in contract compliance. (That revealing oversight will be addressed in Chapter 16.) But the *addition* to the approved list of minorities was also intriguing—and like the omission of women, it was unnoticed by the national press. It suggested some implicit but unexamined, and indeed highly problematical, assumptions in the evolving and inherently expanding, group-rights approach to EEO. What was new in the revised plan under the Nixon administration, as opposed to the original plan under the Johnson administration, was the inclusion of Cubans.

Adding Cubans made some immediate political sense for a Republican administration, in that the large bloc of Cuban Americans who had fled Castro's revolution was overwhelmingly Republican.[26] And the Cubans were indubitably of Spanish heritage. But the Cuban émigrés had fled Castro's violent revolution in 1959 largely because they represented the propertied, professional, conservative classes who had the most to lose. Hence as an American ethnic group of quite recent origin, they had found economic success with unusual speed.[27] If the chief rationale for affirmative action was to compensate for historic discrimination against a disadvantaged minority, then the Cubans' had little credible claim—except perhaps against Fidel Castro's government—and it is not clear that the proud and upwardly mobile Cubans ever pressed such a claim.[28] Thus the mere possession of a Spanish surname seemed a fanciful reason to single out for special protection such a relatively prosperous, Republican group. If the rationale of the Philadelphia Plan was to remedy past discrimination against minority groups, then the Cuban inclusion raised troublesome questions about the relationship between ends and means, about the linkage between actual disadvantage and mere group membership, and about the relative primacy of individual and group rights. Yet the questions raised by Cuban exceptionalism seem not to have been seriously explored by the Labor Department, or questioned by the public during the hectic events of 1969.

For America in the summer of 1969, "minorities" essentially meant blacks, especially in cities like Philadelphia, which had so recently been put to the torch. It certainly did not seem to mean women—although the snowballing feminist movement was beginning to press the question. Logically it seemed to include Asians, who were indubitably a racial minority with a history of severe discrimination and deprivation, especially in the West—although Asians, like Cubans, did not seem to be organizing to press the matter.[29] Hispanic Americans were specifically listed as ethnic minorities, although confusion over their appropriate rubric (Spanish Speaking, Spanish Surnamed, Hispanic "heritage," Chicano) suggested the primitive conceptual level of nascent group-rights theory by the end of the 1960s. Such difficult questions of definition, rationale, and consistency would surface later, when the basic black mold was pretty much set according to a proportional model. But by the summer recess of Congress in 1969, Shultz had revised the Philadelphia Plan and with his President's backing, Shultz was ready to do battle to defend it. He did not have to wait long.

The Interagency Legal War Over the Revised Philadelphia Plan: The Silberman Brief and Racial Preferences

On August 5, Comptroller Staats sent Shultz a 17-page formal opinion that the revised Philadelphia Plan was *still* illegal. Both sides had been braced and indeed well rehearsed for this renewed confrontation. The previous January, Staats had begun to receive complaints from the contractors' associations that his November 1968 directive outlawing the negotiations over manning tables was being widely ignored by federal EEO and contract compliance officials. The complaining builders cited as a particularly flagrant example the defiance of Staats's ruling by an EEO specialist at HEW, who had announced at pre-bid meetings in New York City in late January that he was going to demand manning tables in construction contracts for Beaver College and Children's Hospital in Philadelphia.[30] On February 4, Dirksen had inserted into the *Congressional Record* two articles from *Barron's* attacking the racial quotas demanded by the EEOC and OFCC, and Dirksen asked the Senate Subcommittee on Separation of Powers to look into the matter.[31] On February 7, Republican Paul Fannin of Arizona introduced a bill that would suspend Johnson's executive order of 1965 and make Title VII and the EEOC the sole means of enforcement and remedy in equal employment. Staats soon was being pressed to demand compliance with his 1968 ruling by Democrat John McClellan of Arkansas, chairman of the Senate Committee on Government Operations.

In response, during March and April Staats began to press the agencies for assurances that they were enforcing their legal obligations.[32] By early May Secretary Finch had come around to an agreement that HEW would comply.[33] GAO lawyers concluded that the matter was successfully resolved in their favor and that "such action closes the file."[34] They were, indeed, temporarily right about HEW and the tractable Finch. But they had underestimated the determination of the Philadelphia Plan's aroused constituency, and also that of Secretary Shultz and the Labor Department.

The grassroots counterattack against the GAO began appropriately enough in Philadelphia, and was sponsored by the area's Democratic congressmen. The complaining Pennsylvanians were led by Representative William A. Barrett, chairman of the housing subcommittee of the House Banking and Currency Committee, who summoned the GAO staff to meet with an upset Philadelphia delegation. Meeting on April 29, the group included staff from the GAO, the OFCC, and the congressmen's offices, plus reporters for the Philadelphia *Bulletin* and *Inquirer,* and black representatives from the Philadelphia Human Relations Commission and the Urban Coalition.[35] The Urban Coalition delegate demanded that the Philadelphia Plan be reinstated, even though he admitted that it was "in effect, a quota system." The Human Relations Commission delegate predicted a race riot if the plan was not promptly reinstituted.[36] Soon it became clear to the GAO that in response to such external pressures as well as to the determination of Shultz and his staff, the OFCC again was gearing up the Philadelphia Plan. In June, Senators McClellan and Fannin queried Staats about the legality of the new Fletcher-Shultz version, and this in turn led to meetings of senior staff from GAO, Labor, and Justice, in which Labor Solicitor Laurence H. Silberman announced his intention to prepare a legal brief to defend the revised plan.[37]

On July 16, Labor released Silberman's 44-page brief.[38] In it the Labor Solicitor dealt only secondarily with the explicit objection to post-award negotiations that had led the Comptroller to strike down the first Philadelphia Plan. To meet that objection, Silberman argued that the revised plan's call for minority hiring goals in the form of percentage ranges provided ample guidelines for contractors to draw up their *own* affirmative action plans *prior* to bidding, and that this change should satisfy Staats's procedural objection of 1968. Silberman was correct. When Staats received the revised plan on August 5, he agreed that the revised plan's requirement of a numerical range "is apparently designed to meet, and reasonably satisfies, the requirement for specificity."[39] But Silberman also assumed that this wasn't Staats's main objection, or at least that it wasn't the main objection of the powerful congressional forces that were rallying behind the Comptroller. So the Labor Solicitor concentrated his brief on defending the President's authority against an anticipated argument that the numerical requirements of the manning table device would violate the Dirksen amendment in Title VII, specifically its ban on minority quotas and preferential treatment on account of racial imbalance. Silberman was right again.

Silberman's main problem, then, in defending the OFCC's broad interpretation of its Title VI and executive order authority, was with Dirksen's ban on racial preferences in Title VII. So he pinned his case for maximum presidential authority on constitutional, not statutory, authority. He relied especially on the 5th Amendment's due process clause, which the Warren Court had used in 1954 to justify ordering school desegregation in the District of

Columbia (while relying on the 14th Amendment's equal protection clause to order school desegregation in the states). Silberman pointed out that a presidential program of affirmative action had existed under Kennedy's executive order *prior* to the passage of the Civil Rights Act of 1964. Thus he argued that the President's power flowed from his constitutional obligation rather than from mere statutory law. But even if the new plan *were* subject to Title VII, Silberman hedged, its racial job targets would still be permissible because its provisions were not expressly *forbidden* by that title. Indeed, Silberman said, the Senate had defeated an amendment proposed in 1964 by Senator Tower (Amendment #962) which would have made Title VII the *exclusive* federal remedy for employment discrimination, thus ousting the President from the field.

Silberman concluded that by rejecting Tower's exclusionary amendment, Congress in 1964 had effectively ratified the President's broad EEO program of executive authority. Furthermore, Silberman argued, Congress had implicitly endorsed that broad interpretation when it had subsequently approved appropriations for the OFCC.[40] Indeed, Silberman's broad reasoning created a virtual carte blanche under which any executive requirement of affirmative action to remedy past discrimination might avoid conflict with the Civil Rights Act, "since that is one of the primary purposes of the act."[41] Silberman's claim for presidential discretion was so broad, in fact, that it later embarrassed him, and he publicly repudiated his own brief, including its implicit defense of racial hiring quotas.[42]

Comptroller Staats and Reverse Discrimination

As Silberman guessed, it was the racial hiring quotas that Staats fundamentally objected to, not the question of timing or specificity of guidelines for bidders. On August 5, Staats ruled that "by making race or national origin a determinative factor in employment," the revised Philadelphia Plan violated Title VII.[43] This was an ambitious reach for the Comptroller General under the GAO's audit authority, especially since it promised to set him against the Attorney General in interpreting a matter of law that seemed considerably removed from the GAO's traditional procurement function. To bolster his case Staats quoted from Sec. 703(j)'s prohibition against preferential treatment on account of any racial imbalance. He also quoted Hubert Humphrey's pledge that "Contrary to the allegations of some opponents of this title, there is nothing in it that will give any power to the Commission or to any court to require hiring, firing, or promotion of employees in order to meet a racial 'quota' or to achieve a certain racial balance."[44] Further, Staats cited the agreement of the civil rights bill's liberal Senate floor managers, Clifford Case and Joseph Clark, that "any deliberate attempt to maintain a racial balance" would violate Title VII, which "simply eliminates consideration of color from the decision to hire or promote." Staats objected to Silberman's long recitation of recent case law concerning the desegregation of schools, voting, and public accommodations because these did not deprive any member of a majority group of his rights through "reverse discrimination." Furthermore, Staats wrote, the judicial relief ordered in those cases was directed at the discriminators themselves, whereas in the Philadelphia Plan the contractors were "more the victims than the instigators of the past discriminatory practices of the unions." Thus the plan was essentially and unfairly "a blanket administrative mandate for remedial action to be taken by all contractors in an attempt to cure the evils resulting from labor union activities."[45]

Curiously, Staats's long letter did not bother to counter Silberman's indirect argument that by rejecting the Tower amendment, Congress had evinced its intention not to restrict the President's executive order program. GAO lawyers, in their earlier discussion with Silberman and with attorney David Rose from Justice, had attacked this argument, which hinged on the earlier rejection by Congress of another amendment.[46] This was Section 711(b) of the original House civil rights bill in 1963, which was advanced by chairman Celler in his Judiciary Committee to provide the executive anti-discrimination program with specific statutory authorization.[47] Celler and the committee's liberal majority had argued that 711(b) was needed to prevent revocation or crippling of EEO programs by future presidents who might be less committed to civil rights reform than Lyndon Johnson.

Congressman Richard Poff of Virginia had objected, on behalf of the opponents, that 711(b) would grant the President "blanket and unlimited authority." Its language authorized the President "to take such action as may be appropriate to prevent the commiting or continuing of an unlawful employment practice by a person in connection with the performance of a contract with an agent or instrumentality of the United States." So Celler agreed to delete the amendment from the civil rights bill.[48] Thus the GAO lawyers argued that Congress in 1964 struck a compromise that only supported the kind of presidential authority *then* in existence, such as the programs of the PCEEO under the Kennedy executive order, none of which contained any quota-like provisions for numerical goals or timetables. But in rejecting 711(b) while accepting 703(j), the GAO lawyers argued, Congress had rejected the imposition of quotas or preferential hiring requirements in any future government program. In his letter of rejection to Shultz, Staats merely assumed this logic without elaborating its historical circumstances and relied upon the anti-quota language of Section 703(j). Staats also agreed that Congress might decide to amend Title VII in the future so as to permit a revised Philadelphia Plan. But until Congress so acted, he ruled, the GAO would not honor contracts signed under the revised plan.

Silberman's brief contained one minor trump card, however, that gave Staats difficulty. On July 2, 1969, the Ohio Supreme Court in *Weiner v. Cuyahoga Community College District* interpreted the Civil Rights Act of 1964 to permit the award of a federally assisted construction contract to a second low bidder, when the original low bidder had failed to submit an affirmative action program with the manning tables required under the Cleveland Plan.[49] In reply to Silberman's citation of *Weiner,* Staats observed that this was only a 5–2 decision of a state court based on conflicting opinions in the lower federal courts, and hence it was not a controlling precedent for the validity of the revised Philadelphia Plan. Staats was, after all, taking on the Labor Department and the civil rights establishment, and probably the Justice Department and the entire presidential administration as well. He could ill afford mounting opposition from the federal courts, and thus he had to act quickly to seal his victory. What Staats needed was to rally his natural constituency, which consisted of the congressional conservatives of both parties, the government contractors, the Chamber of Commerce and the National Association of Manufacturers, and, ironically, most of organized labor. Meanwhile, Shultz and the administration seemed determined to implement their plan despite the Comptroller's objection.

Subsequent events broke quickly. The day following Staats's new ruling, Shultz held a press conference and called the Comptroller's opinion an attack upon the entire affirmative action concept that would "destroy all reason for the existence of the Executive Order and the OFCC."[50] The Comptroller was an agent of Congress, Shultz said, not the executive branch. His opinion was not solicited by the Department of Labor, was not supported by the Department of Justice, and was based upon his abstract interpretation of a law unrelated to his proper field of procurement. Two days later, on August 8, the President issued an executive order calling for each agency head to maintain a "continuing affirmative program" of EEO in federal employment, under the oversight of the Civil Service Commission. Actually, Nixon's order was redundant, because Johnson's order of 1965 (as amended to include sex in 1967) already required this. But by issuing his own order to supersede Johnson's on federal employment, and by adding that the affirmative programs "must be an integral part of every aspect of personnel policy and practice in the employment, development, advancement, and treatment" of civilian federal employees, Nixon formally stamped the effort with his imprimatur.[51]

Then on September 22, Attorney General Mitchell sent Shultz a 20-page brief that disagreed with Staats's ruling.[52] Mitchell drew on the Supreme Court's recent *Gaston County* decision on voting discrimination to argue that "the obligation of nondiscrimination, whether imposed by statute or by the Constitution, does not require and, in some circumstances, *may not permit* obliviousness or indifference to the racial consequences of alternative courses of action which involve the application of outwardly neutral criteria [emphasis added]."[53] The Republican Attorney General, whose brief sounded strikingly like the arguments of Joseph Rauh, the Leadership Conference, and the Americans for Democratic Action, conceded that a contractor was free to insist on his customary union referrals. But since he was also at least theoretically free to obtain his employees from non-union sources, the government was "free to bargain for his assurance to do so," and the contractor had "no right to contract with the Government on his own terms."[54]

Mitchell's opinion displayed an inner tension, however. He relied on the modern executive's broad procurement authority to attach virtually any requirements it wished to government contracts with minimal explanation. But this conflicted with the government's traditional practice, reaching back to the King's obligations under Magna Carta, to provide for due process when acting against alleged wrongdoing, such as against racial discrimination. This was one reason why Title VI was so short and Title VII was so long. The former simply relied on the government's procurement authority, and thus didn't require "findings" of wrongdoing. But because Title VII dealt with harmful behavior and provided for damages and remedies, it was quasi-judicial in character and included court-like due process. Mitchell's brief, however, sought at least some cover for both its constitutional and statutory flanks. So despite its main reliance on the President's constitutional authority, it genuflected to the Civil Rights Act, specifically by upholding the Labor Department's findings in the August 1969 hearings of systematic racial discrimination in Philadelphia. These OFCC findings presumably had justified the remedial "special measures" that the OFCC then required in Philadelphia's six specified construction trades.

Spokesmen for organized labor, however, were quick to point out that in its haste to declare and implement the revised Philadelphia Plan, the Department of Labor had reversed the normal order of proceeding. Labor had held the hearings in Philadelphia only *after* it had first announced the policy that the hearings would ratify ex post facto.[55] Fletcher had announced the new order on June 27. He then gave notice of the public hearings on August 16, and held them in Philadelphia on August 26 and 27. Testimony from invited witnesses at the Labor Department's hearings in Philadelphia was virtually unanimous in support of the plan that the Labor Department had already announced. Fletcher finally spelled out the plan's detailed requirements on September 23, the day *after* the Attorney General sent Shultz his legal approval. As one critic wryly observed, "The Department of Labor decreed discrimination first and found it later."[56] Even Silberman's associate solicitor at Labor, James E. Jones, Jr., admitted to "administrative sloppiness."[57] A strong defender of the plan, Jones conceded that "It would have been more orderly to have held all the hearings prior to the issuance of the original order." But Shultz and his Labor colleagues were moved by a sense of urgency. There were threats of renewed racial violence across the land on the one hand, and on the other, a gathering storm of conservative opposition in the Congress.

Showdown in Congress

That summer violent black protest broke out in late July in Chicago and Pittsburgh. Long threatened and almost routinely expected since the Watts explosion of August 1965, the black violence this time was different from the previous summer contagions. Whereas before the rioting and arson had been spontaneous and chaotic, by 1969, there were organized demands for specific objectives, with most of the violence held in bargaining reserve. In Chicago, job protests launched by a coalition of black neighborhood organizations shut down twenty-three South Side construction projects involving $85 million in contracts. They thereby idled 1,200 construction workers, half of them black.[58] The demonstrations in Pittsburgh were more violent than in Chicago, but were similarly organized and focused on job discrimination in construction. One clash in Pittsburgh in late August left 50 black protestors and 12 policemen injured. Only four of the city's 25 building-trades unions had black membership exceeding 2 percent, although Pittsburgh's metropolitan population was 23 percent black. Racial violence over jobs also occurred in Seattle, and black coalitions announced job protest drives for New York, Cleveland, Detroit, Milwaukee, and Boston.[59]

We know from hindsight that violent black protest was dying out in 1969, and that the summer of 1968 was the last of the "long hot summers."[60] But from the perspective of the participants in 1969, in the wake of the murders of King and Kennedy and the Kerner Report's condemnation of pervasive white racism as chiefly responsible for the nation's turmoil, it was a not unreasonable expectation, during Nixon's first year in office, that the chaos would continue unless and until public policy produced demonstrable payoff for urban blacks demanding jobs.[61] Indeed, the massively destructive rioting following King's killing the summer before had seemed to accelerate, not derail, the passage of the fair housing law of 1968. Not surprisingly, then, in 1969 the more radical civil rights leaders sought to heat up the urban confrontations toward a symbolic showdown over the Philadelphia Plan.

For Secretary Shultz, a pressing case in point was the explicit threat of racial turmoil in Boston. When Shultz announced the full implementation of the Phildelphia Plan on September 23, he alluded to the "exceedingly complex" problems of the troubled cities like Pittsburgh, and explained that the plan that was designed for Philadelphia "will not fit snugly into every other city."[62] Shultz called instead for tailoring each solution to the "many parties and intricate issues" peculiar to each city." "There is no solution," Shultz said, "like a hometown solution," where the local community could devise their own plan rather than have it imposed by Washington. But in Boston, the spiritual leader of the civil rights activists was in no mood for the delays that such tailoring might require. Father Robert F. Drinan, S.J., was dean of the Boston College Law School, and also chairman of the Massachusetts advisory committee to the U.S. Commission on Civil Rights. In September, Drinan demanded that the Labor Department immediately install the Philadelphia Plan in Boston.[63] Arthur Fletcher replied to Drinan that Labor's legal and political strategy was to consolidate its test case in Philadelphia alone, while taking on the GAO in Congress. In reply Drinan lashed out at Fletcher: "If serious disorders break out in the near future," Drinan warned, "the Department of Labor can be cited as the proximate cause of such disorders."[64] "I cannot predict what might happen," Drinan warned Shultz, "when the black community of Boston learns that you have broken your promises."[65]

Meanwhile, opposition to Shultz and the revised Philadelphia Plan was building in Congress. On August 7, Dirksen wrote Shultz to remind him that as the author of Section 703(j) banning racial quotas, he felt duty-bound to reassert the intent of Congress against the Philadelphia Plan's imposition of racial quotas. He therefore planned to notify the chairmen of the House and Senate appropriations committees of the "imperative" that no funds should be appropriated or "be available for the use or liquidation of contracts issued or conditioned upon any minority labor plan which is in violation of the 1964 Act."[66] But one month later, Dirksen was dead, consumed by lung cancer and felled by heart failure. In an abrupt and fateful transition for Dirksen's midwestern brand of Republican conservativism, he was replaced as Senate minority leader by Pennsylvania's Hugh Scott, who had been elected party whip the previous January. Scott was up for re-election in 1970, and he was courting the state's civil rights forces by supporting the plan that Dirksen had so strongly opposed.[67] So the mantle of Senate leadership in opposing the Philadelphia Plan fell by default to North Carolina's Sam Ervin. As the third senior Democrat (to Mississippi's James Eastland and John McClellan of Arkansas) on the Senate Judiciary Committee, Ervin chaired the guardian subcommittee on Separation of Powers.

The subcommittee was an excellent instrument for Ervin's purposes. It appeared, on its face, to be something of a conservative hanging jury. It consisted of Ervin, McClellan, Democrat Quentin N. Burdick of North Dakota, and Republican Roman L. Hruska of Nebraska. Burdick was usually aligned with his party's liberal leadership, but McClellan shared Ervin's views on civil rights, and Hruska was a crusty Plains conservative who generally voted with Dirksen. Hruska was to achieve a certain notoriety the following year, when he advanced his own novel theory of proportional representation by supporting Nixon's nomination of federal judge G. Harrold Carswell to the Supreme Court, defending Carswell against charges of mediocrity by arguing that mediocre Americans also deserved appropriate representation on the high bench.[68] But only Senators Ervin and McClellan participated in the hearings of October 27 and 28. The committee's chief counsel and staff director, Rufus Edmisten, who later served as deputy chief counsel for Ervin's Senate Watergate hearings, was joined by Chicago law professor Philip B. Kurland as the committee's chief consultant.

Ervin's chief goal for his subcommittee, however, was not to vote out a bill for the parent Judiciary Committee. Rather, it was to sponsor a public debate. The formal purpose of the hearing was to discuss S. 931, Fannin's bill to suspend the use of Johnson's 1965 executive order and make Title VII the sole means of EEO enforcement and remedy.[69] But Fannin's bill, which if taken seriously would have snarled the Senate in a protracted civil rights fight that few senators desired, was merely a stalking horse for a simpler and cleaner instrument. That turned out to be an appropriations bill rider that eventually was sponsored by Democrat Robert C. Byrd of West Virginia. Of the eleven persons testifying at Ervin's Philadelphia Plan hearings, seven opposed the plan. These included Fannin, Democratic Representative Roman C. Pucinski of Chicago, spokesmen for the AFL-CIO and for several contractor associations, and Comptroller General Staats.[70] The opponents generally argued that the OFCC's "goals and timetables" were euphemisms for racial quotas with deadlines. To demonstrate their point, they pounced on the plan's curious device of a *maximum* range for minority hiring. As the director of the Philadelphia contractors association asked, "Except for camouflage, what purpose is served by establishing a range with a maximum figure? If an employer's efforts to recruit minority workers should bear such fruit as to exceed the maximum figure in the range, will the employer be declared in noncompliance for recruiting too many minority workers?"[71]

The plan was defended by Senators Brooke and Javits, with Professor Kurland making few dents in Javits's defense of the plan, and with the customary senatorial deference blunting the testimony. But the plan's chief supporting witnesses, who were aggressively grilled by Ervin, were Jerris Leonard, the assistant attorney general for civil rights, and Secretary Shultz. Leonard's support was weak. He claimed no specific statutory authority for the plan, not even Title VI, and he defended its requirements as "intended to implement what we believe is probably a constitutional obligation, but certainly a moral obligation" of the federal government.[72] This was "not to participate in and support unlawful discrimination by those doing business with employers who discriminate in employment." Ervin attacked that vulnerable circumlocution, quoting the anti-discrimination language from Johnson's executive order (which in turn had been taken almost verbatim from Kennedy's executive order) as well as from Title VII, including the last's elaborate explications of quota-denial by Senators Humphrey, Case, and Clark. Ervin then asked Leonard whether all that clear English language in Sec. 703(j) didn't "express the congressional intent that there will be no preferential hiring of persons on the basis of race?" "Senator," Leonard unhappily conceded, "that is what it says."

Leonard countered, however, with the practical argument that "[w]here you have had a pattern and practice of exclusion of minority group members, if you apply this language literally, you would never have any employment of minority group members." He cited the example of Chicago, where he claimed that fewer than 4,000 of the 95,000 members of 17 construction trades unions were black, and explained that "what we are talking about is catching up." Ervin pounced again: "Do you contend that there has been discrimination in reverse in the past and that therefore you are going to practice discrimination in reverse in the future to overcome that past discrimination?" No, Leonard replied, the Labor Department couldn't be doing that, because the Philadelphia Plan contained wording that specifically prohibited reverse discrimination.

And so it went, for two days. Ervin demonstrated remarkable devotion and fidelity to the text and spirit of a civil rights act that he had done his utmost in 1963 and 1964 to destroy. But when he got to Shultz, he met a more convinced witness than Leonard, whose heart didn't seem to be behind his testimony. Whereas Leonard had tried to avoid the dangers of Title VII, Shultz took them on directly by taking issue with the Clark-Case answer to one of Dirksen's rhetorical questions of 1964. Dirksen had then asked whether an employer would be guilty of discrimination under Title VII if he was under contract to a union hiring hall that sent him only white males. Dirksen said he sought this clarification for the employer "to protect him from endless prosecution under the authority of this title."[73] The Clark-Case answer of 1964 had been: "If the hiring hall discriminates against Negroes, and sends him only whites, he is not guilty of discrimination—but the hiring hall would be." But of course Labor's OFCC had no direct EEO authority over the unions. That was their brute problem. The EEOC had Title VII jurisdiction over unions, but no real power. The OFCC had potentially vast contractual power, but no jurisdiction over unions.

The Justice Department, however, had both jurisdiction and power over unions, at least on paper. Section 706 of Title VII gave Justice the authority to bring pattern-or-practice lawsuits against unions (much as Section 707 authorized pattern-or-practice lawsuits against employers). But the Justice Department during Lyndon Johnson's administration had never exercised this authority. A Republican Justice Department, however, was less likely to be so solicitous toward organized labor. Later that same week, in fact, the Justice Department not so coincidentally brought the Nixon administration's first anti-discrimination suits under Section 706, directed against five construction unions in Seattle.[74] Ervin found the Attorney General's 706 authority to be an ample tool provided by Congress for attacking discrimination in unions. But Shultz argued that reliance on the pattern-or-practice authority of Justice alone would leave the OFCC virtually helpless against building contractors who placed the blame for their too-white workforce on their unions. So Shultz agreed with Ervin that it "may be true technically" that the employer is not guilty of discrimination when the unions never send any blacks to his hiring gate. But "the result in terms of equal access to his employment" is discriminatory. To Shultz, such a result was intolerable. It represented the difference between ineffective passive nondiscrimination and results-oriented affirmative action.[75] This was why Shultz had revived and revised the Philadelphia Plan, which the *Congressional Quarterly* referred to without euphemism as a "nonnegotiable quota system."[76]

At the conclusion of the Ervin hearings, subcommittee counsel Rufus Edmisten asked Staats what he thought should be done to resolve the dispute. "I suggest the appropriations route," Staats replied. "A provision in appropriation bills could provide a more recent and more explicit policy statement by the Congress."[77] Staats's main interest was GAO authority, not civil rights policy. Thus the Fannin bill, even in the unlikely event that it passed, offered him far less than would a simple appropriations rider through which Congress could confirm the final authority over federal expenditures of its faithful fiscal watchdog, the Comptroller General. Thus much of Staats's testimony was a legal brief supporting the GAO's position that (1) Congress had passed no statutory provision to authorize the Philadelphia Plan's approach to contractual procurement; and (2) the inclusion in any contract of terms, not specifically authorized by law, which tended to lessen competition or increase the probable cost to government, was illegal.[78]

But the Comptroller faced a difficult legal and political problem in trying to enforce his stand alone. This was because in disallowing contract payments, the GAO was procedurally obliged not to prevent payments from being made for services rendered, but rather to attempt to recover them from the contractor or from agency accounts. This meant that in disputed cases, the GAO had to try to collect the disallowed expenditures by going to court. And in court the Comptroller by law was represented not by GAO counsel, but by the Attorney General! As part of the legislative branch, the GAO had no way on its own to bring the case into the courts for litigation. Only the Attorney General had authority by law to bring such suits or to defend the government in suits brought by private parties. So in the bureaucratic politics of Washington, the dispute over the Philadelphia Plan was many-layered, and disguised multiple agendas beneath a symbolic moral dispute over racial equality.

That fall Staats had been coordinating his moves not only with Ervin, but also with the congressional leadership, especially the chairs of the appropriations committees.[79] Staats reminded congressional leaders that the expansive executive branch had challenged GAO authority in 1966 in an Air Force case, then again in an AEC case, and as recently as the previous January when Attorney General Ramsey Clark had told the Secretary of the Air Force to ignore the Comptroller General. Now, Staats said, the Labor Department was being joined by the departments of Defense, HEW, and Transportation in publicly ignoring both the GAO and the clear stipulations of Congress, both in standard procurement law and in the Civil Rights Act of 1964.[80] Thus the big question was "not whether the Attorney General or the Comptroller General is correctly interpreting the law, but whether the agent of the Congress or the Executive has the final authority to decide."

Early in December, Staats wrote to Robert Byrd, who as a senior Democrat on the Senate Appropriations Committee chaired an obscure subcommittee on deficiencies and supplementals, and also to George H. Mahon, chairman of the House Appropriations Committee, and requested a rider to clarify his case.[81] On December 15, Byrd obtained the unanimous consent of his subcommittee to add Staats's rider to a minor supplemental appropriation for damage caused by Hurricane Camille. The heart of the rider was a simple but powerful statement. Drafted by the GAO's general counsel, it asserted that no congressional appropriation "shall be available to finance, either directly or indirectly or through any Federal aid or grant, any contract or agreement which the Comptroller General of the U.S. holds to be in contravention of any Federal statute."[82]

On December 18, Byrd's rider dominated a spirited Senate debate. The alarmed White House held a news briefing where Shultz attacked the rider as potentially a "great tragedy."[83] The rider had surfaced so quickly, with Congress rushing its crowded agenda toward the Christmas adjournment, that its supporters were able to cast the issue as essentially a contest between the statutory authority of Congress and an overreaching executive. The Nixon administration rallied its Republican minority to defend a Republican presidency. But the Democratic-controlled Senate defeated an attempt by Javits to kill the rider, and then rejected attempts by Javits and Griffin to water it down. After Byrd accepted an amendment providing for judicial review of the Comptroller's rulings, the Senate passed the Byrd rider by a vote of 52 to 37.

This sent the bill to a conference committee, where disagreement by House conferees sent the Senate amendment to the full House. Meeting with the Republican congressional leaders on December 22, Nixon emphasized the importance of exploiting the Philadelphia Plan to split the Democratic constituency and drive a wedge between the civil rights groups and organized labor. The Byrd rider, Nixon explained, would force the civil rights people to take a stand—for labor *or* for civil rights. The Nixon administration's "line" would be: "The Democrats are token oriented—We are job oriented."[84] The Philadelphia Plan would even the score against labor, Nixon said, for the administration's humiliating defeat over Judge Haynsworth.

Later that day the President issued a statement opposing this backdoor attempt through an appropriations rider to vest the Comptroller General with "a new quasi-judicial role," and in the process "kill the 'Philadelphia Plan' effort of this administration to open up the building trades to nonwhite citizens."[85] Nixon threatened to veto the rider, and he mounted a major White House lobbying campaign to save his administration's major civil rights initiative.[86] In response, Staats worked closely not only with the contractor associations, but also with the head lobbyist for the AFL-CIO, Andrew Biemiller, in lobbying the full House.[87] The contest confounded normal alignments, as labor joined the southern Democrats in opposition, and the Nixon administration joined the civil rights liberals. Responding to White House pressure, House Republican leader Gerald R. Ford made the vote a litmus test of party loyalty, one soured by traditional conservative scruples, but considerably sweetened by the enraged opposition of organized labor.

The result was a major victory for Nixon—and especially for Shultz. On December 22 the House rejected the Senate rider by a vote of 156 to 208, with the crucial difference coming from Republicans supporting the President by a solid margin of 124 to 41. A majority of Democrats supported the Byrd rider, with the core supporters being southern Democrats (61 to 6). But enough northern and western Democrats stayed loyal to labor to reduce the vote of nonsouthern Democrats against the rider to 78, and thereby produce a Democratic split of 115–84 in favor of the rider.[88] In response to the House vote, the Senate later that day decided on a roll-call vote of 39 to 29 to drop its now hopeless proposal, and both houses quickly adjourned until January 19, 1970.

A Philadelphia Plan for All Government Contractors

Nixon's successful stand behind Shultz and the Philadelphia Plan had ironically split the Southern Strategy's core coalition of Republicans and conservative Democrats. In doing so it had shielded the permanent government in the executive branch from congressional interference with the expansive plans of the OFCC, and ultimately of the EEOC as well. Technically, the legislative battle had ended in a standoff. This left the original law intact, including the anti-quota provisions of Section 703(j). Critics of affirmative action quotas would emphasize this legally correct interpretation.[89] But the conservatives had clearly taken a major beating in Congress. They were soon to take another one in the courts. On June 9, the Supreme Court declined to review *Weiner v. Cuyahoga Community College,* thus leaving intact the ruling of the Ohio courts that the Civil Rights Act of 1964 did not prevent the college district from rejecting a low bidder whose affirmative action plan had failed to satisfy the OFCC.

Then on March 13, 1970, a federal district judge in Pennsylvania granted a Justice Department motion to dismiss a suit filed against the Philadelphia Plan on January 6 by an association representing eighty contractors. In *Contractors Association of Eastern Pennsylvania v. Secretary of Labor,* the contractors raised the Staats objections in court, claiming that the Philadelphia Plan constituted illegal and unconstitutional executive action because it was unauthorized by Congress, and because its racial quotas violated the Civil Rights Act of 1964.[90] Judge Charles R. Weiner ruled, however, that the Philadelphia Plan was needed to end an "unpalatable" employment practice that "has fostered and perpetuated a system that has effectively maintained a segregated class. That concept, if I may use the strong language it deserves, is repugnant, unworthy, and contrary to present national policy."[91] But the ultimate and humiliating proof of defeat for Staats and Ervin was revealed early in 1970 when the Labor Department issued a new set of rules that would extend the Philadelphia Plan's model of proportional representation by race and selected ethnicity (but not yet by sex) in employment to basically *all* of the activities and facilities of *all* federal contractors—which by Arthur Fletcher's estimate covered from one-third to one-half of all U.S. workers. While the battle over the revised Philadelphia Plan was being waged during the summer and fall of 1969, the Labor Department had maintained that the plan's controversial requirement for racial percentages was uniquely tailored to the construction industry. It was targeted only against a minority of construction trades like Philadelphia's six skilled crafts, which included less than 10 percent of the city's construction workers. These target trades, the Labor Department explained, were identified only after an empirical process of hearings and findings by department oficials. Thus the revised Philadelphia Plan was defended during the autumn's congressional battle as construction-specific and narrowly tailored by due process to fit the empirical findings.

Speaking on behalf of organized labor, George Meany rejected these "findings" as based not on proper exploratory hearings *prior* to an order, but rather on loaded *post hoc* hearings designed to confirm the order. The Labor Department's data, Meany charged, were based on a key survey that "consisted of a memorandum by one government employee quoting the 'conservative estimate' of another government employee."[92] Meany made his case after the battle had been lost. But even in victory, the Philadelphia Plan's unusual requirements had never been suggested for the nation's *non*-construction workers. Accordingly, at Ervin's Senate hearings, Jerris Leonard had emphasized the contrast between the uniquely intermittent, hiring-hall selection of construction workers, and "the usual industrial situation, where the employer has a more or less fixed group or complement of employees, [and where] the requirements of the Executive order ordinarily can be met if that employer engages in an affirmative recruiting program."[93] Leonard's normative example was "going into the minority areas in an attempt to seek out qualified minority group employees as new hires."

On January 15, however, Ervin announced that the Labor Department had lied to him. He had discovered that at the same time the department was assuring Congress that its Philadelphia Plan would require only "good faith efforts" of construction contractors and would not set minority quotas, it had secretly issued a new order extending racial quotas to *all* federal contractors. Ervin quoted from the text of the Labor Department's Order No. 4 of November 20, 1969, signed by OFCC director John Wilks. Its language seemed to apply to all government contracts the Philadelphia Plan's model of proportional representation: "The rate of minority applicants recruited should approximate or equal the ratio of minorities to the applicant population in each location."[94] "Unlike the Philadelphia Plan," Ervin said, "Order No. 4 makes no pretense of requiring good faith efforts to raise the percentage of minority group employees in federal contract work. The order makes such hiring flatly mandatory." Like the revised Philadelphia Plan, however, Order No. 4 had been preceeded by no hearings. Neither had it been published in the *Federal Register* for review and comment. "Unrevealed to the Congress or the public until now," Ervin charged, Order No. 4 "requires the imposition of flat quotas. It was an integral part of the Philadelphia Plan. In its scope, it makes the announced Philadelphia Plan look like small potatoes."[95]

This time, Shultz kept his head down. Wilks had to explain to the press that he had made a procedural mistake. He had signed an order that should have been circulated as a draft. He said the purpose of the order was to make mandatory the voluntary guidelines of Lyndon Johnson's Plans for Progress program, which the Nixon administration had discontinued.[96] But order No. 4 applied to all contractors employing more than fifty employees and holding federal contracts worth at least $50,000, whereas the cutoff for the Philadelphia Plan model in construction only had been $500,000. Moreover, Order No. 4 did not set specific goals or percentage ranges, like the Philadelphia Plan. This was because as a practical matter it was impossible for the Labor Department to attempt to survey the minority role in manpower availability for 250,000 contractors employing 20 million workers in all realms of economic activity. So Order No. 4 simply required contractors to set their own "significant, measurable, and attainable" hiring goals and objectives, Wilks explained. *But,* he added, these goals "would have to equal the minority ratio of the local applicant population." Failure to satisfy the OFCC could lead a contractor to be dropped as "a reasonable bidder" until the "inadequacy" of his minority employment profile was "remedied." The *Wall Street Journal* reported the explanation of the customary "knowledgeable source" that the OFCC had made the previously voluntary Plans for Progress guidelines mandatory for all contractors to offset criticism that Labor had "singled out" the construction industry with the Philadelphia Plan.[97]

In the past, the shock of an exposé like Ervin's trump against the red-faced OFCC would have sent the over-reaching agency into retreat before congressional umbrage. But the turn of the decade into 1970 seemed to mark a socio-political watershed. Despite the fragmenting impact of Black Power and ghetto riots and the election of Nixon, the civil rights constituency was emerging from the 1960s with a momentum that increasingly was generated from with*in* as well as outside the citadels of government. Neither Meany's attack on the Labor Department's post-hoc findings nor Ervin's exposé seemed appreciably to slow or even significantly modify the consolidation of the OFCC's victory of December. In the upstart OFCC's triumph over the venerable GAO, the executive branch was united, the Congress was divided, and the courts appeared to be supportive. Under these conditions the permanent government had traditionally sought to expand its jurisdiction, authority, and budget. What was new by 1970 was the growing insider role of the civil rights lobby within the executive establishment, and especially within the subgovernment of regulatory administration. The arcane, technical nature of the modern administrative state yielded enormous advantages to the new social regulators. Despite Sam Ervin's undeniable flair, and George Meany's earthy prose, it is hard to capture and sustain great public indignation over whether something called Order No. 4 was published in the *Federal Register*.

On February 3, Secretary Shultz issued the reworked Order No. 4 to a scarcely breathless public. Yet students of public policy and public administration are increasingly aware that out of such bureaucratic boilerplate, which purport to be merely instrumental measures of administrative implementation, can come fundamental shifts in public policy.[98] The detailed new Order No. 4 required contractors to file an affirmative action program within 120 days of signing a contract. An acceptable program first required an analysis of all major job categories to identify any "underutilization" of blacks, Spanish Surnamed Americans, American Indians, and Orientals within each category. Such underutilization was defined as "having fewer minorities in a particular job class than would *reasonably be expected by their availability* [emphasis added]."[99] The order explained that contractors should determine this by considering such factors as "the minimum population of the labor area surrounding the facility," and "the percentage of the minority work force as compared with the total work force in the immediate labor area."[100] That done, the contractor must then design "specific goals and timetables" to "correct any identifiable deficiencies." That failing, to the satisfaction of the OFCC, the contractor would cease to do business with the federal government.

The Labor Department followed up crisply on its December victory and on its largely unchanged reissue of Order No. 4. On February 9, Shultz designated 19 major cities as targets for speedy adoption of OFCC-approved

Philadelphia-type plans in construction.[101] He warned that failure to work out an acceptable "hometown" solution would result in the imposition of a tough Philadelphia model by federal authorities in Washington.[102] By June, hometown solutions had been approved for Boston and Denver, but the OFCC imposed it own stiff plan on Washington, D.C.[103] Also in June, the Labor Department proposed new rules in the apprenticeship program to replace the ability rankings and oral interviews permitted by the original Wirtz guidelines of 1963.[104] The goal of the new rules was to produce a minority enrollment roughly proportionate to the number of minorities in the local population of apprenticeable age.

Finally, that June, President Nixon appointed George Shultz as the first director of the new Office of Management and Budget. As William Safire observed, "Nixon considered Shultz *un homme sérieux*—a man whose presence carried the authority of intellect and command, who could offer sensible and sometimes startling conclusions in a way that Nixon admired."[105] To White House aid Stephen Hess, "Shultz made it on brains."[106] As White House reporter Dan Rather concluded in his post-Watergate analysis of Nixon's top dozen lieutenants, the ex-Marine-turned-academic, Shultz, was "the biggest surprise of all," and "the only one who truly prospered at every step along the way."[107] Replacing Shultz as Labor Secretary was Under Secretary James D. Hodgson, a former vice president for industrial relations at Lockheed who had been a prime supporter of the Philadelphia Plan.

The Nixon administration, then, having fought so vigorously and successfully for such an unlikely Republican initiative as the proportional hiring model of the Revised Philadelphia Plan, and then having extended its new requirement of racial and selected ethnic proportionality to all federal contractors, received for its pains from the civil rights community what amounted to an ungrateful spit in the eye. On June 30, Herbert Hill told the 61st annual convention of the NAACP that "The Nixon administration is destroying the Philadelphia Plan."[108] Hill said the "abandonment of the Philadelphia Plan" was a "pay-off to the building-trades unions for their support of the war in Indochina. The federal government has capitulated to the 'hard hats.'" The "racist building trades unions," Hill explained, had "succeeded in administratively nullifying the Philadelphia Plan" by substituting the "fraud" of the hometown solutions, which were "a meaningless hodge-podge of quackery and deception, of doubletalk and doublethink."[109]

Even allowing for Hill's customary acerbity and also for the revivalist traditions of the NAACP's convention rhetoric, the public savaging of Nixon's most singular civil rights commitment seemed to mock its political wisdom as well as its moral vision. The Philadelphia Plan had always seemed alien to Nixon's Republican, even Dirksenian instincts for equal procedures and voluntarism. Nixon's early embrace of Shultz's unexpected proposition for Philadelphia had always ill consorted with the main thrust of his administration's admittedly not very coherent policies in civil rights. That thrust was better represented by Nixon's policy on voting rights, and later, by his uncertain response to the rising feminist demand for the Equal Rights

Amendment. But in seeking to further the interests of formerly excluded groups, especially blacks and women, by asserting the primacy of equal individual rights, the Nixon administration also ran afoul of the political whiplash that virtually any civil rights proposal was bound to generate at the beginning of the 1970s. Contrary to Ehrlichman's optimistic assessment of the Administration's civil rights balancing act, there *was* no "sweet and reasonable middle" when dealing with a zero-sum issue like affirmative action versus reverse discrimination. As a disgusted Nixon muttered to his senior lieutenants in the summer of 1970, "the NAACP would say my rhetoric was poor if I gave the Sermon on the Mount."[110]

CHAPTER XIV

Nixon, Congress, and Voting Rights

Extending the Voting Rights Act

The Nixon administration's curious embrace of a group-rights policy in the Philadelphia Plan was initially successful in driving a wedge between labor and civil rights groups. But this came to be seen as a liability when the White House went courting the hard-hat ethnic vote for 1972. Furthermore, the Philadelphia Plan puzzled and angered conservatives in both parties while also managing to draw attacks from the NAACP. Both the OMBE and the Philadelphia Plan were implicitly premised on compensatory theory, which for different reasons drew attacks from both the left and right. A less controversial path was available: to defend the more consistent logic of nondiscrimination theory, as Lyndon Johnson had done. Now that this classic EEO model was established national policy, and stamped with a Republican imprimatur by the celebrated Dirksen compromises, its defense became a conserving posture. Most congressional conservatives, to be sure, had fought to prevent its passage in the 1950s and the early 1960s. But once the nondiscrimination model became law in 1964, even intractable conservative opponents like Sam Ervin had proven remarkably quick to rally to its standard.

Furthermore the race problem, like the logic of nondiscrimination, was properly national, not merely regional. Anchored in the midwestern heartland, the congressional Republicans of the 1960s had been eager to restrict the reach of Lyndon Johnson's civil rights laws to the South, where the problems of desegregating restaurants, hotels, schools, and ballot boxes seemed regionally confined and defended by Democratic politicians. Hard-eyed political logic argued that Republicans should support the mass enfranchisement of southern blacks, whose liberal-redistributionist politics would drive southern white voters toward the Republican party. Southern whites especially resented the regional application of the Voting Rights Act, and Nixon gave voice to that grievance by calling for an even-handed reform that would equally affect the individual voting rights of the citizens in all fifty states.

The battle over extending the Voting Rights Act of 1965 featured an iron-ical transposition of roles, with the civil rights coalition in Congress defend-ing the status quo, and the Nixon administration arguing for broad new fed-eral jurisdiction and authority in policing voting rights. Thus in hearings on voting rights extension held by House Judiciary Subcommittee No. 5 during May-July 1969, chairman Emanuel Celler in effect played the stubborn con-servative and Attorney General John Mitchell the expansive liberal. But con-fusion and resentment over the apparent role reversal mirrored a broader confusion over the relationship between means and ends in civil rights policy, especially within the Nixon administration. Ostensibly the main questions in dispute were whether to allow the 1965 law to expire at the end of its sta-tutory five-year run; to extend its triggering formula and hence the southern regional coverage and preclearance requirements; or to make its application national. Resentful white southerners who demanded a common national standard in the name of fairness were also voicing a grudging tribute to the success of the law. But the evolution of the dispute reveals the twisted con-tours beneath it.

Congressional proponents of extension conceded in the hearings on re-newal that when the Voting Rights Act was passed in 1965, "Congress ex-pected that within a 5-year period Negroes would have gained sufficient vot-ing power in the States affected so that special federal protection would not be needed."[1] This assumption was consistent with the arguments of the John-son administration and congressional leaders in 1965 that two of the law's nineteen sections should therefore be temporary. The first of these was the triggering formula in Section 4, which applied only to those states where a literacy test was required for voter registration *and* where less than 50 per-cent of the voting age population was registered or had voted in the 1964 presidential election. Second was the requirement in Section 5 that these states (or the affected counties within them) could not implement any new election laws until they either (1) obtained permission from the federal district court for the District of Columbia (so as to avoid federal judges in the South who might share Senator Eastland's views on race relations) by proving that the new provisions did not have the purpose *or effect* of racial discrimination; or (2) submitted their new election procedures to the Attorney General and per-suaded him not to object within 60 days.[2]

Thus seventeen of the nineteen sections of the 1965 law were indeed per-manent and applied throughout the United States. There was little disagree-ment about the wisdom of this. But sections 4 and 5 expired on August 5, 1970. These two were designed by Congress to be temporary because they radically imposed federal authority on some states but not on others. They suspended literacy test laws in seven southern states (Alabama, Georgia, Lou-isiana, Mississippi, South Carolina, Virginia, and 39 counties in North Car-olina) while not suspending them in fourteen nonsouthern states that also required literacy tests or a similar device. Furthermore, they required the prior permission of unelected federal officials for any changes in local electoral laws.[3] This was an extraordinary interference on a selective basis with traditional state prerogatives. It had been justified in the eyes of Congress (and the Supreme Court) by palpable violations that were no longer tolerable to the national polity. But Congress approved this radical short-circuiting on the assumption that normal federal-state relationships would resume at the expiration of the "5-year cooling-off period."[4]

The results of the law had been spectacular. By the summer of 1969, more than 800,000 new black voters had been registered in the covered states.[5] Federal voting examiners had been sent to 64 counties and parishes in five states (Alabama, Georgia, Louisiana, Mississippi, and South Carolina), where black registration had subsequently increased from 29 percent of the voting-age population to 56 percent. Black registration in Mississippi jumped from 7 percent to 60 percent of the eligible population (while white registration, always high, soared to a stunning 92 percent). The Justice Department had filed 19 voting rights suits under the act and had reviewed 345 proposed changes in election laws, disapproving only ten under Section 5's preclearance provision. Conservatives cited this record in arguing that the law had performed successfully, and that thus the two exceptional sections for triggering and preclearance should terminate as the law's proponents had originally planned. Congressional scholar Gary Orfield, a strong supporter of extension, agreed that "Most resistance to the laws was based not on the voting issue but on the drastic changes in state and local powers imposed by the law. . . . [the] grant of veto power over state legislation to an appointed federal official was unprecedented and strongly resented."[6]

Conservative opponents had argued in 1965 that Section 4's triggering formula was both arbitrary and irrational. It included Louisiana, for example, where 47.3 percent of the eligible population had voted in the 1964 presidential election, but not Hawaii, where 52 percent had voted, or even Texas, which had produced only a 44 percent turnout, but lacked a literacy test. The 1965 law had been extraordinarily effective, to be sure, mainly because the conservatives had lost that argument. But by 1969 they were applauding the act's efficacy while disapproving its theory. They further claimed that it was wrong to demand extension in 1970 on the basis of the old 1964 election data, when the 1968 presidential election had produced such improved turnout that five of the seven originally covered states had passed the 50 percent triggering score.[7] Opponents of extension thus concluded that the 1965 law was ill-advised in the first place, but that judged by its own terms it had been a signal success. According to those same terms, then, the two exceptional sections were properly temporary, and hence required no further congressional action. The law had been successful as originally designed, they said, and hence its two temporary provisions should be allowed to expire— as originally intended.

In reply, proponents of extension pointed to evidence of continued southern resistance and predicted that the covered jurisdictions would quickly reimpose the old racially discriminatory provisions. They pointed to a 1968 report of the Civil Rights Commission, *Political Participation*, that described continued harassment of Negroes attempting to register and new attempts to block black office-holding in the Deep South.[8] Several federal court decisions had documented further evidence of attempts by white officials to dilute the impact of black electoral strength.[9] To liberal Democrats it seemed imperative that the Nixon administration not be allowed to redeem its southern strategy by dismantling the federal apparatus that had so dramatically enfranchised the mass of southern blacks, who after all constituted the core of liberal Democratic strength in the South. As a consequence one of Attorney General Ramsey Clark's last official acts had been to send the Speaker of the House on January 15, 1969, a proposal simply to extend the Voting Rights Act of 1965 for five additional years. One week later, when the first session of the 91st Congress convened, Emanuel Celler introduced an identical bill, H.R. 4249, and scheduled hearings to begin in May. The civil rights coalition hoped that this early pressure would flush Nixon out on the matter, and avoid the delay that tended to work on the side of expiration, especially in light of the omnipresent possibility of Senate filibuster.

The Quest for a Coherent White House Policy

True to its early form, the Nixon administration had no plan. During Nixon's first month in office the Leadership Conference had probed the higher reaches of the new administration, using the good offices of Senator Brooke and Pat Moynihan, and finding the Attorney General's office undecided and torn by conflicting pressures.[10] On February 18, Nixon directed Mitchell to study the matter of voting-rights extension and recommend a policy by March 7.[11] But the internal debate at Justice ran well into May without apparent resolution. Anchoring one bloc of opinion were the forty staff attorneys in the Civil Rights Division, most of them permanent civil service lawyers hired during the expansionist days of the Johnson administration. They tended to share the liberal views of former Attorney General Ramsey Clark, and many had cut their teeth on field work in the South dealing with the most troublesome local officials.[12] Not surprisingly they urged their new boss, Assistant Attorney General Jerris Leonard, to support a straight five-year extension.[13] Leonard agreed, and on February 27 he recommended to Mitchell a simple extension "at a minimum."[14] Leonard warned that after August 1970, some of the covered states and counties would likely combine restoration of the literacy tests with a general requirement for voter reregistration, and thereby disfranchise substantial numbers of blacks who had been newly enfranchised by the Voting Rights Act of 1965. Such a straight extension of the voting rights act, however, would not redeem President Nixon's opposition to regional legislation. So Arthur Burns constructed a proposal that would deregionalize the Voting Rights Act by discontinuing the triggering and preclearance provisions that were so offensive to conservatives.

Given the demonstrable effectiveness of the 1965 law in achieving its goal of equal access to the voting booth for southern blacks, the key to the extension question seemed to Burns to concern politics more than practicality: in any revised law, what would replace the controversial but effective sections 4 and 5? How would the Administration counter charges that it had sold out to the South, where some white officials would likely respond to the expiration of sections 4 and 5 by reimposing racist restrictions? Burns rejected one obvious solution: simply extending the ban on literacy tests to cover the entire nation—a proposal that Leonard and the Civil Rights Division were pressing on Attorney General Mitchell. Burns objected to this expediency on the conservative grounds that state literacy tests in theory served a rational public purpose, were not inherently evil, and fell within the legitimate and historic purview of state responsibilities in the federal system.[15] Furthermore, such tests had not been shown by evidence to be used discriminatorily in the 14 *non*southern states that required them.[16]

Instead, Burns called for replacing sections 4 and 5 with four new provisions. First, no person who registered to vote in any state during the five-year span of the original Voting Rights Act could subsequently be denied the right to vote because of failure to pass a literacy or similar test. This proviso would apply nationwide, but was obviously designed to protect newly enfranchised black voters in the South, and thus to protect Republicans from attacks for breaking the faith. Second, to similarly protect potential *new* registrants from racial discrimination, Burns proposed that completion of the sixth grade or of six months of honorable service in the U.S. armed forces should exempt voter applicants from any literacy or similar test in any state. Third, Burns proposed that the Attorney General should retain his present authority to ask a court to suspend literacy tests "if they were being used with a discriminatory purpose or effect." In addition the Attorney General would be empowered to assign federal voting examiners and election observers to any county in the United States.

Finally, Burns proposed that a special presidential advisory commission be appointed to study the impact of corrupt election practices on the right to vote. The Civil Rights Commission had concentrated its voting studies on discrimination against black voters in the South, virtually all of whom were Democrats. But Republicans had complained for a decade about how John Kennedy had stolen the presidential election from Richard Nixon in 1960, when Democratic machines in cities like Chicago voted their loyal and occasionally deceased supplicants early and often. Illiterates seemed rarely to vote Republican, Burns knew, and he reminded Nixon that by following his plan, the administration would redeem its pledges, honor the high principles behind those pledges, and "avoid paying homage to illiteracy."[17]

Late in May the Justice Department floated a proposal based on Burns's sixth-grade strategy, and Bryce Harlow's nose-counters in legislative liaison sought out the reactions of Republican leaders on the Hill.[18] Harlow's staff posed both the sixth-grade approach and its implicit alternative, a national ban on literacy tests. The congressional reactions were mixed. Republican leaders in the House strongly preferred an administration proposal that was national rather than regional in scope, and all were enthusiastic about the proposed anti-fraud provisions. But Gerald Ford seemed to speak for the majority in expressing "great apprehension" about banning all literacy tests, which he feared would be divisive within Republican ranks.[19] Bill Timmons summarized the Republican leaders' reactions to Harlow: "Don't do it. Will be infringement on what is left of states' rights. . . . Illiterates seldom vote Republican." Lamar Alexander reported to Harlow that the apprehensive Republican congressmen supported instead the sixth-grade proposal, which they thought would "breeze through" the House. But if a national ban on literacy tests "is what the President wants," Alexander concluded, "they will go to the well" with him.[20]

In the Senate, Minority Leader Dirksen also opposed the literacy-test ban, and preferred the administration's sixth-grade option. But what Dirksen really preferred was simple extension of the 1965 act, in which he felt considerable pride of authorship. McCulloch felt the same way in the House. During the negotiations over the civil rights bills of 1964 and 1965, both Republican leaders had quietly excluded northern districts like their own from coverage. Preserving the Voting Rights Act through simple extension therefore appealed as a conserving gesture to Republican leaders like Dirksen and McCulloch, partly because it would leave their constituencies untouched. When this form of situational conservativism combined with the more ideologically conservative defense of states rights, which objected to the enlargement of Washington authority over traditional state responsibilities, the administration's nationalizing initiative faced stiff odds even within the President's own party. Senator John Tower of Texas, for example, told Harlow that a national ban on literacy tests would be a "drastic change" that could be "suicide for the Republican party."[21]

On the other hand Dirksen's chief lieutenant, Hugh Scott, did not like the sixth-grade approach. But Scott was up for re-election in Pennsylvania and he was more interested in wooing black voters than in defending state-rights principles. He told Lamar Alexander that he supported abolishing all literacy tests, and he thought the sixth-grade proposal would get "40 or better Democratic votes and could therefore pass the Senate."[22] But what might have been a welcome nationalization and liberalization of standards if advanced by Lyndon Johnson Democrats was inherently suspect when advanced by Nixon Republicans. Alexander summarized what was abundantly reported in the newspapers: "Civil rights groups have attacked the first Justice Department proposal as a 'southern bill' and a step backward, and they indicate they will make a big issue out of it." By the end of May, the Administration's indecision was becoming embarrassing, as the unresolved debate repeatedly forced Mitchell to postpone his testimony before Celler's committee.[23] Then, on June 2, the Supreme Court delivered a surprising opinion that weighed in decisively against Burns and the sixth-grade option by reinforcing Leonard's argument for a nationwide ban on literacy tests.

The *Gaston County* Decision and the Heavy Hand of History

On Monday, June 2, 1969, Associate Justice John Marshall Harlan, Jr., read his majority opinion in the Court's 7–1 decision in *Gaston County v. United States*.[24] In a decision that was to have a profound effect on civil rights law, especially as it bore on the compensatory implications of affirmative action, Harlan ruled that the impartial enforcement of literacy tests would serve only to perpetuate inequalities in a social environment characterized by *a history of* unequal educational opportunity. The North Carolina county had sought to reinstitute its literacy test. County officials first restructured the test to be scrupulously impartial both on its face and in its administration, then they sought a "bail-out" judgment from the federal courts as provided for under Section 4(a) of the Voting Rights Act. But the federal district court for the District of Columbia had found that while the new literacy test was indeed impartially designed and administered, it was nevertheless ineluctably coupled with the county's history of segregated and inferior schooling. The court held that the new test therefore deprived the county's blacks of the equal voting rights in violation of the Voting Rights Act.

In upholding the lower court decision, Justice Harlan concluded from the 1965 law's legislative history that a main reason why Congress adopted the provision to suspend tests concerned "the potential effect of unequal educational opportunities on the exercise of the franchise." Gaston County's racially segregated schools had "deprived its black residents of equal educational opportunity," Harlan said, "which in turn deprived them of an equal chance to pass the test."[25] Harlan was reasoning from an incipient theory of historical lag that predated the Voting Rights Act of 1965. In 1964 the Supreme Court had upheld a district court's decree that Louisiana's new, uniformly applied, and objective "citizenship test" for voter registration could not be implemented until Louisiana ordered re-registration for all voters. The Court reasoned that Louisiana's history of black disfranchisement conferred upon the federal courts "not merely the power but the duty to render a decree which will so far as possible *eliminate the discriminatory effects of the past* as well as bar like discrimination in the future."[26] The surface logic of this nascent historical theory seemed plausible, especially in light of Louisiana's manifest historical culpability. But its boundless implications remained unexplored.

In *Gaston County*, Harlan applied the historical theory to the Voting Rights Act, and concluded that the county's school desegregation had occurred too recently to provide an equal chance to pass the enfranchisement test for the already adult black population. But he suggested no helpful criteria whereby future public officials or courts might determine whether or when the heavy hand of history may have sufficiently passed to permit impartial laws to be impartially applied. That pregnant question was not before the Court. And Harlan spoke only of Gaston County and her Jim Crow past. But the Justice Department was quick to see how the logic of *Gaston County* affected the debate over voting rights extension.

On the same day the Supreme Court announced its decision in *Gaston,* John Mitchell and Jerris Leonard had met in Ehrlichman's office with senior White House officials to argue that the sixth-grade strategy be replaced by a national ban on literacy tests. One week later Leonard sent Ehrlichman a five-page brief explaining the post-*Gaston* reasoning that had led the Justice Department to oppose the Burns approach.[27] Leonard found inescapable the conclusion that even if *all* the presently covered states successfully obtained Section 4(a) approval to reinstate equitably administered literacy tests after the 1965 act expired in August 1970, "the Attorney General would have the duty to reopen the case and seek a reimposition of the suspension of the literacy tests."[28] This was because all the covered southern states shared a history of de jure school segregation. *Gaston County* had gutted the voting act's bail-out provision, on the apparent theory that one could not bail out from one's history. Furthermore, Leonard reasoned that the renewed suspension would also apply to all other tests or devices which might have a racially discriminatory purpose *or effect.* Thus Leonard concluded that under the *Gaston County* doctrine, the ban on such tests would necessarily continue into the forseeable future even if Congress enacted *no* new legislation. Thus the Justice Department's earlier proposal to replace a renewal of the literacy test ban with a sixth-grade provision "has lost its validity." So had the earlier assumption that the failure to pass any new legislation at all would enable the President to claim that he was "keeping his commitment against regional legislation," because he nevertheless "would be violating it in fact."

There was theoretically a third possibility, Leonard said. This was to ask for outright repeal of the 1965 law's regionally discriminatory ban on literacy tests. But he conceded that this was defensible neither on political nor moral grounds, because "if successful it would permit a reimposition of discriminatory tests; and it would seriously alienate the Negro community, with immeasurable consequences."[29] "Even if such a measure could pass in Congress," Leonard said, "it would in all probability be struck down by the courts; for the broad language and the reasoning of the *Gaston County* case lead me to the conclusion that the Supreme Court would reach the same result on the basis of the Fifteenth Amendment." The statutory ban on southern literacy tests seemed, like history itself, to have evolved into a one-way street, which Congress could not repeal and the federal courts would not allow to expire.

Leonard did not pursue his argument to the conclusion it invited—that the *Gaston County* doctrine necessarily now included the heretofore uncovered states, like Texas, with a history of de jure school segregation. Such logic would extend Section 4's trigger to include all or portions of the nineteen southern and border states and the District of Columbia, where segregated schools were either required or permitted when the Supreme Court issued the *Brown* decision in 1954. This in effect would only compound the South's resentment at Washington's regional enforcement policies. So Leonard concluded that "The only course available to the President consistent with his position against regional legislation is to propose a nationwide ban on literacy tests." "If this were done in permanent form," Leonard added, seeing an implicit benefit, "it would permit a revision of Section 5, so that the Act would have no regional application."

The Liberal Sociology of Attorney General John N. Mitchell

On June 26, Attorney General Mitchell finally presented his voting-rights proposal before Celler's Subcommittee No. 5, having five times postponed his scheduled appearances while the Administration struggled with its strategic dilemma. Mitchell announced that the Administration found unacceptable a mere five-year extension of the 1965 law, as proposed in Celler's H.R. 4249. Instead he proposed a five-part package of amendments. The centerpiece was a nationwide ban on literacy tests until January 1, 1974. The package also included a new nationwide restriction on state residency requirements for presidential elections, new authority for the Justice Department to send voting examiners and observers anywhere in the country, and similar new nationwide authority for the Attorney General to initiate suits in federal courts and ask for a freeze on discriminatory voting laws. Finally, the Attorney General asked for statutory authorization of a presidentially appointed advisory commission to study voting discrimination "and other corrupt practices."[30] As Steven Lawson acknowledged in his comprehensive study of black voting rights, "The Nixon administration clearly had challenged the liberals in a vulnerable area."[31]

Seizing the reformer's initiative, Mitchell cast Celler and his liberal majority on Judiciary in a conservative role as defenders of the status quo. Mitchell's testimony implied that the Celler subcommittee's defensive posture was lacking in imagination, was less effective than the Administration's more national vision, and was no longer even fair. To improve the defective status quo, Mitchell pursued the logic of *Gaston County* considerably further than had Harlan's regionally circumscribed opinion. Four million poorly educated blacks had fled the South since 1940, Mitchell said. They had spread themselves throughout those 14 nonsouthern states, like New York and California, where literacy tests were also required and where the migrating Negroes' inferior education would continue to deny them an equal chance to pass the tests. Mitchell associated his views with those of liberal organizations like the American Civil Liberties Union, the Leadership Conference on Civil Rights, and the NAACP, which in recent years had urged that such literacy tests be abandoned. The staff director of the Civil Rights Commission, Howard A. Glickstein, had urged such a policy in earlier testimony before the same subcommittee (the commission's chairman, Theodore M. Hesburgh, had written President Nixon the previous March 28 to urge "banning the use of literacy tests nationwide").[32]

Then Mitchell brought the force of his fairness argument home to the Democratic committee chairman from Brooklyn. Not only did the formula of 1965 no longer make much sense in the South in light of subsequent events—

e.g., in all seven southern states by then in excess of 50 percent of eligible blacks were registered. Indeed by 1969 in uncovered states like Texas and Florida, there were more counties where fewer than half the eligible blacks registered or voted than there were in Louisiana and Alabama, which were covered. Worse, "a higher percentage of voting-age Negroes went to the polls in the Deep South than in Watts or Washington."[33] "Little more than one-third of the voting-age Negro population cast 1968 ballots in Manhattan, the Bronx, or Brooklyn, New York City," Mitchell said, "and this amounted to only one-half the local white turnout." Mitchell's four million migrating southern blacks had carried their history with them, and deposited its burden on Mr. Celler's doorstep.

The Attorney General concluded his testimony with what amounted to an indictment from the liberal left of the irrationality and inequities that would follow from a mere five-year extension of the 1965 act, thereby leaving "the undereducated ghetto Negro as today's forgotten man in voting rights legislation."[34] "A higher percentage of Negroes voted in South Carolina and Mississippi," Mitchell said, "where literacy tests are suspended, than in Watts or Harlem, where literacy tests are enforced." In his peroration Nixon's Attorney General stood foursquare with the NAACP, the ACLU, and the Civil Rights Commission:

> I want to encourage black people to vote. I want to encourage Mexican-American and Puerto Rican citizens to vote. I especially believe that minority citizens, who may feel alienated from our society, should be given every opportunity to participate in our electoral processes.
> I want to encourage our Negro citizens to take out their alienations at the ballot box, and not elsewhere.[35]

Not surprisingly, Chairman Celler and his northern Democratic colleagues resented the Attorney General's suggestions of liberal hypocrisy and double standards. When Mitchell at one point argued that requiring literacy tests for Negroes and Puerto Ricans in northern ghettos, which would clearly include riot-scarred communities like Bedford-Stuyvesant and Ocean Hill-Brownsville in Celler's Brooklyn district, provided a "psychological barrier for people without education," Celler snorted, "It is very difficult for us to legislate on the basis of indefinite psychology."[36] Celler replied testily to Mitchell that the "sickness" of conscious voting discrimination was in the southern states, not in states like New York. To Celler, Mitchell's argument was "like saying because you have a flood in Mississippi you have to build a dam in Idaho." Democrat Donald Edwards, the liberal Californian who in the next Congress would chair Judiciary's new subcommittee for civil rights oversight, pointed out to Mitchell that the Justice Department had no evidence of voting discrimination in Watts or Harlem, or even any complaints of such treatment. Celler stopped short of explicating the message this implied—that public officials in such urban Democratic districts generally wanted larger, not smaller turnouts of racial and ethnic minorities, and their eagerness to count these votes lay behind the Attorney General's proposal for a special commission to study voting fraud and corruption in such notorious Democratic bastions as Cook County, Illinois.

Edwards pressed on Mitchell the committee's tactical complaint over the dangers of delay. He charged that the Administration's complicated new package, which contained items like voter fraud provisions which the liberals awkwardly claimed were extraneous to the Voting Rights Act, would cause a delay that might prevent any action before the August 1970 deadline. Edwards' questions reflected the political suspicion of the panel's majority that Nixon and Mitchell were not serious about enforcing voting rights and didn't care if the 1965 act expired. When Edwards asked Mitchell if the proposed new nationwide coverage would lead him to request more enforcement funds for the civil rights division, Mitchell replied that "the limited amount of litigation which would be required under our proposal would not require additional personnel or an additional budgetary appropriation."[37] This attitude reinforced liberal fears that the Mitchell Justice Department wanted to revert to the discredited method of case-by-case litigation that had characterized the voting-rights provisions of the civil rights acts of 1957, 1960, and 1964, and that had so demonstrably failed to enfranchise significant numbers of black voters. But the debate over geographic distinctions, which appeared to give the Administration a kind of moral edge as custodians of a more national vision, had obscured important differences over function and process in federal-state relations. That is, arguments over Section 4's triggering formula had obscured the Administration's quiet determination to terminate Section 5's preclearance provision.

The Tarbaby Logic of Preclearance

Section 5 had not been regarded by the architects of the Voting Rights Act as a mainstay of its enforcement machinery. Rather, the law's massive enfranchisement of southern blacks had derived almost entirely from its triggering formula in Section 4 and the dispatch of federal examiners and observers authorized in Section 3—together with the effective voter registration drives conducted by the civil rights organizations. Mitchell told Celler's subcommittee that only 345 changes in electoral laws had been submitted to the Justice Department for Section 5 preclearance since 1965, and only ten of them had been disapproved (the majority by his own administration). But he admitted that "where local officials have passed discriminatory laws, generally they have not been submitted to the Department of Justice." In these cases the department's Civil Rights Division had to ferret them out by going to court, a process it had found necessary in only seven cases.[38]

Mitchell's testimony and the assumptions behind it, however, were not consistent with the major current study of the problem: the Civil Rights Commission's 1968 report, *Political Participation*.[39] Most of the report's pages described willful violations of both the letter and the spirit of the Voting Rights Act by Black Belt counties, especially in Mississippi and Alabama, where a burst of diehard legislative creativity had produced nakedly racist circumlocutions. These included many of the shopworn devices to prevent black registration and voting itself. But most of them only hastened the dispatch of federal observers and examiners, and hence were generally self-defeating. More creative, and somewhat more subtle, were attempts to minimize or prevent the election of blacks to office. Such racist manipulation of otherwise permissible changes included switching from district to at-large elections, switching from elective to appointive office, extending the terms of elective offices and even abolishing them, increasing the minimum qualifications and filing fees for office, and preventing newly elected officials from obtaining required bonds. It was, overall, a shabby parade of trickery, and a fairly obvious one.

It was also, however, a highly localized counterrevolution, one that was closely watched and sharply challenged. The commission's report revealed a vigorous network of private observers whose challenges had generated a wave of test cases that were welling up toward a sympathetic Supreme Court. The report concentrated on the tactics that diluted the impact of the new black vote. In doing so it seemed to concede that the relative dilution of various voting blocs or political groups was inherent in drawing up rules to govern political competition—that is, southern Democratic legislators had been penalizing Republicans for generations, and rural legislators had penalized city dwellers and suburbanites. More importantly and more immediately, the report acknowledged in its conclusion that the "dramatic increase" in black political participation in such a short period represented "unprecedented progress."[40] White officeholders and candidates were manifesting a greater responsiveness to black needs and concerns, the commission agreed; there had been a decline in open appeals to racism by candidates and officials; and "contrary to the dire predictions of violent reaction," progress had taken place quietly and without major conflict.

The report's catalogue of offenses was substantial, especially in Mississippi and Alabama. But those two states contained three-quarters of the 38 counties named in the commission's complaints, out of a total of 593 southern counties or equivalent jurisdictions covered by the Voting Rights Act. *Political Participation* thus concentrated on approximately 6 percent of the covered counties, and implicitly acknowledged that nine out of ten jurisdictions were responding well to the new law. The attempts at local evasion had drawn court challenges under Section 4 that the report followed closely. But it said very little about Section 5. So in Attorney General Mitchell's testimony before the Judiciary subcommittee, the demonstrable success of recent black enfranchisement, coupled with Section 5's relative invisibility, tended to deflect attention from the fact that the Administration's plan would scuttle Section 5.

To southern congressmen, however, Section 5's minimal impact, and by presumption its minimal need, made it all the more intolerable as a standing insult not only to their region and to traditional state-rights doctrine, but also to bedrock principles of law. Congressman Richard H. Poff of Virginia, the ranking southern Republican on the Judiciary Committee, resented the promiscuity of sections 4 and 5. Both sections applied to his own Virginia. Yet the Civil Rights Commission itself had concluded as early as 1961 that blacks encountered "no significant racially motivated impediments to voting" in Virginia, and since 1965 the Justice Department had sent not a single federal examiner or observer into any precinct in Virginia.[41] Yet none of Virginia's more than 2,000 precincts could even switch from paper ballots to machine voting without applying for permission from a Washington agency.

In theory the 1965 law provided for bail-out from such subservient status through an escape-clause mechanism. But as a practical matter this seemed quite impossible to Poff. The recent *Gaston County* decision appeared to have ordained that electoral jurisdictions could not control the sins of the past in their own history. And the extraordinary language of Section 5, especially as it was being interpreted by the federal courts, seemed to Poff to guarantee that covered jurisdictions could not successfully vouch for their *future* innocence. Poff explained:

> Every lawyer knows that it is practically impossible to marshal the evidence to prove an absolute negative. The burden is particularly onerous when the negative is "not guilty." For Virginia to succeed under the escape clause would require probative evidence, both verbal and written, from 765 general and precinct registrars in 2,031 precincts throughout the State. Such evidence would have to be marshaled for every precinct in every election, National, State, and local, both general and primary, conducted during the 5-year period preceding the institution of the lawsuit.

With respect to each precinct in each election, Poff said, the evidence must establish a prima facie case that incidents of discrimination have been "few in number" and that they have been "promptly and effectively" corrected. It must furthermore establish that "the continuing effect" of past discrimination "has been eliminated"—whatever that might mean—and that "there is no reasonable probability of their recurrence in the future." "That is tantamount," Poff concluded, "to requiring the accused to prove both past innocence and future innocence."

Poff's complaint echoed the 1965 argument of his colleague and senior Republican on the Judiciary Committee, William McCulloch. Both Gerald Ford and McCulloch had objected then to the triggering formula and its ban on literacy tests. They had called instead in their Republican substitute bill for a sixth-grade provision to escape literacy tests, plus nationwide authority for the Attorney General to assign federal registrars to any county where he had received at least twenty-five substantiated complaints of racial disfranchisement.[42] Moreover, the McCulloch-Ford plan of 1965 would have freed the affected areas immediately after they complied with the directives of the federal examiners. But by 1969 McCulloch had changed his mind. After the House had rejected his Republican substitute bill in 1965, he had supported the Johnson administration's Voting Rights Act, thereby endowing it among Republicans with his considerable prestige. By 1969 McCulloch, like Dirksen, felt protective instincts through pride of co-authorship, and he shared in a broad congressional consensus that the liberal voting reforms of 1965 had worked and didn't need fixing.

Thus when the Attorney General on July 1 made a second appearance before Celler's subcommittee to explore further and defend the proposals he had introduced on June 26, McCulloch attacked him. The provisions in the Administration bill, McCulloch charged, "sweep broadly into those areas where the need is least and retreat from those areas where the need is greatest."[43] The Nixon bill "creates a remedy for which there is no wrong," McCulloch said, "and leaves grievous wrongs without adequate remedy." McCulloch's attack was unusual in that it concentrated on defending Section 5. The Administration bill, he charged, would repeal the preclearance provision "in the face of spellbinding evidence of unflagging Southern dedication to the cause of creating an ever-more sophisticated legal machinery for discriminating against the black voter."

McCulloch found three particular advantages in Section 5. The first was that it shifted the burden of proof from the complainant to the political jurisdiction to show that the new electoral procedure did not have the purpose or would not have the effect of discrimination. McCulloch was pleased because "Section 5 strips away the presumption of the legality that so often cloaked imaginative and clever schemes," and thus acted as a deterrent to prevent their multiplication in the South. McCulloch's deterrent logic was plausible, and was consistent with the minimalist pattern of Section 5 use. But the same minimalist pattern of objection suggested that the stripping away of presumptions of legality would primarily affect the innocent majority of officials—those presiding over 94 percent of the covered southern counties—whom the Civil Rights Commission had commended in *Political Participation* for their good faith efforts in implementing the law. McCulloch's syllogism seemed to link virtually all of southern electoral officialdom in the public mind with a kind of blanket, "unflagging Southern dedication" to perpetrate racial evil. This danger had been anticipated in 1965 by Solicitor General Archibald Cox, who had argued that Section 5 was "totally unnecessary and probably unconstitutional," as well as "exceedingly dangerous."[44] Justice Hugo Black had agreed, arguing that having a federal court rule on a state law before it was implemented was to render an advisory opinion without a trial. But Black's argument was a dissent over Section 5, which was rejected by Earl Warren's majority opinion in *South Carolina v. Katzenbach* (1966).

McCulloch's third point called attention to a much more recent decision of the Supreme Court, *Allen v. State Board of Elections,* a Section 5 case that was decided just the previous March.[45] What McCulloch found appealing in *Allen* was its agreement that private citizens could police the local jurisdictions by bringing suit in federal district courts. Prior to *Allen,* McCulloch said, "neither the State governments nor the Federal Government seemed to show much interest in observing the command of Section 5."[46] Deploring this "mutual apathy," McCulloch applauded *Allen* as meaning that "at long last after 4 years section 5 will become effective." McCulloch's brief allusion to *Allen* was more prophetic than he knew, because the import of *Allen* was to be vastly broader than its relatively minor ruling on private rights to sue in local district courts. But because the *Allen* decision did not reach the merits of any claims of voter discrimination, it was quickly overshadowed by *Gaston County*, which so dramatically did. *Gaston* tipped the Nixon-Mitchell strategy toward a collision course with the civil rights coalition in Congress, where a growing consensus approved the success of the law. Outside the resentful southern delegation there was little enthusiasm for revising the formula that had brought that success.

The Voting Rights Act of 1970

The resolution of the debate of 1969–70 over voting-rights extension demonstrated the instability of the Nixon administration's strategic leverage in a Congress dominated by the opposition party. Bryce Harlow had early concluded that such partisan odds dictated a strategy of seeking ad hoc, "floating coalitions" in Congress. This meant building upon a core bloc of Republican loyalists, then searching for allies either by reaching to the left or to the right, depending upon the issues.[47] Such coalition-building worked best in the House, as in January 1970 when Nixon rallied conservative Democrats to help sustain his veto of a HEW appropriation he called budget-busting and inflationary, or when he forged an alliance with liberal Democrats the following April to pass his radical Family Assistance Plan.

But the floating coalition strategy was most vulnerable in the Senate. There Nixon's problem was the substantial bloc of Republican senators from urban-industrial states whose moderate-to-liberal politics blunted their party loyalty. They constituted about a third of the GOP's 43 senators—men like Edward Brooke of Massachusetts, Clifford Case of New Jersey, Jacob Javits and Charles Goodell of New York, Mark Hatfield and Robert Packwood of Oregon, John Sherman Cooper of Kentucky, Charles Mathias of Maryland, Charles Percy of Illinois, and Hugh Scott and Richard Schweiker of Pennsylvania. It was this bloc of Republican "progressives" whose independence had pushed Dirksen in 1968 to reverse himself on the fair housing bill. When Dirksen suddenly died in September 1969, the minority leader's post was won by Hugh Scott, with damaging consequences for Nixon's legislative agenda, especially in civil rights. As a result of these cumulative forces in the Senate, Bryce Harlow's floating coalitions often floated in reverse, with his not very loyal Republican bloc coalescing with the hostile Democrats. Such a bipartisan coalition on the left had defeated the Administration in November 1969 by rejecting Nixon's nomination to the Supreme Court of Clement Haynsworth.[48] Because Celler's Judiciary Committee had forced a House-first strategy on the administration by beginning hearings on voting-rights extension early in 1969, the contest over whose voting-law coalition would "float" was first joined in the House.

In mid-July 1969 a bipartisan, Celler-McCulloch coalition in House Judiciary voted 28–7 to extend the 1965 law for five years with no amendments. But in the fall the White House rallied the conservative coalition of midwestern Republicans and southern Democrats behind the Administration's bill, and on December 11 in a floor vote they substituted it for the Judiciary Committee bill by the slim margin of 208 to 204.[49] The Nixon bill's common national standard drew the support of 129 Republicans and 79 Democrats, most of the latter from the southern and border states. James J. Delaney, a New York Democrat and member of the Rules Committee, defended the President's call for a suffrage measure that applied uniformly throughout the nation—for applying the law "in simple language to every State and not just a particular few in any area."[50] Clarence Mitchell called the vote "a cataclysmic defeat for civil rights engineered by the President."[51] In emotional language that was out of character, the NAACP lobbyist charged that "the Klan was on the floor of Congress today, waiting to lynch Negroes at the polls." The House then voted 234–179 to pass the amended bill, H.R. 4249, and send it to the Senate to consider in the second session of the 91st Congress in 1970.[52]

When H.R. 4249 came over to the Senate the Democratic leadership gave it the customary routing to the Judiciary Committee, where desultory hearings were expected in Ervin's subcommittee. But to avoid the strong likelihood of committee burial, the leadership instructed Judiciary to return the bill by March 1. There on the Senate floor a bipartisan majority of Scott-led Republicans and northern Democrats reversed the House action and resubstituted the five-year extension of the 1965 law. Also added was a controversial provision sponsored by Majority Leader Mike Mansfield to enfranchise eighteen-year-olds, a move that Nixon opposed as a statutory enfranchisement that was constitutionally impermissible. On March 13 the Senate's radically revised voting-rights bill was returned to the House.[53] The eighteen-year-old provision was difficult for politicians to oppose while so many young draftees were dying in Vietnam, and this helped change the political arithmetic on voting-rights extension in the House. On June 17 the House voted 224–183 to accept the Senate bill and thereby avoid a conference in which Senator Eastland's Judiciary Committee would have been heavily represented. Despite overwhelming warnings by Nixon's advisers that enfranchising eighteen-year-olds required a constitutional amendment, and that the youth vote, like the southern black vote, would go overwhelmingly to his Democratic opponent in 1972, Nixon signed the bill into law on June 22.[54]

Ironically, the new law not only retained the original, regionally-targeted, 1964-based formulas for preclearance and triggering that the Nixon administration had sought to remove. It also added a new triggering formula for the rest of the country, which identified jurisdictions in which less than half of the minority voting-age populations were registered on November 1, 1968, or had voted in the 1968 elections. This extended coverage to three districts in Alaska; to Apache County, Arizona; Imperial County, California; Elmore County, Idaho; Wheeler County, Oregon; and Bronx, King's (Brooklyn) and New York (Manhattan) counties in New York. It also suspended the use of literacy tests in *all* states until August 1975, and set a 30-day residence requirement for voting in presidential elections. Congress had bought the Nixon-Mitchell nationalizing rationale, but rejected the trade-offs behind the rationale that would have removed the regional stigma from the South. In the process Congress had secured the enfranchisement of millions of new black and youthful voters whose demographic characteristics spelled millions of Democratic ballots.

To complete the irony of the Nixon administration's backfired strategy in voting rights, the Attorney General decided for internal and administrative reasons to reorganize the Civil Rights Division of the Justice Department. Since its creation in 1957 the CRD had been organized into sections corresponding to geographic regions, with each section handling the full range of civil rights matters arising within its area.[55] This was a common form of internal organization among the federal mission agencies, and it offered the advantage of facilitating both intra-agency coordination between Washington headquarters and the field, and also interagency cooperation between different agencies serving the same region. But new administrations often brought with them a wave of internal reorganizations, as politically appointed new agency heads sought to stamp their imprimatur on bureaucracies that seemed firmly fixed to traditional policies.[56] This produced pendulum swings of reorganization, typically from a structure governed by geographical or hierarchical level to a functional organization and back, and in September 1969 Attorney General Mitchell and Jerris Leonard reorganized the CRD by function. This in turn produced a new Voting Rights Section within the CRD, and as Steven Lawson observed, it thereby "furnished an unforeseen opportunity for disgruntled attorneys to move into it."[57] "The government's suffrage lawyers soon seized the chance to enforce section five," Lawson wrote, and thus "the departmental reorganization had the unanticipated consequence of producing an experienced team of attorneys dedicated to furthering compliance with section five."

"Iron Triangles" and the Civil Rights Bureaucracy

The career lawyer-bureaucrats in the Civil Rights Division who were "disgruntled" over the new Nixon-Mitchell policies were firming up the third anchor of an "iron triangle." Among students of bureaucratic politics, the Washington environment of power and money was famous—or infamous—for encouraging the formation of symbiotic, triangular relationships between constituent groups who organized and lobbied for new programs (or lobbied to increase their size and share of old ones); the executive agencies that ran these programs and benefited from their growth; and the congressional committees that funded them, and whose members profited politically from the contributions and votes of the target constituent groups.[58] The agricultural extension networks and the Army Corps of Engineers were classic examples of America's pluralist tendency toward such triangular back-scratching, and the high-budget programs of the Great Society spawned a wave of new entrants in areas like the war on poverty, federal aid to education, and health care.[59]

Thus the lawyers in the CRD's new Voting Rights Section were following the lead of a new but rapidly growing cluster of subagencies where an increasingly formidable civil rights bureaucracy was being knit together.[60] These organizations included the surviving grand-daddy of them all, the U.S. Commission on Civil Rights, plus the Civil Rights Division of the Department of Justice—both of them created by the Civil Rights Act of 1957. The Civil Rights Act of 1964 led to the creation in 1965 of the EEOC and the OFCC in Labor, the Office of Civil Rights to enforce Title VI in HEW and its equivalents in Defense and HUD, and by 1972 to similar offices in all eleven cabinet departments. The OMBE was created in Commerce in 1969, followed later that year by the Voting Rights Section in Justice.[61] This federal network was also reaching out to parallel networks in state and local government, as for example in the EEOC's state-grant liaison with local EEO agencies. Somewhat more slowly, the network included parallel structures in the larger houses of private commerce and industry.[62]

The clientele base of the federal civil rights bureaucracy was the organized minority community, consisting at first almost exclusively of blacks, then slowly including Hispanics. By 1970, Asians and women were rather weakly linked to the network, and the organized minority groups of the future, like the elderly, the physically and mentally handicapped, and homosexuals, were not yet highly mobilized. Presiding over the clientele base for minorities in the private sector were the great civil rights organizations—the NAACP, the National Urban League, the fading SCLC (King was dead, leaving no strong successor, and SNCC and CORE had self-destructed), the broad-based and effective Leadership Conference on Civil Rights and its components, including the AFL-CIO and the giant foundations like Ford and Rockefeller.

Historically in Congress, because minorities had been a weak or even shunned constituency and the powerful committee chairmanships had been dominated by white southern Democrats, few committees had shown much solicitude for minority interests. The few who did included the House committees on Education and Labor and the Judiciary, and in the Senate the Labor and Public Welfare Committee.[63] By 1970, however, there had been a proliferation of new subcommittees, partly in response to the expanding programs and constituencies of the Great Society.[64] This had produced new constituency-focused, spin-off subcommittees like Representative Edwards's Judiciary subcommittee No. 4 on Civil Rights Oversight.[65] In 1971 the thirteen black members of the House formed the Black Caucus, and they sent the Nixon White House on a merry spin for about a year, until it concluded that the game was unwinnable.[66]

In such a rapidly changing environment, with its accompaniment of ghetto riots, assassinations, violent radical protest, intensifying war in Vietnam, and rising campus disorders, and faced with an overwhelmingly Democratic Congress, it is small wonder that the Nixon administration found it so difficult to govern effectively or to forge a coherent legislative program. But much of the incoherence was internal to the Administration. It had no consistent theory of civil rights, and especially during the crucial first year, no one seemed to be in charge. By the fall of 1969 the executive committee of the Leadership Conference had understandably, but quite wrongly, overestimated the coherence of the Nixon administration's civil rights policies by perceiving in them a "confusion [that] is planned"—a "trememdous amount of deception that appears to be practiced upon the public," and that essentially amounted to a "hoax" on the American people.[67] Moynihan was much closer to the mark, however, when he wrote President Nixon, in his famous and widely leaked "Benign Neglect" memorandum of February 1970:

> During the past year, intense efforts have been made by the Administration to develop programs that will be of help to the blacks. I dare say, as much or more time and attention goes into this effort in this administration than any in history. But little has come of it. There has been a great deal of political ineptness in some departments, and you have been the loser.[68]

Moynihan's first suggestion was that Nixon gather together his senior officials and try to pull together a coherent policy—because "There really is a need for a more coherent Administration approach to a number of issues."[69] But lacking that coherence, the Nixon White House left a vacuum to be filled by other forces in a process that worsened the contradictions.

Nixon paid a heavy political price for his incoherent policies. His support for Shultz's Philadelphia Plan not only alienated many white ethnic workers whose votes Nixon was attempting to dislodge from their weakened Democratic loyalties, but also raised expectations among blacks that easily turned to disappointment and cynicism. His OMBE earned a similar fate on a lesser scale. His voting-rights policy angered the civil rights coalition while failing to appease the southern whites whose loyalties it would repay. In the process it enfranchised millions of new black and youthful voters whose votes seemed certain to reward the Democrats. Worst of all, Nixon's failed nomination of Haynsworth and Carswell reinforced the widely held suspicion that he was "Tricky Dick," a disingenuous man with shifty eyes and a plastic smile, whose false rhetoric intoned a litany of uniting America while he sought to pack the Supreme Court with second-rate reactionaries like G. Harrold Carswell.

Meanwhile, he lost most of his civil rights battles. He had failed to reap much profit even from those he won—such as opposing the Whitten amendment and battling Elmer Staats and Sam Ervin on the Philadelphia Plan, or directing Shultz to smooth the desegregation of southern school districts through his effective network of state and community committees. Nixon's tendencies toward policy contradiction and confusion in combination with a Democratic Congress created an environment in which iron triangles tended to prosper. These now included the rapidly growing and impressively organizing civil rights bureaucracy, their strengthening allies among the congressional committees, and their effectively organized constituency base under the umbrella of the Leadership Conference. Yet the archetypal triads had generally excluded one component of government that was crucial in accounting for the modern string of civil rights victories. This was the federal courts.

In the traditional triangles, such as the vast rivers and harbors enterprise that centered on the Army Corps of Engineers, the federal courts functioned only as a distant and neutral fourth party, much as did the GAO, both acting as contractual arbiter and referee.[70] But in the federal government's new "growth" fields of social regulation like civil rights and environmental policy, recourse to the sympathetic federal courts was increasingly successful. It was also more necessary for the civil rights groups. This was because the iron-triangle model held limits for minority groups seeking expanded job opportunities, rather than merely program benefits tailored specifically for their clienteles. The civil rights clienteles could capture the new offices and sub-agencies, and by and large they set about doing so with telling effect, much as had the farmers and veterans and similar groups before them. But for minority groups in an enforcement era that was increasingly driven by the models of proportional representation and equal results, the third leg of the triangle—the congressional leg—became a problem.

Congress had proliferated subcommittees to foster programs for veterans, farmers, small business entrepreneurs, the elderly, non-English speakers, the physically handicapped, the hungry—and such groups found many friends *and few enemies* in Congress. But minority groups seeking to displace non-minorities from jobs according to a controversial model of proportional representation, and doing so through a network of bureaucratic intrusion backed by federal coercion, could count on stiff opposition from congressional conservatives, whose ranks included many senior committee chairmen. So for the minority groups and their EEO agencies, the third leg of the triangle was less likely to be found in Congress. Instead, they increasingly found it in the federal courts.

CHAPTER XV

The "Color-blind" Constitution
and the Federal Courts

*The Strange Commonality of **Plessy** and **Brown:** Social Science,
History, and Revolution by Judicial Footnote*

In 1964, when the President and Congress joined the Supreme Court in pro-
claiming that racial discrimination violated national policy, the majority coalition supporting the Civil Rights Act seemed to share two broad conclusions about the decade since *Brown v. Board of Education* (1954). One was
that the *Brown* decision had overturned the separate-but-equal doctrine of
Plessy v. Ferguson (1896) and had enshrined in its place the visionary dissent
of Justice John Marshall Harlan. "Our Constitution is color-blind," Harlan
had proclaimed, "and neither knows nor tolerates classes among its citizens."[1] in 1954 the *New York Times* greeted Warren's ruling in *Brown* by
observing that Harlan's lonely dissent in *Plessy*, which for half a century had
been "a voice crying into the wilderness," had at last been transformed by
Brown into "the law of the land."[2]

A second staple of consensus in 1964, more firmly held by liberals than
moderates, was that *Brown II*, the Supreme Court's implementing decision
of 1955, had been a shameful retreat from the moral grandeur of *Brown I*.
The Warren Court's charge to the lower federal judges to pursue school desegregation "with all deliberate speed" had led to a decade of southern intransigence marked by the barest tokenism. If the *New York Times* represented a fair barometer of cosmopolitan convictions about such matters, then
enlightened opinion in 1964 agreed that *Brown I* had struck a mighty blow
for the color-blind Constitution and that *Brown II* had wilted from the challenge. Time and events, however, have since turned this interpretation on its
head.

The first assumption, that *Brown I* had overruled *Plessy*, was of course
true to the extent that *Brown* held unconstitutional the racially segregated
schooling that *Plessy* had approved. But *Brown I* shared with *Plessy* a legal
reasoning that was rooted in a common social logic, and that set both decisions apart from Harlan's abstract principle of color-blindness in a republi-

can Constitution.[3] Earl Warren's opinion in *Brown* left Harlan's famous dissent in *Plessy* curiously unmentioned. It seemed a missed historical opportunity that the *Times* should celebrate Harlan's sweet vindication but the Supreme Court itself in *Brown* did not. Both Justice Henry Billings Brown in his *Plessy* opinion in 1896 and Earl Warren in *Brown* fifty-eight years later argued that the "facts" of social science should heavily determine a claim that state-enforced racial separation could be constitutional—much as social facts might determine the wisdom of separating citizens by age or sex in certain public policies. Despite this common premise in constitutional logic, Brown and Warren disagreed on the nature of the social and scientific facts in dispute. Speaking for the majority in *Plessy*, Justice Brown had stated: "We consider the underlying fallacy of the plaintiff's argument to consist in the assumption that the enforced separation of the two races stamps the colored race with a badge of inferiority. If this be so, it is not by reason of anything found in the act, but solely because the colored race chooses to put that construction upon it."[4] To Justice Brown the stigma attached by Negroes to their relegation to nonwhite railroad cars was strictly in the eye of the beholder, and reflected folk attitudes toward questions of social relations that laws were powerless to efface.[5]

In his opinion in *Brown*, however, Warren drew an opposite conclusion from the facts of state segregation and social perception. "To separate [school children] solely because of their race," Warren said, "generates a feeling of inferiority as to their status in the community that may affect their hearts and minds in a way unlikely to be undone."[6] In both *Plessy* and *Brown*, then, the key to violating the equal protection clause appeared to be not racially different treatment or the denial of benefits on racial grounds, or even the intent of the legislators in framing the segregation laws. Rather, it lay in the subjective, psychological attribute of social "stigma."[7] The majority in *Plessy* held that the Negro plaintiffs wrongly inferred social stigma from the state's separate-but-equal policy. Although Warren had couched his contrary psychological judgment in the cautionary subjunctive "may," he concluded for the unanimous Court that "[s]eparate educational facilities are inherently unequal."[8] "Whatever may have been the extent of psychological knowledge at the time of *Plessy v. Ferguson*," Warren held, the trial court's finding in *Brown* is "amply supported by modern authority," and "any language in *Plessy v. Ferguson* contrary to this finding is rejected." Warren then appended his controversial Footnote Eleven on "modern authority," which cited seven studies by social scientists, beginning with the doll studies of social psychologist Kenneth B. Clark and including Gunnar Myrdal's *An American Dilemma*.[9]

Warren has been roundly criticized for his reliance in *Brown* on the same constitutional logic and transient social science that had underpinned *Plessy*, in effect substituting the egalitarian psychology of the 1950s for the racist sociology of the 1890s.[10] But the stigma-centered sociology of *Brown* and its companion case for the District of Columbia, *Bolling v. Sharpe*, allowed Warren to avoid the absolute prohibition against racial distinctions by governments that was offered by Harlan's radical attack on *Plessy*.[11] This careful formula

of sidestepping Harlan's doctrine of color-blindness was crucial to the Warren Court's lonely nurture of tokenism during the first decade, and then to
its accelerated demands for racial busing during the second decade—but for
opposite reasons. If the chief offense of segregation lay in *perceptions* of stigma
rather than in racially triggered state *behavior,* then the latter was merely
instrumental to the former, and only the federal bench could make the finely
calibrated judgments about what state behavior was acceptable and when it
became unacceptable. Such a flexible formula could strategically accommodate quite gradualist and tokenist school policies during *Brown*'s first decade,
when judicial authority risked concerted defiance. Once the President and
Congress had joined the issue, however, the stigma-based logic of *Brown*
could empower the federal judiciary to command social policy in ways that
constitutional color-blindness could never have permitted.

Because *Brown* had stopped short of holding unconstitutional all racial
classifications by government as impermissibly color-conscious per se, then,
the Court was able to maximize its newly claimed authority and jurisdiction
while minimizing limitations on its discretion in enforcing its decree and af-
fording remedies. Schools in the border areas, like Delaware and Kansas,
could be pressed to desegregate promptly, while school systems in the Deep
South need not be unduly prodded into early defiance. The enormous strategic significance of this posture at the time was unclear, and in all likelihood
its long-range, radical potential for empowering the federal bench was less
appreciated than its more immediate and conservative potential for avoiding
the social chaos and judicial humiliation that could flow from the courts'
failure to enforce immediate color-blindness throughout the nation.[12]

It was ironical that the egalitarian social science of *Brown* would lead not
to equal treatment but instead to color-conscious state policies. But the Supreme Court's modern deference to social science was paralleled by a deference to historical experience—in this case also adumbrated by judicial footnote, and leading the Court in the same ironic direction of unequal treatment.
In *Bolling,* Warren argued that some government classifications based on race
might be legitimate (for authority he cited the Japanese-American Relocation
Cases from World War II, which had upheld a racial classification as reasonably related to the government's conduct of the war), as were many state
classifications based on sex and age. But "classifications based solely on race,"
Warren added, "must be scrutinized with particular care, since they are contrary to our traditions and hence constitutionally suspect."[13] The special judicial scrutiny that Warren called for in *Bolling* referred to a recent development in constitutional law: the notion of "protected classes."

At its inception during the era of World War II, the concept had nothing
to do with immutable attributes like race or national origin. Rather it derived
from a post-New Deal doctrine of "preferred rights" that we associate with
justices like Frank Murphy, William O. Douglas, and Wiley Rutledge, who
argued in the 1940s for stiffer constitutional standards to defend such "preferred" freedoms as speech and religion. These justices were reacting as public men to a war of world conquest by fascist totalitarianism, with its horri-

fying revelations of Holocaust. As jurists they were reacting also against the older constitutional tradition of judges like Oliver Wendell Holmes, Louis Brandeis, Benjamin Cardozo, and Felix Frankfurter, who preached a doctrine of judicial self-restraint by the undemocratic courts, especially in deference to legislative majorities. In the 1960s, Warren himself would lead a second surge of judicial activism, based on the preferred-rights evolution of the 1940s under Chief Justice Harlan Fiske Stone, to advance the specially protected claims of racial minorities and the criminally accused.

In 1954, however, the *Brown* decision owed much of its unique future promise not only to Warren's deference to the authority of modern social science in Footnote Eleven but also to yet another judicial footnote. This was Justice Stone's then obscure but subsequently celebrated Footnote Four, which was squirreled away in 1938 in the otherwise quite unremarkable case of *United States v. Carolene Products* of 1938.[14] In *Carolene Products,* Stone had agreed with the New Deal court's newly ascendant Holmesian liberalism that the Court should customarily yield to the legislative will of the majority, especially in matters of economic regulation. But Stone speculated in his footnote that there "may be narrower scope" for granting the benefit of the doubt to a law that threatens to deprive "discrete and insular minorities" of fundamental freedoms guaranteed by the Bill of Rights.[15]

Some freedoms, then, were to be cherished more than other freedoms; some rights deserved a stouter defense. Although the Constitution itself nowhere hinted at the notion of a hierarchy of rights, it did not seem unreasonable to speculate whether protection of free speech might somehow outrank in intrinsic importance our constitutional protection against, say, excessive bail. Liberal critics of judicial activism had historically argued that appointed and tenured justices were not free to write such speculations into the Constitution. But from Stone's seed of a footnote in 1938 grew the modern judicial doctrine that accords special judicial scrutiny to protect the class interests of certain Court-selected, "discrete and insular minorities." The Stone Court had primarily in mind the threatened minorities of the 1940s—i.e., the exponents of unpopular political and ideological beliefs, whose chief shield was the 1st Amendment guarantee of freedom of expression.[16] But by the 1960s, minorities of belief were yielding to more immutable minorities of race and national origin, whose shield was the 14th Amendment's guarantee of equal protection.[17] John Hart Ely has defended the post-*Carolene Products* incursions of special judicial scrutiny against the Holmesian norm of self-restraint by arguing that past discrimination had stripped the protected classes, especially southern blacks, of their representation in the democratic process. In this view, strict constructionist judges who normally deferred to legislative policies were nonetheless obliged to extend special protection to minorities whose voices were underrepresented in the legislatures.[18] More difficult, and less answered, were the questions of whether such special constitutional status for protected groups was permanent, and if not, what criteria should be employed to return the nation eventually to a condition of unitary citizenship.

This chapter explores the federal courts' increasingly tangled use of *historical* assumptions and analogies in interpreting the Constitution and civil rights statutes. Natural rights doctrine held that human rights were equal and immutable, and most were broadly captured in the founding Constitution. Subsequent amendments recognized and repaired glaring omissions, such as those including blacks and women in the polity. But history remained a one-way street of discovery, a Whiggish road to expanding liberty where individual freedoms, once discovered, were equally and permanently applied. The Stone Court, however, discovered that some rights were more important than others. And the Warren Court discovered that some citizens, because of their minority group membership, required more equal protection than others. Past discrimination, it appeared, had bequeathed such an unfair burden to certain minority groups that mere anti-discrimination would not bring justice. Classic liberalism's injunction to stop doing evil could not seem to repair the collective damage quickly enough to bring equal opportunity to these historically victimized citizens in their lifetimes.

Historical lag thus seemed to the Warren and Burger courts to require a compensatory theory of preferential discrimination. First in the schools, then in housing, voting, and employment, the federal courts grappled with the heavy hand of history. Their emerging compensatory theory and their proportional model of equitable relief was thus forged in the crucible of black protest. By 1970, the civil rights coalition was displacing the original formula of equal treatment for individuals with a formula of proportionally equal results for groups. In 1954, however, it was classic egalitarianism that seemed so radical. The newly evolving doctrine of protected classes was immature. But no such novel theory was required to remedy the discrimination suffered by the Linda Browns of America. What was lacking in 1954 was the judicial determination to provide traditional relief in the form of court orders to stop the racial discrimination. In 1954 Harlan's vision of the color-blind Constitution still appeared too radical and sweeping. Thus the *Brown* court in all likelihood shied away from Harlan's terrible swift sword primarily because the price of unanimity on such explosive social rulings was deliberate ambiguity over the immediacy and extent of its impact.[19] That is, caution was in order, else all hell might break loose.

The Schizophrenia of **Brown II**: Protected Classes and Equitable Relief

The NAACP's Thurgood Marshall, when arguing before the Supreme Court in 1955 for relief for his victorious plaintiffs, put the point with elegant simplicity: "The only thing that the Court is dealing with," Marshall explained, is "whether or not race can be used."[20] "What we want from the Court," he said, "is the striking down of race." But having won *Brown I*, Marshall's plaintiffs would win no direct relief from the court in *Brown II*. Linda Brown would graduate from all-black schools in Topeka. The dismaying story of tokenism under a decade of *Brown* II is familiar, but the obvious ironies of Linda Brown's fate are compounded by more hidden ironies.

When the Court's implementation decision was announced on May 31, 1955, the white South responded with a sigh of relief that the NAACP's call for "majestic instancy" had been rejected, and that instead the five cases were remanded to the local federal district courts.[21] In a cautious gesture that was designed in part to forestall a violent southern reaction against the Court's call for a social revolution in *Brown I,* the Court in *Brown II* asked not for admission to public schools on a nondiscriminatory basis, but only for a "prompt and reasonable start toward full compliance."[22] Warren's opinion borrowed the phrase "all deliberate speed" from an obscure opinion by Holmes. But the enigmatic modifier "deliberate" (if not, indeed, the full oxymoron) was inserted at the insistence of Felix Frankfurter, who feared a disastrous southern backlash against a Court that could not enforce its decree.[23] This sealed the bargain between Warren and the more conservative justices, with the unanimous decision in *Brown I* balanced by a constitutionally novel remedy in *Brown II* that would deny Linda Brown her individual relief, but would buy time for a gradual enforcement by local federal judges who would be provided no clear judicial guidelines or deadlines.[24]

Brown II, then, *was* a conservative decision, as the relieved white South and the disappointed civil rights community immediately perceived. Or at least it so appeared in 1955 and for a decade thereafter. Looking back from 1963, Columbia law professor Louis Lusky (who as Stone's law clerk in 1938 had drafted Footnote Four) concluded that "Conceptually, the 'deliberate speed' formula is impossible to justify."[25] Since its beginning judicial review was grounded in the judicial duty to give a *litigant* his rights under the Constitution, Lusky said. "But the apparently successful plaintiff in the *Brown* case got no more than a promise that, some time in the indefinite future, other people would be given the rights which the Court said he had." Legal scholar Lino Graglia, a conservative critic who regarded the twin *Brown* decisions as a prelude to "disaster," called *Brown II*'s denial of relief to Linda Brown and her co-plaintiffs "the racist element of *Brown II.*"[26]

By imbedding in its implementation decree in *Brown II* a presumption against Harlan's color-blind Constitution, however, the Court ironically linked enforcement to class actions that would ultimately require *more,* not less, government classification by race. In *Brown II,* Warren announced that in fashioning decrees to relieve plaintiffs in school desegregation cases, the federal judiciary as "Courts of equity" would be guided by the "traditional attributes of equity power."[27] Few Americans knew what that meant—and with good reason. The tradition of equity jurisprudence, which in the United States had long since blurred into the judicial discretion of the civil code, nonetheless had an ancient lineage that was distinct from the law. Extending from Aristotle's *Rhetoric* through Bacon, Coke, Hobbes, and Blackstone to the Founders of the American Republic, it had remained remarkably unchanged. Designed to temper the harsh rigor of the law in particular cases, the equity power functioned as an *exception* to the general rule of law. It offered to individuals a judicial means of relief from "hard bargains" in cases of fraud, accident, mistake, or trust, and generally providing a means of easing the suffering from "unjust and partial laws."[28] The English legal tradition had differentiated equitable and legal pleadings, gradually expanding the courts of law while confining the equity power to the courts of chancery, where a last resort was available when the normal processes of the common law had been exhausted. At the Philadelphia Convention the American Founders agreed that the "higher law" of a written constitution required a unified federal judiciary, and in Article III they broke with England's dual tradition and combined all cases in law and equity under the federal bench. During the next century and a half the codification movement in the civil code gradually abolished the formal distinctions between equity and legal procedures, but left intact equity's historic substantive meaning of relief for individuals from "hard bargains" in the law.

By 1955 the equity jurisprudence practiced by the federal courts still remained basically what it had been for two millennia, bound by the Roman maxim: *Aequitas sequitur legem*—"Equity follows the law." But the structural and procedural constraints that would restrain this necessary but potentially dangerous power of judicial discretion were gone. *Brown I* had enunciated in nascent form a subjective, psychological right to equal protection for minorities against the stigma attached to victims of discrimination. But Warren's discussion of the equity tradition in *Brown II* had summoned a conservative usage. The lower courts needed the equity power's "practical flexibility," Warren said, to balance the public interest against the plaintiffs' personal interests. This might mean, however, that Linda Brown's right to prompt relief must yield to some larger but vaguely defined public interest in balancing the claims of her protected class against the need for public order. Normal judicial relief could have directed the segregated school in Oliver Brown's neighborhood to admit his daughter Linda. The tradition of equitable remedy summoned up from the mists of history by Earl Warren in *Brown II* offered no relief for Linda Brown's "hard bargain." It held potential as a mighty engine, however, of judicial discretion for a future day.

From *Brown II* to Busing

Brown II thus had the immediate and conservative effect of remanding the enforcement of the school desegregation cases back to the federal district and circuit judges in the South. These jurists included the *Fifty-Eight Lonely Men* in the Fourth and Fifth judicial circuits (none was female), pillars of their local civic establishments whose cautious decisions during the subsequent decade were described by J.W. Peltason in 1961 as virtual socio-legal inevitabilities.[29] The lamentable but not surprising result was that after a full decade of court-ordered desegregation, only 2.3 *percent* of southern black schoolchildren were enrolled with white children in desegregated schools. In light of high black birth rates, this meant that more black schoolchildren were attending segregated southern schools in 1964 than in 1954. Many federal judges in the South showed uncommon courage in the face of repeated threats to their persons and families.[30] But overall the extraordinary deference of the federal courts to white southern sensitivities amounted essentially to a decade of default. Even when southern defiance triggered the anger of the Court, as had Governor Orval Faubus's at Little Rock in 1957, the Supreme Court rallied less to enforce the commands of school desegregation than to buttress its own prestige as the final interpreter of the Constitution.[31]

With the passage of the Civil Rights Act of 1964, however, the patience of the federal courts quickly grew thin. Within the next few years the courts fundamentally transformed the ambiguously enforced duty of local governments to desegregate their schools into an ill-defined duty to integrate them.[32] The transition began most notably with *United States v. Jefferson County Board of Education* in 1966, in the Fifth Circuit Court under the leadership of Judge John Minor Wisdom.[33] A Republican appointed by Eisenhower, Wisdom was exasperated with a decade-old pattern of compliance through pupil-placement and "freedom of choice" plans that had often very carefully taken race into account in order to guarantee the barest tokenism.[34] In *Jefferson*, Wisdom ordered the school districts not merely to desegregate by abandoning the racially triggered transfers of the free-choice plans, but also to "undo the harm" of their Jim Crow past by racially balancing their school populations according to the new HEW guidelines. For Wisdom, the remedy must speak not to individuals but to the entire race, and the courts were required to hold citizens of the present accountable for the sins of the distant past, even when they were yet unborn. This would require "the organized undoing of the effects of past segregation."[35]

J. Harvie Wilkinson called *Jefferson* the crucial turning point which transformed the face of desegregation law: "To speak thus is to thrust law to the forefront of social change, to adopt an admirable, if impossible, goal."[36] "Relief to the class, as opposed to the individual relief practiced heretofore," Wilkinson said, "initiated the idea of compensatory justice, which later would influence the Supreme Court on such imposing issues as student busing and affirmative action programs." It was a radical change. Yet the courts slipped into the compensatory model while inching their way along a frustrating spectrum of implementation, without apparent awareness of the indefinable and perhaps insatiable burdens it would eventually place both on the courts and on society. For requiring present society to "undo" the effects of the past was to require a remedy for which there was no rational definition, and hence no principled termination point.

Jefferson signified the transition in 1966 from the piecemeal approach of *Brown II* to a positive obligation imposed by the courts on school boards to plan affirmatively and comprehensively for school integration. Far from being blind to color, school superintendents would henceforth be increasingly required to use race as a lodestar for educational decision-making—not only in student assignment, but also in faculty placement, transportation policy, and school curriculum. Wisdom had voiced in *Jefferson* his hope that reliance on the HEW guidelines would help extricate the courts from the burdensome school controversies.[37] But *Jefferson* had the opposite effect, enmeshing the federal courts ever more deeply in school policies on student transportation, school finance, student remediation, teacher training, and testing.

In the spring of 1968, the Supreme Court itself rose above the circuit fray to reject a freedom of choice plan. The dispute arose in New Kent County in rural eastern Virginia, where traditional countywide school busing had produced no white students at all in the county's all-black Watkins school, which enrolled 85 percent of the county's black students. Assistant U.S. Solicitor Louis Claiborne scoffed at the county's elaborate freedom-of-choice plan as a transparent apparatus to avoid sending white children to predominantly black schools and to minimize black attendance at predominantly white schools—which it was, and did. Claiborne argued for an "old-fashioned, traditional concept of neighborhood schools." In the traditionally bused and residentially integrated rural South, this would produce immediate and significant school integration. In *Green* the Supreme Court shifted the burden of achieving integration from the backs of black children and their parents to the southern school systems. *Green* charged school boards with the responsibility to develop an affirmative plan that, in the words of Justice William Brennan, "promises realistically to work *now*."[38]

Jefferson and *Green* represented a leap in enforcement requirements that reflected more than accumulating dissatisfaction with the glacial creep of school desegregation. It reflected in addition the positive incentive provided by the Civil Rights Act of 1964, and especially by the guidelines developed by HEW in response to titles IV (on school desegregation) and VI (on federally assisted programs). These guidelines seemed to promise a basis for order, uniformity, and specificity in school desegregation that had been sorely missing from *Brown II*. They offered as well the sanction of alleged educational expertise at HEW to back the writ of mere judges. So in affirming *Jefferson*, the Fifth Circuit insisted that "courts in this circuit should give great weight to future HEW guidelines."[39]

By the late 1960s the federal courts were rapidly developing a one-two punch to replace the feckless ambiguities of *Brown II*. The first was provided by the damning shadow of history—the need to "undo" the baleful effects of the past. Its leverage would be the courts' expansive new powers of equitable relief. But equity jurisprudence offered no substantive model to define the appropriate remedy for the protected class. The equity power provided the judges a rationale for extraordinary interventions, but it provided no substantive guidance to the nature of the relief. This was provided by an evolving, proportional theory of compensatory justice, together with its necessary handmaiden, statistics. Together they allowed impatient judges to set numerical goals when the school officials would not.

The compensatory theory, however, was inchoate. It was ushered in by history, and crept into judicial logic inductively. It evolved in response to practical imperatives that were theoretically on a level with Lyndon Johnson's common-sense metaphor, advanced in his Howard University speech of 1965, about the unfair footrace between the healthy and the hobbled. The courts' evolving model of compensatory justice in fact owed most of its analytical power to historical analogy itself—specifically to the demonstrable reality of historical lag in group achievement. This was, however, a relatively new use for history. The federal courts in the late 1960s were also exercising an *old* use for history. This was the practice of the bench functioning as judicial historian, summoning the muse Clio to reconstruct legislative history and thereby divine legislative intent. In 1968 in the field of housing discrimination, this produced a bizarre result. The case was *Jones v. Mayer,* and it signalled the intention of the mature Warren Court to read into legislative history the policy preferences of the Court's majority.[40]

The Uses and Abuses of History: Trumping the Fair Housing Act of 1968

When Congress passed the landmark open housing bill in April 1968, the surprising victory capped a two-year battle in which both sponsors and opponents had assumed that no existing federal statute prevented private discrimination in the housing market. Politically, the controversial bill's passage had hinged on congressional negotiations over a compromise formula to protect the intimacy of Mrs. Murphy's boardinghouse. Then on June 17, 1968, the Supreme Court ruled in a 7–2 decision that Mrs. Murphy had in fact been covered by congressional law *for more than a century.* Her boardinghouse was therefore *still* covered, despite the recent labors of Congress. The lawmakers, still reeling from the wave of ghetto riots that followed King's assassination, were understandably astounded.[41]

In *Jones v. Alfred H. Mayer Co.,* Justice Potter Stewart found for the majority that the Civil Rights Act of 1866 had barred all racial discrimination in the purchase or rental of property, whether public or private.[42] The 1866 law had provided that "All citizens of the United States shall have the same right, in every State and Territory, as is enjoyed by white citizens thereof to inherit, purchase, lease, sell, hold, and convey real and personal property." Its legislative history reveals a determination to protect blacks against state-imposed disabilities in contractual relations, not in such private arenas as housing. The lawsuit filed a century later was originally brought in 1965 by Joseph Lee Jones, a black resident of St. Louis, against the developer of a suburban subdivision. With no claim available under Missouri law, Jones's lawyer appealed to the federal statute of 1866. But the lower courts all dismissed because the 1866 law applied only to state action and did not reach private refusals to sell.

The Supreme Court had agreed to hear the Jones case in December 1967, when the open housing bill in Congress appeared to have almost no chance of passing. When the surprising congressional turnaround allowed President

Johnson to sign the bill into law on April 10, 1968, most observers expected the Court to dismiss its earlier writ of certiorari, as it often did in such cases, and defer to the new law, which would doubtless produce its own legal challenges for the courts to resolve.

Instead Justice Stewart, a moderately conservative Republican who had been appointed by Eisenhower, wrote an opinion that was based on the broadest possible reading of the 13th Amendment and the 1866 statute. The *Jones* decision in effect substituted the Court's estimate of congressional policies that were arguably intended in 1866 for those just passed after protracted debate in 1968. Stewart found that the 1866 statute was a valid exercise of the power of Congress to enforce the 13th Amendment by "abolishing all badges and incidents of slavery in the United States." "[W]hen racial discrimination herds men into ghettos and makes their ability to buy property turn on the color of their skin," Stewart wrote, "then it too is a relic of slavery." [43]

The implications of *Jones* were profound and immediate. Congress in its 1968 law had covered only residential realty. But the Court ruled that businesses should be included as well, and also personal property in addition to real property. Congress had just provided for staged implementation through 1970, much as it had in the Civil Rights Act of 1964, and enforcement was structured through new administrative machinery established at HUD. But the Court had suddenly provided for immediate relief through lawsuits. Congress had hammered out another of its typical compromises over coverage, providing permanent exemptions for Mrs. Murphy's boardinghouse and religious institutions and the like. The Supreme Court's new rule, however, allowed no exemptions at all as far as race was concerned. Even the most sympathetic students of civil rights law agreed that this was "a revolutionary reading of the statute." [44]

John Marshall Harlan, joined by Byron White, wrote a lengthy dissent. Harlan argued that the 1866 Civil Rights Act had not been intended by its sponsors to reach private discriminatory conduct, and this view found strong support from constitutional historians. In the sixth volume of his *History of the Supreme Court of the United States,* Charles Fairman sought to "disembarrass the field of history" from the mangling it had received in *Jones,* where "the Court appears to have had no feeling for the truth of history, but only to have read it through the glass of the Court's own purpose." [45] "It allowed itself to believe impossible things," Fairman said, "as though the dawning enlightenment of 1968 could be ascribed to the Congress of a century ago." In his dissent Harlan argued that even if the historical debate over legislative intent was inconclusive, the recent passage of a national fair housing law "so diminishes the public importance of this case that by far the wisest course would be for this Court to refrain from decision and to dismiss the writ as improvidently granted." [46] By 1968, however, with the cities still burning, the Warren Court had the bit of impatient social justice firmly in its teeth. [47]

The Judicial Use of Compensatory Theory: From School Desegregation to Voting Rights

History had its selective uses in judicial construction, as *Jones* had shown. History could thus occasionally be the friend of social justice—especially when the courts of the "Second Reconstruction" mined the legislative lode of the original Reconstruction following the Civil War. But in the eyes of the Warren Court, history was for the most part the custodian of our legacy of inequality, and hence the enemy of social justice. Particularly in the school desegregation cases, the federal courts had begun to view history as an unfair legacy of accumulated inequities that mere nondiscrimination was insufficient to correct. A history of de jure segregation seemed to establish not only the geographic distribution of the problem, but also, and far more subtlety, the nature and extent of the Court-imposed remedy.

In litigation over the Voting Rights Act, the Supreme Court found in *Gaston County* in June 1969 that the blameless conduct of public officials was overwhelmed by the accumulated sins of the past. The hands of government in Gaston County, North Carolina were procedurally clean in 1969. But historically they were deeply stained by traditions and institutions of racial discrimination. *Gaston* meant that after June 1969, no jurisdiction with a Jim Crow past (which conceivably could also include the Bedford-Stuyvesants of America with their southern black immigrants, and perhaps the Hispanic barrios and the western Chinatowns with their own unique versions of Jim Crow) could escape the critical scrutiny of the Civil Rights Division of the Justice Department. Although the Voting Rights Act appeared on its face to be premised on classic nondiscrimination theory, compensatory theory was implicit in *Gaston* in the general sense that black voters from segregated schools would require special federal protection until the harmful effects of historical discrimination had somehow been undone. But the Voting Rights Act's seeds of compensatory theory lay less in the triggering provision in Section 4, which had been the focus of *Gaston,* than in the preclearance provision in Section 5.

Gaston had been preceded in the spring of 1969 by another crucial voting-rights decision, one in which ironically the author of *Gaston,* Justice Harlan, found himself in sharp dissent. The case was *Allen v. State Board of Elections,* decided by the Court that March.[48] The focus of *Allen* was the meaning and reach of Section 5. The case came to the Court in the form of four appeals, three of them coming from Mississippi as challenges to a batch of laws passed by Mississippi's legislature in 1966. The new laws were designed to blunt the impact of the previous year's Voting Rights Act on the political power of black Mississippians by switching from district to at-large elections for county supervisors, moving from elected to appointed superintendents of education in eleven heavily black counties, and making more burdensome the requirements for independent candidates to qualify for public office.[49] In all three cases the three-judge federal trial court in Mississippi had dismissed the complaints on the grounds that they did not come within the purview of Section 5, which dealt only with voter registration procedures and equal access to the voting booth. The lead case, *Allen,* was from Virginia, where a lower federal court had held that Virginia's write-in provision to assist illiterate voters served a rational purpose and was not administered discriminatorily. It thus did not violate Section 4, and the suit was dismissed. But the Supreme Court, citing the interest of "judicial economy," nonetheless heard all four cases together in appeal as challenges to Section 5's preclearance provision.

The main substantive question in *Allen* was whether Section 5 reached beyond the voting booth to include such broad electoral policies as forms of districting and whether offices were appointive or elective—questions that bore not on access to the voting booth but on the relative political power reflected in the outcome of balloting. The three Mississippi cases carried the naked imprint of racial discrimination, and thus offered the Warren Court an opportunity to expand the scope of federal anti-discrimination statutes from the bench. The Virginia case, on the other hand, represented a harmless and even benign confusion, and thus offered the Court a balanced way to rule on the scope of Section 5 without the necessity of finding discrimination when the trial courts had not.

Chief Justice Warren's opinion in *Allen* observed that Congress in 1965 had been "well aware of the extraordinary effect the Act might have on federal-state relationships and the orderly operation of state government."[50] The "unusual" and in some respects "severe" requirements of Section 5, Warren said, were applied by the language of the law to "any voting qualification or prerequisite to voting, or standard, practice, or procedure with respect to voting." The 1965 law defined "voting" to mean "all action necessary to make a vote effective in any primary, special, or general election." This included registration, casting a ballot, and "having such ballot counted properly and included in the appropriate totals of votes cast."[51] Warren then read the broadest possible meaning into the language of 1965, arguing that it was aimed at the "subtle as well as the obvious state regulations which have the effect of denying citizens their right to vote because of race."

In his broad interpretation of "the basic purposes of the act," as distinct from its specific language, Warren leaned heavily on the equal protection logic of the reapportionment cases. In *Reynolds v. Sims,* the "one-man, one-vote" decision of 1964, Warren had both defended and expanded the Supreme Court's radical plunge in *Baker v. Carr* (1962) into the heretofore forbidden "political thicket" of legislative apportionment. *Reynolds* propounded a new requirement of mathematical equality in the population of legislative districts, and Warren defended the new doctrine by alluding to the unfairness of voter "dilution." "The fact that an individual lives here or there," Warren observed, "is not a legitimate reason for overweighting or diluting the efficacy of his vote."[52] The concept of the dilution of power was inherently relative and enormously complex, in that all apportionment and districting decisions necessarily favored some groups over others. But *Reynolds* posited the concept without defining it. Both *Baker* and *Reynolds* had been

14th Amendment cases, with the equal protection clause providing the rationale for the federal courts to command state legislators and local governments to reapportion their districts to the federal judges' satisfaction. Then, in *Allen,* Warren extended the "dilution" logic of the reapportionment cases to apply to voting-rights law and racial discrimination. "The right to vote can be affected by a dilution of voting power as well as by an absolute prohibition on casting a ballot," Warren said.[53] "Voters who are members of a racial minority might well be in the majority in one district, but in a decided minority in the county as a whole. This type of change could therefore nullify their ability to elect the candidate of their choice just as would prohibiting some of them from voting."

Two justices dissented from Warren's holding in *Allen*. Hugo Black essentially repeated his dissent in *South Carolina v. Katzenbach,* where he had argued that Section 5 violated the Constitution by giving federal courts and agencies the power to hold up the passage of state laws until they were approved in Washington. To Black, this was "reminiscent of old Reconstruction days," when the southern states were treated like "conquered provinces," and state officials were obliged to come to Washington "with hat in hand begging for permission to change their laws."[54] But the old Alabamian, who would toss out Section 5 altogether (and who volunteered in his dissent a readiness to throw out the Mississippi laws *after* they had normally been put into effect and *then* challenged in court) was crying out alone. Far more damaging to the majority's ruling was the challenge of Justice Harlan.

Namesake grandson of the famous dissenter in *Plessy,* Harlan was born only three years after the *Plessy* decision. Like Felix Frankfurter, with whom he served during the first seven years of his sixteen years on the high bench, Harlan placed great value on adherence to precedent, maintaining the balance of federalism, and judicial deference to the legislative branch as the proper source of social policy. Unlike Earl Warren, who was described by his biographer, G. Edward White, as "openly humanitarian and just as openly anti-professional, almost contemptuous of the niceties of a legal argument when fundamental American beliefs called out to be affirmed," Harlan adhered to the methodological canons of "process liberalism."[55] A chain-smoker who was nearly blind during his last seven years on the Supreme Court (he retired with cancer in September 1971—the same month Hugh Black died—and died three months later), Harlan was a deliberate worker and a prolific explainer. He valued the unanimity that had marked the Warren Court's desegregation decisions. But his commitment to judicial detachment and disinterestedness, self-restraint for the elitist judiciary, and careful search for "principled adjudication" established Harlan as the chief critic of the Warren majority's impatient egalitarianism.[56] Unlike the absolutist Black, Harlan firmly supported Section 5. But unlike Warren, he understood its function as part of a complex regulatory scheme that the majority was ignoring.

To Harlan, the heart of the Voting Rights Act was Section 4's triggering formula, which suspended all literacy tests and similar devices for a least five years, and opened the door to federal observers and referees to enfranchise voteless black citizens. Thus Section 5's requirement for preclearance "was designed solely to implement" and "assure the effectiveness of the dramatic step that Congress had taken in [Section] 4's substantive commands." A state might consistently violate Section 5 and still escape it, so long as it complied with Section 4—but not vice versa. Furthermore, in prescribing relief Section 5 stated that "no person shall be denied the right to vote *for failure to comply* with such qualification, prerequisite, standard, practice, or procedure" (emphasis supplied by Harlan in his quotation from the statute). Congress thus was concerned with changes in electoral procedure with which voters could *comply*. But changing from a district to an at-large election did not require voters to comply with anything at all. All of the stages of the electoral process enumerated in Section 5 concerned the *procedures* by which votes were processed and finally counted, Harlan said, not the amount of political power that blacks or other groups might subsequently derive from exercising the franchise.

Harlan's view of the Voting Rights Act's regulatory context was reinforced by its legislative history. In hearings before the House Judiciary subcommittee in 1965, Democrat James C. Corman of California had asked Assistant Attorney General Burke Marshall whether the Administration's bill shouldn't reach beyond the question of who can vote and "address itself to the qualifications for running for public office as well as the problem of registration?" "The problem that the bill was aimed at," Marshall had replied, "was the problem of registration, Congressman. If there is a problem of another sort, I would like to see it corrected, but that is not what we were trying to deal with in the bill."[57] To Harlan the majority decision, then, was permitting the "tail to wag the dog" by construing Section 5 in isolation "to require a revolutionary innovation in American government that goes far beyond" the congressional premise in Section 4 that "once Negroes had gained free access to the ballot box, state governments would then be suitably responsive to their voice, and federal intervention would not be justified." The new ruling would require prior federal approval of all state laws "that could arguably have an impact on Negro voting power, even though the manner in which the election is conducted remains unchanged."[58] This would require the federal courts in the future "to determine whether various systems of representation favor or disfavor the Negro voter."

How would judges determine this? How was a court to decide, Harlan asked, whether an at-large system was to be preferred over a district system? "Under one system, Negroes have *some* influence in the election of *all* officers; under the other, minority groups have *more* influence in the selection of *fewer* officers. If courts cannot intelligently compare such alternatives," Harlan said, "it should not be readily inferred that Congress has required them to undertake the task." Moreover, such a construction would logically extend well beyond states and counties triggered by Section 4, to encompass areas "which in the past permitted Negroes to vote freely, but which arguably have limited minority voting power by adopting a system in which various legislative bodies are elected on an at-large basis." Thus the statute, Harlan wrote, as newly interpreted by the Court "deals with a problem that is national in scope. I find it especially difficult to believe that Congress would single out a handful of states as requiring stricter federal supervision concerning their treatment of a problem that may well be just as serious in parts of the North as it is in the South."[59]

Warren's majority opinion had avoided the vulnerable areas that Harlan attacked. Relying instead on his preferred analysis of "the basic purpose of the Act" beyond its text, Warren stretched Section 5 to include all electoral arrangements which might dilute the future impact of black voting power.[60] The metaphorical logic of dilution had led Warren to the reapportionment precedents in *Baker* and especially *Reynolds,* which were grounded in the 14th Amendment's equal protection clause. But in his dissent Harlan lectured Warren on his constitutional rationale: "This is a statute we are interpreting, not a broad constitutional provision whose contours must be defined by this court," Harlan replied. "And the fact is that Congress consciously *refused* to base [Section] 5 of the Voting Rights Act on its powers under the Fourteenth Amendment, upon which the reapportionment cases are grounded."[61] Harlan therefore concluded that the Supreme Court's extension of the equal protection clause into state reapportionment decisions in *Baker* and *Reynolds* (in both of which Harlan had dissented), with its reference to the dilution of the political power of disadvantaged groups, could not be further extended to a congressional statute based on the 15th Amendment. The relevant precedent was not *Reynolds,* Harlan said, but *Gomillion v. Lightfoot* (1960), an opinion by Justice Frankfurter (whose mantle of judicial restraint Harlan carried, often and in dwindling company, on the Warren Court of the 1960s) that had struck down a gerrymander against black voters in Tuskeegee, Alabama.[62]

Harlan's suggestion in *Allen* that the application of Section 5 was inherently national in scope was reinforced by his own majority opinion in *Gaston* three months later, when Harlan had in effect ruled for the Court that the South's history of Jim Crow could not be logically quarantined *either in space or in time.* Thus Congress might continue to believe that it was passing or renewing regional laws with specified time limits. But the expansionist logic of the Warren Court was shaping legislative policy far beyond the confines either of statutory language or of the will of congressional majorities. Even Harlan in his *Allen* dissent, holding aloft the tattered banner of Felix Frankfurter and judicial restraint, was swept along by the Warren Court's sea change of judicial intervention in social policy. By 1970 the Supreme Court lacked Earl Warren. But the momentum of its judicial logic was accelerating. When Nixon appointed Warren Burger Chief Justice in the fall of 1969, the salient civil rights issues of school desegregation, housing discrimination, and voting rights had already reached the high bench and had been similarly decided. Only the major area of job discrimination and equal employment law remained unclarified by the Supreme Court.

Swann, Griggs, *and Racial Proportionality*

The voting-rights suits received little public attention—unlike the school de-segregation cases, which tended to be closely watched for signs of judicial trends. The national media devoted inordinate attention to Nixon's Southern Strategy, even though his posturing over school desegregation and busing disguised a pattern of quiet and effective cooperation by executive agencies with the toughening judicial mandate in the South. When, in the fall of 1969, Nixon's Justice Department and HEW petitioned the federal courts for fur-ther delay in desegregating 33 Mississippi school districts, the Supreme Court in *Alexander v. Holmes County* issued a sharply worded, two-page per cur-iam order that *all* school districts must abandon dual school systems "at once" and operate "now and hereafter" unitary systems.[63] Virtually all com-mentaries on the inside workings of the Nixon administration on civil rights policy, from Ehrlichman's *Witness to Power* to Woodward and Armstrong's *The Brethren,* agree that the President and his strategists were content with *Alexander.* The Nixon administration had asked for "reasonable" delay, but the Supreme Court had insisted instead on immediate integration, and thus the courts and not the White House should be held responsible for the con-sequences.[64]

By 1971 the Supreme Court had built its equitable powers of remedy into a plenary instrument. When that year in the *Swann* case the Supreme Court upheld federal district judge James B. McMillan's order for a massive pro-gram of busing in North Carolina's 550-square-mile Charlotte-Mecklenburg school system, the Nixon administration distanced itself from the racial bal-ancing. Solicitor General Erwin Griswold argued that total integration was not a constitutional mandate, so the remedy required only the disestablish-ment of dual school systems, not "racial balance or integration of every all-white, all-Negro, or predominantly Negro school."[65] Griswold pointed out that Congress in Title IV had carefully defined what desegregation was and what it was not: Desegregation meant "the assignment of students to public schools and within such schools without regard to their race, color, religion, or national origin," but it "shall *not* mean the assignment of students to public schools in order to overcome racial imbalance."[66] Title IV had further stipulated that in authorizing the Attorney General to file desegregation suits,

> . . . nothing herein shall empower any official or court of the United States to issue any order seeking to achieve a racial balance in any school by re-quiring transportation of pupils or students from one school to another or from one school district to another in order to achieve such racial balance, or otherwise enlarge the existing power of the court to insure compliance with constitutional standards.

Speaking for a unanimous court, Chief Justice Burger found the federal courts basically unrestrained by this language—or at least unrestrained by its prohibitions when dealing with public schools in the South. Burger explained in *Swann* that because the courts' equitable powers of remedy were historic as well as broad, Congress could not have intended to *reduce* them in 1964. Rather it had intended to restrict their expansion to affect de facto segregation *outside* the South. Burger's notion that Congress did not intend for its definition of desegregation in Title IV to apply to southern schools was bizarre.[67] But with the troublesome statutory language of 1964 thus summarily disposed of, the Supreme Court in *Swann* approved Judge McMillan's order that the school system bus an additional 13,300 children to achieve racial balance. As a practical matter the plan ordered in *Swann* was sensible and worked well, chiefly because Charlotte's city and county school systems were already combined in an unusual form of metropolitan organization, and because only 29 percent of the system's 84,000 pupils were black.[68] But as a matter of statutory interpretation, *Swann* demonstrated that the Court's new claims to remedial powers through equity were virtually unlimited. As Burger himself expressed it—with unconscious irony —"in seeking to define the scope of remedial power or the limits on remedial power of courts in an area as sensitive as we deal with here, words are poor instruments to convey the sense of basic fairness inherent in equity."[69] Or more precisely, the words of Congress were poor instruments when pitted against the Supreme Court's newly asserted powers of historic equity.

By comparison with the high public visibility of the courts' school desegregation cases, however, the judicial evolution of equal employment law remained obscure. Court challenges over the new equal employment provisions of 1964, whether contained in or administratively spawned by titles VI and VII, tended to hinge on arcane technicalities of labor and contract law, like seniority and bumping rights and bidding protocols.[70] These important early test cases often involved Title VI and the bureaucratic intricacies of contract compliance rather than Title VII—as when the Third Circuit Court of Appeals upheld the Philadelphia Plan in 1971, and in response the Supreme Court denied certiorari.[71] As the job discrimination suits bubbled up slowly from the state circuits and the federal districts, they basically did not command the public attention that their importance deserved. But this dramatically changed when the Supreme Court's *Griggs* decision burst like a bombshell in 1971. *Griggs v. Duke Power Co.* signified the Supreme Court's transition from the equal treatment standards of employment discrimination that underpinned the Civil Rights Act of 1964, to the equal results standards of a new body of "disparate impact" case law that normatively rested on a model of proportional representation in the workplace.[72]

Griggs was a class action brought by thirteen of the 14 Negro workers among the 95 employees at the Duke Power Company's Dan River steam station near Draper, North Carolina. Prior to 1965 the company had restricted blacks to the plant's labor department, where the highest paying jobs paid less than the lowest paying jobs in the four all-white operating departments. These were coal handling, an "outside" department but more desirable than the shovel work in the labor department, and the "inside" departments of operations, maintenance, and laboratory and test. In 1955, Duke had sought to upgrade the quality of its workforce by requiring a high school education for initial assignment to any department except labor (which therefore meant for all "white" departments), and also for transfer from coal handling to the three higher-paying inside departments. On July 2, 1965, the day Title VII became effective, the company ceased restricting blacks to the labor department, and extended to that department the high school requirement for transfer into better job categories, but not for initial employment. Incumbent white employees hired before 1955 and lacking high school diplomas, however, were exempt from the new diploma requirement.[73]

Also in July 1965, the company began requiring, for placement in any department other than labor, that all new employees must achieve acceptable scores on two professionally developed general aptitude tests. Then, in September 1965, Duke began to permit incumbent employees, white or black, who lacked a high school education to qualify for upward transfer by passing the two tests. At trial the company conceded that the pencil-and-paper tests did not measure the ability to learn to perform a particular job or category of jobs. But the Duke management explained that the new higher standards were evenly applied to both races in an attempt to upgrade the workforce in a dangerous workplace. Thus the company had practiced systematic racial discrimination prior to the effective date of Title VII, but thereafter had applied all requirements equally, with the exception of the diploma requirement for the grandfathered employees hired prior to 1955 in one of the four more desirable departments—all of whom were necessarily white.

The results, however, were far from equal. The 1960 census showed that only 34 percent of white males in North Carolina had graduated from high school. Worse, the graduation rate for black males was only 12 percent. The EEOC in its amicus brief cited test results in a similar case in which 58 percent of whites had passed the tests compared with only 6 percent of blacks. Moreover, although the new requirements were racially neutral on their face, Duke's Jim Crow past meant that the only employees exempted were white. When Griggs and his twelve black colleagues sought relief in the federal district court, they lost.[74] The trial court found that although Duke Power had followed a policy of overt racial discrimination prior to 1965, such conduct had since ceased. Because Title VII was intended to be prospective only in its application—beginning on July 2, 1965—the impact of prior inequities, the trial court ruled, was beyond the reach of remedies authorized by the Civil Rights Act.[75]

Griggs's NAACP lawyers then appealed to the Fourth Circuit Court of Appeals in Richmond. In 1970 the appeals court agreed with the trial court that because there was no showing of discriminatory purpose on the part of Duke Power in adopting the diploma and test requirements, and because the new standards had been applied evenly to blacks and whites alike, there was no violation of Title VII.[76] But the appeals court rejected the district court's finding that residual discrimination arising from employment practices prior to the Civil Rights Act was insulated from remedial action. This retroactive view was consistent with the logic of the historical theory as applied in voting rights cases like *Gaston,* although it was not consistent with the language of

Title VII. The appeals court limited its retroactivity to an equal treatment basis, however, holding that only those plaintiffs hired before 1965 had been discriminated against, because similarly situated whites had not been subject to the same requirement.[77]

After reading the appeals court decision, John Pemberton, deputy general counsel for the EEOC, drafted a letter of collegial advice to Jack Greenberg, who as director–counsel of the NAACP Legal Defense Fund was the chief lawyer for the *Griggs* plaintiffs. Pemberton urged Greenberg to accept the limited victory from the Fourth Circuit and not risk appealing to the Supreme Court.[78] Pemberton read the appeals court decision to say that Title VII prohibits an employment test "only if it is adopted with an affirmative desire to discriminate or is administered in a discriminatory fashion."[79] According to Title VII, the job-relatedness of a professionally developed test did not matter, but only the intentions surrounding its use. This struck Pemberton as a clear articulation of the traditionally narrow, "bad motive" approach, and hence as "tragic insofar as it will allow all sorts of totally invalid tests and educational requirements to screen out blacks."

But Pemberton, a self-described "plaintiff's lawyer" who had directed the American Civil Liberties Union in the 1960s, found *Duke Power* to be a vulnerable case on several counts for the plaintiffs to press on the new Burger Court. The tests given by Duke (the Wonderlic and Otis) were widely used in industrial and commercial employment, were selected by a professional psychologist who had testified at trial on their behalf, and were precisely the kinds of tests Congress had sought to protect when adopting the Tower amendment in 1964. Also, the decision as it stood would affect only four of the 13 plaintiffs, since the remainder could either avoid the tests under the Fourth Circuit ruling, or had since been promoted by Duke anyway (including all plaintiffs with high school diplomas). Finally, the steam plant jobs sounded fairly complex and dangerous to Pemberton, thus indicating legitimate business needs for high standards. Moreover, Duke Power had agreed to waive the tests for any employee who could earn a high school diploma or its equivalent, and the company had begun a program to subsidize such efforts through adult education. "All of this means," Pemberton reluctantly concluded, "that the record in the case presents a most unappealing situation for finding tests unlawful. We are therefore reduced to making hypothetical arguments about what might happen in other cases—which is simply not a powerful litigating posture."[80]

Greenberg applied to the Supreme Court for a writ of certiorari in *Griggs,* anyway. Pemberton joined in with a strong amicus brief from the EEOC—to join the amicus brief of the United Steelworkers (which balanced an opposing brief from the U.S. Chamber of Commerce) and also that of Solicitor General Griswold and Assistant Attorney General Jerris Leonard.[81] The plaintiffs' chief leverage on appeal in *Griggs* was the dissent of Judge Simon Sobeloff of the Fourth Circuit, himself a former U.S. Solicitor General, who had argued the original *Brown* case before the Warren Court in 1954. Sobeloff's dissent had argued for a disparate impact standard that would disregard intent, and in doing so he quoted the 1968 language of *Quarles v. Phillip Morris, Inc.* that "Congress did not intend to freeze an entire generation of Negro employees into discriminatory patterns that existed before the act."[82] Sobeloff also cited the guidelines on employment testing issued by the EEOC in August 1966, and argued that the federal courts should give deference to the agency charged with administering the act, much as the courts had come to defer customarily to the more expert policy rulings of the other federal regulatory agencies.[83]

The EEOC's 1966 guidelines required that any test that rejected blacks at a higher rate than whites must be statistically validated with full documentation by employers, and done so separately for blacks and whites ("differentially validated"). In August 1970 the EEOC issued a more comprehensive set of guidelines, the professed goal of which was identical rejection rates for minority and non-minority job applicants.[84] Thus the NAACP and the U.S. Solicitor General could argue before the Supreme Court that modern psychometric science informed the guidelines of the agency established by Congress to administer Title VII and that Judge Sobeloff had called for the customary judicial deference to the EEOC's rules and regulations. Moreover the logic of the historical theory, as demonstrated in *Gaston County,* suggested that a history of inferior segregated schooling had made fair competition impossible between blacks and whites on voting literacy and employment tests alike. Thus while discriminatory motives on the part of the Duke Power Company were not only historically demonstrable and subsequently probable, they were also irrelevant. What mattered was not discriminatory motive but the racial *effect* of employment provisions—whether their impact fell disparately upon blacks and whites, whether their results were racially proportionate. Clearly the diploma and test requirements at Duke Power did not produce such results.

Chief Justice Burger and the Legacy of Earl Warren: The Supreme Court as Historical Revisionist

When Chief Justice Burger first received the *Griggs* petition for certiorari, he saw no merit in the appeal and placed it on the "dead list."[85] A law-and-order conservative who disapproved of his predecessor's liberal crusade on behalf of the rights of the accused, Burger nevertheless had a moderate record in civil rights cases.[86] Hugo Black, himself a passionate exponent of school desegregation, called Burger's civil rights record "decent." The new Chief Justice thought the Fourth Circuit court's compromise in *Griggs* was sensible. But Justice Brennan, who had recused himself from *Griggs* because he had once represented the Duke Power Company, thought his former client ought to lose the case. So Brennan persuaded Justice Stewart to request its full discussion at the judicial conference. The conference then voted to grant review, and oral arguments were heard in December 1970.[87] The Chief Justice, after several failed efforts either to derail the case or to build a coalition that would affirm the Fourth Circuit decision, eventually decided to vote with the majority. As the new Chief Justice he was anxious to appear to be the Court's leader, like his strong-willed predecessor, and voting with the majority would allow him to assign the decision to himself.[88]

Writing for a unanimous Court (of eight) on March 8, 1971, and reversing the Fourth Circuit, Burger sought to soften the decision's impact by inserting disclamatory dicta. He insisted that the Civil Rights Act "does not command that any person be hired simply because he was formerly the subject of discrimination, or because he is a member of a minority group."[89] "Discriminatory preference for any group, minority or majority," Burger proclaimed, "is precisely and only what Congress has proscribed." But Burger acknowledged the power of the historical theory, which once again was presented by the segregated history of North Carolina, as it had been in *Gaston*. This train of judicial logic, Burger explained, required the Court to take into account "the posture and condition of the job-seeker." *Griggs*, like *Gaston*, was a case requiring statutory rather than constitutional interpretation. So Burger claimed that the Supreme Court's decision was mandated by Congress. "Under the [1964 Civil Rights] Act," Burger said, "practices, procedures, or tests neutral on their face, and even neutral in terms of intent, cannot be maintained if they operate to 'freeze' the status quo of prior discriminatory employment practices." The Chief Justice thus borrowed from Judge Sobeloff's 1970 dissent in *Griggs*, which in turn had borrowed from the trial judge's 1968 opinion in *Quarles*: "The Act proscribes not only overt discrimination," Burger paraphrased, "but also practices that are fair in form, but discriminatory in operation." The opinion found no fault with either the district or the appeals courts in finding no discriminatory intent on the part of the power company. "[B]ut good intent or absence of discriminatory intent," Burger said, "does not redeem employment procedures or testing mechanisms that operate as 'built-in headwinds' for minority groups and are unrelated to measuring job capability." Burger even acknowledged that Duke Power had indicated some measure of benign intent by offering to finance two-thirds of the cost of tuition for high school training for its workers. "But Congress directed the thrust of the Act to the *consequences* of employment practices," Burger explained, "not simply the motivation."[90]

Burger's interpretation in 1971 of the legislative intent of Congress in the Civil Rights Act would have been greeted with disbelief in 1964. As the bill's floor co-manager that spring with Joseph Clark, Clifford Case had told the Senate that "Title VII clearly would not permit even a Federal court to rule out the use of particular tests by employers."[91] "[N]o court could read title VII as requiring an employer to lower or change the occupational qualifications he sets for his employees," Case had explained, "simply because fewer Negroes than whites are able to meet them." The subsequent Clark-Case memorandum had assured the Senate that Title VII required no employers to abandon bona fide qualification tests where, "because of differences in background and education, members of some groups are able to perform better on these tests than members of other groups. An employer may set his qualifications as high as he likes, he may test to determine which applicants have these qualifications, and he may hire, assign, and promote on the basis of test performance."[92] Clark told the Senate that he personally disagreed with Title VII's intent approach and equal-treatment criterion, and preferred instead a results-centered definition. But Clark's job was to explain and defend the compromise bill's language and meaning, which the Administration and the congressional leadership wanted passed.

The Duke Power Company in its court defense had appealed to the protection of Title VII's Section 703(h), the amendment added by Senator Tower to protect the employee tests used by Motorola, and the district and appeals courts had agreed. Burger's opinion, however, emphasized a codicil in the Tower amendment's language—that such tests, including their administration or action upon their results, must not be "designed, intended or used to discriminate because of race, color, religion, sex or national origin." Burger chose to emphasize not the conditional phrase "because of" race or color, but rather the verb phrase "or used." And Burger interpreted "used" to mean unintentionally as well as intentionally. He then turned to the EEOC guidelines on testing, as updated and elaborated in August 1970, after the district and appeals courts had already acted on *Griggs*.[93] The "touchstone" of the case, Burger said, was "business necessity." This was a term Congress had never used and Burger did not define or further explain. Burger then addressed the legislative history of Title VII, and especially Section 703(h), in two paragraphs that Donald Horowitz, in *The Courts and Social Policy*, called "halting and embarrassed."[94] Burger argued that the concept of business necessity must have been implied in Tower's language in 703(h), which in its original version had read that the protected tests must be "designed to determine or predict whether such individual is suitable or trainable with respect to his employment, promotion, or transfer."[95] This allowed Burger to conclude for the Court that the administrative interpretation of Title VII by "the enforcing agency is entitled to great deference." "Since the Act and its legislative history support the Commission's construction," Burger reasoned, "this affords good reason to treat the [EEOC] guidelines as expressing the will of Congress."[96]

In 1964, Senator Dirksen had anticipated such broader readings of the scope of government's power over business in Title VII. So he offered an amendment to stipulate in Section 706(g) that violation required a finding that "respondent has intentionally engaged in or is intentionally engaged in an unlawful employment practice." This would avoid, Dirksen explained, a situation wherein "[a]ccidental, inadvertent, heedless, unintended acts could subject an employer to charges under the present language."[97] Five years later, the EEOC explained in its official *Administrative History* of 1969 that the record of Congress in passing the Civil Rights Act "establishes that the use of professionally developed ability tests would not be considered discriminatory."[98] The EEOC acknowledged that under the "traditional meaning" which was the "common definition of Title VII," an act of discrimination "must be one of intent in the state of mind of the actor." But in the interim the agency's rapidly evolving enforcement policy had come to "disregard intent as crucial to the finding of an unlawful employment practice," and instead to emphasize forms of employer behavior "which prove to have a demonstrable racial effect without clear and convincing business motive." Because "the courts cannot assume as a matter of statutory construction that Congress meant to accomplish an empty act by the [Dirksen] amendment," the EEOC *Administrative History* explained, then "the Commission and the courts will be in disagreement." "Eventually this will call for reconsideration of the amendment by Congress," the text conceded, "or the reconsideration of its interpretation by the Commission."[99] In the aftermath of *Griggs*, however, neither had to occur, because the Supreme Court itself had interpreted the Tower amendment on tests and the Dirksen amendment on intent into meaninglessness.

By 1971, Dirksen was dead, the Congress remained paralyzed over revising Title VII, and the historical logic of the Warren Court had achieved a snowballing momentum that swept the reluctant Chief Justice Burger before it. As Gary Bryner observed in the *Political Science Quarterly*, "[w]hile the court's *Griggs* ruling is in agreement with the EEOC and OFCCP guidelines, it conflicts with the working and legislative history of title VII."[100] "Here, the court seems to be primarily concerned with consistency in discrimination cases rather than adherence to legislative intent," Bryner concluded. In *Griggs* the Court extended the logic of discriminatory effect that *Gaston* had inferred from the historical theory and largely dismissed, as it did in *Swann*, the statutory restrictions of the Civil Rights Act.[101] "The effect of [Burger's] opinion," the *New York Times* reported, "was to approve the guidelines issued by the Equal Employment Opportunity Commission."[102] By calling for "great deference" to the EEOC's policy guidelines, *Griggs* lent blanket judicial approval to the agency's reinterpretation of its legislative charter. Alfred Blumrosen reflected the EEOC's delight in the broad triumph that Burger's opinion had brought them: "*Griggs* redefines discrimination in terms of consequences rather than motive, effect rather than purpose."[103] It was a tour de force of administrative ingenuity, determination, and luck. The "poor, enfeebled thing," in growing partnership with the federal courts, had come a long way from its defanged origins in the legislative compromise of 1964.

The Logic of Compensatory Theory in the "Rights Revolution"

The NAACP and the EEOC had a strong moral case in *Griggs*, and Duke Power made a highly vulnerable target. Duke had long co-existed amicably and profitably with racist regimes in Carolina. When finally forced by Congress to end its Jim Crow policies, it had immediately slapped on high school diploma requirements and intelligence and ability tests. The region's poor black workers had been taught for generations in miserable segregated schools and confined to the dirtiest and lowest-paying job category at the power utility. Then when the old system was destroyed by federal law, Duke greeted the new spirit of equal access by creating new hiring and promotions criteria in the form of educational attainments that the vast majority of black workers in Carolina lacked, and written tests that half of the high school graduates in the country would fail (the *New York Times* reported that none of the black laborers at Duke Power had passed the tests).[104]

Lawyers at the NAACP and the EEOC also advanced a challenging legal theory to transform a judicial legacy of "mere" nondiscrimination that they saw as ineffective. The evolving historical theory had logical underpinnings that permitted its partisans both on and before the bench to draw at least plausible inferences from the "broad purpose" and historical context of Title VII. Thus Burger reasoned in *Griggs* that the "markedly disproportionate" rates of black and white advancement at Duke Power must have occurred "because of" the inferior education received by segregated blacks.[105] This causal linkage by historical inference allowed the *Griggs* holding to be reduced to a shorthand formula: "If sufficient disparate impact is present, intention to discriminate need not be proved."[106] Such historical logic reinterpreted statutory law according to the vision of the plaintiffs' lawyers and the judges as to what the legislators *should* have intended, or *might* well have intended in the broadest construction. In light of what the legislators of 1964 had *said* they meant, however, the *Griggs* decision contravened the twofold legislative intention, which was to forbid preferential hiring on a racial basis, and "to allow an employer's bona fide use of professionally developed tests despite their disparate impact on culturally disadvantaged minorities."[107]

The historical or compensatory theory, on which the Supreme Court under both Warren and Burger was basing its increasing judicial activism in ever-widening fields of social policy, had evolved out of *Brown*'s original concern for black school children. As long as the desegregation campaign was based on or appeared to be based on the classic liberal model of nondiscrimination—that government must do no harm to citizens because of immutable characteristics like race, and should not permit most private organizations to do the same—the particular constituency of the victim group seemed to matter little. When such equal-treatment protection was extended from schools to voting rights, the nondiscrimination logic extended easily to Hispanics, Asians, and American Indians, whose claims of unfair treatment at the polls were matters of historical record. Women of course had been nationally enfranchised in 1920, and by nature didn't live in identifiable areas the way racial and ethnic groups did. As the historical logic of compensatory theory spread to the employment field, however, the model of proportional representation on which it was based increasingly ran afoul in the 1970s of three new problematical areas that the original *Brown* model, based as it was on black school children, was ill-designed to accommodate.

The first problem area was illustrated by the tensions within *Griggs*, where an implicit model of proportional representation in the workplace clashed with Burger's assurance that discriminatory preference for any group, minority or majority, was "precisely and only what Congress has proscribed." The proportional interpretation of Title VII, together with the Philadelphia Plan approach to Title VI, led rather quickly to cries of "reverse discrimination" and to counterclaims by white workers that produced the controversial *We-ber* decision of 1979.[108]

The second problem area also involved claims of reverse discrimination, but in the white-collar field of admission to professional schools. There the blue-collar origins of *Griggs* and the Philadelphia Plan collided with the complex cultural variables that governed paths to professional success. The American Jewish community, with its long historical memory of the evils of racial and ethnic quotas, and its more recent memory of vindication through the virtues of classically liberal nondiscrimination, divided sharply over the logic of proportional representation by race and ethnicity. These developments fueled a "neoconservative" movement that strained the historic black-Jewish alliance, and in the 1970s focused a national debate on the *DeFunis* and *Bakke* cases.[109]

The third problem area involved the role of women in the burgeoning "rights revolution." The anti-ERA wing of the women's movement, which dominated the feminist reform impulses of the victorious Democratic party as it entered the 1960s, was challenged at mid-decade by the "second-wave feminists" led by the National Organization of Women, who pressed a race-sex analogy in their drive for the ERA. The defensive battalions of Esther Peterson sought to disarm the growing ERA challenge by claiming that women were equally protected by the 5th Amendment and the 14th Amendment, and that the federal courts needed only to look to find them there. As sound as this presumption was in theory, however, in practice the courts remained blind—even as they were making wholesale discoveries of new black rights in the Constitution and in a century of congressional statutes.

By March 1971, when the Supreme Court announced the *Griggs* decision, the federal courts had found precious little in the way of equal protection for women. In *Goesaert v. Cleary* (1948), the first U.S. Supreme Court case to consider the constitutionality of discrimination against women under the equal protection clause, the Court had upheld a Michigan law prohibiting any woman except the wife or daughter of the male owner of a licensed liquor establishment from working as a bartender.[110] In 1961 in *Floyd v. Florida*, the Supreme Court searched the 14th Amendment again, this time to determine whether a woman convicted of killing her husband with a baseball bat in an altercation could claim protection from an all-male jury.[111] Florida law protected women from being pulled by jury duty away from home and maternal obligations by providing that they could not be called unless they voluntarily registered for jury duty—which, not surprisingly, few did. The Warren Court upheld the Florida law as a rational state classification, explaining that "[d]espite the enlightened emancipation of women from the restrictions and protections of bygone years, and their entry into many parts of community life formerly considered to be reserved to men, woman is still regarded as the center of home and family life."[112] However lightly the Warren Court may have treated legislative intent in statutory cases concerning race relations, it showed a conservative respect for the strict construction of constitutional precedents where women's rights were concerned. Faced with such a sustained judicial disregard for women's claims to the same constitutional protections that blacks enjoyed—and especially black males—the feminists forged within their own ranks a new bipartisan consensus for the ERA. Then they descended on the new Nixon administration in a phalanx.

CHAPTER XVI

Women, the Nixon Administration, and the Equal Rights Amendment

The Semi-Feminist Legacy of the Democratic Party

When Richard Nixon was sworn in as President in January 1969, there seemed to be little reason to expect that women's rights would become a major issue of public policy in his administration. Nixon had routinely endorsed the proposed Equal Rights Amendment as a presidential candidate in 1968, as he had done in 1960 (as Lyndon Johnson had also done as senator, and as President Eisenhower in 1956 and Senator Kennedy in 1960 had appeared to the American public to have done).[1] The ERA's chief historic constituency lay among the Republican-inclined women's professional and business clubs, and Republican presidential platforms had regularly endorsed the ERA since 1940.[2] But when by 1953 it had become clear that the ERA stood virtually no chance of passing the House and could only pass the Senate with the nullifying Hayden rider attached, the ERA cause lapsed into the status of a routine political abstraction. No congressional hearings had been held on the amendment since 1956.[3]

Moreover, the Democrats remained split by the issue, or at least by 1968 they appeared to be. John F. Kennedy as representative and senator had mirrored his labor constituency's opposition to the ERA (or support of it with the Hayden rider, which amounted to the same thing), and only the bowdlerization and release by Democratic committeewoman Emma Guffey Miller of a letter she received from Kennedy had made him appear to accept the ERA.[4] The AFL-CIO Democrats and the Women's Bureau coalition had kept the ERA out of the Democratic platforms in both 1960 and 1964.[5] In response to urgings by Esther Peterson and Arthur Goldberg, Kennedy in December 1961 appointed his President's Council on the Status of Women, and in its final report of October 1963, *American Women,* the commission concluded that the ERA "need not now be sought" because a more immediate source of constitutional protection for women lay in the 5th and 14th amendments.[6]

In response, Kennedy, on November 1, issued Executive Order 11126, which created two follow-up groups, the Citizens' Advisory Council on the Status of Women (CACSW) and the Interdepartmental Committee on the Status of Women (ICSW). Because the Kennedy administration could legitimately claim credit for the Equal Pay Act of 1963, and the Johnson administration could claim credit (with luck) for including sex within Title VII's ban on employment discrimination, the ERA seemed safely defused, at least as a divisive issue that might split the Democratic coalition. But beneath the surface of policy continuity, tensions were building and a new national network of feminist women was being created, one that by 1970 would surround and enfold Peterson and her ERA-resisters.[7]

Initially, the Kennedy-Johnson administration's two new instruments, the status-of-women council with its national scope, and the federal government's interagency committee, were held in close rein by the Women's Bureau network. When President Johnson early in 1965 appointed Peterson as his special assistant for consumer affairs, she retained her assistant secretaryship in Labor but relinquished her directorship of the Women's Bureau. Peterson's replacement at the bureau, Mary Dublin Keyserling, was an economist with strong anti-ERA convictions. As the Johnson administration's senior woman (India Edwards and Liz Carpenter complained that Johnson would be able to name more women to important jobs if Peterson did not get them all), Peterson served as executive chairman of the interdepartmental committee and controlled the staff for both the committee and the council. The committee was easier to control, because its formal membership of seven cabinet secretaries, plus the heads of the Civil Service Commission, OEO, and the EEOC, meant that without strong leadership from the top, it could—and largely did—become a paper organization of deputies standing in at routine meetings.[8]

The council was harder to control, however, because its twenty members were nationally prominent citizens assembled under the chairwomanship of editor Margaret Hickey of the *Ladies' Home Journal*. But full-time staffs often control the agenda of part-time boards, and Hickey was part-time like her colleagues. Thus Peterson was able to focus the council's efforts on such matters of traditional Women's Bureau concern as the minimum wage, unemployment compensation, and defense of state protective labor laws. Catherine East, the executive secretary in the Women's Bureau for both the committee and the council—and something of a feminist subversive for the ERA—complained in 1968 that "Two task forces established by Margaret Hickey in an attempt to circumvent this narrow view—one on relationships with state commissions and one on women in poverty—were aborted by staffing provided by the Women's Bureau."[9] As a result, East claimed, in 1966 "Miss Hickey resigned because of the frustrations involved."

In June 1966, Hickey was replaced as chairwoman of the council by retiring Democratic Senator Maurine Neuberger of Oregon, who concentrated on creating an active state commission in every state. The Women's Bureau had encouraged this drive, which had originated in 1962 with Virginia R. Allan, then president of the National Federation of Business and Professional

Women's Clubs. By February 1967 the drive had created commissions in all fifty states—including Texas, the last holdout.[10] Conferences of state commissions were held in Washington in the summers of 1964 and 1965, with the Women's Bureau in essential control of the agenda. But the watershed conference was in 1966, when debate was heating up over the EEOC collapse on single-sex classified ads, and NOW was formed under the radicalizing whip of Betty Friedan. By the fourth conference, held in Washington in late June 1968 and presided over by Neuberger, the U.S. Government's semi-official women's agenda had been strongly liberalized, but in a Democratic direction that excluded the troublesome ERA. At that meeting the CACSW adopted the reports of four task forces, the most notable of which was Marguerite Rawalt's task force on family law and policy. Its recommendations embraced an expansive feminism, including the "basic human right of a woman to determine her own reproductive life."[11] In a like vein the report called for repeal of all laws penalizing abortion; for egalitarian revisions in marriage law, including alimony and custody rights, no-fault divorce, and equal property settlements; for liberal provision of birth control information and services; and for public funding of day care for children and equal rights for illegitimate children. Three of the task forces called for various government guarantees of income maintenance. Not one mentioned the ERA.

During 1966–68 the Women's Bureau had tried to hold its traditional line in defense of protective legislation, especially on the council's labor and standards task force. But Senator Neuberger was determined to avoid the frustrations of her predecessor in the chair, and she was wise enough in the ways of Washington to know how to do it. She purposely insulated the council staff from the federal bureaucracy by placing her former Senate office manager on it, and by appointing citizen task forces without coopting representation from the federal agencies.[12] As a result, even the labor standards task force produced egalitarian recommendations that states repeal or amend special protective laws that set maximum hours and weight-lifting limits for women, and excluded them from certain occupations. In the summer of 1968 the Washington conference of state commissions continued to duck the ERA issue but otherwise blessed the liberal-feminist task force recommendations. The commissioners were addressed by Vice President Humphrey, secretaries Wirtz of Labor and Cohen of HEW, Coretta Scott King, Mayor Walter Washington of the District of Columbia, and the director of the Women's Job Corps (who was Mayor Washington's wife). The sponsoring CACSW remained technically nonpartisan, or at least bipartisan. But the agenda and the speakers' platform were overwhelmingly liberal-Democratic.

To judge from the tone and substance of the 1968 conference, it seemed fair to conclude that by the end of the Johnson regime the liberal feminists had overcome the protectionists among the Democratic women. But the Democrats had exacted their traditional price, paid in the coinage of philosophical inconsistency, by avoiding the divisiveness of the ERA. Five months later, however, Richard Nixon was elected President. Republican women, who long had nurtured the ERA tradition, now had access to the presidency. But they lacked the continuity and momentum associated with the dominant Democrats in the women's movement. It was a crucial juncture, and the transition in domestic policy was being presided over by Dr. Arthur Burns—no feminist, he.

A Republican Brand of Feminism

The Burns-McCracken transition task forces of late 1968 reflected a Republican commitment to approach the urban crisis indirectly. This meant not through Great Society categorical grants and OEO-style community action programs, but through tax incentives to spur voluntary efforts by the private sector and decentralized efforts by the states to attack the sources of urban decay. Thus the Burns Report to Nixon of January 1969 emphasized the techniques of macroeconomic policy in supporting private and local efforts, and lumped most of the "social issue" under its catch-all Urban Affairs chapter. Only twice were civil rights issues mentioned in the entire 117-page report. One recommendation supported the proposal of Shultz's task force on manpower for an attack on job discrimination against minorities in construction so as to increase the supply of skilled labor (and presumably lower its cost). The second supported the legislative approach of Republican congressman Charles Goodell of New York to arm the EEOC with prosecutorial authority, but deny it the judge-and-jury authority of the cease-and-desist power.[13] The Burns Report did not altogether ignore minority and special "problem" groups. It called for early fulfillment of Nixon's campaign promises to call a conference on the needs of Mexican-Americans, and another on the problems of the aging (including the prompt appointment of a White House Assistant on the Aging), and also new joint Labor-Defense programs for the employment of veterans, "particularly Negroes," when the war in Vietnam was concluded. But there was not one word about women. Nor was there any mention of the ERA.

This was a critical moment of policy formulation. The feminist agenda of the Democrats, with its radicalizing commitment to abortion rights, wholesale changes in family law, and class-based programs for income redistribution, but with its internally imposed silence on ERA, had been swept aside by Nixon's victory. But the regime change created a vacuum that seemed unlikely to be filled by the Republican women's leadership. They had traditionally embraced the ERA, to be sure, but with a quaintly feckless fervor, and their elite origins had historically proved unsympathetic to the class-based demands of the Democratic coalition. Judging from the Burns Report, the newly empowered establishment of Republican males perceived no vacuum at all. There, in limbo, the "woman question" was poised.

But not for long. The previously eclipsed Republican and pro-ERA wing of the feminist movement was immediately heard from. And they had caught the virus. At his press conference on February 6, 1969, Nixon was challenged by Vera R. Glaser, the Washington bureau chief for the North American

Newspaper Alliance. She noted that only three of the Administration's first 200 top-level appointments were women, and asked whether this meant that women were "going to remain a lost sex?"[14] In light of Glaser's subsequent role both in shaping the agenda of the Republican feminists and in flushing the Administration out when it attempted to duck the issues, Nixon's reply was ironical: "Would you be interested in coming into the government?" [*Laughter*]

Unlike most newspaper reporters (on both counts), Glaser was a woman and also a Republican. Indeed, she had been director of public relations for the women's division of the Republican National Committee when Vice President Nixon had kicked off his 1960 presidential campaign. On the heels of the press conference she wrote a story that focused on the President's seeming inability, as symbolized by his all-male Cabinet (unlike President Eisenhower's), to locate qualified women for senior appointments.[15] Catherine East saw Glaser's story, promptly sent her a packet of material on women's issues and feminist proposals from the Democratic status-of-women groups, and met with her to help supply information and apply pressure.[16] East's strategic position at the nerve center of the Democratic women's network provided a kind of informal transition task force, an institutional source of continuity for the Republican women who historically had carried little weight in party or policy councils, and whose role in the Democrat-dominated panels of the Kennedy-Johnson years had been minimal.

Newly armed with East's wealth of data, Glaser wrote a five-part series on the new women's movement that was widely syndicated in the spring of 1969 in the nation's daily newspapers. In April, Glaser and East secured an appointment with Burns and persuaded him to meet with a group of mostly women correspondents. The news briefing was held on May 15, and Burns was asked whether the Administration was planning any initiative to open up opportunities for blacks and women, who seemed to be boxed into the lower federal jobs. "Not as a matter of policy," Burns replied. "On the contrary, we've been trying very hard to place Negroes in government positions, and I'm not aware of any discrimination against the better half of mankind."[17] This answer prompted the following colloquy:

Q. You're not?

A. No, I'm not.

Q. Are you really serious about that?

A. Oh yes. I'm speaking only for myself, and I may be blind.

Q. There isn't anybody in this room among the veteran women correspondents who hasn't run up against real hard-core discrimination.

A. You ought to make more noise about that because to some of us, the idea is abhorrent.

When asked why a country with more than half its population female should be "run practically by a male enclave," Burns conceded that "we don't have very many" women making policy. "If there is a prejudice against women, if there is discrimination against women," he said, "it is of the unconscious variety, and that may be so."

In response to Burns's off-hand invitation to women to "make more noise about" discrimination, Glaser wrote him on May 23 to catalogue the sex discrimination that Burns and his colleagues had failed to notice. Glaser cited the customary data documenting the nation's persistent gender disparities in income, unemployment, and education. But she leaned hardest on the failed federal response. The Supreme Court had *never* held that a state law discriminated on the basis of sex, she said. One-third of the federal workforce was female, but less than 2 percent of the top jobs were held by women. Of more than a thousand political appointments to be filled by the new administration, less than a dozen had thus far gone to women (whereas Johnson had named 4 assistant cabinet secretaries, 6 women ambassadors, and 8 women to federal commissions). The EEOC was forced to rely on the Justice Department for enforcing Title VII, but Justice to date had initiated *no* sex discrimination cases, although they had prosecuted at least forty-five suits charging racial bias.

What should be done? Glaser provided a catalogue. The President should honor his campaign pledge and ask Congress to pass the ERA. He should recommend that Congress provide cease-and-desist authority for the EEOC; add sex to Title VI for OFCC enforcement, especially concerning education; and add sex discrimination to the formal jurisdiction of the Civil Rights Commission. On presidential authority alone, Glaser said, Nixon should appoint a White House assistant for women's issues, direct the Attorney General to begin filing sex discrimination suits under Title VII, and direct Labor Secretary Shultz to issue the OFCC guidelines on sex discrimination that were blocked in October 1968 by organized labor under the Johnson administration.[18]

By the summer of 1969, the momentum of the women's movement among elite whites had raised the public consciousness of a feminist agenda that could not easily be dismissed as radical "bra-burning."[19] But the invisibility of women leaders and women's issues in the Nixon administration was noticed and resented by Republican women who denied being feminists themselves. Representative Florence Dwyer of New Jersey, who was the ranking Republican on the House Government Operations Committee (and whom Glaser told Burns "certainly cannot be categorized as a 'professional feminist'"), had written the President on February 26 to ask that he appoint a White House adviser on women, or create "an independent agency to strengthen women's rights and responsibilities."[20] Dwyer's letter, however, had produced only a routine acknowledgment from the White House. Nixon's only visible early gesture toward women was to invite the Cabinet wives—there being no Cabinet husbands—to sit in on a Cabinet meeting. But Glaser's letter to Burns reported a negative reaction to this gesture that was widely shared among career women and party workers, who resented the symbolic role of women as loyal but silent, wifely observers to male shapers of policy. She told Burns that the resurgent women's rights movement was in a state of ferment, and that women voters now outnumbered men. Awareness of the pervasiveness of sex bias was becoming an "explosive issue," Glaser warned, but one that also created a magnificent opportunity for Nixon to mobilize women in 1972.

On July 8, Congresswoman Dwyer, impatient with the Administration's inaction and angered at Nixon's "empty gesture" with the Cabinet wives, released a sharp memo to the President that was co-signed by three Republican colleagues in the House, Representatives Margaret M. Heckler of Massachusetts, Catherine D. May of Washington, and Charlotte T. Reid of Illinois.[21] "None of us are feminists," Dwyer explained. But in her litany of complaints and her list of recommendations, which paralleled Glaser's but in greater detail, Dwyer reflected the Republican women's traditional class-based resentment against sex discrimination in business and the professions and increasingly in government as well. Of women college graduates, Dwyer said, 20 percent could find "no better employment than clerical, sales or factory jobs." Women accounted for only 7 percent of America's physicians, 1 percent of engineers, 3 percent of lawyers, and 8 percent of scientists. Law and medical schools typically held women to a 7 percent quota, despite women's higher scores on entrance examinations. The Republican congresswomen's blast at their President included a list of seventeen recommendations, with presidential support for ERA hearings listed ninth, still well behind demands for more immediately realizable presidential appointments.[22]

At this point Pat Moynihan, already jousting with Burns for primacy in directing domestic policy, weighed in with one of his distinctive missives to Nixon. It was a political mood-piece that registered the surging feminist anger, even among stalwart Republicans like Florence Dwyer, that Burns and the Republicans' male establishment in general seemed unable to see. As the resident White House liberal, certified Democrat, and unofficial emissary to the left half of the political spectrum, and especially to the disaffected youth, Moynihan wrote Nixon in August that the media's fascination with the racial rhetoric and ideological fratricide between Maoists and Black Panthers at the SDS convention in Chicago that June had obscured a feminist groundswell.[23] In Chicago the Panthers' display of racialist venom against whites had been warmly received by the predominately white SDS audience. But the Panthers' arrogant display of male chauvinism had backfired disastrously. The key to understanding the portent of these bizarre events, Moynihan wrote, was not race or ideology, but rather class. "The blacks are struggling to break out of the lower class matriarchy, and make much of their dominance over women," Moynihan explained. "But their white middle and upper middle class supporters are made up about equally of young males and females, equally well educated, equally well off, and increasingly resentful of the many subtle ways in which women are excluded from the 'serious' things of American life. (By all accounts, the women radicals are the most fearsome of all.)"

Moynihan acknowledged as an "essential fact" that "we" have "educated women for equality in America, but have not really given it to them. Not at all. Inequality is so great that the dominant group either doesn't notice it, or assumes the dominated group likes it that way." He reminded Nixon that at their recent meeting with national educational leaders, Yale's Erik Erikson had pointed out that there were no women present—and Erikson might well have commented on the general absence of women from the centers of power and prestige in American higher education. "It is considered too important for them," Moynihan said. *"They* teach kindergarten." Moynihan was willing to bet that "there are proportionately more women in the Marine Corps than on most University faculties." He concluded by predicting that female equality would be a major cultural and political force of the 1970s, and by recommending that Nixon through his appointments and pronouncements should take advantage of a surging force that was ripe for creative leadership.

Nixon's Unplanned Task Force on Women

The upshot of these growing pressures was a White House decision, consistent with tradition and instinct, to study the matter. So in September a new task force on women's issues was added to the Burns portfolio. As late as August 20, the Burns list of planning task forces on domestic policy for the 1970 legislative and administrative agenda included nothing on women, although it included studies for such specific groups as the aged, the mentally and physically handicapped, prisoners, and small businessmen (including minority business enterprise).[24] But the women's issue was heating up too fast to continue being ignored. Unlike the established and Democratic-dominated CACSW, a new study group on women's issues would offer the virtues of being White House-picked and Republican-controlled. The deliberations and recommendations of the women's task force would thus be filtered during the fall of 1969, along with the other twenty planning groups for the 1970 agenda, through the more cautious Burns, while Moynihan concentrated his efforts on the Family Assistance Plan.

The eleventh-hour creation of the women's task force, together with the White House men's relative ignorance both about women's issues and the identity and capability of women leaders, meant that Vera Glaser and Catherine East would enjoy unusual freedom in shaping the task force's membership and agenda. Indeed Glaser listed East first among her thirteen nominees for the task force. But as a federal employee who was experienced in the world of interagency networks, East was more appropriate and probably more valuable as the group's staff director—a post to which she was subsequently appointed.[25] That appointment was made by Virginia R. Allan, who was also on Glaser's list, and who was invited by Burns to chair the task force. Allan's credentials as a vice president of the Calahan drug store chain in Michigan, and as a former president of the Business and Professional Women's Club of America, made her a strong choice as chairwoman. Allan was also on East's list, as were Evelyn E. Whitlow, a California Republican who was active in the National Association of Women Lawyers; Evelyn Cunningham, a black Republican who directed the Women's Unit for the governor's office in New

York; and Dorothy Haener, a Democrat from the Women's Department of the United Auto Workers, who shared the UAW's maverick position among unions as favoring the ERA.[26]

These nominations allowed Burns's office to avoid other nominees who were prominent Democrats, such as Aileen Hernandez or Betty Friedan. The Burns practice of turning to citizens "outside" the federal establishment kept off the task force such women as Elizabeth J. Kuck, a labor Democrat who had been appointed to the EEOC by Lyndon Johnson in 1968, or Elizabeth Duncan Koontz, another Democrat and former president of the National Education Association, who had recently been brought in to head the Women's Bureau by Secretary Shultz. Glaser herself was an obvious candidate for the task force, and given her unique leverage with the media, she proved to be a crucial member.

By mid-September Burns had selected the 13-member task force, appointed Allan to the chair, and assigned them a broad mission of policy review and recommendation that gave them ample running room.[27] Politically, Burns could scarcely do less. Nixon in 1969 had no women's rights agenda at all, and he could scarcely risk ignoring the rising cry of the Republican women. So visible was the issue, however, and so important was the need to appear to be responding sympathetically to it that the Burns task force on women's issues was elevated to a presidential operation. On October 1, Nixon announced that he was appointing a Presidential Task Force on Women's Rights and Responsibilities. Over the list of the thirteen task force members, the two-sentence release was terse in summarizing the new study group's vague charge: "The task force will review the present status of women in our society and recommend what might be done in the future to further advance their opportunities."[28]

Evelyn Whitlow acted as rapporteur for the orientation session when the group first met in Washington on September 25. Whitlow recorded that Burns opened the meeting by acknowledging that the President was aware that he hadn't done enough to enlarge the opportunities for women, and "it was only recently that he [Burns] had become conscious of certain prejudices through a meeting with VERA GLASER who pointed out negative attitudes of which he had been previously unaware, and which he had more recently observed in others."[29] Congresswoman Dwyer then reported on the meeting that she and representatives Heckler, May, and Reid had held with President Nixon on July 9. The President's response had been to instruct his Cabinet officers when he met with them the following day to make special efforts to appoint and promote more women to senior positions in the agencies. Dwyer told the task force, however, that "our attitude should not be that of feminists. We should join the men and let them know that we are not going to move over any longer."

The task force was then briefed by officials from most, but not all, of the federal agencies whose jurisdictions seemed to concern women in important ways. Oddly, HEW was not represented. But senior women officials presented the perspectives of the EEOC and the Women's Bureau, and two men represented the Labor Department and one the Justice Department. What was most curious, and perhaps most revealing, was the absence from the large Labor delegation of any representative for the OFCC, whose guidelines on sex discrimination had been hanging fire since the previous fall. There they had remained during the lame duck months of the Johnson administration, held hostage to the tension in Democratic regimes between organized labor's protection of seniority and guild-like hierarchies of craft skill in a unionized workforce dominated by white males, and egalitarian threats by minorities and—increasingly now— women. In the fall of 1969 the Democratic regime was long departed, but Labor's new Republican secretary was preoccupied by the racial fight over the Philadelphia Plan.

Leading off the agency orientation was Commissioner Elizabeth Kuck of the EEOC. Kuck emphasized the weak enforcement powers and small budget of the EEOC in the face of rising complaints of sex discrimination, which already accounted for a quarter of all complaints. Yet the commission had accumulated a backlog of 3,000 cases, Kuck said. Given the agency's ponderous, case-by-case method of conciliation, in which the commission was successful in only half of its attempts, the backlog, like the complaints of sex discrimination, was also still growing. Such failures left complainants only two options. One was to bring an expensive and time-consuming private suit for individual relief. The other was to hope that the Attorney General would file a pattern-or-practice suit. But no Attorney General had ever brought such a suit to combat sex discrimination, despite the simultaneous coverage of race and sex in Title VII.

Kuck reported that the EEOC was seeking stronger enforcement powers and that two such bills were pending before Congress. In the Senate Labor and Public Welfare Committee, she explained, S.2453 would give cease-and-desist power to the EEOC, while S.2806, the Administration's bill, would provide prosecutorial authority to the EEOC with enforcement coming through the federal courts. Kuck was politically circumspect about her preference between the two approaches.

Kuck was also circumspect about the commission's embrace of compensatory theory and proportional representation, with its roots in blue-collar discrimination and its implications of minority quotas to compensate for historical lag. The task force she addressed was dominated by educated and affluent women drawn from the Republican party's historic equal rights wing, with its emphasis on nondiscrimination and open access to the male-dominated world of business and the professions. Kuck thus concentrated on the primacy of "nondiscrimination [which] requires that selection be on the basis of individual characteristics," not group characteristics.[30] For this reason the EEOC had very narrowly interpreted the BFOQ claims for exemption to Title VII based on sex, and its guidelines required that help wanted ads could not be listed in columns segregated by sex. Different retirement ages by sex must be phased out, and state laws purporting to protect women must yield to Title VII's command of nondiscrimination. Whitlow's detailed minutes record no mention at all of affirmative action.

Representing the Justice Department was Benjamin Mintz, deputy director of the Civil Rights Division. Mintz was candid to the point of bluntness. His division had only 115 attorneys to handle all Title VII litigation for the entire country. They had brought suit, he said, in 47 Title VII cases, all of them charging racial discrimination. This was because since his division's creation in 1957 the racial situation had been explosive, and the CRD had concentrated first on school desegregation, then on voting. Now the highest priority was assigned to employment discrimination on the basis of race—witness, he said, the current crisis in the construction trades in Chicago. According to Whitlow's minutes, Mintz explained that the Justice Department regarded "discrimination against Negro men [as] a more serious social problem than is discrimination against women and that the prejudices based on race are much more deeply ingrained than are the prejudices based on sex." Mintz "admitted that the Justice Department reviews its policy regularly and that there are no women sitting at the policy level which establishes priorities."[31] Mintz concluded by volunteering that "the Justice Department in making its assessment and in establishing priorities has not made it a practice to consult with the Department of Labor."

The large Labor Department delegation, for its part, produced a curious testimony. Almost all of it avoided the one controversial policy area where imminent change might be expected: the unissued guidelines on sex discrimination by the OFCC. Labor's contract compliance office was something of an institutional ghost at the meeting, although it is not clear that the task force understood this. The OFCC's fight over the Philadelphia Plan would consume the department's energies until the end of the year, when President Nixon's intercession ensured the defeat of Comptroller General Staats and the congressional conservatives. But the Labor Department views had the odd ring, especially to feminist ears, of an agency long captured by a constituency that was profoundly masculine in its orientation and membership, yet one which contained the government's only subagency with a primary concern for women.

The Women's Bureau's deputy director, Mary Hilton, explained to the task force that the ERA was in fact not necessary. "She indicated," Whitlow wrote, "that OFCC 11246 [*sic*] outlaws sex discrimination in Government contracts and said this was administered by the Labor Department."[32] Hilton then added, according to Whitlow, a puzzling disclaimer: although the government was committed to equality in training, Hilton said, "a demand for employees must exist with the result that if employers are unwilling to hire women, the need for training women for those jobs does not exist."

This enigma was clarified by Malcolm R. Lowell, the department's Manpower Administrator. Like the Department of Justice, Lowell said, Labor had placed its priority on racial discrimination. Labor had done this partly because there had been no political pressure on the gender front, and hence no confrontations and negotiations—witness, again, the effect of the protesting black construction workers in northern cities. Thus he elucidated Hilton's point by explaining that while Labor policy was not to discriminate in "manpower" training by sex, the department's program rules and structure dictated that the demand factor controlled the supply factor. If an industry did not hire women, Lowell explained, then there was by definition no demonstrable shortage and thus Labor could not train them. Labor unions, especially in construction, wanted to limit the number of workers entering the trades and many such unions "did not want women."[33]

The implication was clear: the government responded to political pressure, and for the most part women had not applied it. Blacks had. Therefore new life had been breathed into the moribund Philadelphia Plan, while the OFCC guidelines on sex discrimination had been shelved.

Bottling Up the Women's Task Force Report

The task force met in Washington for two days every two weeks through the end of November, with the agenda being fed by the practiced hand of East, and with Glaser needling the Administration through her syndicated stories in such newspapers as the *Detroit Free Press* and the *Miami Herald*.[34] By November 24, the task force had reached a remarkably firm consensus. In this Republican group the old ERA-centered dualism, which had so long divided and compromised the Democratic-dominated status-of-women groups, was absent.[35]

On December 15, Allan sent Nixon the task force's 40-page final report. It contained five recommendations, the first two of which were concerned not with substance but with visibility. First, the President should establish an Office of Women's Rights and Responsibilities, whose director would serve as a Special Assistant reporting directly to the President. Second, he should call a White House conference on women's rights and responsibilities in 1970, the fiftieth anniversary of the ratification of the suffrage amendment and establishment of the Women's Bureau. The third recommendation, however, called substantively for a presidential message to Congress recommending eleven legislative actions, the *first* one being passage of a joint resolution sending the ERA to the states for ratification. The report observed that past opposition to the ERA had been based on its conflict with state protective laws for women. But "[s]ince these laws are disappearing under the impact of Title VII of the Civil Rights Law of 1964 and State fair employment laws, opposition will be much less and may evaporate in the light of information developed at the hearing."[36]

Next came Title VII, which should be amended to transform the private right of 1964 into a public right by empowering the EEOC to enforce the law (the recommendation did not choose between the Administration's prosecutorial approach and the cease-and-desist approach favored by the Democratic leadership of Congress). Congress should also extend Title VII coverage to employees in state and local governments and educational institutions. Other legislative proposals would add gender protection to titles of the 1964 act other than Title VII (e.g., those covering public accommodations and education), add sex discrimination to the jurisdiction of the Civil Rights Commission, and provide government support for child care, including business deductions from the IRS for care of children or disabled dependents. Two of the report's legislative recommendations sought more equitable policies for *men,* by equalizing Social Security benefits that favored widows and wives of disabled workers, and by providing equal fringe benefits for husbands as well as wives of federal employees.

The report had six recommendations for the President himself and his executive branch. Listed first was a demand that the Secretary of Labor issue "immediate" Title VI guidelines on sex discrimination by government contractors. The President should also direct the Attorney General to file sex as well as race discrimination suits under Section 707's pattern-or-practice provision. Finally, the report urged the President to "appoint more women to positions of top responsibility" in the government. This would "achieve a more equitable ratio of men and women," not through preferential quotas for women, but through firm instructions from agency heads "that qualified women receive equal consideration in hiring and promotions."[37]

The equal rights perspective of the Republican women, as distinct from the compensatory emphasis of the EEOC and the OFCC, had dominated the task force. What was needed, it implied, was a level playing field, not special advantages or preferences for women. The final report included a legislative recommendation that the Fair Labor Standards Act be amended to extend coverage of its equal pay provisions to "executive, administrative, and professional employees." Written by professional elites, the report made a good case for such a change. It argued that when the equal pay bill of 1963 was made an amendment to the Fair Labor Standards Act, the exemptions of the FLSA automatically excluded the executive, administrative, and professional employees who were originally included by the equal pay bill's sponsors. Such women were subsequently protected from sex discrimination, to be sure, by Title VII. But unlike the FLSA, Title VII did not permit the identity of a complainant to be withheld from an employer. Thus complaining professional and business women were vulnerable to employer retaliation under Title VII, which in any event provided no enforcement power for the administering agency.

In agreeing to recommend that equal pay protection be extended to women in managerial positions, the task force majority had rejected the proposal of the UAW's Dorothy Haener that the FLSA be amended to cover the very poor as well as the affluent, and thereby require the minimum wage for excluded groups like domestic servants and agricultural workers. Haener's proposal was rejected by the majority not because they were unsympathetic but because Congress had consistently rejected such extensions in the past, and its adoption might overburden the task force report. Nevertheless Haener's blue-collar perspective reflected in mirror image the class orientation of the task force majority, and her dissent alone marred the unanimity of an unusually coherent and forceful document.[38]

After receiving the report in mid-December, Burns wrote Virginia Allan on January 9, 1970, that he hoped she was enjoying some leisure, now that the report "has been filed." [39] Burns thanked Allan and her task force colleagues for completing their difficult task in such good order. But he made no comment on the substance of the report's far-ranging recommendations. The report was "now being reviewed here at the White House," Burns said. But "filed" was the better term, after all.

Flushing Nixon Out on Women's Issues

Not only did the White House not release or comment on the task force report; the staff did not even get around to the routine task of sending its recommendations to the relevant agencies for comment until April. By then, however, the new feminist consensus for ERA was gaining too much momentum and publicity for the Administration to sit much longer on the report and the issues it raised. In late 1969 the League of Women Voters reversed almost a half-century of opposition and endorsed the ERA. On February 7, 1970, the CACSW, of which Virginia Allan was a member, officially embraced the amendment. "Sensing that the time had come to advance the cause of justice and equality for men and women," the CACSW reversed the judgment of Kennedy's commission of 1963. [40] The council also "borrowed" attorney Mary Eastwood from the Justice Department and published her legal brief supporting the ERA. In March, Martha Griffiths entered Eastwood's brief into the *Congressional Record,* and then launched her successful 1970 campaign to force ERA from Celler's Judiciary Committee through a discharge petition. Also that spring, women from NOW began to disrupt the hearings on eighteen-year-old voting by Senator Birch Bayh's subcommittee of Judiciary on constitutional amendments by shouting demands for a hearing on the ERA. [41] Inside the Administration, women like Rita E. Hauser at the United Nations and Patricia Reilly Hitt in HEW began to pepper the White House with memos warning that Democrats like Griffiths and Bayh were beginning to take the lead on the ERA. Hauser and Hitt reminded Ehrlichman and Garment that the percentage of women voting Republican had declined from 51 percent in 1960 to 43 percent in 1968, and they urged Nixon to follow Eisenhower's lead and endorse ERA to the Congress. [42]

On April 10, Burns's assistant Charles Clapp, who handled the women's task force and monitored the issues surrounding it, warned Garment that an army of five hundred of "our most prominent women" was coming to Washington to demand action from the White House. It would include virtually all the Republican national committeewomen, state vice chairwomen, presidents of state federation clubs, prominent Negro and ethnic women, and all appointed and elected Republican women as well. Clapp said he had been told that the senior staff was preparing a statement pretty much confined to only one of the task force's recommendations. So Clapp thought it best to issue the statement and release the embargoed report *after* the women went home. "It obviously is a group with a great deal of clout and if we're not going to say too much it might be better to avoid saying it while they're all convened in one place," Clapp volunteered.[43] "That will avoid going into specifics and yet get them off our back, so to speak." But the women came and stayed on their back. Glaser supplied a copy of the task force report to Marie Anderson, editor of the women's page of the *Miami Herald,* which published it in tabloid form.[44] The bootlegged publishing of an embargoed report to the President on such a hot topic attracted unusual attention to the document, and left the White House with little practical recourse but to release the report.[45]

But the senior White House aides did not want to release the report without first getting from Nixon a clear position on the ERA. For Leonard Garment, whom the press often identified as the staff's resident liberal Republican, that clearly meant Nixon's timely *re*-endorsement. On May 25, Garment sent Ehrlichman for his signature a decision paper for the President, arguing the pros and cons of the ERA and urging an early presidential decision to release with the report.[46] The position paper first summarized the arguments in favor of endorsement, and did so pretty much as the women's task force had presented them. The arguments listed against the amendment were chiefly three. First, it was impossible to predict how the federal courts would interpret the amendment, and the increased litigation could flood the courts with novel constitutional challenges to a vast and intricate body of state statutes on family relationships, property and contract law, education and health and safety. Second, the ERA would almost certainly make unconstitutional the state protective laws. Garment added that the AFL-CIO had opposed the amendment earlier that month in testimony before Senator Bayh's subcommittee, arguing that losing such historic protection would lead to "wholesale exploitation of women workers."[47]

Third, the ERA, with all of its potentially radical and unpredictable implications, might not even be necessary. This was the position taken in 1963 by Kennedy's commission on women—that the courts should more actively discover the principle of equality that is implicit in the 5th and 14th amendments. Garment pointed out that federal courts and the EEOC had already reduced the effect of state protective laws, and federal district courts had recently held that state laws exempting women from jury service and discrimination against women applying to state colleges were unconstitutional under the 14th Amendment. The first federal court decision under Title VII against a state protective law had come in Los Angeles in 1968, and that same year a federal district court in South Carolina had struck down a private company's rule that the travel and physical duties of its commercial representatives limited such jobs to men only.[48] Garment's draft made no mention of the commonplace argument that the ERA would require the military conscription of young women and allow them into combat.

Indeed, Garment clearly favored the ERA and his decision paper was slanted in its favor. His summary of the views of leaders within the Administration, for instance, listed six persons as having gone on record in support of the ERA: the President himself (in 1960 and 1968), the Vice President (in 1968),

Mrs. Nixon (at least according to Connie Stuart at a May 4 press briefing), Secretary Shultz, Patricia Hitt in HEW, and Jacqueline Gutwillig, the Nixon-appointed chairwoman of the CACSW in testimony on May 6 before the Bayh subcommittee. Only Assistant Attorney General William Rehnquist, who as head of the Office of Legal Counsel was the Attorney General's lawyer, was listed in opposition. But Garment quoted Rehnquist in hyperbolic excess: the amendment should be opposed, Rehnquist had urged, because its "overall implication . . . is nothing less than the sharp reduction in importance of the family unit, with the eventual elimination of that unit by no means improbable."[49]

Garment's paper then considered three presidential options beyond re-endorsement, but discouraged all three. Direct opposition to the amendment was discouraged because such a reversal would require a strong argument that seemed politically unavailable. Endorsment of a modified version, such as the Hayden rider, might please most of organized labor, but the legal and political tides seemed to be running against this position, and the Senate Judiciary Committee had rejected it in 1964. Garment alluded briefly to the possibility of additional language that would specify more precisely the effects of the amendment. This approach would be explored in the congressional debate in the coming year. But by neglecting to explore it further, Garment's superficial treatment represented weak staff work and served the President poorly. Garment also discouraged a fourth possibility, which was taking no position pending further review. This "would be preferable to flat opposition," he wrote, but "as an expression of doubt" it would still represent a change in position, and thus only "postpone the date when a definite position would have to be taken, possibly creating some ill will in the interim."[50]

Despite these efforts by the Republican feminists and their allies within the higher councils, Nixon made no endorsement, or rather re-endorsement. Like Brer Fox, he lay low—while the ERA debate shifted in 1970 to the congressional committees and, thanks to Griffiths' triumph in August with her discharge petition, even to the floor of the House.[51] The nearness of the congressional elections of 1970 would preclude any real chance for ERA in the 91st Congress. But its momentum could carry over strongly into 1971. Meanwhile the White House was fastening on any positive action it could safely take to rally the aroused women's support. And once again, Secretary Shultz was prepared to take the initiative on behalf of civil rights. As he had done with the Philadelphia Plan, he would again revive an initiative that had died in the late hours of the Johnson administration.

A Philadelphia Plan for Women?

Shultz was fighting his major public battles over equal employment opportunity one at a time. First priority went to cracking the refractory construction unions with the Philadelphia Plan, a high-risk contest that was narrowly won only in the closing days of 1969. But meanwhile Shultz was preparing for Round Two on women. One of his early moves was to appoint Elizabeth Koontz to head the Women's Bureau. As a Democrat whose professional association with the NEA had kept her largely free of the Democrats' ancient wars over industrial labor standards and protective legislation, Koontz shared Shultz's view that the federal courts and the EEOC were rapidly rendering the question moot by trumping state protective laws with Title VII. During 1969 she was able at least to neutralize her bureau's opposition to the OFCC's proposed guidelines on sex discrimination that had been blocked in 1968.[52]

Meanwhile the EEOC further softened up the opposition with two initiatives in 1969. In March the commission published the results of its national survey on women in the workforce. This showed that women constituted 40 percent of white collar workers, but held only one of every ten management positions and only one of every seven professional positions.[53] Then, on August 19, the EEOC issued a guideline that formally asserted the primacy of Title VII over state protective laws. That same month, which typically was a dead one during the broiling dog days of semi-evacuated Washington, Libby Koontz quietly chaired the OFCC's panel in hearings on the same issues.

By early 1970, Shultz's victory with the Philadelphia Plan allowed him to extend its logic of proportional representation beyond construction to cover *all* areas of government contracting by issuing Order No. 4 on February 3. In the process the compensatory remedies that Labor had originally sought mainly for blacks in Philadelphia were officially and nationally extended to Spanish Surnamed Americans, American Indians, and Orientals. But what about women? The key to the Philadelphia Plan's de facto quota system for racial and ethnic minorities was geographic. The residences of these minorities were unevenly distributed by neighborhood and region, and where their concentration contrasted with their relative invisibility in the workforce, the arithmetic difference provided both evidence of unfairness and a ready target for a federally coerced remedy. Women, however, were evenly distributed everywhere. As the more radical liberationists phrased it, women co-habited with "the enemy." And unlike men, virtually all of whom were assumed as adults to be entering the labor force, the majority of women (60 percent in 1970) were *not* in the labor force.

Moreover, the Philadelphia Plan and Order No. 4 were based on a blue-collar model fashioned by the Labor Department that carried an implicit male bias. Unlike black and Hispanic males, women were not demonstrably clamoring for entry into the construction trades and steel mills—at least not in 1970.[54] The EEOC's burgeoning file of complaints from women mainly charged discrimination not over hiring and firing, but over promotion and equal pay and benefits.[55] The dominant voice of aroused feminism in 1970 came not from the Eleanor Roosevelt and Esther Peterson wing of the Democratic party, but rather from the upper-middle class and professional women of NOW and WEAL, from Republicans like Virginia Allan, Vera Glaser, CACSW chairwoman Jacqueline G. Gutwillig, and otherwise politically mainline GOP congresswomen like Florence Dwyer.[56]

Faced with these pressures, Shultz on June 2 provided his grateful President with a measured response to the rising clamor of his Republican women by signing and sending to the White House the OFCC's new guidelines on sex discrimination. Garment forwarded them to the Oval Office with the recommendation that their prompt release along with the hot-potato task force report could "put the President in the best possible light" by placing the emphasis on "what the President *has done* rather than on what he is being urged to do."[57] The new guidelines were a model of nondiscrimination. They demanded equal treatment for both sexes in recruitment and advertising, job opportunities and fringe benefits, pay and seniority, and physical facilities.[58] They required equal treatment of married and unmarried persons, and barred penalties for childbearing. They stipulated that "An employer must not deny a female employee the right to any job that she is qualified to perform in reliance upon a State 'protective' law."

Thus Secretary Shultz signed and the President approved a Magna Carta commanding equal treatment for women in the workforce. Compared with Title VII as enforced by the puny EEOC, the obscure OFCC through the dimly understood Title VI and the jumble of executive orders reached directly to a third of the U.S. workforce, and set the tone for the rest. Now the visible but weak EEOC and the muscular but obscure OFCC had joined in a bracing one-two punch for nondiscrimination by gender, including men. The hoary, crippling feud over protective legislation was essentially over. And the entire Republican package was consistent with the feminist steamroller that was gathering momentum for the ERA. There was no mention at all of the goals and timetables that had put the sharp teeth into Order No. 4. The OFCC guidelines' brief section on Affirmative Action was limited to urging extra efforts in recruitment, such as including women's colleges in itineraries of recruiting trips, and in training and management trainee programs.[59]

On June 9 the White House announced the guidelines and formally released the women's task force report, now embargoed for half a year (and widely available for the preceding two months), at a press conference that was presided over by Elizabeth Koontz. Glaser led the press correspondents in grilling Koontz closely on the reporters' most pressing concerns. These were not over hiring more women in construction or defense plants, but rather over ERA and equal opportunity for women in the professions, business, and government.[60] Why didn't the President use such an apt occasion for reaffirming his support for the ERA, Glaser asked, now that it was moving into a difficult contest in Congress? Had the President even read the report?, asked the redoubtable Sara McClendon, the dean of the women correspondents who had jousted so prominently with Kennedy and Johnson. Koontz replied awkwardly that the President was on record since 1960 and 1968 in support of the ERA, and White House aide Gerald Warren artfully volunteered that "the President's support for the position for equal rights for women has been stated and is continuing." "For pity's sake, Jerry," McClendon exclaimed, "you can't get away with that. The President said 'I will support it *when I get* to be President.' "[61] "It is not the Equal Rights for Women Amendment,"

Glaser interjected, "it is the Equal Rights Amendment which would do a great deal to help both sexes in this country." Warren replied lamely that "the [President's] support is unflagging. It is as he stated it and there are studies under way as to the implications, as Mrs. Koontz said."

Nixon thus rested most of his case for responding to the women's constituency on Schultz's sex discrimination guidelines, and stood pat on his prepresidential endorsements of ERA, content to allow Congress to absorb most of the controversy. There were occasional other gestures of lesser import, such as appointing Virginia H. Knauer as his special assistant for consumer affairs early in 1969, and later that year appointing Helen D. Bentley to head the Maritime Commission. In 1970 Nixon unobtrusively brought Barbara Franklin from the New York banking community to represent women's interests in the White House. But Franklin functioned with an informal portfolio and without the full title and office, much less the national conference, that the women's task force ranked so high in their long list of recommendations. When Secretary Finch in 1971 sent Nixon a brief report on his efforts to strengthen the role and authority of women at HEW, Nixon replied by noting on the margin that "This is an excellent job. However I seriously doubt if jobs in government for women make many votes from women."[62]

Beyond the modest symbolism of senior appointments there was the announcement, a month after the release of the OFCC's guidelines, that the Justice Department was filing its first pattern-or-practice suit on sex discrimination. The suit was filed in Toledo against Libbey-Owens-Ford and Local 9 of the AFL-CIO's United Glass and Ceramics Workers.[63] The Justice suit, however, implicitly raised again the question of symmetry in the race-sex analogy: if Justice was now belatedly adding sex to race discrimination as targets for pattern-or-practice suits, then why didn't Labor simply extend its racial protections to sex discrimination in its OFCC guidelines? That is, why didn't it extend the Philadelphia Plan to women? The NAACP, at its 61st annual convention in Cincinnati in July 1970, passed a series of resolutions expressing solidarity in the fight for equality between the sexes as well as between the races. But while the NAACP urged the OFCC to expand its new guidelines on sex discrimination to include goals and timetables, its six resolutions on "Equal Rights for Women" managed to make no mention at all of the ERA![64]

The NAACP resolutions reflected the same conflicts within the traditional liberal coalition that had neutralized the entire Leadership Conference on the question of ERA. By 1970, goals and timetables had become a standard item in the black agenda. But quotas for the affected class had never been prominent in the feminist agenda. To the contrary, radical feminism historically had meant radical egalitarianism—it had meant the ERA. But the logic of the race-sex equation, which *had* been prominent in the feminist agenda, would seem to require the extension of goals and timetables to women. The result of these pressures was the silent and invisible construction and almost clandestine release of a fascinating ream of bureaucratic boilerplate whose importance is inversely correlated with the obscurity of its title.

Revised Order No. 4: The Very Perfect Model
of a Modern Regulation

The relation between the obscure Revised Order No. 4 and the grail of the ERA captures one of the great ironies of the civil rights era. The half-century campaign for the Equal Rights Amendment climaxed in a congressional battle that was spectacularly won, only ultimately to fail of ratification in the states and die. Yet in the most crucial year of the ERA's congressional triumph, 1971, the federal bureaucracy secreted a virtually unnoticed and in many ways unfathomable regulation that arguably would have a more direct and powerful impact on women's employment in the decade to come than the crusade for the 26th Amendment on gender equality. The unlikely candidate for this honor was blessed with the magnificently bureaucratic rubric of "Revised Order No. 4."

In the arcane world of regulatory paper that drives the federal colossus, Revised Order No. 4 was issued in the dead hours, eluding all but the most cursory notice of even of the normally hawk-eyed *Wall Street Journal*. In the broadest social and political terms of national discourse, the congressional debate and approval of ERA and the fight for its ratification were indubitably more important than the muted rumble of the civil rights bureaucracy. But the careful obscurity of Revised Order No. 4 disguised an additive process that was linking to the original claimants of black civil rights an expanding series of new petitioning groups. By adding the female majority to the established racial and ethnic minorities, Revised Order No. 4 in effect defined almost three-quarters of the American population as "affected class" minorities whose distribution in the workforce should proportionately reflect their distribution in the population. The implications of such a zero-sum model of proportional representation would send Marco De Funis and Allan Bakke into the federal courts. But the celebrated reverse discrimination suits of the 1970s were directed by white males against quotas and set-asides for racial and ethnic minorities, not for women. This was appropriate, because the main body of feminist leaders was demanding uncompromising egalitarianism rather than compensatory or preferential discrimination. They were united behind the spirit of the ERA, which rejected gender preference. Thus the proportional model for women was a muddled affair that emerged from the compliance bureaucracy far less in response to women's demands than to its own internal pressures for consistency. Its promulgation in 1971 was an almost embarrassed gesture, whispered into the night wind.

The federal archives little reveal the inner decisions of 1971 that essentially completed the evolution of President Kennedy's race-centered executive order of 1961. The process occurred far from the White House. It produced a regulation drawn up by the OFCC that was approved by Secretary of Labor James Hodgson (Shultz having shifted to head the new OMB in the summer of 1971) and that resulted in the publication in the *Federal Register* of a dense notice of proposed rule on the last day of August, when the Labor Day holiday left Washington virtually depopulated.[65] Then on a *Saturday* in early

December, Revised Order No. 4 was officially promulgated to a non-observant nation.[66] The *Wall Street Journal,* normally ever-vigilant for intrusions on entrepreneurial freedom by federal bureaucrats, registered it with a yawn on the fourth column of page 30.[67] The bottom-line result was that the OFCC's Order No. 4, covering *all* contracts and subcontracts over $50,000 outside of construction, now included women as an affected class, *and* required of employers the same written affirmative action plans with detailed goals and timetables to remedy any deficiencies of underutilization at *all* levels of employment. Revised Order No. 4 was thus essentially the original Order No. 4 with women added as an affected class.[68] The ERA, as Glaser and her feminist colleagues tirelessly observed, would perforce apply egalitarian principles equally to men as well as to women. But the new OFCC regulation applied only to "females."

The Civil Rights Act of 1964 had commanded the equality of race and sex by requiring nondiscrimination. But the regulations of the enforcement agencies were concerned less with theoretical abstractions than with concrete deficiencies and remedies. They therefore required the counting of blacks and Hispanics and Orientals and American Indians and now females. The logic of compensatory theory pressed ineluctably toward such a conclusion, for the logic of the historical theory that drove the model demanded the inclusion of the nation's (and the world's) largest group of victims from past discrimination. Thus to deny women the remedies afforded the traditional minorities would violate both the logic and the politics of the race-sex equation. To rule otherwise would seem to acknowledge the theoretical vulnerabilities of compensatory theory and its proportional model in equal employment law.

Revised Order No. 4, however, converted the proportional model into something of an ambiguous abstraction. The Philadelphia Plan of 1969 had calculated discrete quotas from roughly verifiable shares of a potential and unbiased universe of a primarily male, blue-collar workforce in construction. Then in 1970 Order No. 4 had extended this beyond construction to all federal contractors. Revised Order No. 4 now vaguely defined underutilization as "fewer minorities or women than would reasonably be expected by their availability." The Philadelphia Plan had pioneered a measurable formula, based on local demographics, for generating the required numbers and had won approval in the federal courts. But what was the appropriate formula for women? No one knew—least of all the employers who were henceforth required to submit such numbers for the approval of compliance officers. Compliance officials in turn were empowered with wide discretion to enforce numerical requirements for women that they could not define in theory. But in practice the compliance officials could negotiate against a host of employers who were manifestly and historically guilty of discriminating quite massively against women *whatever* the model.[69]

Theoretically, then, Revised Order No. 4 was pregnant with possibilities for mischief as well as for beneficence. It endowed compliance officials in all federal agencies with almost plenary powers for determining what numbers placed contractors "in full compliance," a status that protected contractors from contract cancellation and from disbarment in future bidding. But in 1971 contract compliance was still overwhelmingly a racial question. The federal compliance authorities were now fully armed, and they were serious about moving promptly to rectify the egregious accumulation of employment discrimination by the nation's most visible employers. So they marshalled their new armamentarium against two giant targets. But the differences were more instructive than the surface similarities.

First, the Labor Department successfully attacked the Bethlehem Steel Company for maintaining a unit rather than a plant-wide seniority system, and thereby perpetuating past discrimination against black workers. The action concerned blacks whose transfer out of the traditional low-paying "black" jobs at Bethlehem's huge plant at Sparrows Point, Maryland, into more desirable units had stripped them of their accumulated seniority.[70] Bethlehem's vulnerability was similar to that of Duke Power in reflecting a segregated past. Newly armed with the sanction of *Griggs,* Labor threatened Bethlehem's contracts by waving Executive Order 11246 and Title VI as interpreted by the new OFCC regulations. This produced an out-of-court agreement that provided both retroactive and plantwide seniority for blacks, and also back pay for an affected class of victimized black workers whose identity could be precisely determined.

Then Labor teamed up with the Justice Department and the EEOC to take on the largest private employer in the country: AT&T. The EEOC had started the challenge by raising the novel objection, in an otherwise routine hearing for a telephone rate increase in Virginia, that the increase should not be granted because the telephone company discriminated against women. But the EEOC lacked the legal muscle to go up against such a formidable target, so Labor and Justice joined the effort. The government team sued all 24 operating companies and 700 establishments within the Bell system of AT&T, with sex discrimination as the leading complaint.

There was a crucial difference, however, between Bethlehem Steel and AT&T, one that highlighted the vulnerability of the race-sex analogy. The powerful battery of government attorneys could find no adequate formula for AT&T to determine *which* women were in the affected class. Unlike a specific steel plant in a measurable demographic environment in a border state with a Jim Crow past, AT&T featured a nationwide pattern of traditional sex segregation. Males constituted 98.6 percent of all AT&T craft workers on December 31, 1971, while 96.6 percent of the lower paid office and clerical employees were women.[71] Some specific plants, to be sure, were more vulnerable than others. AT&T's Michigan Bell company, for example, represented a fairly clear violation of the Equal Pay Act. There 500 women classified as "switchroom helper" performed the *same* craft jobs that elsewhere in the Bell system was classified at higher pay as the male craft job of "frameman." But the government was suing *all* of AT&T.

In response, AT&T's lawyers challenged the government to find a formula that could accurately identify the affected class of female employees throughout the AT&T system who continued to suffer the effects of past discrimination. The government simply could not do so.[72] The government's general case charging wholesale sex discrimination was indeed overwhelming. But its specific case, beyond such blatant violations of equal pay as the double standard at Michigan Bell, was unprovable by the population-based formulas that had worked with blacks in generating remedies through numerical goals based on local and regional demographics. On the other hand, the government held in ready reserve the broad-gauged new weapon of *Griggs*. As a result the parties settled out of court in a consent decree that awarded back pay to 13,000 women and 2,000 minority men, and committed AT&T to establishing targets for hiring and promotion to increase the numbers of women and minorities in each job classification (including goals for *males* in the previously all-female job category of telephone operator). AT&T admitted to no guilt in the consent decree, but agreed to a record-breaking settlement. On the other hand the government, while unable to prove its major case by identifying the affected class of women victims for whom remedy was sought, won a landmark multi-million dollar settlement. It acknowledged the expansive principle of a compensatory theory that provided remedies for employment practices that were ostensibly neutral on their face, but that nevertheless had the effect of perpetuating past discrimination.

Passing the Equal Rights Amendment

The AT&T settlement came in 1973. It was delayed in part by a suit filed in federal court by the Communications Workers of America, who claimed a right to be included as a plaintiff in the settlement. The CWA charged that its union bargaining rights were violated by the consent decree, which discriminated against white males and would force the union to discriminate against them also. The court eventually rejected all CWA claims except, ironically, its right to bargain over maternity leave.[73] But the union's counterclaim illustrates the resistance in labor ranks that disassociated most of organized labor from the elite battalions of ERA feminists. The feminists' upper-middle-class perspective on the potential benefits of Title VI and OFCC regulations was captured by Glaser, when she sharply questioned Koontz at her press conference releasing the new sex discrimination guidelines in June 1970. When Koontz described the EEOC's backlog of 600 complaints from women over discriminatory seniority lines and job classifications and fringe benefits, Glaser shifted the focus from blue-collar to professional grievances. What did the government intended to do, Glaser asked Koontz, about the more than 100 complaints "filed against colleges like Harvard, the University of Michigan and the like." "These colleges have more than $3 billion in Federal grants," Glaser said. "Now, is the Department of Labor going to pursue this and will those contacts be cut off?"[74]

Koontz had replied evasively. The Civil Rights Act of 1964 had exempted educational institutions from EEOC scrutiny under Title VII, Koontz explained. She did not elaborate further, but the Labor Department had little experience dealing with educational institutions under Title VI—not even with blacks, much less with women. Feminist anger however, was focused on sex discrimination in business and the professions, including the academic professions. In the early 1970s feminists rallied around the symbolic figure of Bernice Sandler, a member of WEAL and part-time instructor in counseling at the University of Maryland's College Park campus, who had been denied a regular faculty appointment.[75] Like the historic alliance of blacks and Jews in the NAACP, the feminists demanding the ERA wanted mainly to smash quotas, not erect counterquotas.

The women's task force had accumulated evidence of "gross discrimination" against women in education, especially in the higher levels of administration and the professions. In the public schools, 75 percent of elementary school principals were men, as were 96 percent of the junior high principals and 90 percent of the senior high principals. Only 5.9 percent of law students and 8.3 percent of medical students were women. Indeed, for several decades professional women in America had been going *backward*. In 1930 women received 40 percent of all graduate degrees, but by 1966 this had fallen to 34 percent. Fifteen percent of all medical doctorates went to women in 1930. But by 1966 this was down to 12 percent.[76] The task force's recommendations for changes in the basic civil rights law and enforcement practices were quite comprehensive. But all paled before the symbolic egalitarian power of the Equal Rights Amendment.

The dramatic story of congressional approval of the ERA has not been fully told. We have far better studies on how and why ERA failed in ratification than on why it emerged in its final form and how it was approved by Congress.[77] Yet the questions of congressional approval and ratification are closely linked. So committed were ERA partisans to the purity and simplicity of its classically antidiscriminatory text and soul—"Equality of rights under the law shall not be denied by the United States or by any State on account of sex"—that they rejected a range of compromise that, in hindsight's cruel and cheating light of what-might-have-been, may have avoided the ERA's ultimate defeat.

In the fall and winter of 1970 the feminist organizations unanimously rejected the bipartisan compromise suggested by the amendment's Senate co-sponsors, Bayh and Republican Marlow W. Cook of Kentucky, with the weighty support of senators Dole, Griffin, Javits, and Kennedy, to add "sex" to the equal protection clause of the 14th Amendment.[78] When, in January 1971, Martha Griffiths reintroduced the ERA into the new 92nd Congress as H.J. Res. 208, hearings were scheduled for late March and early April by Judiciary Subcommittee No. 4, chaired by ERA-supporter Don Edwards of California. Only two of the subcommittee's eighteen main witnesses opposed the ERA resolution. One was Sam Ervin, who continued to propose amendments exempting women from military conscription and permitting state protective legislation and privacy laws (Griffiths contemptuously referred to the latter as Ervin's "potty argument"). The other was Mary Dublin Keyserling, testifying for the National Consumers League and the National Council of

Catholic Women.[79] The formidable array testifying in favor of ERA included Griffiths, Virginia Allan, Marguerite Rawalt, Aileen Hernandez, Bernice Sandler, Olga M. Madar of the UAW, Norman Dorsen of the ACLU, and Thomas I. Emerson of the Yale Law School.[80]

The surprise witness of the hearings, however, was William Rehnquist. Testifying, on April 1, for the Justice Department, Rehnquist began by affirming, somewhat obliquely, that "President Nixon and this Administration support the goal of establishing equal rights for women."[81] *But,* he added, "opponents of that amendment have raised significant questions which deserve the serious consideration of the Congress." Rehnquist then read a 29-page statement that argued against the need for or the desirability of a constitutional amendment. Instead he discussed the relative merits of H.R. 916, a bill sponsored by Democrat Albert J. Mikva of Chicago and reported by the House Education and Labor Committee.

The ERA was problematical, Rehnquist said, because its ambiguity would require protracted litigation to dispel, as it had with the 14th Amendment. Politically, Rhenquist conceded, the post-Civil War Congress had no realistic alternative to the constitutional amendment process. But the legislative history to date on the ERA already demonstrated broad disagreement on the meaning of its sweeping language even among its supporters. To this ambiguous record would then be added the complex ratifying debates of three-fourths of the states. Moreover, Rehnquist said, recent decisions in the lower federal courts indicated a new willingness to extend 14th Amendment protections to the field of sex discrimination. Furthermore, many of the inequalities suffered by women occurred in the area of private rather than public conduct, and could therefore only be reached by statute. Such a statute, Rehnquist said, was H.R. 916, which responded in detail to the recommendations of the presidential task force on women.

Rehnquist then analyzed the fourteen sections of H.R. 916. They would basically amend the Civil Rights Act of 1964 by adding protections against sex discrimination in most of the titles beyond VII, which of course already contained it. H.R. 916 would also similarly amend the Open Housing Act of 1968 and the Fair Labor Standards Act of 1938. Rehnquist supported the basic statutory approach and many of the specific provisions of H.R. 916. He opposed, however, adding sex discrimination to the Civil Rights Commission's investigatory charge, urging it instead for the Women's Bureau, newly armed with subpoena power. He also opposed simply adding sex to the categories protected by Title VI in all federally assisted programs. Such a blanket approach could have a confusing impact on the Administration's complex new proposals for revenue-sharing and family assistance, he said, and the complications should first be carefully explored by the mission agencies, especially HEW and Treasury.[82] The same caution applied to amending the 1968 fair housing law, Rehnquist said, where HUD had its hands full exploring applications to race.

Rehnquist also disagreed with the women's task force on amending the FLSA to include protection for women in executive, administrative, and professional employment. He supported instead the previous year's testimony by Elizabeth Koontz that Title VII already provided such employees with sufficient protection against sex discrimination. Finally, Rehnquist agreed that it was time to include state and local governments and educational institutions under Civil Rights Act coverage, and also to beef up the enforcement powers of the EEOC. But the Administration was willing to support this, he insisted, *only* if the EEOC were given prosecutorial authority, *not* the cease-and-desist powers provided in H.R. 916.

The feminist leaders were astounded by this demonstration of the Nixon administration's "support" for the ERA. Jacqueline Gutwillig scrawled across the top of Rehnquist's written statement: "This is incredible—to both proponents and opponents. That as fine a lawyer as Mr. Rehnquist could give such testimony is most curious. With friends like this we don't need enemies."[83] She sent her editorialized copy of Rehnquist's testimony to President Nixon the next morning, politely suggesting how puzzled her CACSW colleagues were by such ambiguous support and internally contradictory testimony. But Gutwillig needn't have worried. Despite Ervin's Spanish proverb that "An ounce of mother is worth a pound of priest," and Celler's insistence that "the Fallopian tube has not become vestigal," and also despite the more serious problem of Ervin's waving a 1971 Roper Poll showing wide disagreement among American women over the meaning and desirability of the ERA, the feminist drive in Congress was superbly organized and unstoppable.[84] Celler's Judiciary Committee reported the ERA resolution out after voting 19 to 16 to attach a Hayden-like rider, sponsored by Republican Charles E. Wiggins of California, that exempted military conscription and state protective laws. The full House, however, stripped the Wiggins amendment away by a resounding vote of 265 to 87, and on October 12 sent the ERA resolution to the Senate by the extraordinary margin of 354 to 23.[85]

Just a month after Rehnquist testified, the Supreme Court for the first time struck down a state's gender classification under the 14th Amendment. In *Reed v. Reed,* the unanimous Court overturned an Idaho statute preferring a man over a woman as executor of an intestate estate.[86] This seemed to support, at long last, the arguments of the special-protection Democrats, and also Rehnquist's caution, that the federal courts might obviate the need for ERA by discovering women in the 5th and 14th amendments. Griffiths, however, dismissed the ruling as merely a narrow technicality that would leave the Court unchanged as a "bottleneck" for women's rights. She told the *Washington Post* that "They're just nine old idiots."[87]

Griffiths and her allies sensed imminent victory in Congress, and they were right. On February 29, 1972, the Senate Judiciary Committee, a former bastion of conservatism, steamrolled over Ervin's lonely objections by a vote of 15 to 1.[88] The symbolism of feminist triumph was rich, for in 1963, Marguerite Rawalt had stood alone on President Kennedy's status-of-women commission in calling for the ERA. In mid-March, the full Senate easily rejected all nine of Ervin's qualifying amendments, and on March 22 voted to send the ERA to the states for ratification with an astounding tally of 84 to

8.[89] Only four days earlier, President Nixon had finally weighed in, timidly and late, by sending Hugh Scott a letter reminding the Senate minority leader, not very credibly, that "as a Senator in 1951 I co-sponsored a Resolution incorporating the original amendment," and "in July of 1968 I reaffirmed my support for it as a candidate for the Presidency." "Throughout twenty-one years," Nixon's letter continued, "I have not altered my belief that equal rights for women warrant a Constitutional guarantee—and I therefore continue to favor the enactment of the Constitutional Amendment to achieve this goal."[90] Nixon's expedient timing seemed to confirm the insincerity of his professed conviction—his State of the Union address of the previous January had managed to devote three paragraphs to "Equal Rights for Women" without mentioning the Equal Rights Amendment at all.[91]

The rest, as journalists and politicians (but not historians) say, is history. By the end of 1972, twenty-three states had rushed through ratifications of ERA, only fifteen short of the required thirty-eight. The Supreme Court rather quickly followed *Reed* with a series of decisions that stopped just short of categorizing women, like blacks, as a class whose suspect classification in statutes merited "strict judicial scrutiny." Instead the Court fashioned a middle range of judicial scrutiny to test classifications by sex.[92] Subsequent constitutional litigation over gender distinctions is filled with efforts to define the dimensions of the Court's middle-way scheme.[93] But it is not filled with clarifying litigation to define the precise meaning of the Constitution's new Equal Rights Amendment.

In 1973, the Supreme Court's abortion decision of *Roe v. Wade* galvanized an anti-ERA coalition of conservatives and traditionalist women led by Phyllis Schlafly, and the resulting stalemate ultimately doomed the feminists' congressional triumph of 1972.[94] In the euphoria of victory in 1971–72, however, feminists turned their energies and momentum to the parallel battle over revising the Civil Rights Act of 1964 and related laws in order to end discrimination against women. By adding their formidable new weight to the civil rights coalition during a presidential election year, feminists promised to tip the balance of power decisively against the conservative coalition in Congress. But the dominant feminist commitment to egalitarian principles also promised to complicate the politics of the civil rights coalition, reinforcing class-based distinctions that the race-sex analogy had blurred.

CHAPTER XVII

Race, Sex, and Civil Rights
Enforcement: Culmination 1972

Confronting the Johnson Legacy in Civil Rights Enforcement

In January 1969, senior officials in the departing Johnson administration made two gestures that between them captured the Democrats' challenge to the Nixon administration in civil rights policy. On January 15 the outgoing Attorney General, Ramsey Clark, sent the Vice President and the Speaker of the House a bill entitled the "Equal Employment Opportunity Enforcement Act." Clark's "Enforcement" bill was familiar. Its recent origins traced back at least to the recommendations of the Kerner Commission, and it had been requested by President Johnson in his civil rights message to Congress of January 24, 1967. Its more distant origins reached back to the Truman proposal of 1948 and the House-passed FEPC bill of 1950.

The enforcement bill's key provision was cease-and-desist authority for the EEOC. Among liberal Democrats this demand had become a test of seriousness in civil rights. It had become the heart of the Johnson administration's Senate bill in 1967 and 1968, which in May 1968 had cleared the Senate Labor and Public Welfare Committee but then died with the expiration of the 90th Congress.[1] It had reached the Senate floor late in the session, with no companion bill realistically available in the House, and it had attracted Senator Dirksen's implacable opposition and threat of filibuster. But the partisan and ideological meaning of Clark's lame-duck submission of 1969 was clear: the Democratic-controlled 91st Congress would move immediately to confront Nixon with a comprehensive civil rights bill designed to round out the liberal agenda that had been half compromised away in order to pass the Civil Rights Act of 1964. This would be an act of completion and fulfillment. It pledged a liberal keeping of the faith, especially with black America, now that the ghetto riots had demonstrated the nationwide scope of racial discrimination, and the unpopular war in Vietnam was devouring disproportionate numbers of the disadvantaged.

Substantively the Democrats' enforcement bill would seek a huge expansion in Title VII coverage and cease-and-desist authority for the EEOC. This would include ending Title VII's exemptions for 10.1 million employees of state and local governments and also for 4.3 million employees of educational institutions. Politically, these structural reforms would capitalize on a growing consensus, at least outside of the most obdurate conservative and business quarters, that the "poor, enfeebled" EEOC lacked the tools necessary to achieve the broad purposes that Congress had ordained in 1964. In an increasingly unstable economy, state and local government was the nation's fastest growing sector of employment. Yet as the newspaper stories and television newscasts continued to attest, the lily-white ranks of the state troopers in black belt states like Alabama and Mississippi continued to hurl defiance in the feeble teeth of the EEOC and the Justice Department as well. Moreover a growing public awareness of discrimination against women by educational institutions, especially the business and professional schools and the colleges and universities, would fuel congressional demand from the growing feminist ranks for a legislative culmination to complete the revolution of 1964.

Clark's lame-duck submission to Congress in January 1969 was paralleled by a recommendation from one of Califano's confidential task forces. On January 2, James Gaither placed in the transition portfolio a recommendation supported by the departments of Justice and Labor, and also by the EEOC, that the OFCC's contract compliance function be shifted from Labor to the EEOC.[2] Secretary Wirtz's unsuccessful initiatives in Philadelphia had angered Labor's trade-union constituency, and within the department the OFCC remained an isolated and alien presence. Moreover, Ramsey Clark's 1968 task force for civil rights was troubled by the government's confused and overlapping structure of EEO enforcement.[3] Thus the organizational issue was closely linked to the enforcement issue. The Democratic leadership of Congress would exploit its head start on the Nixon administration. It would take up the bill Clark had sent over and press it forward in committee hearings, principally in the form of bills sponsored by Harrison Williams in the Senate (S. 2453) and Augustus Hawkins in the House (H.R. 6228). The Nixon White House in fashioning their own bill would be forced to react to the Democrats' initiative, and to do so within a pre-emptive political context in which Democrats would equate lesser measures with opposition to sincere civil rights enforcement. On February 13 Nixon asked Attorney General Mitchell to propose a position for the Administration. His request triggered a policy debate between assistant attorneys general Leonard and Rehnquist over basic ends and means in civil rights policy.

The Debate of the Assistant Attorneys General: Leonard Versus Rehnquist

Jerris Leonard's Civil Rights Division (CRD) reflected the ties of continuity between the career attorneys in Justice who had cut their teeth fighting southern racism, and the Democratic liberals in Congress and the Johnson administration who had worked so closely with the Leadership Conference. Leonard shared his division's consensus that the problem of employment discrimination was simply not being met by the jerry-built structures produced by the great compromise of 1964. Leonard's policy recommendations, which reached Ehrlichman in the White House through Deputy Attorney General Kleindienst, included early reports that he had met with leaders of the Leadership Conference and they had lost confidence in the government's EEO program.[4] The Civil Rights Commission was preparing for summer publication a report documenting widespread discrimination by state and local governments, and the commission would recommend cease-and-desist power for the EEOC.[5] Additionally, the Justice Department's pattern-or-practice suit against the "merit system" in Alabama's all-white highway patrol had documented the persistence of segregationist policies. But the CRD's limited resources precluded lawsuits on a volume basis, Leonard said, and that left most of the burden on the EEOC.

The EEOC, however, was so hobbled by its statutory confinement to retail complaint-processing that the result was a widespread lack of enforcement. During FY 1968 the commission had received 15,000 complaints, but had successfully conciliated only 513 cases. Meanwhile its complaint backlog exceeded 30,000, with processing time averaging 18 months and growing longer. During the same year fewer than 100 private lawsuits had been filed, and the Attorney General had filed only 22 pattern-or-practice suits (only six of which had been referred by the EEOC). Indeed, in four years under Title VII, private parties had won only four cases in contested lawsuits without the Attorney General intervening as a party, and in three of those four the Justice Department had filed an amicus brief. The private enforcement of the private rights recognized in 1964 had brought weak enforcement. Granting cease-and-desist authority to the EEOC, Leonard said, would convert nondiscrimination in the workforce into a public right, and shift the burden of enforcement to public authorities.

Ironically, Leonard added, in employer eyes this weak enforcement profile translated not into satisfaction but rather into charges of government harassment. These employer charges were "understandable," Leonard said, because the government's enforcement programs were duplicative, overlapping, uncoordinated, and inefficient. The majority of industrial and construction firms and unions were covered both by Title VII and Executive Order 11246. This made them subject to enforcement by both the EEOC *and* the OFCC and *by each contracting agency as well,* where independent EEO offices lacked uniform policies and standards. Employers in the private sector complained of receiving uncoordinated inspection visits from compliance officials brandishing the OFCC's regulations under Title VI and Executive Order 11246, and also from the EEOC with its Title VII guidelines. The employers were confused and upset.

In the public sector, local agencies in the states and cities were exempt from the better known strictures of Title VII and the EEOC, Leonard continued. But they were nonetheless subjected to a bewildering variety of sanctions through the indirect chain of federal contracts and grants. The circuitous reach of Title VI mirrored the government's sprawling and incoherent network. For example, the Secretary of Transportation had independent statutory authority of his own. The Pentagon enforced nondiscrimination in the states' National Guard units. The office of civil rights in HEW monitored state and local welfare agencies; a similar office in OEO policed the antipoverty agencies; HUD monitored the cities' EEO programs in urban renewal and public housing. The SBA did the same thing for small businesses, the Labor Department for state employment services and apprenticeship programs, and so forth. As for nondiscrimination in federal employment, which was exempt from Title VII, Civil Service regulations assigned to each federal agency the primary responsibility for its own employment practices. And above all, *no* single federal officer or office had the responsibility for coordinating all this.[6]

What should be done? Leonard believed that the basic approach of the Williams-Hawkins bills was "on the whole satisfactory."[7] It would grant the EEOC cease-and-desist authority, based on the NLRB model. Thirty-two of the 38 state EEO agencies possessed such authority, and it seemed to have worked well for them. The Democrats' bill would also transfer the OFCC's contract compliance authority to the EEOC. By thus adding Title VI's admittedly "drastic" remedies of contract cancellation and debarment to the EEOC's armament, Leonard argued, the EEOC would greatly enhance its leverage in conciliation. Thus mere possession of the big stick should maximize efficiency while minimizing its use—as it had done with the state FEPCs.

Leonard proposed, however, that the administration submit its own bill. It would differ from the Democrats' bill by adding three safeguards to reinforce presidential authority over a potentially overzealous regulatory agency. First, nondiscrimination in state and local governments would be enforced by the Attorney General, not the EEOC, using the Justice Department's pattern-or-practice suits. Second, the Justice Department "would control all litigation to which the [EEOC] is a party or in which it has an interest."[8] Third, the EEOC should have a separate general counsel, appointed by the President, with exclusive authority to issue complaints and to prosecute cases. This was the device that Senator Taft had forced on the NLRB in the Taft-Hartley Act of 1948, in order to separate structurally the prosecutor and judge functions within the regulatory agency.

Leonard agreed that under its present obligations the EEOC "plays the role of advocate in holding public hearings and educational efforts." But the commission also "has the prosecutor's function of determining whether there is 'probable cause' after investigations made by its staff." It was therefore "undesirable to vest in a Commission, which has an obligation to act as prosecutor, the quasi-judicial function of determining the merits" of whether employers had violated their nondiscrimination obligations. Thus an independent general counsel would "meet the criticism of those who allege that both the present EEOC and OFCC have attempted to be both judge and prosecutor." Leonard was referring to Dirksen's objections, which continued the tradition of Taft. An independent counsel patterned after the Taft-Hartley precedent would thereby "allow [the] EEOC to address itself to the merits," Leonard concluded, "without having the obligation to 'back up' its staff, as in the present structure, and thus allows for free exercise of unbiased judicial determination."

It was unclear from Leonard's argument, however, just what the EEOC's independent counsel would be independent *of*. Was it to free the commissioners from a prosecutorial function and mentality so they could make unbiased judicial determinations? If so, what then would become of the commission's advocacy role? The Dirksen compromise had created the EEOC more in the image of the Civil Rights Commission than the NLRB. The young EEOC had evolved accordingly, especially under Chairman Alexander. Deprived at the outset of its prosecutorial role, the commission by 1969 had concentrated its public energies on advocacy for the affected classes, holding hearings throughout the country to generate pressure and publicize the malfeasance of discriminatory employers. It seemed unlikely that an independent general counsel would fundamentally affect that role definition and adversary momentum. Leonard did not elaborate. But William Rehnquist did.

If Jerris Leonard's Lincolnian brand of Republicanism coincided with the liberal consensus among the CRD's career attorneys, William Rehnquist's southwestern brand of Republican conservatism reflected the party's major growth constituency in the southern and western suburbs. As head of the Office of Legal Counsel, Rehnquist was essentially the chief lawyer for Attorney General John Mitchell. As Nixon's 1968 campaign manager, Mitchell was sensitive to the political dimensions of the party's future. Rehnquist had been appointed to his Justice post on the recommendation of Kleindienst, who had worked with Rehnquist in Arizona to nominate and elect Barry Goldwater in 1964, and who had directed field operations for the Nixon campaign in 1968. In March 1969, Rehnquist sent Kleindienst an advisory opinion on civil rights enforcement. He began by agreeing with Leonard on several criticisms of the structure and EEO procedures they had inherited from the Johnson administration.[9]

Litigation under Title VII had been extremely slow, Rehnquist conceded. Despite a trickle of plaintiff victories, the Justice Department's small number of pattern-or-practice suits were tied up in a "logjam" in federal district trial courts and subsequent appeals. Furthermore, private suits had proven such a drain on resources, Rehnquist said, that even with NAACP financing they were clearly not "really a feasible remedy for an *individual*." Many federal courts had been sympathetic or open-minded with respect to Title VII litigation, Rehnquist acknowledged, but several in the Deep South had made it "extremely difficult for plaintiffs to prevail." Title VII was such a specialized field of law that its administration would be more just and efficient if handled by an agency with expertise, rather than "having the law made piecemeal by hundreds of district judges." The EEOC had been criticized because "[i]ts present function of determining reasonable cause calls for giving the benefit of the doubt to the charging party," Rehnquist acknowledged. But perhaps "it cannot be expected to act judicially," he added, "until it performs judicial functions."

On the other hand, Rehnquist observed, there were even stronger reasons for opposing such a strengthening of EEOC authority. At the heart of Rehnquist's negative verdict lay the standard Republican objection to New Dealish regulatory commissions: "Administrative agencies are inferior to courts as finders of fact. They lack objectivity and tend to favor one or another of the groups whose interests are protected by their statute." "This would be particularly true," Rehnquist added, "of the EEOC." Although in most cases an individual plaintiff-employee was at a disadvantage in suing a defendant-employer, he acknowledged, this was offset by the Attorney General's authority to bring pattern-or-practice suits. Rehnquist noted the recent criticisms by senators Dirksen and Fannin of the lack of coordination between the OFCC and EEOC, and more pointedly their charge that "both agencies have gone beyond the intent of the law by encouraging racial quotas."

On March 5 Rehnquist met with Leonard in the office of John Dean, who headed the Justice Department's legislative reference section under Kleindienst, to seek areas of agreement and to clarify disagreements. Leonard and Rehnquist agreed that transferring the OFCC's contract compliance functions to the EEOC would streamline the confused and inefficient EEO process. Such a linear, rationalized process would logically make Title VII, with its detailed definitions and protections, the exclusive federal remedy for employment discrimination. Further, both men shared the Republican conviction that the EEOC, like the NLRB, should have an independent, presidentially appointed general counsel. Finally, they agreed that the NLRB parallel also required that were the EEOC to gain cease-and-desist power, its orders should be enforced through federal *appeals* courts, not the district trial courts. For losers in a hearing decision by a regulatory board to seek vindication through the district courts in a trial *de novo,* they agreed, defeated the main purpose of a quasi-judicial regulatory body, which was to handle most disputes outside the courts through administrative law. But here their agreement ended.[10]

On the central question of cease-and-desist authority for the EEOC, Rehnquist dissented vigorously. He objected that the EEOC was biased in favor of the employee-complainant, and he disliked the agency's development of the group-based "rightful place" doctrine. By this Rehnquist meant the compensatory theory of EEO law, and Dean summarized it as the doctrine that hiring or promotion practices should favor the groups that have suffered from long-standing practices of discrimination. Rehnquist also "objected on political grounds," Dean reported to Kleindienst, that giving the Attorney General (much less the suspect EEOC) authority to sue state and local governments over their employment practices would unacceptably interfere with established patterns of federalism that the Republican party had historically defended. Rehnquist closed his case by observing that he saw no reason for supporting stronger civil rights legislation since Nixon had not been supported by the people the bill would help.

The Republican Judicial Strategy Reaffirmed

At the White House, in response to Nixon's circulation on March 28 of a full range of proposals with request for comment, Counselor Burns as the senior aide for domestic policy weighed in strongly on Rehnquist's side. The domestic agenda he had sent the President in January had recommended no new civil rights legislation, and in it Burns had warned Nixon of the likelihood of Democratic proposals in the 91st Congress to endow the EEOC with NLRB enforcement powers—under which "the Commission would operate as investigator, prosecutor, judge, and jury."[11] When, in April 1969, the White House scrambled to pull together some sort of domestic policy, Burns voiced the opposition of the business establishment to cease-and-desist authority for the EEOC. "The words cease-and-desist and N.L.R.B. are inflammatory words to most businessmen," Burns said. "They find the N.L.R.B. and its activities among the worst in the federal government and in many instances, they are absolutely right in this evaluation."[12] He had a better, Republican remedy: "A proposal that was first advanced by then Congressman Charles Goodell last year and subsequently picked up and refined by Senator Dirksen," Burns said, "would equip the E.E.O.C. with the right to go into federal district court to enforce the law in the event its conciliation and mediation procedures fail. This procedure would guarantee a fair trial in a court of law with all the rules of evidence applying."

The federal *district* courts, Burns had said, not the appeals courts, as in the NLRB model, where the courts reviewed the record forwarded by the regulatory board and customarily granted great deference to the expertise of the regulatory agency. A non-jury trial *de novo* in the district courts would provide all of the formidable protections of the adversary system to the battery of lawyers that substantial employers could bring to the contest. But even Rehnquist challenged the wisdom and efficiency of following a regulatory commission hearing and ruling with a full new trial in federal district court. None other than Senator Dirksen, however, father of the grand compromise of 1964, had approved this as the remedy most consistent with the Republican philosophy in the Congress. For Burns, the political judgment was clear: "The worst possible result would be to create another N.L.R.B. in this very emotional and complicated area," he warned. "If the plan to furnish the E.E.O.C. with cease-and-desist power is adopted, this is exactly what would happen." Burns also regarded Leonard's proposal to give the Attorney General pattern-or-practice authority to sue state and local governments as "a massive federal intrusion into local affairs" that would take "a long step in the direction of destroying our present federal system." Burns predicted, not very opaquely, that such a move would prompt "massive resistance from certain areas of the country."[13]

By early August the Administration's working group under Ehrlichman's direction had hammered out its proposal.[14] At its core was the Republicans' vintage judicial strategy of maximizing the role of adversary proceedings in court so as to minimize the judgmental discretion of New Dealish regulatory agencies. Thus in the Administration's bill the EEOC would seek remedies not through cease-and-desist authority, but rather by bringing suits in federal district courts. The EEOC's new prosecutorial authority, moreover, would be tightly circumscribed. The commission could sue only if its attempts at mediation and voluntary compliance failed. A recommendation to sue would have to be made not by the commissioners but by an independent general counsel appointed by the President. The suit would be brought by EEOC trial lawyers. But should the EEOC wish to appeal a district court decision, it would "forward this recommendation to the Justice Department for review and approval." Further, the OFCC would remain in the Labor Department, where it would continue to handle affirmative action programs under Title VI. By the summer of 1969, Shultz had committed himself to the Philadelphia Plan, with a major enforcement role envisioned for Labor through the OFCC. Also unchanged would be continued jurisdiction by the Civil Service Commission over the EEO program for federal employees. Finally, unlike the Democratic proposals, the Administration bill would not extend coverage to the employees of state and local governments and educational institutions.

The Cease-and-Desist Model for the EEOC

At the hearings held by the labor subcommittee of the Senate Committee on Labor and Public Welfare during August and September 1969, subcommittee chairman Harrison Williams introduced the bipartisan Williams-Javits bill. Designated S. 2453, it was co-sponsored by senators Hart, Kennedy, Scott, and thirty other senators, most of them liberal Democrats. The bill not only would grant cease-and-desist authority to the EEOC but, after three years, it would also transfer to the EEOC the Attorney General's authority to file pattern-or-practice suits.[15] Coverage under Title VII would be broadened in four ways. First, a phased-in expansion would newly cover employers or unions with *eight* or more employees or members, as opposed to the current law's lower limit of twenty-five. This would extend coverage to an estimated additional 9.5 million employees and 90,000 employers. Second, coverage would be extended to an additional 10.1 million employees working in 81,000 units of state and local government. Third, approximately 4.3 million employees of 130,000 educational institutions, mostly teachers, would be newly included. Fourth, EEO responsibility for the federal government's 2.5 million civilian workers would be shifted from the Civil Service Commission to the EEOC.[16]

Although Congress would pass no enforcement bill in the 91st Congress, the debates of August and September 1969 previewed the range of dispute that led toward the Equal Employment Opportunity Act of 1972. The coalition behind the reformist push of northern Democrats and liberal Republicans was led by the familiar civil rights organizations and figures of the 1960s, including Clarence Mitchell, Jack Greenberg, Joe Rauh. The line-up of lobbyists and organizations reinforced the customary notion that civil rights meant black civil rights. They were supported on the hearing panel by the kind of liberal senators in both parties who were elected from urban industrial states, and who had sought assignment to a committee with a liberal constituency like Labor and Public Welfare. The leading Senate supporters were Williams of New Jersey, Javits of New York, Gaylord Nelson of Wisconsin, Walter Mondale of Minnesota, Alan Cranston of California, and Richard Schweiker of Pennsylvania.

A prominent new element, however, was provided by representatives from women's organizations. Testimony and supporting statements for the committee's liberal bill came from the National Federation of Business and Professional Women's Clubs, the League of Women Voters, the United Universalist Women's Federation, WEAL, and the General Federation of Women's Clubs. With the black civil rights organizations still smarting from the poor public relations environment produced by black nationalist rhetoric and by ghetto burning and looting, the political value of the new bipartisan feminist push was incalculable.

At the kickoff hearings in August 1969, *no* spokesmen for civil rights organizations supported the Nixon administration's bill, S. 2806. Introduced by Senator Winston L. Prouty of Vermont, the Administration's bill also drew no support from the women's groups. On the other hand, testimony against the need for any additional enforcement authority at all was provided by the Southern States Industrial Council, the Associated General Contractors of America, and the National Association of Manufacturers. By bracketing the Administration on the right, these groups left Nixon's court-enforcement approach in what Ehrlichman hoped would be viewed as the "sweet and reasonable middle." On the EEOC itself, however, three of the five sitting commissioners testified *against* the President's proposal. Together they represented the three main affected classes: former chairman Clifford Alexander was black, commissioners Vincent Ximines was Hispanic, and Elizabeth Kuck was female. A fourth commissioner, Luther Holcomb, was a white male southerner. The White House elected not to bring Holcomb's soft Texas drawl to the hearing room to defend the Nixon proposal, so Holcomb merely submitted a brief written statement preferring court enforcement to cease-and-desist authority.[17]

This pattern of opposition left EEOC chairman William H. Brown III a heavy burden of support. But the weight of credibility placed on Brown's testimony was doubly burdensome because Brown had used the White House press conference of May 6, which had announced his presidential appointment as the commission's new chairman, to demand cease-and-desist authority for the EEOC.[18] The minutes of Brown's first meeting as chairman, on May 12, 1969, record that "The chairman expressed his strong opposition to the district court review, the possibility of *de novo* hearings and the independent counsel."[19] But, Brown by August, had been brought around to the Administration's position by Ehrlichman and his senior colleagues, including a brief session with the President himself.[20] The result was a spirited performance by Brown as lead-off witness on August 11, in which he informed the incredulous committee that *he* had persuaded President Nixon and his senior advisers—not the other way around—of the superiority of court enforcement over cease-and-desist authority.

The EEOC: Quasi-judicial Regulator or Advocate and Prosecutor?

On August 11, Brown told the Senate Labor subcommittee that "I now regard as preferable" the Administration's court-enforcement approach "since it embodies a mechanism more conducive to enforcing the law than merely administering it."[21] "The cease and desist approach would inhibit such an attitude," Brown said, "for it carries with it a presumption of quasi-judicial neutrality toward the problem title VII seeks to correct." "An active enforcement stance, which I think absolutely necessary," he said, "would thus be at odds with the Commission's own machinery." As recently as July 25, however, Brown had written Senator Ralph Yarborough, chairman of the full Labor and Public Welfare Committee, that the EEOC supported Senator Williams's bill and its cease-and-desist provision. So at the hearing Williams asked Brown to please explain "how this remarkable change has come to you." Brown's letter to Yarborough had rested Brown's support for cease-and-desist authority partly on the traditional appeal to social equity, policy consistency, and White House sincerity—i.e., that the other regulatory agencies possessed it. He had also appealed philosophically to the need to transform the private right of 1964 into a public right, and hence to shift the burden of enforcement from the victims of discrimination to the government. But Brown in August told the committee that he had since waged a campaign within the highest councils of the Administration to construct an even *stronger* bill than Senator Williams's bill. The result was the Administration's new proposal, now Senator Prouty's bill, S. 2806.

The Administration's bill was philosophically stronger, Brown explained, because it was now clear to him that cease-and-desist authority implied quasi-judicial *neutrality* about employment discrimination. The EEOC, however, must be an aggressive advocate, not a neutral judge. As a practical matter, moreover, the traditional cease-and-desist authority hadn't seemed to work very effectively during the past decade in the state EEO agencies that possessed it. Minority unemployment in such states had not basically been reduced, and in many of them it had grown. Furthermore, the court-enforced approach would take effect much faster and at less cost than the cease-and-desist approach, Brown said, which would require a huge expansion of staff. To begin court litigation immediately under the Administration's proposal, the EEOC would need to add to its present staff of 650 only fifty new attorneys in the first year and twenty-five more in the second. But to switch to the cease-and-desist model of large regulatory agencies like the NLRB would require adding 125 "personal" attorneys to service the now staffless and part-time EEOC commissioners (each NLRB member was full-time and was assigned 22–25 such attorneys, Brown said). It would further require adding 130 hearing examiners and a like number of court stenographers and clerical staff, all to total perhaps 400–500 new highly trained and expensive person-nel.[22] Senator Schweiker asked Brown: "Is this your own idea or has anyone in the administration asked you to take this position?" Brown replied: "Let me make that very clear again, because this is my proposal, and I had to sell the idea to a number of different people from the White House on down. . . . As a matter of fact, they asked me to come up with a stronger proposal and this is what I have done."

Whether anyone on the Senate panel actually believed Brown's claim to such solo policy innovation is doubtful. But in his testimony Brown had made two assertions, rather in passing and without further elaboration, that gleamed with insight. The first was Brown's rejection of "quasi-judicial neutrality" for the EEOC. This concept lay at the heart of the American regulatory tradition. By rejecting it for civil rights enforcement, Brown was reflecting a consensus within the commission that the EEOC was properly an advocate of the vic-tims of discrimination, not a neutral judge of their claims. This moral stance worked to shape the commission as an adversary party, like the Civil Rights Commission, not as a quasi-judicial party with claims to third-party neutral-ity.

Brown's second insight was historical. "The cease and desist legislation came about at a period of time in history back in the 1930s," Brown told Schweiker, "when most of the courts were hostile. This is not presently true."[23] Brown's history wasn't quite right—the cease-and-desist device as a regula-tory instrument was derived primarily from the progressive era prior to World War I. The courts did indeed remain hostile to economic regulation through the 1930s, and partly as a result the NLRB stood as a monument to the New Dealers' determination to regulate economic life through quasi-judicial agen-cies that could avoid the conservative courts. By 1969, however, the Warren Court seemed to have moved from the right of the regulatory agencies to their left. Brown sensed the gap between this new reality and the political habits and instincts inherited from the past.

Throughout the thirty years of partisan debate over the federal role in civil rights enforcement since Franklin Roosevelt's FEPC, the political parties had fallen into ossified, knee-jerk patterns of commitment and rhetoric, re-sponding to the NLRB model with fossilized reactions of New Deal vintage. The national Democrats carried the Truman banner, demanding cease-and-desist authority as their litmus test of sincerity on civil rights. The Republi-cans rallied to Senator Taft's standard, mustering their judicial strategy to bridle the NLRB and hobble the EEOC. In the intervening decades the GOP's judicial strategy had produced a stabilizing accommodation in economic reg-ulation. But, in civil rights policy, the circumstances that attended the judicial strategy's birth had evolved so far and so fast that it had reached a point of imminent backfire. By 1969, both sides seemed to be betting on the wrong horse.

In economic policy, the judicial strategy over the years had sufficiently dulled the cutting edge of New Dealish economic regulation that Republican politicians and business interests had learned to live with and even prosper under the modern regulatory state. Both the federal courts and the regulatory agencies adjusted to incremental reforms like the Administrative Procedures Act of 1946 and the Taft-Hartley law of 1947, which gave large employers access to and leverage within the regulatory system. Organized labor had gained its legitimacy, to such an extent that the reactionary Goldwater candidacy had rallied the Chamber of Commerce to his banner far more than the National Association of Manufacturers, whose large enterprises had quietly come to value the stability and predictability that institutionalized labor relations conferred.

In civil rights enforcement, however, the Republicans' judicial strategy had been outpaced by events. If not a museum piece by 1970, it had at least become a strategic anachronism. The judicial activism of the Warren Court had created a momentum that the Burger Court would confirm far more than it would challenge.[24] Writing in 1970, Rutgers law professor Alfred Blumrosen claimed that after working at the EEOC during its first year, 1965–66, "I came to the conclusion that it would not help—but would positively harm— the drive to end employment discrimination if the Commission were given that additional statutory power which the liberals believed so important in 1965."[25] Like William Brown, Blumrosen sensed dangers of cooptation in the NLRB model, and boundless new opportunities in the accelerating interventions of the federal courts. "I believed a *more powerful* Commission would become a captive of those interests which were to be regulated," Blumrosen explained, "while the existing weak institution enabled civil rights groups to use the federal courts which are favorable to their demands."

The dominant liberal community, however, clung to its demand for cease-and-desist authority to empower the allegedly toothless EEOC. Luther Holcomb, the last leaf on the tree from the EEOC's brief founding era of internally structured, Madisonian balance of interests, disagreed with the strident demands of fellow commissioner Clifford Alexander for cease-and-desist authority. But Holcomb understood the symbolic power with which such a weapon "would lend dignity, it would lend status" to the frustrated EEOC.[26]

The Revolt of the AFL-CIO: 1970

When Secretary Shultz testified before the Senate Labor subcommittee on behalf of the Administration's bill in August 1969, he made a strong case against pulling the OFCC out of the Labor Department. Shultz argued that Labor was the appropriate agency for developing coordinated manpower programs, and asked the Senate to give him a fair chance to prove that his new Philadelphia Plan could produce effective results.[27] Coming from the new Republican Secretary of Labor and backed by the Administration, this commitment in 1969 won the support of most congressional liberals and black civil rights leaders, and both the Senate and House Labor committees shelved the proposal to move the OFCC to the EEOC. Congress remained wary of such interagency shop-swapping, which tended to threaten the stability that underpinned the established triangular coalitions.

By the winter of 1970, however, the AFL-CIO, which anchored the constituency linking the Labor Department to the congressional committees, had been twice bloodied. First came the Nixon-Shultz victory in the Christmas eve battle over the Philadelphia Plan. Then on its heels came the OFCC's promulgation of Order No. 4 to extend proportional hiring requirements by race and selected ethnicity to all federal contract work. Organized labor's response was an internal vow to purge the cancerous OFCC from the Labor Department and merge it with the EEOC.

Political commentators widely interpreted this demand as mainly a punitive response to the Philadelphia Plan, one that would actually weaken civil rights enforcement by overwhelming the small and fragile EEOC. They noted that Title VII's ban on racial quotas might nullify the OFCC's Philadelphia Plan following transfer to the EEOC. But union spokesmen defended the OFCC transfer as a rationalizing measure to strengthen civil rights enforcement by placing its fragmented components under one roof.[28] When in late late July both the House and Senate Labor subcommittees rejected the Administration's civil rights bills and reported out similar cease-and-desist bills that did not include moving the OFCC to the EEOC, the AFL-CIO cashed in its chips with the Leadership Conference and demanded support for expelling the OFCC from the Labor Department.[29] The reluctant conference swallowed labor's demand as the price of maintaining the civil rights coalition's crucial solidarity. "I just want a bill," Joe Rauh explained wearily, "more than I want a fight [with labor]."[30] The *New York Times* quoted an unnamed civil rights leader's remark that the coalition was "on the brink of disaster" before the conference agreed to back labor's stand. In the same story the *Times* quoted an unidentified EEOC official's explanation that labor's demand was "very sad," but that "without labor's support, there just isn't a chance."

The anonymous EEOC official was half right—at least in the 91st Congress. Labor was not strong enough by itself to persuade the congressional civil rights coalition to break the faith that Shultz had kept. Pro-labor senators were also pro-civil rights senators, and *not one* was willing to risk getting caught in the whiplash by introducing labor's amendment, especially not so late in an election year.[31] On September 30, the full Senate by a vote of 41–26 rejected an amendment offered by Republican Peter H. Dominick of Colorado to substitute the Administration's bill for the Democratic leadership's cease-and-desist bill.[32] The following day the Senate voted 37–30 to reject Ervin's amendment to exempt state and local governments, and then the Senate voted 47–24 to pass the bill. Although the House had passed a similar bill granting cease-and-desist power to the EEOC as early as 1966, the Senate had never before done so. But the appearance of a breakthrough for the cease-and-desist solution was deceptive.

While the AFL-CIO had been too weak to prevail in the Senate, it was strong enough, especially as the congressional elections of November 1970 approached and the 91st Congress hurried its major bills toward the December recess, to block the parallel civil rights bill in the House. There the full Labor and Education Committee had reported its cease-and-desist bill out on August 21 and dispatched it to the Rules Committee, where it was delivered to the inhospitable hands of chairman William M. Colmer, Democrat of Mississippi.[33] In comparison with his predecessor, Judge Smith, Colmer had been a reasonably loyal lieutenant for the Democratic leadership. But Colmer naturally opposed the cease-and-desist bill. He found abundant allies among his committee's Republicans, fellow southern Democrats, plus the unusual backing of organized labor, and above all the clock.[34] Rauh had defended the Leadership Conference's bitter bargain with labor by explaining that "[t]his bill puts the lily-white Mississippi State Patrol under the jurisdiction of the commission. This is worth fighting for."[35] Colmer agreed on the combatworthiness of the cause, but from the perspective of (white) Mississippi; he called the bill "vicious," and refused to schedule any hearings.[36]

The congressional stalemate while the calendar ran out placed the EEOC's chairman in a quandary. Brown pleaded with the Administration to accept the Senate judgment and support the cease-and-desist approach as the only one likely to pass. "My own position would become impossible," Brown told Garment, "should the Administration react in the negative."[37] But the Administration had grown disenchanted with Brown. The disfavor flowed less from Brown's public dissent—indeed, Ehrlichman privately defended Brown's loyalty and sympathized with his dilemma—than from his failure as a manager. In December 1969 an inspection report by the Civil Service Commission had cited an "unbelievable mess" in the EEOC's personnel management.[38] A follow-up study by the Budget Bureau (which that summer became the OMB) had drawn a devastating picture of unordered priorities, undisciplined management, irregular appointments and prolonged vacancies in supergrade slots, and an environment characterized by high rumor and low morale.[39] The OMB identified the major impediment to improvement as Chairman Brown, who resisted delegating management authority to the commission's senior staff, and refused to share policy authority with the other commissioners.[40] Because the EEOC was regarded as too fragile, volatile, and politically sensitive to shape up by the traditional method of imposing a reorganization from above, the OMB urged the White House to consider appointing Brown to a federal judgeship and finding another chairman.[41] Ironically, the OMB staff preferred as the "most logical solution in the last analysis" the transfer of the EEOC to the Secretary of Labor![42]

But the symbol and role of Chairman Brown and the EEOC were too sensitive in the fall of 1970 for the White House to implement what OMB associate director Arnold R. Weber called the "more drastic remedies . . . i.e., a change in management or transfer to the Department of Labor."[43] So Brown was allowed to lobby on the Hill for the cease-and-desist bills even though the Administration opposed them. William Gifford, special assistant to Shultz at the OMB, wrote Shultz in December that "[w]e have secretly sought to tie [the House bill] up [in Rules] because of the Administration's position against the cease and desist orders." Brown, however, remained "unaware that the Administration sought to stall the bill."[44] By December the not-so-secret stalling was easy, and the 91st Congress expired, like the two Congresses before it, with no civil rights enforcement bill.

Surprise Reversal in the House: 1971

The renewed EEO enforcement drive got off to a fast start in the 92nd Congress. Representative Augustus Hawkins introduced his cease-and-desist bill (H.R. 1746) early in January, the House Labor subcommittee held hearings on it in March, and on April 7 the subcommittee recommended the bill to the full committee.[45] On June 4 the bill was sent to the Rules Committee, which no longer was in a position to block the determination of both party leaderships that *some* form of EEO enforcement bill must be sent to the Senate before the end of the first session.[46]

The Hawkins bill was similar to the Williams-Javits bill of the previous fall. It would give cease-and-desist power to the EEOC, and extend its EEO coverage to (1) worker units of eight or larger, (2) employees of state and local governments, (3) employees of federal agencies, and (4) educational institutions. But unlike the Williams bill, it would also transfer the OFCC to the EEOC. This solidified the unity of the black-labor coalition behind the Hawkins bill.[47] But the cease-and-desist and OFCC-transfer provisions also solidified the Republicans against it. Lobbyists for the U.S. Chamber of Commerce and the National Association of Manufacturers weighed in on the side of the Republicans' court-enforced substitute bill (H.R. 9247), which was offered on the House floor by John T. Erlenborn of Illinois.[48]

Erlenborn's substitute was essentially the Nixon administration bill. At its heart was the provision for compliance through EEOC suits in federal trial courts. It made no additions to the coverage provided by the 1964 law. But it established Title VII as the exclusive remedy for unlawful employment practices and it prohibited class action suits.[49] It also provided that only the Attorney General could intervene in private suits or conduct EEO litigation in the federal courts of appeal and the Supreme Court. Erlenborn's substitute bill had earlier been rejected by a vote of 19–14 in the Education and Labor Committee. But the membership of that committee, like its Senate counterpart, the Labor and Public Welfare Committee, was considerably more liberal than the full body, and for this reason both committees suffered reversals on floor votes more often than most standing committees. When the floor vote on the Hawkins bill came due in mid-September, the Nixon administration rallied the southern Democrats to the classic conservative coalition, aiming to substitute the Erlenborn for the Hawkins bill.[50]

The ensuing debate, however, was unusually confused. It was no longer clear whether cease-and-desist or court trials was the more aggressive enforcement policy, since both sides were making strong arguments based on institutional arrangements that had never been tried. Even more confusing was the anomaly that the Republicans found themselves wooing conservative Democrats by offering a cowbird's egg: the Philadelphia Plan's racial quotas. The southern Democrats wanted no enforcement bill at all. So they tended to back the Erlenborn substitute, whose judicial approach appeared less immediately threatening than an empowered EEOC.

But the Republican leadership responded by making three basic arguments for their bill, only the first of which the southern conservatives found very appealing, and the last of which the southerners, and many conservative Republicans as well, found most unappealing indeed.[51] The Republicans' most unifying argument was to oppose extending coverage to employees in state and local governments. House Republicans generally shared the view of Arthur Burns that this represented an unprecedented and unacceptable intrusion by a federal agency into state and local affairs. This objection easily rallied the Republicans' traditional coalition partners among the southern Democrats. But it did not deal with the heart of the Hawkins bill.

The second argument did. It held that court enforcement provided *more* effective EEO protection than the cease-and-desist approach. It would be more effective, the Republican leaders claimed, partly because it was faster (by 1971 the EEOC already had a 21-month backlog of approximately 25,000 complaints, and NLRB cases under cease-and-desist authority were averaging 630 days to resolve). Furthermore it would not jam the trial courts, they claimed, in the areas where most complaints originated (the top five complaint states were all southern—Texas, Louisiana, Florida, Alabama, and Tennessee) because the median time completion for non-jury trials in federal courts in those districts was only eight months. Court enforcement was also more effective, the Republicans said, because it was fairer and more thorough. This was because the rules of civil procedure in federal trials, where a *preponderance* of evidence was necessary to convict, provided broad discovery of evidence backed by the judges' contempt power. The more limited and cumbersome use of subpoena authority by regulatory boards with cease-and-desist power, on the other hand, was narrowly designed to provide only enough "substantial" evidence to find reasonable cause for administrative relief, not the "preponderance" of judicial evidence that was traditionally necessary to convict.[52] These arguments, however, stirred little enthusiasm among the southerner congressmen, whose fondness for the federal judiciary was weak.

The Republicans' third argument against the Hawkins bill was that it would transfer the OFCC to the EEOC, and thereby threaten the Nixon-Shultz Philadelphia Plan. The minority report of the Education and Labor Committee was careful not to mention the controversial Philadelphia Plan per se. Instead it spoke more generally about the OFCC's "commendable progress" in providing equal employment, and about the basic functional distinction between contract compliance and the regulatory function of processing EEO complaints. But the "bugaboo" of racial quotas raised by the OFCC transfer only added to the confusion of the debate. On the central question of enforcement power, the Republicans were rallying to a judicial strategy that liberal scholars were beginning to support, especially in the new light of *Griggs*. Such arguments elicited no enthusiasm at all among the southerners. But in the scrambled politics of the Hawkins-Erlenborn contest, the southerners found agreement in unlikely quarters.

Women's groups in particular did not share in the Republicans' new appreciation for the reformist legacy of the Warren Court. "If there is any group that should not be willing to trust their rights to the federal courts," Martha Griffiths declared on the House floor, "it is the women. They have never won."[53] The *Wall Street Journal*, citing the legal arguments of Albert Blumrosen of Rutgers, commented that a "knee-jerk reaction by liberals is blinding them to the potential inherent in the GOP counterproposal."[54] The AFL-CIO was widely suspected of seeking the OFCC's transfer in the "cynical belief," the *New York Times* editorialized, that the overburdened commission would be less effective than the existing office.[55] Presumably such a transfer would weaken or destroy the Philadelphia Plan, since the EEOC's Title VII explicitly banned racial ratios and quotas in hiring. But Democrat Edith Green of Orgeon saw it the other way around—that the "racial quotas" of the Philadelphia Plan would pollute the EEOC, placing it "in the intolerable situation of being compelled to select between contradictory expressions of Congressional intent."[56] Because neither enforcement method had ever been tried before by the federal government in civil rights policy, all arguments had some rational plausibility. But they remained abstract, and hence were less useful in convincing opponents or the undecided than in rationalizing one's own policy stance.

On September 15, with the vote scheduled for the following day, the floor manager for the Hawkins bill, Democrat John H. Dent of Pennsylvania, sensed danger. So he offered a package of sweetener compromises that included an amendment to bar the EEOC from imposing quotas or requiring preferential treatment of minority group citizens.[57] Representative Hawkins himself said he would accept such an amendment to get his bill passed, but he declared that the anti-quota ban was redundant because it was already in the federal civil rights laws. Erlenborn, however, attacked the anti-quota amendment, with which Dent was wooing Edith Green and her centrist Democratic allies, as "a blatant attempt to undercut the authority of the OFCC for the Philadelphia Plan." Thus the Democratic floor leader of the liberal Hawkins bill was proposing anti-quota amendments that were opposed by the Republican bill's floor leader, whose stance accomplished the rare trick of offending both organized labor and southern Democrats. In such confused circumstances, a consensus seemed impossible, and a decisive resolution seemed at best improbable.

On September 16, the House by the narrow margin of 202–197 rejected the Hawkins bill and substituted for it the Erlenborn bill—which was to say, the Nixon bill.[58] Thus the Nixon bill suddenly became H.R. 1746, and was sent to the Senate.

The Senate Debate and the Shadow of the Philadelphia Plan

In the fall of 1971 the Senate received the Erlenborn bill from the House and processed it through the Labor and Public Welfare Committee, thereby setting the stage for the real debate on the Senate floor early in 1972. Senator Williams's Labor subcommittee held rather pro-forma hearings early in October, where the now familiar and largely partisan arguments over cease-and-desist authority and transferring the OFCC were rehearsed again.[59] The membership of the full committee was so geographically skewed that no southerner sat among its seventeen members. Not only was there no Sam Ervin, there was no representative of Dirksen's brand of midwestern Republican conservatism either—other than perhaps freshman Republican Robert Taft, Jr., of Ohio, whose vague centrism was a faint shadow of his father's conservative faith. Indeed, the conservative side of the political spectrum seemed virtually unrepresented on the committee. Thus through a kind of geo-ideological default, the committee debate focused almost exclusively on which civil rights bill—the Williams-Hawkins bill (S. 2515) or the Dominick-Erlenborn bill (S. 2617, which was essentially the same as H.R. 1746)—promised to be the fastest and most effective enforcement bill.

During the October hearings Undersecretary of Labor Laurence H. Silberman testified in support of Labor's contract compliance program and against transferring the OFCC to the EEOC. Yet Silberman was asked not one question hinting at the sensitive area of racial preferences. Because the Republicans loyally supported the Nixon administration, the liberal Democrats approved of the Philadelphia Plan, and the southern Democrats were missing, the Senate committee managed to avoid discussing the most controversial question at the heart of the bill. Silberman himself raised the affirmative action question, however, in an indirect way when he emphasized the fundamental incompatibility between the functions performed by the EEOC and the OFCC. The OFCC program "is not enforcement minded," Silberman explained, because it didn't have to be. "[W]e have such tremendous sanctions that every time we go to use it," he said, "the contractor falls into compliance." "I am absolutely convinced that if you transfer this affirmative action program to EEOC it won't be affirmative action any more," he concluded. "It will be swallowed up by non-discrimination and it would be a lot more passive."[60]

In emphasizing the fundamental distinction between the EEOC's enforcement role against discrimination, which was weak, and the OFCC's non-enforcement role, which was paradoxically very powerful, Silberman explained that government contractors were *not* assumed by the OFCC to be guilty of employment discrimination. Rather, they were merely required to conform to the contract stipulations that were "incidental" to the government's traditionally broad procurement authority. This was the main justification cited by the federal trial court that in January 1970 upheld the Philadelphia Plan in *Contractors Association of Eastern Pennsylvania v. Secretary of Labor*.[61] In response to charges that the Philadelphia Plan required racial quotas and thus violated the Civil Rights Act, the Labor Department and spokesmen for the Nixon administration distinguished between "fixed and rigid" quotas and deadlines on the one hand and flexible goals and timetables on the other. The federal courts, however, conceded that "clearly the Philadelphia Plan is color-conscious"; they regarded its racially compensatory thrust as the heart of the matter, and otherwise ignored semantic distinctions.[62] The appeals court in *Contractors Association* explained that the Philadelphia Plan was based on Labor Department "findings" that minorities were underrepresented in the construction workforce in Philadelphia "due to the exclusionary practices of the unions representing the six trades."[63] The contractors had never challenged the truth of this assertion. Thus the Philadelphia Plan had properly sought to identify and remedy the effects of past discriminatory practices, the appeals court held.

The OFCC's hearing in September of 1969 and its attendant "finding" of discrimination by six Philadelphia craft unions were therefore crucial, because they allowed the courts to dismiss Title VII's ban on racial preference as limited to Title VII only, and entirely without bearing on Title VI or executive order programs like the Philadelphia Plan.[64] The OFCC finding thereby justified the imposition of racially preferential hiring goals as a remedy for demonstrated past discrimination—in Philadelphia. Silberman later reported that the Philadelphia Plan's close linkage between discrimination and remedy had been "decisive" in persuading doubtful Justice Department officials and, subsequently, several federal judges as to its constitutionality.[65] Once the Labor Department had won its point in Philadelphia, however, it never bothered with the procedure again. The hearings and findings in Philadelphia in 1969 would suffice for the rest of the country. As Silberman put it, "with [the Philadelphia Plan] battle won, we went on to spread construction plans across the country like Johnny Appleseed." "We no longer even purported to base these orders," Silberman admitted, "on findings of discrimination."

When the Third Circuit upheld the Philadelphia Plan in April 1971, it emphasized that the plan applied to construction contracts only. "This choice is significant," the court said, "for it demonstrates that the Presidents were not attempting by the Executive Order program merely to impose their notions of desirable social legislation on the states wholesale."[66] Yet a year *before* the Third Circuit's ruling, the OFCC's Order No. 4 had extended the Philadelphia Plan's scheme of numerical goals and timetables, proportionally based on minority population ratios, beyond construction to cover government contractors in all areas of service and supply throughout the country. In the meantime the Nixon administration had transformed the narrow model the courts had approved for Philadelphia construction unions into a regulatory instrument of extraordinary reach, yet one whose contours were difficult to perceive and whose impact was hard for ordinary citizens to assess or understand. Because voters could understand the connection between cabinet-level orders and racial busing of their schoolchildren, and then register at the polls their opinion about the stewardship of the President and his party, Nixon took pains to lay the blame at the doorstep of the courts. But the social impact of the Philadelphia Plan was sufficiently obscure, and was felt most immediately by Democratic trade unions, that Nixon had less to fear from voter backlash. In a pinch, such as an election year, a sharp dose of antibusing rhetoric could more than compensate for any electoral damage caused by the confusing Philadelphia Plan.

The Leadership Compromise of 1972

The second session of the 92nd Congress, when it convened in January 1972, shared a bipartisan consensus that the equal employment principles of 1964 were sound but the enforcement mechanism was not. Congress was weary of seven years of debate over enforcement. The *New York Times* captured that consensus when it observed in January that "[t]he question of giving enforcement teeth to the commission is one of the last of the pure civil rights issues in Congress that began with the Civil Rights Act of 1964 and continued through the Voting Rights Act of 1965 and the Open Housing Act of 1968."[67] Between January 16 and February 15, the debate on the Senate floor concentrated on whether to adopt the congressional leadership's Williams-Hawkins model of cease-and-desist authority, or the Nixon administration's court-enforced model as embodied in the Dominick-Erlenborn bill. In 1969 the new Nixon administration had enjoyed using the Philadelphia Plan to split the liberal coalition of labor, blacks, and feminists. But this strategy had also split the conservative coalition by driving a wedge between the moderate Republicans and the conservative southern Democrats. The trick for the Administration in 1972 was to enlist southern support to prevent passage of the cease-and-desist bill, and then to obtain liberal Democratic acquiescence in a court-enforced bill that would leave the contract compliance program intact in the Labor Department.

In this delicate task the Administration was ultimately successful.[68] During the January 16–February 15 debate in the Senate the Administration's Dominick bill was thrice rejected. But the Senate's Democratic leadership was unable to close debate and force through the cease-and-desist bill. The Administration profited from such a stalemate, for time sweetened the congressional leadership's appetite for compromise. In 1972, however, unlike 1964, the bipartisan coalition of conservatives was so split by Republican loyalty to the Nixon-Shultz Philadelphia Plan that southern attempts to reassert the comprehensiveness of Title VII's ban on racial quotas and ratios—in effect, legislatively to reverse *Contractors Association*—were blocked by the Nixon administration. Sam Ervin was still there and leading his battalion. But Everett Dirksen was not.

Ervin was an ideologue who lacked the late minority leader's sense of timing and instinct for negotiation. The North Carolinian fired his blunderbuss early, often, and usually ineffectively. He tended to rush forward amendments that had not been thought through, amendments whose supporters had not been courted and counted in advance. Such skills were little necessary on the Watergate committee of 1973, where Ervin's constitutional scruples were essential, and his constitutional homilies were the stuff of folk heroism. Temperamentally unsuited to play Dirksen's role as conservative broker and moderator in the second half of the second phase of the great civil rights reform of 1960–72, Ervin squandered his opportunities in the showdown on the Senate floor in 1972. But the Nixon forces did not.

In the floor battle over amendments that began on January 20, Nixon and his congressional allies demonstrated that they narrowly lacked the strength to substitute their Dominick bill for the Williams bill. On January 24 the Senate for the first time considered the Dominick substitution and rejected it by a roll-call vote of 41–43.[69] On the same day the Senate agreed 40–37 to permit the EEOC to try its own cases in the federal appeals courts as well as in the district trial courts. Although the Administration had lost on both votes, the close margin provided a bedrock of strength with which to wear the Democratic leadership down. Thereafter, loyalist Republicans generally provided the margins to pass compromises acceptable to the Administration while rejecting the southern forces under Ervin and James B. Allen of Alabama. Thus, on January 26, the Senate agreed that in exchange for the Administration's support for extending Title VII coverage to state and local governments, the Attorney General rather than the EEOC would conduct all litigation in suits brought against state and local governments. Then on January 31, when Ervin sought to undo that compromise by excluding coverage of state and local governments altogether, he was rebuffed by a bipartisan majority of 16–59. A subsequent attempt by Ervin to exclude coverage of educational and religious institutions failed by a similarly decisive vote of 25–55 on February 1 (although Ervin was later able to win an amendment excluding religious institutions alone).

The proposal to expand federal power over state and local governments in job discrimination was the main target of southern opposition in 1972.[70] This position, like opposition to lunch counter desegregation in 1964, was theoretically defensible on principled grounds of federalism. But its public relations were similarly atrocious, for they conjured up images of defending George Wallace's lily-white Alabama state patrol. If northern blacks provoked images of arsonists and rioters, southern blacks had not been part of the mass ghetto rioting, and they remained demonstrably excluded from any fair vision of public service in the Deep South—certainly including Ervin's North Carolina and Allen's Alabama. Nevertheless Senator Ervin, still smarting from his defeat over the Philadelphia Plan in 1969, launched a series of amendments that he argued with a sense of moral outrage that seemed ill-suited to the evil practices they would appear to defend. The result was a boomerang effect that rallied the Republicans around their President.

On January 28, Ervin warmed to his task by denouncing the Williams-Javits bill as a "ridiculous" bill designed "to rob all American citizens of basic liberties."[71] It represented, he said, nothing less than "the greatest threat of tyrannical power ever presented to the American Congress throughout the history of this Nation." Like H.L. Mencken, Ervin was a master of hyperbolic overkill, whose broadsides at his prime target managed to insult and alienate numerous secondary targets. For Mencken, this was a fine strategy for selling his deliciously outrageous opinions. But, in 1972, Ervin needed to attract allies in a Congress that was intent on bringing its seven-year debate to a close by finding a reasonable middle ground for compromise. Ervin refused to acknowledge this manifest reasonableness. He had already alienated feminist leaders, many of whom were prominent Republicans, by leading the attack on the ERA in a similar mocking spirit. Thus he produced a double-barreled slur, although probably unintentionally, when he told the Senate that the history of the EEOC "down to this date shows that virtually all the *men* who have been appointed to serve on this Commission were men who were psychologically incapable of holding the scales of justice evenly"—men who were so biased that they would be disqualified to serve on a jury passing merely on questions of fact, and who certainly should never sit jointly as the prosecutor and the judge.[72] Commissioners Aileen Hernandez, Elizabeth Kuck, and, by 1972, Ethel Bent Walsh thus failed to qualify even for Ervin's wholesale psychological disqualification.

In response to Ervin's string of assaults, Javits responded with reasoned moderation. In one of his amendments Ervin charged, for example, that requiring merely "substantial" rather than "a preponderance" of evidence to make a judgment of discrimination was a "crowning prostitution" because "substantial" meant merely "a scintilla, a perceptage point beyond being purely imaginary." In reply Javits politely devastated this legally indefensible argument—which Ervin, the simple country lawyer from the Harvard law school, well knew was nonsense—by citing a half-century of regulatory law, including opinions by that apostle of judicial restraint, Felix Frankfurter, and also by recounting his own experience as New York attorney general in dealing with its highly regarded and prudent State Commission on Anti-Discrimination.[73] Ervin stubbornly pressed his amendment on evidence to a vote, and was badly beaten, 22–43.

In response to this beating Ervin wisely postponed his next scheduled amendment, which was to require full *jury* trials in district court proceedings. Javits could anticipate little difficulty in dispatching such a cumbersome apparatus as jury trials in a bill designed to improve the efficiency of administrative procedures for a regulatory agency that was mired in backlog (the Senate ultimately obliged Javits, rejecting jury trials 30–56 in Ervin's futile, last-ditch fusillade of February 22). Ervin's concern for a preponderance standard was based on a strong argument that he had successfully advanced in the Senate before, and Javits and Williams had agreed to it in their 1970 bill. In 1972, however, Ervin was attaching it to ponderous appeals requirements for judicial *re*hearings of EEOC judgments in trials *de novo*, like his jury trial proposal. In rejecting it the Senate was essentially rejecting the old southern tactic of requiring elaborate, expensive, time-consuming procedures when blacks sought to exercise basic civil rights like voting or applying for jobs on the Alabama state patrol. But having won that particular amendment contest, and more importantly the larger point, Javits and Williams then readily agreed to Ervin's preponderance test as an evidential standard appropriate for EEOC judgments themselves—as they had in 1970.

Having fired too many of his arrows in scattershot fashion, Ervin (together with Alabama's Allen) then pressed ahead undaunted, with his amendment to bar reverse discrimination. The amendment, which Ervin cobbled together far too hastily (and twice changed in the process of merely introducing it on the Senate floor), stipulated that "No department, agency, or officer of the United States shall require an employer to practice discrimination in reverse by employing persons of a particular race, or a particular religion, or a particular national origin, or a particular sex in either fixed or variable numbers, proportions, percentages, quotas, goals, or ranges."[74] Ervin made it plain that his target was not only the EEOC, which was already bound by the 1964 ban but was beginning to ignore it. He was especially gunning for the racial quotas of the Philadelphia Plan, whose sponsoring OFCC the Senate had already voted (on January 26) to keep in the Department of Labor.

In his haste to ban all agents of reverse discrimination, however, Ervin had included any "officer" of the United States. But *any* officer would necessarily seem to include federal judges or officials implementing judicial decrees. Javits was quick to attack this vulnerability in Ervin's hastily designed bludgeon. Javits pointed out that Ervin's amendment would do more than destroy the Philadelphia Plan. Worse, it threatened to undo the entire corpus of federal court decisions reaching back to 1967, including *Quarles v. Philip Morris* in 1968 and *Contractors Association* in 1970.[75]

Quarles had ruled that the history of segregation in the South's tobacco industry had stripped the *bona fides* from its white-dominated seniority system. In the eyes of the *Quarles* court, southern history had thus rendered largely irrelevant Title VII's intent requirement, and also the exemption for bona fide seniority or merit systems that were not the result of intention to discriminate.[76] The *Quarles* dictum had held that Congress in 1964 could not have intended to freeze blacks in a permanent historical lag of disadvantage, and Javits placed Ervin and his supporters in the posture of heartless freezers, most of them with southern accents. Ervin's amendment was therefore not conservative but radical, Javits implied, because "it would torpedo orders of courts to correct a history of unjust discrimination in employment on racial or color grounds." Ervin's amendment, Williams said, would therefore "deprive even the federal courts of any power to remedy clearly proven cases of discrimination."[77]

Javits then announced that the Secretary of Labor had authorized him to express the Administration's opposition to Ervin's amendment. When the votes were tallied, a familiar pattern emerged. Ervin could count on the support only of three small groups: his own hard core of southern Democrats; a corporal's guard of like-minded southern Republicans like Thurmond of South Carolina, Tower of Texas, and Edward J. Gurney of Florida; and a fringe of nonsouthern Republican conservatives like Goldwater and Fannin of Arizona (Goldwater missed the January 28 vote), Nebraska's Hruska and Carl T. Curtis, Norris Cotton of New Hampshire, and Milton R. Young of South

Dakota. With most Republicans staying loyal to their own minority leadership and their President, and the nonsouthern Democrats remaining solid behind the Senate majority leadership, Ervin's battalion was reduced to a platoon. His amendment lost, 22–44.

Thus the Nixon administration used Ervin's intransigent bloc to hold off the cease-and-desist bill (Nixon told Haldeman "I will veto the EEOC. Not appealable. Won't discuss it.")[78] while successfully fending off Ervin's threats to the Philadelphia Plan. This tactic ultimately forced the congressional leadership to settle for a court-enforced bill that left the liberal coalition disappointed but nevertheless "moderately pleased" with the final result.[79] Clarence Mitchell told Tom Wicker of the *New York Times* that he was "completely satisfied with the outcome."[80]

The EEO Act of 1972 and the Legacy of "Reverse Discrimination"

The shape of that final result, the Equal Employment Opportunity Act of 1972, was largely determined as early as February 8, when the Democratic leadership admitted defeat on the Williams-Hawkins bill. By February 22 the amendment battles had sorted out the essential compromises, and the Senate voted 73–21 to close debate and then voted 73–16 to send the bill to the House. The House called for a conference and the conferees filed their reports on March 2, which the Senate accepted on March 6 and the House on March 8.[81] In conference the House basically accepted the Senate version, for to do otherwise would invite reopening the debate in the Senate, and therefore risk watching the bill die as time ran out in an election year—as had happened in 1970.[82] The conference agreement deleted the House bill's conservative provisions restricting class actions and establishing Title VII as an exclusive remedy. The Senate accepted the more conservative House language requiring courts to find that respondents "intentionally" engaged in unlawful employment practices, and the House accepted the Senate stipulation that judicial relief may include "any other equitable relief [in addition to reinstatement or hiring with back pay] as the court deems appropriate."[83]

In the end the Administration's court-enforced approach had prevailed, as had its insistence that the OFCC remain in the Labor Department, and that the Civil Service Commission retain jurisdiction over nondiscrimination in federal employment. The Administration was comfortable with the law's other main compromises. These included extending coverage to (1) employers and unions with fifteen or more full-time workers (rather than the Williams bill's eight, but down from the existing threshhold of 25); (2) employees in state and local government (but exempting elected officials and their policy advisers, and requiring the Attorney General to bring suits against government units, not the EEOC); and (3) employees of educational (but not religious) institutions. Also, the Attorney General's pattern-or-practice authority would be transferred over a two-year period to the EEOC. But the President would appoint the commission's general counsel, and the Attorney General would control EEOC appeals to the Supreme Court (but not to the circuit courts of appeal).

As for reverse discrimination, the core compromise of Section 703(j) was of course still embedded in the heart of the Civil Rights Act of 1964. Attacks by congressional conservatives on the Philadelphia Plan in 1969 and 1972 had failed. But under the doctrine of the separation of powers, the defeat of a legislative proposal does not accomplish the statutory enactment of an executive program with which the proposal may deal.[84] The lower federal courts, however, had begun to reason their way around Title VII's prohibition against discrimination, whether invidious or compensatory, and the Supreme Court had agreed. Legal scholars have debated whether the 1972 law fundamentally altered the anti-discriminatory core of the Civil Rights Act of 1964. Supporters of race-conscious remedies have argued that the rejection of the Ervin amendments in 1972 signified congressional acceptance of the OFCC's nationwide imposition of numerical hiring requirements.[85] Critics of compensatory preferences for minorities have countered that the Supreme Court has always viewed quite narrowly such inferences of congressional intent from *non-action,* whether in appropriating funds for otherwise unauthorized activity or in acquiescing in unauthorized executive conduct.[86] The court has been reluctant "to discover in congressional silence an implicit delegation of power," one critic concluded, "particularly where such delegation would effectively upset some explicit statutory scheme or vest the delegatee with extraordinary authority."[87] The Supreme Court itself, in *Bakke* and *Weber* during the Burger years and subsequently under Chief Justice Rehnquist, continued to split narrowly and indecisively when interpreting the hard, zero-sum questions that lie at the heart of civil rights disputes.

As to the more specific circumstances of the Senate debate in 1972, the defeated Ervin proposals were broadly directed toward eliminating the authority of executive agencies to impose compensatory quotas as a *remedy.* But Ervin and Allen's clumsy prohibitions seemed designed to protect Alabama's lily-white highway patrol from judicial challenge, and thus they failed their own intent test in the congressional environment of 1972. Ervin might well have won a ban against the imposition of minority preferences and quotas by administrative agencies without findings of discrimination that were consistent with the Administrative Procedures Act. This requirement for the minimal due process of findings based on two-party evidence was certainly implicit in Title VII, and even in *Contractors Association.* But it was not explicit in congressional language.

Such an amendment might have considerably limited the subsequent administrative imposition from Washington of a proportional model of minority representation in the nation's workforce, backed by the coercion of federal funding that encompassed a third of the American labor force. But Ervin's ambition far overreached the consensus of his colleagues, and as a result the Philadelphia Plan's supporters claimed that Congress in 1972 had at least indirectly approved the compensatory preferences required by the OFCC. This was especially ironical in light of President Nixon's subsequent behavior in his campaign for re-election. For having fought so successfully to defend the Philadelphia Plan and to unleash the EEOC as a prosecutor in the courts, Nixon followed his victory in the Equal Employment Opportunity Act of 1972 by turning up his campaign rhetoric and directing it against the institutions he was empowering and the purposes they were furthering.

Richard Nixon and the Irony of Civil Rights

At Watergate, Sam Ervin would become an American folk hero as defender of the Constitution against the abuse of executive power—a far nobler purpose than defending Jim Crow as a local option of federalism. But Richard Nixon had manifestly defeated Ervin in the civil rights battles of 1969 and 1972. In deflecting civil rights initiatives toward the courts, Nixon's judicial strategy had sought to shift to the federal judiciary the popular discontent of the 1970s with the 1960s' legacy of social experimentation. Thus Nixon's civil rights victories ironically freed him to run for re-election by taking a stand against the school integration orders and minority quota requirements that he had done so much to further. The strategy worked magnificently, at least in the short run—like Nixon's similarly unRepublican imposition of wage and price controls. Sam Ervin would have his revenge in the slightly longer run.

On March 16, 1972, one week following the Administration's congressional victory in the EEOC debate, and two days after George Wallace had decisively won the Democratic presidential primary in Florida—Nixon made a televised address to the nation demanding immediate congressional action to stop school busing for racial balance.[88] He asked for a moratorium on such busing, coupled with an emergency appropriation of $2.5 billion for remedial education in the disadvantaged neighborhood schools where students would perforce remain in the absence of massive busing. The following day, Nixon sent Congress a special legislative message that concentrated on the abuses of busing, but did so in language that seemed equally applicable to his own OFCC and HEW. Objecting to such "arbitrary Federal requirement[s]—whether administrative or judicial," Nixon insisted that "the remedies imposed must be limited to those needed to correct the particular violations that have been found."[89]

In early April, Nixon began sending Ehrlichman and Haldeman detailed memoranda on politics and policy. He directed Ehrlichman to delegate domestic program and Cabinet coordination to Shultz and Kenneth Cole and other White House assistants, and thus free himself and a team of spokesmen to "sell our line" for the re-election campaign.[90] The "gut" issues of the campaign, Nixon said, would be crime, busing, drugs, welfare, inflation— "issues the Democrats hate."[91] By mid-May, however, it was clear that the Democrats controlling Congress had no intention of cooperating with Nixon by acting on his busing moratorium. Nixon was especially disturbed by Wallace's strong showing in the early polls and primaries, particularly in Florida and Michigan.[92] So Nixon directed Ehrlichman and his presidential surrogates to counter the Wallace appeal by "zeroing in on a state and local basis," exploiting the tensions over busing orders in suburban communities like Pontiac, Michigan, a suburb of Detroit where a federal judge had ordered busing across the city school district line.[93]

Accordingly, in the late spring Ehrlichman indoctrinated his cadre of hard-selling spokesmen and toured the television news circuits in such targeted communities as Detroit, Fort Worth, Rochester, Grand Rapids, Buffalo, and Wilmington. Like Nixon's proposed busing moratorium, which was dying in a congressional committee, Ehrlichman's team called for widespread intervention in suits involving court-ordered busing. The proposed intervention was so widespread, in fact, that Ehrlichman aide Ed Morgan asked why they must intervene in support of school boards like the one in Augusta, Georgia, "which even my most conservative southern friends admit is racist."[94]

The rightward shift of Nixon's campaign rhetoric was accelerated in July when Bishop Stephen G. Spottswood, national chairman of the NAACP and keynote speaker at its Detroit convention, declared that "the NAACP considers itself in a state of war against President Nixon and 'his Constitution wreckers.' "[95] Rising Jewish complaints over racial quotas had drawn the Anti-Defamation League and the American Jewish Congress into an acrimonious exchange with black militant leaders, and the NAACP was caught in the middle.[96] On August 4, Philip E. Hoffman, president of the American Jewish Committee, wrote Nixon urging him to "reject categorically the use of quotas and proportional representation in implementing vitally essential affirmative action programs."[97] On August 11, Nixon wrote in reply that "I share the views of the American Jewish Committee in opposing the concepts of quotas and proportional representation."[98]

Nixon's election-year hard line was going to require some clever balancing. Charles Colson, whose political assignment in the White House included wooing ethnic white voters, had early acknowledged the awkwardness of courting hard-hats with an encumbrance like the Philadelphia Plan. The plan's employment quotas had gotten the Administration in deep trouble with the unions, Colson said, but downplaying this "would need to be handled very adroitly because we cannot, of course, publicly undermine what the Department of Labor has been doing."[99] So, in his letter to the American Jewish Committee, Nixon hedged his opposition to quotas: "With respect to affirmative action programs, I agree that numerical goals, although important and useful as tools to measure progress which remedies the effect of past discrimination, must not be allowed to be applied in such a fashion as to, in fact, result in the imposition of quotas." Nor, Nixon said, "should they be predicated upon or directed towards a concept of proportional representation." Nixon added that he was asking his department heads "to review their policies to ensure conformance with these views."[100] The agency heads responded by issuing clarifications that essentially repeated the Civil Service Commission's earlier clarification of May 11, 1971—i.e., "fixed and rigid" quotas were illegal, but requiring good faith efforts to achieve numerical goals and timetables was not.[101] In such communications the model of proportional representation was not mentioned, although this was the substantive heart of the matter—not how flexible the quotas or goals might be.

As the presidential campaign reached the party conventions in Miami, Nixon increased his references to quotas. In his acceptance speech in Miami on August 23, Nixon compared the Republicans' "open convention without dividing Americans into quotas" with the elaborate system of racial and sexual quotas that George McGovern had engineered and profited from at the Democratic convention. "[M]y fellow Americans," Nixon proclaimed, "the way to end discrimination against some is not to begin discrimination against others."[102] He told a radio audience in his Labor Day salute to the work ethic that "quotas are intended to be a short cut to equal opportunity, but in reality they are a dangerous detour away from the traditional value of measuring a person on the basis of ability."[103] The *Wall Street Journal* explained this as an "election year effort to woo the traditionally Democratic union and Jewish vote," and during the summer and fall the White House so dampened the enforcement efforts of the OFCC through an internal reorganization that its new (since the previous February) black director, George Holland, resigned in protest.[104]

In light of the poll results of October and the election returns of November, Nixon's rightward lunge seemed quite unnecessary. The McGovern-Shriver campaign had collapsed, and the economy was thriving and inflation had abated under the temporary impact of Nixon's wage and price controls. The opening to China, the arms control negotiations and atmosphere of detente with Russia, and Henry Kissinger's proclaimed triumph of peace in Indochina, all conspired to win the President a landslide re-election. Within two years, the "real" Richard Nixon would appear to be the culpable architect of Watergate, driven from the presidency in disgrace. But in the long view of continuity in civil rights policy, the real Richard Nixon was not only the demagogue of busing and the hypocrite of quotas during the warm months of 1972. He was also the expedient and successful defender of the Philadelphia Plan, the careful but quiet enforcer of school desegregation in the South, the architect of judicial empowerment for the EEOC.

The contrast between the campaign rhetoric and the governmental reality of 1972 disadvantaged the latter, as rhetoric is wont to do. On February 15, the day after Nixon had invited the congressional foes of busing to the White House to promise a joint fight against it, an assistant to the White House communications director released a progress report on its civil rights achievements—and less than boldly announced: "We have nothing to be ashamed of."[105] The report included the customary statistical aggregations of presidential puffery. But, in historical perspective, it provided a more revealing benchmark, as an indicator of continuity in direction and quiet acceleration in velocity, than the shrill headlines over busing and the artful codes of the quota debate. In its almost apologetically released progress report, the Nixon administration claimed credit for boosting the percentage of nonmilitary ra-

cial and ethnic minorities among federal workers to 19.5 percent; for doubling aid to black colleges and increasing minority business aid by 152 percent.[106] The percentage of black children in all-black schools nationwide had
decreased from 40 percent in 1969 to 12 percent in 1972. For Hispanics,
Nixon had proposed in his special legislative message on busing an "educational bill of rights for Mexican-Americans, Puerto Ricans, Indians, and others who start under language handicaps." This initiative was quickly buried
along with the busing moratorium, but it would be revived and enacted under President Ford.[107]

For women, the ERA was pushed through Congress by the bipartisan
feminist coalition without significant Administration support. But the Administration's efforts in the 92nd Congress did help produce for women in 1972
an important package of education amendments. The public battles over college aid and school busing overshadowed a major feminist victory in Title
IX, which barred federal aid to any educational program practicing sex discrimination. Similar protections had been included in the 1971 training acts
for health manpower and nursing, and also in a widening range of authori-
zations and appropriations including revenue-sharing, public works, economic development, Appalachian redevelopment, and water pollution control.[108] The Revenue Act of 1971 included a new provision for child care
deductions (although Nixon vetoed a child development bill that would have
funded day care for poor families), and in the same year benefits were equalized for married women who were federal employees. Such initiatives were
usually sponsored in Congress by feminists and their allies, but the Nixon
administration generally supported them.

As for the EEOC, the symbol of the seven-year legislative debate that
Nixon had shaped and concluded in 1972, the "poor, enfeebled" agency had
grown from a staff of 359 and a budget of $13.2 million in 1968 to a staff
of 1,640 and a budget of $29.5 million in 1972.[109] For fiscal year 1973 the
Nixon budget for civil rights enforcement increased from $49.9 million to
$66.3 million, providing for a doubling of OFCC compliance checks from
22,500 in 1971 to 52,000 in 1973.[110] In March 1973, *Business Week* reported that despite "loud cries of anguish" from civil rights forces over Nixon's deep budget cuts in social action programs, his budget for fiscal 1974
"almost unnoticed" had basically doubled the 1972 allotments for civil rights
enforcement agencies. The EEOC budget increased 107 percent, the contract
compliance budget for all agencies rose by 66 percent, and the Justice Department's budget for the Office of Civil Rights increased by 67 percent.[111]
As for encouraging blacks and Hispanics in business, despite the controversy
and cynicism that had greeted Nixon's initiative with the Office of Minority
Business Enterprise in 1969, the Republican effort to encourage minority en-
trepreneurship had started small but had grown like Topsy. The budget for
minority procurement set-asides for all agencies had increased from a found-
ling $8.2 million in fiscal 1972 to $242.2 million in fiscal 1974—an increase
of almost 3000 percent. In the modern American administrative state, the

future lay far more in the budget than in campaign speeches and newspaper headlines.

By 1972, the election rhetoric notwithstanding, the President and Congress had finally dropped the other shoe, rounding out the basic structure of civil rights policy and enforcement machinery that John F. Kennedy had begun in 1961 with Executive Order 10925, and which Lyndon Johnson and the 89th Congress had courageously fulfilled. The compromise between the Nixon administration and the 92nd Congress in 1972 had completed a rationalizing process. It was a legislative tidying up that reflected more than it shaped a fundamental shift in authority and power since 1965—one that had been determined more in the federal courts and the agencies of the permanent government than in the White House or the halls of Congress.

Conclusions

The Civil Rights Movement and the Continuitarians

In his memoir *Thinking Back,* historian C. Vann Woodward claimed that his critics on both the left and the right shared a common flaw of historical vision which he labeled "Continuitarianism."[1] By this Woodward meant a reductionist tendency to flatten out the peaks and valleys of historical experience, to average down its variety by collapsing it into a relentless stream of continuity, moving toward the present like a glacier. From the myopic perspective of the present, distant mountains blurred into hillsides, rolling ineluctably toward us to produce what we are. Widely admired as the "dean of southern historians," Woodward had devoted a distinguished career at Johns Hopkins and Yale to demonstrating the *dis*continuity of the South's historical experience. Constant change in the South had created a broad menu of usable pasts, Woodward argued, and this maximized the freedom of choice for present generations to shape the future. "I am not a determinist of any kind," he said.

Discontinuity to Woodward thus meant freedom from history's chains. His brilliantly revisionist *Origins of the New South* (1951) had attacked the old, conservative orthodoxy of continuity between the New South and the Old, with all its Darwinian implications of an inevitable caste system.[2] In *The Strange Career of Jim Crow* (1955) Woodward sharpened his attack on the ideological blindness of conservatives to the reality of change in southern history.[3] But in *Thinking Back* Woodward extended his critique to "the neo-Continuitarians of the left."[4] These were generally younger radical scholars whose skeptical view of the possibilities of liberal reform had reinforced a post-1960s pessimism.[5]

Looking back at the 1960s through the lens of the Reagan era, such neo-Continuitarians were nostalgic about the turbulent decade's social activism. But they tended to flatten the achievements of government reform into an

450

amalgam of tame and feckless neo-New Dealism.[6] Liberal reform was seen as vintage Lyndon Johnson—a bourgeois, pluralist incrementalism with no fundamental cutting edge and no serious redistributionist effect. In 1969, political scientist Theodore Lowi disparagingly called this modern genre of reform "interest-group liberalism." It demoralized and corrupted government, Lowi claimed, because its informal bargaining between group brokers and its plans without standards could not achieve justice. It thus created new structures of privilege in the name of reform, while institutions of popular control atrophied.[7] From that perspective, it could be conceded that perhaps the early civil rights movement had won the initial campaigns of 1964–65. Those, however, were the easy battles, over segregated restaurants and hotels and all-white voting. But, the indictment continued, when the formalistic, procedural rights of individual competition had yielded after 1965 to substantive demands by minorities and the poor for collective equality, the government had lost interest and the old continuity had reasserted itself.

We encounter this interpretation increasingly in contemporary American textbooks, explaining to a new generation of students that the civil rights reforms of the 1960s were essential but not fundamental. The formalistic concern of the reforms of the 1960s with procedural equality was successful in removing the more egregious violations of southern segregation, students are told. But the reforms made little dent in the standard inequalities that were built into the foundations of liberal pluralism's modern welfare state, where reformist credentials disguised a polity that was racist and sexist at its core. One textbook in African-American history, published in 1982 by black historians Mary Frances Berry and John Blassingame, devotes only four pages of the 485–page text to the civil rights movement and reforms of the 1960s. Berry and Blassingame summarize the Civil Rights Act of 1964 and the Voting Rights Act of 1965 in only three sentences each. They then largely dismiss them with the observation that "political action did not improve the overall black condition."[8] In *The Struggle for Black Equality,* Harvard Sitkoff acknowledges the improvements brought by the "Second Reconstruction" of the 1960s in black education and political participation. But Sitkoff cites the reform era's larger economic failure and concludes that a "Third Reconstruction, aiming for economic justice, is imperative . . . ."[9]

Sitkoff's call for a Third Reconstruction has become a common reaction to the perceived failure of the civil rights reforms of the 1960s. Allen Matusow, in his fresh and critical interpretation of the Kennedy-Johnson achievements, concludes that as far as President Johnson was concerned, by 1966 "the civil rights movement was over." Johnson's fair housing bill represented "shirking responsibility while appearing to fulfill it," Matusow said, and his fair employment policies left the enforcement agencies "to drift listlessly on their rudders."[10] Like Sitkoff, Matusow agrees that the reforms of the 1960s brought legal and political equality to southern blacks. But also like Sitkoff, Matusow suggests that social and economic equality were pursued at a procedural level that, like the war on poverty, left substantive structures of inequality intact. Thus a Third Reconstruction will be required ("Perhaps a century hence," Matusow volunteered) to destroy the ghetto walls and integrate the black underclass into the American mainstream.

William Chafe in his analysis of postwar America concedes that "[w]hen the civil rights movement was compatible with the traditional values of individualism and competitiveness in American society, its demands were granted, however reluctantly." But, Chafe continued, "where the movement threatened structural change, questioned deep-seated cultural values, and entered areas that would require a redistribution of political and economic power, resistance set in." In retrospect, Chafe concluded, "the verdict must be primarily negative."[11] From this Continuitarian perspective, sometimes even the post-1965 requirements of affirmative action have been acknowledged only grudgingly, and often dismissively.[12]

Such a presentist and gloomy Continuitarianism fails to recapture the American world of 1960 and the extraordinary distance it had traveled by 1972. Folk singer Joan Baez caught the triumphal mood of discontinuity better, in celebrating "The Night They Tore Old Dixie Down." It took more than a night, but by the end of the 1960s Jim Crow was dead. The black mobilization for civil rights was a revolutionary social movement that utterly destroyed the biracial caste system in the South. As Martin Luther King well understood and eloquently preached, the black liberation also freed southern whites from a systemic corruption and a shared though lesser victimhood that he understood far better than they. In normally stable, centrist America, the civil rights movement stands out from the continuities of workaday government as a rare and stunning achievement of liberation. In only a dozen years the combination of black social movement and government reformers, in Chafe's acknowledgment, had "toppled segregation, destroyed discrimination within the law against the blacks and members of other minority groups, led to the massive increase in the franchise accomplished by the Voting Rights Act of 1965, paved the way for countless legal and political battles to abolish economic discrimination under the 1964 Civil Rights Act, and achieved—albeit a hundred years late—the legal rights to full citizenship deferred at the end of the First Reconstruction."[13]

The Two Black Americas

The statistics of the black social revolution of the 1960s are impressive. Building on the destruction of the biracial caste system, black gains in political participation, education, economic opportunity, and in public opinion demonstrated discontinuities of heartening magnitude. Between 1962 and 1970, black voters in the South increased from 1.5 million to 3.3 million, and black registration in the seven states covered by the Voting Rights Act increased from 30 percent of those eligible to almost 60 percent (while white registration stabilized at around 70%). The number of black elected officials in the South followed suit, increasing from approximately seventy in 1965 to 700 in 1970, and to 1,600 in 1975.[14] By 1970 the South's historic mold of white racial demagoguery had disappeared. Segregationist politicians like Lester Maddox of Georgia and Ross Barnett of Mississippi disappeared with it—and others, like South Carolina's Strom Thurmond and Alabama's George Wallace, successfully adjusted to the new political demography by courting black voters. In education, the percentage of black children attending school with whites in the South was only 2 percent in 1964, but by 1972 the South led the nation with more than 60 percent. Nationally the median number of school years for black citizens increased from 10.7 in 1960 to 12.7 in 1970 (only .4 of one year less than whites). Black college attendance shot up from only 234,000 in 1963 to 1.1 million in 1977—a 500 percent jump, with blacks by the latter decade attending college at the same rate (32 percent) as whites.[15]

Economically, the jump was much less dramatic than in politics and education, and it obscured some long-range secular trends while exaggerating others. Blacks started from far behind, but vastly accelerated their catching-up. The economic expansion of the 1960s joined with targeted government efforts to help reduce the unemployment rate of adult black males by 1969 to 3.8 percent—six percentage points below its 1955–65 average (which was twice the rate for white men).[16] The number of black Americans living in poverty declined from 55 percent in 1959 to less than 34 percent in 1970, and the proportion of black families earning more than $10,000 a year (in constant 1960 dollars) leaped from 13 percent in 1960 to 31 percent in 1976 (it had been 3% in 1947). Income for white families increased 69 percent during the booming 1960s, but for black families the increase was 109 percent. This raised black family income from 48 percent of white earnings in 1960 to 61 percent in 1970. Outside of the South, by 1970, black husband and wife two-income families earned 88 percent of what white two-income families received, and the differences in earning power between black and white women had virtually disappeared.[17]

Black America, however, was not headed for a socio-economic convergence with whites. Rather, American blacks were moving toward a sharp class bifurcation that by the 1980s would build *two* black Americas—a booming black middle class and a devastated black underclass.[18] The economic downturn of 1973–83 reversed many of the gains of poor blacks, while the black middle class continued to profit from the expanded opportunities opened by nondiscrimination and affirmative action programs.[19] By 1986, 45.5 percent of black families exceeded $20,000 in annual income, and 24.3 percent earned between $20,000 and $35,000 (as did 28.5% of white families).[20] A Rand study of 1986 concluded that "the real story of the last forty years has been the emergence of the black middle class," whose growth was "so spectacular that as a group it outnumbers the black poor."[21] Indeed, in the quarter-century since the Civil Rights Act of 1964, the black middle class had so successfully exploited its legacy of nondiscrimination and affirmative action, while the decline of the black underclass had continued and in some ways accelerated, that the class divisions among blacks became larger than those among whites. By 1988 the top fifth of white families ranked by income made 8.6 times as much as the bottom fifth, but among blacks the figure was 13.8 times.[22]

But while the civil rights era had willed a legacy of liberation and achievement for middle-class blacks, the black underclass had collapsed disastrously.[23] Only 10.2 percent of white families earned less than $10,000 annually by 1985, but 30.2 percent of black families did so. The percentage of out-of-wedlock births among blacks had soared from 21.6 in 1960 to 61.4 in 1985. The share of black families headed by women mirrored this catastrophic plunge, rising from 24.4 percent in 1960 to 56.7 percent in 1985. In 1986 the National Urban League's annual report, *The State of Black America,* included an essay by black economist Glen C. Loury entitled "Beyond Civil Rights."[24] Loury concluded that the pathology of the black underclass had taken on a cultural life of its own and could not effectively be reversed by civil rights policies. "With upward of three-fourths of children born out-of-wedlock in some inner-city ghettos," Loury said, "with black high school drop-out rates of better than 40% (measured as the fraction of entering freshmen who do not eventually graduate) in Chicago and Detroit, with 40% of murder victims in the country being blacks killed by other blacks, with fewer black women graduating from college than giving birth while in high school, with black women ages 15–19 being the most fertile population of that age group in the industrialized world, with better than two in five black children dependent on public assistance, and with these phenomena continuing apace notwithstanding two decades of civil rights efforts—it is reasonably clear that civil rights strategies alone cannot hope to bring about full equality."[25]

By aggregating black economic statistics to define an average black trend, critics could deplore a continuity of depressed black performance since the 1960s. But the aggregated data disguised the sharp discontinuity of economic achievement within the black community. Increasingly there were two black Americas, and the reforms of the civil rights era, by breaking the iron grip of the historic correlation between race and poverty, had freed black Americans, like most Americans, to sort themselves out politically according to their class interests.[26]

Racial Attitudes and Public Policy

Thus the reforms of the civil rights era produced mixed results in black economic life, but the explosion of the black middle class was a major break with the past. In the softer realm of public opinion, the era produced a major and heartening discontinuity as well—although it carried a qualitative hedge. It is clear that the civil rights movement coincided with a sea change in white attitudes since World War II, when national survey research centers began systematically collecting baseline data from polling samples.[27] From these longitudinal surveys emerged a consistent and indeed overwhelming picture of growing racial tolerance among white Americans.[28]

A basic barometer was the change in public attitudes toward racially integrated schools. It is a blunt testimony to the distance we have come that in 1942 only 2 percent of southern and 42 percent of northern whites agreed that black and white children should go to the same schools.[29] But by 1972 the North-South regional gap had converged dramatically, with 70 percent of southern and 90 percent of northern whites supporting integrated schools.[30] Similar questions involving public transportation and jobs produced national averages that began in the 45 percent range in 1942 and soared to the 90th percentile by the 1970s, with rapid southern catch-up accounting for most of the convergence. For a nation that fought and won the world war against fascism with a racially segregated army, this was a breathtaking change in race relations.[31]

There are problems, however, with such a self-congratulatory assessment of the triumph of racial toleration in America. One is that it accords poorly with our historical recollection of the fever chart of change during 1960–72. The civil rights era features a sharp break after 1965, with black rioting in northern ghettos, the eruption of Black Power in Mississippi, the assassinations of Martin Luther King and Robert Kennedy, the political resurgence of Wallace and Nixon, and a nationwide backlash against school busing. Another is that it is inconsistent with surveys of black attitudes.[32] Black Americans were not systematically polled until the mid-1960s, and for several years the volatility of racial politics in a period dominated by ghetto riots confused the results of these studies. But the longitudinal survey data of black attitudes since 1964 in some categories present a mirror image of white attitudes. As whites saw more and more "real change" in the position of black people in the country, blacks saw less. As whites increasingly saw black civil rights leaders "trying to push too fast," especially in the 1970s, blacks said they were moving "too slow."[33]

Most problematical, however, has been the gap between agreement with abstract principles and steps to implement the principles. This is a familiar phenomenon in survey research, and it reinforces a skeptical view that respondents to polls tend to supply socially acceptable answers to questions of broad principle while providing more revealing (and often inconsistent) answers to practical questions of application.[34] Accordingly some survey researchers have discounted the striking postwar trends toward racial tolerance, which they claim represent merely "slopes of hypocrisy" that belie the continuity of white racism underneath.[35] Continued patterns of residential and urban school segregation, they argue, are inconsistent with the plunge in racial prejudice shown by the polling data. In a major scholarly study of postwar attitudes and trends published in 1985, Howard Schuman and his colleagues grappled with this problem, especially in regard to the gap between principle and implementation and the leveling off of white support for government intervention after the late 1960s.[36] The problem of increasing surface tolerance with an undertow of racial prejudice is compounded by evidence of the same disjunction among blacks. Support for federal intervention to integrate schools, for example, declined among *both* whites and blacks after 1966–68.[37]

Beneath the rosy slope of progressive liberal tolerance, then, lies evidence of a watershed in the 1966–68 period. This corresponds more intuitively with our historical sense of the flow of events. But what does it mean? Does it reinforce a Continuitarian-of-the-left view that polling evidence of improving racial tolerance has been a mirage? Schuman and his colleagues in *Racial Attitudes in America* are troubled but are unwilling to draw such a pessimistic conclusion from such complex events and indirect measurements. Their questions of implementation, however, probe for two very different sets of principles without recognizing the difference.[38] They define their implementation questions as those dealing with "approval or disapproval of steps the government might take to combat discrimination *or* to reduce racial inequalities in income or status" [emphasis added]. Combating discrimination and reducing inequalities, however, are different goals that may involve radically different policies. Indeed, these two goals may require contradictory policies—which was what the fight over the racial quotas of the Philadelphia Plan was all about.

The Watershed of 1965: Two Phases of the Civil Rights Era

It is this policy watershed, between classic liberalism's core command against discrimination on the one hand, and the new theory of compensatory justice on the other, that was crossed during 1966–68. After 1965 the civil rights era moved from Phase I, when anti-discrimination policy was enacted into federal law, to Phase II, when the problems and politics of implementation produced a shift of administrative and judicial enforcement from a goal of equal treatment to one of equal results. The Phase II shift from nondiscrimination to preferential treatment for minorities produced a stiffening of white resistance that was reflected in the public opinion polls.[39] In 1978, Seymour Martin Lipset and William Schneider reviewed nearly a hundred opinion polls taken over the previous forty years. They concluded that a large majority of white Americans continued to support not only nondiscrimination, but also positive programs to compensate for past discrimination—such as special training programs, head start efforts, financial aid, and community development funds.[40] These majorities continued to support "soft" affirmative action in the sense of President Kennedy's original executive order of 1961. This required special recruiting efforts for minorities but it also included compensatory programs that were positive-sum and not driven by an equal results test.[41] But when compensatory efforts moved in Phase II into the zero-sum game of preferential treatment or racial quotas that would predetermine the results of competition, most Americans drew the line—including majorities of blacks in some polls.[42]

The fundamental shift of goals and means that distinguished Phase I from Phase II, then, was the shift from "soft" to "hard" or from positive-sum to zero-sum affirmative action. The shift was from a goal of equal treatment with positive assistance, such as special recruitment and training efforts, to a goal of equal results or a proportional distribution of benefits among groups. This shift created the tension, captured by the polls, between Phase II's preferential treatment and the American consensus for equal opportunity and against equal results through minority preferences.[43] Historian J. Mills Thornton has emphasized this distinction between the individualist ideal of personal liberty which dominated the civil rights movement through 1968, and the collectivist ideal of material equality that transformed the post-1968 phase.[44] "From the perspective of the ideal of individual liberty," Thornton said, "the Civil Rights Movement ended because, with the Civil Rights Act of 1964, the Voting Rights Act of 1965 and finally the Fair Housing Act of 1968, the movement had achieved its goals." "In these terms," he concluded, "the movement ended for the same reason that World War II ended: the enemies had been defeated."[45] But if the enemy was defeated, what accounts for the turmoil and disarray in Phase II, when the liberal coalition, having at last triumphed over the conservatives on the battlefield, fell to bickering over the terms of the new order?

The answer, voiced by a growing chorus of contemporary critics from within the liberal reform tradition itself, was that in the process of implementing the victories of the anti-discrimination laws of 1964–65, the enforcement agencies and courts and the civil rights coalition behind them fundamentally transformed the original goals themselves. During the 1970s, students of public policy built a substantial literature studying the obscure but crucial process of policy implementation.[46] They concluded that the politics of implementation often transform the original policy goals, sometimes deflecting them in novel and occasionally even contradictory directions. Thus in the process of implementating the anti-discriminatory commands of Phase I of the civil rights reforms of 1964–65, the EEOC and the OFCC, together with their allies in the civil rights coalition and in the agencies and Congress and also increasingly in the federal courts, fundamentally transformed the Phase I goal of equal treatment into the Phase II goal of equal results. This, however, created an internal contradiction because it required the unequal treatment of citizens by race and minority status so as to compensate for the results of past discrimination.

This shift was becoming more apparent by 1970, when Shultz's Philadelphia Plan had survived the attack by congressional conservatives allied with Comptroller General Staats, and the proportional hiring model tailored for the construction industry had been extended through Order No. 4 to apply to all federal contractors and federally assisted programs. In the fall of 1970 the Civil Rights Commission released a 300-page report entitled *The Federal Civil Rights Enforcement Effort*.[47] Harvard sociologist Nathan Glazer reviewed the CRC report for *The Public Interest*—a new journal of public affairs which Glazer co-edited, and which provided a forum for liberal social critics, like his co-author Patrick Moynihan, who were troubled by the trend toward "reverse discrimination."[48] Glazer was struck by the tension between the CRC report's hard-hitting thesis—that there had been a "major break-down" in federal enforcement—and the bulk of the report's evidence. Readers, he said, would be mystified, because the report described a "growing and rapidly increasing array of federal officials carrying out the laws and orders to enforce equality." [49]

The CRC report summarized a formidable battery of new federal statutes and executive orders, most of them added since 1961, and it described the impressive growth of enforcement agencies and their staffs. The OFCC, with only 28 employees in 1967, was budgeted for 173 for fiscal 1973. The EEOC had 570 employees, and was targeted in the Nixon budget for 1,500 by fiscal 1973. Defense had 166 EEO positions, with 171 additional positions scheduled for 1971. Similar increases were set for HEW, HUD, and Justice. Six thousand federal employees were being trained as EEO counselors for their agencies by the CRC itself.

Most striking to Glazer, however, was the absence in the report on civil rights enforcement of a specific concern with *discrimination*. To Glazer the CRC report signified a shift in goals "from *equal opportunity*—which scarcely seems at issue in most of the report—to an attempt to ensure a *full equality* of achievement for minority groups." [50] This was something entirely new: "The CRC report abandons as the measure of success in federal civil rights enforcement the elimination of discrimination," Glazer said. "*Indeed, there is scarcely a reference to any single case of discrimination by anybody in this enormous report*—which indicates the CRC's sense of the present importance of that issue. It uses a new measure—the achievement of full equality of groups." As a result the test was no longer: "Are members of minority groups discriminated against? It is: *Are they found in employment, at every level, in numbers equal to their proportion in the population?*"

Glazer noted that the CRC seemed unaware of any problem in stating, quite blandly, that the definition of "equal opportunity" had become the "actual employment of minorities," although there was no attempt to argue whether or why "equal opportunity" would lead to any such result. The CRC reported that 15 percent of federal employees were black, without noting that only 11 percent of the U.S. population was black. Yet in order to further increase black representation in federal jobs, Glazer observed, the "arithmetic and algebraic components of the Federal Service Entrance Examination . . . have largely been eliminated." Moynihan had warned the graduating class at the New School for Social Research as early as 1968 against the dangers to liberal principles of political demands for proportional representation, which were setting blacks against Jews in the fight for jobs and control in the New York City schools. "Once this process gets legitimated there is no stopping it," Moynihan warned. "If ethnic quotas are to be imposed on American universities and similarly quasi-public institutions, it is Jews who will be almost driven out." [51]

The EEOC as a Model of Affirmative Action

Glazer's 1971 review of the 1970 CRC report had observed that the report was "perhaps most severe" in criticizing the EEOC itself. "It would be interesting to know," he speculated, "to what extent the procedures to ensure merit have been followed in staffing this agency, and whether the CRC believes the procedures it urges for all other agencies of the federal government (not to mention all private industry and other employment) would actually lead to harder working, more committed, and steadier employees than the EEOC—on the CRC's evidence—now seems to have."[52] From its beginning the EEOC had been self-conscious in developing its hiring procedures into a model of affirmative action employment. Commissioner Jackson had explained as early as 1966 that "the Commission's staffing pattern is unlike any other in the history of Federal agencies and departments."[53] "More than 40 percent of *all* Commission employees are Negro," Jackson observed, and more importantly, he added, minority group personnel in the EEOC held *more* than 40 percent of the higher-grade jobs.

The rapid growth of the EEOC provided Commissioner Jackson's unique agency with the means to flesh out its model vision, and thereby provide answers to Glazer's questions. By 1980 the EEOC budget had grown to $125 million and its staff had grown to 3,746 (from a budget of $2.3 million and a staff of 314 in 1966). Because the EEO Act of 1972 placed virtually all government workers in the United States under the obligations of Title VII (although exempting, under separation-of-powers logic, most staff employees in the legislative branch), the EEOC was thereafter charged with enforcing its affirmative action guidelines against itself. But the EEOC had little to fear from charges of hypocrisy, even if it were judged according to a strict model of racial, ethnic, and sexual proportionality in the applicable workforce. This was because in 1971, blacks represented 11 percent of the U.S. population, 12 percent of the U.S. workforce, 15 percent of the federal civilian workforce, but almost half (49%) of the EEOC staff. Women and Hispanics represented 38 and 5.8 percent of the U.S. workforce, respectively, but constituted 47 and 9 percent of the EEOC staff.

Nevertheless the EEOC remained dissatisfied. It sought to construct what Chairman Brown called a "perfect" affirmative action model within the agency. So, as it required of others, the EEOC instituted a series of annual affirmative action plans with ambitious goals and timetables. The 1975 plan explained that whereas most such plans were aimed at "increasing the representation of females and minority groups to approximately their representation in the civilian labor force," the employment situation at the EEOC was "rather unique in that women and most minority groups hold a larger share of the jobs than labor force figures might deem appropriate."[54] The EEOC enforced its own even more ambitious goals firmly, and by 1978 the percentage of its own workforce was 49.1 percent black, 44 percent female (16.6% white female), 11.2 percent Spanish-surnamed—and 20.6 percent white male. Then the following year the agency's 32 district offices, which included ten white male directors, were reorganized as part of a Special Hiring Plan. This process produced nineteen new field office directorships. When they were filled, the number of white male directors had been reduced from ten to two (the others chosen were 5 black males, 4 black females, 4 Hispanic males, and 4 white females).

Not surprisingly, the Special Hiring Plan was regarded by the EEOC's displaced white male regional directors as a special firing (or genetic demotion) plan targeted on their race and sex. A reverse discrimination suit was filed in federal court in 1978, charging that the EEOC had violated the heart of the law it was created to enforce. The federal trial court agreed. In ruling that the EEOC's Special Hiring Plan constituted a violation of Title VII by the guardians of Title VII, the trial judge found that:

> The EEOC claims to have been determined to build an agency that exemplified its congressionally imposed duty to attack discrimination. That determination can hardly be faulted. In its execution, however, the EEOC fell into a now not unfamiliar trap of believing that matches of employee profile and workforce makeup is the legal command. No law known to this court so requires. Such a fixation with numerical matches is fraught with the hazard of standing the noble goal of Title VII on its head by highlighting the status it seeks to blur.[55]

The judge concluded with an editorial obiter dictum: "I do not intend to demean the high-mindedness of the EEOC, despite the close kinship of that quality with high-handedness . . . . [but] in plain language, this is a case of the cow stepping into the bucket."

The War Over Reverse Discrimination

Why did the cow step into the bucket? Why was it that by the early 1970s the legal majesty of the American state once again, as it had in the segregationist era between *Plessy* and *Brown,* ordained that citizens who had wronged no one must be denied important rights and benefits because of genetic attributes like the color of their skin?[56]

Proponents of the equal-results test of affirmative action that evolved during Phase II would reject the judge's rural metaphor. Their arguments against the sufficiency of well-intentioned but "mere" nondiscrimination carried considerable force. From its origin in the Wagner Act to President Kennedy's borrowing for his executive order of 1961, the term "affirmative action" had always required something more than mere even-handedness or nondiscrimination. The "something" remained ambiguous, to be sure. But it ranged from the NLRB's opposition to unfair labor practices within a larger presumption that labor unions were a desirable norm in modern industrial society, to the PCEEO's insistence that employers make special recruiting efforts to hire "underrepresented" minorities. The notion of underrepresentation, with its implicit, ethno-racial standard of proportional distribution, provided the key for the transition from the soft affirmative action of the Kennedy approach to the hard or equal-results affirmative action developed under Nixon. In between, the Johnson administration courageously won the great campaigns of 1964–65, then struggled inconclusively with the complexities of implementation while the Vietnam War consumed its energy and its future.

Johnson's footrace metaphor in the Howard University speech of 1965 captured allegorically the historical core of the compensatory approach: innocent minorities had inherited a crippling legacy, and social justice therefore required a period of preferential discrimination. Johnson never followed through on his speechwriter's rhetoric. But the non-labor wing of the civil rights coalition rather quietly fashioned a results-centered rationale that countered the equal-treatment creed with an essentially historical argument about institutions and culture. They insisted, with de Maupassant, that equality must mean more than the equal right of the rich and the poor to sleep under the public bridges. Liberalism's historic command not to discriminate could not achieve its goal of a color-blind society because the racism of the past had become institutionalized. Current institutions thus might perpetuate discrimination even though no one in those institutions remained personally prejudiced.[57]

The camel's-nose logic of affirmative action, and the historical argument that past discrimination required compensatory remedies, struck a resonant chord on the federal bench. To judges, the doctrine of make-whole relief provided a familiar rationale for compensatory remedies, even in the absence of statutory provisions. The traditional concept of make-whole relief, as in prohibitory orders or the award of money damages, was being driven by civil rights litigation during the 1960s toward a new norm of equitable remedy. Traditional make-whole relief had typically involved a dispute between private parties leading to a judge-ordered, one-time, one-way transfer that retrospectively righted a claimed wrong. But the new equitable remedies sought to balance the class-action interests of multiple parties (some of them absent) rather than to declare specific winners and losers in a bipolar dispute. The school desegregation cases had invited the rapid expansion of the logic of equitable remedy, and the defiance of southern die-hard regimes provided public legitimacy for the growing assertiveness of the federal courts in social policy during Phase II.

In *Contractors Association, Allen, Gaston, Swann, Griggs,* and similar test cases the federal courts embraced the compensatory argument to justify as equitable relief a preferential treatment for minorities in fields as varying as voting, employee and union seniority, education, hiring and promotions, and job testing.[58] Increasingly the judicial decree sought less to judge past behavior than to adjust future behavior. As the courts declared that Congress's overarching substantive purpose (more jobs for minorities) was more compelling than its specific procedural prohibitions (no discrimination in a color-blind society), sympathetic legal scholars argued that the Warren-era judges were correcting a formalist bias of the late 19th century. Anglo-American law had then emphasized fault and blameworthiness, defined punitive damages in tort and criminal prohibitions in terms of malicious "intent,"

and held that courts properly interpreted statutes only by divining and applying their legislative intentions.[59] Thus in Phase II the intent test in statutory
interpretation was replaced by statistical demonstrations of disparate impact
or disproportionate results. Hard affirmative action through preferential remedies was transforming nondiscrimination into modern, no-fault civil rights.

The classic liberals or "neoconservatives" who attacked Phase II's radical
shift during the early 1970s concentrated on the specter of "reverse discrimination," and throughout the 1970s the running dispute centered on a string
of controversial lawsuits brought by white males—most notably those of Marco
DeFunis, Allan Bakke, and Brian Weber.[60] The results were mixed and the
judicial indecision continued through the 1980s. The philosophical dispute
will continue because it is not resolvable through logic and evidence alone.
Historians cannot resolve society's philosophical disputes. But historians now
enjoy a full generation perspective on the events of the 1960s, and with that
wider angle of vision they are obliged to transcend the dispute between lawyers and judges and legal scholars over reverse discrimination. That debate
has been over whether the cow *should* have stepped into the bucket. It remains a vital and unresolved question, but the learned arguments of lawyers
and judges (and philosophers) cannot satisfy the historians' need to understand *why* the civil rights reforms of the 1960s passed so rapidly through two
phases that contained such severe internal tensions and contradictions.

Looked at from this longer and wider perspective, the two-phase transformation of civil rights policy takes on an importance and meaning that
transcends its salience as a morality play over principles that were either betrayed or fully matured. Instead, from the perspective of the 1990s the Phase-
II shift of civil rights policy appears to have been the unwitting cutting edge
of a vast but quiet revolution in the nature of the American state itself.

The Quiet Revolution in the American Regulatory State

We have long associated the 1960s with revolutions manqué—with suburban
guerillas and with commercial breakthroughs in blue jean design, all heralded
by maximum rhetorical exposure and minimum lasting significance. But the
civil rights revolution, which was acutely self-conscious as a legitimate claim-
ant both to revolutionary title and impact, was part of a larger regulatory
shift that lacked a central or self-conscious direction. Indeed, unlike Phase II
of the civil rights revolution, it lacked even the modicum of self-awareness
necessary to signal its presence and galvanize opponents in the time-honored
Madisonian fashion. The phenomenon in question was a deep, national shift
in the American administrative state, beginning around 1960, away from the
consolidation that followed the New Deal and World War II, and toward a
regulatory apparatus that paradoxically combined disaggregation with growth
and with even greater intrusiveness by government.

The shift was quiet, massive, unanticipated, and largely unperceived. Its
components were unconnected by coherent design, yet together they ushered

in a new era of social regulation with profound but imperceived consequences for civil rights policy.[61] The federal budget, which had grown to $92 billion by 1960 but was almost half consumed by defense outlays, doubled by 1970 to reach $195 billion. During the same period, however, the defense share of the budget dropped by one-fifth, even in the face of the Vietnam War. This was the Great Society's "welfare shift." It endowed an enlarged array of social agencies with new domestic programs and funds, and provided them with unprecedented discretion in distributing these funds to new constituencies.[62]

The political need to avoid an explosion of Washington bureaucrats produced a federal bureaucracy that added only 400,000 employees during the 1960s, and then remained relatively static during the 1970s. But this coincided with a genuine bureaucratic explosion in state and local government, where employment increased during the 1960s by 4 million or 40 percent.[63] The relative freeze in federal personnel led to a pattern not only of farming out public services to state and local governments, but also of jobbing them out to a growing army of private contractors. These were the "beltway bandits" whose skill at winning federal contracts and grants extended far beyond the infamous Washington beltway and its spectacular repository of interest-group associations. Accompanying this process was an explosion in federal regulations—between 1969 and 1974 the number of pages in the *Federal Register* sextupled.[64] Finally, and most relevant to our civil rights focus, there occurred a vast mobilization of constituencies, as the Great Society programs and the surprising initiatives of the Nixon administration in environmental and consumer protection and safety spun a mosaic of new "iron triangle" networks across the face of government.

The convergence of these trends had created by the end of the 1960s an administrative apparatus that was sprawling, complex, and disjointed. It linked not only federal, state, and local governments but also a host of private organizations. Its growth coincided with the grass-roots, decentralizing, citizen-participation movement of the 1960s, and the combination created "something new . . . the explicit adoption of citizen participation as an adjunct to bureaucratic decision making. It signified that administrative agencies were now mobilizing and organizing political constituencies on their own."[65] Students of public administration have called the regulatory results of this marked shift of the 1960s the "new social regulation."[66]

Paradigm Shift of the 1960s: The New Social Regulation

The new social regulation emerged from an unplanned and unanticipated confluence of the distinctive grassroots and citizen-lobby movements of the 1960s—the civil rights, antiwar, environmental, consumer protection, and worker health and safety movements of the 1960s. Its common denominator was a demand that citizens be protected from being harmed by private or government behavior. Its means were a combination of executive and legislative guarantees and new social programs that emphasized the rights of consumers and employees over the needs of producers. These new programs were to be policed or run either by the established mission agencies, such as the Office of Education in HEW, the OFCC in Labor and the OMBE in Commerce, or more characteristically by the establishment of independent new agencies, like the EEOC or the Environmental Protection Agency. A chief advantage of new agencies was their freedom from the constraints imposed by established clientele relationships in the old line agencies. Between 1900 and 1964, only *one* such regulatory agency had been established at the federal level with such a primary responsibility: the Food and Drug Administration, created in 1938 out of the old Bureau of Food and Drugs. Between 1964 and 1972, however, *seven* new federal regulatory agencies were created by Congress with social regulation as their mandate. These were the EEOC (1964), the National Transportation Safety Board (1966), the Council on Environmental Quality (1969), the Environmental Protection Agency (1970), the National Highway Traffic Safety Administration (1970), the Occupational Safety and Health Administration (1970), and the Consumer Product Safety Commission (1972).

The new social regulation differed from the traditional norms of economic regulation in several fundamental respects. Economic regulation sought to control such evils as monopoly, collective restraint of trade, price-fixing, and labor abuses by regulating market entry and rates and prices. This was typically pursued through a relatively unified and vertical form of agency surveillance that focused on coherent areas like railroads, the communications and power industries, the stock market, airlines, trucking.[67] Only a relatively modest number of major industrial managers felt its direct impact and dealt with its regulators. But social regulation confronted the reality or prospect of direct physical and economic harm to millions of employees, consumers, and citizens. Its form evolved through a common, three-step pattern that linked the civil rights movement to the movements for environmental, health, safety, and consumer protection, and that created a quite different model of regulation.[68]

The first step in the pattern was often a scandal or disaster that would command national and therefore congressional attention, such as the Thalidomide controversy in 1962, the Santa Barbara oil spill of 1969—or the racial brutality in Birmingham in 1963 and Selma in 1965 (where television played the publicist role of a Rachel Carson or Ralph Nader). Second, Congress would respond to the new urgency by creating a new agency and new programs. Congress would typically pattern the new agencies after existing commissions or boards like the FTC and the NLRB, and new program jurisdictions would be cannibalized out of the turf of existing administrative agencies, like HEW and Commerce.

This approach, however, often produced poor fits. New social programs and their administrative units often coexisted awkwardly within the established relationships of the big mission agencies. Thus the OFCC, as a special and novel project of Secretary Wirtz, was resented by the old-line constituencies in the Labor Department. President Nixon's creation of the OMBE elicited a similar lack of enthusiasm from the established constituencies in small business. Nonetheless, such bureaucratic tensions tended to diminish and sort themselves out in the mission agencies, once the triangular coalitions adjusted to the new realities.

In the regulatory agencies, however, the congressional instinct for shaping the new social agencies according to the old economic model caused major problems. As the first of these new entities in the modern era, the "poor, enfeebled" EEOC got caught up in crossfires over role and function that *no* one understood at the founding. For Congress the standard model of regulation was economic. Its history reached back to the creation of the ICC in 1887—the founding of the regulatory era and its "headless fourth branch of government." The economic model of regulation generally dealt with the challenges that came before it on a retail, case-by-case basis, and used court-like procedures that were basically adjudicatory. Agencies like the FTC and the FPC dealt typically with dichotomous variables, like approving or disapproving a challenged rate, price, merger, or unfair labor practice. For this reason the main club or sanction of existing regulatory agencies took the form of disapproval or cease-and-desist authority. Their trial-like hearing procedures were governed by the judicial model of the Administrative Procedures Act of 1946. Similarly their verdict, in the judicial analogy, was essentially yes or no.

Social regulation, on the other hand, tended to deal not with dichotomous variables but rather with a complex spectrum of risk. It addressed such questions as how much use or discharge of what chemical over what period of time would impermissibly pollute the food or air or water. This form of regulation, however, posed a dilemma for Congress. Pressed to act by a broadly based public outcry, Congress was nevertheless uncomfortable with the *redis*tributive politics and the zero-sum implications of much of the new social regulation (clean air regulations could force millions of old cars off the roads and jack up the price of new ones). So Congress typically responded to the crisis by giving a contradictory set of charges to the new agencies of social regulation. They must make sure that the specific triggering offense never happens again. But they must at the same time cast a wide net to prevent a broad class of potential but still unknown harms. The new agencies must move quickly and decisively. But they must also grant wide access in thorough policy deliberations to the newly mobilized armies of citizen participants. Congress in its founding statutes therefore often spelled out elaborate procedural directives for the new agencies of social regulation. But all the same Congress did not stipulate clear substantive policies with definitions and standards, because Congress lacked both the expertise and the will to do so.[69]

The result, in the third stage, was that the new agencies of social regulation developed implementing strategies that created a new model of regulation. Their mode of enforcement was wholesale rather than retail, and their administrative style was more legislative than adjudicatory. This therefore led not to cease-and-desist orders, but rather to a detailed process of "notice-

and-comment" rule-making. Proposed guidelines, regulations, and standards were published in the *Federal Register*. Comments were received, hearings were held, revised rules were promulgated—often in lengthy and technical detail.

Given contradictory and ambiguous mandates from Congress, and given also a vast and horizontal, layer-cake structure of regulation (covering the entire nation's air, water, worker safety, consumer health), the new regulatory agencies adopted notice-and-comment rule-making as a procedure for two reasons. First, it maximized their political freedom to respond to their organized and insistent new constituencies. Second, producing detailed standards with which producers must comply proved to be a faster, less expensive, and farther-reaching method of enforcement than the retail, case-at-a-time approach of the adjudicatory model. The rule-making approach of the social regulators won ready approval from federal courts, which had learned to defer to the expertise of the economic regulators when confronting highly technical questions of scientific import.[70]

The rule-making rather than the cease-and-desist model of enforcement provided legal advantages to the social regulators as well. In 1971 in the *Griggs* case, the Burger Court deferred to the EEOC's new guidelines on employee testing. But in doing so the Court pushed the substantive result of *Griggs* far beyond the mere testing question at hand, and granted broad approval to the disparate impact version of the equal-results test. Although Congress in the Civil Rights Act of 1964 had required an intent test and had prohibited proportional ratios and quotas, Congress generally resisted reopening the debates of 1964 and rearguing on constant appeal the rulings and guidelines of the new civil rights agencies. By 1972, Congress was anxious to resolve the nagging disputes over jurisdiction and authority in civil rights enforcement so it could free itself from substantive controversies and channel them through the regulatory agencies and, when challenged, the courts.

The regulator's rule-making approach brought a further legal advantage: it blocked noncompliance by complaining regulatees by depriving them of legal defenses that had previously been sufficient. The promulgation of a minimal pollution standard that had to be met, for example, effectively severed the old common-law connections in tort that had tied the nature of the relief to the specific harm done. The technical standards severed the direct link between intent and harm, because evidence of damage was sufficient, and hence proof of the polluter's evil designs need not be adduced. The social regulators' main goal remained the reduction or prevention of future pollution. Hence it was forward-looking, rather than retrospective, as in the standard adjudicatory model. Such regulatory standards tended to sever the old direct link of tort between harm and relief. The agency's prescribed standards, as a forward-looking form of equitable relief, could be justified to prevent future environmental harm. Or, in the case of civil rights, agency standards could compensate for an aggregated accumulation of past harm that required no specific finger-pointing. Thus the relevance of intent was greatly weakened, and social regulation followed a no-fault rationale.

Civil Rights Policy and Social Regulation:
Similarities and Differences

The birth of the new civil rights agencies and offices of the 1960s, including especially the EEOC and the OFCC, marked the national transition to the new era of social regulation. Prior to the 1960s, social regulation had remained a state and local responsibility in the Madisonian scheme. Wide veto power against its centralization under Washington bureaucrats was exercised by the political parties, regional and economic groups, and the congressional seniority system. So the new civil rights agencies had to cut the path, caught up between the old forces and the new. Modeled after the NLRB, yet deprived of the quasi-judicial, cease-and-desist authority that typified the economic model of regulation, the EEOC turned for support to its constituency of newly mobilized minorities, and it gravitated quickly toward the new model of wholesale rule-making. In hindsight, the stubborn opposition of Senator Dirksen and the congressional conservatives to cease-and-desist authority for the EEOC appears to have been short-sighted and self-destructive of their larger interest in preventing another regulatory runaway, like the CIO capture of the early NLRB.[71]

It is one of the great ironies of the civil rights era that the time-tested judicial strategy of the congressional conservatives backfired on them so badly. History, however, cannot reveal whether a straight NLRB or FEPC model could have long insulated the EEOC from the powerful magnet of rule-making that was driving the new wave of social regulation. The power of this shift during the 1960s swept up not only the regulators themselves, but also the Congress. In 1960, a national FEPC was politically inconceivable, and a national Environmental Protection Agency was improbable. But only a decade later, Congress created the EPA, and then strengthened weak federal pollution legislation through far-reaching "amendments" for air in 1970 and for water in 1972. Similar revisions of existing statutes in 1972 greatly extended the reach of federal EEO regulation by race and sex in the previously sacrosanct areas of education and state and local government.[72]

The evils attacked by social regulators conferred upon them a halo effect. The civil rights laws of 1964, 1965, and 1968, with their nationwide bans on discrimination in public facilities and accommodations, and private jobs and housing, became politically unassailable. And all Americans wanted breathable air, drinkable water, cars that did not explode, toys that did not maim. By 1972 a consensus had formed that minimal standards in such matters were essential to protect the new range of public rights. The administratively and scientifically complex decisions required in formulating these standards were regarded as legitimate matters for negotiation and compromise. But their existence and function was not.

There is little evidence, however, that Americans during the administrations of Kennedy, Johnson, and Nixon saw much connection between social regulation in the interest of the environment, consumer products, worker safety, and public health on the one hand, and civil rights on the other. Neither did senior policy-makers in the Presidency or the Congress. With hindsight we can see the convergence of the two regulatory streams on a wholesale model of remedy involving notice-and-comment rule-making and standard-setting. But their differences remain fundamental.

The key to these differences lay in conflicting theories of public rights. For the consumer-environment regulators, all citizens deserved equal protection from the harms of polluted environs or dangerous products and services. The reach of these new protections was determined by statutory and administrative policy rather than by constitutional guarantees. The civil rights regulators, however, diverged from the equal treatment tradition after 1965, and shifted to an equal-results approach based on a model of proportional representation grounded in group rights theory. This required compensatory treatment for such discrete population subgroups as blacks, Hispanics, women, the disabled. It thus set them against the American constitutional tradition of equal individual rights, and as a result the minority preferences required by the civil rights agencies drew them, unlike the other social regulators—the anti-pollution monitors and the safe toy police—into thorny controversies like the constitutional suits over reverse discrimination.

Clientele Capture, the Agencies, and the Courts: From Iron Triangles to Quadrilaterals

A final difference between the civil-rights and the consumer-environment varieties of social regulation involved the old problem of clientele capture. Like the traditional "Big Six" regulatory agencies that concentrated on economic regulation, the new social regulatory agencies were spawned by populistic crusades against various corporate-government abuses.[73] But unlike the traditional regulatory agencies, the new social regulators resisted capture by the enterprises they were created to regulate. Capture of the Civil Aeronautics Board by the airlines, for example, was neither particularly difficult nor surprising. Given the typically vertical structure of economic regulation, the airlines could concentrate on one single-focused agency, with little to fear from the organized opposition of passengers. But because the social regulators held a horizontal, layer-cake jurisdiction over a wide range of enterprises, the prospect of being captured by the regulated industries was greatly reduced.[74] The multifarious producers of water and air pollution, for example, had few other common interests that might unite them and sustain their concerted action.

The result for the consumer-environment regulators was a considerable measure of independence, but purchased at the price of sustained political conflict. The high costs of pollution abatement and vehicle safety and similar measures tended to stiffen the unity of resisting producer associations to consumer-environment regulation. If they could not easily capture OSHA or the EPA, they could lobby fiercely with the White House and in Congress. As a result congressional oversight hearings in the 1970s became adversarial and whiplashed agencies like the EPA and the Consumer Product Safety Commission in the middle.[75]

In civil rights regulation, on the other hand, the regulators were offering employers a broadened applicant pool and workforce at relatively modest costs (heavier costs were levied against union investments in seniority). As a consequence the regulatees were divided; small employers represented by the U.S. Chamber of Commerce continued to resist, but not the larger enterprises of the National Association of Manufacturers. Basically the consumer-environment agencies concentrated their regulation on selected industries to protect the rights of all citizens, while the civil rights agencies regulated all employers to advance the interests of selected groups who had suffered from past discrimination.

Given the divisions among the regulated industries and, for the civil rights constituencies, the growing rewards of mobilization, the old prospect of regulatee-capture was replaced by the increasing likelihood of capture by the organized constituencies of the new social regulation itself. The relative evil of "capture," of course, lies in the eye of the beholder.[76] But the redistributive nature of the new social regualtion, especially in the civil rights area with its competition over jobs, promotions, appointments, and admissions, provided a mighty incentive to the organized constituent groups not only to escape the old law of capture, but to reverse it.[77]

A striking attribute of the civil rights era, especially during Phase II, was the success with which the victim-clients of the new anti-discrimination agencies did most of the capturing. As a newly self-conscious and empowered interest group within the American pluralist system, the civil rights coalition, with black groups in the vanguard and coordinated by the well-financed Leadership Conference on Civil Rights, claimed the dominant client role. The first agency to be so "captured" was the Civil Rights Commission. It had originally been designed, like most national advisory commissions, to reflect a cross-section of the nation through a mixed distribution of party, section, race, sex, and political inclinations. But by 1964 it spoke with one voice that was indistinguishable from that of the Leadership Conference. Then in 1965 came the OFCC in Labor, which was captured at its founding (not unlike its parent department, and Agriculture and Commerce as well). Similarly captured at the founding were the proliferating offices of civil rights within the mission agencies, especially Justice, HEW, Defense, HUD. The EEOC between 1965 and 1968 followed the pattern of the Civil Rights Commission, and for the same reason of convergence of interest and function. In 1969 Nixon added the OMBE in Commerce.

The civil rights coalition then reached outward into the equivalent agencies in state and local governments and ultimately into the parallel echelons of personnel and management throughout the larger institutions of private industry and commerce.[78] In a sequential development, women and Hispanics followed the black initiatives in the latter 1960s, and in the 1970s a snowballing effect added Asians and American Indians, the elderly and the physically and mentally disabled, and drew new claims from lesbians and gays and other groups. Parallel activity swelled the ranks of the environmentalists and consumer advocates.[79] Interest-group lobbying for social regulation probably reached its apotheosis of effectiveness in the mobilizing of the physically handicapped during the 1970s.[80]

The administrative and regulatory agencies in the new fields of social regulation thereby joined their organized constituent groups to form the base of triangular relationships with the appropriate congressional committees. But in the new social regulation, the old iron-triangle paradigm, which historically had described the relatively closed policy worlds of the Army Corps of Engineers and the agricultural interests, no longer seemed to explain how policy was made.[81] What was happening instead was that in the expanding programs of social entitlement, the old triangles were stretching their contours into quadrilaterals. This new, fourth policy intersection represented the increasing intervention of the federal courts.[82] From the perspective of the late 1970s, Martin Shapiro observed how this new quadripartite dance of negotiation had maximized the advantage of the organized constituencies over the administrative agencies themselves in bargaining for expanded benefits:

> Congress or the courts or both give the interest group standing to lobby the agency and the courts. The agency knows that, unless it satisfies the group, the group will sue, thus increasing the cost and delaying the implementation of the proposed policy. The agency knows that unless it anticipates the policy views of the courts, the judges will find some way to reverse or at least delay the proposed policy if a suit takes place. The interest group knows that the cheapest thing to do is to persuade the agency. It also knows that, if properly approached, courts may strengthen its statutory entitlements, thus providing a stronger base for negotiation with the agency.[83]

The old iron triangles were transformed into quadrilaterals primarily by the ability and increasing willingness of the judiciary, in response to the lawsuits of the interest groups as plaintiff-clients, to monitor and reverse the decisions of both the Congress and the administrative agencies in the expanding fields of social legislation.

This new mode of public law litigation represented a radical shift in the role of the courts.[84] The lawsuit was no longer a bilateral dispute between private parties, but rather a grievance about public policy involving multiple parties. The courts' fact-finding was predictive and legislative rather than retrospective and adjudicatory, and the relief was typically an on-going, affirmative decree by a participant-judge. In his enthusiastic analysis of this seismic legal shift, Abram Chayes conceded that the new public law litigation "inevitably becomes an explicitly political forum and the court a visible arm of the political process"—a process in which the trial judge has "passed beyond even the role of legislator and has become a policy planner and manager."[85]

Temporary Remedies and Permanent Politics

By the middle 1970s the expansion of judge-made social policy had threatened, in the eyes of some critics, to supplant the imperial presidency, now crippled by Watergate, with an imperial judiciary.[86] Much of the growing criticism came from conservatives who opposed the substance of the court-ordered policies themselves, such as school busing for racial balance or ethnic hiring quotas.[87] But the new role of the courts in civil rights litigation posed two major problems that were independent of policy preferences. One of these attracted immediate attention and concern from across the spectrum of political philosophy: if the courts were to become increasingly a political forum and a visible arm of the political process, and the judge was to become an active fact-finder, negotiator, planner and manager and also monitor of social policy, how then was the judiciary to maintain its independence, integrity, and prestige as a neutral third-party for settling disputes? Was the federal judiciary, through its quadrilateral involvements and its increasing politicization through the new public law litigation, degenerating into a "jurocracy" in which judicial principles disolved into a relativistic sea of political preferences?[88]

The second problem was not so immediately apparent. It involved the relation between means and ends in constitutional and civil rights history, and especially the link between temporary measures and permanent principles. Wartime or health emergencies, for example, were permissible until the war or epidemic was over. If some kinds of minority preferences were regarded by the federal courts as a necessary but temporary measure in social policy, looking toward a transition to a color-blind Constitution, then what criteria would govern the transition? When the federal courts in 1969 began approving numerical quotas and ratios as compensatory relief for past job discrimination, they generally did so with apologetic disclaimers, looking toward the day when the historic discrimination had been balanced off, and the Constitution could properly assume its color-blind character. But as in the case of the school busing decrees, the courts adduced no clear set of principles according to which compliance might be judged, and in response to which the judicially enjoined institutions might be allowed to return to their normal independent operations. When Chief Justice Warren Burger struck down the Duke Power Company's employee tests in *Griggs* in 1971, for example, he explained that the Civil Rights Act "does not command that any person be hired simply because he was formerly the subject of discrimination, or because he is a member of a minority group."[89] "Discriminatory preference for any group, minority or majority," Burger explained, "is precisely and only what Congress has proscribed." But the long-run practical result of *Griggs* was to mock Burger's disclaimer.

Seven years later, Burger voted with the majority in the *Bakke* case to order the University of California to admit Alan Bakke to medical school. Justice Harry Blackmun voted to uphold the university's minority quota, and thus to exclude Alan Bakke because he was white. But in doing so Blackmun nevertheless insisted that "I yield to no one in my earnest hope that the time will come when an 'affirmative action' program is unnecessary and is, in truth, only a relic of the past."[90] Blackmun explained that the United States was going through a regrettable but necessary stage of "transitional inequality." He hoped, however, that "within a decade at the most," American society "must and will reach a stage of maturity where acting along this line is no longer necessary." "Then persons will be regarded as persons," Blackmun said, "and discrimination of the type we address today will be an ugly feature of history that is instructive but that is behind us."

The following year Eleanor Holmes Norton, who had been appointed by President Carter in 1977 to chair the EEOC, explained that "[t]he affirmative action tool was invented only after years of appalling evidence showed that discrimination immobilized some groups in the work place."[91] "Not only has the tool been limited to such groups," Norton said, "but there is a general consensus in the society that it probably ought to be limited, except in the fairly restricted circumstances where certain white groups have experienced discrimination, and that it ought to be temporary." That same year Justice Blackmun in *Weber* agreed that affirmative action quotas were a "temporary tool for remedying past discrimination without attempting to 'maintain' a previously achieved balance."[92] Yet despite these disclaimers from the highest authorities, everything we know about the normal politics of social regulation points in the opposite direction.[93] In pluralist America, interest groups have historically entrenched themselves in the political infrastructure in defense of their claimed rights and entitlements. The organized beneficiaries of affirmative action programs have entrenched themselves with no less energy than have the beneficiaries of similar group-based entitlements among farmers, veterans, homeowners, rentiers, the elderly.

Thus, during Phase II of the civil rights era, two contradictory forces emerged under the rubric of one implementing strategy. One was an exceptional rationale that minority preferences were a temporary expedient, defensible only as a bridge to the color-blind Constitution of individual rights. The other was the normal pluralist process that embedded the new model of proportional representation, group rights, and equal results deeply within the social and political structure. But within these tensions lay another source of contradiction: the two dominant political blocs in the civil rights campaign, blacks and women, followed paths that hid fundamental differences beneath surface similarities.

Black Rights, Women's Rights, and the Race-Sex Analogy

Feminist leaders of the 1960s shared a belief, based on a historical analogy of shared victimhood from group discrimination, that the war against sex discrimination should be linked to the principles and methods of the black civil rights movement. This after all had been the original model of the 19th century's first-wave feminism, which had close ties to the abolitionist movement before the Civil War and demanded during the Reconstruction that women's equality be linked to black equality. Excluded then, American feminists had continued to play catch-up. Their 19th Amendment was the blacks' 15th; their unenacted ERA was the blacks' 14th. Their eleventh-hour inclusion in the civil rights act designed for blacks in 1964 came from a surprising tactical coup on the House floor that was replete with its own ironies of strange and expedient political alliance. The flagship organization of second-wave feminism, the National Organization of Women, was created in 1966 to demand that the new EEOC fight sex discrimination with the same zeal and weapons the agency was devoting to racial discrimination.[94]

But in order for the second-wave feminist leaders to exploit the race-sex analogy by demanding the same remedies accorded to blacks, they first had to resolve their historic internal dispute. For generations American feminism's historic fault line had separated cultural feminists who defended the special-protection legacy of the liberal Democratic establishment and egalitarian feminists who demanded the ERA. This class-based split had set their liberation model apart from the black model. Generations of racial oppression had so flattened the black class pyramid that the dominant stream of black protest had been classically egalitarian. The flagship organization of black liberation in the 20th century had been the NAACP, and Thurgood Marshall's brief in *Brown v. Board of Education* had echoed the NAACP's consistent, equal-treatment credo: the Constitution was color-blind.

Women, on the other hand, were evenly distributed throughout the social pyramid, and hence they mirrored its class divisions. This had endowed American feminism with its pronounced dualism, and much of the feminist energy of the early and middle 1960s was devoted to defusing the ancient quarrel between the senior protectionists of the Women's Bureau coalition and the ERA coalition of the future—the Rawalt-Friedan-Griffiths Democrats and the Glaser-Allan-Gutwillig Republicans, joined in a bipartisan phalanx for the ERA.[95] The second-wave feminist shift from protectionist dominance at the beginning of the 1960s to ERA dominance by 1970 was thus a class-related transition. Power had shifted from the Democrats' traditional blue-and pink-collar constituency to a bipartisan coalition of business and professional elites. The shift was fueled primarily by demographic and economic trends and by the self-selectivity of an affluent and highly educated national leadership.

The federal courts, to be sure, played their role in providing incentives, as they had with black civil rights—but in an opposite direction. On the one hand, beginning in 1968 the lower courts began to strike down the state protection laws on Title VII grounds, thereby undercutting the foundations of the Women's Bureau coalition. On the other hand the federal courts continued to fail to find women in the 5th and 14th amendments to the Constitution. It was not until 1971, in *Reed v. Reed,* that the Supreme Court sustained a feminist challenge on 14th Amendment grounds, and by then it was too late to deflect the ERA juggernaut in Washington by arguing that a constitutional amendment was now unnecessary. By 1972 mention of the federal courts at a NAACP convention drew predictable and instinctive applause. But at NOW conventions during the first half-dozen years, mention of the federal bench drew hoots of derision.

The year 1965 was a watershed for women in a dual sense. It marked the opening for business of the EEOC and launched the implementation process for the liberal reforms of 1964–65, thereby providing a catalyst for the egalitarians of NOW to rout the special protectionists and unify the second-wave feminists under the banner of the ERA. Thus for egalitarian feminists, 1965 reversed the old momentum, sending special protectionists into decline and reviving the ERA for congressional passage after a half-century in protectionist limbo. Yet, at the same time, 1965 marked the beginning of the transition from Phase I to Phase II of the civil rights era, when the cutting edge of the black civil rights movement began to turn against the equal-treatment doctrine as a consistent, coherent, and sufficient theory of liberation. Just as the women's movement was at long last solidifying behind the equal-treatment creed of the ERA, the black civil rights movement was beginning to abandon it.

At the level of rhetoric, feminist leaders continued throughout the decade to salute the logic of the race-sex analogy. They were consistent, for example, in insisting that women be included in the Title VI goals and timetables. But analogizing race and sex discrimination has been analytically helpful only to a point, and the analogy has been made to carry too much weight.[96] By 1972, the heart of feminist leadership identified more with the Myra Bradwell tradition of aspirations for professional school than with Rosie and her riveting.[97] In 1960 the dominant feminist leadership was Democratic and ideologically rooted in a working-class constituency. The women of the Women's Bureau coalition were determined on the one hand to preserve special legislation that protected group rights, and on the other to press for new advances through amendments to the New Deal's fair labor-standards formula that would enlarge the protections of this class. But by 1972, when the ERA was sent by Congress to the states, the cutting edge of second-wave feminism was bipartisan, rooted in the middle- and upper-class culture of business and the professions, driven by classic liberalism's defense of individual rather than group rights, and united to attack barriers to women's advancement through Title VII and the ERA.

Thus midway through the era, the two great engines of the civil rights movement passed metaphorically like trains in the night. The women were finally moving in unity under elite leadership and the ERA banner toward a common standard of citizen rights. But the black movement, riven by ghetto riots and cries of Black Power, and caught up in the quadrilateral momentum for racial preferences, shifted its emphasis toward racial difference.

The Irony and the Triumph of the Civil Rights Era

Like all periods of profound social transformation, the civil rights era produced changes with unintended and ironical consequences. The strategic role reversal of the black civil rights movement and second-wave feminism was one of them. Another was the boomerang of the congressional conservatives' judicial strategy, which further empowered a federal judiciary that was abandoning the conservative instincts and judicial restraints upon which the strategy had been historically premised. This led to a triumph of nondiscrimination in 1964–65 that with surprising rapidity added requirements of minority preferences, invoking in the process the heavy hand of history to justify its compensatory logic.

All this culminated under the putatively conservative regime of Richard Nixon. His administration was marked by public disputes over busing and Supreme Court nominations that increased racial polarization. Yet hidden beneath its divisive public rhetoric was a quiet and surprisingly effective process of desegregating southern schools and public institutions. It was also under Nixon that the permanent government invented and consolidated the machinery that would nationally enforce Phase II's decrees of hard affirmative action. The Nixon irony is sharpened by the contrast between the relative moral indifference that characterized the political calculus of Nixon's domestic policy decisions, and the passion that drove President Johnson to strike at the jugular of America's caste system. If there is an unanticipated hero in this story about the entrepreneurs and manipulators of federal policy, it is Lyndon Johnson.

A further irony of the civil rights era involves the consequences of Phase II's affirmative action effort. What difference has it made? The historical evidence of 1960–72 cannot answer a question that depends on subsequent events. But a generation later, scholars and social critics remain locked in close debate. Sociologist Paul Burstein, in his study of the struggle for equal employment opportunity since the New Deal, describes the "great historical irony" of the epic campaign launched by American blacks: "one of the groups most responsible for the adoption of EEO legislation has not utilized it, while another group that had no articulated interest in the legislation has proved to be its major beneficiary." The former group was American Jewry. The latter was American women, who according to Burstein "were not for the most part involved in the struggle for EEO legislation on their own behalf, [yet they] have proved to be its main beneficiaries. . . ."[98] Burstein's analysis of the economic consequences of the EEO legislation of 1963–72 led him to conclude that while the proportional group share of income earned by nonwhite men held steady after 1964, and the share of white men declined, the rate of increase for nonwhite women increased by half, and that of white women doubled.[99]

A final unintended consequence has become so commonplace that it has lost some of its ironical force. This is the tendency in pluralist America for public programs launched to help the less fortunate in society to fall under the control of highly organized groups and be run in the interest of the middle or more affluent classes.[100] Social scientists ranging from Glen Loury on the right to William Julius Wilson on the left have agreed that the affirmative action programs of Phase II, which were politically sustained by an anti-poverty (and anti-riot) rationale that was clearly race-centered, have mainly served to benefit the rapidly growing black middle class, while leaving the black underclass further isolated and decaying.[101] Affirmative action programs whose benefits are automatically triggered by racial, ethnic, or gender traits have therefore become vulnerable in their social rationale because they conflate an immutable, genetic status with a variable, socio-economic one.[102]

The continuing debate over the effectiveness of affirmative action programs finds its origins in the civil rights era of 1960–72, but not its resolution. Its ironies abound. But ironists flirt dangerously with cynicism, because the ironic stance allows us to comment coyly, armed by hindsight, on the paradoxical contradictions that entrap historical actors and confound their best-laid plans. It is the *intended* consequences of the civil rights era's great social reforms, however, that mark its most fundamental and promising breaks with the continuities of history. In 1960, most American blacks still confronted the formal as well as informal symbols of daily humiliation that marked a caste society. Southern whites, the chief but not the exclusive offenders, were also—but unknowingly—victimized economically and brutalized psychologically by the daily corruptions of caste dominance.

Similarly in the America of 1960, males dominated the worlds of power and status so overwhelmingly that most women and girls could realistically aspire to an independent status and achievement, beyond the rewards of housewifery and motherhood, little loftier than that of airline stewardess. Even that was reserved for white women only—young, slim, and single. While the percentage of women who worked outside the home continued to rise in 1960, their horizons for reward in income and status had in many ways declined from levels reached earlier in the century. Most American men and boys took for granted a world of male norms in which women were auxiliaries, and in which the creative talents, beyond the acknowledged procreative capacities, of half of humankind remained largely untapped. Feminism was at bottom about gender freedom, and everybody had a gender. Martin Luther King understood that the same principle obtained about racial freedom for all of God's children.

The civil rights era was a rare American epiphany. Its mixed legacy now defines the conditions of our civil life, and commands our best efforts to understand.

Essay on Sources

Archival Sources: The Presidency

Because the modern American administrative state has been dominated since Roosevelt and the New Deal by the executive branch, historians of national policy have anchored their research in the presidential library system. The archival core of this study is drawn from the John F. Kennedy Library in Boston, Massachusetts; the Lyndon Baines Johnson Library in Austin, Texas; and the Richard Nixon Presidential Materials Project in Alexandria, Virginia. Documents were also drawn from the Dwight D. Eisenhower Library in Abilene, Kansas, and the Gerald R. Ford Library in Ann Arbor, Michigan. Litigation associated with the Watergate episode has prevented the normal development of a Nixon Library. But the public access and the research services offered by the National Archives staff at the Alexandria repository, which opened its first files to the public late in 1986, follow the standard policies of the presidential libraries.

Beginning in 1961, White House Central Files officials in coordination with the National Archives adopted a standard system of classification for the operational files that would also be used by the presidential libraries. The heart of the presidential libraries lies in the White House Central Files (WHCF), especially the alpha-numeric subject file, which contains approximately sixty primary subject categories and more than 1000 subcategories. For all administrations since the Eisenhower presidency, the key file for students of civil rights policy is Human Rights (HU). The HU file is subdivided into categories like Equality (a general classification that during the period 1960–72 referred mostly to issues in black civil rights), Education, Employment, Housing, Voting, and Women.

For research in civil rights policy, or in domestic social policy generally, other important WHCF files in the post-1960 presidential libraries are Legislation (LE), Judicial-Legal Matters (JL), Education (ED), Federal Aid (FA), Labor-Management Relations (LA), Housing (HS), Local Governments (LG), States (ST), Welfare (WE), and Meetings and Conferences (MC). Especially important is Federal Government— Organizations/Agencies (FG). The FG files include, in addition to the Legislative Branch, the Supreme Court, and cabinet-level departments, a broad array of subcabinet agencies and organizations, both permanent and temporary—such as regulatory boards, advisory commissions, presidential task forces, major interagency committees, and quasi-independent units like the Equal Employment Opportunity Commission (EEOC) and the Commission on Civil Rights.

The second major documentary source in the presidential libraries is the collection of office files produced by the senior presidential aides on the White House staff. These are variously called the aide files, staff member and office files, or the files of internal subunits like the Domestic Council, the National Security Adviser/NSC Staff, the congressional relations or liaison office, and Counsel to the President. Most important are the files of the chief aides, the senior presidential assistants whose titles varied widely and rarely indicated the range of their responsibilities. In the Kennedy administration, the aides most involved with civil rights policy were Theodore Sorensen, Harris Wofford, Lee White, and vice presidential aide George Reedy. In the Johnson White House they were Bill Moyers, Joseph Califano, James Gaither, and Harry McPherson. In the Nixon administration, the key aides were John Ehrlichman, H. R. Haldeman, Arthur Burns, Daniel Patrick Moynihan, Leonard Garment, Kenneth Cole, Charles Colson, and Charles L. Clapp. The WHCF and the aide files are supplemented by collections of donated papers, exit interviews, oral history interviews, audio-visual materials and other federal records.

The aide files are rich in executive-level or policy-related correspondence, memoranda, and reports, and they often contain sensitive and classified documents that are not found in the WHCF (although there is a great deal of duplication, including cross-reference copying by the archivists). The aide files best capture the most important policy debates involving the heads of the mission agencies, the Attorney General, the Budget Director, congressional liaison, special task forces and interagency committees, and also party officials and external interest groups. As the focus of all of this effort, the President dominates the dialogues with his omnipresence, but rarely participates directly—although this was less true of Richard Nixon.

So rich is the combined lode of the WHCF and the aide files that their availability in the presidential libraries reinforces the danger of viewing the policy-making process excessively from the White House perspective. To counter this inherent bias, three additional sources were used within the executive branch to provide a wider perspective. One was the papers of the Bureau of the Budget (BOB—after 1969, the Office of Management and Budget or OMB), which are available through 1972 in the National Archives. As the institutional memory of the Presidency, serving the dual functions of custodian of policy continuity and expert troubleshooter, the BOB/OMB produced a stream of internal reports to the Budget Director that are rich in policy insight and independent criticism.

Second, the files of the mission agencies and their heads similarly document agency views and debates that were screened from the White House. In civil rights policy the chief such source was the Secretary of Labor, whose office files (under secretaries Arthur Goldberg, Willard Wirtz, George P. Shultz, and James Hodgson) are also available through 1972 in the National Archives. The Labor Department's role was unique because the DOL provided the administrative support for the EEO committees headed by vice presidents Nixon and Johnson, and in 1965 Secretary Wirtz created the Office of Federal Contract Compliance Programs (OFCC) to enforce the executive orders and Title VI on contract compliance.

A final source of important documents on civil rights policy in the subpresidency, lodged within the executive branch but partly independent of it, is the EEOC. The minutes of the commission meetings, beginning in July 1965, are available at the EEOC Library in Washington through the 1970s. The names of complaint cases are deleted in the minutes, as required by the privacy provisions of the Civil Rights Act of 1964, but the internal policy debate is well documented.

Congress and the Federal Judiciary

Archival documents that reflect the inner workings of the Congress and the Judiciary are far less accessible for researchers than those generated by the Presidency. Consequently social scientists tend to rely on published documents rather than internal memoranda and reports, and on interviews with sources who often remain anonymous. Historians of the recent American past enjoy early access to the presidential libraries, but the archives of Congress and the federal courts, and also the case files of the Department of Justice, generally remain closed for much longer periods. As partial compensation, Congress publishes its floor debates and a massive volume of committee hearings, and also its briefer committee reports on legislation—although this normally excludes markup sessions and conference committees. Congress has provided, moreover, its own rough equivalent of the Presidency's BOB/OMB, in the form of the General Accounting Office (GAO). For this study the files of the Comptroller General on the Philadelphia Plan provided a unique angle of vision on intergovernmental relations.

As for the federal courts, documents reflecting the closed deliberations of the judiciary are not generally available for the recent period, and the courts have traditionally been jealous of their privacy. The oral arguments and printed decisions are generally available, however, and the discovery process in adversary proceedings in the trial courts can provide documentation not normally available to researchers. Beyond these published sources, this study has relied on the secondary literature, including the extensive discourse in the law reviews (see especially the endnotes for Chapter 15).

Archives of Nongovernmental Organizations

Outside of the federal government, the drive of the major civil rights organizations was coordinated by the Leadership Conference on Civil Rights (LCCR), whose valuable papers are available in the Library of Congress (LC). Also available in the manuscript division of the LC are the papers of the NAACP, which are extensive but give the appearance of having been sanitized prior to accession. The LC also holds the papers of the National Urban League, which has not been mined as extensively as the NAACP papers but is often more revealing, and also the papers of Whitney Young, Jr. Within the LCCR, the split between the black and labor camps, which the Nixon administration sought to exploit, is best documented from the union side through the AFL-CIO's George Meany Memorial Archives in Silver Spring, Maryland. Hispanic issues generally were not a major policy concern of the federal government or of the LCCR during 1960–72, although they rapidly became important shortly thereafter.

Similarly, feminist groups such as the National Organization of Women (NOW) felt excluded from and often antagonistic toward the black-labor alliance represented by the LCCR. For women's issues and feminist politics the presidential libraries and the DOL papers are more helpful than the LC manuscript collections. See especially the Kennedy Library on the President's Commission on the Status of Women; the

Clapp file in the Nixon papers, which includes the President's Task Force on Women's Rights and Responsibilities; the Women's Bureau archives of the Labor Department; and the minutes and hearings of the EEOC.

Published Documents

Students of federal policy rely heavily on standard resources published by the U.S. Government Printing Office (GPO), such as the *Public Papers of the Presidents,* the *Congressional Record* and committee hearings and reports; the Supreme Court's *United States Reports;* and the dry but important *Federal Register.* Among the many reports of the U.S. Commission on Civil Rights, of particular use were *Political Participation* (1968), and *The Federal Civil Rights Enforcement Effort* (1970). The Civil Rights Commission also sponsored the study by Richard P. Nathan, *Jobs and Civil Rights,* which was published by the Brookings Institution in 1969. There were two reports on women's issues by presidentially appointed advisory groups: *American Women:* Report of the President's Commission on the Status of Women (Washington: GPO, 1963); and *A Matter of Simple Justice:* Report of the President's Task Force on Women's Rights and Responsibilities (Washington: GPO, 1970).

The Johnson Library holds detailed reports from the major agencies in its *Administrative History* series. Especially useful were the administrative histories for the departments of Labor and Justice, the Budget Bureau, and the EEOC. Additionally, the DOL Library holds an administrative history covering the agency's Nixon-Ford years.

Commercially published reference works include the Congressional Quarterly's essential *Congressional Quarterly Almanac* and the multi-volume series *Congress and the Nation.* The Bureau of National Affairs, also in Washington, published three useful compilations: *State Fair Employment Laws and their Administration* (1964); *The Civil Rights Act of 1964* (1964); and *The Equal Employment Opportunity Act of 1972* (1973). The *New York Times* and the *Wall Street Journal,* both indexed, provided generally superior news-reporting and also contrasting editorial stands on civil rights policy.

Secondary Sources: Books

Detailed references are supplied by the more than 1600 endnotes in this study. But a few secondary authorities merit special mention. General bibliographies for the civil rights field are plentiful—two recent examples are William H. Chafe, *The Unfinished Journey* (New York: Oxford University Press, 1986), 489–501; and Susan M. Hartman, *From Margin to Mainstream: American Women and Politics Since 1960* (New York: Knopf, 1989), 191–99. An unusually comprehensive guide, Floyd D. Weatherspoon, *Equal Employment Opportunity and Affirmative Action: A Sourcebook* (New York: Garland, 1985), contains 1133 bibliographic entries. In women's history, Patricia Zelman has mined the Kennedy and Johnson libraries in *Women, Work, and National Policy: The Kennedy-Johnson Years* (Ann Arbor: UMI Research Press, 1980). Cynthia Harrison's more recent synthesis, *On Account of Sex: The Politics of Women's Issues 1945–1968* (Berkeley: University of California Press, 1988), draws additionally on important Labor Department records and also the rich collection on women's history in the Schlesinger Library at Radcliffe College.

In black civil rights, the tradition of archivally based, monographic histories for each presidential administration reaches from Franklin Roosevelt to John F. Kennedy—and then stops, with Carl M. Brauer, *John F. Kennedy and the Second Reconstruction* (New York: Columbia University Press, 1977). In the baker's dozen years following 1960, the escalation of civil rights protest, Great Society reform, ghetto riots and Black Power, and Vietnam War and campus turmoil seem to have overwhelmed the monographic tradition. One response of researchers has been to narrow and intensify their focus, as did Charles and Barbara Whalen in *The Longest Debate: Legislative History of the Civil Rights Act of 1964* (Cabin John, Md.: Seven Locks Press, 1985). Another has been to attempt to place civil rights issues in a broader policy context, as did Allen J. Matusow's *The Unraveling of America: A History of Liberalism in the 1960s* (New York: Harper & Row, 1984). A few researchers, like Steven F. Lawson in *Black Ballots: Voting Rights in the South 1944–1969* (New York: Columbia University Press, 1976) and *In Pursuit of Power: Southern Blacks and Electoral Politics 1965–1982* (New York: Columbia University Press, 1985), have reconstructed the evolution of one aspect of civil rights policy through several administrations.

Most historians, it seems, followed the trend of the 1970s toward social history. Interest in national elites was displaced by bottom-up history at the grassroots, and during the 1980s we reaped a harvest of community studies of social movements. For a perceptive review of this literature see Steven F. Lawson, "Martin Luther King, Jr. and the Civil Rights Movement," *The Georgia Historical Quarterly* LXXI (Summer 1987): 243–60. Now that the presidential archives are open through 1980, and we have a generation of perspective between us and the 1960s, historians of federal policy are obliged to return to the mother lode of documents produced by the decision makers—although its mass grows ever more intimidating.

In that task a final obligation is to acknowledge the indispensable scholarship of nonhistorians. Because the reforms of the civil rights era linked law and social policy in so many novel ways, research by students of law, political sociology, public administration, and related disciplines of social analysis are essential. For historians the work of contemporary social analysts is especially important, and the following are recommended (listed in order of publication): Morroe Berger, *Equality by Statute* (Garden City, N.Y.: Doubleday, 1952); Jack Greenberg, *Race Relations and American Law* (New York: Columbia University Press, 1959); Paul H. Norgren and Samuel E. Hill, *Toward Fair Employment* (New York: Columbia University Press, 1964); Michael I. Sovern, *Legal Restraints on Racial Discrimination in Employment* (New York: Twentieth Century Fund, 1966); Ruth P. Morgan, *The President and Civil Rights: Policy-Making by Executive Order* (New York: St. Martin's Press, 1970); Alfred W. Blumrosen, *Black Employment and the Law* (New Brunswick, N.J.: Rutgers University Press, 1971); and Jo Freeman, *The Politics of Women's Liberation* (New York: David McKay, 1975).

The standard law text is Derrick A. Bell, Jr., *Race, Racism and American Law* (rev. ed.; Boston: Little, Brown, 1980). A recent assessment of the impact of EEO policy is Paul Burstein, *Discrimination, Jobs, and Politics: The Struggle for Equal Employment Opportunity in the United States since the New Deal* (Chicago: University of Chicago Press, 1985).

Notes

Introduction

1. During the 1980s historical research shifted from the national civil rights movement, with its focus on politics and law, to local communities where the grassroots mobilization of the social movement for black liberation could be reconstructed. In black civil rights this shift of focus is reflected in Charles W. Eagles, ed., *The Civil Rights Movement in America* (Jackson: University Press of Mississippi, 1986). A perceptive discussion of the grass roots literature is Steven F. Lawson, "Martin Luther King, Jr., and the Civil Rights Movement," *The Georgia Historical Quarterly* LXXI (Summer 1987): 243–60. Historical research on second-wave feminism is entering the archival phase of policy research for the 1960s—see Cynthia Harrison, *On Account of Sex: The Politics of Women's Issues 1945–1968* (Berkeley: University of California Press, 1988).

2. Paul Burstein, *Discrimination, Jobs, and Politics: The Struggle for Equal Employment Opportunity in the United States since the New Deal* (Chicago: University of Chicago Press, 1985), 15.

3. See Jeffrey L. Pressman and Aaron B. Wildavsky, *Implementation* (Berkeley: University of California Press, 1973); Erwin C. Hargrove, *The Missing Link: The Study of the Implementation of Public Policy* (Washington: Urban Institute, 1975); Eugene Bardach, *The Implementation Game: What Happens After a Bill Becomes a Law* (Cambridge: Massachusetts Institute of Technology Press, 1977); George C. Edwards III, *Implementing Public Policy* (Washington: Congressional Quarterly Press, 1980); Robert T. Nakamura and Frank Smallwood, *The Politics of Policy Implementation* (New York: St. Martin's Press, 1980); and Walter Williams et al., *Studying Implementation* (Chatham, N.J.: Chatham House, 1982); Garry D. Brewer and Peter deLeon, *The Foundations of Policy Analysis* (Chicago: Dorsey Press, 1983), 249–317.

4. *Public Papers of the Presidents: Lyndon B. Johnson, 1965* (Washington: U.S. Government Printing Office, 1966), I, 636.

5. In *Jenkins* v. *Collard*, 145 U.S. 546, 560–561 (1891), the Supreme Court ruled that executive orders have "the force of public law," and in *United States* v. *Eaton*, 144 U.S. 677, 688 (1892) the Court held that violating provisions of an executive order may be made a crime punishable by sanctions and penalties, *if* Congress so provides.

6. A brief but lucid introduction to the constitutional and political problems of executive orders in civil rights policy is Ruth P. Morgan, *The President and Civil Rights: Policy-Making by Executive Order* (New York: St. Martin's Press, 1970). The

482

historic constitutional confusion of genre and authority surrounding executive orders is reflected in their slow and uncertain evolution. A numbering system for executive orders was not devised until 1907 and was then applied retroactively. Executive orders were not required to be published and judicially noticed in the *Federal Register* until 1935.

7. *Youngstown Sheet & Tube v. Sawyer*, 343 U.S. 534 (1952).

8. Arthur S. Link, *Woodrow Wilson: The New Freedom* (Princeton: Princeton University Press, 1956), 243–52; Kathleen L. Wolgemuth, "Woodrow Wilson and Federal Segregation," *Journal of Negro History* XLIV (1959): 158–73; and Morton Sosna, "The South in the Saddle: Racial Policies During the Wilson Years," *Wisconsin Magazine of History* 54 (Autumn 1970): 30–49.

9. The Pendleton Act of 1883 had sought to establish civil service appointments by merit and fitness, and one of the first regulations banned religious discrimination. But the civil service rules remained silent on racial discrimination until 1940.

10. In 1939 Attorney General Frank Murphy, anticipating war and anxious to avoid the abuse of civil liberties that had characterized World War I, created within the Criminal Division of the Justice Department a Civil Liberties Unit (which in 1941 was redesignated the Civil Rights Section). Though small and lacking statutory authority, the section built a core of constitutional and statutory argument and procedure that was expanded upon after the section was elevated to divisional status by the Civil Rights Act of 1957. See Robert K. Carr, *Federal Protection of Civil Rights: Quest for a Sword* (Ithaca: Cornell University Press, 1947).

11. Harvard Sitkoff, *A New Deal for Blacks* (New York: Oxford University Press, 1978); John B. Kirby, *Black Americans in the Roosevelt Era* (Knoxville: University of Tennessee Press, 1980). In *Farewell to the Party of Lincoln: Black Politics in the Age of FDR* (Princeton: Princeton University Press, 1983), Nancy J. Weiss argues that the Roosevelt administration's political and symbolic gestures toward Afro-Americans were too minimal and unconvincing to explain the mass switch of black party loyalty, and concludes instead that jobs from the broad recovery effort rather than specific civil rights policies explain the partisan shift.

12. Richard M. Dalfiume, *Desegregation in the U.S. Armed Forces: Fighting on Two Fronts, 1939–1953* (Columbia: University of Missouri Press, 1969), 25–63; President's Committee on Equality of Treatment and Opportunity in the Armed Forces, *Freedom to Serve* (Washington: U.S. Government Printing Office, 1950), 47–48. Black manpower in the U.S. military jumped from 3,640 at the beginning of 1939 to 467,883 by the end of 1942.

13. Louis Ruchames, *Race, Jobs, & Politics: The Story of FEPC* (New York: Columbia University Press, 1953). Ross, *All Manner of Men*, is a fair-minded and bittersweet memoir by the FEPC's last chairman. Louis Kesselman, *The Social Politics of FEPC: A Study in Reform Pressure Movements* (Chapel Hill, University of North Carolina Press, 1948), concentrates on the National Council for a Permanent FEPC and its unsuccessful congressional lobbying from 1943 through 1946. Herbert Garfinkle, *When Negroes March: The March on Washington Movement in the Organizational Politics for FEPC* (Glencoe: Free Press, 1959), traces the MOWM beyond the victory of 1941 through the frustrations and decline of the 1950s.

14. Executive Order No. 8802, 6 *Fed. Reg.* 3109 (1941).

15. Richard Polenberg, *War and Society: The United States, 1941–1945* (Philadelphia: Lippincott, 1972), 113–23; Morgan, *The President and Civil Rights*, 28–59.

16. Ruchames, *Race, Jobs, & Politics*, 47–48.

17. Executive Order No. 9346, 8 *Fed. Reg.* 7183 (1943).

18. Will Maslow, "FEPC—A Case History in Parliamentary Maneuver," *The University of Chicago Law Review* 13 (June 1946): 410–11.

19. Frank P. Huddle, "Fair Employment in Practice," *Editorial Research Reports* I (18 January 1946); Fair Employment Practice Committee, *Final Report* (Washington: U.S. Government Printing Office, 1947).

20. The national virulence of blue-collar racism was illustrated by the streetcar strike in wartime Philadelphia. When the war drained away white males from the higher-paid platform jobs of motorman and conductor, they were replaced by white women rather than by black maintenance workers. In 1944 black protest produced a paralyzing strike by white workers, which was put down only through the intervention of the FBI and 5000 U.S. Army troops.

21. See the memoir by Malcolm Ross, *All Manner of Men* (New York: Reynal & Hitchcock, 1948), especially chapters six and eight.

22. U.S. Congress, House, *To Investigate Executive Agencies,* Hearings before the Special Committee to Investigate Executive Agencies, House of Representatives, 78th Cong., 1st and 2d Sess., on H. Res. 102 (Washington: U.S. Government Printing Office, 1944), Part 2, 1894–2173 ff. The FEPC's chief defender on the Smith Committee was Representative Jerry Voorhis, a California Democrat who in 1946 was defeated for re-election by Republican Richard Nixon.

23. Executive Order No. 9980, 13 *Fed. Reg.* 4311 (1948).

24. Donald R. McCoy and Richard Reutten, *Quest and Response: Minority Rights in the Truman Administration* (Lawrence: University Press of Kansas, 1973), 251–69. By the end of 1951, Truman's FEB reported that only 488 complaints had been filed in the federal agencies, with no discrimination found in 60 percent of the cases. The FEB itself had heard 62 appeals cases, finding discrimination in only thirteen.

25. William C. Berman, *The Politics of Civil Rights in the Truman Administration* (Columbus: Ohio State University Press, 1970), 24–37, 158–65. A Senate vote on cloture failed by a margin of 55 to 33 (a two-thirds vote of 64 in favor was required under Senate Rule 22 to limit debate). All attempts to shut off filibusters against civil rights bills had failed since Senate Rule 22 was first adopted in 1917.

26. Executive Order No. 10308, 16 *Fed. Reg.* 112302 (1951).

27. McCoy and Reutten, *Quest and Response,* 269–81.

28. Executive Order No. 10479, 18 *Fed. Reg.* 4899 (1953). The new committee was urged on the President by White House aide Maxwell Rabb, who had assumed the role of minorities adviser. Rabb in turn had requested a draft proposal from Jacob Seidenberg, the staff director of Truman's contracts committee, who continued as executive director of the Eisenhower committee under the chairmanship of Vice President Richard Nixon. On September 3, 1954, Eisenhower issued Executive Order 10557, which required standardized language in the nondiscrimination clauses of all U.S. government agencies, and was recommended by a contracts subcommittee chaired by Attorney General William Rogers.

29. Jack Greenberg, *Race Relations and American Law* (New York: Columbia University Press, 1959), 160–61.

30. Robert Fredrick Burk, *The Eisenhower Administration and Black Civil Rights* (Knoxville: University of Tennessee Press, 1984), 89–108.

31. Jacob Seidenberg, "The President's Committee on Government Contracts: 1953–1960—An Appraisal," January 1961, Mitchell Papers, Box 130, Dwight David Eisenhower Library (hereafter cited as DDEL).

32. In his thousand-page memoir, Richard Nixon devotes only one sentence to the anti-discrimination efforts he led during the Eisenhower administration. Richard M. Nixon, *The Memoirs of Richard Nixon* (New York: Grosset & Dunlap, 1978), 437.

33. Executive Order No. 10590, 20 *Fed. Reg.* 409 (1955).

34. Alan Ronald Schlundt, "Civil Rights Policies in the Eisenhower Years" (unpublished Ph.D. dissertation in History, Rice University, 1973), especially chap. III, 59–91.

35. Morroe Berger, *Equality by Statute: The Revolution in Civil Rights* (rev. ed.; New York: Farrar, Straus, & Giroux, 1978), chap. four, which discusses the experience of New York state.

36. Berger reports only six cases for the entire country during the decade 1926–

36, and only twenty in the following decade—a third of those originating in New York state. Berger, *Equality by Statute*, 160–61. For an inventory of state and municipal FEP statutes circa 1959, see Greenberg, *Race Relations and American Law*, 372–400. By 1964, 30 states had FEP laws; see *State Fair Employment Laws and their Administration* (Washington: Bureau of National Affairs, 1964).

37. Bernard Schwartz, ed., *The Economic Regulation of Business and Industry: A Legislative History of United States Regulatory Agencies* (New York: Bowker, 1973), vol. IV: 2953–82.

38. These were the Civil Aeronautics Board, the Federal Communications Commission, the Federal Power Commission, the Federal Trade Commission, the Interstate Commerce Commission, and the Securities and Exchange Commission. See David M. Welborn, *Governance of Federal Regulatory Agencies* (Knoxville: University of Tennessee Press, 1977).

39. Francis E. Rourke, *Bureaucracy, Politics, and Public Policy* (Boston: Little, Brown, 1984), 48–90; Bernard Schwartz, *The Professor and the Commissions* (New York: Knopf, 1959), 164–67. The criteria of administrative law required that the regulatory agencies demonstrate only "substantial evidence" in support of their findings of fact, as opposed to a "preponderance of evidence" in civil law and evidence "beyond reasonable doubt" in criminal law.

40. James A. Gross, *The Making of the National Labor Relations Board: A Study in Economics, Politics, and the Law, 1933–1937* (Albany: State University of New York Press, 1981), chap. 1.

41. Alfred W. Blumrosen, *Black Employment and the Law* (New Brunswick, N.J.: Rutgers University Press, 1971), 11–14.

42. Jay Anders Higbee, *Development and Administration of the New York State Law Against Discrimination* (University: University of Alabama Press, 1966). New York provided for a maximum jail sentence of one year and a maximum fine of $500. Most state laws contained occupational exemptions, such as for domestic service and employment by nonprofit religious, educational, and social organizations. Numerical or threshhold exemptions were also common, typically excluding employers with five or fewer employees, as in the case of New York.

43. *State Fair Employment Laws and Their Administration* (Washington: Bureau of National Affairs, 1964), Part I.

44. Paul H. Norgren and Samuel E. Hill, *Toward Fair Employment* (New York: Columbia University Press, 1964). Southern cities with FEP commissions included Atlanta, Louisville, Memphis, Oklahoma City, and Tulsa.

45. Berger, *Equality by Statute*, 170–86. In New York state the annual number of complaints averaged only 300 during the commission's first decade, then rose to a plateau of around 500 after age discrimination was added in 1959.

46. The state and municipal FEPCs shared similar experiences. From its founding in 1948 to 1965, the Philadelphia FEPC averaged 110 complaints a year, found probable cause in 25 percent, and resolved *all* of them through conciliation. See Witherspoon, *Administrative Implementation of Civil Rights*, 107–37.

47. Michael I. Sovern, *Legal Restraints on Racial Discrimination in Employment* (New York: Twentieth Century Fund, 1966), chap. 2; Berger, *Equality by Statute*, 181–204. Berger's classic study gives the FEP commissions generally high marks, especially New York's. He criticizes them only for "execessive caution," but acknowledges the extreme difficulty of measuring their effectiveness against their potential.

48. Ronnie Dugger, *The Politician: The Life and Times of Lyndon Johnson: The Drive for Power, From the Frontier to Master of the Senate* (New York: Norton, 1982), 310 *passim*.

49. Doris Kearns, *Lyndon Johnson and the American Dream* (New York: Harper & Row, 1976), 135–59; George Reedy, *Lyndon B. Johnson: A Memoir* (New York: Andrews and McMeel, 1982), 109–20; Steven F. Lawson, *Black Ballots: Voting Rights in the South, 1944–1969* (New York: Columbia University Press, 1976), 183–202.

50. Rowland Evans and Robert Novak, *Lyndon B. Johnson: The Exercise of Power* (New York: New American Library, 1966), 119–40; Lawson, *Black Ballots,* 227–30.

51. J. W. Anderson, *Eisenhower, Brownell, and the Congress* (University: University of Alabama Press, 1964). See also Lawson, *Black Ballots,* 165–202; and Burk, *Eisenhower and Black Civil Rights,* 204–26.

52. Daniel C. Berman, *A Bill Becomes a Law: The Civil Rights Act of 1960* (New York: Macmillan, 1962); Lawson, *Black Ballots,* 203–49.

53. Berman, *A Bill Becomes a Law,* 135.

54. Transcript, "George Reedy Oral History Interview," 12 December 1968, Lyndon Baines Johnson Library (hereafter cited LBJL), Tape 1, 11.

Chapter I

1. On the sit-in movement of 1960 see Harvard Sitkoff, *The Struggle for Black Equality 1954–1980* (New York: Hill & Wang, 1981), 69–96.

2. Carl M. Brauer, *John F. Kennedy and the Second Reconstruction* (New York: Columbia University Press, 1977), 30.

3. "Statement by the President Upon Signing Order Establishing the President's Committee on Equal Employment Opportunity," 7 March 1961, *Public Papers of the Presidents of the United States: John F. Kennedy, 1961* (Washington: U.S. Government Printing Office, 1962), 150.

4. Executive Order No. 10925, 26 *Fed. Reg.* 1977 (1961), Part III, Subpart A, Section 301.

5. Associated Press, "JFK's Bias Ban Gets Expected Reaction," *Philadelphia News,* 7 March 1961.

6. United Press International, "Ban Race Barriers on Jobs," *Grand Forks* [North Dakota] *Herald,* 7 March 1961.

7. Kennedy's civil rights dilemma was politically well understood at the time. Subsequent scholarship has emphasized his intellectual and emotional distance from the liberal ideology of state intervention that drove the civil rights coalition, at least before the watershed events of 1963. See Brauer, *Kennedy and the Second Reconstruction,* chaps. 1–3; Herbert S. Parmet, *Jack: The Struggles of John F. Kennedy* (New York: Dial, 1980), 188–89, 409–14, 461–78; Parmet, *JFK: The Presidency of John F. Kennedy* (New York: Dial, 1983), 50–56, 249–76; and Allen J. Matusow, *The Unraveling of America: A History of Liberalism in the 1960s* (New York: Harper & Row, 1984), 20–25, 62–95.

8. Brauer, *Kennedy and the Second Reconstruction,* 27.

9. Earl Mazo, *Richard Nixon: A Political and Personal Portrait* (New York: Harper & Row, 1959), 257; Richard M. Nixon, *Six Crises* (New York: Doubleday, 1962), 363.

10. Steven F. Lawson, *Black Ballots* (New York: Columbia University Press, 1976), 252.

11. Stephen E. Ambrose, *Nixon: The Education of a Politician, 1913–1962* (New York: Simon and Schuster, 1987), 641–51.

12. Theodore H. White, *The Making of the President 1960* (New York: Atheneum, 1961), 201–8.

13. *New York Times,* 27 July 1960. The Republican civil rights plank of 1960, moreover, promised legislation in a field the Democrats had studiously ignored—"to end the discriminatory membership practices" in labor unions. See James M. Sundquist, *Politics and Policy: The Eisenhower, Kennedy, and Johnson Years* (Washington: Brookings, 1968), 250–53.

14. Donald B. Johnson and Kirk H. Porter, *National Party Platforms, 1840–1972* (3rd ed.; Urbana: University of Illinois Press, 1966), 599–600.

15. *New York Times*, 8 October 1960.

16. Brauer, *Kennedy and the Second Reconstruction*, 61–81; Sundquist, *Politics and Policy*, 254–59.

17. Brauer, *Kennedy and the Second Reconstruction*, 62–63; *New York Times*, 21 December 1961.

18. Leonard Baker, *The Johnson Eclipse: A President's Vice Presidency* (New York: Macmillan, 1966), chap. 2; Rowland Evans and Robert Novak, *Lyndon B. Johnson: The Exercise of Power* (New York: New American Library, 1966), 305–8; Transcript, "George Reedy Oral History Interview," Tape 5, 22–28, Lyndon Baines Johnson Library (hereafter cited as LBJL).

19. George Reedy, *Lyndon Johnson: A Memoir* (New York: Andrews and McMeel, 1982), 133. Reedy recalls that he saw the message "fleetingly through the courtesy of a fellow staff member, but before I could protest, it was on its way to the White House."

20. Evans and Novak, *Johnson*, 308.

21. The oral history interviews in both the Kennedy and Johnson libraries echo with consensual reports that Johnson, while never close to Jack Kennedy, and while despising Bobby Kennedy, developed a respect for President Kennedy. George Reedy recalled that Johnson and Kennedy "always got along rather well" (Transcript, George Reedy Oral History Interview, 12 December 1968, Tape 2, 22, LBJL). Gerald Siegel, like Reedy a Johnson aide in the Senate, observed that throughout Johnson's miserable exile in the vice presidency, President Kennedy "really went out of his way to try to give him substantive things to do, important things" (Transcript, Gerald W. Siegel Oral History Interview, 26 May 1969, Tape 2, 16, LBJL). Jerry Holleman, a Texas labor leader who served as Assistant Secretary of Labor and executive vice chairman of the new committee under Johnson's chairmanship, volunteered that "As a matter of fact, [Johnson] kind of liked Kennedy" (Transcript, Jerry Holleman Oral History Interview, 19 April 1971, Tape 1, 37, LBJL).

22. Transcript, Hobart Taylor, Jr. Oral History Interview, 6 January 1969, Tape 1, 16, LBJL.

23. The oral history interview of Hobart Taylor, Sr., in the Johnson Library provides rich insights into the fabric and tone of the loose biracial system in post-depression Texas. Within that context the childhood of Hobart, Jr., is depicted as relatively sheltered and privileged. See the transcript, Hobart Taylor, Sr., Oral History Interview, 5 May 1969, LBJL.

24. Taylor, Jr., oral history, Tape 1, 14. James Rowe warned Johnson that as chairman of the equal employment committee "you will become the target of the ADA and the "advanced liberals" because you are not doing anything and also the target of the Southerners every time you try to do something even minor." Rowe to Johnson, 22 December 1960, box 2, Vice Presidential Papers, Civil Rights File (hereafter cited as VPP/CRF), LBJL.

25. Ibid., 12. Taylor claimed that he fancied the alliterative quality of "affirmative action," and likened its potential legal and philosophical content to the Fourteenth Amendment's "equal protection" clause.

26. The National Labor Relations Act, 5 July 1935, *U.S. Statutes at Large*, vol. XLIX, Sec. 9(c), 454.

27. Section 297.2(c) of the New York Law Against Discrimination provided for the state commission, upon a written finding of unlawful discrimination, to order the respondent to "cease and desist from such unlawful discriminatory practice and to take such affirmative action, including (but not limited to) hiring, reinstatement or upgrading of employees, with or without back pay, restoration to membership in any respondent labor organization, admission to or participation in a guidance program, apprenticeship training program, on-the-job training program or other occupational training or retraining program, the extension of full, equal and unsegregated accommodations, advantages, facilities and privileges to all persons, payment of compensa-

tory damages to the person aggrieved by such practice as, in the judgment of the commission, will effectuate the purposes of this article, and including a requirement for report of the manner of compliance." Texts of the state laws may be found in *State Fair Employment Laws and Their Administration* (Washington: Bureau of National Affairs, 1964).

28. Sundquist, *Politics and Policy*, 238–53; Lawson, *Black Ballots*, 203–49.

29. This is the consensus of the literature and the oral histories, and is supported by George Reedy: "Abe Fortas definitely drafted the order." Reedy, letter to the author, 10 November 1985; Transcript, Abe Fortas Oral History Interview, 14 August 1969, 13, LBJL.

30. Reedy, letter to the author, 10 November 1985.

31. Gerald W. Siegel Lyndon B. Johnson, 27 December 1960; Theodore Kheel to Johnson, 29 December 1960, box 2, VPP/CRF, LBJL.

32. "Conversation with Dick Goodwin, White House," n.d.; "A Summary of My Conversations Today on the Executive Order," 4 March 1981, box 8, VPP/CRF, LBJL.

33. *New York Times*, 12 January 1960; President's Committee on Government Contracts, *Final Report* (Washington: U.S. Government Printing Office, 1961).

34. On iron triangles see J. Leiper Freeman, *The Political Process: Executive Bureau-Legislative Committee Relations* (New York: Random House, 1955); Douglass Cater, *Power in Washington* (New York: Vintage, 1964); Hugh Heclo, *A Government of Strangers* (Washington: Brookings, 1977); Heclo, "Issue Networks and the Executive Establishment," in *The New American Political System*, ed. Anthony King (Washington: American Enterprise Institute, 1978), 87–124; and Hugh Davis Graham, *The Uncertain Triumph: Federal Education Policy in the Kennedy and Johnson Years* (Chapel Hill: University of North Carolina Press, 1984), 190–93.

35. William H. White, an admirer of Johnson, conceded at the time that Johnson "privately flinched" at the prospect of chairing the new anti-discrimination committee; White, *The Professional: Lyndon B. Johnson* (Boston: Houghton Mifflin, 1964), 228.

36. Reedy to the Vice President, 10 February 1961, Files of George Reedy, Box 2, LBJL.

37. Reedy to the Vice President, 13 February 1961, VPP/CRF, Box 8, LBJL.

38. Transcript, Nicholas deB. Katzenbach Oral History Interview, 12 November 1968, Tape 1, 3, LBJL. Katzenbach had concluded from his historical and legal study that the Vice President "was a member of the Executive Branch with only this very narrow legislative function, but there were no executive powers that would have been improper for him to exercise."

39. Reedy to the Vice President, 10 February 1961, Reedy files, Box 2, LBJL.

40. Executive Order 10925, Sec. 102 (b) and (e).

41. Johnson to the President, 14 February 1961, VPP/CRF, Box 8, LBJL. Katzenbach was responsible for drafting most of the order's Part II specifying nondiscrimination requirements in government employment, and he insisted on provision for a hearing prior to any contract termination. Katzenbach to Moyers, 20 February 1961, VPP/CRF, Box 8, LBJL.

42. Reedy to the Vice President, 3 March 1961, VPP/CRF, Box 8, LBJL.

43. Michigan and Montana adopted the first state equal-pay laws in 1919. By 1961, 24 states had equal pay laws, but no state included sex in their job discrimination protections. A major reason was the existence in 43 states and the District of Columbia of laws especially protecting women by regulating maximum hours, prohibiting night work, and excluding certain occupations thought to be too strenuous or hazardous. See generally chap. 8; Leo Kanowitz, *Women and the Law* (Albuquerque: University of New Mexico Press, 1969), 102–10; and Morag MacLeod Simchak, "Equal Pay in the United States," *International Labor Review* 103 (June 1971): 541–50.

44. "Statement of the President Upon Signing Order Establishing the President's

Committee on Equal Employment Opportunity," *Public Papers of the Presidents 1961* (Washington: U.S. Government Printing Office, 1961), 150.

45. Executive Order 10925, Sec. 302 (a) and (b).

46. Ronnie Dugger, *The Politician: The Life and Times of Lyndon Johnson: The Drive for Power, from the Frontier to Master of the Senate* (New York: Norton, 1982), 310.

47. Robert A. Caro, *The Years of Lyndon Johnson: The Path to Power* (New York: Alfred Knopf, 1982), interprets Johnson's career as driven by a raw ambition uncluttered by principle and unredeemed by ideology.

48. Arthur M. Schlesinger, Jr., *A Thousand Days* (Boston: Houghton Mifflin, 1965), 933.

49. Reedy oral history, Tape 3, 8.

50. Holleman oral history, Tape 1, 43.

51. Richard Fenno, *The President's Cabinet* (New York: Vintage, 1959); Thomas E. Cronin, *The State of the Presidency* (Boston: Little, Brown, 1980), 253–96.

52. Jacob Seidenberg, "The President's Committee on Government Contracts: 1953–1960—An Appraisal," January 1961, Mitchell Papers, box 130, DDEL.

53. The three southerners were Howard E. Butt, Jr., a businessman and Baptist lay minister from Kerrville, Texas; publisher Silliman Evans, Jr. of the *Nashville Tennessean;* and Atlanta lawyer Robert B. Troutman, Jr. The three non-southerners were gas and electric executive Donald C. Cook of New York City; steel and aluminum executive Edgar F. Kaiser of Oakland, California; and editor Howard Woods of the *St. Louis Argus.*

54. An early push was made by Johnson aides George Reedy, Bill Moyers, and Walter Jenkins to have Reedy appointed as executive vice chairman at a supergrade level of GS-18. The three Johnson aides sought to counter the nomination of Hyman Bookbinder, a veteran legislative lobbyist for organized labor and a civil rights activist who was backed by Meany and Goldberg. Johnson compromised by appointing Holleman, who as first president of the Texas AFL-CIO was closer to Meany and Goldberg than to Johnson, but who was not identified like "Bookie" as a militant ADA liberal on civil rights. The three also urged the appointment of Hobart Taylor as special counsel to the committee. Reedy, Moyers, and Jenkins to the Vice President, 9 March 1961, box 2, VPP/CRF, LBJL. Reedy told Johnson that Bookbinder was a "fairly reasonable man who can still keep all of the civil rights groups quiet with the sole exception of Clarence Mitchell, whom nobody can satisfy these days," but that "Bookie is basically a civil rights type and I think his appointment as executive vice chairman would give you the image of having put the committee completely under labor domination." Reedy to the Vice President, 10 March 1961, box 2, VPP/CRF, LBJL.

55. *New York Times,* 7 March 1961.

56. Seidenberg concluded that the regional offices had been established on the assumption that a grass-roots presence would invite greater complaint traffic, and that they would facilitate faster and more direct communication with the regional office networks of the contracting federal agencies. But both assumptions proved to be wrong. Complaint traffic was not significantly increased, and the contracting agencies insisted on dealing with the contract committee through Washington's chain-of-command hierarchy rather than through parallel regional negotiations. Seidenberg, "Appraisal," 51–55.

57. During the early days of the Kennedy administration, White House aides Fred Dutton, Harris Wofford, and Richard Goodwin were involved in developing executive initiatives in civil rights policy. Dutton argued that a reasonable budget for the new equal employment committee would have each participating agency contribute $50,000 annually, and Wofford argued for $100,000 each. The latter figure totaled approximately $2.8 million, depending on how many contracting agencies one counted. Neither man was apparently familiar with the prorated formula under Eisenhower, which

had allocated 50 percent of the committee's budget to Defense, 17.5 percent each to the AEC and GSA, and 7.5 percent each to Commerce and Labor. The Justice Department provided legal services in lieu of cash, and Labor also provided necessary housekeeping services. The PCEEO adopted this rough model for its budget of $425,000 for its first year, which was less than the comparable annual budget for the New York City Commission on Intergroup Relations ($483,215), the New York State Commission Against Discrimination ($650,000), or the U.S. Commission on Civil Rights ($850,000). Dutton to the President, and Dutton to Goodwin, 9 February 1961, Files of Meyer Feldman, Box 802, John F. Kennedy Library (hereafter cited as JFKL).

58. Holleman oral history, 19 April 1971, 44, LBJL; Brauer, *Kennedy and the Second Reconstruction*, 79–84. Paul H. Norgren and Samuel E. Hill, *Toward Fair Employment* (New York: Columbia University Press, 1964), 149–79, compares the administration (including budgets) and performance of Roosevelt's two FEPCs, the Truman and Eisenhower contracts committees, and the PCEEO.

59. Reedy to the Vice President, 10 April 1961, VPP/CRF, Box 8, LBJL.

Chapter II

1. A study of employment at Lockheed-Marietta by the National Urban League in 1957 had revealed that the 1300 black employees (of a 1957 total of 18,000) were restricted to thirty of the plant's 450 job categories. See Ray Marshall, *The Negro and Organized Labor* (New York: Wiley, 1965), 226–31.

2. *New York Times*, 8 April 1961; Michael I. Sovern, *Legal Restraints on Racial Discrimination in Employment* (New York: Twentieth Century Fund, 1966), 108–11. Lockheed-Marietta's employment had peaked in 1955 with 20,000 workers producing the turboprop Hercules and the small executive JetStar, and only 1400 of these workers had been black. Lockheed's laid-off black workers generally fell at the bottom of the seniority ladder for re-employment rights on new contracts.

3. *New York Times*, 14 May 1961, and editorial 20 May 1961.

4. Roland Evans and Robert Novak, *Lyndon Johnson: The Exercise of Power* (New York: New American Library, 1966), 317; John G. Feild, "Equal Employment Opportunity," *The Anti-Defamation League Bulletin* (May 1962): 14–16.

5. Reedy oral history, 12 December 1968, LBJL; Reedy, letter to the author, 10 November 1985.

6. The author is indebted to Adam Yarmolinsky for sharing a draft of his unpublished memoirs.

7. "Plan for Equal Job Opportunity at Lockheed Aircraft Corporation," *Monthly Labor Review* (July 1961): 748–49.

8. *New York Times*, 5 June 1961, 15 June 1961.

9. Leonard Baker, *The Johnson Eclipse* (New York: Macmillan, 1966), 156; Carl Brauer, *John F. Kennedy and the Second Reconstruction* (New York: Columbia University Press, 1977), 81.

10. Transcript, Harris Wofford Oral History Interview, 29 November 1965, 126, JFKL. Wofford was interviewed by Berl Bernhard, who had been staff director for the U.S. Commission on Civil Rights during the Kennedy administration. The oral history interviews in the Kennedy Library, including those for the Robert F. Kennedy project, were often conducted by participants in the administration or by contemporary journalists. The seven-volume transcript for Robert Kennedy's oral history interviews contains the interviews conducted by John Bartlow Martin and Anthony Lewis in 1964, and Arthur Schlesinger, Jr., early in 1965.

11. Troutman to the Equal Employment Opportunity Committee, 24 May 1961, VPP/CRF, box 2, LBJL.

12. *New York Times*, 22 June 1962.

13. Taylor oral history, 6 January 1969, LBJL.

14. Troutman to the Equal Employment Opportunity Committee, n.d. [August 1961], VPP/CRF, box 2, LBJL.

15. Ibid, 2.

16. *New York Times,* 17 June 1962.

17. Peter Braestrup, "8 Companies Sign Negro Job Pledge," *New York Times,* 12 July 1961. The eight companies were: Boeing, Douglas Aircraft, General Electric, Martin, North American, RCA, United Aircraft, and Western Electric.

18. Peter Braestrup, "12 More Concerns Vow Job Equality," *New York Times,* 25 November 1961.

19. *New York Times,* 22 June 1962.

20. Troutman to the Vice President, 6 December 1961, VPP/CRF, box 1, LBJL. Troutman and his four-man staff had negotiated with twenty-four firms, and Feild and his larger committee staff had dealt with twenty-eight, with increasing assistance from the Department of Defense.

21. Holleman to Dungan, 3 January 1961, Files of Lee White, box 23, JFKL.

22. Transcript, Burke Marshall Oral History Interview, 19–20 January 1970, 23, JFKL.

23. *New York Times,* 25 January 1962; 6 February 1962. The six firms were: Boeing, Douglas, Lockheed, Martin, North American, and RCA.

24. Marshall oral history, 24, JFKL.

25. *New York Times,* 7 February 1962. The House committee's bill had been pushed by Representative James Roosevelt, a former member of the Eisenhower contracts committee whose special subcommittee had canvassed the country the previous fall holding hearings on fair employment practices. Secretary Goldberg had testified in support of the bill "in principle," but it was not expected to get past the Rules Committee, and its chief value was to apply pressure for more aggressive enforcement measures.

26. *New York Times,* 15 February 1962.

27. *New York Times,* 15 February 1962; 11 August 1962; Ray Marshall, *The Negro Worker* (New York: Random House, 1967), 92–119.

28. *New York Times,* 5 April 1962.

29. Troutman to the President, Vice President, and Secretary of Labor, 7 May 1962, VPP/CRF, box 11, LBJL. When Troutman moved his operation from Atlanta back to Washington in April 1962, he moved into a suite across Pennsylvania Avenue from the White House, while Feild's staff worked in the Government Accounting Office Building in midtown Washington. Secretary Goldberg said he approved the continued separation of the two staffs partly to avoid "friction."

30. Peter Braestrup, "U.S. Panel Split Over Negro Jobs," *New York Times,* 18 June 1962.

31. *New York Times,* 22 June 1962. The *Times* editorialized about the split on June 19, and on June 20 published a letter from Lyndon Johnson "refusing to promote a controversy where in my opinion none exists."

32. Brauer, *Kennedy and the Second Reconstruction,* 149–51. Troutman's chief black convert was Benjamin Mays, president of Morehouse College in Atlanta. Mays persuaded the NAACP's Roy Wilkins to grant Troutman a hearing, but the Urban League's new executive director, Whitney Young, refused to see Troutman. This reflected an unusual role reversal of the two organizations, with the low profile, bread-and-butter Urban League, despite its historic ties to the white business community and its funding support from the noncontroversial Community Chest, taking a more radical stance than the historically more aggressive NAACP.

33. *New York Times,* 17 June 1962.

34. *New York Times,* 20 June 1962.

35. *New York Times,* 4 July 1962; *New Republic* 147 (10 September 1962): 6.

36. Secretary Goldberg had asked Holleman to help arrange and fund a lavish

banquet for Vice President Johnson, featuring Mexican food catered out of Texas. Holleman accepted a $1,000 donation from financier-swindler Billie Sol Estes, and when the Estes scandal broke, Holleman was quickly forced to resign. In his oral history, Holleman blames Goldberg's lust for a Supreme Court appointment, which led him to sacrifice a loyal and honest lieutenant in order to protect his own reputation. In Holleman's view his abandonment by Goldberg and Johnson destroyed a career that was headed for deputy secretary, and ultimately for the secretaryship and full cabinet rank. Holleman oral history, 45–54, LBJL. See also Reedy oral history, 19–20, LBJL.

37. Reedy to the Vice President, 2 August 1962, and attached Kheel recommendations, Reedy files, box 5, LBJL; PCEEO, "Minutes of the Fifth Meeting," 22 August 1962, 2–5, VPP/CRF, box 10, LBJL. Peter Braestrup covered the story of the Kheel report in the *New York Times* on 19 August 1962, and the report was subsequently commercially published as Theodore W. Kheel, *Report on the Structure and Operations of the President's Committee on Equal Employment Opportunity* (Englewood Cliffs, N.J.: Prentice-Hall, 1962).

38. Seidenberg, "Appraisal," 217.

39. Taylor oral history, 19, LBJL; Reedy oral history, 18–19, LBJL.

40. The AFL-CIO Civil Rights Committee was distressed at the prospect of Holleman being replaced by Taylor, whom it referred to as the son of a Texas banker, rather than by Feild, whose ties with the AFL-CIO and Secretary Goldberg were close. Boris Shishkin to President Meany, 10 August 1962, box 9, George Meany Memorial Archives; *New York Times*, 5 August 1962.

41. Troutman to the President, 22 August 1962, Files of Theodore Sorensen, box 30, JFKL.

42. Draft report, Troutman to the President and Vice President, "Plans for Progress—One Year's Accomplishments," 30 August 1962, 5, Sorensen files, box 13, JFKL. Troutman noted that "these gains were made by the firms almost without support from Negro leaders in finding Negro applicants for the new jobs."

43. Kennedy to the Vice President, 22 August 1962, Sorensen files, box 13, JFKL.

44. By August 1963, Plans for Progress had grown to 111 companies. But its public profile had been greatly reduced, and the PCEEO had created a new advisory council chaired by George William (Bill) Miller, president of Textron.

45. *New York Times*, 27 July 1962.

46. *New York Times*, 18 April 1962; Sovern, *Legal Restraints*, 112–13.

47. PCEEO, "Complaint Report as of March 30, 1962," attached to memorandum, Percy H. Williams to Feild, 2 April 1961, Reedy files, box 5, LBJL.

48. "Report by President's Committee on Equal Employment Opportunity," *Monthly Labor Review* 85 (June 1962): 652–54. The internal breakdown of complaint processing from government contractors revealed that of 704 complaints filed by the end of 1961, the PCEEO staff had fully processed 386. Of these 96 were dismissed for lack of jurisdiction and 75 were dismissed because the complainant failed to provide sufficient information for investigation. Of the 215 remaining, 76 were dismissed as without cause. This left 139 cases where the committee required "corrective action" in the complainant's interest to be negotiated with the contract employer.

49. PCEEO, *Report to the President* (Washington: U.S. Government Printing Office, 1963). By 1963 the PCEEO reported requiring corrective action in 44 percent of contract and 27 percent of federal employment complaints.

50. [HEW] Assistant Secretary [Quigley] to Wofford, 12 May 1961, Wofford files, box 12, JFKL.

51. *New York Times*, 23 August 1961.

52. Reynolds to Wofford, 13 October 1961, Wofford files, box 13, JFKL.

53. *U.S. News & World Report* 52 (5 March 1962): 83–85. Blacks filled one out of every four federal jobs in the nation's capital itself.

54. Memorandum E-13, Holleman to Heads of Federal Agencies, 4 May 1962, box 7, VPP, LBJL.

55. Johnson to [Federal Agency Heads], 15 March 1963, box 7, VPP, LBJL.

56. Taylor to the Vice President, 2 April 1963, box 7, VPP, LBJL.

57. Department of Labor, *The Economic Situation of Negroes in the United States,* Bulletin S-3 (Washington: U.S. Government Printing Office, 1962), 8.

58. U.S. Commission on Civil Rights, *Report: 1959* (Washington: U.S. Government Printing Office, 1959); Lawson, *Black Ballots* (New York: Columbia University Press, 1976), 227–30.

59. *1961 United States Commission on Civil Rights Report:* Vol. l, *Voting;* Vol. 2, *Education;* Vol. 3, *Employment;* Vol. 4, *Housing;* Vol. 5, *Justice* (Washington: U.S. Government Printing Office, 1961).

60. White had joined the staff of Senator Kennedy's small business subcommittee in 1957. As a generalist White House aide, like Max Rabb in the Eisenhower White House, he claimed no special prior expertise in civil rights policy. Wofford describes his "penstroke" role in *Of Kennedys and Kings: Making Sense of the Sixties* (New York: Farrar, Straus and Giroux, 1980), 124–77.

61. Katzenbach to White, 20 October 1961, Sorensen files, box 30, JFKL.

62. Foster Rhea Dulles, *The Civil Rights Commission: 1957–1965* (East Lansing: Michigan State University Press, 1968), 132–45; *New York Times,* 14 October 1962.

63. Minutes, Civil Rights Subcabinet Group, 14 April 1961, Wofford files, box 14, JFKL; William L. Taylor to Lee White, 19 March 1962, White files, Box 25, JFKL. When Kennedy reorganized the "bowl of mush" State Department in the fall of 1961, Dutton was dispatched from the White House to take charge of State's congressional relations. First Wofford and then White inherited the leadership of the subcabinet group, which gradually lost its effectiveness, like the Cabinet it mirrored, as the PCEEO became the primary point of executive coordination on the one hand, and the subcabinet group's members grew more protective of their agencies on the other.

64. Wofford to Sorensen, Goodwin, Dutton, and [Burke] Marshall, 11 September 1961, White files, box 20, JFKL.

65. Ibid. White added that his recommendation had general agreement from Justice, Defense, HEW, Housing, VA, the Civil Rights Commission, and the White House staff.

66. White's policy recommendations to Kennedy of November 13 included *reactivating* the Committee on Equality of Treatment and Opportunity in the Armed Services—Truman's old Fahy Committee—but as a study committee without any enforcement functions. Its chief proponent in the Defense Department was Adam Yarmolinsky, and Kennedy obliged by appointing a new committee under the chairmanship of Gerhard Gesell. But Yarmolinsky's subsequent zeal in attacking off-base segregation so irritated southern conservatives in Congress that Congressman Phil Landrum essentially blacklisted Yarmolinsky from senior appointment in the war on poverty.

67. Katzenbach to White, 1 December 1961, White files, box 20, JFKL.

68. Brauer, *Kennedy and the Second Reconstruction,* 126–51.

69. Transcript, Robert Kennedy and Burke Marshall Oral History Interview (with Anthony Lewis), 4 December 1964, Vol. 6, Tape 6, 699, JFKL.

70. Arthur M. Schlesinger, Jr., *Robert Kennedy and His Times* (Boston: Houghton Mifflin, 1978), 334–35.

71. Marshall oral history, 29 May 1964, 56–57, JFKL.

72. For a discussion of Kennedy's self-inflicted wound on housing desegregation, see, generally, Brauer, *Kennedy and the Second Reconstruction,* 85–88; Schlesinger, *Robert Kennedy,* 334–35; Herbert S. Parmet, *JFK: The Presidency of John F. Kennedy* (New York: Dial, 1983), 258–59; and Allen J. Matusow, *The Unraveling of America* (New York: Harper & Row, 1984), 68–69.

73. President Kennedy signed Executive Order 11063 for nondiscrimination in housing on November 20, 1962, and announced it at a news conference the same day. The order took effect immediately, and covered federally owned or operated housing; public housing and housing in federally subsidized urban renewal projects; housing built with federal loans, such as for the elderly and college students; and housing insured by the FHA and the VA. But it did not cover conventional loans and mortgages by financial institutions regulated by the FDIC and similar agencies, and it did not cover existing housing. The order established a President's Committee on Equal Opportunity in Housing to oversee its administration and coordinate its enforcement.

74. Executive Order 11114, *28 Federal Register* 6485 (22 June 1963).

75. Quoted in Arthur M. Schlesinger, Jr., *A Thousand Days: John F. Kennedy in the White House* (Boston: Houghton Mifflin, 1965), 646.

76. Reedy oral history, 12 December 1968, 5–6, LBJL.

77. Doris Kearn, *Lyndon Johnson and the American Dream* (New York: Harper & Row, 1976), 160–69.

78. Holleman oral history, 19 April 1971, 61, LBJL.

79. Wofford oral history, 22 May 1968, Tape 3, 126, JFKL.

80. "Plans for Progress: Atlanta Survey," a Special Report of the Southern Regional Council (January 1963), 5, 15.

81. Kennedy to the Vice President, 21 February 1963, Sorensen files, box 30, JFKL; Leslie W. Dunbar to Hobart Taylor, 29 May 1963, White files, box 23, JFKL.

82. *New York Times,* 17 April 1963. A *Fortune* survey in July 1963 of Plans for Progress signers in twelve cities revealed the same pattern. See Charles E. Silberman, "The Businessman and the Negro," *Fortune* 68 (September 1963): 97–99.

83. *New York Times,* 1 March 1963.

84. Taylor oral history, 6 January 1969, 18–19, LBJL.

85. Brauer, *Kennedy and the Second Reconstruction,* 216–18; Attorney General to John A. Hannah, 15 December 1962; Hannah to the Attorney General, 2 January 1963; "Resolution of the United States Commission on Civil Rights," 30 March 1963; press release, "Text of Letter from the President to the Chairman, United States Commission on Civil Rights, Dr. John A. Hannah, April 19, 1963, all in Marshall files, box 30, JFKL.

86. Taylor oral history, 6 January 1969, 19–20, LBJL.

87. Kennedy and Marshall oral history, 4 December 1964, 702–4, JFKL. The Kennedy statement about Hobart Taylor is deleted in the oral history transcript, and is quoted here from Schlesinger, *Robert Kennedy,* 337.

88. John F. Kennedy, *Public Papers of the Presidents,* 1963, 221–30; *New York Times,* 1 March 1963. Kennedy's voting rights proposal included provisions to expedite voting suits in the courts, to ensure uniformity in applying local tests of voter qualifications, and renewed support for stipulating a six-grade presumption of literacy.

89. Berl I. Bernhard and William L. Taylor to Lee C. White, 21 February 1963; Bernhard and Taylor to Wilfred Rommel, 19 February 1963, Sorensen files, box 30, JFKL.

90. Brauer, *Kennedy and the Second Reconstruction,* 212–13; *New York Times,* 29 March 1963. The six Democratic senators were Clark, Douglas, Engle, Hart, Humphrey, and Williams. The eight Republicn senators were Beall, Case, Fong, Javits, Keating, Kuchel, Saltonstall, and Scott.

91. Taylor Branch, *Parting the Waters: America in the King Years 1954–63* (New York: Simon and Schuster, 1988), 673–*passim;* David Garrow, *Bearing the Cross: Martin Luther King, Jr., and the Southern Christian Leadership Conference* (New York: William Morrow, 1986), 231–86; Adam Fairclough, *To Redeem the Soul of America: The Southern Christian Leadership Conference and Martin Luther King, Jr.* (Athens: University of Georgia Press, 1987), 111–40.

92. Schlesinger, *Robert Kennedy,* 355–60. Kennedy and Marshall had met earlier in the day in New York urging resistant owners of national chain stores to desegregate their lunch counters in the South.

93. Transcript, Burke Marshall Oral History Interview, 28 October 1968, 12, LBJL.

94. McPherson to the Vice President, 18 February 1963, box 7, VPP, LBJL.

95. PCEEO, Minutes of the Seventh Meeting, 29 May 1963, box 10, VPP, LBJL.

96. Transcript, John Macy Oral History Interview, 18 March 1969, 18, LBJL; Marshall oral history, 19–20 January 1970, 24, JFKL.

97. Transcript, telephone conversation of Johnson with Sorensen, 3 June 1963, C-66-1, 14, LBJL.

98. PCEEO, Transcript of Eighth Meeting, 18 July 1963, 33–38, LBJL; Schlesinger, *Robert Kennedy,* 360–61.

99. Transcript, Jack Conway Oral History Interview, 11 April 1972, JFKL.

100. Kennedy and Marshall oral history, 4 December 1964, 706, JFKL.

101. Ibid., 708.

102. Marshall oral history, 19–20 June 1970, 25, JFKL.

103. Lee White Oral History Interview, 28 September 1970, Tape 1, 16, LBJL.

104. Marshall oral history, 28 October 1968, 13, LBJL.

105. Kennedy and Marshall oral history, 4 December 1964, 707, JFKL.

Chapter III

1. Carl M. Brauer, *John F. Kennedy and the Second Reconstruction* (New York: Columbia University Press, 1977), 259–60.

2. Harvard Sitkoff, *The Struggle for Black Equality 1954–1980* (New York: Hill and Wang, 1981), 129–51.

3. On the role of Martin Luther King in revising his nonviolent strategy after the Albany collapse, and then inviting a violent confrontation in Birmingham, see David J. Garrow, *Protest at Selma: Martin Luther King, Jr. and the Voting Rights Act of 1965* (New Haven: Yale University Press, 1978), 2–4, 220–27. See also Garrow, *Bearing the Cross: Martin Luther King, Jr. and the Southern Christian Leadership Conference* (New York: Random House, 1986); and Adam Fairclough, *To Redeem the Soul of America: The Southern Christian Leadership Conference and Martin Luther King, Jr.* (Athens: University of Georgia Press, 1987), 51–55, 86–90, 100–109, 118–39.

4. The racial violence in Birmingham was echoed by black protests involving short-lived violence in eight cities, including Chicago, Philadelphia, Savannah, and Cambridge, Maryland. It marked the rapid transition from the old pattern of *community* riots featuring white aggression, such as in the classic Detroit riot of 1943, to the *commodity* pattern of black protest violence against white property and police authority, as in Watts in 1965 and Detroit in 1967. See Hugh Davis Graham, "On Riots and Riot Commissions: Civil Disorders in the 1960s," *The Public Historian* 2 (Summer 1980): 7–27; Morris Janowitz, "Collective Racial Violence: A Contemporary History," *Violence in America,* ed. Hugh Davis Graham and Ted Robert Gurr (Beverly Hills: Sage, 1979), 261–86; and Richard Maxwell Brown, *Strain of Violence* (New York: Oxford, 1975), 228–35.

5. Brauer, *Kennedy and the Second Reconstruction,* 181–204.

6. John F. Kennedy, "Radio and Television Report to the American People on Civil Rights," June 11, 1963, *Public Papers of the Presidents: John F. Kennedy, 1963* (Washington: U.S. Government Printing Office, 1964), 468–71.

7. Brauer, *Kennedy and the Second Reconstruction,* 259–78; Arthur M. Schlesinger, Jr., *Robert Kennedy and His Times* (Boston: Houghton Mifflin, 1978), 372–74; Robert Kennedy and Burke Marshall oral history, 4 December 1964, 773–94,

JFKL; White oral history, 28 September 1970, 10–11, LBJL; Sorensen oral history, 3 May 1964, Tape 5, 133–34, JFKL.

8. Transcript, Norbert Schlei Oral History Interview, 20–21 February 1968, 44, JFKL.

9. President Kennedy's extraordinary schedule of meetings during the summer of 1963 included the following: 100 business leaders on June 4, the U.S. Conference of Mayors on June 9, 300 labor leaders on June 13, national civil rights leaders on June 14 (including King, Randolph, Young, and Farmer), 250 religious leaders on June 17, a bipartisan group of governors on June 18, 244 lawyers on June 21, 100 heads of women's organizations on July 9, and coming full circle to the Business Council on July 11.

10. White oral history, 28 September 1970, 12–13, LBJL.

11. Kennedy and Marshall oral history, 4 December 1964, 778–79, JFKL.

12. Schlei oral history, 20–21 February 1968, 47–49, JFKL.

13. Schlei to the Attorney General, 4 June 1963, Attorney General files, box 11, JFKL.

14. Transcript, telephone conversation of Johnson with Sorensen, 3 June 1963, 8, C-66-1, LBJL. Johnson recorded the conversation on his dictaphone.

15. Ibid.

16. Ibid., 6.

17. Sorensen to the Attorney General, 31 May 1963, Attorney General files, box 11, JFKL.

18. Sorensen to the Attorney General, 3 June 1963, Attorney General files, box 11, JFKL.

19. Kennedy, *Public Papers*, 11 June 1963, 468–69. The address was hurriedly drafted by Sorensen on the heels of the afternoon's successful resolution of crisis with Governor Wallace in Alabama, and it represented a triumph of timing and grace under difficult circumstances. The urgency of its message was reinforced the evening it was delivered by the sniper murder of civil rights leader Medgar Evers in Mississippi.

20. In October of 1960, however, Senator Kennedy had used almost identical language while campaigning in Wisconsin.

21. Kennedy and Marshall oral history, 4 December 1964, 777, JFKL.

22. Kennedy, "Special Message to the Congress on Civil Rights and Job Opportunities," *Public Papers*, 19 June 1963, 486. Kennedy also cited municipal ordinances barring segregated public accommodations in Washington, D.C.; Wilmington, Del.; Louisville, Ky.; El Paso, Tex.; Kansas City, Mo.; and St. Louis, Mo.

23. Solicitor General Archibald Cox had proposed that Kennedy establish a national conciliation commission by executive order to provide third-party mediation when local communications broke down (or had never been established). Cox feared that the Justice Department would be swamped by the kind of negotiations that increasingly had occupied the time of Burke Marshall and the Attorney General, and he also found inherent role incompatibilities between law enforcement and mediation. Cox to the President, 17 June 1963, Attorney General files, box 9, JFKL. Vice President Johnson also pressed for some sort of federal conciliation service, as he had in 1957 and again in 1959. But Senate liberals then viewed Johnson's conciliation device as a substitute for stronger measures rather than an accompaniment to them, and Johnson had abandoned the effort.

24. *The Civil Rights Cases*, 109 U.S. 3 (1883).

25. Schlei oral history, 20–21 February 1968, 45–46, JFKL.

26. "Memorandum Concerning Administration's Civil Rights Bill as Background for Meeting, Hotel Roosevelt, July 2, 1963," National Office Headquarters File, box 5, National Urban League Papers, Library of Congress (hereafter cited as NUL Papers).

27. "Labor Views on Administration Civil Rights Package," attached to Andrew

J. Biemiller to Kenneth O'Donnell, 10 June 1963, Civil Rights file, box 9, Meany Archives.

28. Meany testimony before the Senate Labor and Public Welfare Committee, 25 July 1963, quoted in *AFL-CIO News* press release, n.d., box 9, Meany Archives.

29. Supported mainly by labor funds, the Leadership Conference opened its permanent Washington office in the summer of 1963 specifically to lobby for the bill's passage. Its bi-weekly reports to cooperating organizations began on July 25, 1963, and were sent by Arnold Aronson from the Industrial Union Department of the AFL-CIO in the Mills Building on 17th Street. See the LCCR Papers, Library of Congress.

30. Executive Committee Meeting, 10 June 1963; Minutes, Board of Directors Meeting, 9 September 1963, NAACP Papers, Library of Congress.

31. National Urban League, "Memorandum Concerning Administration's Civil Rights Bill," 2 July 1963, 8, NUL Papers, LC.

32. Williams to Sorensen, with attached memo to the President, 15 June 1963, Attorney General files, box 9, JFKL.

33. See Hugh Davis Graham, *The Uncertain Triumph: Federal Education Policy in the Kennedy and Johnson Years* (Chapel Hill: University of North Carolina Press, 1984), especially chs. 1–2.

34. Kennedy to Secretaries Wirtz and Celebrezze, 4 June 1963, Sorensen files, box 30, JFKL.

35. Wirtz to the President, 10 June 1963, Sorensen files, box 30, JFKL.

36. "Statement by the President on Equal Employment Opportunity in Federal Apprenticeship and Construction Programs," *Public Papers*, 4 June 1963, 439.

37. Executive Order No. 11114, 28 *Fed. Reg.* 6485 (1963). In his oral history interview for the Johnson Library, Lee White reveals his embarrassment at forgetting to tell Vice President Johnson that the President had finally decided to issue the order banning discrimination in federally supported construction projects. "Believe it or not," White confessed, "I checked with every damned guy in government, I think, except Johnson! . . . And I've never seen a more surprised, disappointed, and annoyed guy than Johnson when the President of the United States issues an executive order, changing the jurisdiction of his committee. There's no goddamned rational explanation for it, except in my mind he wasn't part of the base-touching apparatus." White oral history, 28 September 1970, 19, LBJL.

38. "Letters to the Secretary of Defense and to the Chairman, Committee on Equal Opportunity in the Armed Forces, in Response to the Committee's Report," 22 June 1963, *Public Papers: Kennedy*, 495–96.

39. White oral history, 26 May 1964, 163–64, JFKL; Brauer, *Kennedy and the Second Reconstruction*, 285; *New York Times*, 7 August 1963.

40. President's Committee on Equal Opportunity in the Armed Forces, "Initial Report: Equality of Treatment and Opportunity for Negro Military Personnel Stationed Within the United States," 13 June 1963. Gesell was subsequently appointed by President Johnson to a federal judgeship.

41. "McNamara Anti-Discrimination Directive Stirs Controversy," *Congressional Quarterly Almanac: 1963* (Washington: Congressional Quarterly, 1964), 366–68; *Wall Street Journal*, 14 June 1963; *New York Times*, 29 July 1963. McNamara appointed Alfred B. Fitt, who for two years had served as Deputy Under Secretary of the Army for Manpower, to head the new Pentagon office for civil rights. Stephen N. Schulman, later to head the EEOC under Johnson, joined the new office to monitor the antidiscrimination program among the 1,030,000 civilian employees of the Defense Department and also the employees of defense contractors.

42. *Congressional Quarterly Almanac: 1963*, 368. Vinson's bill received no further action, nor was it designed to.

43. Brauer, *Kennedy and the Second Reconstruction*, 283–84; White oral history, 26 May 1964, 156, JFKL.

44. Katzenbach to the Attorney General, 29 June 1963, Attorney General file, box 11, JFKL.

45. Mansfield to the President, 18 June 1963, Sorensen file, box 30, JFKL; Allen J. Matusow, *The Unraveling of America* (New York: Harper & Row, 1984), 91–92; Brauer, *Kennedy and the Second Reconstruction*, 265–66.

46. Charles and Barbara Whalen, *The Longest Debate: A Legislative History of the 1964 Civil Rights Act* (Cabin John, Md.: Seven Locks Press, 1985). A study of the Congress-makes-a-law genre that concentrates on legislative politics rather than on the substance of the civil rights issues, *The Longest Debate* was written by former Republican congressman Charles Whalen and his wife. Rep. Whalen represented Ohio's third district from 1967 to 1979, and while *The Longest Debate* heavily emphasizes the role of Whalen's fellow House Republican from Ohio, McCulloch (who retired from the House in 1973), it is based on extensive research in the relevant congressional papers and presidential libraries, as well as on numerous interviews.

47. Kennedy and Marshall oral history, 22 December 1964, Part II, Tape 1, 875–76.

48. Ibid., 879.

49. U.S. Congress, House, Hearings Before Subcommittee No. 5 of the Committee on the Judiciary, 88th Cong., 1st sess., 26 June 1963, 69–79.

50. Subcommittee No. 5 was originally the special antitrust subcommittee of Judiciary that had traditionally attracted congressmen of populistic and liberal persuasions. See James L. Sundquist, *Politics and Policy* (Washington: Brookings, 1968), 264–65.

51. The Democratic members of Subcommittee No. 5 were, in order of seniority, Celler (N.Y.), Peter Rodino (N.J.), Byron Rogers (Colo.), Harold Donohue (Mass.), Jack Brooks (Tex.), Herman Toll (Pa.), and Robert Kastenmeier (Wisc.). The Republicans were McCulloch (Ohio), William Miller (N.Y.), George Meader (Mich.), and William Cramer (Fla.). The full Judiciary Committee contained 21 Democrats and 14 Republicans, the latter including four relatively new liberal Republicans. These were John Lindsay from Manhattan's blue-stocking district; Charles Mathias from the affluent Washington suburbs of Montgomery County, Maryland; William Cahill from Camden, New Jersey; and Clark MacGregor from Minneapolis, Minnesota.

52. Hearings, Subcommittee No. 5, 1394. Celler had opened the hearing with the announcement that 168 bills filled his committee's hopper. Merely printing them required 906 pages in the first volume of printed hearings.

53. Ibid., 1415. See the Whalens' discussion, *The Longest Debate*, 2–9.

54. As Dirksen explained to the Judiciary Committee, "in good conscience and as the minority leader for my party—I could sponsor all but title II in the administration bill, because the titles relating to voting rights, enforcement of school desegregation cases, the community relations office, the withholding of Federal funds, guarantees, and insurance where segregation was permitted, the extension of the Civil Rights Commission, and other items were clearly in conformity with the pledges made to the Nation by the Republican Party platform in 1960." Hearings, Senate Judiciary Committee, 88th Cong., 1st sess., 16 July 1963, *Civil Rights—The President's Program, 1963*, 24.

55. Included in the testimony were 27 statements from the customary panoply of interested and affected organizations, including appeals from individuals ranging from state-rights theoretician John Satterfield from Yazoo City, Mississippi, to the grand old man of American democratic socialism, Norman Thomas.

56. Brief of Professor Paul A. Freund, Hearings, Senate Commerce Committee, 88th Cong., 1st sess., part 1 (1963), 1183–90. Supporting legal briefs were also submitted by law professors Robert Rhodes of Notre Dame and Herbert Wechsler of Columbia.

57. Ibid., 1187.

58. Hearing testimony of Attorney General Robert Kennedy before the Senate Judiciary Committee, 88th Cong., 1st sess., 18 July 1963, 93 *passim.*

59. Katzenbach to the Attorney General, 19 August 1963, Attorney General files, box 11, JFKL.

60. Paul R. Clancy, *Just a Country Lawyer: A Biography of Senator Sam Ervin* (Bloomington: Indiana University Press, 1974), 78–88, 142–55.

61. Ervin's autobiography, *Preserving the Constitution* (Charlottesville: Michie Co., 1984), is a disappointing rehash of his public rhetoric.

62. Hearing, Senate Judiciary Committee, 88th Cong., 1st sess., 23 August 1963, 380.

63. Powell's 31–member committee contained only two southerners (three counting Perkins—all Democrats), and Roosevelt's seven-member subcommittee contained none. Like most such constituency-based authorization committees, most members of Education and Labor had requested their committee assignment because they were sympathetic to its clientele. They were drawn to the committee because they supported organized labor, liberal positions on civil rights, and federal aid to education. Among the Democrats, these included such prominent urban congressmen as Powell, Roosevelt, John Brademas (Ind.), and Roman Pucinski (Ill.); among Republicans, it included Peter Frelinghuysen (N.J.), Albert Quie (Minn.), and Charles Goodell (N.Y.). Roosevelt's subcommittee included himself, Pucinski, John Dent (Pa.), and Neal Smith (Iowa) on the Democratic side, and Republicans Goodell, William Ayres (Ohio), and Edgar Hiestand (Calif.).

64. Hearings before the Special Subcommittee on Labor, House Committee on Education and Labor, 87th Cong., 1st sess., October 23 and 24 (Chicago), 26 and 27 (Los Angeles), and November 3 and 4 (New York), 1961.

65. Hearings before the Special House Subcommittee on Labor, 87th Cong., 2nd sess., January 15–19 and 24, 1962.

66. The sales manager explained that the company had tried to expand sales to Negroes in the early 1950s. But when so many black households defaulted on their payments, the company had thenceforth discouraged such sales by avoiding Negro neighborhoods (and salesmen) and by requiring identification of "color" on the back of the credit applications. Because the salesmen were contract rather than salaried employees, they were not effectively covered by New York's fair employment laws.

67. Article II, Section 2 of the Constitution makes the administration of government employment an executive function. Relying on this provision, the federal courts had generally refused to interfere with discretionary functions of federal administrative agencies. For example, *United States v. McLean*, 95 U.S. 750 (1877), 753: "But courts cannot perform executive duties, or treat them as performed when they are neglected. They cannot enforce rights which are dependent for their existence upon a prior performance by an executive officer of certain duties he has failed to perform."

68. Testimony of Will Maslow, Hearings, Special House Subcommittee on Labor, 87th Cong., 2d sess., 3 November 1961, 568–73.

69. The star chamber was denounced by Alabama congressman Armistead I. Selden, Jr., a Democrat, who was joined in opposition testimony by fellow Democratic representatives Gillis Long and Joe Waggoner of Louisiana and from representatives Albert Watson and William Jennings Bryan Dorn of South Carolina, as well as by South Carolina's attorney general. Dorn argued most forcefully that he was proud of Clemson University's recent and peaceful desegregation, but that the cry for FEPC was rewarding increasing racial agitation, and that "the situation grows gradually worse with the federal efforts to solve the problem." Hearings, House General Subcommittee on Labor, 88th Cong., 1st sess., 11 July 1963, 1573, 2336–37. By 1963 Roosevelt's subcommittee had grown to 10 members, including a black congressman, Augustus Hawkins of California, who claimed authorship of California's fair employment commission.

70. This imbalance is documented and discussed in Paul Burstein, *Discrimination, Jobs, and Politics: The Struggle for Equal Employment Opportunity in the United States since the New Deal* (Chicago: University of Chicago Press, 1985), 97–124.

71. H.R. 405, Sec. 5(a), in Hearings, General House Subcommittee on Labor, 88th Cong., 1st sess., 3–11).

72. Audiotape log 108.2, 28 August 1963, JFKL; Whalens, *The Longest Debate*, 22–28.

73. Katzenbach to Celler, 13 August 1963, Attorney General file, box 11, JFKL.

Chapter IV

1. William Julius Wilson, *The Declining Significance of Race* (Chicago: University of Chicago Press, 1978), 88–99; Reynolds Farley, *Blacks and Whites: Narrowing the Gap?* (Cambridge: Harvard University Press, 1984), 56–81; Frank Levy, *Dollars and Dreams: The Changing American Income Distribution* (New York: Russell Sage, 1987), 56–57.

2. Emmett S. Redford and Marlan Blissett, *Organizing the Executive Branch: The Johnson Presidency* (Chicago: University of Chicago Press, 1981), 152–56, 180–83.

3. Transcript, Samuel V. Merrick Oral History Interview, 17 October 1966, 56, JFKL.

4. See, for example, the testimony of Secretary Wirtz, Hearings, Subcommittee on Labor of the House Committee on Education and Labor, 88th Cong., 1st. sess., 6 June 1963, 443–57; John D. Pomfret, "Economic Factors Underlie Negro Discontent," *New York Times*, 18 August 1963.

5. Wirtz testimony, Hearings, General Subcommittee on Labor, House Committee on Education and Labor, 88th Cong., 1st sess., 6 June 1963, 445.

6. Herman P. Miller, *How Our Income Is Divided* (Washington: U.S. Government Printing Office, 1963); Miller, *Rich Man, Poor Man* (New York: Crowell, 1964), 84–124.

7. Quoted in the *New York Times*, 18 August 1963. Miller's sobering study was formally released by Secretary of Commerce Luther Hodges at a hearing of the Senate Subcommittee on Manpower on 31 July 1963. See Miller testimony, Hearings, Subcommittee on Employment and Manpower, Senate Committee on Labor and Public Welfare, 88th Cong., 1st sess., 31 July 1963, 321–88.

8. Quoted in the *New York Times*, 28 June 1963.

9. Ray Marshall, *The Negro and Organized Labor* (New York: Wiley, 1965), 242–44.

10. *Steele v. Louisville and Nashville Railroad Co.*, 323 U.S. 192 (1944).

11. In testifying for a stronger federal anti-discrimination law to provide "genuine punishment" against segregated unions, George Meany told James Roosevelt's House subcommittee in 1962 that the AFL-CIO was a democratic federation whose only weapon beyond persuasion was expulsion. This was useful against corrupt union leaders, as in the AFL-CIO's expulsion of the Teamsters in 1957, but not against what Meany called a "corrupt bottom." Hearing testimony, General Subcommittee on Labor, House Committee on Education and Labor, 87th Cong., 2nd sess., 24 January 1962, 994.

12. *New York Times*, 6 February 1964; Independent Metal Workers and Hughes Tool Co., Case No. 23–CB–429, 147 NLRB No. 166 (1964).

13. *New York Times*, 3 July 1963; *The Crisis* (August–September 1963), 397.

14. *New York Times*, 7 July 1963. The new notion of compensatory demands for racial job quotas were widely discussed at the NAACP convention, but they found no place on the formal agenda or report. Prior to 1962 there had been almost no public discussion of the concept of "benign quotas." See Daniel W. Dodson, "Can Intergroup Quotas Be Benign?" *Journal of Intergroup Relations* (Autumn 1960), 12–17;

and Oscar Cohen, "The Case for Benign Quotas in Housing," *Phylon* (Spring 1960), 20–29.

15. Farmer testimony, Hearings, General Subcommittee on Labor, House Committee on Education and Labor, 87th Cong., 2nd sess., 26 June 1962, Pt. III, 2224.

16. August Meier and Elliott Rudwick, *CORE: A Study in the Civil Rights Movement, 1942–1968* (New York: Oxford University Press, 1973).

17. Charles E. Silberman, "The Businessman and the Negro," *Fortune* 68 (September 1963): 186, 191.

18. Transcript, Sealtest Meeting, 23 January and 13 February, 1963, minutes provided from the research files of Elliott Rudwick through the courtesy of August Meier. The Sealtest agreement excluded contractual union obligations. At Sealtest's request, it was not publicly announced, and Sealtest declined to place job listings in the *Amsterdam News* because such ads would be "discriminating."

19. Ibid., 191–92. CORE's regional dualism was awkwardly racial, with southern chapters remaining majority black, and those in the North and West often predominantly white—until the radicalization and Black Power movement of 1964–65.

20. James Farmer, *Lay Bare the Heart: An Autobiography of the Civil Rights Movement* (New York: Arbor House, 1985).

21. *New York Times*, 25 June 1963. Dukes was joined in demanding quota hiring by another black cleric, the Rev. Dr. Gardner C. Taylor, a former member of New York City's board of education.

22. *New York Times*, 3 July, 7 July 1963. Congressman Powell told the *Times* that he had led successful quota demands for Negro bus drivers in Harlem in the 1930s.

23. White House press conference, 20 August 1963, *Public Papers of the Presidents: John F. Kennedy* (Washington: U.S. Government Printing Office, 1964), 633–34; "Kennedy Opposes Quotas for Jobs on Basis of Race," *New York Times*, 21 August 1963.

24. Quoted by Senator Ervin, Hearings, Senate Judiciary Committee, 88th Cong., 1st sess., *Civil Rights—The President's Program, 1963*, 413.

25. Ibid., 413–141.

26. Ibid., 415.

27. Ibid., 416.

28. Ervin's biographer observed that the senator was privately embarrassed by the "miserable" southern record on civil rights, but would not admit it in public. Instead, Ervin either defended the entire South under the blanket of state-rights, or conceded that there were problems in Alabama and Mississippi, but not in North Carolina, where he maintained against all evidence that there was no racial disfranchisement. Paul R. Clancy, *Just a Country Lawyer: A Biography of Senator Sam Ervin* (Bloomington: Indiana University Press, 1974), 187–88.

29. Hearings, Subcommittee No. 5, House Judiciary Committee, 88th Cong., 1st sess., 26 June 1963, 2239–40.

30. Bayard Rustin, quoted in the *New York Times*, 15 December 1963.

31. *Wall Street Journal*, 5 September 1963.

32. *New York Times*, 9 November 1963.

33. *New York Times*, 15 November 1963.

34. *The Crisis* (May 1963), 291; (June–July 1963), 345. Wilkins's autobiography (written "with" Tom Mathews), *Standing Fast* (New York: Viking, 1982), reveals little about the policy dialogue within the NAACP power structure. The NAACP papers in the Library of Congress—unlike the National Urban League papers—reveal little internal debate over ends and means during the turmoil of the middle and late 1960s, this despite the vigorous public debates over ghetto rioting, Black Power, and reverse discrimination. This suggests either that no such internal debate occurred, which is unlikely, or that the manuscript collection was sanitized prior to donation.

35. *New York Times*, 12 September 1962. See Nancy J. Weiss, "Whitney M.

Young, Jr.: Committing the Power Structure to the Cause of Civil Rights," in John Hope Franklin and August Meier, eds., *Black Leaders of the Twentieth Century* (Urbana: University of Illinois Press, 1982), 331–58.

36. "Draft of a Position Statement on 'Special Consideration to Close the Gap,'" 21 January 1963, box 43, Series V, Headquarters File, National Urban League Papers, Library of Congress.

37. The Veterans' Preference Act of 1944 gave service-disabled veterans an automatic award of 10 extra points on federal civil service exams, and non-disabled veterans received 5 points. If the veterans' augmented total scores equaled the normal passing score of 70, the names of disabled veterans were moved to the top of the eligibility list, and the names of nondisabled veterans were moved ahead of non-veterans with the same rating.

38. Wendell G. Freeland to Guichard Parris, n.d., box 7, NUL Papers.

39. Lawrence Lowman to Guichard Parris, 5 February 1963, box 7, NUL Papers.

40. "A Statement urging a crash program of special effort to close the gap between the conditions of Negro and white citizens," by the Board of Trustees of the National Urban League, 9 June 1963, box 43, NUL Papers.

41. "The National Urban League's 'Marshall Plan' Proposal," 19 July 1963, box 38, NUL Papers.

42. Whitney Young, Jr., "A Marshall Plan for America," Speech file, box 42, NUL Papers.

43. *Wall Street Journal*, 5 September 1963.

44. *New York Times*, 9 October 1963; "Is Equality Unfair?" *America*, ·12 October 1963, 412–13.

45. "Should There Be 'Compensation' for Negroes," *New York Times Magazine*, 6 October 1963.

46. The following year McGraw-Hill published *To Be Equal* under Young's authorship, although the book was ghost-written by Lester J. Brooks. The book offered a ten-point program for Young's domestic Marshall Plan. Noting that Truman's rehabilitation of Europe had cost $17 billion between 1947 and 1951, Young called for a five-year commitment of $5 billion, and a phase-out of the program "as the need for it diminishes over the next decade." The ten points included the customary categories of effort: better schooling, access to entrace and technical jobs, management training, open housing, improved health and welfare, black role models on public boards and commissions. But the plan remained vague, and called for hiring and promoting "qualified Negroes" on an "equal basis," with blacks preferred only if equally qualified with competing whites.

47. Testimony of Edward E. Goshes on H.R. 10144 before the House Committee on Education and Labor, 21 February 1962, *Congressional Quarterly Almanac, 1962* (Washington: Congressional Quarterly Press, 1963), 412.

48. For a discussion of "iron triangles" in federal education policy, see Hugh Davis Graham, *The Uncertain Triumph* (Chapel Hill: University of North Carolina Press, 1984), 190–93.

49. *New York Times*, 27 July 1963.

50. U.S. Department of Labor press release, 6 June 1963; Circular 64–7, Bureau of Apprenticeship and Training, 17 July 1963.

51. Quoted in Marshall, *The Negro and Organized Labor*, 234–35.

52. *New York Times*, 20 October 1963; 18 December 1963.

53. *New York Times*, 9 October 1963.

54. *New York Times*, 13 December 1963; "Preferential Hiring," *Newsweek*, 16 December 1963, 69.

55. *New York Times*, 13 December 1963.

56. Howard Schuman, Charlotte Steeh, and Lawrence Bobo, *Racial Attitudes in America* (Cambridge: Harvard University Press, 1985), 1–42, 135–38, 193–214.

57. The standard study of fair employment in the 1960s is Michael I. Sovern,

Legal Restraints on Racial Discrimination in Employment (New York: Twentieth Century Fund, 1966). For the economic debate on the causes of unemployment and the effect of equal employment laws, see W. H. Landes, "The Effect of State Fair Employment Laws on the Economic Position of Nonwhites," *American Economic Review* 56 (1967): 578–90; M. H. Liggett, "The Efficacy of State Fair Employment Practices Commissions," *Industrial and Labor Relations Review* 22 (July 1969): 559–67; Dale L. Hiestand, *Discrimination in Employment* (Ann Arbor: Institute of Labor and Industrial Review, 1970); and Ray Marshall and Charles B. Knapp, *Employment Discrimination: The Impact of Legal and Administrative Remedies* (New York: Praeger, 1978).

58. Farmer, *Lay Bare the Heart,* 221.

59. Herbert Hill, "Twenty Years of State Fair Employment Practice Commissions," *Buffalo Law Review* 14 (Fall 1964): 22–69.

60. Paul H. Norgren and Samuel E. Hill, *Toward Fair Employment* (New York: Columbia University Press, 1964), 230; on the New York commission, see 116–30.

61. Morroe Berger, *Equality by Statute* (New York: Columbia University Press, 1952).

62. Alfred W. Blumrosen, "Anti-Discrimination in Action in New Jersey: A Law-Sociology Study," *Rutgers Law Review* 19 (1965):187–200.

63. Blumrosen, *Black Employment and the Law* (New Brunswick: Rutgers University Press, 1971), 53–54.

64. Leon H. Mayhew, *Law and Equal Opportunity: A Study of the Massachusetts Commission Against Discrimination* (Cambridge: Harvard University Press, 1968).

65. Hill, "Twenty Years of State FEPC," 47.

66. Mayhew, *Law and Equal Opportunity,* 198.

67. Ibid., 197.

68. Hill, "Twenty Years of State FEPC," 33.

Chapter V

1. For a sprightly account of the journey of H.R. 7152 through the House Judiciary Committee, see Charles and Barbara Whalen, *The Longest Debate* (Cabin John, Md.: Seven Locks Press, 1985), 29–70.

2. *Civil Rights, 1963* Report of the U.S. Commission on Civil Rights, 71–92; *New York Times,* 1 October 1963. On October 6 the Senate Labor Subcommittee approved such a bill, with FEP enforcement lodged in the Labor Department.

3. Proceedings and *Minutes,* Subcommittee No. 5, 11–12 September 1963; *Longest Debate,* 31–34; *New York Times,* 30 October 1963.

4. Celler had been heavily lobbied by such LCCR leaders as Clarence Mitchell, Joseph Rauh, Andrew Biemiller, Walter Fauntroy, and Arnold Aronson, and he had pledged to a NAACP convention in Washington on August 7 that he would add both a strong Title III and a FEPC provision to H.R. 7152.

5. Celler and McCulloch had each added another new title. Celler's would provide for prompt appeals to higher federal courts when unsympathetic federal district judges (in the South) would transfer voting challenges back to state courts. McCulloch's amendment, interestingly, directed the Census Bureau to gather voting statistics specifically by race.

6. Quoted in the *Wall Street Journal,* 7 November 1963.

7. Arthur M. Schlesinger, Jr., *Robert Kennedy* (Boston: Houghton Mifflin, 1978), 436.

8. Kennedy and Marshall oral history, 22 December 1964, vol. I, 10–11, JFKL.

9. Marshall oral history, 19–20 January 1970, 33–34, JFKL.

10. Marshall and Kennedy oral history, vol. 7, 22 December 1964, 884–86.

11. Ibid.; Katzenbach oral history, 12 November 1968, 17–19, LBJL.

12. Whalens, *Longest Debate,* 45–46.

13. Attorney General Robert F. Kennedy testimony, Hearings before the House Judiciary Committee, 88th Cong., 1st sess., on H.R. 7152, as Amended by Subcommittee No. 5, 15–16 October 1963, 2656.

14. Ibid., 2658.

15. The Republicans were, in addition to Robert Griffin of Michigan, Carroll D. Kearns (Pa.), Peter H.B. Frelinghuysen, Jr. (N.J.), Peter A. Garland (Maine), Charles E. Goodell (N.Y.), Albert H. Quie (Minn.), and William H. Ayers (Ohio).

16. "Equal Employment Opportunity Act of 1962," H. Rept. No. 1370, 87th Cong., 2d sess., 21 February 1962, 6–7.

17. Matthew A. Crenson and Francis E. Rourke, "By Way of Conclusion: American Bureaucracy since World War II," in Louis Galambos, ed., *The American Administrative State* (Baltimore: Johns Hopkins University Press, 1987), 139–46.

18. Louis Ruchames, *Race, Jobs, and Politics* (New York: Columbia University Press, 1953), 131 passim; Herbert Garfinkel, *When Negroes March* (Glencoe: Free Press, 1959), 77–89.

19. James O. Freedman, *Crisis and Legitimacy: The Administrative Process and American Government* (Cambridge: Cambridge University Press, 1978), 137–49.

20. James Roosevelt's memorandum of 21 February 1962 was cited in the statement by Griffin and Frelinghuysen in "Equal Employment Opportunity Act of 1963," H. Rept. No. 570, 88th Cong., 1st sess., 22 July 1963, 15–17. Rep. Roosevelt's memorandum in turn quoted his father from S. Doc. No. 8, 77th Cong., 1st sess., 206 (1941).

21. Statement by Griffin and Frelinghuysen, H. Rept. No. 570, 15.

22. Griffin and Frelinghuysen objected, however, to the Democrats' last-minute change in 1962 that allowed an EEOC commission member as well as the aggrieved party to file a complaint. This would make the EEOC a "self-starting agency with the power to conduct fishing and harassing expeditions," they said, and would unwisely "couple broad investigative power with an unlimited right to file charges."

23. *Wall Street Journal,* 12 July 1963.

24. *Wall Street Journal,* 26 October 1963.

25. *New York Times,* 29 October 1963. Lewis reported the essential elements of the compromise as it affected Title VII. But this was an exception to the broad-brush coverage of the equal employment provisions provided by the national media generally, which concentrated on the disputes over titles II and III.

26. The turmoil was triggered by a surprise motion by Rep. Arch Moore (R.-W.Va.) to report out the liberal subcommittee bill. The frantic nature of the negotiations that followed are reflected in the committee's emergency postponements and the bizarre maneuvers to capture the vote of Chicago Democrat Roland Libonati, whose erratic behavior as a member of Mayor Richard Daley's unforgiving political machine ensured his early political rertirement. See Whalens, *Longest Debate,* 47–66.

27. Audiotape log 116.7, 23 October 1963, JFKL; Whalens, *Longest Debate,* 50–54.

28. Southerners voting against the leadership compromise and for the motion to report out the liberal subcommittee bill were Democrats Robert T. Ashmore (S.C.), John V. Dowdy (Tex.), Elijah L. Forrester (Ga.), William M. Tuck (Va.), and Basil L. Whitener (N.C.); and Republicans William C. Cramer (Fla.) and Richard H. Poff (Va.).

29. Anthony Lewis, "Civil Rights Compact," *New York Times,* 30 October 1963; "Legislative History of H.R. 7152," Burke Marshall file, box 27–28, n.d., JFKL.

30. Quoted in *New York Times,* 17 October 1963. Roy Wilkens issued a statement on behalf of the Leadership Conference in which he called the Judiciary vote "no cause for rejoicing," and deplored the action of the Administration and the Re-

publican House leadership in "securing the defeat of the subcommittee bill on civil rights." *New York Times,* 30 October 1963.

31. The scholarly literature of 1963 was dominated by Morroe Berger, *Equality by Statute* (Garden City, N.Y.: Doubleday, 1952), which judged the pioneering New York commission to have been procedurally conservative and conciliatory, yet successful in protecting minorities from discriminatory acts. The following year two more critical studies were published: Paul H. Norgren and Samuel E. Hill, *Toward Fair Employment* (New York: Columbia University Press, 1964), especially 93–113 on the state FEP experience; and Herbert Hill, "Twenty Years of State Employment Practice Commissions," *Buffalo Law Review* 14 (Fall 1964): 22–39, which condemned the ineffectiveness of the state commissions.

32. Lyndon B. Johnson, *Public Papers of the Presidents, 1963–64* (Washington: U.S. Government Printing Office, 1965), I, 8–10.

33. Whalens, *Longest Debate,* 84–86.

34. Colmer, quoted in *New York Times,* 10 January 1964; Smith, quoted in Whalens, *Longest Debate,* 110.

35. 110 *Congressional Record,* 7 February 1964, Pt. 2, 2462–2513.

36. *Congressional Quarterly Almanac,* 1950, 375–84. Rogers's House colleague, Florida Democrat Charles E. Bennett, accomplished a similar feat by winning voice-vote approval of an amendment barring discrimination because of physical disability. In 1950, such "friendly" amendments were widely understood as conscious parodies of serious legislative purpose.

37. Donald Allen Robinson, "Two Movements in Pursuit of Equal Employment Opportunity," *Signs* 4 (1978–79): 413–17.

38. Susan D. Becker, *The Origins of the Equal Rights Amendment: American Feminism Between the Wars* (Westport, Conn.: Greenwood, 1981), 197–234.

39. Carl M. Brauer, "Women Activists, Southern Conservatives, and the Prohibition of Sex Discrimination in Title VII of the 1964 Civil Rights Act," *Journal of Southern History* XLIX (February 1983): 37–56.

40. A fresh and balanced assessment is Cynthia Harrison, *On Account of Sex: The Politics of Women's Issues 1945–1968* (Berkeley: University of California Press, 1988).

41. Cynthia Harrison, "A 'New Frontier' for Women: The Public Policy of the Kennedy Administration," *The Journal of American History* LXVII (December 1980): 630–46; Harrison, *On Account of Sex,* 109–65.

42. Peterson's pivotal role is reconstructed in Patricia G. Zelman, *Women, Work, and National Policy: The Kennedy-Johnson Years* (Ann Arbor: UMI Research Press, 1980). See also Chapter 8. In the House floor debate on H.R. 7152 during February 1964, Celler quoted Assistant Secretary Peterson's recommendation against adding sex discrimination to Title VII. See 110 *Congressional Record,* 8 February 1964, Pt. 2, 2485.

43. Quoted in Brauer, "Women Activists," 43.

44. Emily George, *Martha Griffiths* (Washington: University Press of America, 1982), 148–52; Harrison, *On Account of Sex,* 176–82.

45. Quoted in *Congressional Quarterly Almanac,* 1964, 348.

46. *New York Times,* 9 February 1964. Also supporting the Smith amendment, in addition to Griffiths, May, and St. George, were congresswomen Frances P. Bolton (R.-Ohio) and Edna F. Kelly (D.-N.Y.).

47. *New York Times,* 16 January 1964.

48. *AFL-CIO News,* 31 January 1964, box 8, Meany Archives.

49. "Comments on Senator Lister Hill's Criticism of Civil Rights Bill," attached to Biemiller to Officers of National and International Unions, 31 January 1964, box 9, Meany Archives.

50. Ibid., 2.

51. Ibid., 3. Biemiller's interpretation was standard writ among liberals in 1964. Labor writer A.H. Raskin explained to readers of the *Reporter* that the civil rights law was "explicit in disavowing any requirement that unions or management grant preferential treatment to rectify racial imbalance or to establish quotas in line with community race ratios." A.H. Raskin, "Civil Rights: The Law and the Unions," *Reporter* 31 (10 September 1964): 23.

52. Lyndon Johnson as quoted by Robert Kennedy, Kennedy and Marshall oral history interview with Anthony Lewis, 22 December 1964, Vol. 7, Pt. II, 871, JFKL.

53. Ibid., 872.

54. Kennedy oral history interview with John Bartlow Martin, 30 April 1964, Vol. 3, Tape VI-A, 470, JFKL.

55. Kennedy and Marshall oral history, 22 December 1964, Vol. 7, Pt. II, 873, JFKL.

56. *New York Times*, 2 February 1964; *Public Papers: Johnson*, 1963–64, Vol. I, 259.

57. See Rowland Evans and Robert Novak, *Lyndon Johnson: The Exercise of Power* (New York: New American Library, 1966), 376–84; Jim F. Heath, *Decade of Disillusionment: The Kennedy-Johnson Years* (Bloomington: University of Illinois Press, 1975), 169–76; Steven F. Lawson, "Civil Rights," in Robert A. Divine, ed., *Exploring the Johnson Years* (Austin: University of Texas Press, 1981), 100–103; Whalens, *The Longest Debate*, 149–57; Allen J. Matusow, *Unraveling of America* (New York: Harper & Row, 1984), 209–12.

58. See generally Whalens, *Longest Debate*, 124–217; Matusow, *The Unraveling of America*, 60–96; James L. Sundquist, *Politics and Policy* (Washington: Brookings Institution, 1968), 259–70.

59. Meier and Rudwick, *CORE*, 192–202.

60. Carl Solberg, *Hubert Humphrey: A Biography* (New York: Norton, 1984), 221–27. Solberg is uncritical in his praise of Humphrey's effective organizational role in winning the battle of the filibuster. But he captures in the ebullient Minnesotan's own words the single-mindedness of his flattering courtship of Everett Dirksen: "I would have kissed Dirksen's ass on the Capitol steps" (224). Mansfield and Humphrey kept the Senate in session both mornings and afternoons as well as on weekends, but they avoided the all-night sessions that threatened the health of the elderly and frail senators, and that also threatened compromise by fraying senatorial nerves.

61. Katzenbach to O'Brien, 9 March 1964, Files of Harry MacPherson, box 21, LBJL.

62. E.W. Kenworthy, "Rights Foes Divided on Dirksen Move," *New York Times*, 17 April 1964. The southern senators debated the relative merits of supporting Dirksen's moderately weakening amendments or supporting the liberals' strengthening moves, but could not agree on a consistent strategy. They wasted most of their energy on marginal jury-trial amendments while Dirksen whittled away their potential centrist allies and left them isolated and impotent.

63. Neil MacNeil, *Dirksen: Portrait of a Public Man* (New York: World, 1970), 223–38. MacNeil provides an extensive and admiring description of Dirksen's artful political manipulations without discussing the substance of any of the issues on which Dirksen was negotiating.

64. *New York Times*, 21 March 1964.

65. *New York Times*, 2 April 1964.

66. *New York Times*, 21 March 1964.

67. Quoted in *Congressional Quarterly Almanac*, 1951, 381.

68. *Wall Street Journal*, 8 April 1964.

69. "Memorandum Describing Changes in H.R. 7152 Embodied in Amendment No. 656 Offered by Senators Dirksen, Mansfield, Humphrey, and Kuchel," Adminis-

trative History, U.S. Department of Justice, Vol. VII, Civil Rights Division, Tab c.1, n.d., LBJL.

70. The Justice Department administrative history summarized the clarification Dirksen demanded in Title VI, which authorized cutting off federal funds to programs found to be discriminating, as follows: "The changes in Title VI are designed to clarify what has always been the intention of the bill: that is, to insure that *only the part of a program or activity in which discrimination is found* is curtailed and that *discrimination in one program or part of a program cannot* result in funds being withheld from other programs or from other parts of the same program within a state [emphasis added]." Litigation over this original understanding led the Supreme Court to reaffirm it in the *Grove City College* decision of 1984. Congress in 1988 amended Title VII to reverse *Grove City* and permit broad cut-off of federal funding.

71. The documents in the Kennedy and Johnson libraries capture little of the active pulse of these negotiations, other than retrospectively through summaries and oral history recollections. In such fast-breaking, ad hoc negotiations, oral communications precluded any extensive documentary record. The Whalens interviewed many of the staff participants in reconstructing the negotiations in *The Longest Debate*, 174–89.

72. *New York Times*, 8 May 1963.
73. *New York Times*, 22 April 1964.
74. Whalens, *Longest Debate*, 189.
75. *New York Times*, 21 April 1964.
76. Herbert Hill, "Twenty Years of State Fair Employment Practice Commissions," 22–69.
77. *New York Times*, 27 April 1963.
78. Illinois FEPC Charge No. 63x-127. On the Motorola case see generally the *New York Times*, March 13, May 29, July 19, November 3, 20, 22, and 26, 1964; Michael Sovern, *Legal Restraints on Racial Discrimination in Employment* (New York: Twentieth Century Fund, 1966), 73, n36; Ray Marshall, *The Negro and Organized Labor* (New York: Wiley, 1965), 252.
79. 110 *Congressional Record*, Pt. 11, 13 June 1964, 13724. The Motorola case is published in the *Congressional Record* at 13492.
80. See *The Civil Rights Act of 1964* (Washington: Bureau of National Affairs, 1964), 335, 346.
81. See *Wall Street Journal*, 3 March 1965; *New York Times*, 22 November 1964, 25 March 1966; 57 *Labor Relations Report* (1964), 264; *Motorola, Inc. v. Illinois F.E.P.C.*, 51 CCH Lab. Cas. par. 51, 323 (Cook County Circuit Court, 1965). Motorola appealed the commission's decision to the Cook County circuit court, which in March 1965 upheld the Illinois FEPC's finding of discrimination against Motorola, but reversed the commission's $1,000 order for punitive damages. In March 1966 the Illinois supreme court reversed the circuit court in favor of Motorola.
82. Humphrey, quoted in 110 *Congressional Record*, pt. 5, 30 March 1964, 6549.
83. Clark and Case, quoted in 110 *Congressional Record*, pt. 6, 8 April 1964, 7213. In 1964, state fair employment statutes specifically prohibited quota systems in Michigan, Ohio, Pennsylvania, and Rhode Island.
84. Ibid.
85. Humphrey, quoted in 110 *Congressional Record*, pt. 10, 4 June 1964, 12723.
86. *New York Times*, 8 May 1964. The congressional debates and editorial discussions of 1964 seldom mentioned sex discrimination, which seemed an afterthought and perhaps a nuisance appendage in a great national debate about racial policy.
87. 110 *Congressional Record*, pt. 10, 4 June 1964, 12723–24.
88. Matusow, *The Unraveling of America*, 96.
89. V.O. Key, Jr., *Southern Politics in State and Nation* (New York: Knopf, 1949); Numan V. Bartley and Hugh D. Graham, *Southern Politics and the Second Reconstruction* (Baltimore: Johns Hopkins University Press, 1975).

Chapter VI

1. For a recent assessment of Johnson's task force operation, see Hugh Davis Graham, *The Uncertain Triumph: Federal Education Policy in the Kennedy and Johnson Years* (Chapel Hill: University of North Carolina Press, 1984), ch. 3.

2. *Public Papers of the Presidents: Lyndon B. Johnson, 1963–1964* (Washington: U.S. Government Printing Office, 1965), I, 705.

3. Hugh Davis Graham, "Short-Circuiting the Bureaucracy in the Great Society: Policy Origins in Education," *Presidential Studies Quarterly* XII (Summer 1982): 407–20.

4. Gordon and Heller to Moyers, 30 May 1964, box 11, EX LE 2, WHCF, LBJL.

5. Moyers to Gardner Ackley et al., 6 July 1964, box 361, EX FG 600, WHCF, LBJL.

6. Lyndon Baines Johnson, *The Vantage Point* (New York: Holt, Rinehart & Winston, 1971), 322–46; Paul H. Conkin, *Big Daddy of the Pedernales* (Boston: Twayne, 1986), 208–42; Marshall Kaplan and Peggy Ciciti, eds., *The Great Society and Its Legacy* (Durham: Duke University Press, 1986).

7. "Task Force Issue Paper: CIVIL RIGHTS," 17 June 1964, box 3, Lee White files, LBJL.

8. Robert W. Kopp, "Management's Concern with Recent Civil Rights Legislation," *Labor Law Journal* 16 (February 1965): 67–86.

9. Ibid., 67.

10. Seidman to the Director, 28 July 1964, Bureau of the Budget papers (hereafter cited as BOB), Record Group 51, series 60.26, EEOC file, box 107, National Archives, Washington, D.C. (hereafter cited as NARA).

11. King Carr to Jack Buchanek, 25 June 1964; Buchanek to Seidman, n.d., box 106, BOB/EEOC 6802–30, NARA.

12. Ibid.

13. Executive Order 11141, 29 *Federal Register* 2477 (12 February 1964). Johnson was responding to a recommendation by President Kennedy's Council on Aging, which delivered its report to Johnson on 16 December 1963. See Johnson's remarks upon releasing the report to the public in *Public Papers of the Presidents: Johnson, 1963–64*, I, 273–74.

14. Seidman to the Director, 28 July 1964, BOB/EEOC, NARA. Unlike the term "Negro," which was dropped from government usage around 1966–67, the term "chairman," which was often statutory, persisted in standard usage into the 1970s.

15. Focke to the Director, 30 July 1964, box 92, BOB series 61.1a, NARA.

16. "Guide for Issuance of Regulations Under Title VI of the Civil Rights Act of 1964," prepared by Norbert A. Schlei, Assistant Attorney General, Office of Legal Counsel, 3 July 1964; Schlei to James L. Morrison, 24 July 1964, box 28, Marshall files, JFKL.

17. Carey to the Director, 31 July 1964, box 92, BOB series 61.1a, NARA.

18. White to the Subcabinet Group, 3 September 1964, Ex FG 600/S, box 3, WHCF, LBJL. By September of 1964 the group listed 35 members, and included among its more active members in civil rights such men (*all* members were men) as Hobart Taylor (PCEEO), Burke Marshall (Justice), John Macy (Civil Service Commission), Peter Libassi and William Taylor (Civil Rights Commission), Stephen Shulman (Defense), Louis Martin (Democratic National Committee), James Quigley (HEW), James Reynolds (Labor), LeRoy Collins (Community Relations Service), and Clifford Alexander (White House staff).

19. "A Civil Rights Program: Points for Discussion," n.d., box 3, White files, LBJL; author interview with Lee White, 16 June 1986, Washington, D.C.

20. Hughes to the Director, 17 November 1964, box 92, BOB series 61.1a, Box 92, NARA.

21. Johnson to Humphrey, 2 December 1964, box 403, EX FG 731, WHCF, LBJL.

22. Humphrey to the President, 4 January 1965, box 4, White files, LBJL.

23. *Report to the President* on the Coordination of Civil Rights Activities in the Federal Government, submitted by the Vice President-elect, 4 January 1965, box 403, EX FG 731, 15, WHCF, LBJL.

24. Moyers to the President, 3 February 1965, box 4, White files, LBJL.

25. "Letter to the Vice President Upon Establishing the President's Council on Equal Opportunity," 6 February 1965, *Public Papers of the Presidents: Johnson, 1965,* I, 152–53; Executive Order 11197, 30 *Federal Register* 1721 (5 February 1965).

26. Johnson, *The Vantage Point,* 161.

27. The popular vote was 43,128,958 for Johnson-Humphrey to 27,176,873 for Goldwater-Miller, providing a Democratic margin of 61.4 percent of the two-party vote that approximated Franklin Roosevelt's landslide of 1936. See Bernard Cosman, *Five States for Goldwater* (University: University of Alabama Press, 1966); and Numan V. Bartley and Hugh D. Graham, *Southern Politics and the Second Reconstruction* (Baltimore: Johns Hopkins University Press, 1975), ch. 4.

28. J. Morgan Kousser, "The Undermining of the First Reconstruction: Lessons for the Second," in *Minority Vote Dilution,* ed. Chandler Davidson (Washington: Howard University Press, 1984), 27–46.

29. Lee Holt, *The Summer That Didn't End* (New York: Morrow, 1965); Doug McAdam, *Freedom Summer* (New York: Oxford University Press, 1988).

30. Katzenbach testimony, *Hearings on Voting Rights,* House Judiciary Committee, 89th Cong., 1st sess. (Washington: U.S. Government Printing Office, 1965), 9.

31. Katzenbach to the President, 18 December 1964, Administrative History of the Department of Justice, vol. VII, Civil Rights, part Xa., Documentary Supplement, Tab B.2.a., 2, LBJL.

32. David J. Garrow, *Bearing the Cross: Martin Luther King, Jr., and the Southern Christian Leadership Conference* (New York: Random House, 1986), 357–430. See also Garrow, *Protest at Selma: Martin Luther King and the Voting Rights Act of 1965* (New Haven: Yale University Press, 1978).

33. Adam Fairclough in *To Redeem the Soul of America: The Southern Christian Leadership Conference and Martin Luther King, Jr.* (Athens: University of Georgia Press, 1987), 225–52, finds less philosophical contradiction in King's post-Albany strategy.

34. Black registration as a percentage of eligible voters in the other southern states in 1964, in ascending rank order, were Louisiana 31.7%, South Carolina 37.3%; Virginia 38.3%, Arkansas 40.4%, Georgia 44%; North Carolina 46.8%, Florida 51.2%, Arkansas 54.4%, Texas 57.7%, and Tennessee 69.4%. For the eleven former Confederate states in 1964 the average level of registration for eligible blacks was 43.3%, whereas the average for whites was 73.2%.

35. King's carefully laid scenario was almost wrecked when shortly before his protest began, Selma's new mayor appointed a shrewd segregationist, Wilson Baker, as director of public safety. Baker, however, could not duplicate Laurie Prichett's control in Albany because Baker could not control the volatile Sheriff Clark. But King's demonstrators could. As Baker later observed, the protesters voted Clark "an honorary member of SNCC, SCLC, CORE, the N-Double A-C-P. . . .And from then on they played him just like an expert playing a violin." Interview of Wilson Baker in Howell Raines, *My Heart Is Rested* (New York: Putnam's, 1977), 200; Garrow, *Protest at Selma,* 220–24; Allen J. Matusow, *The Unraveling of America* (New York: Harper & Row, 1984), 180–87.

36. Steven F. Lawson, *Black Ballots: Voting Rights in the South, 1944–1969* (New York: Columbia University Press, 1976), 308–22.

37. *Public Papers of the Presidents: Johnson, 1965,* I, 5.

38. Rauh to Katzenbach, 12 February 1965, Justice Department Administrative History, Civil Rights, Doc. Supp., LBJL.

39. Cox to the Attorney General, 23 February 1965; Greene and Lindenbaum, Draft Act "To Enforce the Fifteenth Amendment to the Constitution of the United States," 5 March 1965; Claiborne to Cox, 29 March 1964; Justice Department Administrative History, Civil Rights, Doc. Supp., LBJL.

40. *Congressional Quarterly Almanac, 1965,* 533–65.

41. Cox to the Attorney General, 23 March 1965, Justice Department Administrative History, Civil Rights, Doc. Supp., LBJL.

42. Howard Ball, Dale Krane, and Thomas P. Lauth, *Compromised Compliance: Implementation of the 1965 Voting Rights Act* (Westport, Conn.: Greenwood Press, 1982), 64–94; Abigail M. Thernstrom, *Whose Votes Count?* (Cambridge: Harvard University Press, 1987), 24–27.

43. This was the dissenting argument of Justice Hugo Black when the Supreme Court upheld the Voting Rights Act in *South Carolina v. Katzenbach,* 383 U.S. 301 (1966). Black's main contention was that Section 5 violated the fundamental compact of federalism by treating the states as "little more than *conquered provinces,*" requiring them "to *beg* federal authorities" for approval of local laws before they could implement them. His dissent argued that to have a federal court rule on a state law before it was implemented was to render an advisory opinion without a trial, whereas the Constitution gave to federal courts jurisdiction to hear only concrete state cases on appeal.

44. *Heart of Atlanta Motel v. United States,* 379 U.S. 241 (1964).

45. *Louisiana v. United States,* 380 U.S. 145 (1965); *United States v. Mississippi,* 380 U.S. 128 (1965). The following year, in *Harper v. Virginia State Board of Elections,* 383 U.S. 663 (1966), the Supreme Court invalidated the state requirement of poll tax payment as a prerequisite for voting in state and local elections.

46. *Public Papers of the Presidents: Johnson, 1965,* I, 282–83. Johnson entitled his speech "The American Promise," but it would be remembered by his invocation of the Movement song, "We Shall Overcome."

47. No single monographic study of the 1965 voting law exists that is equivalent to the Congress-makes-a-law studies of the civil rights acts of 1957, 1960, or 1964. But Steven F. Lawson's longitudinal, two-volume study reconstructs the evolution of the full policy process in voting-rights law, including its implementation. Lawson's *Black Ballots* covers voting rights in the South from 1944 to 1969, and its sequel, *In Pursuit of Power* (New York: Columbia University Press, 1985), covers 1965–82.

48. The literature on the Voting Rights Act has been dominated by political scientists and legal scholars who have concentrated on its implementation and litigation. See for example Harrel P. Rogers and Charles S. Bullock III, *Law and Social Change: Civil Rights Laws and Their Consequences* (New York: McGraw-Hill, 1972); U.S. Commission on Civil Rights, *The Voting Rights Act: Ten Years After* (Washington: U.S. Government Printing Office, 1975); Ball, Kane, and Lauth, *Compromised Compliance;* and Thernstrom, *Whose Votes Count?* Steven Lawson describes the drafting and enactment of the 1965 law in *Black Ballots,* 307–21.

49. The nine liberal senators on the Judiciary Committee were Democrats Burch Bayh of Indiana, Quentin Burdick of North Dakota, Philip Hart of Michigan, Edward Kennedy of Massachusetts, Edward Long of Missouri, and Joseph Tydings of Maryland; and Republicans Hiram Fong of Hawaii, Hugh Scott of Pennsylvania, and Jacob Javits of New York. Voting against the bill in committee were chairman James Eastland of Mississippi, John McClellan of Arkansas, Sam Ervin of North Carolina, and Olin Johnston of South Carolina—all Democrats. Senator Johnston was absent with illness, but was recorded against the bill; he died on April 18.

50. J. Morgan Kousser, *The Shaping of Southern Politics: Suffrage Restriction and the Establishment of the One-Party South, 1880–1910* (New Haven: Yale University Press, 1974).

51. In the House Judiciary hearings, chairman Celler opposed a similar amendment by arguing that the target of the bill was voting discrimination, and the New York law requiring English literacy, which had served generations of immigrants well as an inducement to citizenship and a doorway to its responsibilities, was not administered discriminatorily. But senatorial courtesy prevailed in the Senate, and the Senate prevailed in conference—despite Dirksen's misgivings. On November 15, 1965 a three-judge federal court in the District of Columbia invalidated the "American flag" provision as "entirely dissociated" from the anti-discriminatory intent of the law and an unconstitutional intrusion by Congress under the inappropriate color of the 15th Amendment. But the following year the Supreme Court reversed, holding in *Katzenbach v. Morgan,* 384 U.S. 641 (1966), that the "American flag" provision was permissible under congressional authority to enforce the equal protection clause of the 14th Amendment.

52. Note [Larry F. Amerine], "The Voting Rights Act of 1965," *University of Texas Law Review* 4 (July 1966): 1411–16.

53. Daniel P. Moynihan, *Family and Nation* (New York: Harcourt, Brace, Jovanovich, 1986), 30.

54. *Public Papers of the Presidents: Johnson, 1965,* I, 636.

55. As Moynihan recalled, Johnson accepted the draft speech without change, requiring only that it be read first to the principal civil rights leaders—King, Wilkins, and Young—whose judgment he trusted. James Farmer of CORE and John Lewis of SNCC had been excluded from the President's circle following their refusal to agree to his proposed moratorium on civil rights demonstrations following the enactment of the Civil Rights Act of 1964. The Goodwin redraft of Moynihan's original text was modestly edited for style—e.g., Horace Busby's objections removed the potentially objectionable term "assimilation" as a civil rights goal. But the reference to "equality as a fact and as a result" in the preceding paragraph drew no editorial comment.

56. Douglas Kiker, "Johnson Spoke for History," *New York Herald Tribune,* 6 June 1965.

57. Lee Rainwater and William L. Yancey, *The Moynihan Report and the Politics of Controversy* (Cambridge: MIT Press, 1967).

Chapter VII

1. Author interview with Lee White, 16 June 1986, Washington, D.C.

2. James Rowe to the President, 22 January 1965; William Taylor to Bill Moyers, 28 January 1965, box 380, EX FG 655, WHCF, LBJL.

3. Macy to the President, 4 December 1964, and 11 December 1964, box 774, Macy files, LBJL. Macy informed Johnson in April 1965 that White preferred to remain in his White House post.

4. *New York Times,* 3 May 1965; *St. Louis Post-Dispatch,* 13 May 1965; Clark to the President, 7 May 1965, box 382, EX FG 655, WHCF, LBJL.

5. Macy to the President, 29 April 1965, box 380, EX FG 655, WHCF, LBJL.

6. Moyers to the President, 29 April 1965, box 774, Macy files, LBJL. Moyers described Graham as "a World War II pilot hero, an engineer and inventor, and a $40,000-a-year corporate executive in the computer field. . . . He is a tough-minded, fair, and talented young man who is also very close to me and would be 'our' man on the FEPC."

7. *New York Times,* 11 May 1965. Early in the Kennedy administration, Secretary McNamara had objected to giving Roosevelt a senior appointment at Defense, and similar objections from within the Kennedy organization had blocked his appointment as Secretary of Commerce. Roosevelt, unlike his mother, Eleanor, had supported Kennedy's presidential candidacy, and Kennedy subsequently sought to reward

him with a senior post in Washington as a launching pad for his New York political ambitions. But like so many of the Roosevelts' children, Franklin, Jr., seemed to lack a steadiness of character and purpose to match his name and ambition. See Arthur Schlesinger, Jr., *Robert Kennedy and His Times* (Boston: Houghton Mifflin, 1978), 241, 813. On the burdens of childhood in the Roosevelt family, see Hugh Davis Graham, "The Paradox of Eleanor Roosevelt: Alcoholism's Child," *Virginia Quarterly Review* 63 (Spring 1987): 210–30.

8. *New York Times,* 3 June 1965. Neither Roosevelt nor Hernandez were suggested on the Macy-Moyers lists.

9. *New York Times,* 13 August 1965. Roosevelt also missed the planning for the White House Conference on Equal Employment Opportunity, which was set for August 19–20.

10. Representative William Brock, quoted in the *Congressional Record,* 18 August 1965, attached to Moyers to FDR, Jr., 21 August 1965, box 380, EX FG 652, WHCF, LBJL.

11. Ibid. On August 3, John Macy sent Marvin Watson a similar complaint about Roosevelt's frequent absences.

12. The Vice President to Hobart Taylor, 8 February 1965, box 106, Series 60.26, BOB, NARA.

13. William Kolberg to John Stewart, 17 February 1965, box 107; Kolberg to Hirst Sutton, 17 February 1965, box 106, Series 60.26, BOB, NARA. Kolberg told Sutton that "the financial affairs of the P.C.E.E.O. are a lot more questionable than I had any idea." The quarterly billings were resented by the mission agencies, and Kolberg conceded that "we must steel ourselves" once again for yet another billing.

14. Humphrey inherited the Vice President's statutory chairmanship of the Aeronautics and Space Council, but he could expect to exert little substantive influence on its established policies or staff, which Johnson had guided and would of course protect. Humphrey's other council chairmanships were largely showcase and quite modest: Youth Opportunity, Indian Opportunity, Marine Sciences, Recreation and Natural Beauty, plus a presiding function on the President's Council on Economic Opportunity.

15. The Vice President to Department Secretaries, 10 March 1965, box 107, Series 60.26, BOB, NA.

16. The Vice President to the President, 9 August 1965, box 4, White files, LBJL.

17. Hubert Horatio Humphrey, *The Education of a Public Man* (New York: Doubleday, 1976), 407.

18. White to the President, 12 August 1965, box 4, White files, LBJL.

19. Lee Rainwater and William L. Yancey, *The Moynihan Report and the Politics of Controversy* (Cambridge: MIT Press, 1967). In response to the controversy, Johnson appointed Chicago businessman Benjamin W. Heineman to defuse it and control the damage. The fall White House conference was downgraded to a planning session, and the conference was postponed until 1966. Heineman could not defuse the controversy, but his efforts to control the damage were partly successful, as 2,500 carefully screened delegates convened at Washington's Sheraton-Park Hotel in June 1966 to approve prearranged recommendations, and to defeat a resolution condemning the President's policy on Vietnam.

20. The Vice President to Watson, 26 August 1965, box 4, White files, LBJL.

21. Katzenbach to the President, 20 September 1965, box 403, EX FG 731, WHCF, LBJL.

22. Katzenbach's assumptions were called by John Doar, assistant attorney general for the Civil Rights Division, the "Norbert Schlei approach," which basically viewed the federal government as far too complex for highly centralized administration. Doar contrasted Schlei's approach to the "LeRoy Collins approach," which called for centralized civil rights enforcement under a coordinating czar. See Allan Wolk,

The Presidency and Black Civil Rights (Rutherford, N.J.: Fairleigh Dickinson Press, 1971), 178–89.

23. Robert Kennedy and Burke Marshall oral history interview with Anthony Lewis, 4 December 1964, Vol. 6, Tape VI, 719–25, JFKL; Berl Bernhard oral history interview, 17 June 1968, 23 July 1968, *passim*, JFKL. As staff director for the Civil Rights Commission, Bernhard admitted that the commission's staff investigations on Mississippi included "just wild, undocumented, hysterical material," which the staff then liberally leaked to the press.

24. White to the President, 20 September 1965, box 403, EX FG 731, WHCF, LBJL.

25. Transcript, White House News Conference, 24 September 1965, box 3, EX HU 2, WHCF, LBJL; Lyndon Johnson, "Memorandum on Reassignment of Civil Rights Functions," *Public Papers of the Presidents, 1965,* II, 1017–19 (the recommendations attributed to Humphrey were appended thereto). Although former CRS director Collins had recently moved to the deputy secretary post in Commerce, he was reported to be angry at the shift of the service to Justice on such short notice and with almost no consultation, and the complaints of the threatened CRS staff were echoed in the press. This likely explains at least in part why dismayed civil rights leaders concentrated their attack on the marginal CRS move as well as the defrocking of Humphrey, and said little about the far more important shift of contract compliance to Labor rather than to the EEOC.

26. Humphrey, *Education of a Public Man,* 408.

27. Author interview with Lee C. White, 16 June 1986, Washington, D.C.

28. Executive Order 11247, 30 *Federal Register* 12327 (1965).

29. John Herbers, "Rights Groups Fear Easing of U.S. Enforcement Role," *New York Times,* 17 October 1965; Christopher Pyle and Richard Morgan, "Johnson's Civil Rights Shake-up," *New Leader,* 11 October 1965; Whitney Young, Jr., "Rights Program Downgraded," *New York Herald Tribune,* 19 October 1955; Allen J. Matusow, *The Unraveling of America* (New York: Harper & Row, 1984), 209–16.

30. Carl Solberg, *Hubert Humphrey: A Biography* (New York: Norton, 1984), 276–78.

31. Richard K. Berg, "Equal Employment Opportunity Under the Civil Rights Act of 1964," *Brooklyn Law Review* 31 (1964): 62–97.

32. Ibid., 64–67, 85–88. When the provision for private suits was originally inserted in the House bill, it was regarded as a rather marginal precaution by liberals against failure of the commissioners to bring suit. The public right was subsequently reinforced by provision for the Attorney General to bring "pattern or practice" suits, leaving the EEOC with the authority to file commissioner complaints.

33. Ibid., 85. See also Comment, "Enforcement of Fair Employment Under the Civil Rights Act of 1964," *Chicago Law Review* 32 (1965): 430–70.

34. *Wall Street Journal,* 28 May 1965.

35. Ibid.

36. EEOC Minutes, meeting #4, 14 July 1965, EEOC Archive, Washington, D.C.

37. "NAACP Files First Major Employment Complaints Under Title VII of the Civil Rights Act," NAACP press release, 14 July 1965, NAACP Papers.

38. Hill was to make a NAACP crusade out of complaint-generation. His memorandum of January 1967 to the NAACP's state conference presidents provided detailed instructions for adding to the NAACP's 2,000 signed and notarized affidavits of complaint to the EEOC. Hill to State Conference Presidents, 2 January 1967, box 362, Series B, Group III, NAACP Papers.

39. *Wall Street Journal,* 15 September 1965. Respondent companies included such southern plants of nationally based firms as Chevrolet in Atlanta, Kaiser in Baton Rouge, and Union Carbide in Eden, North Carolina.

40. *New York Times,* 14 December 1965. In its first annual report in 1966 the

EEOC summarized the complaint patterns. The commission investigated only 43 percent of the nearly 9,000 complaints received. Twenty-three percent were rejected for lack of jurisdiction, 16 percent required more information, 11 percent were referred to state commissions, and 7 percent remained pending. About half originated in the South, 82 percent were directed to employers, and 60 percent charged racial discrimination. The big surprise was that 37 percent charged sex discrimination. Only 1 percent involved Hispanic Americans, and only 2 percent involved religion.

41. Roosevelt to Schultze, n.d., box 107, Series 60.26, BOB, NARA.

42. Cornelius J. Peck, "The Equal Employment Opportunity Commission: Developments in the Administrative Process 1965–1975," *Washington Law Review* 51 (1976): 831–65.

43. U.S. Equal Employment Opportunity Commission, *Legislative History of Titles VII and XI of Civil Rights Act of 1964* (Washington: U.S. Government Printing Office, n.d.), 3006.

44. Ibid., 3189.

45. Ibid., 3005.

46. Ibid., 3019.

47. Ibid., 3004.

48. Alfred W. Blumrosen, "Administrative Creativity: The First Year of the Equal Employment Opportunity Commission," in Blumrosen, *Black Employment and the Law* (New Brunswick: Rutgers University Press, 1971), 68–69.

49. Ibid., 74–79.

50. EEOC Minutes, meeting #26, 2, 22 September 1965.

51. EEOC Minutes, meeting #34, 4, 7 October 1965.

52. Ibid., 3.

53. Blumrosen, *Black Employment*, 72.

54. Berg, "Equal Employment," 89.

55. Michael I. Sovern, *Legal Restraints on Racial Discrimination in Employment* (New York: Twentieth Century Fund, 1966), 88.

56. *EEOC Administrative History*, Vol. I, 36.

57. *New York Times*, 26 November 1965; *EEOC Administrative History*, Vol. 1, 15–17, 59–67, LBJL. Form EEO-1 was required annually from employers, EEO-2 from the apprenticeship committees, EEO-3 from unions, and EEO-4 from state employment agencies.

58. *EEOC Administrative History*, 20.

59. Ibid., 67.

60. Blumrosen, "Administrative Creativity," 51–101; author interview with Richard Graham, 12 June 1986, Washington, D.C.

61. The affirmative action panel was chaired, ironically, by Donald Thomas of Plans for Progress. The other chairmen and their panels were: John Doar, assistant attorney general for civil rights, on patterns of discrimination; Alfred Blumrosen for complaint procedures; Richard Berg for sex discrimination; Charles Markham, EEOC research director, for reports and record-keeping; Vincent Macaluso of the PCEEO on apprenticeship and training; and Percy Williams, also of the PCEEO, on referral, hiring, promotion, and firing.

62. Transcript, White House Conference on Equal Employment Opportunity, 20 August 1965, 2, EEOC Library.

63. Ibid., 29.

64. 110 *Congressional Record*, pt. 6, 8 April 1964, 7213.

65. Transcript, White House Conference on EEO, 81, EEOC Library.

66. Ibid., 83–84.

67. *Wall Street Journal*, 25 March 1966.

68. Ibid.

69. The ambivalence of the civil rights organizations about racial identification in

records was reflected in the NAACP response to the Wirtz announcement. Clarence Mitchell was on record with his denunciation of such proposals as "scurrilous," and even Herbert Hill called it "potentially dangerous." But the LDF's director-counsel, Jack Greenberg, called it a "useful principle to apply across the board." *New York Times,* 19 May 1966.

70. McPherson to the President, 28 April 1966, box 381, EX FG 655, WHCF, LBJL. Johnson finally relented on the exchange of letters, but he severely edited the draft "Dear Frank" letter that McPherson sent him, calling it "Too long, too fulsome, and too formal," and striking out references to Roosevelt's "distinguished" service and to Johnson's "regard and affection of an old friend of the family." Johnson to Roosevelt, draft (undated), initialed 13 May 1966, box 381, EX FG 655, WHCF, LBJL.

71. Macy to the President, 11 July 1966, box 774, Macy files, LBJL. Holcomb wrote Johnson on June 21 that he preferred not to be considered for the chairmanship, and Alfred Hernandez, president of the League of United Latin American Citizens (LULAC) wrote Senator Yarborough that Holcomb's appointment would "set back the cause of Mexican-Americans of Texas and the Southwest 20 years." Holcomb was given to writing sycophantic letters to Johnson, typically pledging, for example, that "Wherever I go, whatever I do, I strive to make a conscientious effort to handle myself in a way that will be pleasing to you." Holcomb to the President, 28 July 1967, box 381, EX FG 655, WHCF, LBJL.

72. Macy to the President, 20 April 1966, box 744, Macy files, LBJL.

73. Commissioner Graham's father was Jewish, but this was not generally known, and the commission's white liberal Republican with the Scottish name and the Ivy League degree presumably represented the standard WASP pedigree.

74. Moyers to Franklin D. Roosevelt, Jr., 17 May 1965, Box 4, White files, LBJL.

75. James C. Falcon to Macy, 28 July 1966, box 774, Macy files, LBJL. Falcon closed his summary of the empty-handed search by observing that "15 names have been presented to the President. Five of these have been asked to serve and are not available. One is available, but the president isn't interested. Oh, shucks."

76. Macy to the President, 27 April 1966, box 381, EX FG 655, WHCF, LBJL. Graham shared a belief, widespread among feminists, that his reappointment was blocked in response to his activities on behalf of women. Graham interview by the author, 12 June 1986, Washington, D.C.

77. O. Jack Buchanek to William D. Carey, 9 August 1966, box 181, Series 61.1a, BOB, NARA. A "prime example" of this scramble for political favor, Buchanek wrote, was that the New York field office was being filled with the patronage appointees of Adam Clayton Powell.

78. Macy to the President, 22 August 1966, box 381, EX FG 655, WHCF, LBJL. The other nominees were, in order, H. Rex Lee (governor of Samoa), David E. Feller (former Goldberg law partner), N. Thompson Powers (former counsel to the PCEEO, assistant to Wirtz, and acting staff director of the EEOC), and Malcolm Andersen (tax counsel for Mobil Oil and board member of the National Urban League).

79. Notes, telephone conversation, Macy to the President, 17 August 1966, box 774, Macy files, LBJL.

80. Congressional Quarterly, *Revolution in Civil Rights,* 3rd ed. (Washington: Congressional Quarterly Press, 1967), 62.

81. *Congress and the Nation, 1965–1968* (Washington: Congressional Quarterly Press, 1969), III, 5.

82. Transcript, Farmer oral history interview, 20 July 1971, 18–19, LBJL.

83. The new partisan lineup for the 90th Congress still heavily favored the Democrats, with margins of 64–36 in the Senate and 248–187 in the House. But the 1966 elections gave Republicans control of half of the governorships.

84. Buchanek to Carey, 9 August 1966; Schultze to Holcomb, 13 August 1966;

Buchanek to the Director, 24 October 1966, box 106, EEOC 6806–15, Series 60.26, BOB, NARA.

85. Buchanek to the Director, 12 August 1966, box 106, Series 61.1a, BOB, NARA.

Chapter VIII

1. See generally Aileen Kraditor, *Ideas of the Woman Suffrage Movement, 1890–1920* (New York: Columbia University Press, 1965); William O'Neill, *Everyone Was Brave: A History of Femimism in America* (New York: Times Books, 1969); William H. Chafe, *The American Woman: Her Changing Social, Economic, and Political Role, 1920–1970* (New York: Oxford University Press, 1972); Jo Freeman, *The Politics of Women's Liberation: A Case Study of an Emerging Social Movement and Its Relation to the Policy Process* (New York: David McKay, 1975); and Carl N. Degler, *At Odds: Women and the Family in America from the Revolution to the Present* (New York: Oxford University Press, 1980).

2. Susan D. Becker, *The Origins of the Equal Rights Amendment: American Feminism Between the Wars* (Westport, Conn.: Greenwood, 1981); Susan Ware, *Beyond Suffrage: Women in the New Deal* (Cambridge: Harvard University Press, 1981). On the postwar years, see Leila Rupp and Verta Taylor, *Survival in the Doldrums: The American Women's Rights Movement, 1945 to the 1960s* (New York: Oxford University Press, 1987).

3. Recent case studies have probed with greater subtlety and some revisionist implications the dichotomies of Kraditor and O'Neill. See Steven M. Buechler, *The Tranrormation of the Woman Suffrage Movement: The Case of Illinois, 1850–1920* (New Brunswick: Rutgers University Press, 1986); and Felice D. Gordon, *After Winning: The Legacy of the New Jersey Suffragists, 1920–1947* (New Brunswick: Rutgers University Press, 1986).

4. The loose coalition supporting women's protective legislation included the League of Women Voters, the Women's Trade Union League, the National Consumer's League, the YWCA, the national councils of Catholic and of Jewish women, and the Women's Christian Temperance Union. See Becker, "They Are Reformers—We Are Feminists," *Origins of ERA*, 197–234; Nancy Schron Dye, *As Equals and As Sisters: Feminism, the Labor Movement, and the Women's Trade Union League of New York* (Columbia: University of Missouri Press, 1980); Judith Sealander, *As Minority Becomes Majority: Federal Reaction to the Phenomenon of Women in the Workplace, 1920–1963* (Westport, Conn.: Greenwood, 1983).

5. In *Adkins v. Children's Hospital*, 261 U.S. 525 (1923), the Court majority struck down a District of Columbia minimum wage law for women, thereby appearing to threaten the entire corpus of protective legislation that had been successfully defended by Louis Brandeis in his famous defense of "sociological jurisprudence" in *Muller v. Oregon*, 208 U.S. 412 (1908). The *Adkins* Court's Darwinian egalitarianism was not vigorously pursued, however, and in 1937 the Supreme Court formally overturned *Adkins* in *West Coast Hotel Co. v. Parrish*, 300 U.S. 379 (1937).

6. Patricia G. Zelman, *Women, Work, and National Policy: The Kennedy-Johnson Years* (Ann Arbor: UMI Research Press, 1980). See also Cynthia Harrison, "A 'New Frontier' for Women: The Public Policy of the Kennedy Administration," *The Journal of American History* LXVII (December 1980): 630–46.

7. Carl M. Brauer, "Women Activists, Southern Conservatives, and the Prohibition of Sex Discrimination in Title VII of the 1964 Civil Rights Act," *Journal of Southern History* XLIV (February 1983): 37–56.

8. Cynthia Harrison, *On Account of Sex: The Politics of Women's Issues 1945–1968* (Berkeley: University of California Press, 1988).

9. Ibid., 109–65.

10. In 1944 a coalition of labor, consumer, civic, and religious women's groups,

led by Mary Anderson and the Women's Bureau and supported by the AFL, CIO, and the ACLU, formed the National Committee to Defeat the UnEqual Rights Amendment (NCDURA) to oppose the ERA and preserve the existing labor standards' laws protecting women. In 1947 the Women's Bureau coalition transformed the negative NCDURA into the positive National Committee on the Status of Women. The latter disbanded in 1950 when the Hayden rider to the ERA proved effective in killing the amendment by exempting women's protective legislation. See Marguerite Rawalt, "The Equal Rights Amendment," *Women in Washington,* ed. Irene Tinker (Beverly Hills: Sage, 1983), 49–78.

11. Zelman, *Women: Kennedy-Johnson Years,* 30–32; Esther Peterson, "The Kennedy Commission," *Women in Washington,* 21–34; Harrison, *On Account of Sex,* 89–105.

12. *American Women: Report of the President's Commission on the Status of Women* (Washington: U.S. Government Printing Office, 1963), 45; Esther Peterson, "The Kennedy Commission," *Women in Washington,* 21–34. But see *On Account of Sex,* 136–37, for Harrison's argument that by isolating and sidestepping rather than rejecting the ERA, the PCSW broke the postwar stalemate over ERA and permitted a unified agenda that was no longer paralyzed by a dispute over constitutional equality.

13. George Meany to James Roosevelt, 29 December 1962; Andrew Biemiller to Bayard Rustin, 23 April 1970, Equal Rights file, box 9, Meany Archives.

14. Peterson's network consisted of newly formed state and city commissions on the status of women, the national Citizens' Advisory Council on the Status of Women (CAC), and the Interdepartmental Committee on the Status of Women (ICSW). The ICSW, which was composed of cabinet-level government officials, and the CAC, composed of leading private citizens, were created by Kennedy's Executive Order 11126 of 1 November 1963 to carry on the work of the expired PCSW.

15. Harrison, *On Account of Sex,* 12–13.

16. Biemiller to Tom Harris, 19 January 1960, Equal Rights file, box 17, Meany Archives.

17. Meany to James Roosevelt, 29 December 1962, Equal Rights file, box 17, Meany Archives.

18. Reedy to the Vice President, 3 March 1961, box 8, VPP/CRF, WHCF, LBJL; Powers to the Secretary [Wirtz], 7 September 1965, NN 370–108, RG 174, Records of the Secretary of Labor, NARA.

19. Humphrey to the President, 9 August 1965, box 4, White files, LBJL.

20. Moynihan to Moyers, 21 January 1965, box 43, EX HU 2–1, WHCF, LBJL. See Zelman, *Women: Kennedy-Johnson,* 85–87.

21. Wirtz to the President, 4 May 1965, box 31, NN 370–108, RG 174, Labor, NARA.

22. Moynihan to Moyers, 21 January 1965, box 45, GEN HU 2–1, WHCF, LBJL.

23. Lee Rainwater and William L. Yancey, *The Moynihan Report and the Politics of Controversy* (Cambridge: MIT Press, 1967).

24. Ibid., 184–87. Murray had urged the PCSW to support the sex amendment to Title VII, but she had opposed the ERA.

25. The Moynihan Report has similarly been criticized for a false historical grounding in the myth that slavery had destroyed the stability of the black family. Yet this was precisely the lesson taught by the historical scholarship that was available to Moynihan. His report spurred the revisionist history of scholars like Herbert G. Gutman, whose *The Black Family in Slavery and Freedom, 1750–1925* (New York: Pantheon, 1976) documented the remarkable cohesion of black families in the rural South through slavery and into the early 20th century. Gutman associated the breakdown of black family stability with northward migration and urbanization, which eroded the communal resilience and religious strength of the southern black peasantry.

26. *Wall Street Journal,* 22 June 1965.

27. *New York Times,* 3 July 1965.

28. John Herbers, "For Instance, Can She Pitch for the Mets?" *New York Times,* 20 August 1965.

29. *New York Times,* 21 August 1965.

30. Zelman, *Women: Kennedy-Johnson,* 73–108.

31. In their oral history interviews for the Johnson Library, both Esther Peterson and Mary Keyserling credit Vice President Johnson more than President Kennedy for pushing the equal pay bill and the status of women commission within the early administration. Keyserling recalls Johnson announcing at his first Cabinet meeting in January 1964 that "The day is over when top jobs are reserved for men." On March 4 of that year he explained to the Women's National Press Club, in a passage that was steeped in the tradition of social feminism, that "Women have a willingness of heart; moreover, they have an instinct for rightness that is as important to decision-making as numbers or logic." Transcript, Mary Dublin Keyserling oral history interview, 22 October 1968, Tape I, 7, LBJL.

32. Harrison, *On Account of Sex,* 185–91.

33. Frances Reissman Cousens, *Public Civil Rights Agencies and Fair Employment* (New York: Praeger, 1969), 13.

34. Ibid., 92–96. ICSW chairman Wirtz appointed a three-woman ICSW subcommittee to draft a policy paper. But the group promptly split, mirroring the larger world of activist women. One side was led by Women's Bureau director Mary Keyserling, who had replaced Peterson as Bureau head and represented the Bureau's commitment to protective legislation. The other was led by Mary Eastwood of Justice and Evelyn Harrison of the Civil Service Commission, who argued that employers should not be required to comply with protective laws that in practice prevented the advancement of women workers. See Harrison, *On Account of Sex,* 175–76, 185–91.

35. Aileen C. Hernandez, *E.E.O.C. and the Women's Movement 1965–1975* (Newark: Rutgers University Law School, 1975), 6–7.

36. *New York Times,* 27 July 1965.

37. In 1965, 43 states and the District of Columbia had laws setting maximum daily or weekly hours for women, and 24 of these plus the federal district set maximum hours of 8 hours a day, 48 hours a week, or both, for women in one or more occupations. Twenty-one states prohibited or regulated night work for women. The only states without such protectionist laws were Alabama, Alaska, Florida, Hawaii, Indiana, Iowa, Minnesota, and West Virginia—an oddly disparate list of dissenters from the long protectionist trend.

38. Wirtz to Roosevelt, 9 August 1965, box 31, NN 370–108, RG 174, Labor, NARA; Harrison, *On Account of Sex,* 195–86.

39. *New York Times,* 2 July 1965.

40. *The Civil Rights Act of 1964* (Washington: Bureau of National Affairs, 1964), 59–60. Vermont's law against sex discrimination was limited to equal pay.

41. *Wall Street Journal,* 29 July 1965. In generally reporting the agreement of employers in Wisconsin that the Wisconsin Industrial Commission was effectively and fairly enforcing the new ban on sex discrimination, the *Journal's* reporter, Joseph D. Mathewson, identified the official in charge as "Virginia Huebner, the attractive, red-haired director of the commission's fair employment practices division."

42. Wirtz to Roosevelt, 9 August 1965, box 31, RG 174, Labor, NARA.

43. *New York Times,* 20 August 1965.

44. *Wall Street Journal,* 17 September 1965.

45. EEOC Minutes, meeting #25, 21 September 1965, 2.

46. *New York Times,* 19 August 1965. Thirteen of the study committee's 17 members were men, and ten represented newspapers or advertising agencies.

47. EEOC Minutes, meeting #25, 21 September 1965, 1.

48. EEOC Minutes, meeting #26, 22 September 1965, 3.

49. *Wall Street Journal,* 13 October 1965.

50. Title 29 CFR Sec. 1604.1–3, 22 November 1965; *EEOC Newsletter* (December 1965), 2; *EEOC Administrative History,* 238–40.

51. *Wall Street Journal,* 23 November 1965.

52. Edelsberg, quoted in the *Washington Post,* 23 November 1965.

53. Title 29 CFR Sec. 1604.1, 23 November 1965; *EEOC Administrative History,* 243–45. In its first annual report, the EEOC observed that "At first glance Title VII appears to have the effect of superseding all state protective legislation. However, there is nothing in the legislative history of Title VII to indicate that Congress intended any such far-reaching result." EEOC, *First Annual Report* (Washington: U.S. Government Printing Office, 1966), 43.

54. Holcomb complained that the Dallas school superintendent was using the city's Negro school principals as "stooges" to reject most applications from Negro parents so they would "feel their children are not qualified." Holcomb to Walter Jenkins, 21 August 1961, box 1, VPP/CRF, LBJL.

55. Holcomb to Watson, 16 December 1965, Confidential file, box 17, HU 2–1, WHCF, LBJL.

56. Holcomb to Watson, 25 March 1966, box 380, EX FG 655, WHCF, LBJL.

57. EEOC Minutes, meeting #84, 24 March 1966, 3.

58. Author interview with Richard Graham, 12 June 1986, Washington, D.C.

59. EEOC Minutes, meeting #86, 30 March 1966, 2; text of Newport News conciliation agreement, quoted by Commissioner Graham in ibid., 4.

60. Ibid., 7.

61. EEOC Minutes, meeting #92, 19 April 1966, 6–8; meeting #93, 20 April 1966, 5.

62. Author telephone interview with Aileen Hernandez, 9 October 1987.

63. 31 *Federal Register* 6414, 28 April 1966; *EEOC Administrative History,* 241–42; *Wall Street Journal,* 29 April 1966.

64. Peterson, letter to the editor, *New York Times,* 3 September 1965.

65. Zelman, *Women: Kennedy-Johnson,* 101–2; author interview with Richard Graham, 12 September 1986, Washington, D.C.

66. Catherine East, "Newer Commissions," *Women in Washington,* 35–44; Harrison, *On Account of Sex,* 185–87.

67. Pauli Murray and Mary O. Eastwood, "Jane Crow and the Law: Sex Discrimination and Title VII," *George Washington Law Review* 34 (December 1965): 232–56.

68. Hernandez, *EEOC,* 10–13.

69. Griffiths to Holcomb, 19 May 1966, box 3, EX HU 2, WHCF, LBJL.

70. Holcomb to Griffiths, 1 June 1966, reprinted in 110 *Congressional Record,* Pt. 10, 89th Cong., 2d Sess. (20 June 1966), 13690.

71. Griffiths's biographer reports that the House speech was "carefully prepared by [Phineas] Indritz, who had moved from the staff of Griffiths's Government Operations Committee to become chief counsel of the House subcommittee on Conservation and National Resources. Emily George, *Martha Griffiths* (Washington: University Press of America, 1982), 152.

72. Ibid., 13689. In quoting Edelsberg, Griffiths cited an article reporting on NYU's Annual Conference on Labor in the *Race Relations Law Reporter,* 35 April 1966, 253–55.

73. 110 *Congressional Record,* Pt. 10, 20 June 1966, 13690.

74. Ibid., 13693.

75. For an apt survey of the issues posed in this literature, see James Q. Wilson, "The Rise of the Bureaucratic State," *The Public Interest* 41 (Fall 1975): 77–103.

76. 110 *Congressional Record,* Pt. 10, 20 June 1966, 13693, 13694.

77. Betty Friedan, *It Changed My Life: Writings on the Women's Movement* (New York: Random House, 1976), 75–86; author interview with Richard Graham, 12 June 1986, Washington, D.C.

78. Friedan, *It Changed My Life,* 80; author interview with Sonia Pressman Fuentes, 9 October 1986, Potomac, Maryland; memorandum, Pressman to Friedan, n.d. (summer 1967), courtesy of Sonia Pressman Fuentes.

79. Zelman, *Women: Kennedy-Johnson,* 104–7; Harrison, *On Account of Sex,* 192–96.

80. *Des Moines Tribune,* 14 November 1966. An early EEOC ruling on whether sex was a BFOQ for the position of flight attendant was delayed by a temporary restraining order on the commission issued by the federal district court for the District of Columbia. At issue was whether participation in the gender guidelines decisions in the fall of 1966 by Commissioner Hernandez was compromised by her prior announcement that following her resignation she would accept the vice presidential position with NOW. Shulman to the President, 30 December 1966, box 122, Confidential file—Agency Reports: EEOC, WHCF, LBJL.

81. Macy to the President, 12 October 1966, box 381, EX FG 655, WHCF, LBJL.

82. Author interview with Richard Graham, 12 June 1986, Washington, D.C. The Hispanic walkout on Graham, who was accompanied by staff director Herman Edelberg, is described in EEOC Minutes, meeting #85, 29 March 1966.

83. Califano to the President, 7 June 1967, box 381, EX FG 655, WHCF, LBJL.

84. Ibid. Johnson appointed Ximenes to chair a new Interagency Committee on Mexican American Affairs. Reporting to Ximenes on this committee, at least on paper, were secretaries Orville Freeman, John Gardner, Robert Weaver, Williard Wirtz, and OEO director R. Sargent Shriver. As these single-interest gestures proliferated, agency heads routinely assigned minority deputies to attend such "cabinet-level" committee meetings.

85. Graham shared with Hernandez, and to a somewhat lesser extent with Jackson, twin frustrations born of ideology and bureaucracy. On the first count, they urged a more aggressive expansion of commission power to attack race and sex discrimination than the chairman and vice-chairman and staff deemed prudent. On the second, they were caught in the coils of the structural and functional compromise that gave the EEOC birth. In an executive agency where the staff reported to the chairman, the commissioners seemed administratively marginal. This led them to press for greater staff support for research to bring commissioners' charges so internal as well as outside forces could shape the EEOC's agenda. The detailed minutes of the commission's tense meetings of November 18 and 23, 1965, capture the new agency's internal frustrations, and Richard Graham's intensity in pressing for commissioner staff and authority may have contributed to his not being reappointed.

86. Shulman to the President, 9 February 1967, box 122, Confidential file—Agency Reports: EEOC, WHCF, LBJL.

87. A joint appeal to defend the state protection laws was sent to Wirtz in the fall of 1966 by the ACLU, the Americans for Democratic Action, American Nurses Association, National Association of Social Workers, National Consumers League, the national councils of Catholic, Jewish, and Negro women, the National Farmers Union, United Church Women, and the YWCA.

88. Olya Margotin to J. Francis Pollhaus, 23 December 1966, box 68, EEOC 1966 file, District of Columbia Office, NAACP, LC.

89. Jo Freeman, *The Politics of Women's Liberation* (New York: McKay, 1975), 177–90.

90. EEOC, *First Annual Report* (Washington: U.S. Government Printing Office, 1966), 6. Not surprisingly, the majority of these complaints came from women, the only significant exception being that 35 of 175 complaints on hiring came from men seeking to enter such traditionally female jobs as nursing. In subsequent years the women's share of EEOC complaints would settle at a level of one-quarter.

91. Nathan, *Jobs and Civil Rights,* 51, Table 2–4.

92. *New York Times,* 10 August 1966.

93. Donald Allen Robinson, "Two Movements in Pursuit of Equal Employment Opportunity," *Signs* 4 (1978–79): 413–33.

94. EEOC, Statement Adopted by the Commissioners at Meeting #123, 19 August 1966; Hernandez, *EEOC,* 15–20.

95. See Freeman's discussion of NOW and the smaller women's groups in *Politics of Women's Liberation,* 44–146.

96. *Wall Street Journal,* 22 May 1967.

97. Transcript, Keyserling oral history, 31 October 1968, 19–33, LBJL.

98. Shulman to the President, 6 February 1967, box 122, Confidential file—Agency Reports: EEOC, WHCF, LBJL.

99. *EEOC Administrative History,* 119–21; Shulman to the President, 4 May 1967, box 123, Confidential files—Agency Reports:—EEOC, WHCF, LBJL; Shulman testimony concerning S. 1308, Subcommittee on Labor and Public Welfare, Senate Committee on Employment, Manpower and Poverty, 4 May 1967.

100. For a discussion of court rulings on the EEOC guidelines of 1965, 1969, and 1972, see Kenneth M. Davidson, Ruth Bader Ginsburg, and Herma Hill Kay, *Texts, Cases and Materials on Sex-based Discrimination* (St. Paul, Minn.: West, 1974), 1001–5.

101. EEOC Minutes, meeting #119, 4 August 1966. As acting chairman of a five-member commission with only three sitting commissioners, Holcomb was able to block any motion, as he did on August 4 when Hernandez and Jackson moved that the term "salesman" be ruled sexually discriminatory.

102. EEOC, *Third Annual Report* (Washington: Government Printing Office, 1969), 14–16; *Fourth Annual Report,* 15–17.

103. *Wall Street Journal,* 6 August 1968.

104. EEOC Minutes, meeting #126, 1 September 1966; Hernandez, *EEOC,* 20–28.

105. EEOC Minutes, meeting #180, 13 February 1968.

106. Title 29 CFR Sec. 1604.31 (February 1968).

107. Freeman, *Politics of Women's Liberation,* 236–37.

Chapter IX

1. The mantle of the Great Society was so comprehensive and its programs so complex that the literature is fragmented. For an analysis of its legacy after twenty years, see Marshall Kaplan and Peggy Cuciti, eds., *The Great Society and Its Legacy: Twenty Years of U.S. Social Policy* (Durham: Duke University Press, 1986).

2. Only three of the 23 chapters in Johnson's official memoir, *The Vantage Point* (New York: Holt, Rinehart and Winston, 1971), deal with the domestic programs of the Great Society. These are chapters 4 on the war on poverty, 9 on education and health, and 15 on urban policy and consumer and environmental protection. Two recent but brief assessments of Johnson's domestic achievements are Vaughn Davis Bornet, *The Presidency of Lyndon Johnson* (Lawrence: University of Kansas Press, 1983), ch. 10; and Paul K. Conkin, *Big Daddy of the Pedernales* (Boston: Twayne, 1986), ch. 9.

3. The best broad survey of federal initiatives against poverty is James T. Patterson, *America's Struggle Against Poverty, 1900–1980* (Cambridge: Harvard University Press, 1981). Allen J. Matusow is critical of Johnson's anti-poverty war in *The Unraveling of America* (New York: Harper & Row, 1984), 217–71. More sympathetic is Charles R. Morris's account, *A Time of Passion: America 1960–1980* (New York: Harper & Row, 1984), 89–98.

4. Doug McAdam, *Political Process and the Development of Black Insurgency 1930–1970* (Chicago: University of Chicago Press, 1982), 181–229.

5. Such a combined business-labor department had briefly existed from 1903 to

1913, but the mutually hostile constituencies had demanded divorce. Its reconstitution in 1967 was recommended by Johnson's outside task force on government organization, chaired by Dean Donald K. Price of the Harvard Graduate School of Public Administration, to combat the parochialism of constituency-captured agencies. The agency merger appealed to Johnson's sense of efficiency and consensus, but his proposal astounded many veteran observers, not the least of them being the Budget Bureau's top expert on government reorganization, Harold Seidman. See the detailed account in Emmette S. Redford and Marlan Blissett, *Organizing the Executive Branch: The Johnson Presidency* (Chicago: University of Chicago Press, 1981), 142–56.

6. Stephen N. Shulman to the President, 9 February 1967, box 122, Confidential file, agency reports: EEOC, WHCF, LBJL.

7. The EEOC minutes during 1965–66 reveal a deepening split between two factions. The dominant faction consisted of chairman Roosevelt, vice chairman Holcomb, and staff director Edelsberg. Their regulatory model was hierarchical and emphasized linear administration under a single executive, like a cabinet agency. The dissenting faction consisted of commissioners Graham, Hernandez, and Jackson. Their quasi-judicial model emphasized their collective role in assessing probable cause in complaint processing, in filing commissioners' charges, and in recommending suits for the Justice Department. This led to a running clash between the commissioners' staff assistants and the main staff under Edelsberg. The split faded somewhat after 1966, when Graham and Hernandez had left and the agency de-emphasized its "retail" role in complaint processing.

8. *EEOC Administrative History,* LBJL, 1969, Ch. IV; James P. Gammon, "Uphill Bias Fight: After Faltering Start, Agency Readies Attack on Job Discrimination," *Wall Street Journal,* 12 April 1967.

9. Richard P. Nathan, *Jobs and Civil Rights* (Washington: Brookings, 1969), 13–32, 66–69; Alfred W. Blumrosen, *Black Employment and the Law* (New Brunswick: Rutgers University Press, 1971), 9–20.

10. Quoted in *Wall Street Journal,* 12 April 1967.

11. Shulman to the President, 9 February 1967, box 122, Confidential file, agency reports: EEOC, WHCF, LBJL.

12. Nathan, *Jobs and Civil Rights,* 74–85.

13. *Wall Street Journal,* 12 April 1967.

14. Department of Justice, *Annual Report of the Attorney General of the United States* (Washington: U.S. Government Printing Office, 1966), 184–85. The other areas of jurisdiction included desegregating public accommodations and facilities under titles II and III of the Civil Rights Act of 1964, and coordinating enforcement of Title VI on fund cut-off.

15. Shulman to the President, 4 May 1967, box 66, EX LE/HU 2, WHCF, LBJL.

16. Macy to the President, 11 May 1967, box 774, Macy files, LBJL.

17. Ibid.

18. Macy to the President, box 774, Macy files, LBJL.

19. The black nominees included Vivian W. Henderson, president of Clark College in Atlanta; Theodore A. Jones, a Chicago businessman and protégé of Mayor Daley; Franklin H. Williams, ambassador to Ghana; Labor assistant secretary George L.P. Weaver; and Edward Sylvester, director of the OFCC.

20. White nominees on Macy's lists included former Civil Rights Commission staff director Berl I. Bernhard; Inland Steel vice president William G. Caples; former assistant attorney general Norbert Schlei, who had just lost a race for lieutenant governor in California; San Francisco lawyer William K. Coblentz; and Army general counsel Albert B. Fitt.

21. Macy to the President, 19 May 1967, box 774, Macy files, LBJL.

22. On Morrow, see Robert Frederic Burk, *The Eisenhower Administration and Black Civil Rights* (Knoxville: University of Tennessee Press, 1981), 77–88; on Brown,

see Leon E. Panetta and Peter Gall, *Bring Us Together: The Nixon Team and the Civil Rights Retreat* (Philadelphia: Lippincott, 1971), 202–3.

23. Transcript, Clifford L. Alexander, Jr. oral history interview, 1 November 1971, 1–15, LBJL.

24. Macy to the President, 10 June 1967, box 774, Macy files, LBJL.

25. James C. Falcon to Macy, 21 June 1967; Macy to the President, 24 June 1967, box 774, Macy files, LBJL.

26. Blumrosen, *Black Employment,* 51–60.

27. *EEOC Administrative History,* 104–28; *Wall Street Journal,* 11 June 1968. For a sympathetic description of the early EEOC that relies heavily on Nathan's *Jobs and Civil Rights,* see Duane Lockhard, *Toward Equal Opportunity* (New York: Macmillan, 1968).

28. Nathan, *Jobs and Civil Rights,* 21–22.

29. *Wall Street Journal,* 20 February 1960.

30. *New York Times,* 5 May 1967; *Wall Street Journal,* 20 July 1967.

31. *Wall Street Journal,* 13 January 1966.

32. EEOC Minutes, meeting #131, 1 November 1966.

33. Donald D. Osborne, "Negro Employment in the Textile Industries of North and South Carolina," Report submitted to the EEOC, 21 November 1966; Nathan, *Jobs and Civil Rights,* 28–29.

34. EEOC Minutes, meeting #133, 16 November 1966; meeting 134, 23 November 1966.

35. *EEOC Administrative History,* 129–45. The ill will reflected in the textile hearings led to several 707 suits that the EEOC recommended to Justice.

36. EEOC Research Report 1967–20, "Employment Patterns in the Drug Industry, 1966," 29 September 1967. The drug hearing was co-sponsored by the Food and Drug Administration, as an added inducement to private industry to participate. EEOC survey data involving 398 drug companies with 133,735 employees showed that only 5.2 percent were black—2.1 percent of them working in clerical, .6 percent in sales, and 1.8 percent in other white-collar jobs. *Wall Street Journal,* 9 October 1967.

37. Blumrosen, *Black Employment,* 102–37; *New York Times,* 7 August 1967.

38. *EEOC Administrative History,* 146–54.

39. EEOC Minutes, meeting #150, 5 April 1967; meeting #178, 20 December 1967.

40. Nathan, *Jobs and Civil Rights,* 30–31.

41. Blumrosen, *Black Employment,* 70–71.

42. Blumrosen, "The Newport News Agreement," *Black Employment,* 328–77. Following the Newport News agreement, a violent strike broke out at the shipyard—the first strike in the company's 81-year history. Congressional conservatives led by Senator Paul Fanin of Arizona attacked the EEOC for unsettling labor relations in the nation's defense industry by requiring racial quotas in apprenticeships and promotions to make up for past bias. See "Official Bias: A Note on the Equal Employment Opportunity Commission," *Barron's National Business and Financial Weekly,* 17 July 1967, 1 *passim.*

43. *EEOC Administrative History,* 119–25.

44. Pressman to Duncan, 31 May 1966, provided to the author by Sonia Pressman Fuentes.

45. Author interview with Sonia Pressman Fuentes, 10 October 1966, Potomac, Maryland.

46. *Holland v. Edwards,* N.Y. Ct. App., 307 N.Y. 38 (1954), 119 N.E. 2d 581, 34 LRRM 2018.

47. *Consolidated Steel Corporation,* 11 LA 891 (1948); *The Civil Rights Act of 1964* (Washington: Bureau of National Affairs, 1964), 74.

48. *Rath Packing Company,* 24 LA 444 (1955); *The Civil Rights Act of 1964,* 74.

49. *Strauder v. West Virginia,* 100 U.S. 303, 25 L. Ed. 644 (1880); *Ex parte Virginia,* 100 U.S. 339 (1880).

50. *Norris v. Alabama,* 294 U.S. 587, 55 S. Ct. 579 (1935); *Hernandez v. Texas,* 347 U.S. 476, 74 S. Ct. 667 (1954).

51. *Phelps Dodge Corp.,* 32 NLRB 338 (1941); *F.W. Woolworth Co.,* 25 NLRB 1362 (1952).

52. Note, "Enforcement of Fair Employment Under the Civil Rights Act of 1964," *University of Chicago Law Review* 32 (1965): 430.

53. Pressman to Duncan, 31 May 1966, 4.

54. *Swain v. Alabama,* 380 U.S. 202 (1965). The rape conviction of Swain, an escaped convict, followed testimony describing a particularly gruesome and prolonged assault, occurring in daylight in a rural setting when the victim was caring for a four-year-old child.

55. *Swain v. Alabama,* 208; Derrick A. Bell, Jr., *Race, Racism and American Law* (2d ed.; Boston: Little, Brown, 1980), 250–53.

56. Pressman to Duncan, 31 May 1966, 6; EEOC Minutes, meeting #109, 23 June 1966.

57. *United States v. New York, New Haven, and Hartford Rail Road,* 355 U.S. 253 (1957).

58. Pressman to Duncan, 31 May 1966, 8.

59. Ibid., note 19, viii.

60. Blumrosen, *Black Employment,* 172–80; Harrell R. Rodgers, Jr., and Charles S. Bulloch III, *Law and Social Change: Civil Rights and Their Consequences* (New York: McGraw-Hill, 1972), 113–37.

61. *EEOC Administrative History,* 248.

62. 110 *Congressional Record* 5 June 1964, 8194.

63. Samuel C. Jackson, "EEOC vs. Discrimination, Inc.," *The Crisis* (January 1968): 16–17.

64. Ibid., 17.

65. *The Civil Rights Act of 1964,* 119.

66. Jackson, "EEOC vs. Discrimination," 17.

67. *EEOC Administrative History,* 249.

68. *New York Times,* 20 July 1968.

69. *EEOC Administrative History,* 17; EEOC Minutes, meeting #101, 25 May 1966; meeting #118, 11 August 1966; meeting #124, 24 August 1966. The minutes kept by Marie D. Wilson, secretary to the commission, from July 1965 through October 1966 were detailed and often verbatim. Beginning with the 127th meeting on 2 October 1966, Chairman Shulman prepared only summary minutes, a practice continued by a series of staff directors from March 1967 through the fall of 1969, when Marie Wilson's more comprehensive minutes resumed.

70. The Legal Defense Fund's suit against Monsanto's chemical plant in El Dorado, Arkansas, illustrates the problems typical of segregated racial lines of seniority. The plant employed 469 hourly workers in three departments. All 80 blacks worked in the labor department at unskilled tasks, with a top hourly wage of $2.94. All whites worked in either the operating or maintenance departments, running or maintaining the plant's machinery. The lowest white hourly wage was $2.96, and the highest $3.69. Monsanto's contract with the Oil, Chemical, and Atomic Workers International Union permitted blacks to transfer to a white department, but at the price of losing all seniority and risking layoff. This was no idle threat, as employment at El Dorado had been declining for more than a decade owing to automation. Black workers, represented by the LDF, sued both Monsanto and the OCAW union. *Wall Street Journal,* 5 January 1967.

71. On seniority in EEO law, see William B. Gold, *Black Workers in White Unions: Job Discrimination in the United States* (Ithaca: Cornell University Press, 1977), 67–98. Gould was a consultant to the EEOC in 1966 when the commission formulated

its legal strategy for attacking race discrimination through seniority. His advice to the EEOC is captured in Gould, "Employment Security, Seniority, and Race: The Role of Title VII of the Civil Rights Act of 1964," *Howard Law Journal* 13 (1967): 1–50.

72. *New York Times,* 20 July 1968.

73. *Whitfield v. United Steelworkers,* 263 F. 2d 546, 551 (5th Cir.) *cert. denied,* 360 U.S. 902 (1959).

74. EEOC *Administrative History,* 249–50; Gould, *Black Workers,* 71–74.

75. *Wall Street Journal,* 5 February 1967.

76. Blumrosen, "Seniority and Equal Employment Opportunity," *Black Employment,* 159–217.

Chapter X

1. *Congress and the Nation, 1965–1968* (Washington: Congressional Quarterly Press, 1969), 373–74; *New York Times,* 29 April 1966.

2. Franklin D. Roosevelt, Jr., to Whitney Young, 20 July 1965, box 42, Headquarters Administration file, Urban League papers.

3. Deputy Attorney General Ramsey Clark to Charles L. Schultze, 21 March 1966; Wilfred Rommel to McPherson, 22 April 1966; Rommel to Califano, 22 April 1966, box 66, EX LE/HU 2, WHCF, LBJL. Rommel was assistant director of the Budget Bureau's division of Legislative Reference.

4. Rommel to McPherson, 22 April 1966, box 66, EX LE/HU 2, WHCF, LBJL.

5. Katzenbach to Califano, 9 October 1965, box 65, EX LE HU, WHCF, LBJL.

6. Katzenbach to Califano, 13 December 1965, box 21, Files of Harry McPherson, LBJL.

7. Katzenbach to Califano, 9 October 1965, box 66, EX LE/HU 2, WHCF, LBJL.

8. Lyndon B. Johnson, "Annual Message to the Congress on the State of the Union," 12 January 1966, *Public Papers of the Presidents: Lyndon B. Johnson, 1966,* I, 3; Johnson, "Special Message to the Congress Proposing Further Legislation to Strengthen Civil Rights," 28 April 1966, ibid., I, 462.

9. John Herbers, "House Approves Enforcement Powers on Job Rights," *New York Times,* 28 April 1966.

10. Katzenbach to Califano, 9 October 1965, box 66, EX LE/HU 2, WHCF, LBJL.

11. See Gary Orfield, *The Reconstruction of Southern Education: The Schools and the 1964 Civil Rights Act* (New York: Wiley, 1969), ch. 4.

12. In his May 4 testimony before Subcommittee No. 5 of House Judiciary, Katzenbach had reminded Congress that in the Civil Rights Act of 1866 (42 U.S.C. 1982) Congress had declared: "All citizens of the United States shall have the same right, in every State and Territory, as is enjoyed by white citizens thereof to inherit, purchase, lease, sell, hold, and convey real and personal property." But since World War II, Congress on five occasions had rejected proposals for a statutory ban on discrimination in federally assisted public housing—once when considering the Housing Act of 1949, and on four subsequent occasions in the 1950s.

13. See George W. Grier, "The Negro Ghettos and Federal Housing Policy," *Law and Contemporary Problems* 32 (Summer 1967): 550–60. The FHA's 1936 *Underwriting Manual* instructed appraisers that residential properties insured by the FHA should be protected from "adverse influences" that included "the infiltration of business or industrial uses, lower class occupancy, and inharmonious racial groups."

14. David J. Garrow, *Bearing the Cross: Martin Luther King, Jr. and the Southern Christian Leadership Conference* (New York: Random House, 1968), 491–526.

15. Monroe W. Karmin, "Segregation in Housing Seen Persisting Despite New Efforts to Curb It," *Wall Street Journal,* 13 June 1966. Karmin reported that for the entire country during the first three and a half years under the Kennedy housing order, the FHA had moved only 30 black families into houses constructed by builders it had

found guilty after investigating disscrimination complaints. Nineteen additional minority families had won their complaints but had decided not to move into all-white developments. Only 13 builders (five of whom were later reinstated) and no lenders were declared ineligible, on grounds of racial discrimination, to participate in FHA mortgage insurance programs. Slightly more effective had been the FHA's mortgage foreclosure program. Because this did not pit the agency directly against its home-building constituency in a painful role conflict, it allowed the agency to open to racial minorities more than 450 previously all-white neighborhoods since 1963.

16. A rare exception was the NAACP's chief Washington lobbyist, Clarence Mitchell, who argued that it was time for the President to seek statutory support from Congress.

17. Author interview with Lee White, 16 June 1986; David L. Lawrence to the President, 10 November 1965, box 47, EX LE HU 2-2, WHCF, LBJL.

18. Califano to the President, 20 November 1965, box 65, EX HU 2-2, WHCF, LBJL.

19. Katzenbach to Califano, 13 December 1965, box 66, EX LE/HU 2, WHCF, LBJL.

20. *Congress and the Nation, 1965–1968,* 365–74; *Wall Street Journal,* 28 April, 10 August 1966.

21. Supporting passage were 183 Democrats and 76 Republicans, and opposing were 95 Democrats and 62 Republicans. There were 169 northern Democrats voting aye and 17 against, while 78 southern Democrats opposed passage and 14 favored it.

22. *New York Times,* 29 April 1966.

23. *Wall Street Journal,* 20 September 1966.

24. The Senate's vote of September 19 was on a cloture motion, and it fell considerably short of the required two-thirds majority (of 93 senators voting) at 52 to 41. Thus the Senate never cast a final vote on H.R. 14765, and the House-passed omnibus bill died upon adjournment.

25. On the Califano task force operation, see Hugh Davis Graham, *The Uncertain Triumph* (Chapel Hill: University of North Carolina Press, 1984), ch. 6; Nancy Kegan Smith, "Presidential Task Force Operations During the Johnson Administration," *Presidential Studies Quarterly* XV (Spring 1985): 320–29.

26. Bureau of the Budget, *Administrative History,* Part III, "The Bureau's Contribution to Program Development," I: 101–2, LBJL.

27. Califano to Katzenbach, 5 September 1966, box 12, Task Forces file, WHCF, LBJL.

28. These included the White House (McPherson and Clifford Alexander), the Budget Bureau (executive assistant director William Carey), HEW (assistant secretary Lisle Carter and his assistants, Peter Libassi and Derrick Bell), and Labor (OFCC director Edward Sylvester), with 16 of the 34 members coming from Justice itself.

29. Clark to Califano, 30 November 1966, box 12, Task Force file, WHCF, LBJL.

30. Most were either carrot or stick proposals, calling for example either for increased federal coercion (e.g., through the federal licensing power, labor injunctions, NLRB rulings), or for new civil rights strings on established programs (e.g., education grants and loans, rural and vocational and state employment programs). Pollack's task force and Attorney General Clark rejected most of them as either premature, politically infeasible, or offering potential benefits that were too marginal for the price of policing established client relationships.

31. Report of the 1966 Interagency Task Force on Civil Rights, 30 November 1966, 5–8, box 12, Task Force file, WHCF, LBJL.

32. Ibid., 9–10.

33. Ibid., 2.

34. Office of Legal Counsel on Employment Proposal No. 2, Background Notebook, nonpaginated, box 12, Task Forces file, WHCF, LBJL.

35. Office of Legal Counsel need assessment on Employment Proposal No. 3, ibid.

36. *Public Papers: Johnson, 1967,* I, 188–89. Johnson added minor proposals to renew the Civil Rights Commission for five years (it was scheduled to expire in 1968), and to increase substantially the funding for the Community Relations Service.

37. J. Francis Poulhaus, "Analysis of S. 1026," box 304, Group III, Series B, General Office file—Civil Rights, NAACP papers.

38. Clark to Califano, n.d., box 21, Task Forces file, WHCF, LBJL; Eugene Tryck to the [Budget Bureau] Director, 14 November 1967, box 5, Series 60.5, Task Force on Civil Rights, BOB/NARA.

39. David L. Rose to Clark, 16 November 1967, box 21, Task Forces file, WHCF, LBJL; Pfleger to Frey, 28 November 1967, box 5, Series 60.5, Task Force on Civil Rights, BOB/NARA. Attending the White House meeting on November 24 were Califano, Secretary Wirtz, EEOC Chairman Clifford Alexander, Housing under secretary Robert Wood, Peter Libassi from HEW, Treasury assistant secretary Robert Wallace, and from Justice, Ramsey Clark, John Doar, David Rose, Roger Wilkins, and Stephen Pollack.

40. Clark to Califano, n.d., 28 November 1967, box 66, EX LE/HU 2, WHCF, LBJL.

41. Even Clark agreed that the FDIC could not legally impose new nondiscrimination requirements on 14,000 banks merely because their deposits were federally insured. The participants did agree, however, to explore further the federal government's potential leverage over the 12,000 banks holding federal deposits, and also over savings and loan associations regulated and insured by the Home Loan Bank Board.

42. Pfleger to Frey, 28 November 1967, box 66, EX/LE HU 2, WHCF, LBJL.

43. *Congress and the Nation,* Vol. II, 1965–1968, 378; *Wall Street Journal,* 25 January 1968.

44. *Public Papers of the Presidents: Johnson, 1968,* I, 24 January 1968, 55–62; *New York Times,* 25 January 1968. The quotation is from Johnson's proposal for open housing of 15 February 1967. In the State of the Union message on January 17, 1968, Johnson had made the same request quite briefly, but had concentrated on the Vietnam war and the problem of rising domestic violence.

45. *Congressional Quarterly Almanac,* 1968, 152–62.

46. *Wall Street Journal,* 19 February 1968.

47. Michael Foley, *The New Senate: Liberal Influence on a Conservative Institution 1959–1972* (New Haven, Conn.: Yale University Press, 1980), 196–200; Arlen J. Large, "Federal Power and 'Flexible' Senators," *Wall Street Journal,* 13 March 1968. Large argued that the election of 1966 had made the Javits-Scott-Case-Morton wing a Republican majority in the Senate, and that Dirksen was leading only a hard core of aging conservatives. Dirksen's son-in-law, Howard Baker, was widely credited with the major role in persuading Dirksen to embrace an open housing compromise as a unity symbol for the 1968 presidential elections, as well as a guarantor of Dirksen's continued leadership.

48. *Congressional Quarterly Almanac,* 1968, 159; 114 *Congressional Record,* 26 February 1968, Pt. 4, 4574. In the Dirksen version, like the Mondale bill (and therefore like the Johnson bill of 1967), and also like the EEOC's original coverage under Title VI, coverage would be phased in over a three-year period.

49. The Senate amendments included, oddly but unanimously (81–0), Sam Ervin's "Indian Bill of Rights." The negative votes on the bill itself, like those against cloture, were overwhelmingly southern. Only three of 29 Republican senators voted against the bill (including John Tower of Texas, paired against it), and only two of 20 southern Democrats voted for it (Ross Bass of Tennessee and Ralph Yarborough of Texas).

50. The key vote on April 10 was a 229–195 procedural vote to prevent sending the bill to conference. A majority of Republicans and most southern Democrats voted unsuccessfully for a conference.

51. "Effective Lobbying Put Bill Across," *Congress and the Nation*, Vol. II, 1965–68, 386–88.

52. *Wall Street Journal*, 1 March 1968, 12 April 1968.

53. The best contemporary exposition of this theme is Large, "Federal Power and 'Flexible' Senators," *Wall Street Journal*, 13 March 1968.

54. *Report of the National Advisory Commission on Civil Disorders* (Washington: U.S. Government Printing Office, 1968); see also Hugh Davis Graham, "On Riots and Riot Commissions: Civil Disorders in the 1960s," *The Public Historian* 2 (Summer 1980), 7–27.

55. Dirksen, quoted in *Congressional Quarterly Almanac*, 1968, 159.

56. James W. Button, *Black Violence: Political Impact of the 1960s Riots* (Princeton: Princeton University Press, 1978), 78–79.

57. Harry McPherson, *A Political Education* (Boston: Little, Brown, 1972), 369–70.

58. Scott McKinney and Ann B. Schnare's *Trends in Residential Segregation By Race: 1960–1980* (Washington: Urban Institute, 1986) concludes that the continuing trend in the 1960s of intensifying patterns of racial segregation in housing was reversed in the 1970s.

59. 114 *Congressional Record*, 28 February 1968, Pt. 4, 4574. In the 1968 housing title, the Secretary of HUD, like the EEOC in Title VII cases, was required to defer to state and local agencies with appropriate jurisdiction, and to protect the confidentiality of the hearing process.

60. See generally 114 *Congressional Record*, 26 February–4 March 1968, Pt. 4, 4570–4980.

61. Ervin, quoted in the *Congressional Quarterly Almanac*, 1968, 157.

62. U.S. Commission on Civil Rights, *The Federal Fair Housing Enforcement Effort* (Washington: U.S. Government Printing Office, 1979), 70–73.

63. Hill, quoted in Karmin, "Segregation in Housing," *Wall Street Journal*, 13 June 1968.

64. *Wall Street Journal*, 12 April 1968.

65. Edward L. Holmgren, quoted in Allen J. Matusow, *The Unraveling of America* (New York: Harper & Row, 1984), 208.

66. Douglas S. Massey and Nancy A. Denton, "Trends in the Residential Segregation of Blacks, Hispanics, and Asians: 1970–1980," *American Sociological Review* 52 (December 1987): 802–25. See also John M. Goering, ed., *Housing Desegregation and Federal Policy* (Chapel Hill: University of North Carolina Press, 1986), 256–63.

67. For a critical assessment of "Affirmative Action in Housing" from the perspective of the middle 1970s, see Nathan Glazer, *Affirmative Discrimination: Ethnic Inequality and Public Policy* (New York: Basic Books, 1975), 130–67.

68. *Report of the National Advisory Commission on Civil Disorders*, 256–63.

Chapter XI

1. *New York Times*, May 31, June 1, 9, and 11, 1963.

2. Report, AFL-CIO Human Relations Committee to Executive Board, Philadelphia AFL-CIO Council, 27 February 1963, box 9, Meany Archives.

3. *Philadelphia Sunday Bulletin*, 3 March 1963. The report noted that racial segregation was so deeply entrenched in the city's AFL craft unions that two all-Negro locals, the motion picture projectionists and the musicians, refused to merge with their larger all-white counterparts even though their seniority had been guaranteed. Nationally, the generally smaller black locals in such racially segregated pairings often feared

that their white counterparts would outvote the black members, who thus would have less chance of holding office. By late 1967, the national AFL-CIO's desegregation drive had reduced the number of all-black locals from approximately 500 in 1960 to fewer than 170 in 1967. The Railway Clerks had merged 135 of their 205 black locals with white units, and the Musicians had merged 18 of 34. *Wall Street Journal*, 19 September 1967.

4. The gulf between national rhetoric and local reality for the trade union movement was demonstrated when William F. Schnitzler, chairman of the national AFL-CIO's civil rights committee, vowed in the spring of 1963 that "We are going to bust that thing." But in Philadelphia, the business manager of Steamfitters Local 420 announced the withdrawal of his 2200-member local from Philadelphia's Building and Construction Trades Council in objection to a negotiated agreement that would add six blacks to a disputed school construction project—one Negro plumber, one steamfitter, two electrical apprentices, and two sheet metal workers. *New York Times*, 1 June 1963.

5. White to the President, 3 June 1963, box 20, Lee White files, JFKL.

6. Hirst Sutton to the Director, 14 June 1966, box 6, series 60.5, BOB/NARA; Cabinet press release, 25 August 1966.

7. William B. Gould, *Black Workers in White Unions: Job Discrimination in the United States* (Ithaca: Cornell University Press, 1977), 172–88.

8. "The Federal Role in Apprenticeship," attached to George P. Shultz to Arthur Burns, 3 July 1969, NN 372-119, Secretary files, Department of Labor/NARA.

9. Ray Marshall et al., *Employment Discrimination* (New York: Praeger, 1978), 26–61. The skilled journeymen in the building trades enjoyed no golden security, however. Despite generally high hourly wages, employment was intermittent and yearly income was low. Median annual earnings for construction in all crafts in 1965 were $5867 compared with $7002 in basic steel, and many "full time" construction workers averaged only 1100 to 1400 hours a year. When the economy was strong, many electricians, carpenters, and other craftsmen left construction for manufacturing, utilities, and other more dependable sources of employment.

10. *Wall Street Journal*, 17 March 1970, 25 June 1970.

11. Michael Sovern, *Legal Restraints on Racial Discrimination in Employment* (New York: Twentieth Century Fund, 1966), 177–204; Marshall, *Employment Discrimination*, 26–61.

12. David H. Rosenbloom, *Federal Equal Employment Opportunity: Politics and Public Personnel Administration* (New York: Praeger, 1977).

13. Hirst Sutton to the Director, 14 June 1966, box 6, series 60.5, BOB/NARA; Sutton to James Gaither, 16 January 1967, attached to Gaither to McPherson and Califano, 19 January 1967, box 3, Task Forces file, WHCF, LBJL; William D. Carey to McPherson, 25 July 1967, box 208, series 61.1a, RG #51, BOB/NARA. A uniform complaint by the Budget Bureau was lack of enforcement staff in the agencies, and this in turn reflected both congressional hostility to aggressive Title VI enforcement, and the administration's budget squeeze from Vietnam inflation. In FY 1967 a Labor Department staff of 16 compliance officers found some discriminatory practices in 75 percent of the programs they investigated, but were able to investigate only 6 percent of the department's 10,000 programs.

14. Department of Labor, *Administrative History*, Vol. I, 165–77, LBJL; Richard Nathan, *Jobs and Civil Rights* (Washington: Brookings, 1969), 101–7.

15. Paul Nitze to Wirtz, 19 March 1968, box 54, NN 371-116, RG #174, Labor/NARA. Nitze referred to the 1966 correspondence in his 1968 reiteration of the Pentagon's earlier objections.

16. Ibid. Wirtz had his deputy, Under Secretary of Labor James J. Reynolds, reply to Nitze. Reynolds argued that the OFCC regulations provided for a national security exemption through which the head of the agency could waive the EEO requirements. Exercising the waiver provision, however, would imply that the Secretary of Defense

was recognizing the contractual authority of the OFCC, and this the Pentagon was unwilling to yield.

17. *New York Times*, 30 March 1966; David L. Rose to McPherson, 29 May 1967, box 5, Series 60.5, BOB/NARA.

18. Nathan, *Jobs and Civil Rights*, 105–6.

19. *Wall Street Journal*, 7 February 1966. The PCEEO had been criticized by civil rights groups for never cancelling a contract during its four-and-a-half years of enforcement effort, and for imposing penalties on only 22 companies. Most of these sanctions involved delay of contract award until compliant behavior could be certified. Such sanctions proved to be quite effective enforcement tools, but they permitted critics to charge the opposite, and to equate lack of contract cancellation with lack of enforcement.

20. As of 1970, OFCC sanctions had led to no cancelled contracts, and the OFCC had instituted debarment proceedings against only seven companies; only three of these led to a formal hearing, and none of them resulted in debarment. *Federal Civil Rights Enforcement Effort* (Washington: U.S. Commission on Civil Rights, 1970), 68–69.

21. *Wall Street Journal*, 9 November 1966; Sylvester to the Secretary [Wirtz], 8 May 1967, box 62, NN 371-20, RG #174, Labor/NARA. Other targets included the National Lead company and Crane Plumbing in New York City, Jacksonville shipyards in Florida, Newport News Shipbuilding and Dry Dock, Allen-Bradley of Milwaukee, B&P Motor Express of Pittsburgh, and Pullman, Inc. of Bessemer, Alabama.

22. Sylvester to the Secretary, 24 March 1967, box 61, NN 371-20, Labor/NARA; Alfred W. Blumrosen, *Black Employment and the Law* (New Brunswick: Rutgers University Press, 1971), 337–407.

23. *New York Times*, 25 and 26 May 1968.

24. Nathan, *Jobs and Civil Rights*, 108–9; Department of Labor, *Administrative History*, Vol. II, 345–58, LBJL.

25. Sylvester to Heads of All Agencies, 3 May 1966, box 60, NN 371-20, RG #174, Labor/NARA.

26. Nathan, *Jobs and Civil Rights*, 108.

27. James E. Jones, Jr., "The Bugaboo of Employment Quotas," *Wisconsin Law Review* (1970): 344–67. Jones was the Labor Department's associate solicitor for civil rights from May 1967 to September 1969, where his division provided the legal justification for the Philadelphia Plan.

28. *New York Times*, 5 February 1966; Marshall, *Employment Discrimination*, 28–34.

29. Jones, "Bugaboo," 346; Lance Jay Robbins, "Federal Remedies for Employment Discrimination in the Construction Industry," *California Law Review* 60 (June 1972): 1196–1224.

30. *Hearings Before the U.S. Commission on Civil Rights*, held in San Francisco, Calif., May 1–3, 1967, and Oakland, Calif., May 4–6, 1967.

31. *Wall Street Journal*, 14 December 1966.

32. Quoted in Jones, "Bugaboo," 346.

33. Sylvester to Wirtz, 20 March 1967, box 61, NN 371-20, Labor/NARA.

34. *Wall Street Journal*, 10 August 1967; Jones, "Bugaboo," 346–47.

35. Wirtz to Heads of All Agencies, 4 May 1967, box 62, NN 371-20, Labor/NARA.

36. Nathan, *Jobs and Civil Rights*, 107.

37. Ibid., 108.

38. Sylvester to Wirtz, 14 June 1967, box 62, NN 371-20, Labor/NARA.

39. Represented on the Philadelphia FEB were the departments of Defense, Justice, Labor, HUD, the GSA and Post Office, plus the Community Relations Service, the EEOC, and the OFCC.

40. Sylvester to Heads of All Agencies, 22 September 1967; Macy to Chairmen

of Federal Executive Boards, 30 October 1967, Reel 28, Executive Records, Department of Labor, LBJL.

41. Warren P. Phelan to Members, Federal Executive Board for Philadelphia, 27 October 1967, Reel 28, Executive Records, Labor/NARA.

42. The Bureau of Apprenticeship and Training was a relatively small operation by federal agency standards. Eighty-five percent of the bureau's 500 staff members were located in the field, where they worked intimately with building contractors and local union officials in screening applicants for "indenture" into apprenticeships.

43. The first test case was *Ethridge v. Rhodes,* 268 F. Supp. 83 (S.D. Ohio 1967), and was brought by NAACP lawyers to block a state contract to build a medical science building at the campus of Ohio State University in Columbus. The low bidder had a hiring-hall contract with an all-white union, and had obtained a waiver from the Governor's 1966 executive order requiring an anti-discrimination assurance. In granting the injunction on constitutional grounds, the federal district court held that the 14th Amendment barred the state from participating in a racially discriminatory practice, even though the "active" discrimination was not by the state or the contractor, but rather by a tertiary party, the union. The district court's decision was not appealed. See Jones, "Bugaboo," 350–54. The Cleveland Plan itself was challenged and sustained in *Weiner v. Cuyahoga Community College,* 44 Ohio 2d 468, 238 N.E. 2d 839 (C.P. 1968). The *Weiner* case is discussed in Chapter 14.

44. Wirtz, quoted in Nathan, *Jobs and Civil Rights,* 110–11; *Wall Street Journal,* 16 October 1967.

45. Letters, Macy to Representative Odin Langen (R-Minn.) and Senators Quentin Burdick (D-ND), George McGovern (D-SD), and Milton Young (R-ND), 22 January 1968, Reel 28, Executive Records, Labor/NARA.

46. Vincent G. Macaluso to William E. Dunn, 22 November 1967; Dunn to Macaluso, 27 November 1967, Philadelphia Plan file, General Accounting Office. The GAO files were made available through the courtesy of Ronald Wartow, managing attorney at GAO, and are hereafter cited as PP/GAO.

47. Dunn to Wirtz, 18 December 1967, PP/GAO.

48. Wirtz to Dunn, 28 December 1967; Dunn to Presidents, Association of General Contractors, 3 January 1968, PP/GAO.

49. This reality was briefly glimpsed by the American public during the litigation over the Gramm-Rudman-Hollings budget reduction act during 1985–86, when the GAO was disqualified by the Supreme Court as a neutral, third-party arbiter of budget disputes, owing to its primary allegiance to Congress. *New York Times,* 12 January 1987.

50. Owen Birnbaum, "Government Contracts: The Role of the Comptroller General," *American Bar Association Journal* 42 (1956): 433–41; A.I. Baer, "The General Accounting Office: The Federal Government's Auditor," *American Bar Association Journal* 47 (1961): 359–72.

51. Comment, "*S&E Contractors* and the GAO Role in Government Contract Disputes: A Funny Thing Happened on the Way to Finality," *Virginia Law Review* 55 (1969), especially 765–76 on the Wunderlich Act of 1954 and its aftermath.

52. Clark's opinion of 16 January 1969 held that "Congress did not intend to set the GAO up as an additional layer of administrative appeal for contractors on disputes clause questions." Opinion of the Attorney General, *Government Contracts Report* 8 (January 16, 1969), #82,490 at 88,002.

53. Charles W. Gasque, Jr. to J. Edward Welch, 29 November 1967, PP/GAO.

54. W.A. Lewis to Marshall H. Lynn, 7 December 1967, PP/GAO.

55. Melvin E. Miller to J. Edward Welch, 11 December 1967, PP/GAO.

56. *Wall Street Journal,* 29 March 1968. The construction unions had accepted Labor Department funding in 1962 for a special Outreach program to recruit minority apprentices, the numbers of which they claimed had doubled by 1966. But Herbert Hill countered that this represented an increase from only 2.5 percent in 1960 to 4.4

percent in 1966, and that fewer than 4 percent of these apprentices were in the highly skilled crafts.

57. *Wall Street Journal,* 17 June 1968.

58. Cramer to Staats, 8 April 1968, PP/GAO. Cramer's specific query concerned federal highway aid under the jurisdiction of the Public Works committee, and this had been involved in the urban renewal contracts in Cleveland.

59. Wilbur R. Allen to Office of Legislative Liaison, 17 May 1968, PP/GAO.

60. Draft, [Assistant Comptroller] Frank H. Weitzel to Staats, 15 May 1968, PP/GAO; Comptroller General Opinion B-163026, Staats to Cramer, copy to Wirtz, PP/GAO.

61. Cramer to Staats, 8 October 1968, PP/GAO. Cramer attached copies of complaints he had received in June from the Philadelphia General Building Contractors Association and the Association of General Contractors of America.

62. Staats to Cramer, 18 November 1968, Philadelphia Plan Document #3, Department of Labor Library (hereafter cited as PPD/DOLL). The DOL collection at the GAO contains 25 documents, beginning with the Comptroller's ruling against the Cleveland Plan on 22 May 1968, and ending with President Nixon's statement of 22 December 1969.

63. Commission on Civil Rights, *Federal Civil Rights Enforcement Effort* (1970), 170–71; Peter G. Nash, "Affirmative Action Under Executive Order 11246," *New York Law Review* 46 (1971): 232–34; Robert P. Schuwerk, "The Philadelphia Plan: A Study of the Dynamics of Executive Power," *University of Chicago Law Review* 39 (1972): 739–40.

64. Califano to Clark, 7 October 1968, box 26; Rose to Califano, 24 October 1968, box 3; Gaither to Frey, 2 January 1969, box 3, Task Forces file, LBJL.

65. *New York Times,* 6 December 1968.

Chapter XII

1. John D. Ehrlichman, *Witness to Power: The Nixon Years* (New York: Simon & Schuster, 1982); Raymond Price, *With Nixon* (New York: Viking, 1977); and William Safire, *Before the Fall: An Inside View of the Pre-Watergate White House* (New York: Doubleday, 1975).

2. The tangle of legal, legislative, and political circumstances that blocked public access to the Nixon presidential papers until late 1986 is described by James J. Hastings, acting director of the Nixon Presidential Materials Project of the National Archives, in "Status of the Nixon Presidential Materials," paper presented at the Nixon conference at Hofstra University, 21 November 1987.

3. See Joan Hoff-Wilson's introduction to *Nixon Without Watergate: A Presidency Reconsidered,* forthcoming from Basic Books. Hoff-Wilson examines Nixon's policies in five areas of domestic reform: welfare, the environment, civil rights, economics, and government reorganization. Her discussion of civil rights concentrates on school desegregation, women's rights, and Indian policy.

4. A succinct overview of the Nixon presidency with bibliography is Joan Hoff-Wilson, "Richard M. Nixon, 1969–1974," *The American Presidents,* III (Pasadena: Salem Press, 1986), 751–85. A critical journalistic account is Jonathan Schell, *The Time of Illusion* (New York: Vintage, 1976).

5. Otis L. Graham, Jr., *Toward a Planned Society: From Roosevelt to Nixon* (New York: Oxford University Press, 1976), 188–263. See also David E. Wilson, *The National Planning Idea in U.S. Public Policy: Five Alternative Approaches* (Boulder, Colo.: Westview Press, 1980), 64–93; and Peri E. Arnold, *Making the Managerial Presidency: Comprehensive Reorganization Planning, 1905–1980* (Princeton: Princeton University Press, 1986), 272–302. For a critical, insider view of Nixon's "New Federalism" that concentrates on his attempt to control the bureaucracy during his

second term, see Richard P. Nathan, *The Plot That Failed: Nixon and the Administrative Presidency* (New York: Wiley, 1975).

6. Daniel Patrick Moynihan, *The Politics of a Guaranteed Income: The Nixon Administration and the Family Assistance Plan* (New York: Random House, 1973). On Nixon's radical proposal for welfare reform, see also Vincent J. Burke and Vee Burke, *Nixon's Good Deed: Welfare Reform* (New York: Columbia University Press, 1974); and M. Kenneth Bowler, *The Nixon Guaranteed Income Proposal* (Cambridge, Mass.: Ballinger, 1974).

7. John C. Whitaker, *Striking A Balance: Environment and Natural Resources Policy in the Nixon-Ford Years* (Washington: American Enterprise Institute, 1976).

8. Herbert Stein, *Presidential Economics: The Making of Economic Policy from Roosevelt to Reagan and Beyond* (New York: Simon & Schuster, 1984), 133–207.

9. Rowland Evans, Jr. and Robert D. Novak, *Nixon in the White House: The Frustration of Power* (New York: Random House, 1971), 133–76, 11.

10. Richard M. Nixon, *Six Crises* (New York: Doubleday, 1962), 391.

11. Quoted from Nixon's Lincoln Day speech in Cincinnati, in Jules Witcover, *The Resurrection of Richard Nixon* (New York: Putnam, 1970), 71–72.

12. A. James Reichley, *Conservatives in an Age of Change: The Nixon and Ford Administrations* (Washington: Brookings, 1981), 175–76.

13. Stephen E. Ambrose, *Nixon: The Education of a Politician, 1913–1962* (New York: Simon & Schuster, 1987), 269.

14. Witcover, *Resurrection,* 343–44.

15. Author interview with John D. Ehrlichman, 5 August 1986, Santa Fe, New Mexico; Joe McGuiniss, *The Selling of the President, 1968* (New York: Trident, 1969), 199–201; Lewis Chester, Godfrey Hodgson, and Bruce Page, *An American Melodrama: The Presidential Campaign of 1968* (New York: Viking, 1969), 462–63.

16. Reichley, *Conservatives,* 176. At the Miami convention in 1968 Nixon received 228 of the 310 votes cast by southern Republican delegates, with 60 of the remainder going to Reagan.

17. Richard M. Scammon and Ben J. Wattenberg, *The Real Majority* (New York: Coward-McCann, 1970), 45–58.

18. Quoted in Chester et al., *American Melodrama,* 624–25.

19. The strongest conservative ideologues among Nixon's senior staff were his "southern" counselor Harry Dent and speechwriter Patrick Buchanan on the White House staff; Richard Kleindienst and William Rehnquist at Justice, and Robert Mardian at HEW. These were balanced against the more moderate to liberal orientations of Daniel P. Moynihan and Leonard Garment in the White House, and in the Cabinet, Robert Finch at HEW, George Romney at HUD, John Volpe at Transportation, and George Shultz at Labor. Neither Mitchell, Haldeman, nor Ehrlichman seems to have had a highly developed ideological orientation when they joined the administration. See Carl M. Brauer, *Presidential Transitions: Eisenhower through Reagan* (New York: Oxford University Press, 1986), 121–69; Reichley, *Conservatives,* 59–78. A contemporary journalistic account is Dan Rather and Gary Paul Gates, *The Palace Guard* (New York: Harper & Row, 1974).

20. Nixon's claims to centrist prudence are illustrated in his description of the moderate path he followed on civil rights policy between the extremist demands of civil rights activists and die-hard conservatives in *RN: The Memoirs of Richard Nixon* (New York: Grossett and Dunlap, 1978), 435–44.

21. The Burns papers in the Gerald R. Ford Library in Ann Arbor, Michigan, contain files on 21 task forces. But the Burns report to Nixon of January 18, 1969, mentions only twelve (although the report contained 18 topical chapters), and not all of the twelve appear to be listed in the Burns papers in the Ford Library. In August 1969 Ehrlichman sent the Domestic Plans Group a list of 16 new task forces that would report that fall to Burns. They included such single-issue topics as the mentally handicapped, prison rehabilitation, the physically handicapped, the aged, and minor-

ity business enterprise. Many were not completed when Burns left to head the Federal Reserve System that October, and their planning initiatives lapsed. Ehrlichman to Domestic Plans Group, 20 August 1969, box 18 A6, Burns Papers, Gerald R. Ford Library (hereafter cited as GRFL).

22. Burns to the President, 18 January 1969, hereafter cited as the Burns Report. It was supplied to the author through the courtesy of A. James Reichley. Prior to Burns's death in 1988, most of the Burns correspondence remained closed in both the Nixon and Ford papers. Assisting Burns in shaping the report's final policy recommendations were McCracken, Alan Greenspan, and Martin Anderson.

23. Reichley, *Conservatives*, 75–77.

24. Burns Report, 81.

25. Ehrlichman interview, 5 August 1986.

26. Quoted from the "Rights" issues file in Tom Cole to Martin Anderson, 2 June 1969, box 18 A6, Burns Papers, GRFL.

27. Congressional Quarterly, *Nixon: The First Year of His Presidency* (Washington: Congressional Quarterly Press, 1970), 68.

28. Ehrlichman, *Witness to Power*, 207–62; Evans and Novak, *Nixon in the White House*, 37–74; Safire, *Before the Fall*, 112–17, 463–78; Reichley, *Conservatives*, 59–78.

29. *Public Papers of the Presidents: Richard M. Nixon, 1969* (Washington: U.S. Government Printing Office, 1971), 1–4.

30. Ambrose, *Nixon*, 252–53.

31. Burns Report, 70.

32. Brauer, *Presidential Transitions*, 137–41, 156–58; Paul Charles Light, *The President's Agenda: Domestic Policy Choice from Kennedy to Carter* (Baltimore: Johns Hopkins University Press, 1982), 44–45.

33. Nixon to Ehrlichman, 13 March 1969, box 31, Ehrlichman files, White House Special Files (WHSF), Richard M. Nixon Presidential Materials Project, National Archives and Records Administration (hereafter cited as RMNP).

34. Harlow to the President, 15 March 1969, box 31, Ehrlichman files, WHSF, RMNP.

35. Haldeman to Ehrlichman, 19 March 1969, box 31, Ehrlichman files, WHSF, RMNP.

36. Alexander P. Butterfield to Haldeman, 7 April 1969, box 31, Ehrlichman files, WHSF, RMNP.

37. Burns to Haldeman and Ehrlichman, 18 March, 1969; Ehrlichman to Burns, 20 March 1969; Burns to Ehrlichman, 27 March 1969, box 31, Ehrlichman files, WHSF, RMNP; Brauer, *Presidential Transitions*, 137–41.

38. *Public Papers: Nixon, 1969*, 284–88; *New York Times*, 15 April 1969; Brauer, *Presidential Transitions*, 156–58.

39. The Ash task force led in the spring of 1970 to the creation of a policy-centered Domestic Council, consciously modeled after the National Security Council, with administrative affairs centered in the new Office of Management and Budget, formerly the Bureau of the Budget. On Reorganization Plan #2, see Otis Graham, *Toward A Planned Society*, 204–13; Arnold, *Managerial Presidency*, 282–86.

40. *Public Papers: Nixon, 1969*, 284; *New York Times*, 15 April 1969. The following day Nixon proposed anti-inflationary budget revisions that reduced HUD's enforcement program for open housing by $4 million. His national address of August 8 on domestic programs concentrated on legislative proposals for tax and welfare reform, revenue-sharing, and manpower training, and contained no mention of civil rights priorities.

41. Arnold, *Managerial Presidency*, 274–77; Stephen Hess, *Organizing the Presidency* (Washington: Brookings, 1976), 112–15; John Osborne, *The Nixon Watch* (New York: Liveright, 1970), 28.

42. Evans and Novak, *Nixon in the White House*, 92.

43. Burns to Ehrlichman and Haldeman, 5 June 1969; Moynihan to the President, 10 June 1969; Moynihan to Ehrlichman, 16 June 1969; Ehrlichman to Burns, 23 June 1969, box 31, Ehrlichman file, WHSF, RMNP.

44. Ehrlichman, *Witness to Power*, 247; Brauer, *Transitions*, 140–41.

45. In a luncheon conversation reported to Ehrlichman by staff assistant John C. Whitaker, Agnew explained that Nixon had asked him to be the key domestic man when the ticket was formed in Miami. His role in the campaign had tarnished his image, Agnew said, but by the summer of 1969 his image had changed and he was ready to assume the role of "domestic chief honcho." Whitaker to Ehrlichman, 25 June 1969, box 31, Ehrlichman files, WHSF, RMNP.

46. Reichley, *Conservatives*, 138; Brauer, *Transitions*, 133–35.

47. White House press release, 4 November 1969, box 1, EX HU-2, WHCF, RMNP.

48. Safire, *Before the Fall*, 468.

49. Ehrlichman, *Witness to Power*, 248.

50. Moynihan to the President, 19 March 1969, box 1, EX HU-2 Equality, WHCF, RMNP.

51. Daniel Patrick Moynihan, "Report to the President," 19 March 1969, over-size attachment No. 564, EX HU-2 Equality, RNMP.

52. Ibid., 17.

53. Moynihan to Jensen, 20 March 1969, box 2, EX HU-2, RMNP.

54. Moynihan Report to the President, 19 March 1969, 19.

55. Moynihan to the President, 10 March 1969, box 1, EX HU-2, RMNP.

56. Moynihan Report to the President, 19 March 1969, 17–18.

57. Ibid., 24, 25.

58. Ibid., 26–27.

59. Ibid., 28.

60. Daniel P. Moynihan, *The Politics of a Guaranteed Income: The Nixon Administration and the Family Assistance Plan* (New York: Random House, 1973), 61–113. Moynihan argued that by not repudiating the social goals of his Democratic predecessors, Nixon accepted them by default. Thus Nixon's planning task forces implicitly accepted the continuity of a liberal agenda, and this continued momentum precluded any fundamental change in course. To Moynihan, this was "the central fact of his Administration."

61. Daniel Patrick Moynihan, *Maximum Feasible Misunderstanding* (New York: Free Press, 1969).

62. Moynihan's occasional missives to Nixon were not all reflective gems. In July 1969, Moynihan wrote Nixon that because the current "crisis of confidence, the erosion of authority—had to be raised to the highest levels of policy concern," then perhaps "a constitutional convention was the one way to do it." The republic had not had a constitutional convention since the Founding, and to thus casually propose one was astounding. "On reflection," Moynihan admitted, "I concluded that you would probably think me out of my mind." Moynihan to the President, 17 July 1969, box 31, Ehrlichman files, WHSF, RMNP.

63. Chester E. Finn, Jr., *Education and the Presidency* (Lexington, Mass.: D.C. Heath, 1977), 25–39, 65–75.

64. Moynihan Report to the President, 19 March 1969.

65. Reichley, *Conservatives*, 138.

66. Ibid., 139.

67. *Public Papers: Nixon, 1969*, 197–98.

68. John H. Bunzel, *The American Small Businessman* (New York: Knopf, 1962), 48–51; William A. Brock, *The Economics of Small Business: Their Role and Regulation in the United States* (New York: Holmes and Meier, 1986).

69. *Congress and the Nation, 1965–1968* (Washington: Congressional Quarterly Press, 1969), 293–97.

70. *Public Papers 1969*, 197–98; *Wall Street Journal*, 6 March 1969.

71. *Wall Street Journal*, 10 July 1969.

72. Roeser had initially proposed that nine major civil rights organizations be enlisted to help run OMBE programs, with each assuming jurisdiction over a special line of business opportunity—e.g., CORE would be assigned gas station franchises, the NAACP would get housing and contracting, the Rev. Leon Sullivan's Opportunities Industrial Corporation in Philadelphia would get shopping centers, and so forth. Although most assignments would go to black groups, Roeser suggested that the Puerto Rican forum might be assigned construction work training. But the civil rights groups resisted being assigned tasks by the government, and Sandoval resented the entire OMBE/black capitalism intrusion on his SBA operations. *Wall Street Journal*, 10 July 1969.

73. Garment to Ken Cole, 20 April 1970, box 2, EX HU-2, RMNP. Ink had performed a similar rescue operation for the Johnson White House in 1965, when the breakthrough in federal aid to education required restructuring the Office of Education in HEW. But in 1969–70 Garment held off, saying he was "especially aware of the sensitivities involved."

74. *Wall Street Journal*, 10 July 1969.

75. Stans to Mayo, 6 August 1969, File TI-6/2, Series 69.1, BOB/NARA.

76. Economics, Science, and Technical Division (Balmat) to the Director, 1 August 1969, File TI-6/2, Series 69.1, BOB/NARA.

77. Nixon to Heads of Departments and Agencies, 5 December 1969, box 2, EX HU-2, WHCF, RMNP.

78. Garment to Ken Cole, 20 April 1970, box 2, EX HU-2, RMNP. Garment also observed that with only two more months to go in the fiscal year, the Administration's goal of $42 million in Section 8(a) procurement set-asides had been missed by $34 million.

79. *Wall Street Journal*, 8 January 1970.

80. Senate Report 91–1343, 10 November 1970.

81. *National Journal* 52 (26 December 1970): 2820–25. The *National Journal* exposed disarray and bickering in the OMBE program and discord at the top between Sam Wyly, the white Dallas electronics executive who headed the advisory council, and Secretary Stans, to whom Wyly refused to report. In response to the *National Journal* article, White House aide Kenneth Cole wrote Ehrlichman: "The lesson is that we have done a superb job of undercutting a Cabinet officer. We have allowed ourselves to be pushed around by Wyley [*sic*] at a time when we probably could have accomplished just as much by forcing Wiley to work through the system established by the Executive Order—a reporting relationship to Secretary Stans." Cole concluded that "we have been had by both Stans and Wiley. We've been had by Stans because we allowed him to put out his report on OMBE's accomplishments during the last year and Wiley because eventually his report has reached the public domain complete with charges and counter-charges by the two organizations." Cole to Ehrlichman, 31 December 1970, box 3, EX HU-2, RMNP.

82. *National Journal* 52 (20 December 1970): 2824.

83. F. Chris Garcia, ed., *Latinos and the Political System* (Notre Dame: University of Notre Dame Press, 1988), 7–118; L. H. Gann and Peter J. Duignan, *The Hispanics in the United States* (Boulder, Colo.: Westview Press, 1986), 69–128; Pastora San Juan Cafferty and William C. McCready, eds., *Hispanics in the United States* (New Brunswick, N.J.: Transaction, 1985); Alejandro Portes and Robert L. Bach, *Latin Journey: Cuban and Mexican Immigrants in the United States* (Berkeley: University of California Press, 1985).

84. *Congressional Quarterly Almanac: 1969* (Washington: Congressional Quarterly, 1970), 846. Congress passed Public Law 91-181 on December 10 and President Nixon signed it on December 30.

85. *Mexican Americans and the Administration of Justice in the Southwest*

(Washington: U.S. Commission on Civil Rights, 29 April 1970); Howard A. Glickstein to Moynihan, 27 April 1970, box 2, EX HU 2, WHCF, RMNP; *New York Times,* 31 May, 3 June 1970.

86. Ehrlichman to the President, 23 April 1970, box 2, Ex Hu 2, WHCF, RMNP.

87. Flanigan to Harry Flemming, 15 July 1970, box 2, EX HU 2, WHCF, RMNP.

88. Chapin to Garment, 31 December 1970, box 3, EX HU 2, WHCF, RMNP.

89. Ehrlichman to the President, 24 November 1970, box 2, EX HU 2, CHCF, RMNP. After early 1969 Nixon generally rejected proposals to meet with black groups, such as the congressional Black Caucus and black civil rights leaders, journalists, or mayors. "The President feels he has met enough with members of this minority group," Flanigan explained to Hugh Sloan in September 1970. When Garment, who was Jewish, noted that he was preparing a separate memo on recruiting Jewish groups that he would send the President through Haldeman, Nixon scrawled on the margin: "No—Don't bother."

90. In 1971 Colson sent Ehrlichman a 52-page confidential memorandum on American Hispanics that was thoughtfully prepared and argued for substantive program initiatives as well as political gestures. Colson's report observed that estimates of the Spanish-speaking population ranged from 9 to 20 million. Only in 1970 did the Census attempt to identify the Hispanic population, the report claimed, by adding a 5 percent sample survey (while blacks got a 100 percent sample) and questionnaires printed in English. Colson to Ehrlichman, 20 December 1971, box 4, EX HU 2, WHCF, RMNP.

91. *Congressional Quarterly Almanac: 1971* (Washington: Congressional Quarterly, 1972), 654–55.

92. Russell Dean to Clark MacGregor, 16 June 1971, box 4, EX HU 2, WHCF, RMNP.

93. Badillo and Roybal to the President, 9 August 1971; George Grassmuck to Dave Parker, 5 October 1971, box 4, EX HU 2, WHCF, RMNP.

94. *Wall Street Journal,* 7 February 1969.

95. *New York Times,* 15 March 1969.

96. *New York Times,* 26 March 1969, 10 April 1969. On March 29, Nixon released a mollifying memorandum to all department heads, reaffirming "my own official and personal endorsement of a strong policy of equal opportunity within the Federal Government." On May 7 he replaced Alexander as EEOC chairman with William H. Brown III, a black Republican lawyer from Philadelphia whose appointment to the commission by Johnson was awaiting Senate confirmation.

97. *Wall Street Journal,* 2 May 1969. Nathan had chaired a Nixon transition task force on welfare, and had subsequently joined the Nixon administration as assistant director in the Budget Bureau for human resources programs. His report repeated the recommendation of Ramsey Clark's interagency task force of 1968 that the ineffectual and relatively invisible OFCC be transferred to the EEOC.

98. *Wall Street Journal,* 7 July 1969.

99. Ehrlichman interview, 5 August 1986. As a partial counterbalance, the Nixon administration did ultimately oppose an amendment that would forbid the termination of federal funds to any school district which had a freedom-of-choice plan. The amendment was attached to the Labor-HEW appropriations bill in the summer of 1969 by Rep. Jamie L. Whitten, a Mississippi Democrat. The Administration initially took no stand on it despite pleas from House Republican leaders, and it was accepted by the full House in July. But in the fall Secretary Finch announced the Administration's opposition to the amendment, and it was defeated in the Senate on December 17.

100. Reichley, *Conservatives,* 183; Gary Orfield, *Must We Bus? Segregated Schools and National Policy* (Washington: Brookings, 1978), 320.

101. For accounts of the effective role of George Shultz, then director of the new Office of Management and Budget, in establishing and coordinating the network of

biracial citizen committees in seven southern states beginning in early 1970, see Nixon, *Memoirs,* 439–43; Ehrlichman, *Witness,* 230–35; Reichley, *Conservatives,* 187–89.

102. See Wilkinson, *From Brown to Bakke,* 193–249. The key decision was *Milliken v. Bradley,* 418 U.S. 717 (1974), in which a 5–4 Supreme Court majority overturned a district court ruling that joined 53 of Detroit's 85 outlying suburban school districts in a metropolitan busing plan to integrate Detroit's schools. Chief Justice Warren Burger wrote the majority opinion, joined by justices Harry Blackmun, Lewis Powell, and William Rehnquist— all Nixon appointees—and Eisenhower appointee Potter Stewart.

103. Reichley, *Conservatives,* 189–90.

Chapter XIII

1. John D. Ehrlichman, *Witness to Power* (New York: Simon & Schuster, 1982), 222–23.

2. *Wall Street Journal,* 31 December 1968.

3. George P. Shultz, "Priorities in Policy and Research for Industrial Relations," *Proceedings of the Twenty-First Annual Winter Meeting,* ed. Gerald G. Somers (Madison, Wisc.: Industrial Relations Research Association, 1969), 1–13.

4. George P. Shultz and Arnold R. Weber, *Strategies for the Displaced Worker* (New York: Harper & Row, 1966). In this study of job retraining for meatpacking workers displaced by automation, Shultz and Weber (who accompanied Shultz to the Labor Department and then to the OMB) denied a need for "bold new ideas," and argued instead for fine-tuning federal programs, such as the retraining efforts under the Manpower Development and Training Act of 1962, to deal with individual workers and their particular problems.

5. George P. Shultz and John R. Coleman, *Labor Problems* (New York: McGraw-Hill, 1959); Reichley, *Conservatives in an Age of Change* (Washington: Brookings, 1981), 74–75.

6. Author interview with John Ehrlichman, 5 August 1986, Santa Fe, New Mexico. The Davis-Bacon Act of 1931 required builders using government funds to pay their workers no less than the prevailing wage level in the work area. Under prodding from organized labor, postwar Congresses had expanded Davis-Bacon coverage to include the federally-assisted construction of airports and hospitals (1946), public schools (1950) defense installations (1951), interstate highways (1956), and area redevelopment and water polution projects (1961).

7. Author interview with William J. Kilberg, Washington, D.C., 17 September 1986. Upon graduation from Harvard Law School, Kilberg became a White House Fellow, where he began working with Shultz in June 1969. In August 1970 he joined Shultz's senior staff in the office of the Solicitor General, and he served as Solicitor from April 1973 through January 1977.

8. Nixon's characterization of Shultz is summarized in Notes of Meetings with the President, 1969–1973, Files of John D. Ehrlichman, 29 December 1969, box 3, WHSF, RMNP.

9. Ehrlichman interview, 5 August 1986.

10. Mayo to Dr. Arthur Burns, 13 February 1969, File RG-1, Series 69.1, BOB/NARA. At the same time Herbert Hill, who was denouncing the OFCC for caving in to the Defense Department and not enforcing EEO requirements in textile contracts, called for the OFCC to be transferred to the Justice Department. Hill to William H. Oliver, 26 March 1969, box 433, Series B, group III, NAACP Papers.

11. Shultz reviewed a draft of this chapter while serving as Secretary of State in the Reagan administration in March 1988. He reported to the author that he could

not recall with sufficient precision the timing and circumstances of early 1969 that led to the Labor Department's commitment to the revised Philadelphia Plan.

12. The textile firms were Burlington Industries, Dan River Mills, and J.P. Stevens Company.

13. *Wall Street Journal,* 28 March 1969.

14. Ehrlichman interview, 5 August 1986.

15. Ehrlichman, *Witness to Power,* 228–29.

16. Richard M. Nixon, *RN: Memoirs of Richard Nixon* (New York: Grossett & Dunlap, 1978), 437.

17. Ehrlichman interview, 5 August 1986.

18. *Department of Labor Administrative History*: U.S. Department of Labor During the Nixon/Ford Administrations, Vol. II, 38–53, DOLL.

19. Title 41 C.F.R., 60-1.40.

20. "Remarks by Assistant Secretary of Labor Arthur Fletcher at Signing of Philadelphia Plan, Philadelphia, Pennsylvania, June 27, 1969; Department of Labor *News,* 27 June 1969, PPD/DOLL.

21. Fletcher to Heads of All Agencies, 27 June 1969, PPD/DOLL. The revised plan applied to all contracts in excess of $500,000, but provided for exemption on certain national security grounds.

22. DOL *News,* 16 August, 23 September 1969; Fletcher and Wilks to Heads of All Agencies, 23 September 1969, PPD/DOLL. Originally, the roofers and water proofers were included among the seven trades targeted in the June 27 revised plan. But the August hearings led to their being dropped from the September 23 list. Similarly, the operating engineers had been included in the FEB's original Philadelphia Plan of 1967. But progress in minority hiring by Local 543 had led to their being removed by the OFCC from the revised plan of June 1969. Thus the six Philadelphia trades covered by the OFCC's September 23 implementation order were the iron workers, steamfitters, sheetmetal workers, electricians, elevator construction workers, and plumbers and pipefitters.

23. Contractors could demonstrate good faith effort by notifying community organizations of their worker needs, maintaining an active file of minority workers, and availing themselves of relevant training programs. But contractors whose efforts failed to satisfy the OFCC faced the immediate sanction of disqualification from the bidding, and the ultimate sanction of being struck from the procurement list of potential contractors.

24. Appendix to the June 27 plan, 3. The appendix was a sample affirmative action form for contractors to submit with their bids.

25. See Nathan Glazer, *Affirmative Discrimination: Ethnic Inequality and Public Policy* (New York: Basic Books, 1975), 46–51; Herbert Hammerman, " 'Affirmative Action Stalemate': A Second Perspective," *The Public Interest* 93 (Fall 1988): 130–34.

26. Thomas D. Boswell and James R. Curtis, *The Cuban-American Experience* (Totowa, N.J.: Rowman & Allanheld, 1983), 168–79; Maria de los Angeles Torres, "From Exiles to Minorities: The Politics of Cuban-Americans," *Latinos and the Political System,* ed. F. Chris Garcia (Notre Dame: University of Notre Dame Press, 1988), 81–98.

27. Alesandro Portes and Robert L. Bach, *Latin Journey: Cuban and Mexican Immigrants in the United States* (Berkeley: University of California Press, 1985), 200–39; Boswell and Curtis, *Cuban-American Experience,* 112–13.

28. Patrick Lee Gallagher, *The Cuban Exile: A Socio-Political Analysis* (New York: Arno Press, 1980), 121–45, 167–74.

29. The Office of Management and Budget during the Nixon administration incorporated Indians, Pakistanis, and Pacific Islanders into the Asian category; Eskimos and Aleuts were included as American Indians.

30. Harry P. Taylor [Executive Director, General Building Construction Association] to Arthur F. Hintze [Director of Government Relations, Association of General Contractors of America], 31 January 1969, PP/GAO, complaining about the defiance of EEO specialist Donald Burstein of HEW.

31. The *Barron's* articles, by Shirley Scheibla, were "Gentlemen's Agreement?— Government is Making Business Its Unwilling Partner in Bias," on 23 December 1969, and "Fairness by Fiat— Some Employees These Days Are More Equal Than Others," on 20 January 1970.

32. Staats to Finch, 25 February 1969; Allen to Miller, 18 March 1969; Staats to Moynihan, 11 April 1969, PP/GAO.

33. Finch to Staats, 7 May 1969, PP/GAO.

34. Allen to Miller, 19 May 1969, PP/GAO.

35. Miller to Office of Legislative Liaison, 2 May 1969, PP/GAO. The participating congressmen included, in addition to Barrett, Reps. James A. Byrne, Joshua Eilberg, and William J. Green, all Democrats.

36. A month later the Human Relations Commission delegate, Clarence Farmer, wrote Staats to warn that withdrawal of the plan promised to "cause tensions which may erupt into demonstrations or civil disorders." Farmer to Staats, 11 June 1969, PP/GAO.

37. Wilbur R. Allen to the file (B-163026), 3 July 1969, PP/GAO; Laurence H. Silberman to Jerris Leonard, 24 June 1969, PPD/DOLL.

38. Laurence H. Silberman, "Authority Under Executive Order 11246," 16 July 1969, PPD/DOLL; DOL *News*, 16 July 1969.

39. Staats to Secretary Shultz, 5 August 1969, 6, PP/GAO.

40. Silberman brief, 12–16.

41. Ibid.; Allen to the file, 3 July 1969, PP/GAO.

42. Laurence H. Silberman, "The Road to Racial Quotas," *Wall Street Journal*, 11 August 1977. Silberman explained that in defending the legality of the original Philadelphia Plan in 1969, the Labor Department had been careful to find that the Philadelphia construction unions had engaged in discrimination, and thus the plan was justified as a remedy to correct past discrimination. This argument had been decisive, Silberman said, in persuading the Justice Department's chief counsel, assistant attorney general William Rehnquist, and later several federal judges of its constitutionality.

43. Staats to Secretary Shultz, 5 August 1969, 2, PP/GAO; 49 Comp. Gen. 59 (1969).

44. Ibid., 9.

45. Ibid., 14, 15.

46. Allen to Miller, 19 May 1969; Allen to file (B-163026), 3 July 1969, PP/GAO. The GAO attorneys were Wilbur R. Allen, Melvin E. Miller, and General Counsel J. Edward Welch.

47. H.R. Rep. No. 914, 88th Cong., 1st Sess., pt. 2 (Judiciary Committee, 1963).

48. See Robert D. Schuwerk, "The Philadelphia Plan: A Study in the Dynamics of Executive Power," *University of Chicago Law Review* 39 (1972): 732–39. For an argument that supports preferential racial quotas as remedies for past discrimination, and that interprets Sections 703(a) and 703(j) as prohibiting racial discrimination *only* against minorities, see Paul Marcus, "The Philadelphia Plan and Strict Racial Quotas on Federal Contracts," *UCLA Law Review* 17 (1970): 817–36. See also William J. Kilberg, "Current Civil Rights Problems in the Collective Bargaining Process: The Bethlehem and AT&T Experiences," *Vanderbilt Law Review* 27 (1974): 81–113.

49. *Weiner v. Cuyahoga Community College District*, 238 N.E. 2d 839 (Ohio Com. Pl. 1968), 19 Ohio St. 2d (Ohio Sup. Ct., 2 July 1969). The low bidder was Reliance Mechanical Contractors, which submitted an affirmative action plan that was "subject to the availability and referral to RMC, Inc. of qualified journeymen and apprentices from Pipefitters Local No. 120," with which RMC had an exclusive hiring-

hall contract. The community college rejected the bid in favor of the second lowest bidder, whose plan pledged that "we will have Negro representatives in all crafts on this project."

50. "Statement by Secretary Shultz," DOL *News,* 6 August 1969.

51. Executive Order 11478, 8 August 1969.

52. Mitchell to Shultz, 22 September 1969, PPD/DOLL; *New York Times,* 24 September 1969.

53. *Gaston County v. United States,* 395 U.S. 285 (1969).

54. Mitchell to Shultz, 22 September 1969, 14–15, PPD/DOLL.

55. On September 22, 1969, President C.J. Haggerty told the 55th convention of the AFL-CIO's Building and Construction Trades that "We are 100 percent opposed to a quota system, whether it be called the Philadelphia Plan or whatever." "Statement of Policy on Equal Employment Opportunity," 22 September 1969, Atlantic City, New Jersey, Meany Archives. The 8th annual convention of the AFL-CIO that October adopted a civil rights resolution asserting that "Nixon officials are covering this retreat in civil rights enforcement by trying to make a whipping boy of unions, especially those in the building trades . . . .The Administration's so-called 'Philadelphia Plan' sets up unsound procedures used in no other industry, segment of the labor market or in government itself."

56. Schuwerk, "The Philadelphia Plan," 741, n100.

57. Jones, "Bugaboo," 388, n249.

58. *New York Times,* 7 August 1969.

59. *Wall Street Journal,* 26 September 1969.

60. Hugh Davis Graham, "On Riots and Riot Commissions: Civil Disorders in the 1960s," *The Public Historian* 2 (Summer 1980): 7–27.

61. James W. Button, *Black Violence: Political Impact of the 1960s Riots* (Princeton: Princeton University Press, 1978).

62. DOL *News,* 23 September 1969, PPD/DOLL. The OFCC had announced its intention to implement the plan in nine other U.S. cities, including Boston. But in response to the direct challenge of the Comptroller General, Shultz confined his defense of the plan on September 23 to the specifics of Philadelphia, and sought to minimize the threat of Washington imposing a cookie-cutter mold throughout the country.

63. Drinan to Shultz, 25 September 1969, box 68, NN 372-119, DOL/NARA.

64. Drinan to Fletcher, 14 October 1969, box 68, NN 372-119, DOL/NARA.

65. Drinan to Shultz, 14 October 1969, box 68, NN 372-119, DOL/NARA. Drinan had enlisted the support of Speaker McCormack and Senators Brooke and Kennedy in demanding a Philadelphia Plan for Boston.

66. Dirksen to Shultz, 7 August 1969, box 68, NN 372-119, DOL/NARA.

67. Scott led the fight against the Nixon proposals in voting rights and school desegregation, and he voted against Nixon's Supreme Court nomination of Clement Haynsworth.

68. *Congress and the Nation: 1969–1972,* Vol III (Washington: Congressional Quarterly, 1973), 295–96.

69. A bill identical to Fannin's was introduced in the House on November 18 as H.R. 14840 by Republican John M. Ashbrook of Ohio.

70. Subcommittee on Separation of Powers of the Senate Committee on the Judiciary, *Hearings: The Philadelphia Plan and S. 931,* 91st Cong., 1st Sess., Oct. 27–28, 1969.

71. Testimony of Harry P. Taylor, Executive Director of the General Building Contractors Association, Inc., of Philadelphia. Ibid., 70.

72. Testimony of Jerris Leonard, ibid., 92.

73. *The Civil Rights Act of 1964* (Washington: Bureau of National Affairs, 1964), 332. Because the Dirksen compromise had already stripped the EEOC of prosecutorial power, Dirksen was necessarily referring to private litigation or prosecution by the

Attorney General under Section. 706. He clearly did not seem to contemplate, in 1964, the likelihood that an administrative agency of the executive branch, outside of the EEOC, might run afoul of Title VII when enforcing minority hiring quotas against employers.

74. *Wall Street Journal,* 10 November 1969.

75. Testimony of George P. Shultz, Philadelphia Plan Hearings, 130–31.

76. Congressional Quarterly, *Nixon: First Year of His Presidency* (Washington: Congressional Quarterly Press, 1970), 51.

77. Testimony of Elmer B. Staats, Philadelphia Plan Hearings, 154.

78. Ibid., 143–49.

79. Staats to Ervin, 8 October 1969; Staats to the Speaker, Senate President, and Appropriations Chairs, n.d., PP/GAO.

80. Miller to Welch, 1 December 1969, PP/GAO.

81. Staats to Byrd, 2 December 1969; Staats to Mahon, 2 December 1969, PP/GAO.

82. Sec. 904, "The Philadelphia Plan," an amendment to H.R. 15209, Supplemental Appropriation for 1970, S, Rept. 91-616, 19. As amended and passed by the Senate, it was renumbered Sec. 1004.

83. *Wall Street Journal,* 19 December 1969.

84. Ehrlichman notes, 22 December 1969, Meetings with the President, RMNP. Nixon spiced the meeting with a George Meany joke that Ehrlichman recorded in dialect: George Meany to Shultz: "When I was a plumber, it never occoid to me to have niggers in the union."

85. *Public Papers: Nixon, 1969,* 1038–39.

86. *Washington Evening Star,* 23 December 1969. Nixon's statement asked Congress to explore a way to provide for prompt court review of legal disagreements between the executive branch and "the chief watchdog of the Congress," with the Comptroller having his own counsel rather than the Attorney General to represent him in such cases. But by December 22, only an up-or-down vote was possible.

87. Staats to Biemiller, 22 December 1969, Meany Archives.

88. *Wall Street Journal,* 23 December 1969; *Congress and the Nation, 1969–1972,* 6a. The unusual ideological divisions over the rider found such leading House liberals as Democrats James O'Hara of Michigan and Chet Holifield of California voting with the southern conservatives. In the Senate, Majority Leader Mike Mansfield also supported the rider.

89. Herman Belz, *Affirmative Action from Kennedy to Reagan: Redefining American Equality* (Washington: Washington Legal Foundation, 1984), 5.

90. *Contractors Association of Eastern Pennsylvania v. Secretary of Labor,* 311 F. Supp. 1002 (E.D. Pa. 1970); *Wall Street Journal,* 13 January 1970; Kilberg, "Current Civil Rights Problems," 86–89. The decision was affirmed on appeal by the Third Circuit Court of Appeals in April 1971; on October 12, 1971 the Supreme Court denied certiorari and the district court's decision stood affirmed.

91. *Wall Street Journal,* 16 March 1970.

92. Meany delivered organized labor's most persuasive attack on the Philadelphia Plan only after the battle had already been lost. In a speech to the National Press Club in Washington on January 12, 1970, Meany disputed the manpower survey numbers that Fletcher had used in his order of June 25, 1969. Meany said, for example, that there was not one iron workers union in Philadelphia with only 12 minorities out of a total membership of 850, as Fletcher's survey had claimed, but rather there were five iron worker locals with 3,575 members, of whom 690 or 19 percent were from minority groups. George Meany, "Labor and the Philadelphia Plan," 12 January 1970, AFL-CIO Archives; *Wall Street Journal,* 13 January 1970.

93. Leonard testimony, Philadelphia Plan hearings, 92–93.

94. *New York Times,* 16 January 1970.

95. Ervin was quoted in an Associated Press wire story, 15 January 1970.

96. *Wall Street Journal,* 16 January 1970. The Nixon administration arranged for Plans for Progress to be absorbed by the private National Alliance of Businessmen.

97. *Wall Street Journal,* 16 January 1970.

98. See Jeffrey L. Pressman and Aaron Wildavsky, *Implementation* (Berkeley: University of California Press, 1973); Erwin C. Hargrove, *The Missing Link: The Study of the Implementation of Public Policy* (Washington: Urban Institute, 1975); Eugene Bardach, *The Implementation Game: What Happens After a Bill Becomes a Law* (Cambridge: MIT Press, 1977); George C. Edwards III, *Implementing Public Policy* (Washington: Congressional Quarterly Press, 1980); Robert T. Nakamura and Frank Smallwood, *The Politics of Policy Implementation* (New York: St. Martin's Press, 1980); and Walter Williams et al., *Studying Implementation* (Chatham, N.J.: Chatham House, 1982).

99. 41 C.F.R. Part 60-2 (1970). This language represented a modification of the original Order No. 4's definition of underutilization as having fewer minorities "than their availability in the community." The looser definition was important because the Labor Department was struggling with the challenge of designing OFCC guidelines for sex discrimination—a category of job discrimination that unlike racial and ethnic discrimination, did not lend itself to a direct extrapolation from residential patterns or community representation to workforce representation.

100. 41 C.F.R. Part 60-2 (1970); *Wall Street Journal,* 4 February 1970; *New York Times,* 4 February 1970; Peter G. Nash, "Affirmative Action Under Executive Order 11246," *New York Law Review* 46 (1971):235.

101. *Wall Street Journal,* 10 February 1970. Shultz designated six of the 19 cities as "priority" areas: Boston, Detroit, Atlanta, Los Angeles, Seattle, and Newark.

102. Like the imposed Philadelphia models, the hometown plans required the negotiation of minority hiring goals and timetables on a proportional model. They tended to have less sharply escalating goals, but their coverage was broader, and included all construction contractors in a metropolitan area, not just federal contractors. Hometown plans emphasized training and upgrading programs more than plans imposed by the Labor Department. But whereas imposed plans often necessarily emphasized recruiting nonunion minority labor, hometown plans emphasized getting minorities into the unions. See Patricia A. Cooper, "Equal Employment Opportunity and the Department of Labor: Executive Order 11246 and the Equal Pay Act During the Nixon Administration," mimeograph, U.S. Department of Labor, April 4, 1975, 12–18, DOLL.

103. *Wall Street Journal,* 2 June 1970. Mandatory plans were also imposed on Chicago (after a hometown solution collapsed), San Francisco, Atlanta, St. Louis, Newark, and Camden. By May 1971 the OFCC staff had expanded from 26 to 119, but it was badly overextended by the city-by-city negotiations. By 1974 there were 62 hometown plans covering the nation's major metropolitan areas, and only six imposed ones. See Augustus J. Jones, Jr., *Law, Bureaucracy and Politics: The Implementation of Title VI of the Civil Rights Act of 1964* (Washington: University Press of America, 1982), 203–44.

104. *Wall Street Journal,* 25 June 1970. The new regulations provided instead for random selection of apprentices from a "pool of eligibles" who needed to pass only minimum qualifications, and shifted the burden of proof in apprenticeship selection from the applicant to the sponsor who rejected him.

105. Safire, *Before the Fall,* 468.

106. Quoted in Reichley, *Conservatives,* 140. Hess said of Shultz that "He was barely known to the president at the time he was appointed, but he soon began to make a strong impression at meetings. His style was to hang back until others had presented their views, and then to cut through to the heart of the subject, summarizing what others had said, and giving his own opinion."

107. Rather and Gates, *The Palace Guard,* 46.

108. Address by Herbert Hill, National Labor Director of the NAACP, 61st Annual Convention, 30 June 1970, NAACP/LC.

109. Within a year the Labor Department was to concede that the hometown solutions were slow to produce the desired results, and in response the OFCC in 1971 imposed in San Francisco, St. Louis, and Atlanta the same kind of mandatory racial quotas in construction that had been imposed in 1970 in Philadelphia and Washington. *Wall Street Journal,* 22 February 1971, 6 May 1971, 4 June 1971, and 5 September 1971.

110. Ehrlichman notes, Meetings with the President, 4 August 1970.

Chapter XIV

1. House Report 91–397, 28 July 1969, 5.

2. Lorn S. Foster, "Section 5 of the Voting Rights Act: Implementation of an Administrative Remedy," *Publius* 16 (Fall 1986): 17–28.

3. The triggering formula in Section 4 of the 1965 law also captured Yuma County in Arizona and Honolulu County in Hawaii. It had originally identified the state of Alaska and several remote western counties as well, but according to Attorney General Mitchell, "they sought and obtained judgments [from the Justice Department] indicating that their tests had not been used discriminatorily." Testimony of John N. Mitchell, House Judiciary Committee, Subcommittee No. 5, *Hearings on Voting Rights Extension,* 91st Cong., 1st sess., 26 June 1969, 220.

4. House Report 89-439, 1 June 1965, 15. In 1966 the Supreme Court upheld sections 4 and 5 of the 1965 act in *South Carolina v. Katzenbach,* 383 U.S. 301, holding that "In acceptable legislative fashion, Congress chose to limit its attention to the geographic areas where immediate action seemed necessary."

5. Steven F. Lawson, *Black Ballots: Voting Rights in the South, 1944–1969* (New York: Columbia University Press, 1976), 334–39; Abigail M. Thernstrom, *Whose Votes Count?* (Cambridge: Harvard University Press, 1987), 17–18.

6. Gary Orfield, *Congressional Power: Congress and Social Change* (New York: Harcourt Brace Jovanovich, 1975), 96.

7. Dissenting Views of Hon. Walter Flowers, of Alabama, House Report No. 91-397, 22.

8. U.S. Commission on Civil Rights, *Political Participation* (Washington: U.S. Commission on Civil Rights, 1968), 21–84; Steven F. Lawson, *In Pursuit of Power: Southern Blacks and Electoral Politics, 1965–1982* (New York: Columbia University Press, 1985), 128–31.

9. *Smith v. Paris,* 257 F. Supp. 901 (M.D. Ala., 1966); *U.S. v. Shannon* (N.D. Miss., Civ. Action No. DC-6928-K; *Hadnott v. Amos,* 394 U.S. 358 (1969).

10. Lamar Alexander to Harlow, 5 February 1969, box 19, EX HU 2-4 Voting, RMNP; Moynihan to Ehrlichman, 29 May 1969, box 19, EX HU 2-4, RMNP; "Memorandum on Meeting with Attorney General John N. Mitchell, 18 February 1969, LCCR Papers.

11. Nixon to Mitchell, 18 February 1969, box 19, EX HU 2-4, RMNP.

12. Howard Ball, Dale Krane, and Thomas P. Lauth, *Compromised Compliance: Implementation of the 1965 Voting Rights Act* (Westport, Conn.: Greenwood Press, 1982), 74–75.

13. Lawson, *Pursuit of Power,* 328n; John W. Dean III to Ehrlichman, 28 May 1969, box 19, EX HU 2-4, RMNP.

14. Leonard to Mitchell, 27 February 1969, box 72, Files of John W. Dean III, WHSF, RMNP.

15. Burns to the President, 17 June 1969, box 19, EX HU 2-4, RMNP.

16. The non-southern states requiring literacy tests in 1969 were Alaska, Arizona,

California, Connecticut, Delaware, Hawaii, Maine, Massachusetts, New Hampshire, New York, Oregon, Washington, and Wyoming. The 14th state, Idaho, had a "good character" requirement which qualified as a "test or device" within the meaning of Section 4(c) of the Voting Rights Act. Literacy tests were first required in Connecticut in 1855 and in Massachusetts in 1857, and reflected nativist reactions to heavy Irish immigration. Similarly, the West Coast literacy tests generally followed in the wake of heavy Asian immigration.

17. Burns to the President, 17 June 1969, 3.

18. Ken Cole to Dr. Arthur Burns, D.P. Moynihan, John Ehrlichman, Wilf Rommel, Bryce Harlow, and James Keogh, 29 May 1969, box 18 A6, Burns Papers, GRFL.

19. Bill Timmons to Bryce Harlow, 2 June 1969, box 19, EX HU 2-4, RMNP. The Republican leadership polled by Harlow included John B. Anderson, Leslie C. Arends, William C. Cramer, Gerald R. Ford, Richard H. Poff, John J. Rhoads, H. Allen Smith, Robert Taft, Jr., and Robert C. Wilson. It curiously did not include William McCulloch, whom Hugh Scott claimed could carry 40 to 50 votes with him.

20. Alexander to Harlow, 2 June 1969, box 19, EX HU 2-4, RMNP.

21. Dale Grubb to Harlow, 2 June 1969, box 19, EX HU 2-4, RMNP.

22. Alexander to Harlow, 3 June 1969, box 19, EX HU 2-4, RMNP.

23. Dean to Mitchell, 26 May 1969, box 72, Dean files, WHSF, RMNP; Dean to Cole, 28 May 1969, box 18 A6, Burns Papers, GRFL.

24. 395 U.S. 285 (1969).

25. Ibid. at 289, 291.

26. *Louisiana v. United States,* 380 U.S. 145 (1964) at 154.

27. Leonard to Ehrlichman, 9 June 1969, box 19, EX HU 2-4, WHCF, RMNP.

28. Ibid., 3.

29. Ibid., 4.

30. Testimony of Attorney General John N. Mitchell, *Hearings on Voting Rights Extension,* 26 June 1969, 1st sess., 91st Cong., 218–45.

31. Lawson, *Pursuit of Power,* 138.

32. Hesburgh to the President, 28 March 1969, box 19, EX HU 2-4, RMNP.

33. Mitchell testimony, *Hearings,* 26 June 1969, 227.

34. Ibid., 1 July 1969, 278.

35. Ibid., 279.

36. Ibid., 26 June 1969, 296.

37. Ibid., 244.

38. Ibid., 220–21.

39. U.S. Commission on Civil Rights, *Political Participation* (Washington: U.S. Government Printing Office, 1968).

40. Ibid., 177.

41. "Separate Views of Hon. Richard H. Poff," House Report No. 91-397, 28 July 1969, 14–15. Poff conceded that historically, Virginia "sorrowfully has not always been innocent of racial discrimination."

42. Lawson, *Black Ballots,* 319.

43. Statement of William McCulloch, *Hearings,* 1 July 1969, 269.

44. Cox to the Attorney General, 23 March 1965, Justice Department Administrative History, Civil Rights, Documentary Supplement, LBJL. Once the Voting Rights Act was passed and judicially affirmed by the Supreme Court, Cox defended the power of Congress under the 14th and 15th amendments to secure and broaden voting rights as virtually plenary. See Cox, "Constitutional Abjudication and the Power of Human Rights," *Harvard Law Review* 80 (1966): 91–103; Cox, *The Court and the Constitution* (Boston: Houghton Mifflin, 1987), 80–81.

45. *Allen v. State Board of Elections,* 393 U.S. 544 (1969).

46. McCulloch statement, *Hearings,* 1 July 1969, 271.

47. Reichley, *Conservatives,* 85–97.

48. In December 1970 a floating coalition of the right in the Senate rejected the

Nixon-Moynihan bill on welfare reform that a left-coalition had passed in the House. See generally Michael Foley, *The New Senate* (New Haven: Yale University Press, 1980), 63–78.

49. *Congress and the Nation, 1969–1972* (Washington: Congressional Quarterly Press, 1973), 494–99.

50. Delaney quoted in House Rules Committee, *Hearings to Extend the Voting Rights Act of 1965 with Respect to the Discriminatory Use of Tests and Devices,* 91st Cong., 1st sess., 1969, 20.

51. *Wall Street Journal,* 12 December 1969.

52. Lawson, *Pursuit of Power,* 139–57.

53. The Senate voted 51–22 to substitute the five-year extension, and 64–17 to add the 18-year-old vote. The revised bill was then passed on a 64–12 roll-call vote. The consensus of constitutional scholars supported Nixon's view that the Constitution did not permit Congress to lower the voting age by statute.

54. Rowland Evans and Robert D. Novak, *Nixon in the White House* (New York: Random House, 1971), 129–31. Nixon's constitutional reservation about enfranchising 18-year-olds by statute was partially sustained by the Supreme Court in *Oregon v. Mitchell,* 400 U.S. 112 (1970), which narrowly held that the law's enfranchisement of 18-year-olds was valid for presidential and congressional elections but not for state and local ones. In response to the constitutional uncertainty, Congress rushed a constitutional amendment through both houses and sent it to the states in March 1971. On June 30 the 26th Amendment was ratified by the requisite 38th state in the record time of three months and seven days.

55. Ball et al., *Compromised Compliance,* 74–77.

56. For a discussion of these reorganizational imperatives, see Hugh Heclo, *A Government of Strangers: Executive Politics in Washington* (Washington: Brookings, 1977); Hugh Davis Graham, *The Uncertain Triumph* (Chapel Hill: University of North Carolina Press, 1984), chap. 4.

57. Lawson, *Pursuit of Power,* 162–63.

58. When Douglass Cater left his post as national affairs editor for *The Reporter* to join Lyndon Johnson's White House staff in May 1964, he had just completed *Power in Washington* (New York: Vintage, 1964), a perceptive work of political journalism about the triangular networks that so often frustrated President Kennedy. Political scientist J. Leiper Freeman published a scholarly analysis of the same phenomenon in *The Political Process* (New York: Random House, 1955).

59. For a discussion of such triangular relationships in education policy during the Kennedy-Johnson years, see Hugh Davis Graham, "Short-Circuiting the Bureaucracy in the Great Society: Policy Origins in Education," *Presidential Studies Quarterly* XII (Summer 1982): 407–20. A sophisticated analysis of the broader phenomenon is Heclo, *A Government of Strangers;* and Heclo, "Issue Networks and the Executive Establishment," in *The New American Political System,* ed. Anthony King (Washington: American Enterprise Institute, 1978), 87–124.

60. Hanes Walton, Jr., *When the Marching Stopped: The Politics of Civil Rights Regulatory Agencies* (Albany: State University of New York Press, 1988), 28–42.

61. The new agencies in the civil rights network tended to be disproportionately staffed by members of constituent groups from the protected classes. When Leon Panetta took over the Office of Civil Rights at HEW in 1969, the OCR staff of 278 included 129 blacks, 15 Spanish-surnamed Americans, 2 American Indians, 1 Asian American, and 131 classified as "other." Panetta and Gall, *Bring Us Together* (Philadelphia: Lippincott, 1971), 128.

62. Nathan Glazer, *Affirmative Discrimination: Ethnic Inequality and Public Policy* (New York: Basic Books, 1975), 211–16.

63. Richard F. Fenno, Jr., *Congressmen in Committees* (Boston: Little, Brown, 1973); Samuel C. Patterson, "The Semi-Sovereign Congress," in King, ed., *New American Political System,* 125–78; David E. Price, "Congressional Committees in

the Policy Process," in Lawrence C. Dodd and Bruce I. Oppenheimer, eds., *Congress Reconsidered* (2d ed.; Washington: Congressional Quarterly Press, 1981), 156–85.

64. By the 93rd Congress there were 57 standing and special committees and 288 subcommittees, which together provided a total of 345 work units, including their chairmanships and burgeoning staffs, for only 535 legislators. See Roger H. Davidson, "Two Avenues of Change: House and Senate Committee Reorganization," in Dodd and Oppenheimer, eds., *Congress Reconsidered*, 107–33.

65. The Edwards subcommittee began its first hearing on May 26, 1971, to investigate charges that Attorney General Mitchell and the Justice Department were not vigorously enforcing the Voting Rights Act. U.S. Congress, House, Civil Rights Oversight Subcommittee of the Committee on the Judiciary, *Hearings*, 92nd Cong., 1st sess., 1971.

66. In response to a series of public attacks by the Black Caucus, Nixon invited them to the White House on March 25, 1971, to discuss his administration's response to their sixty "demands." Most of these centered on economic issues like welfare reform and job discrimination, but they ranged broadly to include areas like African policy. On May 18 Nixon sent each caucus member a 115-page report that was prepared by the OMB under the direction of Shultz and that summarized the Administration's efforts. When the Black Caucus attacks continued, the White House abandoned the effort. Nixon to Congressman [Charles C.] Diggs, 18 May 1971, box 41, Series 69-1, OMB/NARA; *New York Times*, 21 May 1971; Ehrlichman interview, 5 August 1986.

67. Minutes, Executive Committee, Leadership Conference on Civil Rights, 10 September 1969, LCCR Papers.

68. Moynihan to the President, reprinted in the *Wall Street Journal*, 3 March 1970.

69. Moynihan's second suggestion was "benign neglect," which he defined as continuing to pay close attention to black progress, but also directing greater attention than in the past to Indians, Mexican-Americans, and Puerto Ricans, while "racial rhetoric fades" and extremists are denied their national forum for "martyrdom, heroics, [and] histrionics."

70. Arthur Maas, *Muddy Waters: The Army Engineers and the Nation's Rivers* (Cambridge: Harvard University Press, 1951).

Chapter XV

1. *Plessy v. Ferguson*, 163 U.S. 537 (1896) at 559.

2. *New York Times*, 23 May 1954.

3. For a fresh interpretation see Charles A. Lofgren, *The Plessy Case* (New York: Oxford University Press, 1987).

4. *Plessy*, 163 U.S. at 551. In his dissent Harlan took judicial notice of the obvious: "Everyone knows that the statute in question had its origin in the purpose, not so much to exclude white persons from railroad cars occupied by blacks, as to exclude colored people from coaches occupied by or assigned to white persons."

5. C. Vann Woodward, "The Case of the Louisiana Traveler," *Quarrels That Shaped the Constitution*, ed. John A. Garraty (New York: Harper & Row, 1966), 145–58.

6. *Brown v. Board of Education of Topeka*, 347 U.S. 483 (1954) at 494–95.

7. For a critical discussion of the evolution of an implicit "stigma" doctrine, see Edward J. Erler, "Sowing the Wind: Judicial Oligarchy and the Legacy of *Brown v. Board of Education*," *Harvard Journal of Law and Public Policy* 8 (1985): 399–426, esp. 409–13.

8. In *Brown* the Court held narrowly that racial segregation was unconstitutional only in public education. But within a short period the Court invalidated state laws

requiring segregated public golf courses, beaches, buses, and parks. This was done by per curiam decisions citing *Brown* as the authority, although the extension of the educational rationale of *Brown* to such disparate activities would seem to have merited a constitutional explication. The *Plessy* ruling in transportation was most directly attacked in *Gayle v. Browder*, 352 U.S. 903 (1956), which rejected bus segregation in Montgomery, Alabama. See Catherine A. Barnes, *Journey from Jim Crow: The Desegregation of Southern Transit* (New York: Columbia University Press, 1983).

9. Richard Kluger, *Simple Justice* (New York: Knopf, 1976), 315–45; David W. Southern, *Gunnar Myrdal and Black and White Relations* (Baton Rouge: Louisiana State University Press, 1987), 127–50.

10. Lino A. Graglia, *Disaster by Decree: The Supreme Court Decisions on Race and the Schools* (Ithaca: Cornell University Press, 1976), 26–32.

11. *Bolling v. Sharpe*, 347 U.S. 497 (1954). *Bolling* extended the stigma-based logic of *Brown* from the 14th Amendment's equal protection clause to the 5th Amendment's due process clause, thereby endowing both with a substantive meaning grounded in social psychology. See the criticisms of Wallace Mendelson, "From Warren to Burger: The Rise and Decline of Substantive Equal Protection," *American Political Science Review* 66 (December 1972): 1226–33; and Erler, "Sowing the Wind," 399–404.

12. David M. O'Brien, *Storm Center: The Supreme Court in American Politics* (New York: Norton, 1986), 233–34. On Warren's bargaining over enforcement to achieve unanimity, see D. Hutchinson, "Unanimity and Desegregation: Decisionmaking in the Supreme Court, 1948–1958," *Georgetown Law Journal* 68 (1979): 1–43.

13. *Bolling*, 347 U.S. at 499.

14. *United States v. Carolene Products Co.*, 304 U.S. 144, 152 n.4 (1938).

15. For an account of the authorship of footnote 4, see Alpheus Mason, *Harlan Fiske Stone: Pillar of the Law* (New York: Viking, 1956), 513–15. See also Louis Lusky, "Footnote Redux: A *Carolene Products* Reminiscence," *Columbia Law Review* 82 (1982): 1093–1123, for the reflections of Stone's law clerk in 1938.

16. See Robert M. Cover, "The Origins of Judicial Activism in the Protection of Minorities," *Yale Law Journal* 91 (June 1982): 1287–1316.

17. Government classifications by race became judicially suspect during the 1960s, although explicit denomination of specific minority groups by the federal courts was largely a phenomenon of the 1970s. In practice, the triggering of strict judicial scrutiny of a law using "suspect" classifications was virtually certain to invalidate the law. During the 1970s race, national origin, and alienage became firmly established as class or group characteristics of a "discrete and insular minority"; gender evolved as a special halfway case (see Chapter 16); and the courts have generally refused to so denominate illegitimate children, the poor, or conscientious objectors.

18. John Hart Ely, *Democracy and Distrust: A Theory of Judicial Review* (Cambridge: Harvard University Press, 1980). On *Carolene Products*, see 75–77.

19. O'Brien, *Storm Center*, 287–95; Kluger, *Simple Justice*, 722–46.

20. 347 U.S. 483, Brief for Appellants on Reargument, 15.

21. Albert P. Blaustein and Clarence Clyde Ferguson, Jr., *Desegregation and the Law* (2d ed.; New York: Random House, 1962), 158–79; Hugh Davis Graham, *Crisis in Print: Desegregation and the Press in Tennessee* (Nashville: Vanderbilt University Press, 1967), 72–90.

22. *Brown v. Board of Education*, 349 U.S. 294 (1955).

23. See the reminiscence of Frankfurter's self-described "law clerk for life," Philip Ulman [interviewed by Norman Silber], "The Solicitor General's Office, Justice Frankfurter, and Civil Rights Legislation, 1946–1960: An Oral History," *Harvard Law Review* 100 (February 1987): 817–52.

24. Stephen L. Wasby, Anthony A. D'Amato, and Rosemary Metrailer, *Desegregation from Brown to Alexander* (Carbondale: Southern Illinois University Press, 1977), 108–30.

25. Louis Lusky, "Racial Discrimination and the Federal Law: A Problem in Nullification," *Columbia Law Review* 63 (1963): 1172, n37.

26. Graglia, *Disaster by Decree,* 36.

27. *Brown v. Board of Education of Topeka,* 349 U.S. at 300.

28. For a critique of the post-*Brown* abuse of the equity power, see Gary L. McDowell, *Equity and the Constitution* (Chicago: University of Chicago Press, 1982). For a contrary view, see Owen M. Fiss, *The Civil Rights Injunction* (Bloomington: Indiana University Press, 1978).

29. J. W. Peltason, *Fifty-Eight Lonely Men* (New York: Harcourt, Brace & World, 1961).

30. Jack Bass, *Unlikely Heroes* (New York: Simon & Schuster, 1981).

31. In *Cooper v. Aaron,* 358 U.S. 1 (1958), the unanimous court denied Little Rock's petition for delay, and lectured Governor Faubus severely on the supremacy of the Supreme Court's authority "to say what the law is." Yet even in *Cooper* the angry Court nonetheless agreed that under certain circumstances district courts might still "conclude that justification existed for not requiring the present nonsegregated admission of all qualified Negro children."

32. The distinction between desegregation and integration was most forcefully advanced by Judge John J. Parker of the Fourth Circuit, who concluded, in *Briggs v. Elliott,* 132 F. Supp. 776 (E.D.S.C.1955), that the Constitution "does not require integration. . . . It merely forbids the use of governmental power to enforce segregation." Parker had approved freedom-of-choice plans in North Carolina, a process that in 1957 had introduced three black pupils into white schools in Charlotte. But by 1960 this tiny opening wedge of three had declined to one. Such tokenism was a common experience under freedom of choice plans, and it led to the Fifth Circuit's attack on them in the middle 1960s and then to Supreme Court's rejection of them in 1968.

33. *United States v. Jefferson County Board of Education,* 372 F.2d 836 (5th Cir. 1966); J. Harvie Wilkinson III, *From Brown to Bakke* (New York: Oxford University Press, 1979), 78–192.

34. Graham, *Crisis,* 120–23, 170–87; Raymond Wolters, *The Burden of Brown: Thirty Years of School Desegregation* (Knoxville: University of Tennessee Press, 1984), 150–55.

35. *Jefferson,* 372 F. 2d at 866.

36. Wilkinson, *Brown to Bakke,* 112.

37. Wilkinson, *Brown to Bakke,* 114. The role of Wisdom and his Fifth Circuit colleagues in *Jefferson* is described sympathetically in Bass, *Unlikely Heroes,* 297–310.

38. *Green v. County School Board of New Kent County,* 391 U.S. 430 (1968) at 439.

39. Wilkinson, *Brown to Bakke,* 107, 327 n.171.

40. *Jones v. Alfred H. Mayer Co.,* 392 U.S. 409 (1968).

41. *Wall Street Journal,* 18 June 1968.

42. Derrick A. Bell, Jr., *Race, Racism and American Law* (2d ed.; Boston: Little, Brown, 1980), 497–508. Justice Stewart's decision was drafted by his law clerk, Laurence Tribe.

43. *Jones,* 392 U.S. at 412.

44. Bell, *Racism and American Law,* 498.

45. Charles Fairman, *Reconstruction and Reunion, 1864–1888* (New York: Macmillan, 1971), 1258. In "Jones v. Mayer: Clio, Bemused and Confused Muse," Gerhard Casper called *Jones* a combination of "creation by authoritative revelation and 'law-office' history." *Supreme Court Review* 89 (1968), 100.

46. *Jones,* 392 U.S. at 450.

47. For a defense of "dynamic" judicial interpretation of statutes, see William N. Eskridge, Jr., "Dynamic Statutory Interpretation," *Pennsylvania Law Review* 135 (1987):

1479–1555. Eskridge argues that statutes that are old and ambiguous, like the 1866 law reinterpreted in *Jones,* are appropriate candidates for judicial modernization, but that statutory text and legislative intent should be judicially respected as constraints on judge-made law.

48. *Allen v. State Board of Elections,* 393 U.S. 544 (1969).

49. The three cases from Mississippi were: *Fairley v. Patterson, Attorney General of Mississippi,* 282 F. Supp. 164 (1967); *Burton v. Patterson,* 281 F. Supp. 918 (1967); and *Whitley v. Williams, Governor of Mississippi,* originally brought as *Whitley v. Johnson, Governor of Mississippi,* 260 F. Supp. 630 (1966).

50. *Allen,* 393 U.S. at 563.

51. Ibid., 563–64.

52. *Reynolds v. Sims,* 377 U.S. 533 (1964). For discussions of vote dilution that generally support Warren's broad interpretation, see Chandler Davidson, ed., *Minority Vote Dilution* (Washington: Howard University Press, 1985); and Howard Ball, "Racial Vote Dilution: Impact of the Reagan DOJ and the Burger Court on the Voting Rights Act," *Publius* 16 (Fall 1986): 29–48.

53. *Allen,* 393 U.S. at 569.

54. Ibid., 595, 596.

55. G. Edward White, *The American Judicial Tradition* (New York: Oxford University Press, 1988), 341.

56. J. Harvie Wilkinson, "Justice John Marshall Harlan and the Values of Federalism," *Virginia Law Review* 57 (1971): 1185–99.

57. *Allen,* 393 U.S. at 564.

58. Ibid., 585, 583.

59. Ibid., 586.

60. As G. Edward White conceded in his admiring biography of Warren, the Chief Justice "equated judicial lawmaking with neither the dictates of reason, as embodied in established precedent or doctrine, nor the demands imposed by an institutional theory of the judge's role, nor the alleged 'command' of the constitutional text, but rather with his own reconstruction of the ethical structure of the Constitution." G. Edward White, *Earl Warren: A Public Life* (New York: Oxford University Press, 1982), 359.

61. *Allen,* 393 U.S. at 588.

62. *Gomillion v. Lightfoot,* 364 U.S. 339 (1960).

63. *Alexander v. Holmes County,* 396 U.S. 1218 (1969), 396 U.S. 19 (1969); Wasby et al., *Desegregation from Brown to Alexander,* 298–425.

64. John Ehrlichman, *Witness to Power* (New York: Simon & Schuster, 1982); Ehrlichman interview with the author, 5 August 1986; Bob Woodward and Scott Armstrong, *The Brethren: Inside the Supreme Court* (New York: Simon & Schuster, 1979), 60.

65. Jonathan Spivak, "Supreme Court to Take Up Desegregation, Pitting Nixon Policies Against the Liberals," *New York Times,* 12 October 1970.

66. 42 U.S.C. 2000 (c), emphasis added.

67. Burger's political reading of the play of congressional motives during 1963–64 was in part correct, especially regarding the determination of key moderates like McCulloch and Dirksen to direct the Civil Rights Act mostly against racial segregation in the South. In 1974 in *Milliken v. Bradley,* 418 U.S. 717, Burger in speaking for a 5–4 majority reversed a lower court order for massive racial busing across school district lines in metropolitan Detroit. The Supreme Court's hostility toward local control of schools in the South was reversed in *Milliken* to protect de facto patterns in the North.

68. Wilkinson, *Brown to Bakke,* 131–51.

69. *Swann,* 402 U.S. at 31.

70. Note, "Civil Rights—Employment Discrimination—Preferential Minority Treatment as an Appropriate Remedy Under Section 703(j) of Title VII," *Tennessee*

Law Review 42 (1974): 397–405. The first federal court decision under the Civil Rights Act of 1964 to require racially preferential treatment was a Fifth Circuit decision in 1969 that required an all-white labor union to admit four blacks and make subsequent job referrals on a one-to-one basis, one white and one black. *Local 53, Asbestos Workers v. Vogler*, 407 F.2d 1047 (5th Cir. 1969).

71. *Contractors Association of Eastern Pennsylvania v. Secretary of Labor*, 442 F.2d 159 (3rd Cir. 1971), *cert. denied*, 404 U.S. 854 (1971).

72. *Griggs. v. Duke Power Company*, 401 U.S. 424 (1971).

73. Ibid., 426–28.

74. *Griggs v. Duke Power Co.*, 292 F. Supp. 243 (1968).

75. Bell, *Racism and American Law*, 619–23.

76. *Griggs v. Duke Power Co.*, 420 F.2d 1225 (4th Cir. 1970).

77. Alfred W. Blumrosen, "Strangers in Paradise: *Griggs v. Duke Power Co.* and the Concept of Employment Discrimination," *Michigan Law Review* 71 (November 1972): 59–110.

78. Author interview with John deJ. Pemberton, Jr., 3 March 1985, San Francisco, California.

79. Pemberton to Greenberg (not sent), 27 January 1970, supplied to the author by courtesy of John deJ. Pemberton, Jr.

80. Pemberton's letter argued for concentrating on the back pay case involving the Albemarle Paper Company. But for technical reasons the Albemarle case was delayed and not decided by the Supreme Court until 1975. By then it was overshadowded by *Griggs*, but it built on the new edifice of disparate impact case law that *Griggs* had accelerated.

81. *New York Times*, 18 June 1970.

82. *Quarles v. Phillip Morris, Inc.*, 279 F. Supp. 505 (E.D. Va., 1968).

83. *Griggs*, 420 F.2d at 1239–44.

84. Phil Lyons, "An Agency with a Mind of Its Own: The EEOC's Guidelines on Employment Testing," *New Perspectives* 17 (Fall 1985): 20–25; *Personnel Testing and Equal Employment Opportunity*, eds. Betty R. Anderson and Martha P. Rogers (Washington: U.S. Government Printing Office, 1970).

85. Woodward and Armstrong, *The Brethren*, 140–41. Woodward and Armstrong based their portrait of the inside workings of the Burger Court primarily on the anonymous evidence provided by former law clerks. Thus *The Brethren*, which is essentially undocumented, must be used with caution.

86. G. Edward White, "The Burger Court," *American Judicial Tradition*, 424–34.

87. Pressure to maintain unaniminity in the desegregation cases blurred the customary dichotomies between liberal and conservative arguments on the Court and extracted a price in internal tension and contradiction, as it did in *Griggs*. See Dennis J. Hutchinson, "Unanimity and Desegregation: Decisionmaking in the Supreme Court, 1948–1958," *Georgetown Law Journal* 68 (1979): 1–96.

88. *The Brethren* paints an unflattering picture of Burger as a manipulative Chief who abused the voting process in conference in order to place himself in the majority and thereby assign opinions. Nina Totenberg, a Court reporter for the *National Observer*, described an example of this behavior in the *Swann* case in "Behind the Marble, Beneath the Robes," *New York Times Magazine*, 16 March 1975, esp. 63–64.

89. *Griggs*, 401 U.S. at 430–31.

90. Ibid., 432; *Wall Street Journal*, 9 March 1971.

91. 110 *Congressional Record* 6415 (1964).

92. Ibid., 7213.

93. Equal Employment Opportunity Commission, "Guidelines on Employee Selection Procedures," 29 CFR @1607, 35 *Federal Register* 12333 (1 August 1970).

94. Donald L. Horowitz, *The Courts and Social Policy* (Washington: Brookings, 1977), 15.

95. 110 *Congressional Record* 13492 (1964). For interpretations that support Burger's view, see George Cooper and Richard B. Sobel, "Seniority and Testing Under Fair Employment Laws: A General Approach to Objective Criteria of Hiring and Promotion," *Harvard Law Review* 82 (1969): 1598–1679; Blumrosen, "Strangers in Paradise," 59–110.

96. *Griggs,* 401 U.S. at 434.

97. 110 *Congressional Record* 8194 (1964).

98. EEOC *Administrative History,* Vol. I, LBJL, 1969, 17.

99. Ibid., 249.

100. Gary Bryner, "Congress, Courts, and Agencies: Equal Employment and the Limits of Policy Implementation," *Political Science Quarterly* 96 (Fall 1981): 411–30.

101. Hugh Steven Wilson, "A Second Look at *Griggs v. Duke Power Company*: Ruminations on Testing, Discrimination, and the Role of the Federal Courts," *Virginia Law Review* 63 (1972): 844–74, esp. 852–58. But see also Eskridge, "Dynamic Statutory Interpretation," esp. 1506–11, for criticism of the "intentionalist" doctrine as incoherent.

102. *New York Times,* 9 March 1971.

103. Blumrosen, "Strangers in Paradise," 62.

104. *New York Times,* 9 March 1971.

105. *Griggs,* 401 U.S. at 429.

106. Michael Brody, "Congress, the President, and Federal Equal Employment Policymaking: A Problem in Separation of Powers," *Boston University Law Review* 60 (1980): 239–305, esp. 260–63.

107. Note, "Business Necessity Under Title VII of the Civil Rights Act of 1964: A No-Alternative Approach," *Yale Law Journal* 84 (November 1974): 98–119; Horowitz, *Courts and Social Policy,* 15.

108. *United Steelworkers of America v. Weber,* 443 U.S. 193 (1979).

109. *DeFunis v. Odegard,* 416 U.S. 312 (1974); *Regents of the University of California v. Bakke,* 438 U.S. 265 (1978); Robert M. O'Neil, *Discrimination Against Discrimination: Preferential Admissions and the DeFunis Case* (Bloomington: Indiana University Press, 1975); Timothy J. O'Neill, *Bakke and the Politics of Equality* (Middletown, Conn.: Wesleyan University Press, 1985).

110. *Goesaert v. Cleary,* 335 U.S. 464 (1948).

111. *Floyd v. Florida,* 368 U.S. 57 (1961).

112. Ibid., 61–62.

Chapter XVI

1. For a description of the ambiguous circumstances surrounding the Eisenhower, Kennedy, and Johnson positions on the ERA, see Cynthia Harrison, *On Account of Race* (Berkeley: University of California Press, 1988), 33–38, 116–19, 182–83.

2. A concise history of the ERA is Marguerite Rawalt, "The Equal Rights Amendment," *Women in Washington,* ed. Irene Tinker (Beverly Hills: Sage, 1983), 49–78. See also Amelia R. Fry, "Alice Paul and the ERA," *Rights of Passage,* ed. Joan Hoff-Wilson (Bloomington: Indiana University Press, 1986), 8–24.

3. Janet K. Boles, *The Politics of the Equal Rights Amendment* (New York: Longman, 1979), 37–40.

4. Esther Peterson, "The Kennedy Commission," *Women in Washington,* 23; Harrison, *On Account of Sex,* 116–19.

5. Throughout the 1950s Representative Emanuel Celler regularly sponsored bills to establish a commission to study the legal status of women, while keeping ERA bills bottled up in his Judiciary Committee. In 1961 Celler's status-of-women bill carried his own version of the Hayden rider: it called for equal treatment of women "except

such as reasonably justified by differences in physical structure, biological, or social function." Celler's proposed commission, which would report to Congress, was preempted by President Kennedy's own status-of-women commission. Rawalt, "The Equal Rights Amendment," *Women in Washington,* 52–58.

6. *American Women*: Report of the President's Commission on the Status of Women (Washington: U.S. Government Printing Office, 1963), 45.

7. Patricia G. Zelman, *Women, Work, and National Policy: The Kennedy-Johnson Years* (Ann Arbor: UMI Research Press, 1980), 23–38.

8. During 1965–66 EEOC commissioner Richard Graham led an active ICSW subcommittee on women and the media that grappled with the problem of sex-segregated classified ads. But when President Johnson allowed Graham's one-year EEOC appointment to expire, the ICSW reverted to its natural torpor. See the summary report of Catherine East to Secretary-designate George P. Shultz, 30 December 1968, box 39, Task Force on Women's Rights and Responsibilities, Charles L. Clapp files, Staff Member Office Files (SMOF), RMNP.

9. Ibid., 3. In 1961 East had been transferred from the Civil Service Commission to the staff of the PCSW, and from 1963 through 1975 she served as executive secretary for both the CACSW and the ICSW.

10. Catherine East, "Newer Commissions," *Women in Washington,* 35–36.

11. Press release, Citizen's Advisory Council on the Status of Women, July 1968, box 39, Clapp file, RMNP.

12. East to Shultz, 30 December 1968, box 39, Clapp file, RMNP.

13. Burns Report, 18 January 1969, 81, 82–83.

14. The President's News Conference of February 6, 1969, *Public Papers of the Presidents: Richard M. Nixon, 1969* (Washington: U.S. Goverment Printing Office, 1971), 75.

15. Quoting the public Nixon on women is like quoting Lincoln on blacks. On April 16, 1969, Nixon told the 17th annual Republican Women's Conference not only how pleased he was with the women he had thus far appointed, but also "how proud I am of the women in this administration who do not hold office, but who hold the hands of their husbands who do hold office. . . . we have one of the finest groups of Cabinet wives I have ever seen." But the following day he told the League of Women Voters on their 50th anniversary that he sensed "a tremendously escalating role of women in politics in the United States." He pointed out that the prime ministers of India, Ceylon, and Israel were women, that certainly in the next 50 years "we shall see a woman President—maybe sooner than you think," and that "in America a woman can and should be able to do any political job that a man can do." *Public Papers: Nixon, 1969,* 291, 297.

16. Jo Freeman, *The Politics of Women's Liberation* (New York: David McKay, 1975), 205–6.

17. Partial transcript, enclosed with Glaser to Burns, 23 May 1969, box 39, Clapp file, RMNP.

18. Glaser to Burns, 23 May 1969, Clapp file, box 39, RMNP.

19. Bra-burning as a media symbol for radical feminist kookiness was resented by feminists as symptomatic of a generalized sexist refusal to take the movement and its social critique seriously. Feminists pointed out that unlike draft card burning, the bra-burning was largely a media concoction, drawn from the feminist demonstration against the 1968 Miss America Contest, where bras and girdles and false eyelashes were thrown into a "freedom trash can." In response, many women's liberation leaders refused to speak to male reporters. Freeman, *Politics of Women's Liberation,* 111–13; Joanna Foley Martin, "Confessions of a Non Bra-Burner," *Chicago Journalism Review* 4 (July 1971): 11–13.

20. Dwyer's letter to Nixon is summarized in Glaser to Burns, 23 May 1969.

21. Dwyer to the President, 8 July 1969, Clapp files, box 39, RMNP.

22. Dwyer's first-listed recommendations concentrated on senior appointments for

women, including a Special Assistant to the President for Women's Rights and Responsibilities and an Office of Women's Rights and Responsibilities for the new presidential assistant to direct and staff.

23. Moynihan to the President, 20 August 1969, box 21, EX HU 2-5, WHCF, RMNP.

24. Ehrlichman to Domestic Plans Group, 20 August 1969, box 31, Domestic Plans Group, Ehrlichman files, RMNP.

25. Clapp memorandum for the file, 20 August 1969, box 39, Clapp file, SMOF, RMNP.

26. Clapp memorandum for the file, 25 September 1969, box 39, Clapp file, SMOF, RMNP.

27. Burns to Allan, 12 September 1969; Allan to Burns, 30 September 1969, box A26, Women's Rights, Task Forces file, Burns Papers, Gerald R. Ford Library (GRFL).

28. White House Press release, 1 October 1969, RMNP. The task force members were, in addition to Allan, Cunningham, Glaser, Haener, and Whitlow: Elizabeth Athanasakos, a Fort Lauderdale judge; Ann R. Blackman, president of a real estate brokerage in Winchester, Mass.; P. Dee Boersma, a graduate student at Ohio State University; Sister Ann Ida Gannon, president of Mundelein College in Chicago; Patricia [Laddie] F. Hutar, president of a public relations firm in Chicago; Katherine B. Massenburg, chair of the Maryland Commission on the Status of Women; William C. Mercer, vice president of personnel relations for AT&T; and Alan Simpson, president of Vassar College. The White House announcement reflected the custom of identifying the marital status of females. It listed "Miss" Allan as "chairman," and also supplied the marital identifications as Miss or Mrs. for all of the women except "Judge" Athanasakos (five of the women were identified as being unmarried).

29. Minutes of First Meeting, September 25–26, 1969, attached to Whitlow to Task Force Members, 11 October 1969, box 39, Clapp files, RMNP.

30. Kuck, paraphrased in ibid., 2–3.

31. Mintz, paraphrased in ibid., 7.

32. Hilton, paraphrased in ibid., 3.

33. Lowell, paraphrased in ibid., 4.

34. Vera Glaser, "The Cabinet's Done Nothing," *Miami Herald,* 9 October 1969.

35. Summary of Recommendations Adopted by the President's Task Force on November 23–24, 1969, box 39, Clapp file, RMNP.

36. Report of the President's Task Force on Women's Rights and Responsibilities, December 1969, box A26, Task Force file, Burns Papers, GRFL.

37. Allan to the President, 15 December 1969, Women's Task Force Report, vii, 13.

38. "Minority Views of Dorothy Haener on Extension of the Fair Labor Standards Act," Women's Task Force Report, 38–39.

39. Burns to Allan, 9 January 1970, box A26, Burns file, GRFL.

40. Citizens Advisory Council on the Status of Women, *Women in 1970* (Washington: U.S. Government Printing Office, 1970), 2.

41. Freeman, *Politics of Women's Liberation,* 212–21.

42. Hauser to John Mitchell, Rogers B. Morton, Ehrlichman and Garment, 29 October 1969; Hauser to William Safire, 2 March 1970; Hitt to Ehrlichman, 15 April 1970, box 21, EX HU 2-5, WHCF, RMNP.

43. Clapp to Garment, 16 April 1970, box 39, Clapp file, RMNP.

44. Glaser, "Female Elite of U.S.," *Detroit Free Press,* 13 April 1970; *New York Times,* 23 April 1970; East, "Newer Commissions," *Women in Washington,* 38.

45. The President's Task Force on Women's Rights and Responsibilities, *A Matter of Simple Justice* (Washington: U.S. Government Printing Office, April 1970).

46. Garment to Ehrlichman, 25 May 1970, box 21, EX HU 2-5, WHCF, RMNP.

47. Garment noted that the United Auto Workers had voted a month earlier at

their annual convention to support the amendment. But most leaders of organized labor opposed the ERA, and this in turn had neutralized the Leadership Conference on Civil Rights. The LCCR had admitted NOW to its membership in March 1967 over the objections of Andrew Biemiller (in April 1968 it had also admitted the League of Women Voters), and in January 1970 NOW's Lucy Komisar asked the conference to endorse the ERA. But the LCCR's executive committee concluded at its April 29 meeting that because no consensus could be reached, the conference would take no position, and its constituent organizations were free to take positions of their own. Marvin Caplan to Washington Representatives, 3 June 1970, LCCR papers, LC.

48. *Edwards v. North American Rockwell Corp.*, 291 F. Supp 199 (1968); *Cheatwood v. South Central Bell Tel. & Tel. Co.*, 303 F. Supp 754 (1968).

49. Garment to Ehrlichman, Memorandum for the President, 25 May 1970, 7–8.

50. Ibid., 5.

51. Rawalt, "The Equal Rights Amendment," *Women in Washington*, 62–65; George, *Griffiths*, 170–72. On 10 August 1970, having forced the ERA resolution out of Celler's Judiciary Committee with a discharge petition bearing the requisite 217 signatures, Griffiths joined 331 of her House colleagues in voting for ERA to only 22 opposed.

52. Freeman, *Politics of Women's Liberation*, 209.

53. The three-volume study was based on data from 123 cities, reflecting employment patterns by race and sex in 60 industries covering all 50 states. See *Hearings* on the Utilization of Minority and Women Workers in Certain Major Industries, Los Angeles, California, March 12–14, 1969; Sonia Pressman, "Job Discrimination and the Black Woman," *The Crisis* (March 1970): 103–8.

54. Subsequent survey data indicate that the proportion of women in blue-collar jobs has remained essentially constant since 1970. The Bureau of Labor Statistics reported in 1987 that between 1970 and 1986, the female percentage of operators, fabricators, and laborers declined from 25.9 to 25.4 percent, and the women's proportion of workers in precision production, craft, and repair increased only from 7.3 to 8.6 percent. Substantial increases came in managerial and business specialties (33.9 to 43.4%) and professional specialties like law, teaching, and writing (44.3 to 49.4%). But the largest category remained that of administrative support, including clerical (73.2 to 80.4%) and service occupations (60.4 to 62.6%). *New York Times*, 17 July 1987.

55. Equal Employment Opportunity Commission, *Sixth Annual Report* (Washington: U.S. Government Printing Office, 1972), 11. The 3,597 complaints of sex discrimination received by the EEOC in 1970 represented 25 percent of all complaints, a proportion that remained constant in 1971 as complaints of sex discrimination rose to 5,820 of 22,920 charges.

56. On the class basis of American feminist leadership circa 1970, see William Henry Chafe, *The American Woman: Her Changing Social, Economic, and Political Roles, 1920–1970* (New York: Oxford, 1972), 226–44.

57. Garment to Cole, 3 June 1970, box 21, EX HU 2-5, WHCF, RMNP.

58. Title 41, Ch. 60, Part 60-20, "Sex Discrimination Guidelines," U.S. Department of Labor, attached to Garment to Cole, 3 June 1970.

59. *Washington Post*, 10 June 1970.

60. Transcript, White House press conference, 9 June 1970; *Wall Street Journal*, 10 June 1970.

61. Transcript, Koontz press conference, 9 June 1070, 10.

62. Finch to the President, 15 April 1971, box 10, President's Office Files—Handwriting, RMNP.

63. Press release, Department of Justice, 20 July 1970; *Wall Street Journal*, 21 July 1970.

64. NAACP, 61st Annual Convention Resolutions, Cincinnati, Ohio, July 1970, box 17, NAACP papers, LC.

65. 36 *Federal Register* 17444, 31 August 1971.

66. 36 *Federal Register* 23152, 4 December 1971.

67. *Wall Street Journal*, 3 December 1971.

68. The order omitted for women the base criterion in Order No. 4 for determining the availability of workers, which was the size of the minority population "of the labor area surrounding the facility." For women the revised order repaired to such less precisely definable criteria as the percentage of the female *workforce* in the immediate labor area and the size of the female *unemployment force* in the local area.

69. In June 1973 the OFCC in effect acknowledged that the proportional model of equal employment law was not universally applicable by publishing "Guidelines on Discrimination Because of Religion or National Origin," 38 *Federal Register* 28967, 19 January 1973. For these two groups, which were not covered by Revised Order No. 4, the OFCC acknowledged the impossibility of enumerating all the varied religious and ethnic groups and the impossibility of determining their availability in order to formulate goals and timetables for them.

70. William J. Kilberg, "Current Civil Rights Problems in the Collective Bargaining Process: The Bethlehem and AT&T Experiences," *Vanderbilt Law Review* 27 (January 1974): 90–92.

71. Ibid., 97; Phyllis A. Wallace, ed., *Equal Opportunity and the AT&T Case* (Cambridge: MIT Press, 1976).

72. "Bias Charges in Hiring: AT&T Fights Back," *U.S. News and World Report* 73 (14 August 1972): 66–68; author interview with William Kilberg, 15 September 1986, Washington, D.C.

73. *Daily Labor Report*, No. 29, F-1 (February 12, 1973).

74. Transcript, White House press conference, 9 June 1970, 4.

75. Freeman, *Politics of Women's Liberation*, 194–202.

76. *Report of the Presidential Task Force on Women's Rights and Responsibilities* (1970), 8, 30.

77. Boles, *The Politics of the Equal Rights Amendment*, concentrates on the ratification process in the state legislatures prior to 1979. Similarly, Gilbert Y. Steiner, *Constitutional Inequality: The Political Fortunes of the Equal Rights Amendment* (Washington: Brookings, 1985), focuses on the critical role of such unanticipated complications as the abortion ruling and Watergate. See also Hoff-Wilson (ed.), *Rights of Passage*, 39–92. Jane J. Mansbridge, *Why We Lost the ERA* (Chicago: University of Illinois Press, 1986), is a fair-minded postmortem by an ERA supporter whose analysis is similar to the political sociology of Boles. Mary Frances Berry, *Why ERA Failed* (Bloomington: Indiana University Press, 1986), is a deftly drawn historical analysis by a lawyer-historian and member of the Civil Rights Commission. Berry concentrates on the lessons of the past provided by the success of the woman's suffrage amendment and especially on the failure between the wars of the child labor amendment, where its supporters won consensus in Washington but fatally failed to build a parallel consensus in the states.

78. 116 *Congressional Record* 18075–78, 91st Cong., 2d sess., 17 October 1970; George, *Griffiths*, 173–81; *Congressional Quarterly Almanac*, 1970, 706–10. In most accounts of congressional passage of the ERA, which are written by ERA supporters, the Bayh-Cook compromise proposal of 1970 is quickly dismissed, much as the ERA lobby dismissed it at the time. Even Berry, whose historical post-mortem thoughtfully explored the unlearned lessons of the failed child labor amendment, ignores the Bayh-Cook compromise as an alternative strategy with greater potential for ratification.

79. *Equal Rights for Men and Women*, Hearings Before Subcommittee No. 4 of the House Judiciary Committee, 92nd Cong., 1st sess., March 24, 25, 31 and April 1, 2, and 5, 1971. This 724–page volume provides unusually full documentation of contemporary arguments and evidence concerning ERA. In addition to the 18 main witnesses, several witnesses made brief appearances, including Andrew Biemiller for

the AFL-CIO and Ruth Miller for the Amalgamated Clothing Workers. Both opposed the ERA and favored H.R. 916, a comprehensive statutory approach to gender equality reported by the House Education and Labor Committee.

80. Emerson was the most prominent legal exponent of the ERA. His most widely cited exposition, written with Barbara A. Brown, Gail Falk, and Ann E. Freedman, was "The Equal Rights Amendment: A Constitutional Basis for Equal Rights for Women," *Yale Law Journal* 80 (April 1971): 871–985. The leading opponents among law scholars were Paul A. Freud of Harvard, who testified against the ERA in Senate committee hearings on the amendment in 1945, and Philip B. Kurland of the University of Chicago. See Freud, "The Equal Rights Amendment Is Not the Way," *Harvard Civil Rights—Civil Liberties Law Review* 6 (March 1971): 234–42; and Kurland, "The Equal Rights Amendment: Some Problems of Construction," *Harvard Civil Rights—Civil Liberties Law Review* 6 (March 1971): 242–50.

81. Transcript, Statement of William H. Rehnquist, Assistant Attorney General, Office of Legal Counsel, on H.J. Res. 208, 1 April 1971, box 21, EX HU 2-5, WHCF, RMNP.

82. Jerris Leonard had taken a similar position when testifying before the House Labor and Education Committee in July 1970 on H.R. 16098, the forerunner of H.R. 916.

83. Gutwillig to the President, 2 April 1971, box 21, EX HU 2-5, WHCF, RMNP.

84. Boles, *Politics of the Equal Rights Amendment*, 103, 139. A succinct congressional perspective is Gary Orfield, *Congressional Power: Congress and Social Change* (New York: Harcourt Brace Jovanovich, 1975), 298–306.

85. *Congressional Quarterly Almanac*, 1971, 656–59.

86. *Reed v. Reed*, 404 U.S. 71 (1971). Chief Justice Burger's opinion held narrowly that states could not pass laws treating men and women differently unless some clear reason was given for doing so.

87. George, *Griffiths*, 179.

88. *Washington Post*, 28 February 1972.

89. *Congressional Quarterly Almanac*, 1972, 199–204.

90. White House press release, Nixon to Scott, 18 March 1972. During the final House debate in the fall of 1971, Nixon aides Garment and Price had drafted a virtually identical letter of support for Nixon to send to Gerald Ford, but Nixon had declined to do so. Ehrlichman to the President, n.d. [September 1972], box 33, Dean file, WHSF, RMNP. Ehrlichman reported that a consensus for sending the Ford letter was shared by Garment, Price, Robert Finch, Donald Rumsfeld, Clark MacGregor, John Dean, and Barbara Franklin.

91. *Public Papers: Nixon, 1972*, 61.

92. The guiding case is *Craig v. Boren*, 429 U.S. 190 (1976). For a critique of the Court's reasoning as "confused" and "muddled" in *Reed, Frontiero v. Richardson, Craig,* and subsequent decisions to sort out the race-sex analogy, see David L. Kirp, Mark G. Yudof, and Marlene Strong Franks, *Gender Justice* (Chicago: University of Chicago Press, 1986), 90–123.

93. See Ruth Ginsburg, "Sex Equality and the Constitution," *Tulane Law Review* 52 (1978): 451–66; Rex E. Lee, *A Lawyer Looks at the Equal Rights Amendment* (Provo: Brigham Young University Press, 1980); and Joan Hoff-Wilson, "The Unfinished Revolution: Changing Legal Status of U.S. Women," *Signs* 13 (Autumn 1987): 16–36.

94. *The Phyllis Schlafly Report*, "What's Wrong With 'Equal Rights' for Women?," February 1972, and "The Right to Be a Woman," November 1972.

Chapter XVII

1. Senate Report No. 1111 on S. 3465, 90th Cong., 2d sess. (May 1968).

2. Gaither to James Frey, 2 January 1969, box 3, Presidential Task Force file, WHCF, LBJL.

3. Gaither Oral History Interview, LBJ Library; author (telephone) interview with James Gaither, 8 July 1982, San Francisco, California.

4. Leonard to Ehrlichman, n.d. [May 1969], box 33, Dean files, WHSF, RMNP.

5. The Civil Rights Commission had first called for a national FEPC with cease-and-desist authority in its report of 1961.

6. Leonard to Ehrlichman, n.d. [May 1969], box 33, Dean files, WHSF, RMNP.

7. Leonard to Kleindienst, 22 February 1969, box 33, Dean files, WHSF, RMNP.

8. Leonard to Ehrlichman, n.d. [May 1969], box 33, Dean files, WHSF, RMNP.

9. Rehnquist to Kleindienst, n.d. [March 1969], box 33, Dean files, WHSF, RMNP.

10. Dean to Kleindienst, n.d. [March 1969], box 33, Dean files, WHSF, RMNP.

11. Burns Report, 82–83.

12. Proposed Message to Congress on "Equal Employment Opportunity," n.d. [April 1969], box A10, Equal Opportunity, Burns papers, GRFL. Burns also opposed transferring the EEO policing authority in federal government employment from the Civil Service Commission to the EEOC.

13. Ibid., 4, 6.

14. Richard T. Burress to Ehrlichman et al., 31 July 1969, box 33, Dean file, WHSF, RMNP. The group included Ehrlichman, Kleindienst, Dean, Leonard, White House aide Robert Brown, EEOC chairman William Brown, and Labor undersecretary James D. Hodgson, but not Rehnquist.

15. Both cease-and-desist and pattern-or-practice authority had been assigned to the EEOC in the Democrats' original House bill of 1963, but had been dropped in the Judiciary Committee negotiations with House Republicans.

16. *Hearings on S. 2453* before the Subcommittee on Labor of the Committee on Labor and Public Welfare, U.S. Senate, 91st Congress, 1st sess., August 11, 12, September 10 and 16, 1969, 3–32.

17. Ibid., 31–32; *Wall Street Journal,* 12 August 1969.

18. Transcript, White House press conference, 6 May 1969; *Wall Street Journal,* 7 May 1969.

19. EEOC Minutes, Meeting #202, 12 May 1969. Tensions between Brown and Alexander grew increasingly disruptive and climaxed at the August 1969 meeting, when Brown ruled Alexander out of order and Alexander stormed out of the meeting. Shortly thereafter Alexander resigned to join the Washington law firm of Arnold and Porter. EEOC Minutes, Meeting #204, 15 August 1969.

20. Author interview with John Ehrlichman, 5 August 1986, Santa Fe, New Mexico.

21. *Hearings on S. 2453,* 41; *New York Times,* 12 August 1969.

22. Ibid., 44–48, 105–8. Alexander countered Brown's argument by testifying that 94 percent of NLRB cases were settled administratively without reaching cease-and-desist procedures.

23. *Hearings on S. 2453,* 106.

24. Vincent Blasi, ed., *The Burger Court: The Counter-Revolution That Wasn't* (New Haven: Yale University Press, 1983), especially Paul Brest, "Race Discrimination," 113–56; Frank Graham Lee, ed., *Neither Conservative Nor Liberal: The Burger Court on Civil Rights and Civil Liberties* (Malabar, Florida: Robert E. Krieger, 1983), especially Jesse Choper, Ray Forrester, Gerald Gunther, and Philip Kurland, "Forum: Equal Protection and the Burger Court," 103–22.

25. Alfred W. Blumrosen, *Black Employment and the Law* (New Brunswick: Rutgers University Press, 1971), 59.

26. Transcript, Luther Holcomb oral history interview, LBJL.

27. Shultz testimony, *Hearings on S. 2453*, 92–97.

28. *New York Times*, 24 September 1970.

29. *Wall Street Journal*, 30 July 1970; *New York Times*, 22 August and 10 October, 1970.

30. Quoted in the *New York Times*, 24 September 1970.

31. *New York Times*, 2 October 1970.

32. *Congressional Quarterly Almanac: 1970* (Washington, D.C.: Congressional Quarterly, 1971), 711–12; *Wall Street Journal*, 1 October 1970.

33. House Report 91-1434, 91st Cong., 1st sess., 21 August 1970, 45.

34. These unpromising circumstances for the bill endowed the congressional deliberations in the fall of 1970 with a desultory quality that was atypical of civil rights debates. The civil rights coalition was hobbled by its internal split, the southern senators made no serious effort to filibuster, and the Nixon administration, having been defeated in committee on its court-enforced approached, quietly encouraged the stalemate in Rules.

35. *New York Times*, 24 September 1970.

36. *New York Times*, 3 December 1970.

37. Brown to Garment, 5 November 1970, box 3, EX HU 2, Equality, WHCF, RMNP.

38. Management Review of the EEOC, Bureau of the Budget, June 1970, Series 69.1, File T2-10/1, EEOC 1970–72, OMB/NARA.

39. Dwight A. Ink to Arnold R. Weber, 27 October 1970, Series 69.1, OMB/NARA.

40. Ink to Weber, n.d., attached to Arnold to Ink, 20 October 1970, Series 69.1, OMB/NARA.

41. Arnold to Frederic V. Mallek, 3 November 1970, Series 69.1, OMB/NARA.

42. Seymour D. Greenstone and Ann C. Macaluso to Ink, 29 December 1970, Series 69.1, OMB/NARA. No record was found in the OMB files or the Nixon presidential papers of Director Shultz's position on this matter.

43. Weber to Ink and Richard P. Nathan, 8 March 1971, Series 69.1, OMB/NARA.

44. William Gifford to the Director, 8 December 1970, Series 69.1, OMB/NARA.

45. *New York Times*, 8 April 1971. H.R. 1746 was co-sponsored by New York Republican Ogden Reid, thus lending it the appearance of bipartisan support. But Reid was the only Republican on the Education and Labor Committee to vote for it.

46. *Wall Street Journal*, 4 May 1971; *New York Times*, 2 June 1971. Nineteen of 20 Democrats on the House Education and Labor Committee supported the Hawkins bill and 11 of 13 Republicans opposed it.

47. Andrew J. Biemiller to State Federations, 30 July 1971, Civil Rights file, Meany Archives. In his Senate testimony Clarence Mitchell complained that "[t]he Labor Department tries to create the impression that only organized labor backs the transfer of the OFCC to the EEOC; that is a conscious effort to conceal the truth. The NAACP backs such a transfer."

48. House Report 238, 92nd Cong., 1st sess., 2 June 1971.

49. In the hearings held by the Senate Labor subcommittee in October 1971, the Rev. Theodore M. Hesburgh testified as chairman of the Civil Rights Commission that the Erlenborn bill's judicial model of enforcement was an advance for the Nixon administration, but one that did not redeem its prohibition of class actions and its provision making Title VII the exclusive federal remedy for job discrimination.

50. *Congressional Quarterly Almanac: 1971* (Washington: Congressional Quarterly, 1972), 644–49.

51. "Minority Views on H.R. 1746," House Report 238, 58–67. The only Democrat on the committee to vote against the Hawkins bill was Romano L. Mazzoli of Kentucky. Thus the Erlenborn-Mazzoli bill, like the Hawkins-Reid bill, were bipartisan in formal co-sponsorship only.

52. Ibid., 62–63.

53. Quoted in *Congressional Quarterly Almanac: 1971*, 648.

54. *Wall Street Journal*, 15 September 1971.

55. *New York Times*, 9 May 1971.

56. "Separate Views of Representative Edith Green of Orgeon," House Report 238, 68.

57. *Congressional Quarterly Almanac: 1971*, 648.

58. The first vote to adopt the Erlenborn substitute came on a recorded teller vote of 200–195, an action that was then challenged and reaffirmed by the crucial roll-call vote of 202–197. A motion to recommit was rejected 130–270, and the House then formally passed the Erlenborn bill, which had become H.R. 1746, by a roll-call vote of 285–106.

59. *Hearings on S. 2512, S. 2617, and H.R. 1746* before the Subcommittee on Labor of the Committee on Labor and Public Welfare, U.S. Senate, 92nd Cong., 1st sess., October 4, 6, and 7, 1971.

60. Ibid., 90–91.

61. *Contractors Association of Eastern Pennsylvania v. Secretary of Labor*, 311 F. Supp. 1002 (E.D. Pa. 1970).

62. *Contractors Association*, 442 F.2d 159 (3rd Circ., 1971) at 173.

63. Ibid., 442 F. 159 at 171. In October 1971 the Supreme Court denied certiorari (*Contractors Association* 404 U.S. 854) and the district court's decision stood affirmed.

64. See Andrew Kahn Blumstein, "Doing Good the Wrong Way: The Case for Delimiting Presidential Power Under Executive Order No. 11246," *Vanderbilt Law Review* 33 (1980): 921–51; Robert P. Schuwerk, "The Philadelphia Plan: A Study in the Dynamics of Executive Power," *University of Chicago Law Review* 39 (Summer 1972): 723–60, 736. Most contemporary law review commentaries tended to defend the Court's logic in *Contractors Association*. See James E. Jones, Jr., "The Bugaboo of Employment Quotas," *Wisconsin Law Review* 40 (1970): 341–55; and Schuwerk, "The Philadelphia Plan," note 8.

65. Laurence H. Silberman, "The Road to Racial Quotas," *Wall Street Journal*, 11 August 1977. One of the administration's chief doubters in 1969, who was successfully persuaded by the discrimination-remedy rationale for the Philadelphia Plan's compensatory quotas, was Assistant Attorney General Rehnquist.

66. *Contractors Association*, 442 F. 2d at 172.

67. *New York Times*, 27 January 1972.

68. See George P. Sape and Thomas J. Hart, "Title VII Reconsidered: The Equal Employment Opportunity Act of 1972," *George Washington Law Review* 40 (1971–72): 824–89. Sape was deputy director of congressional affairs for the EEOC, and Hart was a staff member of the labor subcommittee of the House Committee on Education and Labor.

69. On the Dominick substitution vote, Republicans supported the President 22–13, and Democrats opposed 19–30. But southern Democrats supported the amendment 16–1. The narrowness of the contest was illustrated on February 9 by a tie vote of 33–33, which rejected an Administration-supported amendment proposed by Roman Hruska of Nebraska that would accept the House-passed Erlenborn bill's provisions for EEOC complaints to be the exclusive remedy for persons claiming job discrimination.

70. Sape and Hart, "Title VII Reconsidered," 847.

71. 118 *Congressional Record*, Pt. 2, 92nd Cong., 2d sess., 1656.

72. Ibid., 1657–58 [emphasis added].

73. Ibid., 1659.

74. 118 *Congressional Record*, 92nd Cong., 2d sess., 28 January 1972, 1661–62.

75. *Quarles v. Philip Morris, Inc.*, 279 F. Supp. 505 (E.D. Va. 1968). For a critical analysis of *Quarles* and seniority, see William B. Gould, "Seniority and the Black

Worker: Reflections on *Quarles* and Its Implications," *Texas Law Review* 47 (1969): 1039–60.

76. See Robert Belton, "Title VII of the Civil Rights Act of 1964: A Decade of Private Enforcement and Judicial Development," *St. Louis University Law Journal* 25 (1976): 225–307.

77. Ibid., 1675, 1676.

78. H[arry] R[obbins] Haldeman, Notes from Meetings with the President, 27 January 1972, box 45, Haldeman file, WHSF, RMNP.

79. *Congressional Quarterly Almanac: 1972*, 247; Richard R. Rivers, "In America, What You Do Is What You Are: The Equal Employment Opportunity Act of 1972," *Catholic University Law Review* 22 (Winter 1973): 455–66.

80. *New York Times*, 27 February 1972.

81. Senate Report 681, 6 March 1972; House Report 899, 8 March 1972. The Senate vote on March 6 was 62–10 and the House vote on March 8 was 303–110. With Nixon's signature H.R. 1746 became Public Law 92-261, the Equal Employment Opportunity Act of 1972 (House conferees had agreed to drop the rubric "Enforcement").

82. See generally *The Equal Employment Opportunity Act of 1972* (Washington: Bureau of National Affairs, 1973), a 415–page synopsis of the law's background, congressional evolution, and provisions.

83. Title VII of the Civil Rights Act, Section 706(g), as amended by P.L. 92-261.

84. Schuwerk, "The Philadelphia Plan," 750; Blumstein, "Doing Good the Wrong Way," 946–61; Herman Belz, *Affirmative Action from Kennedy to Reagan* (Washington: Washington Legal Foundation, 1984), 5.

85. Schuwerk, "The Philadelphia Plan," 751–60.

86. *Ex parte Endo*, 323 U.S. 283, 303 n.24 (1944); *Youngstown Steel & Tube Co. v. Sawyer*, 343 U.S. 579 (1952); *SEC v. Sloan*, 436 U.S. 103 (1977); *TVA v. Hill*, 437 U.S. 153 (1977).

87. Michael Brody, "Congress, the President, and Federal Equal Employment Policymaking: A Problem in Separation of Powers," *Boston University Law Review* 60 (1980): 239–305, 299.

88. "Address to the Nation on Equal Opportunity and School Busing," 16 March 1972, *Public Papers: Nixon, 1972*, 425–29; A. James Reichley, *Conservatives in an Age of Change* (Washington: Brookings, 1981), 197–99.

89. *Public Papers: Nixon, 1972*, 436, 437.

90. Nixon to Ehrlichman, 8 April 1972, box 163, Haldeman files, WHSF, RMNP.

91. Ibid.

92. On March 14 Wallace won a plurality of 42 percent of the Democratic presidential primary vote in Florida, more than doubling the vote of his nearest rival, Hubert Humphrey. Wallace then took his campaign northward, placing second in the Wisconsin and Pennsylvania primaries in April and winning more than 40 percent of the vote in Indiana in early May. He also easily won primaries in Alabama, North Carolina, and Tennessee. McGovern and Humphrey mounted last-minute "stop-Wallace" drives in Michigan and Maryland, where primaries were scheduled for May 16. Wallace was shot and permanently crippled while campaigning in Laurel, Maryland, on May 15. He won the Michigan and Maryland primaries the next day, but was forced by his injuries to drop out of the race.

93. Nixon to Ehrlichman, 17 May 1972, box 163, Haldeman files, WHSF, RMNP. In the Detroit school desegregation case, *Milliken v. Bradley*, 418 U.S. 717 (1974), the federal district court in 1972 had joined 53 of Detroit's 85 outlying suburban school districts with the city school system and ordered cross-district busing for 310,000 students. In 1973 the Sixth Circuit Court of Appeals affirmed the busing decree by a vote of 6–3, but in 1974 the Supreme Court reversed by a vote of 5–4, with Burger writing the majority opinion.

94. Ehrlichman to Nixon, 1 June 1972; Morgan to Ehrlichman, 18 May 1972,

box 26, Ehrlichman files, WHSF, RMNP. Ehrlichman entitled his June 1 memo to the President "Selling Our Line: An Interim Report."

95. Transcript, Keynote Address by Bishop Stephen S. Spottswood, 5 July 1972, NAACP Papers, LC; Robert J. Brown to Ehrlichman, 17 July 1972, box 4, EX HU-2 Equality, WHCF, RMNP.

96. Irving Spiegel, "Two Jewish Leaders Score Minority Job Quotas," *New York Times,* 30 June 1972; David A. Schulte to Kenyon C. Burke, 13 July 1972, Anti-Defamation League, Series E, LCCR papers, LC.

97. Philip E. Hoffman to the President, 4 August 1972, box 4, EX HU 4 Equality, WHCF, RMNP.

98. Nixon to Hoffman, 11 August 1972, EX HU-2 Equality, WHCF, RMNP.

99. Colson to Staff Secretary, 26 July 1971, box 3, EX HU-2 Equality, WHCF, RMNP.

100. Robert E. Hampton to Heads of Departments and Agencies, 18 August 1972, attached to James B. Clawson to Tod Hullin, 21 November 1972, box 4, EX HU-2 Equality, WHCF, RMNP.

101. J. Stanley Pottinger to Ehrlichman, 20 November 1972, EX HU-2 Equality, WHCF, RMNP.

102. *Public Papers: Nixon, 1972,* 788.

103. Ibid., 852; *Wall Street Journal,* 5 September 1972.

104. *New York Times,* 14 June 1972. On July 6 at the NAACP convention in Detroit, Herbert Hill savaged the Labor Department for abandoning the Philadelphia Plan model in favor of toothless "hometown" plans.

105. *New York Times,* 16 February 1972.

106. Ibid.

107. In 1974 Congress would pass and President Ford would sign a $585 million authorization for bilingual education, and in 1975 Ford would propose and Congress would enact a far-reaching amendment to the Voting Rights Act to protect language minorities.

108. See Joan Hoff-Wilson, *Nixon Without Watergate* (New York: Basic Books, forthcoming); Freeman, *Politics of Women's Liberation,* 202–5.

109. *Wall Street Journal,* 11 May 1971.

110. Ibid., 25 January 1972.

111. *Business Week,* 24 March 1973, 74–75. Between 1972 and 1974 the EEOC budget rose from $20.8 million (actual) to $43 million, and the Justice Department's budget for the Civil Rights Division increased from $10.7 million to $17.9 million.

Conclusions

1. C. Vann Woodward, *Thinking Back: The Perils of Writing History* (Baton Rouge: Louisiana State University Press, 1986), 59–79.

2. C. Vann Woodward, *Origins of the New South 1877–1913* (Baton Rouge: Louisiana State University Press, 1951).

3. First published by the Oxford University Press in 1955, *The Strange Career of Jim Crow* was republished in revised and expanded editions in 1957, 1966, and 1974. See Howard N. Rabinowitz, "More Than the Woodward Thesis: Assessing *The Strange Career of Jim Crow,*" and Woodward, "*Strange Career* Critics: Long May They Persevere," *Journal of American History* 75 (December 1988): 842–68.

4. Woodward, *Thinking Back,* 75.

5. Woodward's chief targets as "fellow-travelers of the continuity line" on the left were two Marxist historians: Dwight B. Billings, Jr., *Planters and the Making of the "New South"* (Chapel Hill: University of North Carolina Press, 1979), and Jonathan Weiner, *Social Origins of the New South: Alabama, 1860–1885* (Baton Rouge: Louisiana State University Press, 1978). The classic Continuitarian of the left for Wood-

ward was Wilbur Cash, whose *The Mind of the South* (New York: Alfred A. Knopf, 1941) Woodward critically revisited in "W.J. Cash Reconsidered," *New York Review of Books* (4 December 1969): 28–34.

6. See, for example, Marvin E. Gettleman and David Mermelstein, eds., *The Great Society Reader: The Failure of Liberalism* (New York: Vintage Books, 1975).

7. Theodore J. Lowi, *The End of Liberalism* (New York: Norton, 1969), 68–93, 287–97.

8. Mary Frances Berry and John W. Blassingame, *Long Memory: The Black Experience in America* (New York: Oxford University Press, 1982), 194.

9. Harvard Sitkoff, *The Struggle for Black Equality 1954–1980* (New York: Hill & Wang, 1981), 237.

10. Allen J. Matusow, *The Unraveling of America* (New York: Harper & Row, 1984), 198, 206, 211.

11. William H. Chafe, "One Struggle Ends, Another Begins," in Charles W. Eagles, ed., *The Civil Rights Movement in America* (Jackson: University of Mississippi Press, 1986), 147.

12. For a theoretical critique of the ideology of liberal anti-discrimination from the perspective of the critical legal studies, see Kristin Bumiller, *The Civil Rights Society: Antidiscrimination Ideology and the Social Construction of Victims* (Baltimore: Johns Hopkins University Press, 1987).

13. Chafe, "One Struggle," 128.

14. Karl Zinsmeister, "The Kerner Commission—Twenty Years Later: Black Demographics," *Public Opinion* (January-February 1988): 41–44. Between 1968 and 1988 the number of black officials elected nationwide increased six times over, to a total of nearly 7,000. The number of black mayors increased from 48 to 303, and black chief executives were elected in Atlanta, Baltimore, Birmingham, Chicago, Detroit, Los Angeles, New Orleans, Philadelphia, and Washington, D.C.

15. Reynolds Farley, *Blacks and Whites: Narrowing the Gap?* (Cambridge: Harvard University Press, 1984), provides a balanced assessment of trends in black demography in the quarter-century since 1960. See also Reynolds Farley and Walter R. Allen, *The Color Line and the Quality of Life in America* (New York: Russell Sage, 1987).

16. Frank Levy, *Dollars and Dreams: The Changing American Income Distribution* (New York: Russell Sage, 1987), 56–57, 132–41.

17. Farley, *Black and White*, 56–81; William H. Chafe, *The Unfinished Journey* (New York: Oxford University Press, 1986), 432–39; Sitkoff, *Struggle for Black Equality*, 232–37.

18. Farley, *Blacks and Whites*, surveys the literature and the demographic evidence of 1984 and arrives at a cautiously mixed verdict that emphasizes overall black improvement. Consistent black gains in education, occupation, and earnings of the employed marked a burgeoning middle class, but coincided among the underclass with chronologically high unemployment and drop-out rates, persistent residential segregation, increasing urban school segregation, and family deterioration.

19. Levy, *Dollars and Dreams*, 132–41. By 1980 unemployment had risen to 11 percent among white high school dropouts and 21 percent among black dropouts. In large central cities the black unemployment rate often exceeded 40 percent, and responded inelastically to economic trends.

20. The number of blacks in professional, technical, managerial, and administrative positions increased by 57 percent between 1972 and 1982, while the white increase was only 36 percent. Between 1970 and 1980 the number of blacks enrolled full time at American colleges and universities nearly doubled, and blacks recorded a 47 percent increase in home ownership compared with a 30 percent increase for whites. Gary Puckrein, "Moving Up," *Wilson Quarterly* 8 (Spring 1984): 74–87.

21. James P. Smith and Finis R. Welch, *Closing the Gap: Forty Years of Economic Progress for Blacks* (Santa Monica: Rand, 1986), ix. The Rand report noted

that in 1984 constant dollars the typical black male's income increased from $4,500 in 1940 to $19,000 in 1980. Between 1940 and 1980 black male wages increased 52 percent faster than white, moving from 43 percent of the white median to 73 percent.

22. Zinsmeister, "Black Demographics," 44.

23. See William Julius Wilson, *The Declining Significance of Race* (Chicago: University of Chicago Press, 1978), 126–54; Wilson, *The Truly Disadvantaged* (Chicago: University of Chicago Press, 1987), 109–20. Zinsmeister, "Black Demographics," estimated the black underclass in 1988 at 8 million, or 30 percent of the U.S. black population.

24. Glen C. Loury, "Beyond Civil Rights," *The State of Black America 1986* (New York: National Urban League, 1986), 163–74.

25. Ibid., 165. See also Loury, "A New American Dilemma," *New Republic* (31 December 1984): 14–23; Loury, "The Moral Quandary of the Black Community," *The Public Interest* 79 (Winter 1985): 9–22.

26. Thomas Sowell, *The Economics and Politics of Race* (New York: William Morrow, 1983), 199–202. Both Sowell on the right and Williams on the left criticize the racial preferences of affirmative action programs for widening the gap between the black middle class and the underclass.

27. The National Opinion Research Center (NORC) at the University of Chicago began its data bank in 1942. Longitudinal survey data on racial attitudes are also provided by George Gallup's American Institute of Public Opinion (AIPO) and by the Institute of Social Research (ISR) at the University of Michigan.

28. The Gallup Poll questions and the responses by percentage and demographic categories are collected in George H. Gallup, *The Gallup Poll: Public Opinion 1935–1971* (New York: Random House, 1972).

29. Paul B. Sheatsley, "White Attitudes Toward the Negro," *Daedalus* 95 (1966): 217–38.

30. Taylor Garth, Paul B. Sheatsley, and Andrew M. Greeley, "Attitudes Toward Racial Integration," *Scientific American* 238 (June 1978): 42–51.

31. Herbert McCloskey and Alida Brill, *Dimensions of Tolerance: What Americans Believe About Civil Liberties* (New York: Russell Sage Foundation, 1983). While in all surveys tolerance increases with education and socio-economic class (i.e., in 1972 southern whites with greater than 13 years of schooling surpassed in approval of school integration northern whites with less than 11 years), McCloskey and Brill emphasize the generational factor in explaining the postwar growth of American tolerance.

32. Howard Schuman and Shirley Hanchett, *Black Racial Attitudes: Trends and Complexities* (Ann Arbor: Institute of Social Research, 1974). Based primarily on survey research conducted in 1968 for the Kerner Commission, with special attention to Detroit, the Schuman-Hanchett study focuses on rising black alienation.

33. Lawrence Bobo, "Whites' Opposition to Busing: Symbolic Racism or Realistic Group Conflict?" *Journal of Personality and Social Psychology* 45 (1983): 1196–1210.

34. At the height of American concern over domestic communism, sociologist Samuel Stouffer demonstrated the disjunction between principle and practice involving civil liberties in *Communism, Conformity, and Civil Liberties* (New York: Doubleday, 1955). See also James W. Prothro and Charles M. Grigg, "Fundamental Principles of Democracy: Bases of Agreement and Disagreement," *Journal of Politics* 22 (1960): 276–94.

35. David O. Sears, Carl P. Hensler, and Leslie K. Speer, "Whites Opposition to 'Busing': Self-Interest or Symbolic Politics?" *American Political Science Review* 73 (1979): 369–84.

36. Howard Schuman, Charlotte Steeh, and Lawrence Bobo, *Racial Attitudes in America: Trends and Interpretations* (Cambridge: Harvard University Press, 1985), 86–104.

37. Ibid., 147–62. White support for federal intervention to integrate schools peaked at 48 percent in 1966; black support peaked in 1968 at 90 percent, then plummeted 30 percentage points in the following decade.

38. Ibid., 86. See Abigail M. Thernstrom, "How Much Racial Progress?" *The Public Interest* 85 (Fall 1986): 96–100.

39. While in the NORC polls the percentage of Americans favoring school integration rose from 30 in 1942 to 74 in 1970 and to 85 in 1977, the percentage of white parents who would object to their children attending a school where more than half of the children were black also rose—from 39 in 1972 to 45 in 1977.

40. Seymour Martin Lipset and William Schneider, "The Bakke Case: How Would It Be Decided at the Bar of Public Opinion?" *Public Opinion* 2 (April 1978): 38–44.

41. Duke constitutional law scholar William W. Van Alstyne, in "Affirmative Action and Racial Discrimination Under Law: A Preliminary Review," *Selected Affirmative Action Topics* (Washington: U.S. Commission on Civil Rights, 1985), 180–89, identifies five kinds of affirmative action. He argues that the line marking constitutional impermissiblity lies between the fourth kind (selecting among alternative policy choices those likely to be of most significant use to ethnic minority persons, as, for example, deciding to budget $1 million for the public library's reading program rather than for paving downtown streets) and the fifth (indexing individuals by minority status and measuring civil rights and distributing benefits according to that index).

42. A Gallup poll of October 1977, when the Bakke case was receiving wide public attention, found that Americans rejected preferential treatment by a margin of over eight to one (including almost two-thirds of black respondents). "Rarely is public opinion," Gallup observed, "particularly on such a controversial issue, as united as it is over this question." George H. Gallup, *The Gallup Poll: Public Opinion 1972–1977*, vol. II (Wilmington, Del.: Scholarly Resources, 1978), 1059. Gallup's question, however, forced respondents to choose between "preferential treatment" for minorities and women on the one hand, and "ability" as demonstrated by tests on the other. Lipset and Schneider, in "The Bakke Case," 39–41, cite other polls that showed closer divisions depending upon the phrasing of the question, but still demonstrated a consensus against preferential treatment in direct competition for jobs, promotions, and school admissions.

43. The consensus recorded against preferential treatment for minorities has been persistent, although not surprisingly it has been weaker among minorities. A Media General/Associated Press poll of 1988 asked a national sample if they agreed that "blacks and other minorities should receive preference in hiring to make up for past inequalities." The sample disagreed 79 percent to 15, with whites disagreeing 85 to 10, Hispanics disagreeing 64 to 31, and blacks agreeing narrowly by a margin of 44–48. Asked whether they agreed that "blacks and other minorities should receive preference in college admissions to make up for past inequalities," majorities in all three groups disagreed: whites 81–14, Hispanics 55–46, and blacks 47–44 (*The Polling Report*, 22 August 1988).

44. J. Mills Thornton III, commentary on Chafe, "One Struggle Ends, Another Begins," 148–55.

45. Ibid., 149.

46. Jeffrey L. Pressman and Aaron B. Wildavsky, *Implementingation* (Berkeley: University of California Press, 1973); Erwin C. Hargrove, *The Missing Link: The Study of the Implementation of Public Policy* (Washington: Urban Institute, 1975); Eugene Bardach, *The Implementation Game: What Happens After a Bill Becomes a Law* (Cambridge: MIT Press, 1977); Robert T. Nakamura and Frank Smallwood, *The Politics of Policy Implementation* (New York: St. Martin's, 1980); Gary D. Brewer and Peter deLeon, *The Foundations of Policy Analysis* (Chicago: Dorsey, 1983), 249–317.

47. *The Federal Civil Rights Enforcement Effort: A Report to the United States Commission on Civil Rights* (Washington: U.S. Government Printing Office, 1970).

48. Although the label "neo-conservative" did not come into common usage until the middle 1970s, the list of individuals, organizations, and publications associated with the neologism would include the founders of *The Public Interest*—Glazer and Moynihan, Irving Kristol, Daniel Bell, Robert Nisbet, Martin Diamond. A similar group of like-minded intellectuals was associated with *Commentary* and included Norman Podhoretz, Milton Himmelfarb, Walter Laquer, Midge Decter, Paul Seabury, Sidney Hook, Diana Trilling, Edward Shils, Peter Berger, Michael Novak, and Bayard Rustin. A parallel and also overlapping group was associated with the American Enterprise Institute: James Q. Wilson, Seymour Martin Lipset, Richard Scammon, Edward Banfield, Ben Wattenberg. See generally Peter Steinfels, *The Neoconservatives* (New York: Simon & Schuster, 1979).

49. Nathan Glazer, "A Breakdown in Civil Rights Enforcement?" *The Public Interest* (Winter 1971): 107.

50. Ibid., 109. In 1975 Glazer published *Affirmative Discrimination: Ethnic Inequality and Public Policy* (New York: Basic Books, 1975), a book-length critique of the proportional model of affirmative action.

51. Daniel Patrick Moynihan, "The New Racialism," in Moynihan's *Coping: Essays on the Practice of Government* (New York: Random House, 1973), 204–5. Moynihan recalled that the recent enthusiasm of the editors of the *Harvard Crimson* for racial hiring targets on the faculty did not apparently extend to a commensurate goal of racial and ethnic proportionality among the undergraduates, where something like seven out of eight Jewish students would have to leave.

52. Glazer, "Breakdown," 112–13.

53. Confidential memorandum, 9 September 1966, attached to Jackson to Mrs. Cernoria Johnson [director of the Washington bureau of the National Urban League], 28 September 1966, box 53, NUL papers, LC.

54. The citation is from court evidence presented in the case of *Jurgens v. Norton*, a "reverse discrimination" suit brought in federal district court against the EEOC by white male employees of the agency in Texas in 1978. It was decided in 1982 as *Dale H. Jurgens et al., Plaintiffs v. Clarence Thomas and Equal Employment Opportunity Commission, Defendants*, U.S. District Court, Northern District of Texas, Dallas Division. Civil Action No. CA-3-76-1183-6, 9 September 1982, Commerce Clearing House # 33,090, 27,289–324.

55. Quoted in ibid., 27,307.

56. See Douglas Rae et al., *Equalities* (Cambridge: Harvard University Press, 1981), 64–81; Herman Belz, "The Civil War Amendments to the Constitution: The Relevance of Original Intent," *Constitutional Commentary* 5 (Winter 1988): 115–41; Belz, comment on "Equality as a Constitutional Concept," *Maryland Law Review* 47 (Winter 1988): 28–37.

57. In 1971 John Rawls provided a philosophical treatise to justify contractarian obligations for compensating benefits to society's least advantaged members in *A Theory of Justice* (Cambridge, Mass.: Belknap Press, 1971). On the eve of the Bakke decision, Ronald Dworkin published a defense of preferential treatment in *Taking Rights Seriously* (Cambridge: Harvard University Press, 1977), 223–39.

58. Owen Fiss, "The Fate of an Idea Whose Time Has Come: Antidiscrimination Law in the Second Decade after *Brown v. Board of Education*," *University of Chicago Law Review* 41 (1974): 742–70.

59. Kermit L. Hall, *The Magic Mirror: Law in American History* (New York: Oxford University Press, 1989), 187, 297; William N. Eskridge, Jr., "Dynamic Statutory Interpretation," *University of Pennsylvania Law Review* 135 (1987): 1479–1555.

60. See Robert M. O'Neil, *Discriminating Against Discrimination: Professional Admissions and the DeFunis Case* (Bloomington: Indiana University Press, 1975);

Barry R. Gross, ed., *Reverse Discrimination* (Buffalo: Prometheus Press, 1977); Timothy J. O'Neill, *Bakke and the Politics of Equality* (Middletown, Conn.: Wesleyan University Press, 1985); Robert Belton, "Discrimination and Affirmative Action: An Analysis of Competing Theories of Equality and *Weber*," *North Carolina Law Review* 59 (1981): 531–98. A superior exploration of the topic's legal, philosophical, and ethical dimensions is Kent Greenawalt, *Discrimination and Reverse Discrimination* (New York: Knopf, 1983).

61. Matthew A. Crenson and Francis E. Rourke, "By Way of Conclusion: American Bureaucracy since World War II," in *The New American State: Bureaucracies and Policies since World War II,* ed. Louis Galambos (Baltimore: Johns Hopkins University Press, 1987), 137–77.

62. Samuel P. Huntington, "The Democratic Distemper," *The Public Interest* 41 (Fall 1971): 13.

63. During the 1960s the amount of federal financial aid received by the subnational governments increased by 230 percent, from $7 billion in 1960 to more than $23 billion in 1970. Crenson and Rourke, "American Bureaucracy," 148.

64. Hugh Heclo, "Issue Networks and the Executive Establishment," in Anthony King, ed., *The New American Political System* (Washington: American Enterprise Institute, 1978), 89–90.

65. Crenson and Rourke, "American Bureaucracy," 151.

66. David Vogel, "The 'New' Social Regulation in Historical and Comparative Perspective," in Thomas K. McCraw, ed., *Regulation in Perspective* (Cambridge: Harvard University Press, 1981), 155–86.

67. In *The Rise of the Corporate Commonwealth* (New York: Basic Books, 1988), Louis Galambos and Joseph Pratt distinguish between rate-of-return regulation, which exercised vertical, industry-by-industry control in sectors like transportation and utilities, and cross-industry regulation. The latter began in areas like the tariff and antitrust policy, and evolved toward an industry-wide pattern, for example in labor relations and the NLRB, that would characterize the social regulators of the 1960s like the EPA.

68. For a comparison of the new social regulation and the traditional economic regulation that does not include civil rights as a field of regulation, see Frederick R. Anderson, "Human Welfare and the Administered Society: Federal Regulation in the 1970s to Protect Health, Safety, and the Environment," in William N. Rom, ed., *Environmental and Occupational Medicine* (Boston: Little, Brown, 1983), 835–64.

69. Gary C. Bryner, *Bureaucratic Discretion: Law and Policy in Federal Regulatory Agencies* (New York: Permagon, 1987), 19–90.

70. It should be noted that in practice there was more overlap between the social and economic models of regulation than this bi-polar treatment suggests—i.e., all regulatory bodies exercised some measure of rule-making authority, and none performed in a purely zero-sum or redistributionist environment.

71. Howell John Harris, *The Right To Manage: Industrial Relations Policies of American Business in the 1940s* (Madison: University of Wisconsin Press, 1982), 111–27; James A. Gross, *The Reshaping of the National Labor Relations Board: National Labor Policy in Transition 1937–1947* (Albany: State University of New York Press, 1981), 151–71; Harris, "The Snares of Liberalism? Politicians, Bureaucrats, and the Shaping of Federal Labour Relations in the United States, ca. 1915–47," in Steven Tolliday and Jonathan Zeitlin, eds., *Shop Floor Bargaining and the State: Historical and Comparative Perspectives* (Cambridge: Cambridge University Press, 1985), 177–78.

72. Anderson, "Human Welfare and the Administered Society," 850–51.

73. David M. Welborn, *Governance of Federal Regulatory Agencies* (Knoxville: University of Tennessee Press, 1977).

74. Paul Sabatier discusses the new clientele capture, through case studies that emphasize air pollution, as a constituency-based counterweight to regulatory decay in

"Social Movements and Regulatory Decay: Toward a More Adequate—and Less Pessimistic—Theory of Clientele Capture," *Policy Sciences* 6 (1975): 301–42.

75. Bryner, *Bureaucrati Discretion,* 203–18. Bryner's case studies are the EPA, OSHA, the CPSC, and the FDA.

76. Political Scientist James Q. Wilson provides a thoughtful exploration of the "capture" literature and its political implications in his conclusion to James Q. Wilson, ed., *The Politics of Regulation* (New York: Basic Books, 1980), 357–94.

77. In *Regulating Business by Independent Commission* (Princeton: Princeton University Press, 1955), Marver Bernstein developed the pessimistic model of capture and regulatory decay, whereby the regulated industries captured the regulators, and the injured third parties (consumers and taxpayers) typically lapsed into unorganized victimhood. See also Barry M. Mitnick, *The Political Economy of Regulation* (New York: Columbia University Press, 1980), 173–204.

78. Paul Burstein, *Discrimination, Jobs, and Politics* (Chicago: University of Chicago Press, 1985), 97–124; Jeremy Rabkin, "Office for Civil Rights," in Wilson, ed., *Politics of Regulation,* 304–53.

79. Samuel P. Hays, "The Politics of Choice in Regulatory Administrations," in *Regulation in Perspective,* ed. Thomas K. McCraw (Cambridge: Harvard University Press, 1981), 124–54.

80. Richard K. Scotch, *From Good Will to Civil Rights: Transforming Federal Disability Policy* (Philadelphia: Temple University Press, 1984); Edward D. Berkowitz, *Disabled Policy* (Cambridge: Cambridge University Press, 1987).

81. Hugh Heclo has argued that in highly technical policy areas like arms control, energy, health, and monetary policy, the iron triangle has been supplanted by "issue networks," where technocratic policy entrepreneurs ("technopols") determined policy through their expertise. Heclo, "Issue Networks," 98–115.

82. Francis E. Rourke, "Bureaucracy in the American Constitutional Order," *Political Science Quarterly* 102 (Summer 1987): 25–28.

83. Martin M. Shapiro, "The Presidency and the Federal Courts," in *Politics and the Oval Office,* ed. Arnold J. Smeltsner, (San Francisco: Institute for Contemporary Studies, 1981), 151–52.

84. Abram Chayes, "The Role of the Judge in Public Law Litigation," *Harvard Law Review* 89 (May 1976): 1282–84, 1302–04.

85. Ibid., 1304, 1302.

86. Donald L. Horowitz, *The Courts and Social Policy* (Washington: Brookings, 1977), 4–19.

87. Thomas Sowell, "*Weber, Bakke,* and the Presuppositions of 'Affirmative Action,' " *Wayne State Law Review* 26 (1980): 1309–36; Sowell, *Civil Rights: Rhetoric or Reality* (New York: Frederick Morrow, 1984); Herman Belz, *Affirmative Action from Kennedy to Reagan: Redefining American Equality* (Washington: Washington Legal Foundation, 1984).

88. Donald L. Horowitz, *The Jurocracy: Government Lawyers, Agency Programs and Judicial Decisions* (Lexington, Mass.: Lexington Books, 1977).

89. *Griggs v. Duke Power Company,* 401 U.S. 424 (1971) at 430–31.

90. *Regents of the University of California v. Bakke,* 438 U.S. 265 (1978) at 403.

91. *A Conversation with Commissioner Eleanor Holmes Norton* (Washington: American Enterprise Institute, 1979), 21.

92. *United Steelworkers of America v. Weber,* 443 U.S. 193 (1979) at 216.

93. James Q. Wilson, ed., *The Politics of Regulation* (New York: Basic Books, 1980), 357–94.

94. Cynthia Harrison, *On Account of Sex: The Politics of Women's Issues 1945–1968* (Berkeley: University of California Press, 1988), 192–209.

95. Hugh Davis Graham, "Civil Rights and the Irony of the Race-Sex Linkage: 1964–1972," paper delivered at the annual meeting of the Southern Historical Association, New Orleans, Louisiana, 14 November 1987.

96. William H. Chafe has pioneered in exploring the interplay of race, class, and gender in modern American life, most notably in *Women and Equality* (New York: Oxford University Press, 1977).

97. In 1873 the U.S. Supreme Court upheld the denial by Illinois of Myra Bradwell's application for a license to practice law. Justice Bradley's in his concurring opinion observed that the "natural and proper timidity and delicacy which belongs to the female sex evidently unfits it for many of the occupations of civil life." *Bradwell v. State,* 83 U.S. (16 Wallace) 130 (1973).

98. Burstein, *Discrimination, Jobs and Politics,* 191–92.

99. Ibid., 150, 137–38. Burstein points out that by the 1980s the historic white male advantage remained enormous, although the relative group income of white men had declined from 175.5 percent of parity in 1953 to only 156.2 percent in 1978. The earnings of *non*white men as a percentage of 1953–78 parity rose from 86.9 in 1962 to 98.6 in 1974; the percentage for nonwhite women rose from 28.5 in 1955 to 52.1 in 1978; and the percentage for white women rose from 38.5 in 1953 to 55.7 in 1978. Thus nonwhite women started farthest behind and made the greatest gains, but the large bloc of white women made the greatest gains in group share.

100. Conservative and market-oriented critics have argued that government welfare and affirmative action programs cripple their intended beneficiaries by encouraging dependency rather than the ability to compete for real economic benefits. See Thomas Sowell, *Ethnic America* (New York: Basic Books, 1981); Walter E. Williams, *The State Against Blacks* (New York: McGraw-Hill, 1981); and Charles Murray, *Losing Ground* (New York: Basic Books, 1984).

101. Loury, "Moral Quandary," 11–14; Wilson, *The Declining Significance of Race,* 1–23, 122–82; Wilson, *The Truly Disadvantaged,* 109–39, 146–49.

102. This has generally been recognized with Asian Americans, who have been excluded from many nonfederal affirmative action programs, even though they share a long history of severe racial discrimination that provided a chief rationale for compensatory minority preferences in the first place.

Index

Administrative Procedures Act of 1946, 130
administrative state, American, 3, 343, 462, 463
affirmative action, 4, 6, 187, 188, 190, 197-200, 247, 290, 295, 296, 352, 472, 475; constitutional basis for, 331; enforcement by EEOC, 459, 460; Executive Order 10925, 40-42; minority hiring goals, 330; origin of phrase, 28, 33, 34; plans for women, 413; quotas, 341, 343; regulations of 1968, 326
AFL-CIO, 11, 22, 36, 48, 115, 208, 231; effort to move OFCC from Labor Department to EEOC, 431; opposition to ERA, 208, 407; opposition to Philadelphia Plan, 290, 325, 341; position on Title VII, 139-40; segregated locals, 103, 278; support for FEPC, 82, 83
age discrimination, 40, 97, 138, 158
Alexander, Clifford L., Jr., 238, 319
Alexander v. Holmes County, 382
Allen v. State Board of Elections, 359, 360, 377, 378, 381
American Jewish Committee, 446
American Newspaper Publishers Association (ANPA), 216, 218-20
antidiscrimination suits, 338
apprenticeship programs, 114, 115, 280, 281, 344
armed forces, 86
AT&T, 414, 415

Baker v. Carr, 378, 381
Bakke decision, 471
Bay Area Rapid Transit system (BART), 286
benign neglect memorandum, 364
Berg, Richard K., 189, 195
Berger, Morroe, 118
Berry, Mary Frances, 451

Bethlehem Steel Co., 414
BFOQ exceptions, 215, 216
Biemiller, Andrew, 140, 208
bilingual education, 110
Black, Hugo, 379
black attitudes, 455
black capitalism, 313-16
black gains, 452, 453
black middle class, 310, 311
black underclass, 310, 312, 453, 454
black unemployment, 60, 100
black-Jewish split, 310, 311, 391
Blackmun, Harry, 471-72
Blassingame, John, 451
Blumrosen, Alfred W., 118, 119, 194, 195, 220, 221, 431
Bolling decision, 368
boycotts, 105
Brotherhood of Locomotive Firemen and Enginemen, 102
Brown, Henry Billings, 367
Brown, William H., III, 428-30, 433
Brown v. Board of Education, 366-69
Brown II, 371, 372
Bryant, Robert E., 149
Buchanek, Jack, 203, 204
bumping rights, 252
Bureau of Apprenticeship and Training (BAT), 114
Bureau of the Budget, 154, 155, 157-59, 203
Bureau of Labor Statistics, 101, 240
Burger, Warren, 382, 383, 386, 387, 389
Burns, Arthur: controversy with Moynihan, 309, 313; domestic policy task forces, 304, 306; equal employment opportunity enforcement policy, 426; voting rights policy, 349, 350; women's issues, 397, 398, 400, 401, 406
Burns report, 304, 305, 396

Burstein, Paul, 475
busing, 320, 374, 382, 383, 445, 446
Button, James W., 273
Byrd rider, 339, 340

Cabinet Committee on Opportunity for the
  Spanish Speaking (CCOSS), 317, 318
Califano, Joseph, 183, 187, 260, 262, 263
Carey, Gordon, 106
Case, Clifford P., 150, 387
cease-and-desist authority, 21, 98, 467; for
  EEOC, 132, 133, 149, 189, 193, 255-58,
  420-34, 439, 443; NLRB, 33, 34; state
  FEPCs, 22
Celler, Emmanuel: hearings on Civil Rights
  Act of 1964, 89, 98, 99, 125-28, 135,
  137; open housing legislation, 267; voting
  rights legislation, 354, 355
Chafe, William, 452
Chayes, Abram, 470
Citizens Advisory Council on the Status of
  Women (CACSW), 222, 394, 395, 406
Civil Rights Act of 1866, 375, 376
Civil Rights Act of 1957, 34
Civil Rights Act of 1964, 6, 152, 296; Title
  II, 90-93, 128, 146; Title III, 127-129,
  133; Title VI, 82, 83, 140, 147, 156, 160,
  183, 282, 405, 410; Title VII, 133, 136-
  40, 146, 147, 150, 151, 156, 158, 192,
  194, 212-15, 246-51, 383-88, 404, 405,
  421, 422, 424, 427
civil rights bureaucracy, 363
Civil Rights Commission, 63, 64, 66, 157,
  185, 317, 319, 457-59
Civil Rights Division, Department of Justice,
  236, 237, 362, 403, 421, 422
civil rights enforcement, 155, 156, 239; in
  construction industry, 279; Humphrey
  proposal, 181; judicial strategy, 430, 431;
  Katzenbach plan, 184-86
civil rights movement, 3, 4, 450-52
civil rights regulation, 469
Clairborne, Louis F., 168, 169
Clapp, Charles, 406
Clark, Joseph S., 150, 388
Clark, Ramsey, 263, 264, 268, 269
class divisions: among blacks, 276, 310,
  453, 476; in women's movement, 205,
  473
classified advertisements, 214-21, 229
Cleveland Plan, 295, 296
clientele capture, 467, 468
Colmer, William M., 433
commerce clause authority for public accommodations legislation, 80, 81, 91, 92
Communications Workers of America, 415
Community Relations Service, 185
compensatory theory, 4, 5, 105, 110-12,
  117, 352, 390, 444, 461; applied to

women, 413, 415; school desegregation,
  370, 373-75
complainant model. *See* retail model of enforcement
Congress of Racial Equality (CORE), 104-6,
  109, 117
Connor, Eugene T. ("Bull"), 74
construction grants, 65-67
construction industry, 278, 279, 281, 284,
  285, 288
Construction Industry Joint Council, 114
construction trades, 102. *See also* craft
  unions
continuitarianism, 450
contract compliance, 42, 48, 52, 53, 57, 59,
  68, 181, 182, 184, 188, 194, 283-85, 289,
  290, 328, 329, 413, 414
*Contractors Association of Eastern Pennsyl-
  vania v. Secretary of Labor,* 341, 437, 438
contractors' associations, 329
Conyers, John, 261
Corps of Engineers, New Orleans District,
  107
court enforcement of civil rights, 155
court enforcement of equal employment opportunity. *See* judicial strategy
court-ordered policies, 471
Cousens, Frances P., 212
Cox, Archibald, 168, 169
craft unions, 280, 281, 287-89, 294, 323,
  325, 327, 341, 438
Cramer, William, 294, 295
Cuban-Americans, 328

defense contractors, 49-52, 54, 56
Department of Defense, 48, 68, 283, 284
Department of Housing and Urban Development, 285
desegregation, 5, 49, 56, 59, 63, 65, 66, 75,
  133
Dirksen, Everett: amendment to Title VII,
  248, 388; Civil Rights Act of 1964, 141,
  142, 144-47; compromises, 148, 151, 273,
  274; equal employment opportunity en-
  forcement policy, 193, 194; Nixon admin-
  istration, 319, 329, 336, 351; open-
  housing legislation, 271
discriminatory intent, 244, 245, 248, 249
disparate impact. *See* proportional representation model
Doar, John, 320
Domestic Council, 309
Dominick-Erlenborn bill, 437, 439
Duke Power Co. See *Griggs v. Duke Power
  Co.*
Dukes, Nelson, 106
Duncan, Charles T., 244
Dunn, William E., 291
Dutton, Fred, 64
Dwyer, Florence, 398-400

East, Catherine, 394, 396, 400
Eastland, James, 144, 145
Eastwood, Mary, 222
economic discrimination, 100, 101, 120
economic model of regulation, 465
Edelsberg, Herman, 197, 198, 217
Edwards, Donald, 355, 356
effects test, 250, 254
Eisenhower, Dwight D., 16-18
employment qualification tests, 150, 151, 248, 251, 383-86, 390
employment statistics, 240-42, 244, 245
equal employment law, 383
Equal Employment Opportunity Act of 1972, 443-45
Equal Employment Opportunity Commission (EEOC), 97, 98; adoption of wholesale model of enforcement, 250; as affirmative action model, 459, 460; chairman search, 179, 237, 238; complaint backlog, 203, 235, 236, 239; coverage of state and local government, 269; delay in appointing commissioners, 177-79; Dirksen attempts to limit Title VII powers, 146-48; discriminatory intent test, 248; growth during Nixon administration, 448; 1966 guidelines, 386, 388, 389; hearings, 241, 242; Katzenbach recommendations, 257; lack of enforcement powers, 157-60, 234, 253, 276; move to enforce sex anti-discrimination, 228, 230, 231, 402; national reporting system, 193-97, 240; Newport News shipyard settlement, 243; NLRB model, 129, 131, 133, 134; object of feminist ire, 223-25; Pollack task force proposals, 265, 266; proposed expansion of enforcement authority, 421-35; reluctance to enforce sex nondiscrimination provision of Title VII, 212-20; retail model of enforcement, 189-93; role in Humphrey proposal, 182; seniority systems, 252; sex discrimination cases, 402
equal employment opportunity enforcement, 281, 458-60; legislation, 435, 436, 439; policy debate, 420-33
equal impact model. *See* proportional representation model
Equal Rights Amendment (ERA), 136, 137, 206-8, 393-96, 391, 415-19; effect on state protective laws, 406, 407; Nixon administration position, 408, 410, 411; women's task force report recommendations, 403, 404
equitable relief, 370, 374
equitable remedy, 461
equity jurisprudence, 371, 372
Erlenborn bill, 434, 436, 437
Erlichman, John, 309, 446
Ervin, Sam: debate on Equal Employment Opportunity Act of 1972, 439-42, 444;

Equal Rights Amendment, 416, 417; omnibus civil rights bill, 262; Philadelphia Plan, 336-38, 342; Senate Judiciary Committee hearings, 94, 95
ethnic groups, 317
executive authority, 65, 66, 67
executive branch, 7
Executive Order 10925, 27, 28, 35, 40, 41
executive order strategy, 27, 31
executive orders, 8, 9, 16, 157, 158, 181, 188, 259

Fair Employment Board, 4, 15
fair employment policy, 76, 83-85, 265
Fair Employment Practice Committee (FEPC): first, 10, 11; permanent FEPC legislation, 15, 19; proposal for permanent national FEPC, 82, 83, 95; second, 12-14; state and local FEPCs, 19, 21, 22, 95, 96, 117-19, 131, 134, 148-49, 193, 196
fair employment regulation, 129-31, 134, 140
fair housing. *See* open housing policy
Fair Housing Act of 1968, 375
Fair Labor Standards Act, 405
Farmer, James, 104, 108, 109, 117, 190, 203
federal contractors, 290, 291, 293-95, 333, 342, 343
federal contracts, 8, 36, 40-42, 47, 158, 287, 292, 296
federal courts, 370, 372, 373, 426, 461-71
federal employment, 18, 184, 333; of blacks, 60-62, 71, 72; of hispanics, 317
Federal Executive Boards (FEBs), 287, 288
federal grants, 37-40, 63-65, 82, 156, 263
Federal Housing and Home Finance Agency, 107
Feild, John G., 48, 52-54, 58
feminism, 222, 472
feminist agenda, 395, 396, 398
feminist leadership, 474
feminist legislation, 448
feminists, 227, 229-31, 418
filibuster, 143, 151
Fletcher, Arthur A., 326, 327, 334, 335
flight attendants, 232
Floyd v. Florida, 391
Focke, Arthur B., 159
Foley, William, 109
Fortas, Abe, 35
Fourteenth Amendment, 369; and ERA, 416; authority for public accommodations legislation, 80, 81, 87, 92, 93
Freeland, Wendell G., 111
Freund, Paul A., 92
Friedan, Betty, 225, 226

Garment, Leonard, 315
Gaston County v. United States, 333, 352, 353, 360, 377, 381

General Accounting Office (GAO), 291-96, 329-32
Gesell report, 86
Glaser, Vera, 396-401, 410, 415
Glazer, Nathan, 457-59
Goesaert v. Cleary, 391
Goldberg, Arthur, 35
Government Contract Committee, 17, 18
Government Contract Compliance Committee, 16
Graham, Richard, 178, 202, 216, 220, 221, 225, 226
Great Society legislation, 153, 175, 233, 234, 263, 276
Green, Edith, 136, 137, 212
Green decision, 374
Greenberg, Jack, 189, 252, 385
Griffiths, Martha, 137, 223-25, 406, 416, 418
Griggs v. Duke Power Co., 383-90, 466, 471
Griswold, Erwin, 382
Gutwillig, Jacqueline, 418

Haldeman, H. R., 307, 309
Harlan, John Marshall, Jr., 352, 366, 376, 379-81
Harlow, Bryce, 307
Haselden, Kyle, 112, 113
Hawkins bill, 255, 256, 265, 434-36
Hernandez, Aileen C., 178
HEW guidelines, 264, 374
Hill, Herbert, 47-49, 54-56, 117-19, 149, 344
Hill, Lister, 139, 140
Hill, Samuel, 118
Hispanic leaders, 226
Hispanic minorities, 110, 316-18
Holcomb, Luther, 218-20, 223, 431
Holleman, Jerry, 48, 57
House leadership compromise of 1963, 132, 133
House Rules Committee, 134, 135
housing discrimination. *See* open housing policy
Humphrey, Hubert: on sex discrimination, 209; President's Council on Equal Opportunity, 161, 162, 175, 180-83, 185, 186; Senate Majority Whip, 143, 150, 151

indemnification. *See* compensatory theory
institutionalized racism, 120, 191
intent test, 250
interagency committees, 44, 45
Interdepartmental Committee on the Status of Women (ICSW), 212, 222, 394
interest-group lobbying, 469-70
International Association of Machinists (IAM), 47, 48
International Longshoremen's Association (ILA), 103

iron quadrilaterals, 470
iron triangles, 38, 362-65
Ives-Quinn Act, 19

Jackson, Samuel C., 178, 194, 195, 237, 249
Jane Crow, 222
Javits, Jacob, 56
Jim Crow. *See* segregation
Job Corps, 211
job discrimination, 19-21, 96, 98, 148, 150, 190, 244
job discrimination suits, 383-88
Johnson, Lyndon B., 4, 6, 22-24, 31-33, 35, 475; chairman of PCEED, 38, 39, 46, 61, 62; choice of EEOC chairman, 201; civil rights legislation (1966–1968), 256-58, 260, 267; exclusion from policy meetings as vice president, 76-78; executive order to end PCEO, 180, 181; and feminists, 227; Great Society legislation, 233, 234; Howard University commencement address, 174, 175; and Robert Kennedy, 67-73; legislative agenda (1964–1965), 153, 162, 166, 170; as new president, 134; Open Housing Act of 1968, 270, 272; pact with Robert Kennedy to pass Civil Rights Act of 1964, 141, 142; on racial quotas, 110
Jones v. Alfred H. Mayer Co., 375, 376
judicial strategy of equal employment opportunity enforcement, 131, 426, 427, 435, 443, 475
jury discrimination, 245, 246

Katzenbach, Nicholas, 35, 65; interagency task force on civil rights legislation, 256-58, 260; legislative strategy for civil rights bills, 87, 99, 143; plans for civil rights enforcement, 184-86; Voting Rights Act of 1965, 163, 164, 166
Kennedy, John F., 4, 6; civil rights policy, 31; civil rights speech of June 11, 1963, 74, 75; Executive Order 10925, 28, 40; fair employment policy, 83-85; legislative proposals, 65, 69, 70, 79-82; PCEED, 27; Plans for Progress, 58; racial quotas, 106
Kennedy, Robert F., 65; and Emmanuel Celler, 127-29; and Lyndon Johnson, 67, 69-73; and William McCulloch, 88, 89; pact with Lyndon Johnson to pass Civil Rights Act of 1964, 141; testimony on administration civil rights bill of 1963, 93, 95, 106-8
Kerner Commission, 276
Keyserling, Mary Dublin, 394
Kheel, Theodore, 57, 58
King, Martin Luther, Jr., 74, 164, 165, 170, 271-73, 452
Koontz, Elizabeth, 409, 410, 415
Kuck, Elizabeth, 402

labor unions, 36, 37, 40, 102, 115, 278, 280, 323, 338; segregated locals, 103, 285, 286, 288; seniority systems, 252, 253

Leadership Conference on Civil Rights, 15, 19, 95, 103, 133, 363

Legal Defense Fund, 236, 251

legislative strategy: Johnson administration, 143, 144, 171, 260-62; Kennedy administration, 65, 76-78, 87, 90, 125; Nixon administration, 360-62, 439, 440, 443

Leonard, Jerris, 337, 342, 349, 353, 421-24

liberal agenda on civil rights, 83, 84

liberal reform, 116-18, 120, 451, 456

liberalism, 4, 5, 9; male-centered, 209, 210; on racial quotas, 198, 200

Lindsay, John, 88, 89

Lipset, Martin, 456

literacy tests, 167, 168, 347, 350, 351, 353-55, 362

Lockheed-Marietta, 47, 48, 49

Lowell, Malcolm R., 403

Lowi, Theodore, 451

Lowman, Lawrence, 111

Lowry, Glen C., 454

Macy, John, 71, 72, 177, 178, 199, 201, 202, 237, 291

manning tables, 286, 289, 290, 293, 295, 329

Marshall, Burke, 77, 88, 90

Marshall, Thurgood, 370

Marshall Plan for blacks, 111-13

Maslow, Will, 96

Massachusetts Commission Against Discrimination (MCAD), 119

Matusow, Allen, 451

Mayhew, Leon, 118, 119

McClendon, Sara, 410

McCulloch, William, 87, 88, 98, 125, 127, 135, 172, 272, 358, 359

McNamara, Robert, 86, 87

McNutt, Paul V., 11

Meany, George, 83, 110, 208, 341

Merrick, Samuel, 101

Mexican-Americans, 227

Miller, Herman P., 101

minority groups, 328, 329

Mintz, Benjamin, 403

Mitchell, Clarence, 199, 272

Mitchell, John, 319, 320, 333, 354-56

Model Cities Program, 294

Mondale, Walter, 270, 275

Morse, Wayne, 144

mortgages, federally insured, 259

Motorola Corporation, 149

Moyers, Bill, 154, 162, 179, 183, 186

Moynihan, Daniel Patrick, 306, 309-13, 364, 399, 400

Moynihan Report on the Negro family, 183, 209, 210

Moynihan Report to President Nixon, 309-12

Mrs. Murphy exemption, 270, 271, 274, 375, 376

Mrs. Murphy's boardinghouse, 89

Murray, Pauli, 222

Myart, Leon, 149

Nathan report, 319

National Advisory Commission on Civil Disorders, 272

National Association for the Advancement of Colored People (NAACP), 15, 55, 104, 110, 117, 190, 208, 278, 325, 411, 446

National Labor Relations Board (NLRB), 11, 20, 33, 34, 103, 130

National Organization of Women (NOW), 212, 226-28, 391, 406

national reporting system, 193-97, 240, 241

National Urban League, 111-13

National Women's Party (NWP), 136, 137, 206

Negro image, 234

Neuberger, Maurine, 394, 395

New Deal, 9, 10

New Jersey Civil Rights Commission, 118

New York teachers' strike, 310

Newport News Shipbuilding and Drydock Company, 220, 221, 243

Nicklis, John, 116

Nitze, Paul, 283

Nixon, Richard M., 4, 475; civil rights legacy, 447, 448; civil rights policy as president, 305-9, 313, 314, 319-21, 364, 443; civil rights policy as presidential candidate, 302-4; Philadelphia Plan, 322, 325, 340; position on feminist issues, 398, 408, 411, 419; school busing moratorium, 445, 446; vice president, 17, 18, 29, 30, 37

nondiscriminating, 9, 10, 16, 28, 35, 40-42, 51, 52, 346, 370, 390

Norgren, Paul, 18

Norris v. Alabama, 245

Norton, Eleanor Holmes, 472

Office of Federal Contract Compliance Programs (OFCC): attempt to transfer to EEOC, 296, 324-27, 421; enforcement activities, 282-87, 291; enforcement powers, 188, 196, 197; hearings on Philadelphia Plan, 330, 338, 342, 343, 438; sex discrimination, 402, 403, 410

Office of Minority Business Enterprise (OMBE), 314-16

Oil, Chemical and Atomic Workers International Union, 56

omnibus civil rights bill of 1966, 260-62

Open Housing Act of 1968, 270-72

open housing policy, 258, 269, 261-65, 267, 269, 276, 375, 376

opinion surveys, 455, 456

Order No. 4, 342, 343, 409
organized labor. *See* AFL-CIO

pattern-or-practice suits. *See* judicial strategy of equal employment opportunity enforcement
Peabody, Mrs. Malcolm, 145
Pemberton, John, 385
pension rights, 232
"permanent government," 7
Petersen, Esther, 137, 206, 207, 222, 394
pharmaceutical industry, 241
Phelan, Warren P., 287-90
Philadelphia Plan, 287-97; congressional opposition, 335-39, 342, 343; court challenges, 341; GAO opposition, 329-34, 339; legislative debate, 431, 432, 436-38, 442; political liability to Nixon, 346; proportional representation model, 326-28; support by Nixon, 340
Pitney-Bowes, 116
Plans for Progress, 46, 51-59, 68, 196
Plessy v. Ferguson, 366, 367
Poff, Richard, 358
political participation, 348, 356, 357
politics of implementation, 457
poll tax, 171-73
Powell, Adam Clayton, 95, 110, 256
Powell amendment, 76, 82
pre-award programs, 283-86, 289. *See also* Cleveland Plan; Philadelphia Plan
preclearance requirement, 169, 347, 356-59, 362, 380, 381
preferential hiring, 105, 112, 113, 115, 116, 245
preferential treatment, 456, 457
President's Commission on the Status of Women (PCSW), 137, 206, 207
President's Committee on Equal Employment Opportunity (PCEED), 39-46, 51, 53, 55, 57-65, 84, 158-62, 181-84, 279
President's Committee for Equal Opportunity in Housing, 184
President's Committee on Equal Opportunity in the Armed Forces, 86
President's Committee on Government Contracts, 29-31
President's Committee on Government Employment Policy, 18
President's Council on Equal Opportunity, 161, 162, 180-82
presidential authority. *See* executive orders
presidential election of 1964, 162
Presidential Taskforce on Women's Rights and Responsibilities, 400-407
Pressman, Sonia, 244-47
program deregistration, 115
progressivism, 206
Project C, 74
proportional representation model, 120, 224, 321, 327, 472; as basis of affirmative action policy, 390, 446, 447; as basis of compensatory justice, 370, 374; in Philadelphia Plan, 341, 343, 344; model for EEOC, 247, 248
protected classes, 232, 368-70, 374
protectionist laws, 206, 208, 212, 213, 227-31, 395, 404, 407, 409, 410, 446
protest demonstration, 106, 142
public accommodations legislation, 79, 81, 89
public opinion, 202

Quarles v. Phillip Morris, 386, 442
quotas: in apprenticeship programs, 114, 115; congressional debate, 1963, 106-9; Dirksen amendments to Title VII, 147, 150, 151; organized labor opposed to federal imposition of, 139, 140; used against blacks, 102, 103; for women, 411-13

race relations, 309-11
race-sex analogy, 3, 218, 220, 222, 223, 228, 232, 411-14, 472-74
racial prejudice. *See* opinion surveys
racial quotas. *See* quotas
racial statistics, 60-62, 138, 191, 194-200
racial violence, 142; Birmingham (1963), 74; Philadelphia, 278; protest demonstrations, 334, 335; riots, 268, 272, 273; Selma, 164, 165; University of Mississippi (1962), 75; Watts, 175
Randolph, A. Philip, 10
Rauh, Joseph, 167
reapportionment, 378, 379, 381
Reed v. Reed, 418
Reedy, George, 39, 43, 46
Rehnquist, William, 417, 418, 424, 425
retail model of enforcement, 189, 235
Reuther, Walter, 110
reverse discrimination, 5, 6, 107, 108, 112, 120, 332, 337, 390, 391, 442, 444, 460, 462
Revised Order No. 4, 412, 413
Reynolds v. Sims, 378, 381
Rodino, Peter, 108, 109
Roe v. Wade, 419
Roeser, Thomas F., 314, 315
Roosevelt, Franklin D., 9-11
Roosevelt, Franklin D., Jr., 179, 180, 190, 201
Roosevelt, James, 95
rulemaking model of regulation, 465-68
Russell, Richard, 148
Russell amendment, 13, 16
Rustin, Bayard, 109

Scheomann, Peter, 115
Schlei, Norbert, 76, 77
Schneider, William, 456
school desegregation: Clark task force proposals, 269; court-ordered, 266, 371-74,

382, 383; de facto, 81, 145, 258, 261, 262; de jure, 126, 127; HEW guidelines, 264, 319, 320
Schuman, Howard, 455, 456
Sealtest Milk, 105
segregation, 9, 10; social stigma of, 367, 368
Seidman, Harold, 157, 158
seniority rights, 139, 140, 248, 251-53, 414
sex discrimination, 40, 97, 204, 207-31, 397-99, 405; addition to Title VII, 136-39; in professions, 416; pattern-or-practice suits, 411
sex statistics, 228
Shapiro, Martin, 470
Shulman, Stephen N., 202, 227, 235-37
Shultz, George, 309, 344; equal employment opportunity enforcement legislation, 431; equal employment opportunity for women, 409, 410; support for Philadelphia Plan, 322-26, 333, 337, 338
Silberman, Laurence H., 330, 331, 437, 438
Sitkoff, Harvard, 451
Small Business Administration (SBA), 314, 315
Smith, Howard W. ("Judge"), 12, 134-36
Sobeloff, Simon, 385, 386
social policy, 3, 7
social regulation, 463-68
Socony Mobil, 56
Southern Christian Leadership Conference, 74
Southern Regional Council report, 68
Southern strategy of Richard Nixon, 303, 304, 319
Sovern, Michael, 195
sponsorship system, 115
Staats, Elmer, 292, 293, 295, 296, 329, 331-33, 338, 339
Stans, Maurice, 315
state and local government employees, 435, 440, 443
statistical evidence of discrimination, 245-47
Steele decision, 139
Stewart, Potter, 375, 376
Stone, Harlan Fiske, 369
Strauder v. West Virginia, 245
Supreme Court, 80, 170, 246, 320, 352, 366-92, 418, 419, 444
Swain v. Alabama, 246
Swann decision, 382, 383
Sylvester, Edward C., 282, 283, 286, 287

Taft-Hartley Act of 1947, 37
task forces, 153, 154; Clark task force, 268, 269; interagency civil rights task forces, 262-65; Johnson administration, 153, 154; Pollack task force proposals, 263-65
Taylor, Hobart, Jr., 33, 51, 58, 62, 69, 71
textile industry, 241
Thirteenth Amendment, 376
Thornton, J. Mills, 457

tokenism, 368
Tower, John, 150
Tower amendment to Title VII, 388
triggering formula, Voting Rights Act of 1968, 167-69, 171, 172, 347, 379
Troutman, Robert B., Jr., 50-58
Truman, Harry S, 14, 15

underutilization. *See* proportional representation model
United States v. Carolene Products, 369
United States v. Jefferson County Board of Education, 373
Urban Affairs Council, 306

voluntary compliance. *See* Plans for Progress
voluntary model, 34
voter registration, 163, 165, 167, 168
voting examiners, federal, 168, 171-73
voting rights, 65, 69, 70, 127, 133
Voting Rights Act of 1965, 171-73, 320, 321, 377-81; Section 4, *see* triggering formula; Section 5, *see* preclearance requirement
voting rights legislation: effects, 357-59; Gaston County decision, 352, 354, 355; Nixon administration, 347-50; voting rights bill of 1970, 361

Wagner Act of 1935, 33, 34
Wall Street Journal, 275
Wallace, George, 75
want ads. *See* classified advertisements
Warren, Earl, 367-69, 372, 378, 379, 381
Wearever Aluminum Company, 96
Webb, James, 72
Weiner v. Cuyahoga Community College District, 332, 341
White, Byron, 246
White, Lee, 63-65, 76, 86, 160, 177, 178, 185
white attitudes, 454-56
White House Conference on Equal Employment Opportunity, 197, 213
Whitfield decision, 252
wholesale model of enforcement, 239, 243
Wilkins, Roy, 55
Wilks, John, 342
Williams-Hawkins bill, 437
Williams-Javits bill, 434, 440
Wilson, Charles, 197, 198
Wirtz, Willard: apprenticeship programs, 114, 281; black employment, 84, 100-102; Philadelphia Plan, 287; racial statistics, 199, 200; sex discrimination, 209, 213, 214
Wisdom, John Minor, 373
Women's Bureau coalition, 159, 206, 207, 228, 393-95, 403, 473
Women's Equity Action League (WEAL), 212

women's groups, 225, 227, 228, 428, 436
women's movement, 205, 206, 212
Woodward, C. Vann, 450

Ximenes, Vincente T., 226, 227

Yarmolinsky, Adam, 48, 86
Young, Whitney, Jr., 111-13

Zelman, Patricia, 211
zero-sum game, 5, 6, 456

NATIONAL UNIVERSITY
LIBRARY   SAN DIEGO